TEXTILES, IDENTITY AND INNOVATION: DESIGN THE FUTURE

PROCEEDINGS OF THE INTERNATIONAL CONFERENCE ON TEXTILES, IDENTITY AND INNOVATION (D_TEX 2017), LISBON, PORTUGAL, 2–4 NOVEMBER 2017

Textiles, Identity and Innovation: Design the Future

Editors

Gianni Montagna & Cristina Carvalho
CIAUD, Lisbon School of Architecture, Universidade de Lisboa, Portugal

CRC Press is an imprint of the
Taylor & Francis Group, an **informa** business
A BALKEMA BOOK

CRC Press/Balkema is an imprint of the Taylor & Francis Group, an informa business

Typeset by MPS Limited, Chennai, India

Published by: CRC Press/Balkema
Schipholweg 107C, 2316 XC Leiden, The Netherlands
e-mail: Pub.NL@taylorandfrancis.com
www.crcpress.com – www.taylorandfrancis.com

ISBN: 978-1-138-29611-4 (Hbk)
ISBN: 978-1-315-10021-0 (eBook)

Textiles, Identity and Innovation: Design the Future – Montagna & Carvalho (Eds)
© 2019 Taylor & Francis Group, London, ISBN 978-1-138-29611-4

Table of contents

3.3. Teaching, research and education

Textiles, Identity and Innovation: Design the Future – Montagna & Carvalho (Eds)
 ISBN 978-1-138-29611-4

Preface

Textiles, Identity and Innovation: Design the Future (D_TEX 2017) is a selection of the best papers presented at the 1st Textile Design Conference 2017 promoted by the D_Tex Lab through the CIAUD – Research Centre in Architecture, Urbanism and Design of the Lisbon School of Architecture of the University of Lisbon. The conference was held at the Lisbon School of Architecture of the Universidade de Lisboa, Portugal on November 2–4, 2017 under the theme Design the Future.

As coordinators of the D_TEX Lab and effective members of the CIAUD – Research Centre in Architecture, Urbanism and Design of the Lisbon School of Architecture of the University of Lisbon we are very pleased to present this book which provides different points of view about textile materials, identities and innovation in an interdisciplinary crossroads of ideas around the complex universe of Textile Design and Technology.

This book is divided into 3 main sections whose themes are the main subjects of the conference:

Section 1 is denominated Textile. It is related to Weaving and Knitting textile research materials culture, Surface, Digital and Virtual Textiles and ending with research on Textile Products. This section is specifically dedicated to the textile structures and textile constructions, to textile surfaces, digital and virtual textiles and the application of the textile material to specific products and uses.

Section 2: denominated Identity, it is related to communication through textiles. The theme Textile Art proposes ideas about textile as an artistic and aesthetic object, proposing the textile as a form of artistic expression with the user. Sub-section Culture and Craft proposes multiculturalism, customization and symbology as the main axes of the section and the section Marketing and Consumption presents concepts and research on materials and objects of textile material.

Section 3: is presented under the umbrella of Innovation. Technical and Smart Textiles are the main themes for this section. Intelligent fabrics are presented here for applications as different as clothing but also for the performing arts and architecture. The sustainable textiles section presents renewed strategies for the sustainability and reduction of textile wastes and their recovery. The last topic dealt with in this section is Teaching, Research and Education which presents different points of view for the teaching materials and textile design for the new generations.

We hope that this publication will be a starting point for an open and more global conception of textile design and its materials, allowing special characteristics of adaptation to the users and the different areas of application.

The main objective was to promote the linking of different areas of knowledge that can expand and use the textile and fibrous materials for the improvement and development of new products and functionalities.

Gianni Montagna
Cristina Carvalho
Coordinators for D_Tex Lab and effective members of CIAUD of the Lisbon School of Architecture of the Universidade de Lisboa

 ISBN 978-1-138-29611-4

Editor biographies

Gianni Montagna has held a PhD in Design from the Lisbon School of Architecture of the Universidade de Lisboa since 2012, with a thesis on the application of smart systems to clothing, teaches Fashion Design in the first academic year of the Masters in Fashion Design and often teaches other disciplines at the same institution, as in the case of Accessory Design. In the textile development laboratory (D_TEX) of the Faculty of Architecture of Lisbon, he is responsible for the area of Textile Design and is responsible for the discipline of Laboratory III of the PhD in Design as well as being professor of the subject of Textiles for the Architecture of the PhD program in Architecture of the same Faculty. He is the scientific coordinator of the International Congress on Textile Design "D_TEX _ Textile, Identity and Innovation" which aims to bring together the different scientific domains around the textile area. He is a guest lecturer in the Doctoral Program in Fashion Design at the University of Minho in collaboration with the Beira Interior University, where he collaborates in the teaching of "Interfaces Design Fashion and Technology".

Cristina Carvalho, was born on September 9, 1968 in Alpedrinha / Fundão, Portugal, graduated in Textile Engineering (Production Branch) at the University of Beira Interior in 1993. She obtained her Master's Degree in Biochemical Engineering / Biotechnology in 1999 at Instituto Superior Técnico of the Technical University of Lisbon, and her PhD in Biotechnology in 2006 at Instituto Superior Técnico, Technical University of Lisbon. She has been a researcher at the CIAUD of the Faculty of Architecture since 2007 and has been an executive member of the Research Center since 2009. Between 2008 and 2012 she coordinated the Master's Degree in Fashion Design (2nd cycle) and is also a lecturer in the Fashion Design graduation course in the areas of Technologies and Materials and in the Ecodesign, in the Doctoral Degree in Design has been professor within the curricular unit of Laboratory III since 2014 in the Lisbon School of Architecture of the University of Lisbon and Member of the FA Council since 2012.

Textiles, Identity and Innovation: Design the Future – Montagna & Carvalho (Eds)
© 2019 Taylor & Francis Group, London, ISBN 978-1-138-29611-4

Committee members

CONFERENCE CHAIRS

Gianni Montagna
CIAUD, Lisbon School of Architecture, Universidade de Lisboa, Portugal
Cristina Carvalho
CIAUD, Lisbon School of Architecture, Universidade de Lisboa, Portugal

ORGANIZING COMMITTEE

Gianni Montagna, *CIAUD, Lisbon School of Architecture, Universidade de Lisboa, Portugal*
Cristina Carvalho, *CIAUD, Lisbon School of Architecture, Universidade de Lisboa, Portugal*
Filipa Nogueira Pires, *CIAUD, Lisbon School of Architecture, Universidade de Lisboa, Portugal*
Maria Alexandra Luís, *CIAUD, Lisbon School of Architecture, Universidade de Lisboa, Portugal*
Luís Ginja, *AEAULP, CIAUD, Universidade de Lisboa, Portugal, Portugal*
Luís Santos, *CIAUD, Lisbon School of Architecture, Universidade de Lisboa, Portugal*

KEYNOTE SPEAKERS

Anne Marr, *University of the Arts of London, UK*
Colin Gale, *Birmingham City University, UK*
Julianna Abel, *University of Minnesota, USA*
Sonja Andrew, *University of Huddersfield, UK*

INTERNATIONAL SCIENTIFIC COMMITTEE

Ana Couto
CIAUD, Lisbon School of Architecture, Universidade de Lisboa, Portugal

Ana Cristina Broega
Engineering School of the University of Minho – DET, Portugal

Ana Margarida Pires Fernandes
CIAUD, Lisbon School of Architecture, Universidade de Lisboa, Portugal

Bernardo Providência
Engineering School of the University of Minho – DET, Portugal

Carla Morais
CIAUD, Lisbon School of Architecture, Universidade de Lisboa, Portugal

Carlos Figueiredo
CIAUD, Lisbon School of Architecture, Universidade de Lisboa, Portugal

Cristina Carvalho
CIAUD, Lisbon School of Architecture, Universidade de Lisboa, Portugal

Elisabete Rolo
CIAUD, Lisbon School of Architecture, Universidade de Lisboa, Portugal

Gabriela Santos Forman
CIAUD, Lisbon School of Architecture, Universidade de Lisboa, Portugal

Gianni Montagna
CIAUD, Lisbon School of Architecture, Universidade de Lisboa, Portugal

Hazem Abdelfattah
Helwan University, Egypt

Helder Carvalho
Engineering School of the University of Minho – DET, Portugal

Helena Britt
Glasgow School of Art, Scotland

Joana Cunha
Engineering School of the University of Minho – DET, Portugal

Maria João Delgado
CIAUD, Lisbon School of Architecture, Universidade de Lisboa, Portugal

Maria João Ferreira
CHAM – FCSH/Nova – Uac, Portugal

Maria João Pereira Neto
CIAUD, Lisbon School of Architecture, Universidade de Lisboa, Portugal

Maria Pereira
Faculty of Engineering, University of Beira Interior, Portugal

Maria Sacchetti
CIAUD, Lisbon School of Architecture, Universidade de Lisboa, Portugal

Maria Sílvia Barros de Held
University of São Paulo, Brazil

Ningtao Mao
University of Leeds, UK

Regina Sanches
University of São Paulo, Brazil

Renata Pompas
Milan, Italy

Rita Salvado
Faculty of Engineering, University of Beira Interior, Portugal

Rui Miguel
Faculty of Engineering, University of Beira Interior, Portugal

Secil Uğur Yavuz
Unibz – Faculty of Design and Art, Italy

Susana Oliveira
CIAUD, Lisbon School of Architecture, Universidade de Lisboa, Portugal

Theresa Lobo
UNIDCOM/IADE, Portugal

1. Textile

1.1. Weaving and knitting

Textiles, Identity and Innovation: Design the Future – Montagna & Carvalho (Eds)
© 2019 Taylor & Francis Group, London, ISBN 978-1-138-29611-4

The use of textiles in Anne Wilson and Kathrin Stumreich's work

Inês Pereira Guerreiro Jorge

ABSTRACT: In this paper, I will analyze some works by artists Anne Wilson (b. 1949) and Kathrin Stumreich (b. 1976), which incorporate textiles and explore the boundaries between hand and body, manual and digital tools. By combining different weaving techniques and digital tools in the creation of installations, videos, performances and community projects, their pieces oscillate between the materiality of the artifact and the immateriality of the digital image and/or sound, the reflection on handwork and the inclusion of the spectator's body. In these works, the dialogue between craft and technology opposes to the common perception that these two fields are irreconcilable; instead, it insinuates that both establish connections, either in the literal or in the symbolic sense. At a symbolic level, such entanglements may refer to the human biological system or to social relations that are being optimized and transformed by the Internet and social networks.

In this paper, I will analyze works by artists Anne Wilson and Kathrin Stumreich, which incorporate textiles and explore the boundaries between hand and body, manual and digital tools. By combining different weaving techniques and digital tools in the creation of installations, videos, performances and community projects, the pieces conceived by these two artists oscillate between the materiality of the artifact and the immateriality of the digital image and/or sound, the reflection on the history and processes of handwork, and the inclusion of the spectator's body.

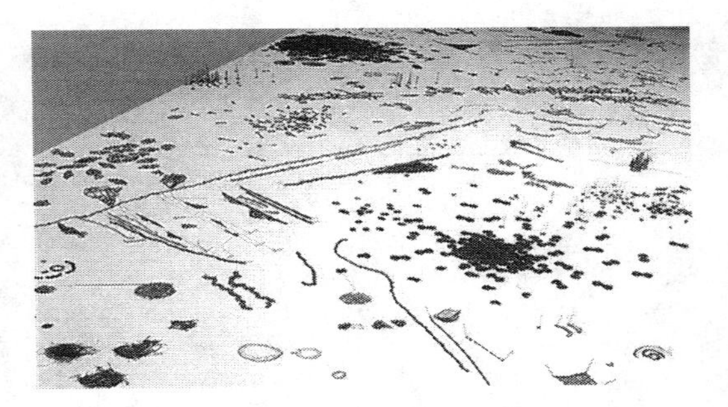

Figure 1. *Topologies,* Anne Wilson, 2002–2008. Lace, thread, cloth and pins on painted wood support, 79 cm high × 188 cm wide × 1.10 m long (overall dimension).

1 THE DICHOTOMY MANUAL/DIGITAL IN SOME INSTALLATIONS BY ANNE WILSON AND KATHRIN STUMREICH

Anne Wilson (b. 1949) has been exploring the dichotomy between manual and digital processes in installations made with organic or poor materials – such as textiles, human hair, pins, wire and glass –, which are subjected to audiovisual recording. Her works often include functional items that thereby lose their utility, while the memory of their lost function is evoked. Moreover, the displacement of these objects – that were once useful – into the art gallery recalls the readymade principle introduced by Marcel Duchamp in 1913. The traditional notion of skill is then subverted through the blend of weaving techniques, digital tools, and randomness, whilst conceptual divides of making/unmaking and creating/destroying are put into question.

Wilson started exploring these notions in 2000, through the creation of abstract landscapes on horizontal platforms that were formed by torn, punctured and stitched tablecloth, such as the installation *Feast*. Later, she filled these surfaces with small constructions of found black lace, which were subsequently scanned, filtered, printed on paper, and once again stitched on the original landscapes, as in the piece *Topologies* (Fig. 1).

Such surfaces resemble tables, whose functional nature leads us to analyze them according to Howard Risatti's book *A Theory of Craft: Function and Aesthetic Expression*. In this book, Risatti (2007) established a taxonomy of objects based on their ability to fulfill the body's physiological needs, i.e. their "applied function" (Risatti, 2007). Such classification divides objects in the following three categories, along with a fourth one related to architecture:

a) Containers – such as jugs, bowls and vases;
b) Covers – such as clothes, blankets and quilts;
c) Supports – such as beds, tables and chairs;
d) Shelters

According to this taxonomy, the bases on which Anne Wilson's scenarios are built fall into the category of "supports". As a verb, "to support" means to serve as a physical or psychological foundation for something or someone.

Figuratively speaking, these supports can stand for dining tables that witness encounters and bear several social codes. In this sense, they incite us to reflect on the traditional notion of family as a symbol of support and protection, as well as on the prevalence of this notion today.

The use of dining tables to represent codes of conduct and social roles is also present in the iconic installation The Dinner Party, developed by Judy Chicago between 1974 and 1979. In this piece, which incorporates crafts such as embroidery, weaving, sewing and china painting, the set of a supper is recreated on a triangular table, where thirty-nine illustrious women who have been overlooked by History are recalled. The table's triangular shape is an allusion not only to femininity, but also to the Last Supper, as thirteen women are placed on each side, as opposed to the thirteen men that were present according to the Gospel accounts.

On the other hand, Wilson's work Feast seems to allude to the analogy between food and sex. This correlation dates back to Sigmund Freud's psychoanalytical studies, which turned sexuality into a key element of the self (Freud 2013 [1901]). Almost one hundred years later, art historian Preziosi (1999) stated the following: "You are not only "what you eat" or what you make, consume, or collect, but you are also, and especially, what you desire." (Preziosi 1999)

The tables comprising both installations can also be interpreted as sewing tables which, in the same way as the work *Dinner Party*, pay tribute to the intricacy and delicacy of manual, domestic and feminine labor. When associated with Risatti's taxonomy, this idea leads us to suggest that these tables represent women as the essential foundation of families, thus challenging traditional Western representations of women as fragile and passive beings. In any case, the insinuation of the hand and the body in these pieces is unquestionable.

The title of the second installation, *Topologies*, carries an additional layer of meaning. Topology is the branch of geometry dealing with "the properties of a figure that are unaffected by continuous distortion, such as stretching or knotting". Therefore, it involves "the study of limits in sets considered as collections of points (...) making a given set a topological space" (Topology 2017). Likewise, in this installation the structural properties of lace are explored through the deconstruction of found black lace and through the creation of large horizontal topographies. As stated in the artist's website, *Topologies* "is a constantly unfolding process of close observation, dissection, and recreation" (Topologies n.d.).

The computer plays a pivotal role in the observation phase, as lace fragments are scanned, filtered, and printed out as paper images. The digital prints once again acquire a physical form through hand stitching, and then are placed in the topography, together with the found and rebuilt lace.

The project *Topologies* contains multiple references, namely regarding connections between material and immaterial systems of relationships – for instance, textile networks and the World Wide Web –, the biological system, microscopic views of cellular structures, and macroscopic views of urban constructions. None of these themes or perspectives is privileged over the other, as in today's information society, in which the information explosion driven by digital information and communication technologies blurs the hierarchies between different contents.

Similarly, artist Kathrin Stumreich (b. 1976) has been exploring the links between weaving and electronics. By manipulating digital tools with aesthetic ends, she celebrates the transforming power of technology within the industry and the arts.

The alliance between craft and digital technology was also examined in the exhibition *The New Materiality: Digital Dialogues at the Boundaries of Contemporary Craft*, held at the Fuller Craft Museum of Massachusetts in 2010. According to the curator Fo Wilson, the show embodied "a wide variety of craft media in objects that all incorporate digital technology, or 'new media', in some manner" (Harrington 2010). Kathrin Stumreich's work exemplifies this type of creations, while reflecting on both craft and technology's ability to create connections, either in a literal or in a figurative sense, as in Anne Wilson's network landscapes.

Furthermore, some of Stumreich's pieces deal with the theme of mobility. In the installation *Der Faden* (Fig. 2), "a string is tied to a starting point and unwinds with the travelling streetcar, wrapping up the city" (Der Faden 2010). The trail of the string is recorded by a piezo (a type of microphone that senses audio vibrations through contact with solid objects) that amplifies the resulting sound and creates a full-scale map of the itinerary. The unravelling sound catches the attention of passers-by, allowing them to produce their own acoustic fantasies (Gustafsson & Falb 2010).

Figure 2. *Der Faden,* Kathrin Stumreich, 2010. Detail of bobbin mounted on a tram.

Such performative character requires the public's action, as the incorporated device incites people to participate in a personalized "tour", which is also rooted in reality. Through the use of technology and motion, *Der Faden* incites us to reflect on the increasing mental and physical passiveness prompted by Western devices of representation – from linear perspective to cinema –, the new possibilities provided by interactive technology, and the blurring of territorial borders caused by the dissemination of Internet.

This installation entails a temporary piece that overcomes the mere object (the bobbin) and its traditional function (to wind the thread for weaving). Accordingly, craft becomes implicated in cutting-edge technology, and therefore in contemporaneity.

2 CRAFT, PERFORMANCE AND COMMUNITY IN SOME WORKS BY ANNE WILSON AND KATHRIN STUMREICH

In Anne Wilson and Kathrin Stumreich's work, the exploration of artistic liminality and hybridism brings about the intersection of the so-called "low" and "high" arts, and particularly of craft and performative arts, fostering a connection between hand and body.

This intersection was equally explored in the creations by John Cage (1912–1992), who produced music for dance pieces and choreographers such as his partner Merce Cunningham, and developed the concept of happening (an ephemeral, theatrical event that challenged the divide between stage and audience, and had neither a defined duration nor a detailed script). In the 1960s, Cage performed a concert with video artist Nam June Paik and, in collaboration with composer Lejaren Hiller, he conceived the multimedia show *HPSCHD*. This performance involved the exploration of randomness, the incorporation of computerized sounds, film screenings, and a projection of drawings (Pritchett 1993).

The works developed by Cage in this period were influenced by Marshal McLuhan's writings on the social effects of media, as well as by R. Buckminster Fuller's thoughts on the power of technology to promote social change. Due to its experimental character and its relations with socialist and utopic ideals, this body of work reached an unprecedented degree of complexity and inspired the emergence of electronic music in the second half of the 20th century.

The synesthetic character of John Cage's pieces is also present in Anne Wilson's *Notations* (2008), whose title refers to a book that was published in 1969 by Cage himself and Fluxus artist Alison Knowles. This installation comprises a notational system formed by audio and visual recordings of Wilson's hands and their repetitive movements, as she crocheted and knitted. The resulting motion sequences were then photographed and captured in twenty musical "scores", which were fixed on the wall. The musical composition by Shawn Decker was inspired by the collected sounds and echoed through space in an aural manner, sacralizing the artist's craftsmanship.

In the meantime, audio-visual recordings of the abovementioned work *Topologies* were integrated in the installations *Errant Behaviors* (2004) and *Mess* (2006), which expanded the meaning of the former, through the investigation of randomness, the absence of a beginning or an end, and the possibility to produce an unlimited number of connections.

In these pieces, images of lace compositions were animated from one frame to the next, replicating the accumulative act of lacing and the "errant behavior" of the hand that the title evokes (Fig. 3). As can be seen in the movie *Modern Times* by Charlie Chaplin, the predictable and flawless behavior of the machine opposed to the one of the hands, whose connection with the mind and the eye induces unique abilities and movements. Likewise, human interactions involve unexpected and uncanny acts that are being increasingly reproduced by technology. Once again, the sound arrangements by Shawn Decker reacted to the images that integrate these works, mixing found and processed sounds and natural, human and artificial rhythms (Errant Behaviors 2010).

The incorporation of craft in performance became a central subject within feminist art in the 1970s and 1980s, with the aim of discussing the women's role in the public and social spheres. In contrast, the association between handwork and performance in some of Kathrin Stumreich and Anne Wilson's pieces alludes to the magical and ritual connotations of craft, as well as to its economic and social role in the past and today.

Between 2008 and 2012, Anne Wilson created three choreographies under the title *Wind-Up: Walking the Warp*, which were presented at the Rhona Hoffman Gallery in Chicago, the Contemporary Arts Museum in Houston, and the Whitworth Art Gallery in Manchester, UK. These performances recreated rituals associated with weaving and resulted in the creation and exhibition of a textile piece.

Acts involved in weaving, such as spinning, counting, winding, and entwining, were interpreted literally, through the movements of bodies in space.

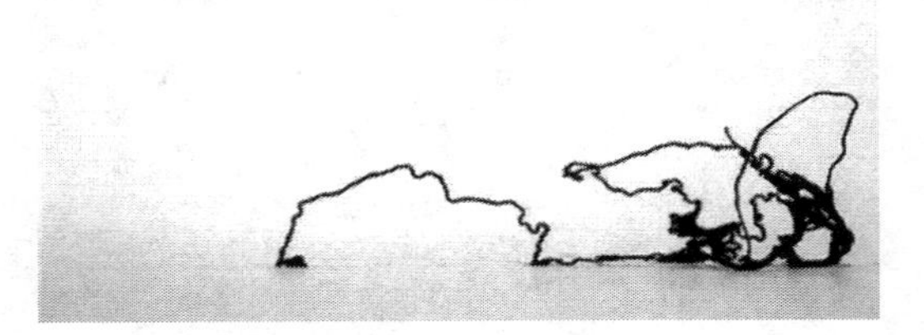

Figure 3. *Errant Behaviors* (still), Anne Wilson, 2004. Video and sound installation. Musical composition by Shawn Decker; animation by Cat Stolen; post-production animation and mastering by Daniel Torrente.

Figure 4. Wind-Up – Walking the Warp, Anne Wilson, 2010. Performance by Hope Stone Dance at the Contemporary Arts Museum in Houston.

The exhibition venues, in turn, were chosen for their historical connections with weaving, and the following de-industrialization and decadence of that industry. Hence, the inclusion of the modal verb "wind-up" in the title alludes both to the conclusion of an action and to the winding involved in textile making.

For the performances exhibited in the USA, the fabric and additional materials were brought from local weaving factories. The performance at the Rhona Hoffman Gallery was visible from outside and originated a sculptural presence, whilst in the Contemporary Arts Museum the two performances were presented in partnership with local group Hope Stone Dance (Fig. 4).

In the UK, the performance was produced in collaboration with choreographer Bridget Fiske, and executed by dancers from The Lowry Center for Advanced Training in Dance. The dancers' costumes were dyed by textile students from Manchester Metropolitan University. Furthermore, the choreographies were inspired by the circulation of steam engines from old cotton industries in Northwest England, while Shawn Decker's musical composition incorporated mechanical rhythms and excerpts from Anglo-Saxon labor songs, which possess a strong cultural identity. In addition, textile fragments from the gallery's Ancient Egypt collection were exhibited, to illustrate the spatial-temporal transversality of the striped-pattern textile.

Equally positioned between weaving, performance, and music, and therefore between materiality and immateriality, Kathrin Stumreich's *Fabricmachine* (2009–2013, Fig. 5) stems from a reflection on musical structures that are present in the texture and composition of textiles. Such work consists of a device with two fabric loops that are driven by a motor. At the same time, light sensors – as if they were needles of a phonograph or of a record player – detect and translate the characteristics of the fabric into sound signals.

The musicality of *Fabricmachine* can be experienced by sliding and altering the position of the sensor on the fabric, or by altering the speed of the motors. Fabric samples thus turn into data storage devices within the interactive installation. Hence, the tone pitch is determined by the quality of the fabric and by the weaving technique; breaks and rhythm are influenced by the seam and length of the fabric; and the musical composition is shaped by the fabric's layout (Fabricmachine 2013).

As in Kathrin Stumreich's former creation *Der Faden* (2010) and in Anne Wilson's *Notations* (2008), *Fabricmachine* makes it possible to modify an organic sound – in this case, the sound of a loom –, and to record it or "print" it on fabric, providing the spectator with a synesthetic experience. Therefore, the device converts an ordinary piece of clothing into an audible machine, overcoming clothing's conventional function of "covering", as defined by Risatti (2007). In this sense, clothing is no longer confined to the ends of protection or camouflage, acquiring the ability to convey emotions.

In the same way as Wilson's performances, Stumreich's piece seems to reflect on, and perhaps offer a response to the collapse of the clothing industry in the West, which failed to compete with the global market, leading many families to unemployment and financial vulnerability. From this viewpoint, *Fabricmachine* illustrates the need to adapt to new demands, by incorporating the latest technologies. On the other hand, whilst being played by the artist in live performances, this device returns the processes and acts inherent in object making to people's daily lives.

Such performative character, along with a communitarian intent, was accomplished by Anne Wilson in *Local Industry* (2010), a work that was presented at the Knoxville Museum of Art during the solo exhibition *Anne Wilson: Wind/Rewind/Weave.* Once again, the exhibition venue was historically linked to weaving. The museum galleries were thus transformed into a factory, while visitors were surrounded by looms and bobbins, becoming "artisans" and participating in a collective action. With the help of hundreds of weavers, the public produced a 19-meter-long piece of fabric that now belongs to the museum collection. When exhibited in a new space, the work is always accompanied by its production archive, to replicate the temporary event.

According to Risatti's taxonomy (2007), the colossal tapestry that resulted from this project transcends its covering function. But does it fulfill this function in a metaphorical sense, as a source of protection or concealment?

We may consider craft as a historically collective trade that brings people together, with a single purpose in mind. Through this initiative, whilst experiencing the dignity and demands of handwork, the public was encouraged to think about the social impact caused by the collapse of textile industries, and about ethical issues regarding global fabric trades. Whereas in *Walking the Warp* (2008–2012) the entanglement of craft in performance aimed to restore memory and national identity, *Local Industry* incited the public to act, accomplishing what Borriaud (2002 [1998]) termed as "relational aesthetics".

Figure 5. *Fabricmachine,* Kathrin Stumreich, 2009-2013. Textile, device with two motorized loops, light sensors.

Indeed, relational art emerged as a reaction to the post-industrial age, during which the manufacturing sector gave way to the service sector, while knowledge economy was supplanted by information society. The project *Local Industry* relates to this artistic movement, as it focuses on "(. . .) human relations and their social context, rather than [on] an independent and private space" (Borriaud 2002 [1998]: 113). Therefore, it ties in with Borriaud's (2002) statement that "(. . .) the most pressing thing is no longer the emancipation of individuals, but the freeing-up of inter-human communications, the dimensional emancipation of existence" (Borriaud 2002 [1998]: 60).

Anne Wilson's and Kathrin Stumreich's performance and communitarian pieces focus on artistic process and social experience, rather than the final product, in line with developments in art and craft after World War II. Likewise, both artists reflect on the increasing immateriality of handwork since the rise of the studio craft movement in the 1950s and 1960s, by subverting the traditional notions of material, object, function, and skill.

3 CONCLUSIONS

In this paper, I have addressed the dichotomies of manual/digital, hand/body in some works by Anne Wilson and Kathrin Stumreich that comprise textiles. In their creations, weaving techniques are incorporated in installations, videos, performances and community projects, in which traditional artisanship is evoked and transformed with the help of digital tools, questioning the alleged rilvalry between craft and technology. The materiality of the artifact is thus entangled in the immateriality of digital images and/or sounds, resulting in a "new materiality" (Harrington 2010).

In the first chapter, I focused on the manual/digital divide whilst analyzing some installations by Wilson and Stumreich. These pieces are constituted by textiles, weaving materials and techniques which are then subverted and transformed through digital tools such as scanners, animation and sensors. Furthermore, the static quality of handmade objects is exceeded through their digital transformation and through their incorporation in textile networks, moving images and mobile installations.

The blend of manual and digital technologies in these creations aroused the idea that both handwork and electronics establish connections, not only at a literal level, but also at a symbolic one. As a result, I explored the works' symbolic meanings, establishing a parallel between textile meshes and social and digital networks.

In the second chapter, I considered the manual/digital dichotomy within an investigation of some performance pieces and community projects by Anne Wilson and Kathrin Stumreich. As in the formerly investigated installations, textiles, weaving materials and techniques were used or served as inspiration, but were then transmuted to performative and participatory projects. At a symbolic level, the association of artisanship and performance in these creations enabled a discussion about craft's magical and ritual meanings, as well as about its past and present role in economy and society.

The hand/body theoretical opposition crosses both chapters. It could be expected that the merge of manual and digital technologies would cause the loss of the hands' traces, or the absence of the body. However, the use of industrial machinery and digital solutions by both artists, along with the incorporation of randomness and error in works that are highly experimental or synesthetic culminates in honoring handwork's imperfections and unexpectedness.

Moreover, Wilson and Stumreich's pieces evoke the body, not only of the artist as (s)he works, but also of the viewer who is immersed in audiovisual, performative, interactive or communitarian projects. According to Shiner (2012), one of the possible ways to clarify the concept of craft is to understand the hand as a "body", that is, as a form of making that involves physicality (a clear example of this idea is the technique of glassblowing).

On the other hand, the presence of the body senses in some of these creations reminds us of the influence of senses on the intellect. In particular, Harrington (2010) stated that the body – and what it feels – is an essential vehicle for mental perception, while Slatman (2005) stressed the following: "Without a body there is no life. Or, to be more precise without a body, there is nothing that touches" (Slatman, 2005). Thus, allusions of the body also entail allusions of tact.

In short, the analyzed pieces show that contemporary artist-artisans can express complex ideas, emotions and meanings whilst incorporating textiles in sophisticated manual and digital technology. Hence, they are capable of combining intellectual and technical skills, i.e. what the ancient Greeks named *techné*, in works that involve handicraft.

Both Anne Wilson and Kathrin Stumreich comment on traditional notions regarding artisanship, such as materiality, objectuality, function, and skill, while creating pieces that are simultaneously material and immaterial; that are based on objects, but mostly focus on processes; that incorporate utility in their content, but do not depend on function; that are skillful, but embrace experimentation.

Their textile creations result from a blend of artisanal and industrial techniques, bring together visual and performative forms of expression, and hold references to the history of handwork and its relationship with industry and fine art.

Therefore, the examined case studies show that the use of textiles whilst merging craft and digital tools generates works that can carry prolific symbolic meanings and discussions concerning contemporary issues.

REFERENCES

Borriaud, N. 2002 [1998]. *Relational Aesthetics.* Paris: Les Presses du Réel.

Der Faden 2010. Retrieved from http://www.kathrinstumreich.com/der-faden/.

Errant Behaviors 2010. Retrieved from http://vimeo.com/12185840.

Fabricmachine 2013. Retrieved from http://www. kathrinstumreich.com/stofftonband/.

Freud, S. 2013 [1901]. *The Psychopathology of Everyday Life.* Worcestershire: Read Books Ltd.

Gustafsson, M. & Falb, H. 2010. Konfrontatioenen Nickelsdorf 2010. Retrieved from http://www.konfrontationen.at/ko10/sound.html.

Harrington, B. A. 2010. "New Materiality" and the sensory power of craft. In Matthew Hebet (ed.), *Furniture Matters: 6-7; September 2010.* North Carolina: The Furniture Society *apud* Wilson, F. 2010, *The New Materiality: Digital Dialogues at the Boundaries of Contemporary Craft.* Massachusetts: Fuller Craft Museum.

Preziosi, D. 1999. Performing modernity: the art of art History. In Amelia Jones & Andrew Stephenson (ed.), *Performing the Body/Performing the Text*: 27–35. London: Routledge.

Pritchett, J. 1993. *The Music of John Cage.* Cambridge: Cambridge University Press.

Risatti, H. 2007. *A Theory of Craft: Function and Aesthetic Expression.* Chapel Hill: University of North Carolina Press.

Shiner, L. 2012. "Blurred Boundaries"? Rethinking the Concept of Craft and its Relation to Art and Design. *Philosophy Compass.* Doi: 10.1111/j.1747-9991.2012.00479.x.

Slatman, J. 2005. The Sense of Life: Husserl and Merleau-Ponty on Touching and Being Touched. *Chiasmi International* 7: 305–325.

Topologies n.d. Retrieved from https://www.annewilsonartist.com/topologies-credits.html.

Topology 2017. *Dictionary.com.* Retrieved from http://www.dictionary.com/browse/topology/.

Textiles, Identity and Innovation: Design the Future – Montagna & Carvalho (Eds)
 ISBN 978-1-138-29611-4

Selection and training of a panel of evaluators for sensory analysis of tactile comfort in Brazil

R.N. Nagamatsu
Faculty of Textile Engineering, University of Minho, Guimarães, Portugal
Faculty of Fashion Design, and Faculty of Textile Engineering, Federal University of Technology Parana, Apucarana, Paraná, Brazil

M.J. Abreu
Faculty of Textile Engineering, University of Minho, Guimarães, Portugal

C.D. Santiago
Faculty of Fashion Design, and Faculty of Textile Engineering, Federal University of Technology Parana, Apucarana, Paraná, Brazil

I. Braga
Faculty of Textile Engineering, University of Minho, Guimarães, Portugal
Faculty of Fashion Design, Federal of University Piauí, Teresina, Piauí, Brazil

ABSTRACT: Increasingly, sensory analysis has been explored by non-food sectors to evaluate different products as a strategy to improve their commercialization. The selection of evaluators is an important step for the sensorial evaluation of products. This paper presents the process of forming a Brazilian tactile sensory panel to evaluate the surface of textile products. Fourteen pre-selected evaluators through triangular tests were trained and monitored for a period of four months. Two monitoring methods selected the evaluators: overall performance was assessed at a significance level of 0.05 using ANOVA; And then line graphs were used to reveal the performance and eventual problems of agreement of the evaluators in the use of the evaluation scale. The results obtained with the use of these methodologies concomitantly for the monitoring were relevant and helped the selection of the evaluators with a smaller margin of error. After the training stage, eleven evaluators were selected to compose the sensory panel based on their power of discrimination, repeatability and agreement with the group in the use of the scale. The aim of this panel is to classify the attributes of the cap textile surface of the Apucarana-Brazil region.

1 INTRODUCTION

Textile sensory analysis has been the focus of studies to measure and compare surfaces in different types of textiles in view of consumer comfort. This method has been adapted from international standards developed for food assessment. Sensory evaluation commonly uses sense perception (sight, smell, hearing, taste, and tactile) to evaluate food products. In textiles, sensory analysis has been investigated mainly by tactile perception (Li & Wang, 2005; Sabir & Doba Kadem, 2016; Bacci, 2012; Chollakup, 2004; Nogueira, 2011; Philippe, et al., 2004; Guest & Spence, 2003).

Tactile sensory comfort in textiles is the result of the amount of tension generated in the material in contact with the skin. Thus, it has a strong relation between the tactile function and the mechanical properties of the fabric. The tactile properties of the sensorial comfort of fabric were standardized through descriptive and psychophysical sensory analysis techniques from the 1990's (Sztandera, et al., 2012). However, Yenket et al. (2007) say that few studies have used trained assessors for tactile material perception. According to Philippe et al. (2004) investigations on sensory analysis emerged in the 1950s with the development of descriptive methods used by the food industry. From the 1970s the complete methodology of descriptive sensory analysis was proposed by Hebert Stone & Joel Sidel, (Spence and Gallace, 2011) becoming a US standard in the 1980s, and is currently an international standard ISO:8586 Sensory analysis – General guidelines for the selection, training and monitoring of selected assessor and expert assessors (Philippe, et al., 2004). These sensory methods generally use the human senses as a measuring tool.

Spence & Gallace, in their research on the importance of multisensory design, have quoted that after the eye, touch is the first sensor to judge the object and is decisive for the final acceptance of a product. They even mention that in the case of textile articles, consumers often even rub the materials on parts of the body where they feel more sensitive, for example on the cheeks, in order to feel their warmth and softness. In addition, they emphasize that the consumer, when touching a poor-quality textile article, brings a more negative evaluation than when compared to a visual evaluation of the product. The study of multisensory processing is revealing some regulations for multisensory perception of objects, this allows the creation of experiences to consumers that potentiate the stimulation of specific classes of sensory receptors in the skin provoking specific sensations in the consumer's mind, such as the sensation of well-being (Spence and Gallace, 2011). In this way, the designer must be attentive to meet the well being needs of the consumer.

In textile products, sensory studies were carried out from the 1980s onwards, with trained or untrained evaluators to make the tactile assessment. Yenket et al., (2007) Some researchers allowed the sensory evaluators to look at the materials during the haptic evaluation, and others Philippe et al. (2004), and Nogueira (2011) blinded the evaluators during the tests using a barrier (evaluation booth) that did not allow the visualization of the textile material. Thus, there has been a growing need in the training and follow-up of specific sensory panels for tactile evaluation of the surface of textile materials. Also, because they are people, it is important to have a careful follow-up to the selection of the final panel of evaluators (Teixeira, 2009). Thus, this work aims to present the process of training and monitoring a tactile sensory panel for evaluation of the textile surface, which may assist the designers in the decision making for development of new textile products.

2 SENSOR PANEL FORMATION

The panel of sensory analysis is formed by selected people called Judges, Tasters or Testers. The stage of selection of people is very important to obtain adequate data for quantitative descriptive analysis (Teixeira, 2009). Recruitment can be done between employees of the companies where the products to be evaluated are developed; By voluntary consumers; Or by the two groups. Selection participants should have time available for both training and sensory evaluation; readiness; Not being allergic or having health problems that affect panel participation; Have good articulation capacity; And have no aversion to the products that will be tested (Esteves, s.d.). Lawless & Civille (2013) add up that for the formation of a textile tactile sensory analysis panel, participants can not present calluses on their hands, deficiencies in the circulation of the hands and fingers, central nervous system disorders, dry or cracked skin.

An important instrument for training and monitoring is the panel leader. He or she should recruit, select, train, and periodically monitor the overall and individual performance of the panel of evaluators through statistical analysis (ABNT, 2016; NF EN ISO, 2014). It is indicated that between 40 and 60 volunteers are recruited to form a trained panel of 7 to 15 evaluators.

Image 1. Evaluators in training session.

The training sessions can determine indicators to gauge the ability of the evaluators in homogeneity, repeatability and reproducibility (ABNT, 2016). These data are relevant for rejecting or accepting an evaluator in the sensory panel. Statistically, when p-value are significant ($p > 0,05$) may be indicative that assessors need more training sessions to improve. The training should be repeated until the results are more homogeneous, or the evaluator can be excluded from the panel. At the end, the selected evaluators will be able to evaluate the intensity of each sensory attribute.

3 MATERIALS AND METHODS

3.1 *Materials*

The research was approved by the Research Ethics Committee involving Human Beings of the Federal Technological University of Paraná, with number CAAE 45651115.5.0000.5547. The training site was a laboratory of the UTFPR with monitored relative humidity and temperature of the air so that the evaluators felt comfortable at a temperature of 22°C (±2°C) and 65% (±5%) of relative humidity (AFNOR, 2014).

In this research seven samples of different textile materials were used. They were cut into a 20 × 20 cm dimension and coded following the standard ISO 6658:2005 Sensory analysis — Methodology — General guidance (ISO, 2004) with random numbers of three digits. Samples were placed behind a visual barrier (booth). They were positioned in different orders and individually to avoid comparison between them (Ellendersen & Wosiacki, 2010). For each sample the evaluator received a card with the 11 attributes with a intensity scale. This disposition was repeated for each evaluator (ABNT, 2016).

Among a group of 43 volunteers, 13 were selected to begin training for the formation of a haptic sensory panel of textile evaluators. The evaluators were previously selected by triangular difference test between two textile samples (ABNT, 2013). This test aimed to pre-select people with haptic acuity, assiduity and interest in participating in the project.

The volunteers are students and employees of the Federal Technological University of Paraná, Apucarana campus, aged between 18 and 50 years. These individuals were trained over a period of four months to meet the requirements of the ISO 8586:2014 (NF EN ISO, 2014).

Therefore, each inexperienced evaluator participated in pre-scheduled sessions at least once a week with a maximum duration of one hour in an air-conditioned laboratory designated to train them. A Lexicon with 11 textile attributes was used: 5 Bipolar (Light–Heavy; Thin–Thick; Cold–Warm; Dry–Humid; Rough–Flat); 3 describing the material surface (Soft, Plushy, Rugged); and 3 describing the properties of the Material (Elastic; Rígid; Falling)(Nagamatsu, et al., 2017).

Bipolar attributes	Surface attributes	Materials attributes
Light-Heavy	Soft	Elastic
Thin-Thick	Plushy	Falling
Cold-Warm	Rugged	Rigid
Rough-Flat		
Dry- Humid		

Chart 1: Lexicon for analysis of Brazilian textile sensory comfort (Nagamatsu, et al., 2017).

These attributes were generated from a list of 299 different terms, and treated statistically until the final formation of the 11 most significant terms. These attributes can be used to define and classify a textile texture to which is possible to specify values that help to categorize information that facilitates the monitoring of performance and the development of products, according to the attributes desired by the consumers.

3.2 *Data analysis*

The performance of the inexperienced panel was evaluated using multivariate statistical methods in accordance with the standards ABNT NBR ISO 11132:2016 Sensory analysis – Identification and selection of descriptors for establishing a sensory profile by a multidimensional approach (ABNT, 2016) e NF EN ISO 8586:2014 Sensory analysis – General guidelines for the selection, training and monitoring of selected assessor and expert assessors (2014). The ANOVA (Analysis of variance) was conducted to evaluate the interactions between samples, assessors and attributes.

The homogeneity of the panel was evaluated according to the interaction between the sample and the evaluator; the repeatability of the panel was considered from the individual repeatability of the evaluators; And for reproducibility of the panel, the interaction between evaluators and sessions was considered (ABNT, 2016).

4 RESULTS AND DISCUSSION

4.1 *Training of evaluators*

Before beginning the training, the evaluators participated in two one-hour meetings to define the attributes and intensity on the scales with examples of textile samples. The references were determined in accordance with the definitions and proposition of the inexperienced panel. It was also decided the gesture of the touch for each attribute.

After the first training session, a meeting was held to discuss the individual difficulties of the assessment. The greatest difficulty was to remember all the gestural of the touch during the session, so a panel containing information about the attribute, the references of the scales and an image of the gesture of the touch was affixed in the cabin.

The 14 pre-selected candidates participated in several sessions. In each session, the 11 attributes of 7 textile samples were evaluated. Evaluator, sample and repetitions tabulated the data of the training sessions. For the multivariate analysis in ANOVA double-factor, it was used repetition data in three sessions (Ellendersen & Wosiacki, 2010). These data were used to gauge the performance of the panel as a whole and of the evaluators individually.

4.2 *Panel performance monitoring*

Panel performance was monitored and data from every three sessions were tabulated and analyzed in ANOVA. Each session was held once a week.

Table 1 shows the first results of significance levels ($p \leq 0.05$) of F_{sample} for each evaluator in relation to each attribute. It was noticed that the evaluators P1 and P12 obtained low performance compared to the others in four attributes. This indicates that the evaluators have low discrimination power therefore requiring more training sessions on these attributes. Table 2 shows the repeatability results. If the evaluator presents a significant p-value at $p \geq 0.05$, it is a sign that the evaluator failed to assign the same score for the same sample in three different sessions. The P4, P6, P11 and P12 evaluators performed well in repeatability of the seven samples in relation to the 11 attributes, however the P5 and P10 evaluators had poor performance necessitating more training sessions. In this first phase it was verified that the evaluators have more difficulty in memorizing the scores assigned to each sample in relation to the attributes.

Table 3 and 4 present the final panel performance after 4 months of training. It is interesting to note that the evaluators who present less power of discrimination are not the same ones that present low repeatability. This makes it difficult to remove evaluators from

Table 1. Significance level to assessors for discrimination of the samples – 1 fase.

ASSESSORS	COLD WARM	FALLING	LIGHT HEAVY	THIN THICK	RIGID	DRY HUMID	ELASTIC	ROUGH FLAT	PLUSH	RUGGED	SOFT
P1	0,000206	0,730357	0,080217	0,002710	0,011675	0,812813	0,024610	0,000337	0,000000	0,168055	0,003811
P2	0,000119	0,000003	0,000001	0,000009	0,002334	0,000008	0,000023	0,000015	0,000506	0,003693	0,000041
P3	0,000000	0,000021	0,000000	0,000000	0,000000	0,003601	0,000000	0,000003	0,000043	0,000003	0,000009
P4	0,000095	0,001719	0,000381	0,000002	0,000000	0,046397	0,000000	0,012564	0,000000	0,000721	0,001827
P5	0,019662	0,000000	0,013058	0,000186	0,000067	0,000755	0,000007	0,003641	0,000000	0,000113	0,000051
P6	0,000000	0,000000	0,000000	0,000000	0,000000	0,000000	0,000000	0,000000	0,000000	0,000000	0,000000
P7	0,000005	0,000000	0,021489	0,000000	0,000000	0,000000	0,000739	0,000000	0,000000	0,000000	0,000000
P8	0,000069	0,032933	0,002341	0,000002	0,000000	0,001510	0,000041	0,000033	0,000000	0,000491	0,000007
P9	0,000029	0,000002	0,000238	0,000000	0,000174	0,000305	0,000001	0,000165	0,000000	0,000001	0,000000
P10	0,000023	0,000000	0,000405	0,000000	0,000609	0,011591	0,033182	0,000000	0,000020	0,000226	0,000193
P11	0,000000	0,000000	0,000000	0,000000	0,000000	0,000000	0,000000	0,000000	0,000000	0,000000	0,000000
P12	0,000673	0,982763	0,244252	0,055034	0,000479	0,000104	0,000001	0,956202	0,000000	0,023648	0,000000
P13	0,000064	0,366788	0,000030	0,000001	0,000000	0,004874	0,000145	0,000609	0,000000	0,000016	0,000016
P14	0,012760	0,000001	0,000091	0,000261	0,001766	0,038472	0,000000	0,000014	0,007347	0,000232	0,000007

Table 2. Significance level to assessors for discrimination of the repetitions – 1 fase.

ASSESSORS	COLD WARM	FALLING	LIGHT HEAVY	THIN THICK	RIGID	DRY HUMID	ELASTIC	ROUGH FLAT	PLUSH	RUGGED	SOFT
P1	0,015625	0,114012	0,110627	0,367565	0,085392	0,784104	0,068813	0,183457	0,117649	0,001178	0,236832
P2	1,000000	0,778773	1,000000	0,039841	0,775571	0,000863	0,003590	0,778773	0,008897	0,980231	0,897528
P3	0,396569	0,022451	0,295926	0,883631	0,286703	0,896150	0,168047	0,082552	0,396569	0,853932	0,761788
P4	0,810140	0,832585	0,479788	0,746215	0,548537	0,916393	0,746215	0,074279	0,396569	0,396569	0,357273
P5	0,000803	0,006755	0,041036	0,329713	0,173405	0,102598	0,883631	0,913099	0,396569	0,000891	0,975759
P6	0,396569	0,396569	0,396569	0,396569	0,396569	0,396569	0,396569	0,396569	0,396569	0,396569	0,396569
P7	0,548537	0,054675	0,001658	0,396569	0,168047	0,000544	0,977056	0,840360	0,396569	0,015052	0,396569
P8	0,301894	0,653403	0,949163	0,840360	1,000000	0,463453	0,543872	0,083457	0,883631	0,332410	0,027197
P9	0,244378	0,706427	0,520935	0,063470	0,631251	0,019681	0,623842	0,840360	0,908462	0,775571	0,109091
P10	0,074968	0,000375	0,000233	0,000863	0,072309	0,150494	0,684495	0,000235	0,917289	0,056279	0,370855
P11	0,396569	0,396569	0,396569	0,396569	0,396569	0,396569	0,396569	0,396569	0,396569	0,396569	0,641050
P12	0,425752	0,493270	0,054465	0,299382	0,960336	0,111350	0,452471	0,418550	0,396569	0,092663	0,531441
P13	0,025170	0,742049	0,872068	0,908462	0,032401	0,481264	0,423146	0,755215	0,396569	0,572042	0,028088
P14	0,127670	0,029909	0,039611	0,377150	0,002956	0,063700	0,334898	0,055357	0,396569	0,289715	0,396569

Table 3. Significance level to assessors for discrimination of the samples – 4 fase.

ASSESSORS	COLD WARM	FALLING	LIGHT HEAVY	THIN THICK	RIGID	DRY HUMID	ELASTIC	ROUGH FLAT	PLUSH	RUGGED	SOFT
P1	0,000000	0,000000	0,000052	0,000000	0,000011	0,000308	0,000000	0,000001	0,000000	0,000014	0,000000
P2	0,000000	0,000000	0,000054	0,000010	0,000000	0,000010	0,000000	0,000000	0,000000	0,000000	0,000000
P3	0,000000	0,000000	0,000017	0,000000	0,000000	0,000002	0,000000	0,000000	0,000000	0,000000	0,000000
P4	0,000000	0,000000	0,000068	0,000694	0,000000	0,011690	0,000000	0,000000	0,000000	0,000000	0,000001
P5	0,000000	0,000000	0,000004	0,000007	0,000000	0,010247	0,000001	0,000011	0,000000	0,000000	0,000000
P6	0,001156	0,000000	0,007170	0,000006	0,000070	0,000090	0,000000	0,000000	0,000000	0,000000	0,000000
P7	0,000000	0,000000	0,000010	0,000010	0,000000	0,000002	0,000000	0,000001	0,000000	0,000000	0,000000
P8	0,000007	0,000018	0,000045	0,000022	0,000002	0,000315	0,000002	0,000000	0,000000	0,000000	0,000000
P9	0,000000	0,000000	0,000000	0,000000	0,000000	0,000175	0,000000	0,000000	0,000000	0,000000	0,000000
P10	0,000000	0,000000	0,000001	0,000004	0,000000	0,025615	0,000003	0,000002	0,000000	0,000001	0,000000
P11	0,000005	0,000000	0,000007	0,000172	0,000004	0,000021	0,000002	0,000000	0,000000	0,000000	0,000000

the panel, and so there is also a need for individual monitoring of the evaluators.

4.3 *Individual monitoring of evaluators*

The individual performance of the evaluators was monitored simultaneously with the overall panel performance monitoring. Therefore, the same averages used to monitor overall panel performance were used. The line graph was used to verify interactions and the individual performance of the evaluator in relation to the panel. The results presented in this work are the average scores of the fresh and warm attribute for seven samples.

Table 4. Significance level to assessors for discrimination of the repetitions – 4 fase

ASSESSORS	COLD WARM	FALLING	LIGHT HEAVY	THIN THICK	RIGID	DRY HUMID	ELASTIC	ROUGH FLAT	PLUSH	RUGGED	SOFT
P1	0,334898	0,364648	0,897528	0,235282	0,695067	0,684495	0,840360	0,361856	0,548537	0,746215	0,087791
P2	0,897528	0,746215	0,561027	0,506631	0,684495	0,396569	0,396569	0,262144	0,108366	0,116057	0,840360
P3	0,334898	0,281435	0,623842	0,531441	0,437916	0,840360	0,281435	0,618625	0,396569	0,746215	0,746215
P4	0,108366	0,908462	0,746215	0,944201	0,787854	0,765895	0,474361	0,641050	0,396569	0,746215	0,709673
P5	0,087791	0,641050	0,182652	1,000000	0,883631	0,924564	0,094858	0,656781	0,168047	0,746215	0,132810
P6	0,100464	0,506631	0,678934	0,323579	0,781885	0,249906	0,235282	0,396569	0,493270	0,897528	0,364648
P7	0,396569	0,803960	0,849671	0,327785	0,168047	0,235282	1,000000	0,746215	0,396569	0,448795	0,396569
P8	0,618625	0,281435	0,313415	0,661029	0,557255	0,535346	0,888664	0,172403	0,349510	0,573753	0,329713
P9	0,235282	0,746215	0,235282	0,235282	0,641050	0,840360	0,262144	0,641050	0,506631	0,235282	1,000000
P10	0,840360	0,094858	0,548537	0,157267	0,281435	0,281435	0,539483	0,599927	0,641050	0,329713	0,061678
P11	0,917289	0,249906	0,746215	0,671212	0,696761	0,147973	0,143088	0,474361	0,396569	0,185368	0,924564

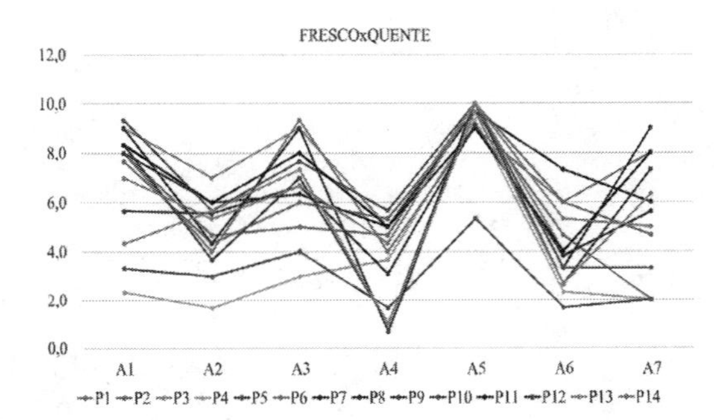

Figure 1. Average scores for fresh and hot attribute of 7 samples – phase 1.

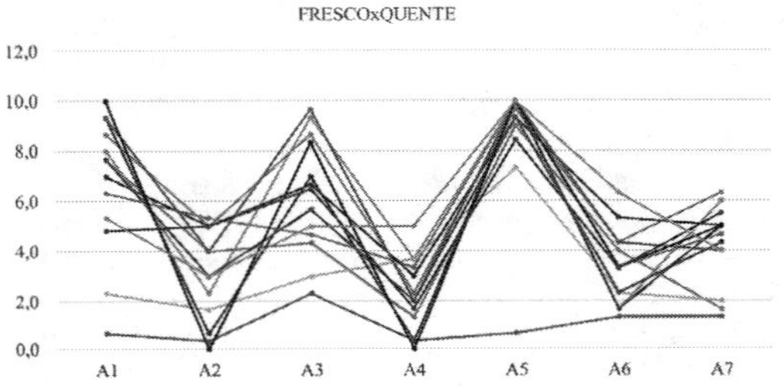

Figure 3. Average scores for fresh and hot attribute of 7 samples -phase 3.

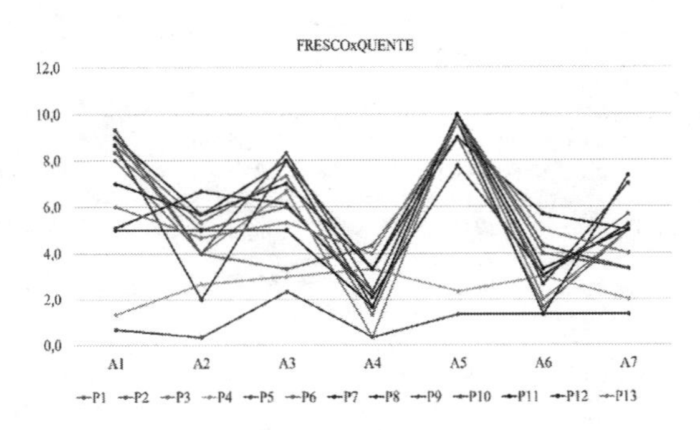

Figure 2. Average scores for fresh and hot attribute of 7 samples -phase 2.

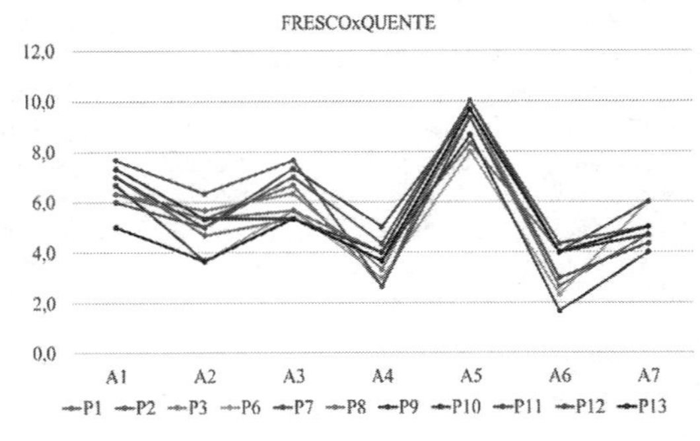

Figure 4. Average scores for fresh and hot attribute of 7 samples -phase 4.

Figure 1 shows the average of the first three sessions, which shows a poor performance, both in the discrimination of the samples and in the use of the scale by the evaluators, except for the sample A5 where only the P5 evaluator does not follow the trend of the panel. Evaluators P4 and P5 rated on average low scores and did not follow the panel score tendency. Differently from what was presented in Tables 1 and 2 where the P4 evaluator showed good overall performance.

Figure 2 shows individual performance after two months of training. In this phase the evaluators present a small improvement in relation to the discrimination between the samples, however the evaluators P4 and P5 still do not agree with respect to the panel. The evaluator P13 requested to be released from the project.

Figure 3 shows an improvement in the performance of the evaluators in samples A4 and A7. In the three monitoring phases of the panel training, the P5 evaluator scored in the same proportion, indicating good repeatability and homogeneity as indicated in Tables 1 and 2. The P4 evaluator, although not agreeing with the panel, also obtained good discrimination of the samples, except for sample A5 in phase 2. At this stage it is noticed that the P4 evaluator inverted the score in the evaluation scale. Thus, at the end of three months the evaluators P4 and P5 were not selected.

Figure 4 shows the last training phase, which shows generally good agreement for all samples.

5 CONCLUSIONS

This study was conceived as a foundation for an applied research in the evaluation of the tactile comfort of caps, where the sensorial evaluation of textile surfaces depends on the performance of well trained professionals to perform sensorial analyzes in a uniform way. This standardization may enable the use of the same techniques among different regions of Brazil, making the sensory analysis data comparable and reproducible

A panel with eleven Brazilian textile tactile evaluators was formed. The selection procedure aimed to measure the ability of the evaluators in relation to discrimination, repeatability and agreement with the group. In this work the methodology used for the recruitment, selection and training of evaluators was presented. During the panel training process, the meaning, the tactile gestural procedure and the references of each attribute in a previously developed lexicon were defined; this helped the evaluators to familiarize themselves with the attributes.

This trained panel will be responsible for quantifying the tactile attributes of the textile surface of caps produced in the city of Apucarana – Brazil. The data obtained from this evaluation will be compared with data obtained in the physical textile tests carried out in the laboratories of the Textile Science and Technology Center of the University of Minho. The expectation of the project is to help the designer explore the textile surface for the development of products that meet the consumer's desirable attributes.

ACKNOWLEDGMENT

This work is financed by FEDER funds through the Competitivity Factors Operational Programme – COMPETE and by national funds through FCT – Foundation for Science and Technology within the scope of the project POCI-01-0145-FEDER-007136.

REFERENCES

ABNT, 2013. *ABNT NBR ISO 4120 – Análise Sensorial - Metodologia – Teste Triangular,* Rio de Janeiro: Associação Brasileira de Normas Técnicas.

ABNT, A. B. d. N. T., 2016. *NBR ISO 11132: Análise sensorial – Metodologia – Guia para monitorar o desempenho de um painel sensorial quantitativo,* Rio de Janeiro: Associação Brasileira de Normas Técnicas.

Bacci, L. e. a., 2012. Sensory Evaluation and Instrumental Measurements to Determine Tactile Properties of Wool Fabrics.. *Textile Reserch Journal,* pp. 1430–1441.

Chollakup, R. e. a., 2004. Tactile Sensory Analysis Apllied to Silk/Cotton Knitted Fabrics. *International Journal of Clothing Science and Technology,* pp. 132–140.

Ellendersen, L. S. N. & Wosiacki, G., 2010. *Análise sensorial descritiva quantitativa: estatística e interpretação.* 1ª ed. Ponta Grossa: Editora UEPG.

Esteves, E., s.d. *Notas sobre a Selecção e Treino dum Painel de Provadores para Análise Sensorial de Produtos Alimentares,* Algave: Universidade do Algave.

Guest, S. & Spence, C., 2003. What Role Does Multisensory Integration Plau in the Visiotactile Perception of Texture?. *International Journal of Psychopysiology,* Volume 50, pp. 63–80.

ISO, 2005. *ISO 6658:2005 – Sensory Analysis – Mothodology – General Guidance,* Genova: International Organization Standardization.

Lawlees, L. & Civille, G., 2013. Developing Lexicons: a review. *Journal of Sensory Studies,* Volume 28, pp. 270–281.

Li, Y. & Wang, Z., 2005. Thermal Sensory Engieneering Design of Textile and Apparel Products. Em: *Environmente Ergonomics: The Ergonomics, and Performance in the Thermal Environmente.* s.l.:Elsevier Science, pp. 473–476.

Nagamatsu, R., Abreu, M. & Santiago, C., 2016. *O Desenvolvimento de Léxico para Análise Sensorial Têxtil Brasileiro.* Buenos Aires, 3 International Fashion and Design Congress, pp. 2112–2117.

Nagamatsu, R. N., Abreu, M. J. A. M. & Santiago, C. D., 2017. *Tactile feeling of textile: a comparative study between textile comfortable attributes of France, Portugal and Brazil.* Aachen, Fiber Society.

NF EN ISO, 2014. *NF EN ISO 8586: Sensory analysis – General guidelines for the selection, training and monitoring of selected assessors and expert assessors.* La Plaine Saint-Denis: Association Française de Normalisation.

Nogueira, C., 2011. *Análise Sensorial de produtos Têxteis,* Guimarães: Universidade do Minho.

Philippe, F., Shacher, L., Adolphe, D. & Catherine, D., 2004. Tactile Feeling: sensory analysis a'pplied to textile goods. *Textile Research Journal,* pp. 1066–1072.

Sabir, E. & Doba Kadem, F., 2016. Comfort and Performance Propieties of Raised and Laminated Denim Fabrics. *Fibres & Textiles,* 24(5), pp. 88–94.

Sztandera, L., Cardello, A., Winterhalter, C. & Schuts, H., 2012. Identification of the Most Significant Comfort Factor for Textiles from Processing Mechanical, Handfeel, Fabric Construction and Perceived Tactile Comfort Data. *Textile Research Journal.*

Teixeira, L. V., 2009. Análise Sensorial na Indústria de Alimentos. *Rev. Inst. Latic. Candido Tostes,* 366(64), pp. 12–21.

Yenket, R., Chambers IV, E. & Gatewood, .., 2007. Color has Littles Effect on Perception of Fabric Handfeel Tactile Properties in Cotton Fabrics. *Journal of Sensory Studies,* pp. 336–352.

Textiles, Identity and Innovation: Design the Future – Montagna & Carvalho (Eds)
 ISBN 978-1-138-29611-4

The interactivity between sales channels on omni-channel retail

Maria José Abreu & C. Daniela Miranda
Department of Textile Engineering, University of Minho, Guimarães, Portugal

SUMMARY: The change in retail, derived from evolution of technology and consumer behavior, leads to a constant upgrading of the strategies employed by retailers on their brands, requiring, in this way, research initiatives on this transformation. This article discusses the importance of interactivity and integration of sales channels in the omni-channel retail, providing a literary review and a contextualization on terms and concepts of this discipline.

1 INTRODUCTION

The walls between the offline world and the online world are collapsing. What is virtual? What is physical? Nowadays, the customer can initiate him purchase on brand's site (through a mobile device for exemple), and raise the product in the physical store (click concept & collect), or he you can start him purchase on the mobile device (outside or inside the store), with the purpose of search products, compare prices, showing the availability in store, the physiological characteristics of the article, among other numerous possibilities that these media offer and finish the purchase in physical store or virtual store.

This opportunity to transition from channel to channel that allows the client to make him buying process and conveniently search, due to technological developments, in particular, mobile devices, social networks and related software-stroke with these applications at the same time, allows for different approaches and strategies for retailers implement your marks, giving rise to new points of contact with the consumer (Agis, 2012). These business strategies have to respond to the progress, not only the technology but also the consumer, who lives and evolves in this change. It's a ubiquitous consumer that wants a complete and unified shopping experience.

Daniel Agis (2012) designates this new technological era as Retail 3.0, where opera creativity of 360° and where all elements work in synergy (Agis, 2012, pp. 83–84). Is in this face of retail that ascends the omni-channel concept which consists in the integration of all sale channels of a brand with a purpose to establishing a relationship with the consumer during 24 hours a day to get in touch with the fashion brand that he want to (Agis, 2012; Harris, 2012).

This article aims to contextualize the reader through the literary review of the omni-channel concept, since there is a shortage of information on this subject, mainly in the retail market. Once, the omni-channel concept is relatively current, your study presents appropriate and relevant (Coelho, 2015).

The article is divided into three chapters. The first introduces the study of generic form, describing the theme of research, syntactically the problem and the importance of it. The second chapter is about the literature review and the contextualization of the omni-channel and the interactivity between the online world and the offline world. Finally, the third chapter concerns the conclusions and future perspectives.

2 THE OMNI-CHANNEL WORLD

2.1 *The evolution of omni-channe retail*

Study the retail's evolution is the best way to contextualize and learn about the omni-channel concept (Harris, 2012).

The Retail 1.0 occurred in the era of pre Internet, where he succeeded the uni-channel concept. At this stage, was used a single channel of relationship between brand and client: a physical store, resulting in a major influence on consumers, providing the establishment of strong ties and offering a familiar and personalized service (Harris, 2012; Agis, 2012, p. 76).

The emergence of the Internet and the introduction of Information and Communication Technologies (ICT) possibility in new approaches to bring the brand to the consumer, amounting, in this way, for Retail 2.0, providing also the development of online channels (Agis, 2012, p. 78). At this stage, retailers had to restructure their strategies, studying what the sales channels and communication that their brands should be present: online channels and/or offline channels. Thus, the uni-channel strategy has evolved into a multi-channel strategy, where the channels operate separately, in other words, retailers work contact points in independent way, allowing customers multiple accesses (Harris, 2012; Verhoef, et al., 2015).

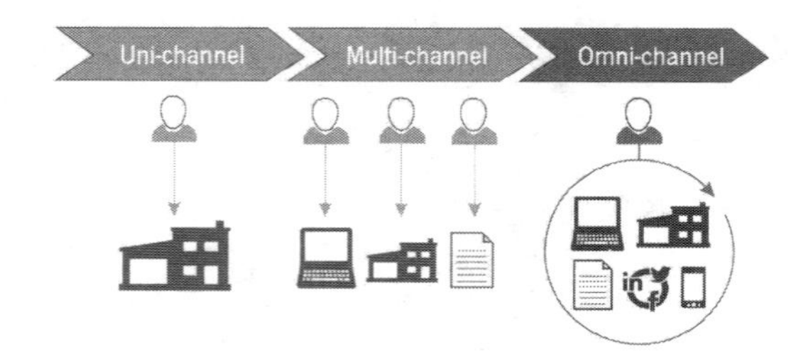

Figure 1. The evolution of the concepts and business strategies Source: Adapted from (Fonseca, 2015).

The retail concept has continued to evolve with the emergence of mobile channels, social media, and software that are directly related to these applications in the integration of these new channels with the channel online and offline channel (Verhoef, et al., 2015). In this way, happens the metamorphosis into the Retail 3.0, as an era where operates creativity of 360°, going through all the stages of production and marketing of products and where all the factors are intertwined in today (Agis, 2012, pp. 83–84), contributing to the mutation of the multi-channel concept to the omni-channel concept where is possible watch the union of all sales and communication channels in a single and unique point of interaction with the consumer, allowing to enjoy a unique and direct experience with the brand through the middle that he wishes to (Harris, 2012; Rigby, 2011).

In comparison to the multi-channel concept, the omni-channel operates more channels and the distinctions between these tend to disappear, turning the "world" in an environment without walls/barriers (Brynjolfsson, et al., 2013; Verhoef, et al., 2015).

In Figure 1 it's possible see the evolution of business strategies in retail and the respective sales and communication channels.

It is necessary to take into consideration that the omni-channel strategy is extremely complex and requires a large investment on the part of retailers, depending on their goals. The omni-channel implementation requires large investments in financial and human resource and also a great understanding of the results that the business model will provide to organisation (Kersmark & Staflund, 2015).

This strategy can be considered a competitive advantage in relation to other strategies used by multi-channel retailers and those who operate exclusively online, once these do not offer an integrated and unified service. Channel optimization offers a full potential by taking the greatest advantage of the physical store and the online store in order to expand and offer consumer experiences, encouraging the purchase and empathy with the brand (Kersmark & Staflund, 2015).

2.2 *Omni-channel concept*

Through an analysis of the semantics of the word omni-channel, which means every channel (multichannel), since the word "omni" is derived from the Latin and means "all"/"universal", while the "channel" means the channel. In addition, all definitions of this concept, have the same foundations and notions of experience versus integration through all channels.

Lazaris (2014) claims that the omni-channel concept was discussed primarily at IDC's Global Retail Insights in 2009 by Parker & Hand and Ortis & Casoli who observed the evolution of multi-channel consumer to a consumer omni-channel, which takes advantage of all channels to contact the tag instead of using them independently (Lazaris & Vrechopoulos, 2014).

In the academic world, Rigby was the first to address the term omni-channel in 2011, claiming to be an integrated experience with the ability to combine the advantages of physical store with the abundant information that an online offers. In this way, allows the retailer interact with the consumer through a huge number of channels (Rigby, 2011; Lazaris & Vrechopoulos, 2014).

Forbes (2014) argues that the omni-channel is the ability of results articulation and information between all channels (Forbes, 2014).

Daniel Agis (2012) underlines that the omni-channel is the "integration of all online and offline channels into a single distribution and marketing strategy" in which your goal is to establish the necessary synergies for a 360° relationship with consumers (Agis, 2012, p. 85). In this way, the union of all brand's channels can establish a conversation and a research and purchase process with their client without barriers.

Daniel Agis (2012) presents six key points on integration between channels (Agis, 2012, p. 86):

- Physical stores are operating terminals for virtual stores;
- Online stores replace the offline stores that has no influence on the image of the brand or operational utility;
- Physical stores can be smaller, optionally without losing commercial operation;
- ICT develop innovative offline formats and operating in Union with the online channel;
- Exclusive online stores open Flagship Stores to enhance his own image;
- Both offline and online platforms, the same brand, help to raise levels of customer traffic from one to the other.

The omni-channel concept offers the retailer the possibility to establish a conversation with the consumers, anywhere in the world, in real time and regardless of the medium used: virtual space (mobile and web devices) or physical space, reinforcing the relationship of trust between the consumer and the brand (Agis, 2012, p. 85),

Today, the consumer interacts with all channels during three distinct stages of the purchase process: pre-purchase, purchase and post-purchase

(Forbes, 2014). Thus, this concept needs to create and manage systems, solutions and operations that work in harmony with the goal of providing a uniform shopping experience, captivating the consumer, whereas the sales and customer support. The objective is to create a platform that offered logistical processes, distribution, database, marketing and communication, to accompany the client 24 hours a day, so that access to the brand the way he want and when he want to (Harris, 2012; Agis, 2012).

On a world scale, a brand that wants to be competitive in the face of their competition requires implementing an omni-channel strategy, which requires a lot of changes in terms of infrastructure and the application of ICT, but will enable the origin of new concepts and strategies in the same (Frazer, 2014).

On application of omni-channel strategy, retailers must take as true goal provide their consumers a unique experience by integrating all channels. Thus, in the implementation of this strategy, another crucial point to consider is all sensory sensations to be explored in the integrated channel marketing experiences. This offers a unique experience for consumers to feel, think, act and interact with the environment, the concept and the history of the brand (Frazer, 2014).

On integration of both channels, retailers have to adapt the best practices for their business, including the price, the shopping experience and building customer relationships. In this new competitive environment there are numerous strategies to achieve success, depending on the product, the level of demand and the target audience of the brand (Brynjolfsson, et al., 2013).

The omni-channel strategy along with the use of mobile devices, one of the main drivers of this phase, allows the consumer have information during the purchase process. Consumer decisions can be molded through the information that brand's channels offer (Brynjolfsson, et al., 2013).

2.3 *Sale channel in omni-channel world*

The channels are points of contact between a brand and they consumer, in other words, are the means by which brand establishes a dialogue with the client, creating an interaction. They are interchangeably used by the consumer during the search and purchase process, facilitating the experience in acquiring the article (Brynjolfsson, et al., 2013).

The omni-channel strategy offers consumers a wide range of channels to access a brand, providing information during the process of researching and purchasing decision. These decisions, for other side, can be molded through the information that the brand channels offer (Brynjolfsson, et al., 2013). For example, the consumer is in a physical store and search on him mobile device the availability of a product in the store and/or in other colours can also find lower prices at other stores.

When developing the strategy of sales channels the retailer has several options to implement on him brand to offer this possibility of interactivity to the client, taking into account him objectives (Verhoef, et al., 2015).

In the omni-channel world barriers between the channels tend to disappear, reinforcing the need for interaction between sales channels and communication channels (Verhoef, et al., 2015).

2.4 *Physical channel (offline)*

Physical stores are locals that provide high quality sensory and experimental level that no other medium is able to offer, being inborn and irreplaceable. Customers want to touch, feel, experiment and live the atmosphere of the store, that is, the physical store must explore the five senses sensory, providing, entertainment and personal experiences (Piotrowicz & Cuthbertson, 2014; Zhang, et al., 2010).

This channel allows retailers a better aproximity with their costumers and also offers customization. In addition, the consumer can purchase the product instantly, if this is available in the store, without having to wait for your reception. He can touch, feel and try out the product before purchase, choosing to pay cash physical which is not possible with the purchase online, this being one of the main obstacles online purchase: the impossibility of physical cash (Rigby, 2011; Zhang, et al., 2010; Bell, et al., 2014).

On the other hand, taking into account the benefits cited, consumers need to spend time, energy and money to visit a physical store and, in some situations, these spaces may not be open in convenient hours (Zhang, et al., 2010; Bell, et al., 2014).

However, the offline stores have to be accessible to consumers and large enough to hold the stock. The retailer has to take the products to acquire certain stores at the right time and, for this purpose, it is necessary to anticipate and meet them client and decide which are the best products for each location/region (Bell, et al., 2014).

In this new era of retailing, it is not expected the end of physical stores but a rehabilitation and redefinition of new strategies. According Stephens (2015), these points of contact with the consumer have the potential to become powerful communication points where retailers can articulate and tell the history of the brand, providing distinctive experiences about the products and the brand. In the final step of the process, the client makes the purchase by the channel that they want (online or offline channels) (Stephens, 2015).

Thus, retailers will change the perspective about their physical stores as a place for the exhibition of brand's products, and will face them as communication points in the history of the brand. They will enjoy the physical store to improve the customer shopping experience, taking advantage of the technological means to offer something truly unique, memorable and differentiator (Stephens, 2015).

In this way, the stores will be invaded by new technologies. Technologies brought by consumers, including mobile devices. Technologies implemented by retailers if direct to customer experiences, such as interactive screens. Technology for the employees who assist the purchasing process, such as tablets. Technologies that help the layout of the product, such as screens and virtual corridors, digital signage, self-service kiosk, vending machines, code QR Code readers that allow consumers to obtain information on the brand and/or product (product knowledge, functionality, availability in the store, prices, among others) (Piotrowicz & Cuthbertson, 2014).

In this perspective, the physical stores will be omni-channel centres, in view of the customer across multiple channels. In this way, the shops will be commercials, as well as to expose social media spaces will emerge allowing the customer to expose in real time their opinion, evaluation and product comparison (Stephens, 2015).

However, it is important to take into consideration that not all consumers are familiar with new technologies, some prefer and feel more comfortable with traditional methods (personal interaction - face to face) at the point of sale, in other words, with traditional cash payment options and personal customer service. In this way, retailers must have a deep knowledge about their target audience and choose which strategies and channels are best for their consumer (Piotrowicz & Cuthbertson, 2014).

2.5 *Virtual channel (online)*

The retailer, through the virtual channel, gives the consumer the possibility to interact with the brand 24 hours a day, where, when and the way him want, allowing the reduction of costs in time and travel (Zhang, et al., 2010).

The purchase in the online world allows consumers to save time during the purchase process, through the easy entry and exit from the site and rapid payment process (Rigby, 2011; Bell, et al., 2014). However, this brings disadvantages, such as the waiting time of the reception of the product and, in some cases, the shipping costs (Bell, et al., 2014).

In addition, the virtual channel presents, as a rule, almost the products collection and provides detailed information about the same compared to the physical channel (Rigby, 2011; Zhang, et al., 2010). As well as the ability to compare prices and offers from brand to brand (Rigby, 2011). In this way, the consumer can conduct research about a particular article and obtain all the necessary information before make the purchase.

The virtual channel also allows the possibility to share in real time the costumer's opinions and get feedback from other customers, friends and family (Zhang, et al., 2010).

On the other hand, this also provides consumers think and reflect on their purchases in the comfort and safety of their homes or at a location of their choice (Zhang, et al., 2010).

2.6 *Mobile channel (use of mobile advise)*

Mobile devices help overcome the barriers between the online world and the offline world as well, the need for integration between channels is evident when the mobile technologies, mainly the consumer's devices are used in the physical store. In omni-channel world, the customers can use their mobile devices inside the store in order to conduct research, compare products, ask for advice, look for alternatives and perform reading QR code (Piotrowicz & Cuthbertson, 2014).

Therefore, mobile devices have been contributing to the fall of barriers between the traditional retail and the Internet, offering multiple points of contact. However, provides information in the online world with the buying process of the offline world, merging the information touch-and-feel in the physical world with the online content in the digital world, allowing greater interaction between the brand and the consumer (Brynjolfsson, et al., 2013).

Consumers want more flexibility of the use of mobile applications, with the pace of life faster, these want to shop where, when and as they wish. In addition, this technology is an aid in the integration of purchase between all channels (Doherty & Ellis-Chadwick, 2010).

Thus, the use of mobile devices is on the rise, offering a more convenient experience omni-channel concept, allowing mobile payments, offering to buy anywhere and circumstance, and the rise of new forms of investment/security (Smith, 2015).

It is important for brands to adjust and adapt their offline and online channels with the mobile device.

2.7 *World online vs. offline world: capital gains of omni-channel world*

In the omni-channel world, the channel offline and the online channel will not become extinct, but rather complement each other, combining and adapting to new strategies, as noted earlier in this article. On the one hand, the online world offers consumers extremely creative and developed skills at the level of communication that no other channels have the ability to offer. On the other hand, the offline provides real, emotional and psychological experiences, that no other offers, with the advantage of exploring all the senses of the human, being an entertainment venue, sharing, self reward and even altruism, that is, when the consumer buys a physical store has the entire shopping experience, achieving direct interaction with product (Agis, 2012, p. 86).

The offline world will not be sufficiently competitive, and it is necessary to create a relationship between all sales channels. The offline channels are supported by online channels, through technology, they are often taken by consumers into the physical store (mobile devices), making these websites to purchase more efficient and innovative. All this will provide dynamic, as noted, a decrease in physical stores without these lose your business value (Agis, 2012, p. 86; Rigby, 2011).

The online world has progressed powerfully in last year's and, on the other hand, the offline world continues to be relevant and indispensable. In this way, Dorman has developed a study that concluded that the two worlds complement and strengthen the brand image, both are extremely relevant to the omni-channel, offering and providing a positive connection to the consumer. There is a synergy between both worlds, namely the offline channel is not falling in the obsolete, due to the explosion of the online world, but rather be a supplement (Dorman, 2013).

Consumers already are omni-channel, they use both online and offline channels in the buying process (Bell, et al., 2014; Piotrowicz & Cuthbertson, 2014). They require a range of choices to select the one that best suits their needs in the purchase rather than an individual strategy of the online world and the offline world as multi-channel world.

In addition, in the physical store and/or virtual store, customers look for increasingly unique and enriching experiences. Want to experience increasingly integrated into the online world invades the physical store, namely, retailers must offer an experience ever more customer-centric, providing in the same virtual world with the real world and vice versa (Smith, 2015).

Consumers want everything to which they have entitled: in the virtual world, availability of all products and information about the same and in the real world, personal and personalized service, possibility to play and experiment with the product and make the purchase as a way to experience and fun. But, taking into account the different experiences and possibilities that both worlds offer and provide, it is necessary to take into account all the different consumer profiles and that each value to their shopping experience differently but all tend to strive for the perfect union between all channels (Rigby, 2011).

In Table 1 are shown the advantages that the online world and the offline world provides to the client.

Table 1. Advantages of the online and offline world.

Advantages	
Online world	Offline world
Detailed information about the product	Purchase as an experience
Consumers have access to opinions and suggestions from other consumers	Variety
Social involvement	Ability and opportunity to touch, test and try the product
Choose more diverse by product	Personal help for employees
Price comparison and special offers	Instant access to products
Get out and enter the site quickly	Possibility to make the exchange more easily
Convenience (access when, where and how you want)	Exploitation of all human senses

Source: Adapted from (Rigby, 2011).

2.8 *The features of omni-channel retail in sales channels*

The omni-channel concept encompasses the relationship between a wide range of different retail areas, since the advertising, communications, marketing, logistics, transport services, distribution, sales tools, promotional activities, performance measures, and work with databases (Agis, 2012).

Retailers need to break down the barriers between the online world and the offline world, unifying and providing services that integrate the sales channels. One of the gains in the fall of walls between channels is "click and collect": buy online and pick up the product on a physical store. This tool offers consumers the best of both worlds, complete information about the price and availability before purchase and immediate compliance, without the need to wait for delivery. Other important distribution activities to instruct are buying on physical store and reception of the product in house, the return in the physical store of a purchase made on the site and free deliveries both in store like home (Piotrowicz & Cuthbertson, 2014).

Promotion activities are about information consistent across all sales channels. Customers are cautious about the equality of information across all channels and it is, therefore, crucial to equal prices, locators of the physical stores on the site, availability of the product in store and technology implementation by retailers if direct to customer experiences and provide information about the product, such as QR codes Code, interactive screens, among other (Piotrowicz & Cuthbertson, 2014; Zhang, et al., 2010)

The purchase and sales tools help consumers during the purchase process both in the online world as in the offline world. In this way, the use of new technologies at the point of sale, such as technology brought by consumers and the technologies, in particular, mobile advise, assist in the process of buying, for example, in obtaining product information, availability, location, among others (Piotrowicz & Cuthbertson, 2014). As well as, the possibility of the consumer show purchase extract (IBM, 2014), after the purchase through the site, the consumer has the possibility of knowing the product and when their long to receive.

Besides, the retailer needs to know their consumer, through their purchase and consumption behaviour, their experience and satisfaction with the brand. Thus, it becomes crucial to get online statistics to meet the client, what he wants in order to provide a perfect experience (Zhang, et al., 2010).

Table 2 presents the features and the pillars of omni-channel having regard to the study by Malin Kersmark and Linda Staflund (2015), however, some changes have been made with the literature studied. In this way, the table presents the strategies and activities to be implemented in sales channels. Refers to the essential points that must be implemented to unite and connect the sales channels.

Table 2. Activities and characteristics of the omni-channel retail.

Activities and characteristics of the omni-channel retail
Distribution and logistics
Click-and-collect (order and pay *online and collect in store)*
Buy in physical store and receive the product at home
Return of the product in the physical store of orders made on the website
Free delivery on physical store
Free home delivery
Sale tools and purchase
Use of mobile technologies on the part of the shop assistants in physical point of sale (e.g., tablets)
Use of technologies within the store (e.g. iterative screens)
Possibility to see the extract from the purchase of a product
Online chat to assist consumers in purchasing process
Promote the consumer experience and the integration between channels
Price consistency across all channels
Location of the physical store on the site of the brand
Online information about the availability of the product in the shop
Detailed information about the product (product characterization, available colors, sizes, among others)
Promote the use of mobile devices to consumers in the physical store (e.g. mobile applications, apps, QR Code)
Data collection and consumer behavior
Getting online statistics about consumer behavior on the Web site

Source: Adapted from (Kersmark & Staflund, 2015).

3 CONCLUSION AND FUTURE PRESPETIVES

In omni-channel strategy all sales channels should work. Each channel features its unique characteristics and so the channels complement each other as a whole, being instrumental in promoting consumer shopping experience. All channels have their features and references that are fundamental in the union's aid.

However, retailers must take into consideration the implications that this strategy can provide to their brand. The omni-channel implementation requires a large investment in human, financial and technological resources, as well as, a wide understanding of own brand and their target audience.

It was found that a brand wishing to remain competitive on a world level, their challenge must be based on the development of strategies that aim to unite all sales channels in order to win, find and offer a seamless shopping experience to the customers, taking into account the online world and the offline world.

Given the topicality of the theme would be appropriate to perform specific research in this area. In this way, it would be relevant to study the necessary strategies a brand would have to implement in order to be considered a promoter of omni-channel concept, that is, what does she need to implement and develop. One could classify a company like implement omni-channel strategy asphalt paving to an unsatisfactory level, satisfactory or excellent? Finally, it would be useful to analyze what are the strategies that should be implemented in business to reach the perfect level of Union between channels.

REFERENCES

Agis, D., 2012. Retail 3.0: *Fututo físico e virtual*. s.l.:ATP.

Bell, D. R. B., Gallino, S. & Moreno, A., 2014. *How to win in an Omni-channel world*. MIT Sloan Management Review, pp. 45–54.

Brynjolfsson, E., Hu and Mohammad S. R, Y. J. & S. R, M., 2013. *Competing in the Age of Omnichannel Retailing*. Summer 2013 Research Feature.

Coelho, S. C. P., 2015. *Desafios do omnichannel na aplicação às empresas nacionais*, s.l.: s.n.

Doherty, N. F. & Ellis-Chadwick, F., 2010. *Internet retailing: the past, the present and the future*. International Journal of Retail & Distribution Management, p. 943–965.

Dorman, A. J., 2013. *Omni-Channel Retail and the New Age Consumer: An Empirical Analysis of Direct-to-Consumer Channel Interaction in the Retail Industry*, s.l.: CMC Senior Theses.

Forbes, F., 2014. *Integração online + offline para entender o consumidor*. [Online] Available at: http://www.varejista.com.br/artigos/tendencias/1208/integracao-online-offline-para-entender-o-consumidor [Accessed 2016 Fevereiro 2016].

Frazer, M., 2014. *Omnichannel Retailing: the merginf of the online and offline environment*. Global Conference on Business and Finance Proceedings, Volume 9, pp. 655–657.

Harris, E., 2012. A Look At Omni-Channel Retailing. [Online] Available at: http://www.innovativeretailtechnologies.com/doc/a-look-at-omni-channel-retailing-0001 [Accessed 26 Outubro 2015].

IBM, 2014. IBM Study: *Consumers Willing to Share Personal Details, Expect Value in Return*. [Online] Available at: http://www-03.ibm.com/press/us/en/pressrelease/42903.wss [Accessed 27 Outubro 2015].

Kersmark, M. & Staflund, L., 2015. *Omni -Channel Retailing: Blurring the lines between online and offline*. s.l.:Jonkoping University.

Lazaris, C., 2014. *Exploring the "Omnichannel" Shopper Behaviour*, s.l.: s.n.

Lazaris, C. & Vrechopoulos, A., 2014. *From Multichannel to "Omnichannel" Retailing*: Review of the Literature and Calls for Reasearch. June.

Piotrowicz, W. & Cuthbertson, R., 2014. *Introduction to the Special Issue: Information Technology in Retail*: Toward Omnichannel Retailing. International Journal of Electronic Commerce, Volume 18, pp. pp. 5–16.

Rigby, D., 2011. *The Future of Shopping*. Harvard Business Review.

Smith, J., 2015. *Looking Back and Ahead: What 2015 can tell us about the 2016 retail environment*. Point B.

Stephens, D., 2015. *The Future of Retail is the End of Wholesale*. [Online] Available at: https://www.businessoffashion.com/articles/opinion/future-retail-end-wholesale [Accessed 12 Junho 2016].

Verhoef, P. C., Kannanb, P. & Inm, J. J., 2015. *From Multi-Channel Retailing to Omni-channel Retailing*. Journal of Retailing.

Zhang, J. et al., 2010. *Crafting Integrated Multichannel Retailing Strategies*. Journal of Interactive Marketing, pp. 168–180.

Textiles, Identity and Innovation: Design the Future – Montagna & Carvalho (Eds)
© 2019 Taylor & Francis Group, London, ISBN 978-1-138-29611-4

Technological evolution in fashion product knitwear companies

G. Montagna
CIAUD, Lisbon School of Architecture, Universidade de Lisboa, Portugal

L. Piccinini
Lisbon School of Architecture, Universidade de Lisboa, Portugal

M.P. Carvalhinha
Faculty of Engineering, University of São Paulo, Brazil

ABSTRACT: This article presents the technological evolution of flat knitting machines and their implications for the creative process of Fashion Clothes Knitwear products (VMMR), walking through what the market denominates 3D printing today. The work investigates the impacts of technological advancements for product development teams, understanding how interdisciplinarity arises out of the need for interaction between designers, modelists and programmers so that the ideas may be adequately applied and turned themselves into products through electronic machine software interfaces.

1 INTRODUCTION

This study describes the technological evolution in the field of Fashion Clothes Knitwear (VMMR), considering the evolution of flat knitting machines from the early 20th century and the eventual change in product development processes related to them.

The VMMR products differentiate themselves from other fashion line products for starting directly from the yarn to semi-finished or finished pieces in high versatility and flexibility processes. The current technological grounds for the production of VMMR pieces involve pieces of equipment capable of producing finished pieces from programs developed by specialized professionals based on references from fashion designers and modelists.

From this viewpoint, the design of fashion product in flat knitting machines is, at present, a complex interdisciplinary area which involves, mainly, three professionals of very distinctive characteristics: the designer, the modelist and the programmer.

The ultimate purpose of this process is to yield pieces with different cost relationships *versus* benefit, which may allow for differentiation of the brands they trade before the final consumer.

These results are reached when there is careful listening and understanding of the target public, of the material and creative resources, whether they are conceptual or technological of development of raw materials, colors and shapes. According to RECH (2002). "The designer is the professional responsible for transforming, in a creative and conscious manner, ideas into shapes, through the combination of the trio: technology, materials plus the social context in the sense of pleasing the human being".

The production technological resources are interconnected with product development resources, so that the creative process is fundamentally interconnected with the technology and the links of the supply chain.

The fashion designer, by expanding its universe of knowledge structures and resources of available machinery also receives valuable information in creation, which, in general, can reach greater differentiation of products. In addition, there is also room for the improvement of the creation process, for the choice of raw materials, both in the part of differentiation in the product conception phase, as well as in the part for the search of technical solutions. Designers have been exploiting unique qualities the knitting industry has to offer, by breaking barriers with yarns and uncommon materials, and playing around with the scale, with natural interaction between art, design and new technologies (SISSONS, 2010).

In this article we are going to investigate and analyze the impact of these technological advancements in product processes and development teams.

2 TECHNOLOGIES INVOLVED IN THE VMMR PRODUCT

The choices in the creation of knitwear pieces in the product development of the pieces start from the choice of the yarns and add the possibility of options in blending of materials to form, as well, the very "fabric" of the piece, its textures and colors.

2.1 *The textile technology subdivisions*

The fabrics are developed after yarns or filaments, obeying a series of techniques, whose result is what we have on our body, a product we wear to cover and protect ourselves (CONTI, 2014).

One of the differences between knitwear product design and orthogonal fabric is that in knitwear companies the choices start out in the research of yarns and fibers that will make them up, since they will affect the tactile and visual characteristics of the knitwear product.

The variety of yarns available is wide and with a large variety of fibers, colors and textures. The advancements in the manufacture of yarns have taken to the development of new and uncommon yarns. By conceiving a knitwear product it is key to have good knowledge of the characteristics and composition of the yarns, and to understand the distinction between one type of yarn and the other, since the fiber composition, the weight, the structure and the finish of a yarn can greatly affect its handling with the behavior of the finished fabric. The yarns are built with short fiber lengths or continuous fiber lengths, which are weaved, twisted or bound between themselves to form a continuum or yarn length. Yarns are made of natural or artificial fibers, or a combination of both (BROWN, 2013):

- Yarns used for the manufacture of clothing are basically divided into orthogonal fabrics and knitted fabrics, as described by Mendes (2010) below.
- Knitted fabrics – The knitting mechanism consists of the formation of yarn loops with the aid of fine and sharp needles. The interlacing and the continuous formation of new loops produce the knitted fabrics
- Orthogonal fabrics – They are articles produced in a loom and formed by the alternative perpendicular interlacing by, at least, two sets of yarns, those of warp or net and those of weft.
- The knitted pieces are still subdivided according to process type and machinery that originate them:
- Weft knitting where flat and circular knitting machines are used.
- Warp knitting, use basically Kettenstuhl and Raschel knitting machines.

This article is concentrated in the technologies around the Fashion Clothes Knitwear.

The circular weft knitting technology is used for finer and lighter pieces, such as socks, underwear and sportswear. Flat knitting is used mainly in coats and fashion clothes, since they allow for greater stitching versatility, volumes and material mixture. In the past years both technologies have evolved and blended in a way to enable for greater quality and economical viability to meet the final consumer.

2.2 *Creative processes of VMMR products*

Even before the technological evolution of flat machines, the VMMR product is related with a different creative process of fashion products coming from other textile formations (flat fabrics, warp knit and knits stemming from circular machines), since they allow for greater variability of shapes, stitches, textures and thinness, colors, materials, all these variations can be combined so as to produce very distinctive results and with relatively small production lots.

Figure 1. Examples of different results achieved from creative works after flat knitting (elaborated by the authors).

In figure 1 are examples of flat knitting works.

In fashion design in flat knitwear companies the possibility of creation is infinite, a field of study which has many techniques and experimentations in sculpting the fabrics in three-dimensional shapes, by adding volume through drapery, pleats and applications. Innovating three-dimensional structures can also be created by applying additional shapes in a background of knitwear fabric. This is a time demanding process, but the results can be impressive. The triangles, the flaps, the knots, the braiding, the frills and the stripes can be previously knitted and then incorporated to the knitting while it is being worked on or applied onto the knitted surface after the conclusion. These structures can be obtained by the contrast of pieces produced with fabrics made mostly with thick yarns and yarns of finer thicknesses, as well as in the exploitation of the physical characteristics of the fiber, for example, working with elastic yarns with excellent stretching properties which add shape and structure which mold the structure of the piece. There are elastic yarns in the market perfect for knitting with other yarns or to embellish the knitted surface (BROWN, 2013).

According to Neves (2010), designers are expected to have perceptive abilities capable of abstracting the demands requested by their clients, the beneficiaries of their project, always seeking innovations and prospecting ways through the intentional reflections performed in the development of new products. The designer has the possibility of interacting through technological supports and guided objectives: documental objectives which follow the technical specification, such as, samples of raw material and texture, template pieces, and all the supporting objects that aid the communication of the project, composing a creative scenario to this context.

The in-depth study in historical design references, engineering, human behavior study and artistic manifestations are sources of inspiration for the development of new products.

According to Sissons (2010), in the case of knitwear, designers have more and more been exploiting the unique qualities the knitwear has to offer, breaking boundaries with yarns and uncommon materials. There is a natural interaction between art, design and new technologies.

The processes of product creation and development permeate some functions within a style department, which traditionally involve the designer and the modelist.

The stylist has the role of imagining the product, presenting it by means of a drawing and detailing desired aspects of texture, weight, fit, materials, etc. The modelist brings the expectations in a three-dimensional manner into flat parts which will be assembled in order to achieve the result expected by the designer. The interaction between these two professionals is what in fact yields the result. According to Mariano (2011):

> *"For all the projecting and esthetic implications, the modeling of the clothing is intrinsically linked to the concept of design. The "doing" which integrates knowledge and sensibility to conceive and configure clothing products. The modeling professional can be considered a designer, since it detects and solves configuration issues, and also drafts the matrixes which enable the standardization and production of clothing in a large scale".*

In this bipolarized design model, the designer is the part that clearly has the role of capturing an esthetic trend and provokes the final user's desire, the modelist is the part whose role is that of translating the esthetic ideals and shapes to patterns which may compose the product, and both get involved in the search for comfort and functionality of the pieces by the user's perspective.

In knitted clothing, questions are presented in relation to the difficult matching modeling and solution by parts of the product development team. The modelists are used to turn the three dimensions of the human body 3D into two dimensions 2D, i.e., modeling planification. However, the knitwear, by its own soft and elastic characteristic, ends up molding itself in direct relation to the structure of the body, or which tends to demand some tests to the solution of the issue of design.

In addition to these two professionals, in flat knitwear, traditionally there was also the role of the rectiliner, who created upon experimenting combinations of yarns and stitches, working as a craftsman who creates upon "doing".

With the evolvement of machinery technology and that of the VMMR-oriented software, both the role of production as well as the duty of design have changed, as described below.

2.3 *VMMR product manufacture technologies*

One of the first technological breakthroughs in VMMR, in the early 20th century, was the production of continuous knitted parts directly from the flat machines in the shape of the pattern of the piece. This system, called fully fashioned, is a knitting reduction technique and, consequently, of construction of different shapes directly in the machine. For enabling the construction of these shapes directly in the machine, it avoids the use of cutting operations of the textile material. After weaving the parts as front body, back and sleeves these are sewn and finished in one sewing step. Still, they presented an evolution by cutting down the production time and simplifying the manufacturing logistics. In the 1960s, the seamless technology was developed, in which pieces are manufactured in the seamless tube system, eliminating other production steps.

The main flat knitting machine manufacturers are Shima Seiki in Japan and Stoll in Germany – both companies working in the avantgarde of knitting technology to develop products from knitting software to standard computerized flat knitting machines and seamless knitting technology. The Japanese flat knitting industry Shima Seiki produces several knitting systems "WHOLERGARMENT" (trademark), of knitwear with feeler gauges (stitch thickness) from very fine to thick. Stoll produces "KNIT AND WEAR" (trade mark) dressing technology using similar systems and a range of feeler gauges, creating fabrics from thin to the thicker, with a more voluminous aspect, giving the appearance of hand-knit. The knitting machine technology is constantly refined and developed for development and assembling time of the pieces, to increase the production speed and enable the machines to knit a greater variety of stitch structures (BROWN, 2013).

According to Montagna, G, (2012),

> *"When it refers to the term Seamless, it is about products made with looms capable of producing pieces of garment in knitted fabric which adopt themselves to the body and reduce the number of sewing, usually reserved for the finish of hems or application of hemming."*

In the late 1990s the seamless technology was adapted, including the use of techniques and tubular and three-dimensional logics, providing many more options of differentiation, to the point of one coat leaving the machine with finished pockets and button holes.

The quality in flat knitted fabric automated technologies also comes from the regularity of the movements, which are precisely the same in all steps, keeping the force and tension, which produces pieces with more regular textures. This advancement is shown

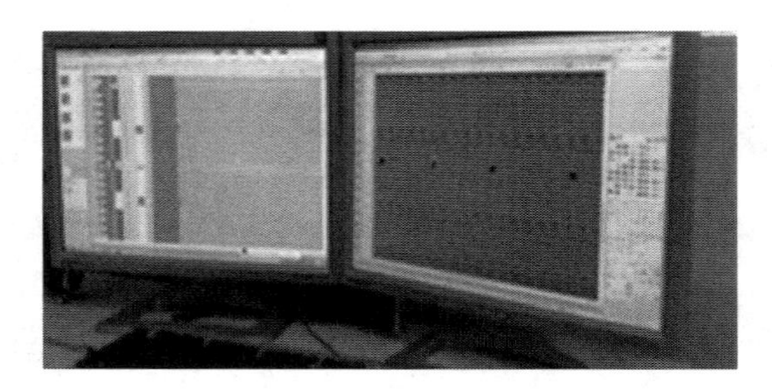

Figure 2. Visualization interface and management of stitches and standards programming. (STOLL, 2014).

in opposition to the technologies prior to automation, in which one person would lead the weaving on the machine, with the force and speed controlled by human movements.

This process for the construction of finished cutless and seamless piece has had significant improvement in the past decade and gained manufacturing speed and quality very similar to the manufacturing in parts, reason for which the industries have more and more been adhering to this new generation. In the manufacture of seamless clothing, its parts of the clothing – front part, back, collars and sleeves, are knitted as tubes, with each component using a separate cone of yarn fed by separate yarn feeders at the same time.

The formation of the piece is worked on by a computer program which, during the knitting process, the components are joined and fused into a piece of clothing. This process results from the production of clothing that is reliable and, as described by Stoll, "a perfect adjustment, providing new freedom of form and design, high quality fabric and the elimination of inconvenient sewing (Brown, 2013)".

The evolution of the VMMR production technology is not only related to the industrial process, but also to the form of how the machine and men relate. When the flat machines stopped being operated by carts, which were moved by human arms, they also stopped having their stitches set by needles repositioned by the operator, and the role of the rectiliner gives room to the flat machine operator.

The electronic machine generation has incorporated software interfaces which receive the production order with all instruction of needle movements which must be followed for each lot to be produced.

These interfaces evolve in a rapid and steady manner. Figure 2 presents a modeling visualization program, stitches and standards available for the programming of the pieces received by the product designer and drawer into patterns by the modelist.

After the development of interfaces for programming the machine, these software have also started to anticipate physical phenomena which could take place during the weaving process of the product, such as retraction, deformation, correction, guide for the finishing process and repair of possible errors.

In several solutions the programmer manages, on the one hand, to work out the commands which define stitches and shapes, and on the other hand, visualize the result in simulations of how the piece will be when ready, weight, fit, among so many others which influence the final visual aspect of the piece. Apart from other technological advancements of the electronic flat looms, machines and programs in the stamping, modeling and cutting areas have also been developed.

In terms of innovation the production of tennis shoes upon flat machines is highlighted. Nike was the pioneer in this innovation, with the Flyknit Technique it has Flywire support. Flywire are yarns that pass through the knitted fabric weftward during the weaving, providing a knitted fabric structure with safer supports. This innovation blends the knitted fabric technique with orthogonal fabric. The knitted fabric is knitted with yarns or Flywire cables in an asymmetric fashion, which involves the foot and provides support. The tennis shoe is knitted with very light yarns in the weft as well as in the warp, with a structure that does not fray to give the sensation of second skin. The Dri-FIT knit is a fabric created to help keep optimum temperature, for better performance in the most varied weather conditions.

Another area in which innovation in clothing is advancing is in intelligent clothing, clothing that has additional active functions relative to the traditional property of the cloth. These functions are obtained through the use of special textiles or electronic devices. These are comprised of intelligent fabrics, textile fabrics capable of feeling and reacting to external stimuli, such as the change of light and temperature. These technological resources introduced in clothing can bring many benefits in the area of health, well-being and the prevention of illnesses, and can present new possibilities for the flat knitted products.

2.4 *New VMMR technologies and 3D printing*

As detailed in the previous item, the production technologies of flat knitted fabric have gone, at the late 20th century, to the development of solutions that produce finished pieces upwards from the yarn, these advancements were associated with the concept of 3D Printing, integrating the flat machines to the edge of the current industrial trends.

The term 3D is originally related to the additive technology, in which material is added layer by layer until it forms the designed object, in an analogous system, but different from the flat machine technologies, which use the yarn as its base:

> *"3D printers (in resin) and milling machines with digital command follow the same technological principle: Guiding the movements of one mechanical device with the aid of a software. The most well-known among them works as a printer, but in three dimensions; passage after passage, a nozzle displaces into three axis and*

Figure 3. Fabric made in knitted fabric with the application of pieces manufactures in 3D resin, Pringle of Scotland, 2015.

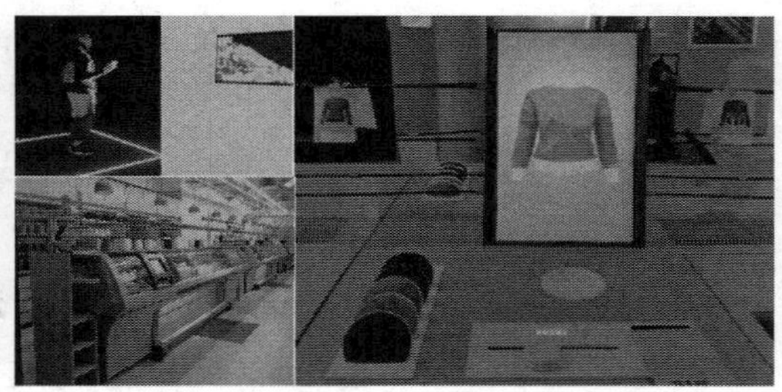

Figure 4. "Adidas Knit For You" store in Berlin, 2017.

overlays layers of material following a digitalized model until it obtains the desired volume." (Soderberg, 2011).

Some initiatives in the clothing industry with 3D printing in resin is also observed, including the complementary form to the traditional applications. One example is the one proposed by the brand *Pringles Of Scotland*, which, in 2015, developed a catalogue with dozens of clothing models with colors and stamps that could be customized online. Specific pieces are then manufactured in 3D printers in resin and attached by hand onto the knitted fabric as presented in figure 3:

The 3D tools existing nowadays facilitate the mass customization as an element of creation of products with innovation and greater worth:

Gilmore and Pine, (2007) identified three different types of customization and it is interesting to notice how the definition of mass customization can be changed to reflect the several possibilities:

- Adaptive Customization – In which there is a customizable standard. The product or service is designed so that the users can change it, by deciding upon some specifications at the moment of purchase or soon after, for a price close to the similar mass product/service.
- Cosmetic Customization – In which a standard product is presented in different form to different clients, i.e., an aesthetic change. This approach is appropriate when the clients use one product in the same way and differ solely in the way they want it to be presented. For example, the product can be showcased in a different way, its attributes and benefits announced in different ways, or the name of the client can be placed in each item.
- Transparent Customization – In which the company provides "unique" goods or services for individual clients, and lets them know explicitly that the products and services have been customized for them.

The most advanced VMMR production technologies can, to an extent, be also considered 3D printing, since, in one single process managed by a piece of equipment, part of the raw material is turned directly into a three-dimensional product, allowing for customizations such as previously described.

Figure 5. Application of "3D printing" for suits at the flagship from Ministry of Supply in Boston, 2017.

One example is the one presented by Adidas at the pop-up store "Adidas Knit For You" in Berlin in the early 2017. In this store, the client enters and faces a rather different structure from a conventional store: a projection room with a body-scanner, a product development table with samples of yarns, knitted fabrics and standards, and an "aquarium" with automatic flat machines and a lean team of finishing production activities (figure 4).

Upon choices of colors and stamps, it is possible to "print" a piece of exclusive flat knitted clothing, tailored, and in a cycle of less than one hour for a process that traditionally takes in the fashion industry more than one year between the design, the production, the distribution logistics and sale.

In an analogous application to Adidas "knit for you", but for the dress-up jackets, the brand Ministry of Supply implanted a flat machine in its flagship in Boston, enabling the client to choose different features of the piece, such as sizes, colors and finishes, which will be printed in 90 minutes and traded for US$ 345 as shown in figure 5.

In the cases of "Adidas" and "Ministry of Supply", some new features are observed: (1) explicit partnership of the retail brand (Adidas / Ministry of Supply) with large machine manufacturing companies (Stoll, Shima Seiki); (2) presence of machine inside the store, enabling the client not only to customize / develop the product, but also to watch the production; (3) recharacterization of a technology consolidated long before, of production of ready pieces, for the use of the term "3D Printing", more aligned with the language of the current customization movements.

3 IMPLICATIONS OF THE TECHNOLOGY FOR PRODUCT DESIGN IN KNITTED FABRIC

In the current technological context, the design of VMMR product is an interdisciplinary area placed between, at least, three important areas of knowledge and, therefore, three professional specialties: the fashion designers, the modelists and the electronic flat machine programmers.

> "*Interdisciplinarity seeks to answer, thus, issues generated by the advancement of the disciplinary modern science, when the latter characterizes itself as a shatterer of the real; a fact which results in the outstanding multiplication of new areas of knowledge*" (Wollner, 2003).

That is, interdisciplinarity is needed to deal with the comprehension of the real world and, in this case, to solve the issue of VMMR product design.

With regard to the performance in the field of design in a broader sense:

"In its higher and more ambitious sense, the design should be conceived as an expanded field which opens itself to several other areas, some closer, others more distant. In this sense, the designer can indeed be an artist, architect, engineer, stylist, marketeer, advertising executive or countless other things. The great importance of the design resides, today, precisely in its ability to build bridges and to forge relationships in an ever shredded world by the specialization and fragmentation of knowledge". (Cardoso, 2011)

In the development of flat knitted fabric product a great complexity related to the variety of possibilities of materials, shapes, colors, textures, among other factors is still observed. Obtaining products with so many variables is a challenge faced by professionals of different academic backgrounds and cultures:

a) Electronic Flat Machine Programmers – they hold technical background and become specialists in advanced technologies.
b) Designer – "professional responsible to transform, in a conscious and creative manner, ideas into shapes, through the combination of the trio: technology, materials plus the social context in the sense of satisfying the human being" (RECH, 2002, p.53).
c) Patternmakers – do the spatial translation, or 3D, of the idea and creativity of the drawings presented by the designers.

The interaction of the designer with the remaining members of the product development team – modelist, the programmer and the finishing of the product – is one of the most important factors in the final result

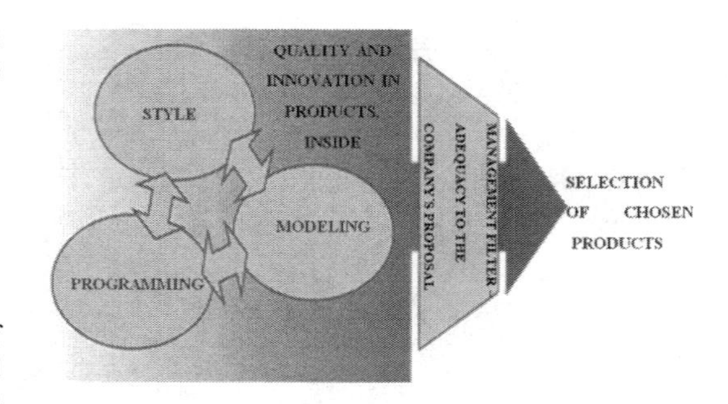

Figure 6. Scheme of product development, Creation team interaction, Piccinini L., 2015.

of the product, however, very often this interaction is limited by the lack of technical knowledge from some of the parties. Maybe this is the strongest reason to defend the systemic integration of the department members of creation (stylist, modelist, programmer), with the guidance of a manager with an external vision, independent, to direct and weigh the several factors that influence the final decision of the collection. In figure 6 below we have one visualization of the interaction of the professionals linked with creation.

The designer conceives the overall idea of the product to be developed, already considering its use and application to the market. The modelist translates the idea into flat shapes which will make up the three-dimensional product; The programmer converts these shapes and textures chosen by the designer into machine language. In this manner, the interaction between the three professionals and the exchange of interdisciplinary knowledge enable the better use of the allowed resources offered by technology and for all the areas involved in the creation. However, important limitations are not observed in terms of technical knowledge of the creation teams with regards to technological elements, despite the effort of machine manufacturers to disseminate this information (Piccinini, 2015).

4 CONCLUDING REMARKS

Innovation is usually an important motivating factor for the differentiation in the market, by performing in different ways in its countless sectors, yielding different results. In the case of knitted fabric clothing, it is oftentimes from the connection of these different pathways that the most interesting results arise in terms of textures, surfaces, properties, aesthetic and functional results.

According to Montagna, G. (2012, p.6):

> *"The significant advancement of science and technology in the different areas, enable, at present, new possibilities in the creation and production of pieces of clothing, giving them potentialities to develop functions far beyond*

those traditionally required and, thus, exponentially increase what is commercially considered added value".

In the case of the VMMR industry, innovation arises, mainly, out of three areas:

- Textile Raw-Material Industry and Spinning – they innovate in the creation of materials with specific properties, seeking to meet new and specific demands;
- Equipment Industry (Flat Knitting Machines) – they innovate in order to differentiate themselves among the competitors, enabling more technical resources for the production and design;
- Fashion Brands and Designers – they innovate seeking to propose novelties in terms of shape, fit, modeling and functionality for the final consumer, reaching a differentiation in the market.
- Specifically dealing with the evolutions in flat knitting machines, one can evaluate that the implementation of new technologies can mean important implications for the distribution of capital and jobs, for the logics of consumption and business models.

In the past years, new technology has also been called 3D printing, following the "vigor" the term has gained with the creation of additive printing technologies. Its main advantages are related with the high quality of the seamless pieces, the short manufacturing time, the elimination of disposables and the ease of customization.

On the one hand, the technological advancements open the way for the reduction of labor costs and make the displacement of industrial production unnecessary to further regions of the globe in search for better labor costs, which can represent an opportunity for local business development.

On the other hand, it requires evolution in the training of the professionals involved in design and production. Product development centers in machine industries have many resources and information available free of cost, however, most of the domestic companies still do not know or do not use these trainings (Piccinini, 2015).

Additionally, the knitted fabric product has become an option for sustainable development, contributing with the growth of the clothing industrial poles. The creation of jobs with the interaction of cutting edge technologies contribute with the local economy, qualifying people in all levels of the industry and improving their life quality. For these reasons, creating and expanding knitted fabric product development locally can be an opportunity to develop and bring a return of social and economic development.

These implications are expected for the future in society and in the economy as new technologies take most part in the clothing sector, but implications for the formation of design teams inside the companies can already be observed.

Since the late 20th century, program-controlled automated machines have been, more and more, incorporated to VMMR companies and, gradually, these pieces of equipment have assumed more resources and more simplified interfaces. This process involves a profound man-machine interaction and, with such, it requires the development of interdisciplinarity in the area of design.

For results that best take profit from technology, it is necessary to develop the ability of interaction between designers, modelists and programmers, in addition to the generation of interdisciplinary knowledge, which is still a great challenge and can represent a great opportunity for the creation and production of VMMR.

ACKNOWLEDGMENTS

This research was only possible with the support of CIAUD – Research Centre in Architecture, Urbanism and Design, Lisbon School of Architecture of the University of Lisbon, and FCT – Foundation for Science and Technology, Portugal.

REFERENCES

Brehim, L. S., Ruthachilling, E. A.: A interface entre as ferramentas tecnológicas promovidas pelo designer na construção de tecido em malha de retilínea. Article, Colóquio da Moda, Caxias do Sul, RG. 2010.

Brown, C.: Knitwear Design, Laurence King publishing Ltd, London, 2013.

Cardoso, R.: Design para um mundo complexo, São Paulo, 2011.

Conti. G M, Curto B., Soldati M. G.: Textile Vivant, Tracks, Experiences and Researches In Textile Design, Milano, La Triennalle, 2014.

Gilmore, P: Architecture as a Product. Digital Technologies in Architecture, 2007.

Jones, S. J.: Fashion design; Manual do estilista, Cosacnaify, São Paulo, 2005

Mariano, M. L. V. Da construção a desconstrução. Master's Degree, Anhembi Morumbi, São Paulo, 2011.

Mendes, D.M. Sacomano, J.B., Fusco, J.P.A.; Rede de Empresas, A cadeia têxtil e as estratégias de manufatura na indústria brasileira do vestuário de Moda. São Paulo. Arte e Ciência, 2010.

Montagna, G; O vestuário inteligente como ferramenta para o design da performance desportiva. Doctoral Thesis in Design, Faculty of Architecture from UTL, Lisbon, 2012.

Neves, B.N.: Estudo de Projeto têxtil em um contexto de design de interação. Master's Degree Dissertaion, Master's Degree with an emphasis in technology and, UFRGS: Porto Alegre, 2010.

Piccinini, L.: Um Estudo do Processo de Desenvolvimento de Produto no Vestuário de Moda na Malharia Retilínea no Brasil, Master's Degree dissertation, EACH-USP, São Paulo, 2015.

Rech, S. R. Qualidade na criação e desenvolvimento de produto de moda nas malharias retilíneas. Master's Degree Dissertaion, Production Engineering. Florianópolis, UFSC, 2002.

Roos, D. B. O setor industrial de malharia de Caxias do Sul. Um estudo de aglomerado de pequenas empresas. Faculdade de ciências Econômicas da Universidade do Rio Grande de Sul, Master's Degree Porto Alegre, 2001.
Sanches, R. A.; et al. Proposta de metodologia para seleção de matérias-primas utilizadas em artigos para vestuário. 5th International Congress on Design Research. Bauru, 2009.
Sissons, S.: Malharia, Fundamentos de design de moda, Porto Alegre, Bookman, 2012.
Soderberg, J.: Impressoras 3-d, a última solução mágica. A ilusória emancipação por meio da tecnologia. Le Monde Brasil, 2013.
Spencer, D.J.: Knitting Technology; A comprehensive handbook and practical guide, Woodhead Publishing Limited, Oxford, UK, 2001.
Stoll, H. P. The History Book, 135 Years, Stool. The Right Way to Knit, ITMA, SHANGAI, 2008.
Textile Vivant. Experiences and researches in textile design. La Triennalle, Giovanni Maria Conti, Barbara del Curto, Maria Grazia Soldati. MILANO, 2014.
Wollner, A.: Design 50 anos, São Paulo, Cosac e Naify, 2003.
WWW.shimaseiki.com, access, November/ 2015

Textiles, Identity and Innovation: Design the Future – Montagna & Carvalho (Eds)
© 2019 Taylor & Francis Group, London, ISBN 978-1-138-29611-4

Pilot study to convert an existing model into a zero waste pattern and cutting system

A. Simões & R. Almendra
University of Lisbon, Lisbon, Portugal

ABSTRACT: The waste in the fashion industry and the almost absence of design practices that encourage the search of sustainable solutions is a wide ranging discussion, and begins to find echoes in alternative approaches to industrial production systems, as reflecting a growing and global concern for the sustainability of the planet. Conventional fashion design methodology has resulted in considerable amount of fabric wasted throughout the garment's production process, and the responsibility for this lies primarily in two factors: the model and the pattern. New approaches to reducing the waste in the fashion industry encompass a number of design concepts for sustainability and shape important advances in the development of new ways of designing garment construction. This paper examines the results of a pilot study aimed at converting a existing man's jacket model, to a Zero Waste pattern cutting system, and what progress and difficulties arise from this in an industrial context.

1 INTRODUCTION

The fabric, being a valuable and sophisticated product due to its process of construction, which includes fibre extraction, spinning, design, weaving or knitting and finishing processes, alongside with the economic investment incorporates material, energy, water and time investments. When fabric is wasted, during the cutting and garments production process, these investments are often lost.

By this we mean that waste is not a problem for fashion and textile industry which only emerges after consumers have worn and got tired of a piece of clothing; there are substantial material and energy losses throughout the entire production chain. One of the most visible moments of waste production occurs during fabric cutting stage, when patterns are placed in the so-called "marker" (markers are a guide used in the cutting process. It is long sheet of bond paper with all of the pattern pieces used to make a style laid out in a configuration intended to reduce fabric waste as much as is possible, including all of the sizes you'd need of a given style. Markers are often made by computer and printed out with a plotter. The marker is laid on top of the fabric layers which cutters then use to cut out all of the pieces at once) or "cutting map" and the 'negative' space around the patterns is wasted.

Fletcher (2014) points out that, on average, between 10 to 20 percent of fabric can become waste at this stage of the production process; the efficiency of its use depends on several factors, including pattern shape, fabric width, sizes combination and the spreading technique used. Other factors limiting cutting efficiency have to do with the intrinsic characteristics of the fabric design, such as whether it is a longitudinal directional fabric (this rule is generally used in man's tailoring regardless of any other characteristic of the fabric), if there is transverse symmetry, the need to match the fabric design (stripe and/or plaid), rapport length, etc. (rapport is a French term meaning 'relation' and it is used to denominate a repetitive design. The elements of a printed design, whether visual or tactile, create a representation focused on the unity, continuity, filling and rhythm of a surface. These elements form the module, which is the smallest area that contains all the visual elements that are part of the image. And when repeated laterally it forms the standard unit (the complete design).

Although this average percentage of 10 to 20 percent of waste per garment may seem minimal, it is a tangible and visible sign of a pattern design method that has long proved to be imperfect, as it reflects an approach and a mind-set that accepts such losses and wastage as an inevitable part of the production cycle. When cutting single garments of large sizes often the efficiency achieved may be even lower.

Through this experiment we intended to develop a pilot study in order to explore the possibility of adapting an existing garment style (more specifically a tailor-made man's jacket), whose pattern and cutting methodology were the traditional ones, applying a Zero Waste concept and goal.

Based on the results obtained, we believe that Zero Waste concept is part of a larger intervention in terms of design for sustainability and that its creative approach may establish new opportunities for design and fashion industry, as it makes clear that it is possible to minimize waste even in preexisting styles.

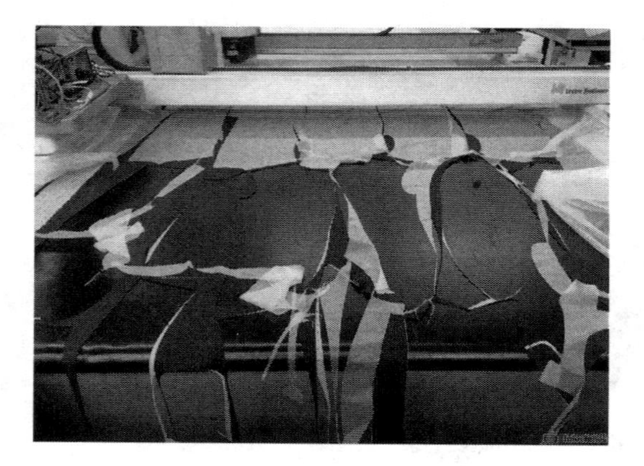

Figure 1. Example of various materials waste (paper, cloth and plastic) resulting from the cutting of a man's jacket in a traditional tailoring manufacturer company (source: author, 17/06/2017).

Figure 2. Chiton was a linen robe worn by both men and women. When extended, it was basically a fabric rectangle (source: The Costumer Manifesto, accessed 17/06/2017).

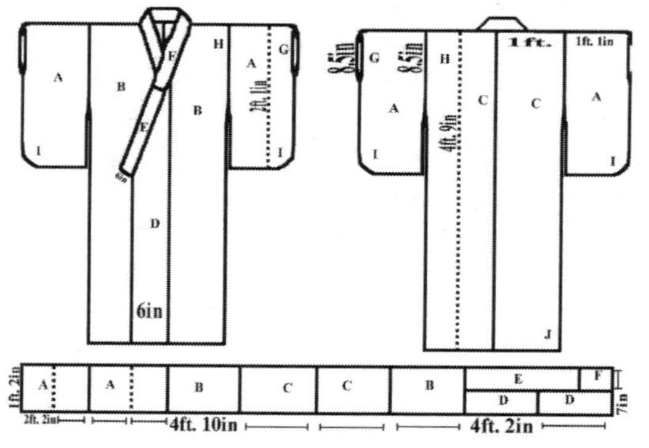

Figure 3. Japanese kimono and its rectangular patterns for fabric cutting (source: http://cosplayqna.tumblr.com, accessed on 17/06/2017).

2 ZERO WASTE FASHION DESIGN: THE CONCEPT AND A FEW HISTORICAL EXAMPLES

The Zero Waste International Alliance defines the Zero Waste philosophy as a pragmatic and visionary goal designed to advise people to imitate sustainable natural cycles, where all the discarded materials are features that others can use. Zero Waste means designing and manage products and processes to reduce the volume and toxicity of waste and materials, preserve and recover all resources, and not burn or bury them. The implementation of Zero Waste design practices, according to this organization, will eliminate all discharges to land, water or air which may pose a threat to planetary health, human, animal or vegetal. The Zero Waste strategy can be defined as a set of practical tools that seek to eliminate the waste and not manage it.

According to Rissanen and Mcquillan (2015), the term Zero Waste applied to the fashion industry emerged especially after 2008, which led many people to think that the expression Zero Waste Fashion Design defines a new phenomenon. Paul Palmer founded the Zero Waste Institute (Yale University, San Francisco, California, USA) in the 1970s and has since published much criticism of the modern waste industry, particularly about recycling. Palmer was one of the first to use the term Zero Waste and in part due to its pioneering work the term was later more easily adapted to fashion. Although the term Zero Waste Fashion Design is new the practice is as old as dressing the body with furs and cloth.

Historically, clothing evolved from simple shapes and during hundreds of years garments were made from cloth rectangles, whereby the waste of fabric was residual or even null. From the ancient Greek tunic to the Japanese kimono there are several samples that we can observe how, from geometric forms, man has covered his body.

Gwilt (2014) says that, when clothing began to adjust to the body, patterns began to include curved lines and, this combination of lines and curves no longer allows its efficient cut: the result is a set of "negative" spaces that end up being wasted.

The Zero Waste Fashion Design concept is based on the principles of design for sustainability, a zero waste approach to fashion and pattern design which purpose is to eliminate, or at least, reduce waste in the initial design stage of clothing products.

This approach attempts to create and/or modify designs for a use of a hundred percent of the fabric and thus, eliminate wastage; that is to say, refers to the planning and implementation of processes through which the entire tissue is used in the garment and has attracted the attention of several international designers to the concept of sustainable fashion.

Nevertheless, in terms of mass production, the awareness of sustainable approaches to fashion and pattern design in the clothing industry, is still at a very early stage.

3 THE PILOT STUDY

This paper as previously said is based on a research project whose purpose was to explore the creative adaptation of the patterns of a pre-existing men's jacket in two Portuguese men's tailoring companies (Dielmar, S.A. and Diniz e Cruz, Lda), by using the concept of Zero Waste Fashion Design.

It was intended to demonstrate that the fashion design process, based on this concept, should be seen as a set of skills under continuous development, and how this creative method of fashion and pattern design can result in a multidisciplinary design optimization (multidisciplinary design optimization allows designers to incorporate all relevant disciplines simultaneously. The optimum of the simultaneous problem is superior to the design found by optimizing each discipline sequentially, since it can exploit the interactions between the disciplines) and not just a methodology for fashion sustainable development.

This innovative experiment also aimed to motivate the two companies involved to think about the issues of environmental impact resulting from their productive activity and to integrate sustainability as part of their business and design philosophy.

The organization of the study was segmented into two distinct moments: firstly, we defined the common criteria and assumptions of the work plan, which comprised the following parameters:

1. Use the same jacket style in both companies (it was defined to use a jacket style without lining, with 2 buttons on the front, 3 patch pockets and two side vents. This style exists in both companies although each of them work with different measurement tables. This factor did not impede the study because the results were compared like for like in the company and not between companies)
2. Use the same fabric composition and useful width (The useful width defined was 148 cm)
3. Keep the external image of the model unchanged
4. Respect the measures and fit of the original style in the new model produced
5. The new model would not result in an undue increase in manufacturing cost
6. An unnecessary complicated construction would not be promoted
7. The results obtained would enhance the optimization of the values previously achieved by both companies.

In the second moment, and following the previous defined organization chart, the CAD (Computer Aided Design. CAD systems automatically calculate the total area of garment pattern pieces placed in a marker so you get the area of the marker that is consumed by garments from CAD system) sector of the companies was asked to carry out a 'marker' based on the pre-existing patterns. Thus, it was possible to obtain fabric metric and linear utilization values as well as the percentage in terms of cutting efficiency for the original patterns.

It was also asked to the Methods-Time department of the company Dielmar to provide the study of manufacturing times of the original jacket as well as the company minute cost of labour for the purpose of calculating costs. This particular request had to do with the fact that only this company was going to produce a prototype for result's demonstration.

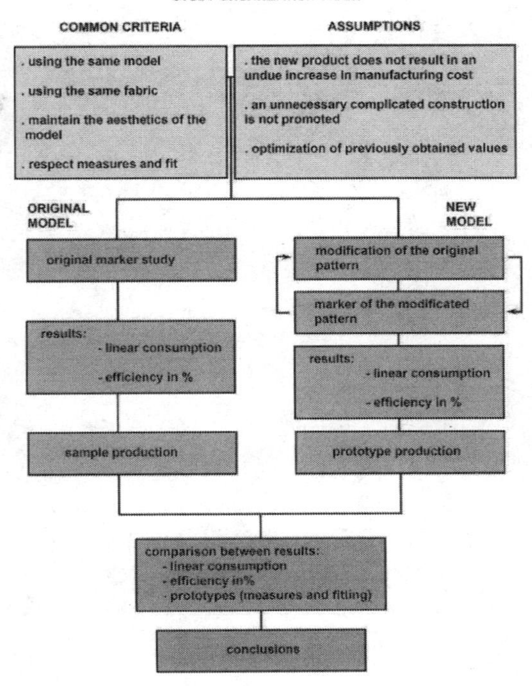

Figure 4. Pilot study organization chart (source: author, 17/06/2017).

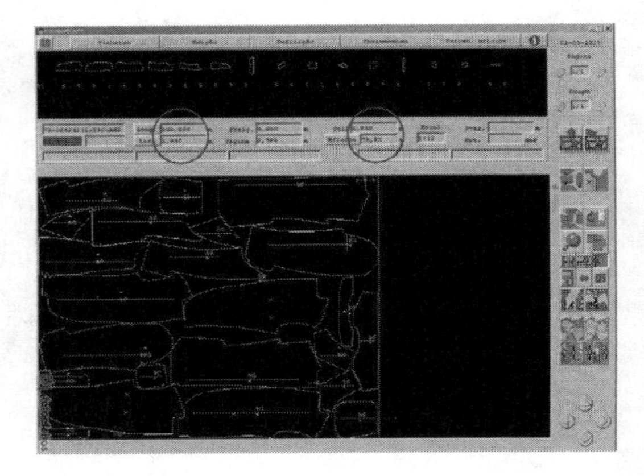

Figure 5. Marker executed with the original patterns in the company Diniz & Cruz (source: author, 17/06/2017).

In the marker executed with the original patterns in Diniz & Cruz we could observe an efficiency in percentage terms of 78.81%, which meant a total waste of fabric of 21,19% and a metric consumption of 185cm fabric. This marker was carried out with Lectra's Diamino program and the patterns were made with Lectra's Modaris program.

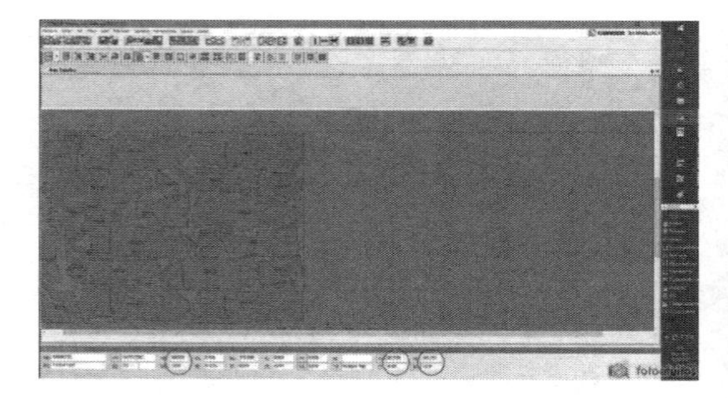

Figure 6. Marker performed with the original patterns in the company Dielmar (source: author, 17/06/2017).

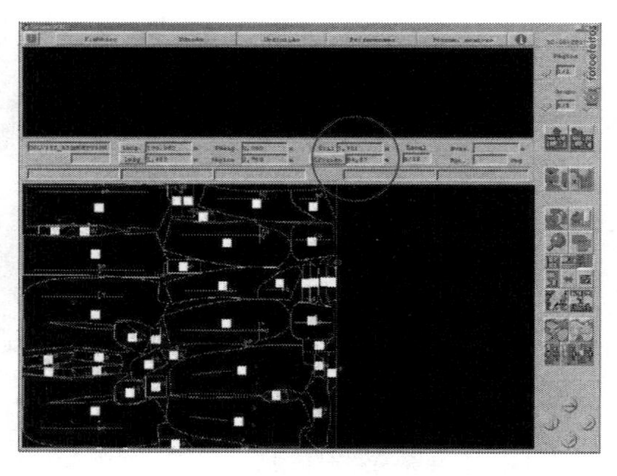

Figure 7. Marker developed with the modified pattern in the company Diniz & Cruz (source: author, 17/06/2017).

In Dielmar the study done with the original patterns presented an efficiency of 82.23%, which performed a 17,77% waste of fabric, and fabric consumption of about 195cm, performed with the Pattern Design System for patterns and Simplex software for the marker.

CAD systems and their effectiveness are constrained by the original logic of its programming; they work with efficiencies within established parameters for an existing industrial pattern system and, as such, have no ability to accommodate completely new concepts of garment construction, which may make it impossible to create innovation.

The next phase was initiated based on this argument and started with a joint meeting with the patterns and CAD designers to explain the purpose of the experiment. We considered that the success of this research would be the greater with the involvement of the entire design team; technology can provide the tools but it is the creative designer's mind that directs its effectiveness.

Being a pioneering project, both for the PhD student and for the designers of the companies involved, and since there is no design methodology formula in fashion design, the study was developed based on the principles of experimentation, measurement and evaluation, which required a restart during the process until a valid and feasible solution was obtained. The knowledge of a series of disciplines that make up the productive processes of a man's jacket, and the deep mastery of the operating method of an industry, proved to be critical to the success of the journey.

In both companies the result of this experiment, as we can observe in figures 7 and 8, points to the fact that, although the initial pattern shape transformation and its modifications can contribute to the best use of the fabric in at least 7,5%, in the Dielmar case since the aims were substantially higher than in Diniz & Cruz, there is space for further improvements in existing styles.

Substantial gains can possibly be achieved if considered style variations through an innovative design, that is to say, modifications in the garment's design and the possible creation of a new style of conventional silhouettes in men's tailoring.

The various constraints imposed on the experiment at the outset were prevented from obtaining more substantial results, first and foremost, and in addition to those mentioned previously, the respect for the 'grain line' (Grain lines are a generally unnoticed aspect of the garment, that is until they are either used in the wrong way and cause fit problem, or used in interesting ways to mould the fabric in different ways to the body. Sometimes they can also be used to turn a print onto a different angle for interesting visual effect. When fashion designers and pattern makers talk about grain lines they are referring to the way that a pattern is cut out when it is laid out on a piece of fabric. Basically, fabric is woven from thread going in two different directions and it is sometimes easiest to remember that fabric is built on tiny squares of threads which criss-cross each other. Grain line correspond to the correct pattern positioning on the fabric for cutting. Changing this rule, in an unthinking way, can create deformed garments and production problems. As our assumption was to keep the piece as close as possible to the original style, this rule could not be changed) in the cutting patterns placement in the new marker.

The comparative study of the efficiency and tissue consumption values obtained Dielmar, reveals that in a primary evaluation the direct gains are substantial. Other secondary materials, such as paper and plastic, not directly incorporated into the model, but used during the cutting process, may yield equally interesting gains.

The objective of the experiment was not to measure or quantify gains associated with cost reduction with waste management, particularly in the handling, storage, transport and landfill; more importantly, and also not possible to quantify, is the value associated with the reduction of the environmental footprint that may result from the implementation of design strategies with the purpose of waste disposal; however, we believe that these gains will be more significant as the implementation of new approaches to the waste issue in the fashion industry is further enhanced.

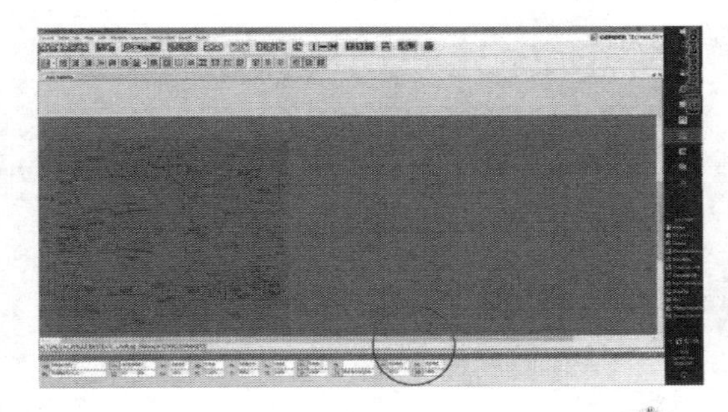

Figure 8. Marker developed with the modified pattern in the company Dielmar (source: author, 17/06/2017).

	EFFICIENCY	CONSUMPTION	PRODUCTION MINUTES	FABRIC (€)	MAKING COST (€)
ORIGINAL JACKET	82,23%	195cm	172,17"	39,00€	44,76€
NEW PROTOTYPE	89,69%	184cm	184,05"	36,80€	47,85€
RESULTS	+7,46%	-11cm	+11,88"	-2,20€	+3,09€

For the purpose of calculating costs, it was considered:
1 - 20,00€ fabric meter
2 - 0,26€ labor minute

Figure 9. Comparative study of costs between models (source: author, 17/06/2017).

Figure 10. Photo of two coats produced: with the original patterns and with the modificated patterns. The outer image of the garments is the same. (source: author, 17/06/2017).

4 CONCLUSIONS

New design approaches to waste reduction that have emerged in recent years involve a few design concepts for sustainability, including methods that explore the reduction of textile waste at the fabric cutting stage. Complete disposal or at least minimal waste should be the main driver of the process.

To solve these challenges we need the different fields to work together on potential solutions, that is, fashion design, pattern design and CAD should work on a 'horizontal' rather than vertical team's organization. It is thus a combined responsibility of the entire team as well as the industry to develop innovative thought processes in relation to the needs and possible results in the search for waste reduction.

This new way of conceiving in fashion design must consider new models of thought that include other type of pattern shapes and markers; the Zero Waste Fashion Design approach offers an excellent creative opportunity when combined with more creative pattern design and cutting solutions and increases the fabric incorporation rate into the model at the design stage without increasing its cost; the excess of post-cut fabric, which would normally be wasted, becomes an integral part of the garment.

Design is about solving problems and the Zero Waste Fashion Design approach sets up an opportunity to think and design in men's tailoring in a holistic way to solve the waste problem. Interdisciplinary cooperation between the areas of fashion design, pattern design and CAD will, in our view, allow us to challenge conventions on what is to fashion design while at the same time contributing to change the perception of what constitutes the creative process in contemporary tailoring art. Imagination will be the limit.

5 FUTURE DEVELOPMENTS

The learning acquired in the process of carrying out this pilot study, and the conclusions we reached, allowed us to delineate some future work that we consider fundamental for the success of our research. In this way, we will continue to collaborate with the two Portuguese companies involved as they represent, among us, the best that is done and produced in male tailoring. We are also in the process of preparing a briefing, through which we will try to sensitize and motivate pattern and CAD designers to the importance of interdisciplinary work and the creation of a system of practices, procedures and rules, a set of circumstances and premises that will have a positive impact on the research objectives. It is also expected a potential future collaboration and exchange of knowledge with other designers whose work is located in the field of Zero Waste Fashion Design.

ACKNOWLEDGMENTS

This research was only possible with the support of CIAUD – Research Centre in Architecture, Urbanism and Design, Lisbon School of Architecture of the University of Lisbon, and FCT – Foundation for Science and Technology, Portugal; Diniz & Cruz,

Lda; Dielmar, S.A.; Mário Batista – Grupo Têxteis e Confecções, Lda and Marzotto Group.

REFERENCES

Fletcher, K., Sustainable Fashion and Textiles: Design Journeys, London: Routledge, 2014.

Gwilt, A., A Practical Guide to Sustainable Fashion, London: Fairchild Books, 2014.

Rissanen, T., What is Fashion Good For? – Part 2, https://timorissanen.com,accessed 06/14/2017.

Rissanen, T. and McQuillan, H., Zero Waste Fashion Design, London: Fairchild Books, 2016.

ZERO WASTE EUROPE, Introducing Zero Waste Europe – The Main Principles, Brussels: https://www.zerowaste europe.eu, accessed on 06/14/2017.

Textiles, Identity and Innovation: Design the Future – Montagna & Carvalho (Eds)
© 2019 Taylor & Francis Group, London, ISBN 978-1-138-29611-4

The textile in seismic architecture: A reinforcement solution

António José Morais, José Afonso & Alexandrino Diogo Basto
CIAUD, Lisbon School of Architecture, University of Lisbon, Lisbon, Portugal

ABSTRACT: The Portuguese government, to put forward the rehabilitation of the oldest buildings, implemented in 2017 a licensing policy that exempted property developers from the need to present the stability project related to the structural recovery of the building to be rehabilitated. This typology of old buildings has usually a poor structural quality, without capacity for seismic resistance. The licensing policy now approved will result in rehabilitated buildings lacking the current structural safety legislation, therefore they do not have the required seismic resistance under the RSA (Safety and Security Regulation).

This is a very serious matter, in two planes. Portugal is a country with high seismic risk, the opportunity is lost to reinforce the old built heritage, adapting it with capacity of seismic resistance capable of withstanding the action of an earthquake; on the other hand, it deceives and harms future purchasers of recovered fires under this licensing policy because they acquire a fire that they think has adequate seismic resistance to an earthquake because it has even been the subject of recovery work and, fact, it does not have seismic resistant capacity. This is a problem of security and civil protection.

This paper presents a possible solution of non-intrusive seismic reinforcement, using the textile fabric as a coating glued to the masonry walls in the building to be recovered. This textile coating thus has two functions, structural reinforcement, and decorative formal element of architecture.

1 INTRODUCTION

There is a class of buildings in Portugal, built in the second half of the 19th century and in the second half of the 20th century, whose seismic resistance is very weak. These buildings were initially constructed of low quality masonry walls and with wooden floors. This type of construction, very common in Lisbon, is known as the "gaioleiro" building, the thickness of its walls only reaches of 0.40 m.

In view of the government's policy of exempting from the obligation to submit the Stability Project in the licensing projects for restoration of old buildings, real estate developers do not feel compelled to introduce structural elements that able "gaioleiros" buildings with adequate seismic resistance (Morais, A. J. 2006). Usually civil engineers to reinforce the old buildings rely on intrusive solutions insertion beams and columns, metallic or reinforced concrete, these solutions are not only expensive but, above all, they pose problems of functionality and space, since it is necessary to "tear" the existing building

In this context of the problem, the solution to develop a project methodology that uses the non-intrusive and, above all, low cost solution. One way of encouraging the real estate developer to reinforce the seismic capacity of the building to be recovered, even if legally not required by law, is to ensure that the cost of this reinforcement operation does not raise the total cost of the building's restoration work. Obviously, otherwise, the developer will not do any structural seismic reinforcement work on the building, because law does not require this. Why spend more money if the law does not oblige you.

To do not increase the final cost, the solution is to gives two functions to a constructive element which is subject to recovery or conservation work. The best element to use this double function is the wall cladding. The coating will necessarily be subject to rehabilitation and/or conservation work. It is a mandatory task, which the real estate developer will have to perform. So if the decorative function does not allow the coating to perform a structural function, the cost of the seismic structural recovery does not weigh on the total cost of the work, since the coating was already included in its budget.

In this paper we present and describe a possible seismic reinforcement solution that adopts this methodology, to be used in this type of buildings – gaioleiro the low-cost buildings, which does not use intrusive solutions, and which takes advantage of the need to repair the coatings, or replace the old for new and suitable coatings, more contemporary and with the aesthetic values of modern architecture.

The possible solution to this methodological objective is given by the textile industry. By adopting as a coating fabric bonding it to the existing walls, we can maintain their integrity and continuity, avoiding

Figure 1. Exterior of the Portuguese Finance Ministry building (Lisbon).

their disintegration, and take advantage of the tensile capacity of these textile materials. This is the solution we propose in this article.

2 THE TEXTILE MATERIAL

The textile fabric is a yarn-based material, made from natural or synthetic fiber. The fabric is manufactured in the textile industry.

Natural fiber fabrics can have three origins; animal origin (wool and silk); mineral origin (asbestos); and vegetable origin (cotton, jute, hemp, flax and sisal).

Synthetic fabrics are man-made fibers, using as raw material chemicals from the petrochemical industry. The most common are polyester (PES), polyamide (PA), acrylic (PAC), polypropylene (PP) and elastomeric polyurethane (PUR), as well as Aramides (Kevlar and Nomex) (Heyse, P. et al 2015). Artificial fabrics come from cellulosic fibers, from materials such as corn and vegetable oils. Among the materials most publicized by the textile industry we have rayon, which is a cellulose fiber, and was the first synthetic fabric produced, known by the most common name of nylon. It is made from a polymer, the polyester, which is derived from alcohols, in chemical terms.

Nowadays polyester fibers, yarns and ropes have application in various economic activities outside the world of fashion. For example, they are nowadays used in reinforcements for tires, fabrics for conveyor belts, seat belts, etc. Polyester fibers are also used as an insulation material and in the filling of pillows, duvets and upholstery.

From the chemical point of view, Polyester is a category of polymers containing the ester group, in its main molecular chain. Although there are many polyesters, the masculine noun "polyester", as a specific material, refers to polyethylene terephthalate (PET).

While synthetic clothing is perceived as having a less natural touch, in relation to clothing made from natural fibers, such as cotton or wool, polyester fabrics have specific advantages over natural fabrics, such as tensile strength. As a result, the polyester fibers are sometimes mixed with natural fiber, to produce a fabric having properties of the synthetic fiber, without losing the touch of the natural fiber.

Figure 2. Preparation of the warp for weaving http://www.textileflowchart.com/2015/04/flow-chart-of-warping-process.html.

Polyesters are also used as raw material for the manufacture of plastic bottles, plastic films, tarpaulins, canoes, paints and varnishes, etc. Unsaturated polyesters reinforced with fiberglass find a wide variety of applications in yacht hulls and in auto parts.

From the chemical point of view, polyester is a synthetic polymer made from terephthalic acid (PTA).

In the solution proposed here, both natural and synthetic fabrics can be used, although synthetic ones have better tensile strength.

2.1 *Textile weaving*

The factory-produced, color-dyed fabrics are sent to various designers and fabric manufacturers and textile applications until they reach the market and into our homes. We usually associate fabrics with cloths, towels, sheets, blankets, curtains, rugs, and clothing, but there is another world of application in architecture. Although the colors and patterns of all these products are dictated by the style of the moment, the clothes are undoubtedly the most influenced items of fashion.

To dye a fabric, it is necessary to subject it to chemical treatment, so that the micro-spaces between the web of the cloths (pores) become larger. In this way, the fabric is better absorbed by the associated pigment in the dyeing process. After this dyeing bath, the fabric is washed and stretched on a metal support.

Nowadays, in the factory, the production process is mechanical and automated, with digital control by computer. Basically, to produce the fabric, two sets of wires are required; one thrown transversely and the other thrown longitudinally; the first undergoes stretching using appropriate equipment.

The second set of wires is launched from a support with longitudinal bars, and, as a rule, each bar launches a single yarn, the entire yarn-to-pattern yarn process being controlled by computer.

3 COLORING THE COLORLESS

The cloth obtained does not yet have color, besides containing still impurities and imperfections, reason they need to be treated previously to be sent to the confection. In effect, the cycle closes with the intervention of the designers. The final confection of the fashion world takes place in parades around the world.

4 TEXTILE ARCHITECTURE

As same one said, "no other building material offers Architecture as much creative freedom as fabrics." The main feature offered by the fabric to the architecture is its flexibility, not from the point of view of structural behavior, but also to provide flexible design solutions capable of adapting to different environments and specific formal spaces. The textile architecture assumes itself as a good alternative to the usual coatings that the construction industry makes available to the architect. The textile architecture brings innovation and creativity to the Architecture Project.

The new developments with fibrous materials, with high tensile strength, together with the development of more accurate lamination techniques, allowed the emergence of so-called architectural membranes. They are widely used in such ephemeral architecture, but may also have permanent character. They make it possible to construct spaces that are much lighter and more functional than the usual conventional solutions.

A new architectural form has developed in recent years based on the textile material. Kennet Snelson, a disciple of Richard Buckminster Füller, experimented with the very flexible textile membranes through which he explored space three-dimensional, innovating an edificatory method that aggregates rigid modular elements connected to cables with flexibility – the tensegrity.

Frei Otto is another pioneer architect in the use of membranes and textile fabrics. He has developed research in the field of Textile Architecture, with his innovation of the tense-structure. In his formation of aeronautical engineer, he realized experiments on tents and membranes, developing methods of calculation pioneering for civil engineering.

The Textile Architecture uses the membrane as the main material for its structures and coverings purposes. These membranes are made up of fiber fabric or mesh, and can receive specific treatment to optimize their performance, being able to cover large spans, with quick and easy assembly and at a lower cost, than solutions in reinforced concrete. Stressed membranes are very light and very thin, and support well the formal drawing with curved surfaces.

The Textile Architecture, whose origin is associated to the tents of the nomadic peoples, is used whenever the time of assembly and disassembly is a critical factor. They are easy to assemble and disassemble, and are easy to carry, because they weigh little. In the Textile Architecture, the tensioned cover is constituted by membranes that soon fulfill two functions, the structural one and the one of seal; result economically. Stressed membranes are made up of fabrics with flexibility, cables, rigid support, anchoring and foundation elements. Structurally they are classified as membrane structures, and they are tensioned. Its main structural feature is flexibility, that is, it changes shape whenever the loading changes position or direction. Initially, the tents were temporary, temporary housing. Nowadays, it is already possible to consider the Textile Architecture as being able to assume a permanent character.

5 ALTERNATIVE RETROFITING SOLUTION TO IMPLEMENT

The architectural design proposed here to facilitate an inexpensive solution to equip old buildings to recover with some seismic resistance, does not resort to the usual intrusive reinforcement solution by inserting a reinforced concrete structure or steel. It is proposed to adopt, as an alternative, the use of layers of textile fabric bonded to existing masonry walls by bonding provided by polymeric materials (Valvona, F. et al 2017). That is, the existing masonry wall is "encased" on both sides by using coatings of textile materials bonded to the masonry wall material through polymers. The polymer material is in this solution the adhesive element that attaches the textile fabric to the existing masonry wall. The textile fabric set associated with the polymeric adhesive agent has the task of maintaining the integrity and cohesion of the existing masonry wall, avoiding its disintegration. It thus has the fundamental purpose is the wall to be continued and solid ensuring a homogeneous behavior, that is, increasing its seismic resistance. Although there is cracking within the existing wall, due to the at breakages, they will be contained, locked, or at least their delayed opening, by the existence of the polymeric sizing membrane and by the textile coating. The polymeric skin, plus the textile fabric, presents tensile strength, blocking the progression of cracking and above all maintaining solid and continuous the whole. The entire wall assembly, now coated and working, is strengthened to maintain its solidity and continuity, thus ensuring the stability of the wall during more cycles of vibration when an earthquake actuates the building (Widder, L. 2017).

The masonry walls of "gaioleiro" buildings possess a good compressive strength, for vertical loads, but very low tensile strength, and reduced ability to handle torsion efforts. Unless these walls are reinforced with materials with capacity to work at tension, we have that the performance of these walls of masonry, when

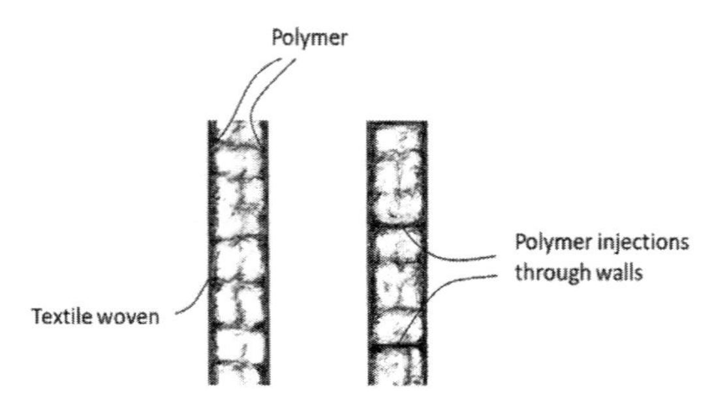

Figure 3. Schematic cut with the proposed solution (Author).

seismic actions occur, presents / displays a very weak seismic behavior. In practical terms, we can say that its seismic resistance is zero. So, if we want to increase the seismic resistance of this type of buildings – the "gaioleiro", it is necessary to add components to the existing masonry wall so that the whole results in a stable system with greater seismic resistance.

The fabric to be used in this reinforcement solution that performs both structural and coating functions, therefore the cost does not increase, it is proposed to coat it with textile fabric based on polyester fibers, which are glued to the material of the wall through of polymers (Bernat-Maso, E. et al 2016). It results in a composite solution, a kind of sandwich, where the viscoelastic reaction of the polymer impregnates the old wall, filling the cracks and cavities of it, acting as micro-anchors, picking up and gluing all the polyester fabric to the existing surface of the masonry wall. It results in a joint structural operation of all components, old and new (Muntasir Billah, A.H.M. & Shahria Alam, M. 2014).

6 BONDING PROCESSES

There are basically three types of bonding: 1) water-based adhesives; 2) solvent-based adhesives; and 3) adhesives that chemically react in contact with air. Despite these differences, all glues have something in common: they take advantage of the adhesive properties that certain polymers have, to keep different elements glued together, that is, united. These polymers are molecules that bind to one another and also to the surfaces of the elements that are intended to be bonded together. This bonding property that polymers produce, we call adhesiveness. The most potent bonding type is called chemical bonding (the third coted). The glue material has a formulation based on chemical compounds, for example cyanoacrylate, which reacts chemically in contact with the humidity of the air, forming polymers. The formed polymer stiffens rapidly, strongly sticking the elements to be bonded. The glue effect is strong. It glues, plastics, ceramics and even metals. This highly adhesive property of chemical adhesives, formulated with cyanoacrylates, allows them to be used in civil construction, restoration work and structural reinforcement.

7 CONCLUSIONS

1. The textile architecture can help the economic recovery of the existing real estate;
2. Although the legislation allows for the licensing of building rehabilitation works without the need to elaborate the stability project, for old buildings, such as the "gaioleiro", it will be necessary for security reasons that the developers will confers some seismic resistance to these buildings;
3. A textile-based solution was presented which together with polymeric glues can seismically reinforce the existing walls and at the same time ensure the formal covering of the building, thus not increasing the cost of the property recovery.

ACKNOWLEDGMENTS

This research was only possible with the support of CIAUD – Research Centre in Architecture, Urbanism and Design, Lisbon School of Architecture of the University of Lisbon, and FCT – Foundation for Science and Technology, Portugal.

REFERENCES

Appleton, J. G. (2005). Reabilitação de Edifícios Gaioleiros. Edições Orion. ISBN: 9789728620059.

Bernat-Maso, E.; Gil, L.; C Escrig, C. (2016). Textile-reinforced rammed earth: Experimental characterisation of flexural strength and thoughness. Construction and Building Materials, 106, pp: 470–479. DOI: 10.1016/j.conbuildmat.2015.12.139.

Heyse, P.; Buyle, G.; Walendy, B.; Beccarelli, P.; Tempesti, A. (2015). MULTITEXCO – High Performance Smart Multifunctional Technical Textiles for the Construction Sector. Procedia Engineering, 114, pp: 11–17. DOI: 10.1016/j.proeng.2015.08.012.

Morais, A. J., (2006). Sistema Pombalino: Caracterização Estrutural e Comportamento sísmico. Artitextos. Faculdade de Arquitectura, Universidade Técnica de Lisboa, Lisboa, 2006, vol. 3, pp. 87–107. ISBN 972-97354-7-6.

Muntasir Billah, A.H.M.; Shahria Alam, M. (2014). Seismic performance evaluation of multi-column bridge bents retrofitted with different alternatives using incremental dynamic analysis. Engineering Structures, 62–63, pp: 105–117. DOI: 10.1016/ j.engstruct.2014.01.005.

Valvona, F.; Toti, J.; Gattulli, V.; Potenza, F. (2017). Effective seismic strengthening and monitoring of a masonry vault by using Glass Fiber Reinforced Cementitious Matrix with embedded Fiber Bragg Grating sensors. Composites Part B: Engineering, 113, pp: 355–370. DOI: 10.1016/j.compositesb.2017.01.024

Widder, L. (2017). Earth eco-building: textile-reinforced earth block construction. Energy Procedia, 122, pp: 757–762. DOI: 10.1016/j.egypro.2017.07.392.

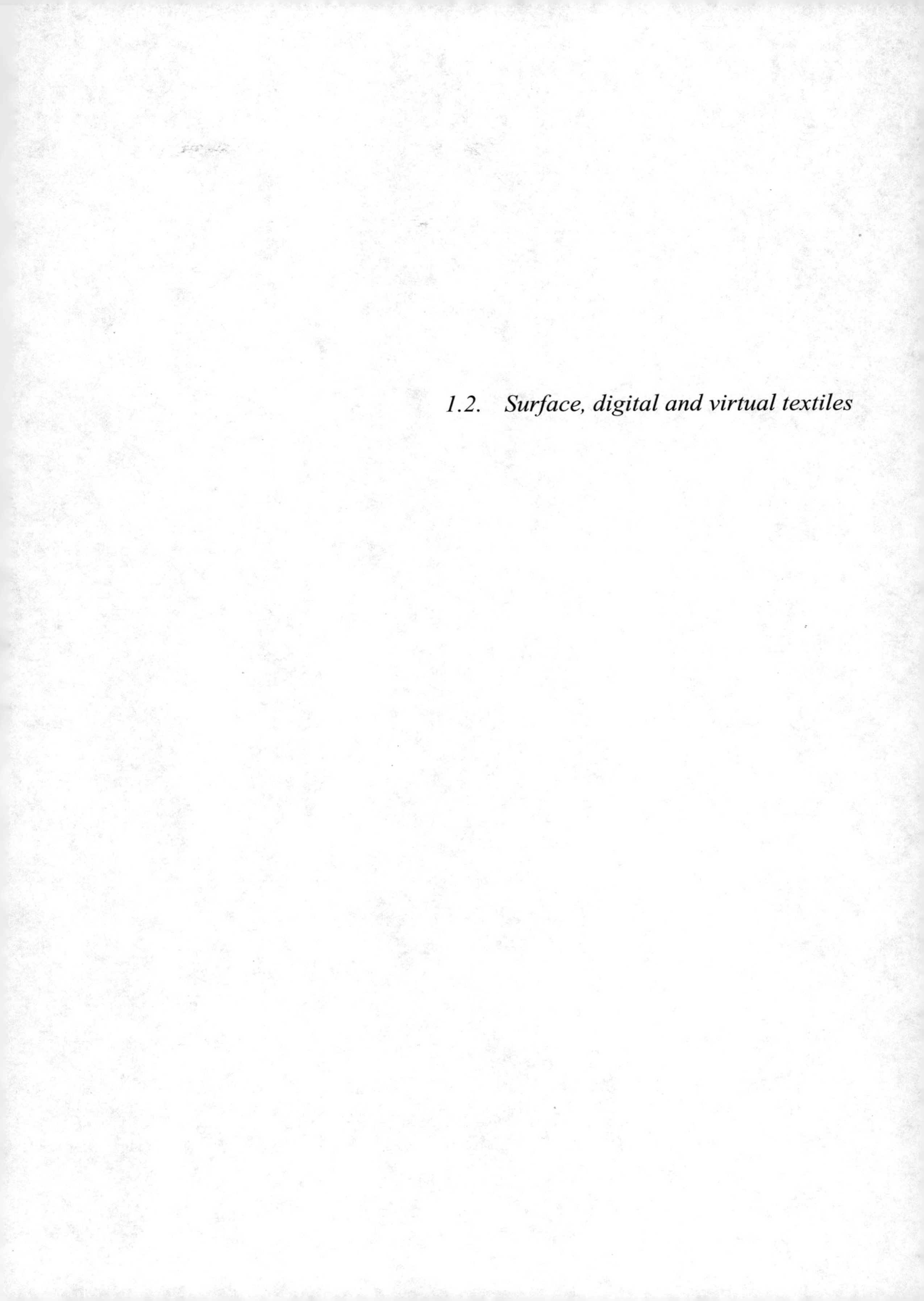

1.2. Surface, digital and virtual textiles

Textiles, Identity and Innovation: Design the Future – Montagna & Carvalho (Eds)
 ISBN 978-1-138-29611-4

An analysis of 3D printed textile structures

E. Grain
Huddersfield University, Huddersfield, UK

ABSTRACT: 3D printing in Fashion is quite a significant innovation which came to realization as part of the zeitgeist of the 2000's. This is a time where technological progress has moved fastest in the last 20 years than it has in the last 100. This exponential growth means that the world in which we live in is changing at a rapid rate and both design and materials will need to alter along with it. Focusing on the areas of technology, design and manufacturing this research aims to look at new textile structures with extensive research undertaken of existing 3D printed textiles and fashion. The main aim is the categorization of all existing 3D printed textile structures in fashion, in order to recognize their properties and benefits.

1 EXPLORING THE TERRITORY

1.1 *Exploration of textile structures*

The design of the structure is the key part of any textile, even before any finish, colour or application is applied, the structure determines the feel, the drape and the look of the whole fabric. As with any other product, textiles have to be constructed with the manufacturing process in mind and therefore are fundamentally limited by the need to consider DFMA criteria (Bingham et al. 2007).

There are 2 main structures in textiles, knit and woven and between those there are many more variations of each which the main principles of structure are kept but the design of the way they are knit or woven are altered slightly to vary the appearance and properties of the final fabric.

Figure 1 shows a jersey knit structure which is different on the front and back: V shapes on the front and arcs on the back/ purl side. This gives the front face a flat plain appearance and the back a slightly arced/looped appearance and it is in the back where the flexibility in the structure lies.

Some of the structures shown here are applied in woven fabrics which is more structured and has less stretch if any at all. These structures were considered but not used going forward in to design development of the designs for this project.

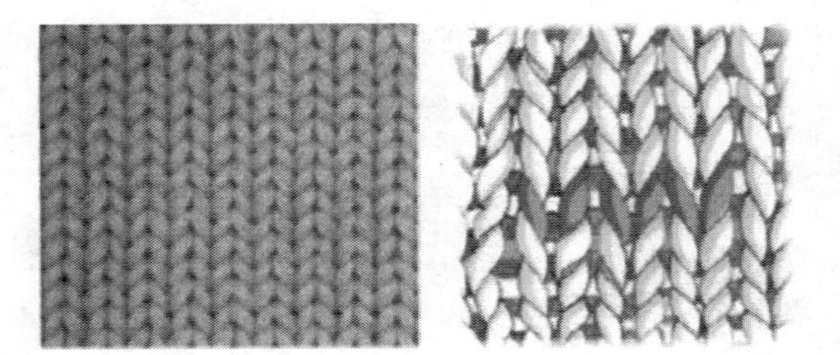

Figure 1. Jersey Knit, plain side. Hallett C & Johnston, A. (2010).

Plain weave is the most basic and the oldest type of weave construction. The warps and wefts crisscross each other at right angles, with each weft thread passing over one warp thread and then under the next warp thread, (Hallett, 2010).

The satin weave is one where the weft yarns are prominent and are woven across 4 warp yarns, then under one, giving it the appearance of a smoother surface and there are more threads laid on top that with a plain weave. This fabric is mainly used in

Figure 2. Jersey Knit, purl side. Hallet C & Johnston, A. (2010).

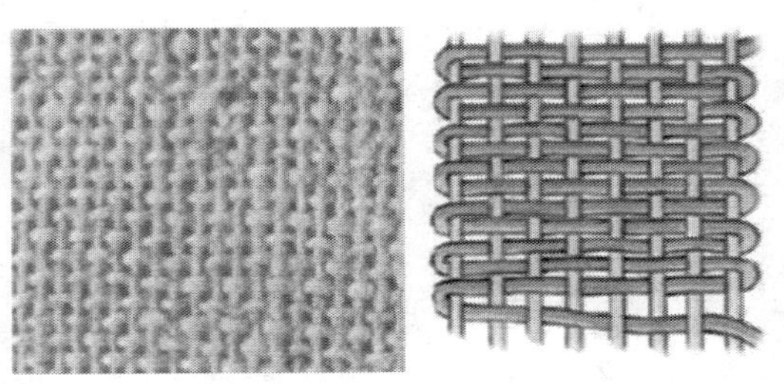

Figure 3. Plain weave Hallett, C & Johnston, A. (2010).

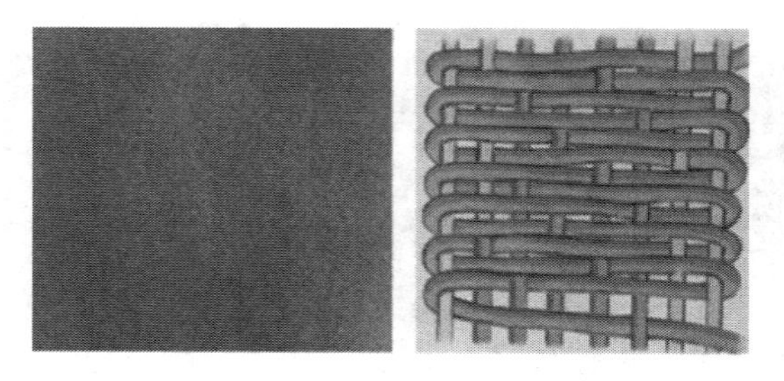

Figure 4. Satin weave Hallett C & Johnston, A. (2010).

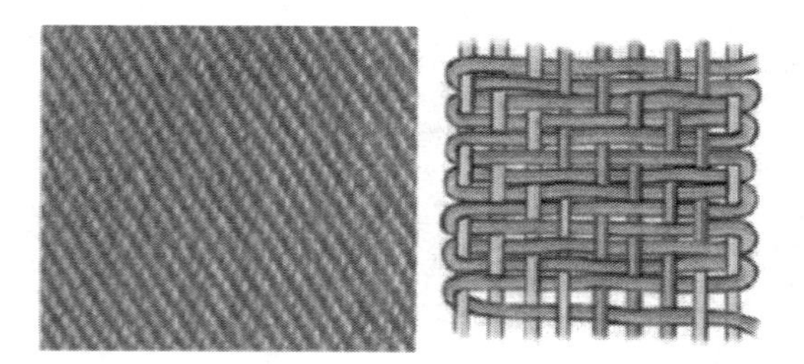

Figure 5. Twill weave Hallett C & Johnston, A. (2010).

Property	Plain	1 × 1 Rib	1 × 1 Purl	Interlock
Appearance	Different on face and back: V-shapes on face, arcs on back	Same on both sides, like face of plain	Same on both sides, like back of plain	Same on both sides, like face of plain
Extensibility				
Lengthwise	Moderate (10–20%)	Moderate	Very high	Moderate
Widthwise	High (30–50%)	Very high (50–100%)	High	Moderate
Area	Moderate-high	High	Very high	Moderate
Thickness and warmth	Thicker and warmer than plain woven made from same yarn	Much thicker and warmer than plain woven	Very much thicker and warmer than plain woven	Very much thicker and warmer than plain woven
Unroving	Either end	Only from end knitted last	Either end	Only from end knitted last
Curling	Tendency to curl	No tendency to curl	No tendency to curl	No tendency to curl
End-uses	Ladies' stockings Fine cardigans Men's and ladies' shirts Dresses Base fabric for coating	Socks Cuffs Waist bands Collars Men's outerwear Knitwear Underwear	Children's clothing Knitwear Thick and heavy Outerwear	Underwear Shirts Suits Trouser suits Sportswear Dresses

Figure 6. Appearance and properties of knit. Hallett C & Johnston, A. (2010).

evening or bridal wear because of its sheen and draping properties.

The twill weave is where the warp and weft are staggered in a diagonal formation to create a chevron like effect in the finished weave. This is the weave used for denim and utility trousers and is quite a strong structure but inflexible.

Different knit structures and properties are shown in this Figure 6. ranging from the plain knit through to interlock knit which is plain knitted on both sides of the fabric. The table notes the properties such as appearance, thickness and warmth, tendency to curl and end uses.

2 3D PRINTED TEXTILE STRUCTURES

Figure 7 the researcher has collated of all 3D printed textile structures from the year 2000 to present day. The infographic identifies the designer, the material used and the process in which it is made and type of structure used in the design.

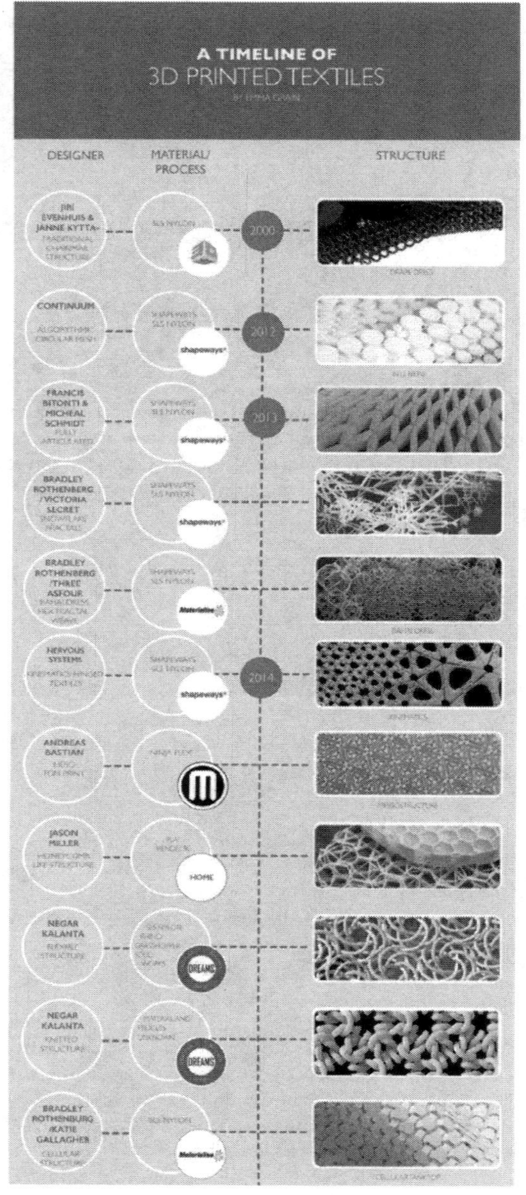

Figure 7. A timeline of 3d printed textiles (Grain, E 2016).

In the following sections the researcher has categorized structures into six main areas: Messo, Linked, Hinged, Flexible, Generative and Hybrid structures.

2.1 *Linked*

Chain mail is the most obvious choice of structure for a hard material, as it has been long functioning well in that role in protective garments since the middle ages. The earliest specimen of interlinked mail was excavated from a 3rd century B.C, Celtic grave in Romania (Williams, 1980).

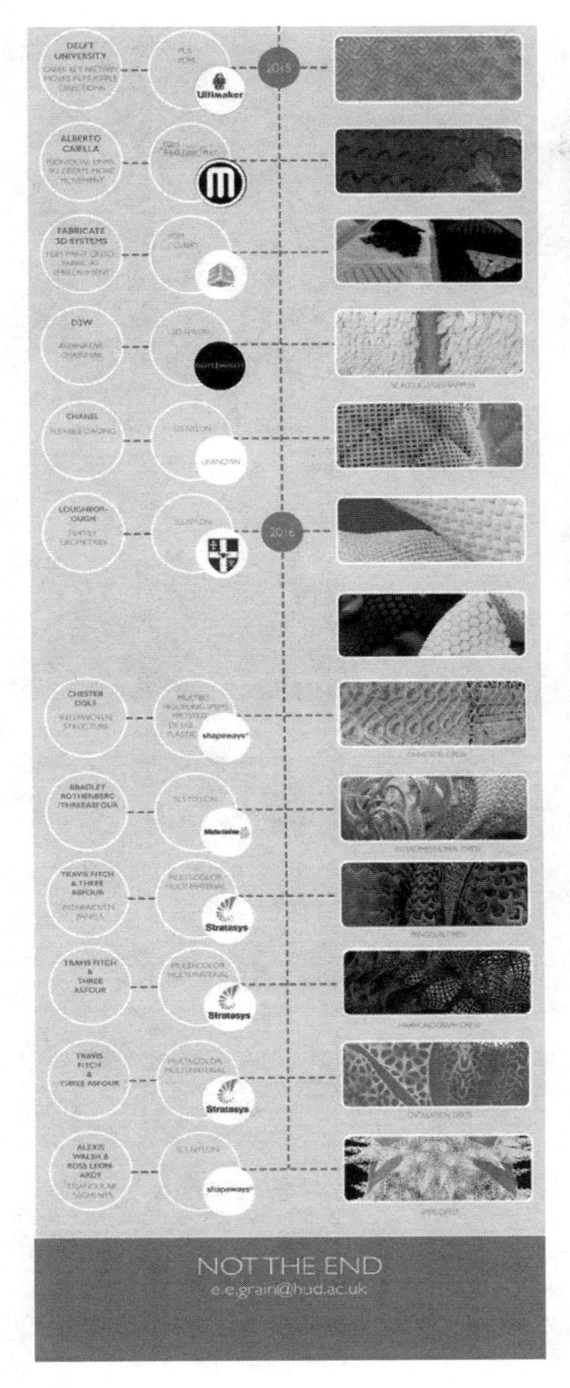

Figure 7. (*continued*)

Jiri Evenhuis, an industrial engineer from Netherlands developed a 3D printable fabric in the late 1990's. In 1999/2000 Jiri collaborated with another Dutch industrial designer, Janne Kyttanen and made the first ever 3D printed dress (Perepelkin, 2013). This chainmail was the first type of 3D printed textile by

Figure 8. Jiri Evenhuis Chainmail dress. Perepelkin (2013).

Figure 9. D2W SLS samples.

Freedom of Creation, which mimicked armor like textile structures from a hard Nylon using SLS. The utilization of such linkable geometry therefore provides the required free movement and drape characteristics inherent in conventional manufactured textiles (Bingham et al. 2007). This dress and a few other items in this collection are on permanent display at the MOMA museum in New York.

2.2 *Mesostructured*

Messo structures are those which have two main properties to allow them to work as they do, those being: synclastic bending and auxetic behavior. Synclastic materials have the fascinating ability to assume compound curvature along two directions allowing the structure to curve around a sphere with no folds or creases. Andreas described this type of structure on his Thingyverse profile where you can download the design to print, he said:

"Auxetic behavior is found in materials with a negative Poisson's ratio, which relates the deformation in

Figure 10. Andreas Bastian mesostructured (Thingiverse 2014).

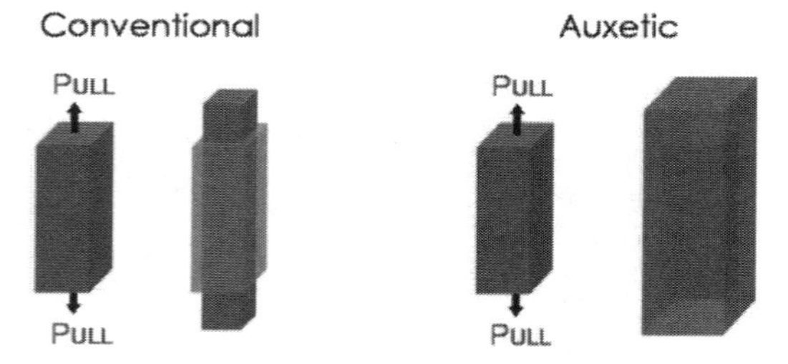

Figure 11. Auxetic structure (Thingiverse 2014).

one direction when the material is stressed in a perpendicular. When compressed in one direction, auxetic materials contract in the other, and when stretched, they expand" (Thingy verse, 2014).

Using these principles and looking at other designers who have done this in their designs such as Delft University with their Greek Key design and also in some on Andreas Zomperelli's designs. Messo Structures appear to be one of the best structures for the materials used in 3D printing because of its stretching properties.

2.3 *Hinged*

The Design Studio Nervous Systems founded in 2007 by Jessica Rosenkrantz and Jesse Louis-Rosenberg, which uses computer simulation to generate designs. One of their most successful design collections is Kinematics. Kinematics is a system of 4D printing that creates complex, foldable forms composed of articulated modules. The hinge mechanisms are 3D printed in-place and work straight out of the machine (Nervous Systems, 2016).

2.4 *Flexible*

Three ASFOUR worked with 3D designer Travis Fitch and Stratasys to create coloured multi material, flexible garments made up of 30 individual pieces for their Spring/Summer 17 collection showed at New York Fashion week.

Figure 12. Kinematics dress (Nervous Systems 2014).

Figure 13. Kinematics hinge (Nervous Systems 2014).

Figure 14. Oscillation dress three as four with Travis Fitch & Stratasys 2016.

Adi Gil of Three ASFOUR explains "3D printing is transformative for designers aiming to take complex designs and realize them as a wearable garment. In the case of 'Oscillation,' Stratasys 3D printing enabled us to visualize 3D patterns as they truly are – complex,

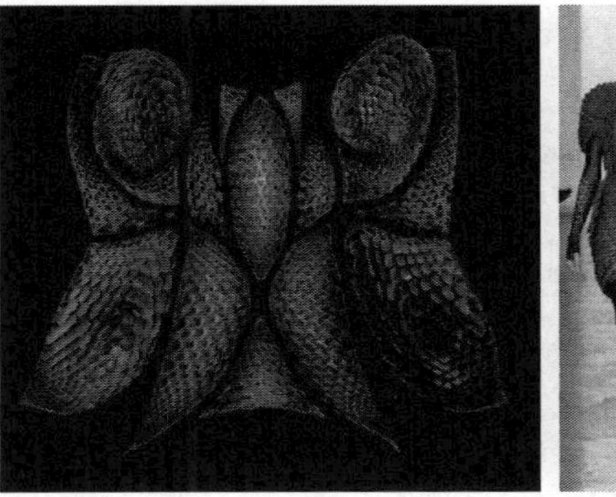

Figure 15. Pangolin dress three as four with Travis Fitch & Stratasys 2016.

Figure 16. Tissue Tessellate by Alessandro Zomparelli (Blender Brussels 2015).

interwoven circles of energy, transforming in shape, color and flexibility as they radiate around the body [...] The stellar parallax of the patterns, the way in which they transform as the viewing angle changes, is only possible through 3D printing" (Swack, 2016).

The Pangolin dress is designed based on a Three-ASFOUR signature piece which then was transformed into its 3D form using an interlocking woven skin using bio mimicry to create the different textures and movement throughout the piece (Stratasys, 2016) The Objet Connex3 3D Printer's precision and ability to vary material properties such as rigidity and color gradation, provided the designers with the geometric control to create nuanced, deliberately placed transformations in the membrane's porosity and flexibility, (Stratasys, 2016).

2.5 *Parametric modelling/generative modelling*

In an interview with Mingjin Lin in 2016 3D Printing industry magazine spoke about her 3D textiles practice. "With parametric modelling you input data rather than shape. That's the reason why I love parametric. It's almost organic, it allows shapes to grow out of data inputs. From nothing comes something. Grown Geometry". (Tovar, 2016). This cellular like design produces shapes unachievable by traditional methods but often by mimicking nature we can find solutions that really work and nature has already done the development for us. The emerging technology uses algorithms to generate every possible permutation of a design solution. The designer simply enters a set of parameters and then chooses the best outcome generated by the software (Howarth, 2017).

Figure 17. Mingjin Lin Inter-fashionality (3dprintingindustry.Com 2016).

Figure 18. Nervous systems V new balance.

Generative design, which mimics the way organisms evolve in the natural world, was the hot topic at Autodesk University – a three-day technology conference held in Las Vegas in November 2016.

New Balance (2016) mentions that through an exclusive collaboration with 3D Systems, they utilized the company's newly-developed laser sintering powder, Dura Form® TPU Elastomer to make significant advancements in the performance of printed parts for running shoes. Within this incredibly flexible midsole, hundreds of small, open cells provide cushioning and structure mimicking shapes from nature to create this highly engineered and high functioning midsole.

Figure 19. Carpace: Alessandro Zomparelli of Design Studio Mhox.

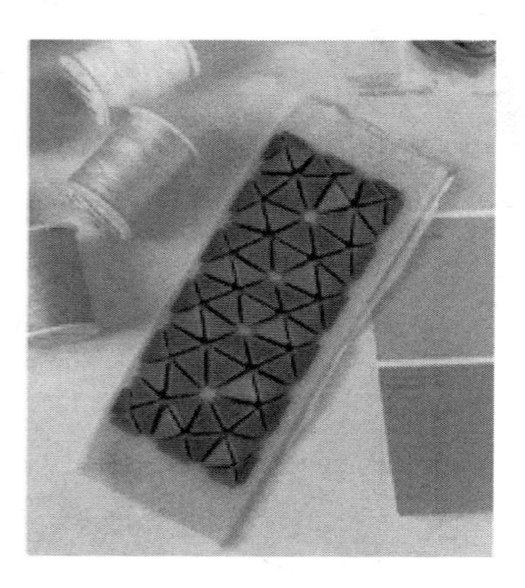

Figure 20. 3d systems fabricate 2015.

2.6 *Hybrid 3d Printed as an Application*

In the recent work of Mark Beecroft 2016 he describes building on his previous research into the possibility of printing knitted textile based structures by Selective Laser Sintering that exhibit the flexibility, stretch and strength of traditional textiles. As shown in Fig 21 the shapes and interlocking loops you'd find on your clothing under a microscope formed the starting point of their research (Alec, 2014). They found that the key to quality and flexibility in the samples would only be achieved from a high spec printer, this means costlier sampling but more effective results. They used the traditional knit and the 3D printing together in samples and called this 'cross fertilization of techniques, old and new'. The 3D print is manually knitted into the textile combining the two methods together as one.

Fabricate by 3D Systems uses a Cubify desktop 3D printing machine which uses FDM technology to print directly onto fabric as seen in Fig 20. After downloading your favorite design, a footprint layer must be printed onto the build plate. The printer then stops in order for you to add mesh fabric before printing the top layers. Once the print is complete, the mesh fabric can be sewn onto clothing or accessories for a customized look (3DPRINTR.COM 2015). Alike 3D Systems using Fabricate and Beecroft in his SLS knit structures, others have used 3D printing either directly onto fabric or joining it to it in some way. Pei et al (2015) has experimented with the adhesion of FDM polymer materials directly onto fabric in a number of decorative designs. This method is more of an embellishment and an addition to existing traditional textiles instead of a replacement for the traditional textile altogether so it always appears in conjunction with fabric rather than being treat as the fabric itself so it adorns the textile and is not worn alone.

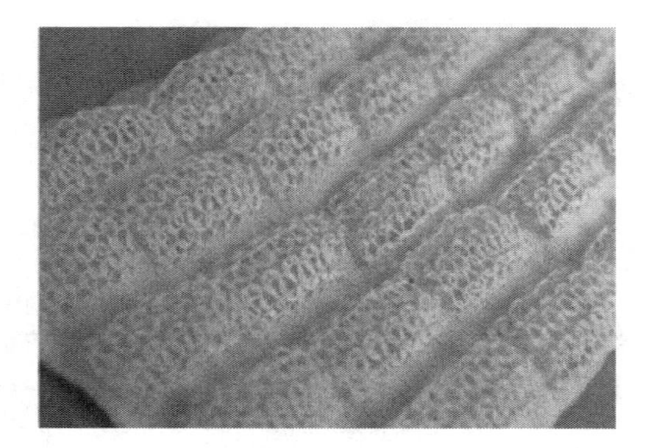

Figure 21. Design by Laura Mcpherson & Mark Beecroft 2014.

2.7 *Analysis of current 3d printed fashion*

The application of 3D printing in fashion, although not commercial and a relatively new application of 3D printing, has been explored by a large number of designers, 3D printing companies and also material manufacturers and students and educators. It is mostly used by students and in high fashion because of the access to 3D modelling, manufacturing and testing. Therefore, generally 3D printing companies and also printing material manufacturers sponsor this technology to demonstrate their potential capabilities.

Figure 22 is produced to categorize the 3D printed fashion into garments that are worn as a whole or adorn to the body as an accessory or attached to an existing traditionally sewn garment. The infographic also shows which manufacturer made the garment, the process and the material used.

2.8 *Adorn or worn?*

To adorn is to decorate or make more beautiful or attractive this has been traditionally the case with jewels or beads or added applications of decoration in the fashion industry. Therefore, the notion of adornment is the item which adorns and in this case it means the

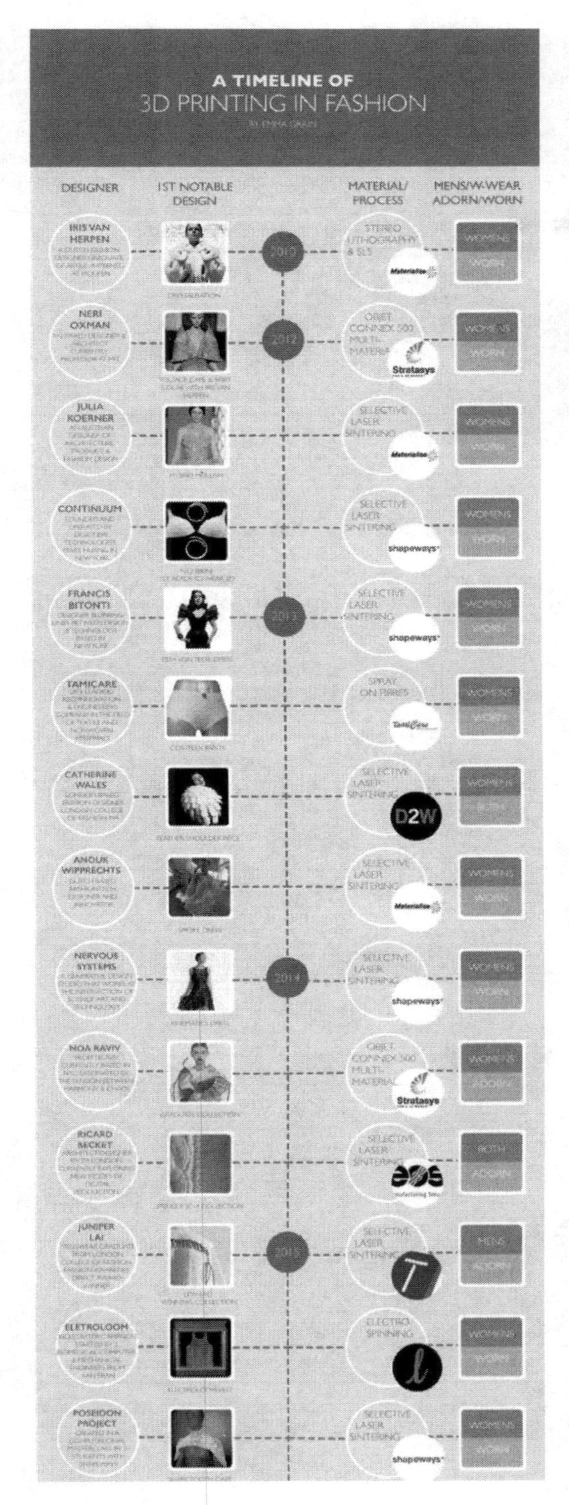

Figure 22. 3d printed fashion timeline (Copyright Grain. E).

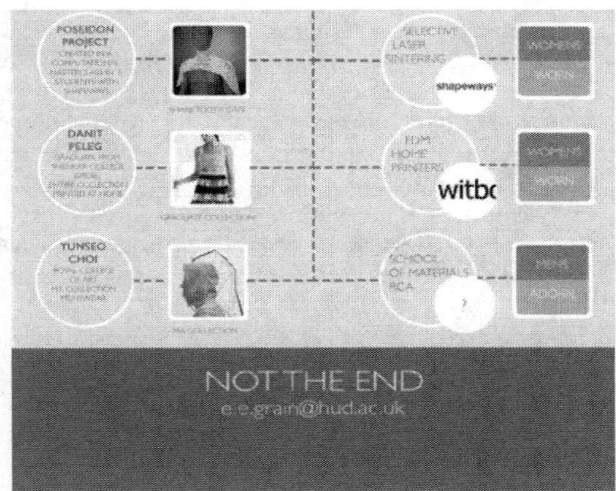

Figure 22. (*continued*)

3D printed structures which are attached to an existing garment or layered on top, acting almost like jewelry or an accessory to the outfit. To wear something or for an item to be worn it has to be on one's body as clothing which has the function to protect or simply to cover up. The researcher has thus identified which of the existing 3D printed fashion designs are either worn or adorn and what this means in relation to how it is designed. For a 3D printed garment to be worn it has to be physically wearable, but for it to be merely adorned it can be attached to something else already wearable or be more sculptural or decorative. In design terms to adorn is much easier at present because of the available materials, but as materials and, manufacturing develop in sophistication, wear-ability will become a more achievable property to attain.

3 CONCLUSION

To conclude, within the research into multiple 3D printed textile structures, it has been found there are 6 main categories, Messo, Linked, Hinged, Flexible, Generative and Hybrid. Each varying in material depending on the properties of the structures. Just like in traditional fashion, certain materials are only used in certain structures whether that be knits, woven or none wovens, each material is selected with the structure in mind and visa-versa. Some of these categories are more suited to being adorned and some more suited to being worn based on whether they are flexible in material or structure. This initial stage of textile design when designing 3D printed fashion is a critical aspect which cannot be ignored if you want the fashion piece to be a garment rather than a sculpture, and it is only with the development of new materials and machines that will allow us to design more flexible and fluid 3D printed textiles with the properties of traditional textiles but with a totally contemporary design and possible new features that traditional fabrics cannot achieve.

Going forward building on this research new textile structures will be designed using the 3D software 3Ds Max and manufactured using multiple 3D printers from FDM to SLS and utilizing the information found in analysing the existing properties of the 3D printed structures that already exist and act as textiles. Already new materials and methods of additive manufacturing are being developed and with interest in the recycled materials element in particular the researcher aims to 3D print new innovative textile structures using recycled polyester garments in a larger project which will be the focus of her PhD.

REFERENCES

Bingham, G. A. (2007). The generation of 3D data for rapid manufactured textiles (PhD Thesis). Loughborough University, Loughborough.

Bingham, G. A., Hague, R. J. M., Tuck, C. J., Long, A. C. Crookston, J. J., & Sherburn, M. N. (2007). Rapid manufac tured textiles. International Journal of Computer Integrated Manufacturing, 20(1), 96–105. doi:10.1080/09511920600690434 (Bingham et al 2007).

Electroloom. (2014). Electroloom. Retrieved from http://www.electroloom.com.

Grain, E. (2016). 3d Printing Fashion With Recycled. Polyester: A Sustainable Journey. In IFFTI, Beijing, Beijing Institute of Fashion and Technology, Retrieved from http://iffti2016.bift.edu.cn.

Hallett, C., & Johnston, A. (2010). Fabric for fashion: A com prehensive guide to natural fibers. London: Laurence King.

Hobson, B. (2015). Using computer algorithms in design is like "sculpting with a new material". Retrieved from https://www.dezeen.com/2015/12/09/video-interview-alessandro-zomparelli-mhox-generative-design-sculpting-new-material-movie/.

Howarth, D. (2015). Adidas combines ocean plastic and 3D printing for eco-friendly trainers. Retrieved from https://www.dezeen.com/2015/12/12/adidas-ocean-plastic-3d-printing-eco-friendly-trainers/.

New Balance. (2016). The future of running is here. Retrieved from http://www.newbalance.com/article?id=4041.

Nervous Systems. (2014). Kinematics Collection. Retrieved from https://n-e-r-v-o-u-s.com/shop/line.php?code=15.

Pei, E., Shen, J., & Watling, J. (2015). Direct 3D printing of polymers onto textiles: Experimental studies and applications. Rapid Prototyping Journal, 21(5), 556–571. doi:10.1108/RPJ-09-2014-0126

PROTEIN (2015). Meet the New York-based fashion designer who wants us to swap craftwork for algorithms. Retrieved from https://www.prote.in/journal/articles/francis-bitonti.

Richard-Beckett.com. (2014). 3D Printed Knitwear - Pringle of Scotland AW 2014. Retrieved from http://www.richard-beckett.com/pringle-3d-fabrics.html.

Stratasys. (2016). Stratasys and three ASFOUR Unveil 3D Printed OSCILLATION Dress as Part of Quantum Vibrations Collection. Retrieved from xhttp://investors.stratasys.com/releasedetail.cfm?ReleaseID=989196.

Swack, M. (2016). New York Fashion Week: New Movement in 3D Printed Fashion By ThreeASFOUR, Travis Fitch and Stratasys [Web log post]. Retrieved from http://blog.stratasys.com/2016/02/16/3d-printed-dresses-new-york-fashion-week/.

Thingiverse. (2014). Mesostructured Cellular Materials: Early Prototypes by Andreas Bastian. Retrieved from http://www.thingiverse.com/thing:289650.

Williams, A. (1980). The manufacture of Mail in Medieval Europe: A Technical Note. Gladius. XV, 105–135.

Yunseo Choi. (2015). Yunseo Choi. Retrieved from http://www.rca.ac.uk/students/yunseo-choi/.

Zomparelli, A. (2016, Dec 17). Retrieved from https://www.instagram.com/p/BOIUukMFmzD/?taken-by=alessandro_zomparelli&hl=en.

Textiles, Identity and Innovation: Design the Future – Montagna & Carvalho (Eds)
© 2019 Taylor & Francis Group, London, ISBN 978-1-138-29611-4

Fashion illustration on the surface design products created by designer Ronaldo Fraga

Rodrigo Bessa & Anna Mae Barbosa
University Anhembim Morumbi, São Paulo, Brazil

ABSTRACT: This article looks into the work of Brazilian illustrator and designer Ronaldo Fraga, whose fashion illustration present on the surface drawings of products aimed to the market is the identity of his work. It also approaches objects from the exhibit "Ronaldo Fraga – Caderno de Roupas, Memórias e Croquis" "Ronaldo Fraga – Clothes notebook, Memories and Sketches", which happened in 2012 at Casa Fiat de Cultura, in the city of Belo Horizonte, Minas Gerais, Brazil. Plus, the article proposes a reflection on this fashion designer's drawings present on his surface design, which makes a dialogue between art and design possible.

1 INTRODUCTION

In Brazil, surface design is a relatively new area of studies. The research and reflection on the field began in the 80s, when its definition and concept resulted from a partnership between Brazilian authors Renata Rubim (2004) and Evelice Anicet Rüthschilling (2008).

SDA, the Surface Design Association, headquartered in The United States, deals only with the textile area on its website "surfacedesign.org", but it is important to point out that Rubim (2004) states that the concept covers not only all specialties from textile design, but also paper, pottery, plastic, rubber and overall drawings or colors on the surface of utensils.

This article presents the fashion illustration by Brazilian fashion designer Ronaldo Fraga, as the main identity and feature contained in his creations for the surface design of several products, which turned out as a book and exhibit "Ronaldo Fraga - Caderno de Roupas, Memórias e Croquis" - "Ronaldo Fraga – Clothes notebook, Memories and Sketches", which happened in 2012 at Casa Fiat de Cultura, in the city of Belo Horizonte, Minas Gerais, Brazil. Besides that, the article proposes a reflection on the relationship between traits of these drawings and Brazilian social and cultural criticism present on this fashion designer's surface design, which makes a dialogue between art and design possible.

2 HISTORICAL CONTEXTO OF THE IMPORTANCE OF DRAWING

In Prehistory, at the Paleolithic stage (up to 10,000 BC), human beings already used graphic representations and expressed several forms of language, messages through surface drawings. Rock art being remarkable as one of the first surface drawings created by humankind in order to convince, orient or inform something. "[. . .] even earlier than writing, men drew" (OLIVEIRA; GARCEZ, 2014).

About 3,500 BC, based on the improvements of drawing, cuneiform writing appeared in Mesopotamia and hieroglyphics in Egypt, containing more than 2,000 symbols, which marked the end of this historical period and expanded the possibilities of the use of drawings. "The invention of writing was an indispensable achievement, from the historical civilizations of Egypt and Mesopotamia" (JANSON, 1996).

Between 3,000 and 2,500 BC, drawing starts being regarded as a sacred status, especially used to decorate the surfaces of temples and tombs within the pyramids of Egypt.

Throughout history, drawing has showed several languages, since Prehistory, in which surface design exposed the rock character in the Paleolithic stage; until the Middle Age, in which drawing reached high levels of representativeness and, after, showed figurative elements of representation, exclusively related to the artist's concepts and beliefs.

Drawing has always been the protagonist of the arts in representing ideas, lifestyles and identity of each historical period. "Surfaces have always supported men's necessity of expressing themselves symbolically. Such ingrained habits can help to understand, by means of a timeline, where the genesis of surface design lies" (RÜTHSCHILING, 2008).

In the fourteenth and fifteenth centuries, drawing had a meaningful change in the history of arts, highlight the works of the Renaissance period, since the representations started to have perspective, unlike what happened, for example, in medieval illustrations. "In the last years of his life, Leonardo da Vinci dedicated himself to his scientific interests more and more. Art and science united for the first time with the development of perspective, by Brunelleschi; Leonardo's works constitute the climax of that trend. [...]" (JANSON, 1996).

With the Industrial Revolution, a chain of changes occurred in Europe in the 18th and 19th centuries, especially the illustrative modality that perfected the industrial design, which was restricted to projections that supported the production process of machinery, and directed professionals towards production. This style of drawing emphasized the perfection of proportions and forms of objects. "[...] Artists with academic degree, for being the only ones majored in drawing, used to be more and more hired by manufacturers in order to create formal concepts and decorations according to the dominant taste [...]" (HESKETT, 2008).

In the late nineteenth and early twentieth centuries, the drawings of Art Noveau influenced the aesthetics of design, architecture, visual arts and fabric.

Two school experiences are remarkable in the twentieth century in which the application of drawing is concerned; the first one took place at Bauhaus State School, and the second one at Moore School of the University of Pennsylvania.

In 1919, in opposition to the first world war, Walter Groupius (1886-1969) founded the Bauhaus State School in Weimar, Germany. The certainty of being able to have the power of art overpower the industry's, joining architecture, art craft, and an art academy, linking to rationalism and functionality to artistic attributes. "Bauhaus was a place in which several vangard branches came together and dealt with the production of typography, advertising, products, painting and architecture." (LUPTON, *et al.*, 2009).

In 1946, in a partnership between Moore School of the University of Pennsylvania and the Ballistics laboratory of the Test Field of Abeeen, the first computer with technological reference that we know in contemporaneity was launched.

As a result of accessibility and the society's need of technology, computers spread slowly, and mainstream digital resources were used more and more to create drawings, improve them and/or combine them to other elements, which determined new possibilities and illustrative identities.

Later to these occurrences, with the improvement of style, the representation became known as technical drawing. "During many centuries, drawing, known today as technical, was not committed to rules and implementing regulations, due to difficulties displaying the volumetric of shapes on flat surfaces. [...]" (TRINDADE, 2012).

Currently, the forms of representation of a drawing are innumerable, because of the range of artistic movements nowadays and the quantity of material and technological resources that can be attributed to an execution. Drawings with an artistic character will limit themselves to absolutely nothing, whereas drawings with a technical character, as a result of technology, demand ever more preciseness, with no space for ambiguities.

In the latest century, humankind witnessed a social, industrial and economic development. Machines became more and more modern, and consequently, society became more and more addicted to this process, especially the surface design for industrial design making.

3 SURFACE DESIGN

In Brazil, the term surface design is a recent concept, used by Brazilian researchers and designers. According to Rüthschilling, (2008), Brazilian surface design appeared in the state of Rio Grande do Sul, as an autonomous knowledge field and professional practice in the 80s, result of historical investigation and market experience whose analysis caused the awareness of the concept. The definition came from a research between Brazilian authors Renata Rubim and Evelice Anicet Rüthschilling.

According to Rubin (2004), the term surface design is little known in Brazil. In the United States it is used to determine a project elaborated by a designer, when the creation on a surface uses color or treatment, being industrial or not.

According to the Surface Design Association (SDA), surface design finds its roots spread in time and space, since the most concrete reference is this foundation in 1977, in the United States of America, made by an association of textile artists, aiming to improve the knowledge, understanding and appreciation of textile items in consumers of art and design, as well as in the general public.

To the SDA, surface design has as characteristics the coloration, patterns and structures of fibers and fabrics. It involves processes of dyeing, printing, painting, embroidery, beautification, quilting, weaving, knitting, felt and papermaking. Surface design is in everyday life, once this design represents the creation of two-dimensional imaging, designed specifically for the generation of standards developed on surfaces in various industrial and craft areas, such as textiles, stationery, ceramics, synthetic materials, among others

Rubim (2004) rules that a Professional in the área of surface design must have in his her profile knowledge about the technological process involving the development of the product, once that is going to lead to problem solutions and productive process management.

Rüthschilling, (2008), agrees with Rubim (2004), when saying that the surface designer must know about techniques and processes, once this knowledge is necessary for better results when it comes to creation and visual effects.

However, there are several design constraints interfering in the work of a surface designer. "[...] such as limitations in the productive process, available Technologies, existing machinery and equipment, target market needs, company's issues and market's" (RÜTHSCHILLING, 2008). All those aspects must be considered and observed by the designer before starting a project.

When it comes to the creative process, the surface design is a receptacle for human expression and possibility for various aesthetic and projective elements in the context of the designer's cultural evolution.

4 RONALDO FRAGA'S FASHION ILLUSTRATIONS ON THE SURFACE DESIGN OF PRODUTS

In contemporaneity, fashion illustrations are more and more present in the design of products, Brazilian and international advertising campaigns from several brands, couture and prêt-à-porter designers. Besides, fashion, artistic or design drawings started being seen in museums and art galleries.

Thus it is fair to say we live a moment of redemption of the value of illustrators' traits, that were remarkable in the history of design in the eighteenth and nineteenth centuries, and became known and recognized by the professional market and final consumers in the twenty-first century.

In addition to artistic expression, drawing is the creative expression present in the surface design of objects. According to Rubin (2004), surface design is "[. . .] any design by a designer, with respect to the treatment and color used on an industrial or non-industrial surface. [. . .]."

The integration between art and fashion design also brings important contributions for the surface design. In this sense, this current study looks into the works of the fashion designer and illustrator Ronaldo Fraga (Fig. 1), known in the Brazilian scenario for drawing the surfaces of several fashion collections, books, household objects, ecobags, among other product. In the latest years, the notoriety of his work has given the designer invitations to exhibit his surface drawings in spaces dedicated to art.

According to Belchior and Ribeiro (2014), Ronaldo Fraga graduated at the Federal University of Minas Gerais, in Brazil, and also studied in New York, at Parson's School. It's important to highlight that his creations have already been exhibited in more than seven countries, and in 2014 he was selected by the Design Museum in London as one of the seven most innovative fashion designer in the world. Throughout his career, he has already participated in Brazilian dance and theater as a set and costume designer, and as a surface designer of several products.

Illustration is present in Ronaldo Fraga's (Fig. 1) professional everyday, and he says in his book "Cadernos de Roupas, Memórias e Croquis" "Clothes notebook, Memories and Sketches" (2013) that he ritually records everything that will be produced for the next fashion show, from the theme to the fabric, from the color chart to the prints, everything is recorded in his books of creative process.

Figure 1. Ronaldo Fraga launching the book "Caderno de roupas, memórias e croquis" - "Ronaldo Fraga – Clothes notebook, Memories and Sketches."

Figure 2. Fashion Illustration on the surface design of produts created by designer Ronaldo Fraga.

From 2013 on, Ronaldo Fraga started signing several items with his fashion illustrations. His fashion drawings didn't only have the purpose of creating a piece of clothing, but also became the surface design of products, such as school supplies, cushions, bed sheets, glasses, cups and an infinity of objects for home (Fig. 2).

In 2014, Ronaldo Fraga illustrated (Fig. 3) another collection for the chain of supermarkets Verdemar, in honor of the 117 years of Belo Horizonte, Minas Gerais state capital, in Brazil. That was the third partnership between the designer and that supermarket chain, as his first two collections were launched in 2009 and 2011.

In addition to the plastic bags (Fig. 3) , the designer from Minas Gerais also designed a line of products for home, for the supermarket chain "Verdemar", with fashion illustrations like the surface design of plates and buckets.

It is noteworthy that Ronaldo Fraga generally uses the fashion illustrations also as the surface design of patterns of some pieces of his collections for adults and for children (Fig. 4) as well, in the line "Ronaldo Fraga para filhotes" "Ronaldo Fraga for puppies".

The trait of the human figure is the main identity of his authorial work as a fashion designer, this need of

Figure 3. Returnable bags of the supermarket chain Verdemar created by Ronaldo Fraga.

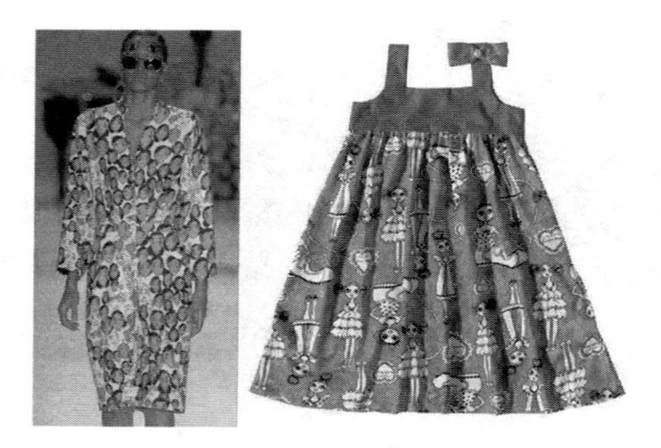

Figure 4. Ronaldo Fraga's surface design on the patterns of dresses for adults and children.

drawing fashion illustrations on the surface design of products, especially the textile design, has been present in his creations since the 90s.

Rüthschilling, (2008, p. 23), comments that "Surface design is a creative activity and technique that is concerned with the creation and development of aesthetic, functional and structural quality, specifically designed for the constitution and/or surface treatments, according to the socio-cultural context and productive processes".

On the catwalk, on April 25, 2016, Ronaldo Fraga, with his summer collection for 2017, promoted once again the fashion illustrations, presenting the main visual identity of his work as designer. Besides the representations of the sketches present in the surface design of the pieces that paraded in the event Sao Paulo Fashion Week – Summer Edition 2017, throughout the show, images of the stylist's hand were displayed on the screen, illustrating the fashion drawings present in the event.

The designer's fashion illustration work is also recognized by curators of art galleries in Brazil. In 2012, the exhibit "Ronaldo Fraga – Caderno de Roupas, Memórias e Croquis" "Ronaldo Fraga – Clothes notebook, Memories and Sketches" took place at Casa Fiat de Cultura in the city of Belo Horizonte, in the state of Minas Gerais, Brazil (Fig. 5). At the facility surface

Figure 5. Exhibit "Ronaldo Fraga – Caderno de Roupas, Memórias e Croquis." - "Ronaldo Fraga – Clothes notebook, Memories and Sketches."

design of dresses, graphic materials, videos, objects and designs of various products were exposed.

The work of art, like the design, results from an open object, that is subject to the different interpretations, according to PIRES et al. (2008, p. 39–40), "[. . .] recreations and re-readings from the user/observer/viewer and the ones who interact with the objects and processes that use digital technologies and virtual and mobility systems, as well as other designers' or creative professionals from the most diverse professional segments".

Rüthschilling (2008), states that surface design is heir of art, in which the freedom of creation is made through the domain of visual language and the author's own logical creativity. This approach is very clear for artists, but misunderstood by other scholars and professionals.

Rubim (2005) points out that surface design is not only an artistic project, because it brings together factors such as reproduction, communication, aesthetic and design, as well as an appropriate methodology and symbolic value.

In this context, Ronaldo Fraga's illustration work calls the attention of Brazilian and international press, for its capacity of social and political criticism, and the need of reflecting about and broadcasting Brazilian culture, when creating drawings on the surfaces of several products. And the creative processes of these objects enable a dialogue between art and design.

5 CONCLUSIONS

The technical fashion sketch is a necessary tool to design for industry. The fashion illustration is the designer's artistic expression, with the purpose of conveying, by means of image, a new behavior, style, habit or way of life. This way, unlike the fashion designer drawing, that aims to represent products that are close to the reality of the prototype, the illustrator has the freedom to make the fashion image in a dialogue with artistic and aesthetic trends in the visual arts.

Acting as a fashion designer or a fashion illustrator is to work with the existing relationships between design and art. First, for the fact that industrial design is related to research, artistic creation, drawing, conception, project, prototype and production; secondly, because both art and design are a reflection of their time and society, from the discourse of an object open and subject to various interpretations, (re) creations or (re) readings.

It is a fact that both art and contemporary design are facets of an increasingly complex reality. This way it is possible to say that fashion illustration dialogues with art and design, once art is its aesthetic reference, techniques, and repertoire in the professionals' creation process.

However, today's fashion illustrations are present in museums and art galleries, as well as other works by various visual artists. Even being exhibited at those environments, fashion illustration is not always regarded as a work of art, either the illustrator seen as an artist.

Considering the fashion illustrations presented in the case study of this article, it is possible to recognize in Ronaldo Fraga's drawings the identity of the traces of his representations, seen by many as an artistic expression and signature of the professional in the area of fashion design. Besides the importance that the fashion designer has in the fashion and visual art scenario in Brazil, it is noteworthy saying that his illustrations are intended to criticize society and politics, referring to Brazilian cultural themes, through surface drawings and his creations in industrial design.

REFERENCES

Casagrande, et al. Mulheres na informática: quais foram as pioneiras? Available at <http://www.academia.edu/5791164/ENIAC_o_primeiro_computador_eletr%C3%B4nico_digital>, accessed on July 20, 2017.

Fraga, Ronaldo. Caderno de roupas, memórias e croquis. 1° ed. Rio de Janeiro: Cabogó, 2003.

Fraga, Ronaldo. Short resume. Available at <http://ronaldofraga.com/blog/>, accessed on May 5, 2016.

Fraga, Ronaldo. Figure 1. Exihibit and Launching the book: "Caderno de roupas, memórias e croquis" - "Ronaldo Fraga – Clothes notebook, Memories and Sketches." Source: http://revistadonna clicrbs. com. br /moda/ ronaldo-fraga-apaixonado-por-desenho-estilista-reune-suas-colecoes-em-livro/, accessed on July 22, 2017.

Fraga, Ronaldo. Figure 2. Fashion Illustration on the surface design of produts. Source: <http://lourdesporlourdes.com.br/?p=7478> , accessed on July 22, 2017.

Fraga, Ronaldo. Figure 3. Returnable bags of the supermarket chain Verdemar. Source: <http://lourdesporlourdes.com.br/?p=7478>, accessed on July 22, 2017.

Fraga, Ronaldo. Figure 4. Ronaldo Fraga's surface design on the patterns of dresses for children. Source: <http://mulher.uol.com.br/moda/álbum/2014/11/03/spfw-inverno-2015-veja-as-apostas-das-marcas-que-participam-do-evento.htm>, Source: <http://www.criancabemvestida.com.br/2014/11/ronaldo-fraga-para-carinhoso.html>, accessed on July 22, 2017.

Fraga, Ronaldo. Figure 5. Exihibit "Ronaldo Fraga – Caderno de roupas, memórias e croquis" – "Ronaldo Fraga – Clothes notebook, Memories and Sketches." Source: <http://guia.uol.com.br/belo-horizonte/noticias/2012/10/09/exposicao-gratuita-em-bh-conta-com-criacoes-e-desenhos-do-estilista-ronaldo-fraga.htm>, accessed on July 22, 2017.

Heskett, John. Design. São Paulo: Ática, 2008.

Janson, H. W.; Janson, Anthony F. *Iniciação* à História da *Arte*/H. W. *Janson*, Anthony F. *Janson*; [tradução Jefferson Luiz Camargo] – 2ª ed. – São Paulo : Martins Fontes, 1996.

LUPTON, Ellen; Miller, J. Abbot (orgs.). Abc da Bauhaus: a Bauhaus e a teoria do design. Ed. Cosac Naify, SP, 2009.

Nunnelly, Carol A. Enciclopédia das técnicas de ilustração de moda. Tradução: Márcia Longarço. Editora Gustavo Gili, Barcelona, Espanha, 2012.

Oliveira, Jô; Garcez, Lucília. Explicando Arte: uma iniciação para entender e apreciar as artes visuais. Rio de Janeiro: Ediouro. 6° Ed, 2004.

Pipes, Alan. Desenho para designers: habilidades de desenho, esboços de conceito, design auxiliado por computador, ilustração, ferramentas e materiais, apresentações, técnicas de produção. Tradução: Marcelo A. Alves. São Paulo: Editora Blucher, 2010.

Pires, Dorotéia Bady (Org.). Design de Moda: olhares diversos. São Paulo: Estação das Letras e Cores, 2008.

Queiroz. João Rodolfo; BOTELHO, Reinaldo (Org.). Ronaldo Fraga. (Coleção Moda Brasileira; 4). São Paulo: Cosac Naify, 2007.

Ribeiro, Rita A. C.; Belchior, Camilo. Design & Arte: entre os limites e as interseções. 1° Ed. Contagem. Do autor, 2014.

Rubim, Renata. Desenhando a superfície. São Paulo: Edições Rosari, 2004.

Rüthschilling, Evelice Anicet. Design de Superfície. Porto Alegre: Ed. Da UFRGS, 2008.

SDA, the Surface Design Association. Mission and History, available at <surfacedesign.org>, accessed on July 22, 2017.

Trinidade, Bernardete. Ambiente híbrido para a aprendizagem dos fundamentos de desenho técnico para as engenharias. Florianópolis, 2002. 118f. Tese de doutoramento em Engenharia de produção – Programa de Pós-graduação em Engenharia de Produção, UFSC, 2002. Available at <https:// repositorio.ufsc.br/handle/123456789/83731>, accessed on July 22, 2017.

Textiles, Identity and Innovation: Design the Future – Montagna & Carvalho (Eds)
© 2019 Taylor & Francis Group, London, ISBN 978-1-138-29611-4

When clothing comfort meets aesthetics

L.L. Matté
Department of Textile Engineering, University of Minho, Portugal
Department of Fashion Design, Federal University of Technology, Brazil

A.C. Broega
Department of Textile Engineering, University of Minho, Portugal

M.E.B. Pinto
Department of Psychology and Psychoanalysis, Londrina State University, Brazil

ABSTRACT: This paper explores the results obtained in the first phase of a PhD research on psychological comfort of clothing. At this initial stage, a variation of the Delphi Method was used to reach for the opinion from the experts about the concept of psychological comfort and which factors would interfere in its perception. The results converge to suggest that aspects related to aesthetics are one of the most important attributes to be considered in the process of assessing the psychological dimension of clothing comfort. Therefore, the present article supports the hypothesis raised by the literary revision, which argues, that the aesthetic dimension has great relevance for the comfort of clothing. It also corroborates that investigations in the psychological comfort of clothing can benefit from considering the importance of aesthetics in the interaction between person-clothing-environment, both in terms of appearance (vision) and in the pleasure promoted by clothing and perceived by the totality of the senses.

1 INTRODUCTION

One of the legacies of modern thought is the separation that splits culture into two mutually exclusive branches: one scientific, quantifiable and "hard", the other aesthetic, evaluative and "soft". While still prevalent, such division has gradually faded out in favor of greater connection and exchange between different areas of knowledge. Another characteristic of contemporaneity is the "aestheticization" of everyday life, in which objects, values, places and relationships, in short, all spheres of life, are surrounded by an aesthetic aura (Flusser 1999; Maffesoli 2012).

Contemporaneity has been marked by this permeability, or fluidity of ideas, which allows "hard" areas such as product engineering, ergonomics, and related human factors, to include more "soft" approaches such as affective ergonomics, pleasure-based human factors and the use of methods such as Kansei Engineering (methodology, developed by Mitsuo Nagamachi), dedicated to understanding the feelings or psychological impressions of a user in relation to a product. The term of Japanese origin encompasses several concepts related to sensitivity, senses, feelings, emotion, affection, intuition and aesthetics (Nagamachi 2002; Helander & Khalid 2005), since they consider the increasing importance of aesthetics, pleasure and emotions in the relation of people to products.

The total experience with a product of clothing, encompasses stimuli and perceptions that go beyond the field of usability, products should also provide enjoyable and perhaps pleasurable experiences. Therefore, not only must the requirements for technical functionality, manufacturing and ergonomics be met, but attention must be given to the entire experience, more specifically to the aesthetics of the form and the quality of the interaction. Dressing, "is no longer merely functional," if at any time it was. The act of dressing increasingly takes on a permanent aesthetic function (Norman 2013; Jordan & Green 2005; Baitello Jr. 2006).

Comfort plays a key role in enhancing product interaction and countless contributions have already been made by the Science of Comfort to promote the well-being of the individual. In the field of clothing comfort, research has focused mainly on the physical-mechanical and/or thermophysiological or even ergonomic comfort, and little regard is given to the more subjective aspects, namely, the aesthetic and emotional aspects, related to the needs Individual or social contexts (Broega 2007). Yet some of the most important researchers in the field recognize that physiological properties "are not the whole story" by stating that, between the most basic perception of suitability and ostentatious conspicuous consumption, there is an important component of self-confidence and being at ease (Fourt & Hollies 1969).

2 CLOTHING COMFORT

In the interface between the human body and its surrounding environment, clothing plays a determinant role in the subjective perception of a user's comfort status. In the scope of clothing, comfort includes the physical, physiological, and social-psychological balance between a person, his clothing and his environment (Branson & Sweeney 1991). This balance is fundamental when it comes to the person-clothing relationship, and the particularities that involve the use of a product that interacts continuously and dynamically with the body, and is often compared to a second skin (Hosseini Ravandi & Valizadeh 2011; Sanches et al., 2015).

Although there is no definitive consensual definition, most researchers argue that comfort is a subjective and multidimensional and experience (Slater 1986, Kamalha et al., 2013, Fan 2009 and Li 2001). Such multidimensionality is illustrated in the work of various comfort theorists who point out to three or more dimensions of comfort, which in turn present themselves in the interrelation between three spheres: person, clothing and environment. Fourt & Hollies, were the pioneers to emphasize the importance of the dynamic person-clothing-environment, and proposing a model of comfort underlying this triad (Fourt & Hollies 1969).

Slater, in defining comfort in general, reveals the complexity of the phenomenon by identifying three dimensions of comfort, namely: physical, physiological and psychological. According to the author, the three factors cannot be separated completely by the considerable number of intersections and points of "common ground" between them (Slater 1977; Slater 1986; Slater 1997).

Building upon the contributions of Fourt & Hollies and Slater, several researchers have been working to specify the dimensions and respective variables of clothing comfort, and to understand the relationships established between them.

The comfort of clothing, for Li (2001), includes 4 main aspects, being those: thermophysiological comfort, sensory comfort, comfort of body movement and comfort of aesthetic appeal, or psycho-aesthetic comfort. The typologies of comfort are defined by the author as follows:

- Thermophysiological comfort: it translates a thermal and moisture state to the surface of the comfortable skin. It involves the transfer of heat and water vapor through textiles or clothing,
- Sensory Comfort ("touch/hand"): set of various neural sensations, when a textile comes into direct contact with the skin,
- Ergonomic comfort: the ability of a garment to "dress well" and to allow freedom of movement of the body,
- Psycho-Aesthetic Comfort: subjective perception of aesthetic evaluation, based on vision, touch, hearing and smell, which contribute to the total well-being of the wearer (Li 2001; Wong & Li 2006; Broega 2007).

The researches on the comfort of clothing, must pay attention to the various dimensions and attributes involved in the perception of comfort, which go beyond physical and physiological parameters, such as; weight, thickness, heat transfer, air permeability, moisture absorption and diffusion, handling and ease of movement; and also consider, parameters related to aesthetics, such as: color, brightness, fashion and style, pondering the culture and the subjectivity of the user. It is to these latter parameters that the studies of the psychological dimension of comfort are dedicated (Hosseini Ravandi & Valizadeh 2011).

3 PSYCHOLOGICAL CLOTHING COMFORT

Psychological comfort is a process of hedonic judgment, whereby the brain forms a subjective perception of sensory stimuli, influenced by many factors. By its subjective nature, psychological comfort is affected by personal idiosyncrasies and, therefore, it is difficult to evaluate. The understanding of which factors contribute to the perception of psychological comfort is fundamental for the establishment of a valid concept, as well as for the definition of attributes to be considered when evaluating this dimension of comfort.

There are many models designed to explain the comfort of clothing that consider the psychological dimension as one of its components. Although they do not intend to present a clear set of attributes to be measured, nor to propose a specific evaluation method, such models contributed to the clarification of the concept of psychological comfort, as well as to elucidate the importance of the role of this aspect in the overall perception of clothing comfort (Slater 1986; Fan 2009).

The comfort model proposed by Branson & Sweeney (1991) is based on the suggestion that the attributes of the person, clothing and environment can be categorized between the physical and sociological dimensions. Such attributes interact by generating physiological / perceptual responses in the individual. Such responses go through a process of "filtering" that involves past experiences and memories, and then are converted in a judgment/evaluation of the comfort of clothing. For the authors, the attributes associated with the social psychological dimension are, as presented in figure 1: Personal Attributes, such as: "state of being"; "self-concept"; "personality"; "body image/catexis"; "values"; "attitudes"; "interests"; "awareness"; "religious beliefs"; "political beliefs". Attributes of Clothing, such as: "fabric-clothing system"; "aesthetics"; "style"; "fashionability"; "appropriateness"; "design"; "color"; "texture"; "body emphasis/de-emphasis". Environmental Attributes, exemplified as: "occasion/situation of wear"; "significant other"; "reference group"; "social

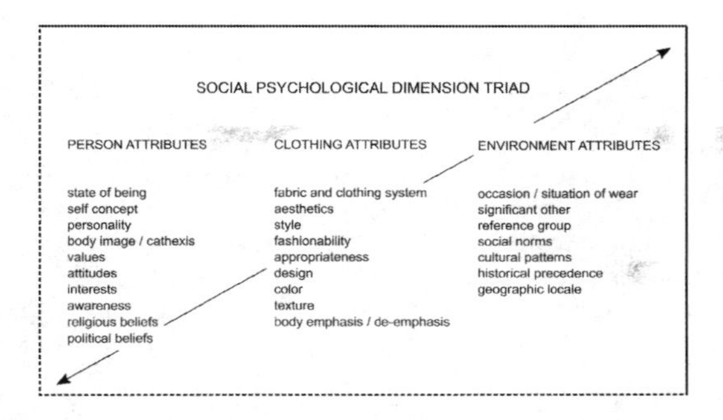

Figure 1. Social-psychological dimension of the garment comfort model proposed by Branson and Sweeney (Branson & Sweeney 1991).

norms"; "cultural patterns"; "historical precedence"; "geographic locale".

According to Fan (2009), psychological comfort occurs when the person has a sense of well-being derived from being secure in relation to his or her own appearance, such factors as aesthetics, cost, performance and social adequacy, some of the factors that contribute to such a sense of well-being. The link between psychological comfort, appearance and aesthetics is argued by Slater (1986) when he states that the "aesthetic clothing" according to the latest fashion gives the wearer mental comfort and a sense of good appearance (Slater 1986, 2001 Roy Choudhury et al., 2011).

For Housseini Ravandi & Valizadeh (2011), aesthetic parameters such as fashion, color harmony and fit are fundamental for psychological comfort. Yu (2011) further clarifies that the "aesthetic values" of color, style, shape, material mix and match, assembly and finishing can influence the appearance of clothing and physical comfort, as well as the sensation of psychological comfort, with intensified experiences and feelings of self-identity.

Among the authors studied here, who considered the psychological (psycho-aesthetic or even social psychological) dimension of the comfort of clothing in their models and definitions, the establishment of the "aesthetic" aspect is consonant as being one important component of the total perception of conformity.

4 AESTHETICS

Aesthetics, is the branch of philosophy dedicated to conceptual and theoretical research on art and aesthetic experience. Its first definition, in the modern sense, was made in the eighteenth century by the German philosopher Alexander Baumgarten in "Aesthetica", and marks its emergence as a philosophical discipline, alongside logic, metaphysics and ethics, concerned initially with the definition of beauty. The term "aesthetics" is rooted in the Greek word aisthesis, aistheton (sensation, sensitive) which can be translated as understanding through sensory perception (Hermann 2005; Hekkert & Leder 2008; Levinson 2009).

Although contemporaneity is characterized as a culture of "visuality," in which the world-consciousness tends to give itself through vision at first, the original Greek aesthetic meant "that which is perceptible through the senses," contemplating, not only the visual form, but the comprehensiveness of the information effects of all senses in human cognition and cognition (Macdonald 2002).

When it comes to "aesthetic perceptions," Burns & Lennon (1994) should consider them as perceptions of aesthetic objects involving sensory stimulation (encompassing the whole of the senses) and the resulting responses to evaluation, preferential or affective to the stimulus. These emotional or affective responses may be, for example, a judgment of the degree to which an object is beautiful or a judgment of the extent to which an object evokes pleasure in the perceiver.

In the field of Design and Consumer Behavior, several authors (Schifferstein & Spence 2008; Hekkert et al. 2003; Blijlevens et al. 2017; Hekkert & Leder 2008; Bell et al. 1991; Veryzer & Hutchinson 1998; Green & Jordan 2002) have already discussed the relations between aesthetics and usability, about the aesthetic properties that would be the source of the purchase preference, and more, about how aesthetics would affect the pleasure and behavior of the user in relation to the product.

In the area of psychology (especially in social psychology), aesthetics were investigated: in studies about the importance of appearance in the formation of first impressions (Burns & Lennon 1993); in the understanding of how cultural esthetic codes interfere in the construction of the appearance and still how the esthetic experience of traditional dress happens (Rudd & Lennon 1994); in how the physical and psychological characteristics of the individual interfere in the aesthetic preferences of clothing (Chattaraman & Rudd 2006); and even how aesthetic properties as colors can function as indicators of sexual interest (Elliot et al., 2013). It is observed that the researches aforementioned, that approach the aesthetics, consider, for the most part, only visual stimuli or solely evaluate, visually perceptible properties.

4.1 *Aesthetics and clothing*

The user's responses to the sensory stimuli of aesthetic products, such as textiles and clothing, occur on several levels simultaneously. In addition, in the consumer decision-making process, the sight and touch of textiles often interact with perceptions from other senses, being the aesthetic appreciation of a textile object, awakened from various stimuli: visual, tactile, olfactory , gustatory, auditory, kinesthetic/proprioceptive (refers to sense of position and body movement) and vestibular (refers to the vestibular system, responsible for the balance and maintenance of body orientation) (Burns & Lennon 1994; Macdonald 2002; Schifferstein & Spence 2008)

Joseph Kupfer, when lecturing on clothing and aesthetic experience, approaches clothing as an aesthetic object and describes, in an enlightening and elegant manner, the complexity of the aesthetic experiences of dress. For the author, clothing can enter the aesthetic experience, both as an object of perception and as an object of clothing. However, the aesthetic complexity of clothing increases when one considers the body in movement. The change or dynamism in the appearance of a dress that flutters in response to the movement or a piece that suffers from the wind, creates new visual perceptions, sometimes auditory, as the individual changes the posture or step (Kupfer 1994).

The aesthetic appreciation of clothing also includes the user's interaction with the garment during use, since clothing is an immediate environment, moving with and against the body, as a second skin. The appearance of clothing can be appreciated by others even better than by the wearer himself, however, the aesthetics of direct interaction is available only to the user (Kupfer 1994).

In addition to visual stimuli, such as the appreciation of color, texture, and even shadows created by garments, some materials provide auditory stimuli (such as the "*froufrou*" of layered skirts or the sound of the friction of synthetic fiber fabrics) or even olfactory stimuli, as the characteristic odor of a piece of skin, and this is one of the most important aspects of the aesthetics of this material (Kupfer 1994).

In addition to textiles, there are no materials that daily offer such rich and meaningful tactile interactions. Besides the action of the textile surface on the skin, there is also the influence of proprioception / kinesthetic on the muscles and joints, especially when the body is in movement, because the pressure that the suit exerts on the body, allows an amplification of the consciousness of Movement and body position. Such contrasting and simultaneous sensations can add complexity to the aesthetic pleasure of interaction with clothing. Clothing as a medium is able to "shape" the person-environment interaction, promoting or hindering the aesthetic experience, and therefore interfering in the perception of clothing comfort (Kupfer 1994; Baitello Jr. 2006).

5 MATERIALS AND METHODS

This research work has as its starting point the results obtained in the first stage of a doctoral research on the psychological comfort of clothing. In this initial phase of the investigation, the Delphi Method was used in order to reach the concept of psychological comfort of clothing and to specify its possible attributes. Originated at the Rand Corporation, Delphi is a widely accepted and validated research method whose purpose is to obtain the most trustworthy consensus of opinion from a panel of experts (Dalkey & Helmer 1962, Sackman 1974, Linstone & Turoff 2002, Linden 2005).

The technique is usually applied presentially, however, to facilitate contact with specialists, many geographically distant (Portugal-Brazil), the questionnaires were sent by e-mail. The panel of experts was selected according to the following criteria: Portuguese-language speakers, researchers whose work is related to textile engineering, fashion design, comfort science, ergonomics, psychology, sociology, etc. Among the 50 experts contacted, 26 participated in the discussion group of concepts and the anonymity of the members was maintained throughout the process.

The open questionnaire submitted to the peer group consisted of four questions related to the psychological comfort of clothing. Although the questions were scripted, experts were encouraged to respond as freely as possible. Responses to the first round were subjected to a qualitative text analysis, with the support of MAXQDA software. The objective was to identify, within the content of the answers, the notions of psychological comfort of clothing that are more recurrent and that could potentially generate consensus.

The analytical process was guided by qualitative data analysis methods (Kuckartz 2014). As a result of the preliminary analysis, a list of 13 categories/clusters was obtained (still to be appreciated and validated by the group of experts), which would be a prospect of possible attributes of the psychological comfort of clothing. For this article, specifically, it was decided to analyze the implications of only one of the 13 categories, "aesthetics". The relationships between aesthetics and clothing comfort were explored in the light of the theoretical framework of several areas such as Psychology, Aesthetics, Design and the Science of Comfort.

6 RESULTS AND DISCUSSION

The great majority of the interviewed experts pointed to aesthetics as a factor influencing the perception of the psychological comfort of clothing. Among the 13 categories generated from the analysis of the answers of the specialists, the "aesthetics" category was one of the most expressive, with 92.31% of specialists referring to aesthetics and related ideas (second only to the category "psychological state" cited by 100% of the specialists), to refer to the formation of the concept of psychological comfort, as well as the parameters to be considered, in a process of evaluation of the psychological comfort of clothing. Table 1 presents excerpts from the answers from the experts who make up the "aesthetics" cluster.

In the creation of categories, the term "aesthetics" was coined as an "umbrella" contemplating from the meanings attributed in Greek origin – to the notions that over time were added to the word, endowing it with meanings related to Experiences, pleasures, judgments, evaluations, pro-realities among other nouns to which the "aesthetic" qualifier can be attached. As for example: aesthetic properties of clothing, color, shape, pattern (in the sense of textile surface design); beauty;

Table 1. Selection of part of the experts' answers referring to aesthetics.

Experts code	Answers excerpts
B	*"Clothes that provide the sensation of well-being that also passes through the composition of beautiful forms, which is pleasing to the eye"*
C	*"A sense of well-being provided by many factors such as color [...], shape / design of the piece"*
I	*"Aesthetics (current values of fashion, attractive to the senses: visual, haptic / touch, smell, sound)"* *"Visual and textural harmony"*
K	*"... from the visual point of view, she (clothing) needs, speaking in fashion-clothing, to express what I need to speak within fashion language"* *"... visually have the result on the body that I imagined."*
N	*"Psychological comfort is the consumer's sense of using clothing regardless of technology or functionality, knowing that the aesthetic aspect of clothing is one of the most eye-catching features of consumers."*
S	*"Thus, sensory experience with an apparel artifact gives rise to an aesthetic experience that, in turn, promotes associations with the cognitive universe of the user"* *"Visual attraction"*
T	*"... well-being that I feel just looking at a piece of clothing ..."*
ZB	*"Patterns and prints that indicate comfort through visuality."*

Concepts of "aesthetic perception" from the definition of Burns and Lennon (1994), including aesthetic preferences and "taste"; And also, from a relationship established by some experts and authors studied (that related directly aesthetic and fashion), fashion trends and style;

In the process of literature review, it was found that there are a number of publications, in the most diverse areas, relating to aesthetics and their ramifications. The aforementioned theories that deal with aesthetics do not refer directly to comfort. However, they transpose into spheres that connect directly or indirectly to the perception of comfort, if we consider comfort a state of balance and well-being perceived by the senses and aesthetics as pleasure or gratification through the senses.

In the Science of Comfort, authors such as Slater, Fan (2009), Li (2001), Yu (2011) and Branson & Sweeney (1991), to name a few, have pointed to aesthetics, or aesthetic properties), as parameters of psychological comfort. Li (2001), when specifying his proposal of dimensions of comfort and proposes psycho-aesthetic comfort as one of them, describes it as "the subjective perception of aesthetic evaluation" considering the influence of 4 senses (vision, hearing, smell and touch).

From the publications of Kupfer (1994), Macdonald (2002) and Burns & Lennon (1994), it was also possible to identify that other senses as proprioception and kinesthetic also interfere in aesthetic perception of clothing and potentially affect the psychological aspect of comfort. However, the idea of aesthetics is still, even in the literature, closely linked to visual perceptions and their affective (hedonic) responses.

Recalling what Macdonald (2002) says about the predominance of vision, it was observed that most of the specialists emphasized aspects perceptible to vision, as can be seen in table 1.

7 FINAL REMARKS

During the process of analysis of the answers obtained, a high number of aesthetic-related sentences were identified, which were equivalent to the content found in the reviewed literature. The aestheticians attributed to the aesthetics the value of influential element of the psychological comfort of clothing, just as the authors studied.

When confronting the literature, it was observed that, when it comes to "aesthetics", the term is approached from a perspective of appearance, considering the characteristics observable by vision, the same can be verified in the frequent mention of the specialists to visuality.

Just as product engineering and ergonomics have opened up for investigations that consider subjective environments such as emotions, pleasure, and the role of aesthetics in the relationships between artifacts and people, the science of comfort can also benefit from such approaches in investigating, in the domains of psychological comfort, the role of aesthetics and affection in the perception of comfort.

Although the question of the relationship between comfort and pleasure is under debate and requires theoretical deepening, it is believed that if this relationship exists, it is above all the aesthetic aspect, the factor that gives comfort a positive valence. For future reflection, we hypothesize that if comfort can be more than a state of equilibrium and neutrality, a pleasant state of harmony, it is the aesthetics that enables it.

ACKNOWLEDGMENTS

This work is supported by FEDER funds through the Competitivity Factors Operational Program – COMPETE and by national funds through FCT – Foundation for Science and Technology within the scope of the project POCI-01-0145-FEDER-007136.

The first author would also like to gratefully acknowledge the support from the Araucaria Foundation of Paraná State and the Federal University of Technology, specially, the Fashion Design Department and the Office of Research and Graduate Studies.

REFERENCES

Baitello Jr., N., 2006. Entrevista AntennaWeb Norval Baitello Junior.

Bell, S.S., Holbrook, M.B. & Solomon, M.R., 1991. Combining Esthetic and Social Value to Explain Preferences for Product Styles with the Incorporation of Personality and Ensemble Effects. *Journal of Social Behavior & Personality*, 6(6), 243–274.

Blijlevens, J. et al., 2017. The Aesthetic Pleasure in Design Scale: The development of a scale to measure aesthetic pleasure for designed artifacts. *Psychology of Aesthetics, Creativity, and the Arts*, 11(1), 86–98.

Branson, D.H. & Sweeney, M., 1991. Conceptualization and measurement of clothing comfort: Toward a metatheory. In S. B. Kaiser & M. L. Damhorst, eds. *Critical linkages in textiles and clothing subject matter: Theory, method and practice*. Monumet, CO, 94–105.

Broega, A.C. da L., 2007. Contribuição para a Definição de Padrões de Conforto de Tecidos Finos de Lã. , 205.

Burns, L.D. & Lennon, S.J., 1994. The look and the feel: methods for measuring aesthetic perceptions of textiles and apparel. In M. R. DeLong & A. M. Fiore, eds. *Aesthetics of Textiles and Clothing: advancing multi-disciplinary perspectives*. Monument, CO: International Textiles and Apparel Association, 120–130.

Burns, L.D. & Lennon, S.J.S., 1993. Effect of Clothing on the Use of Person Information Categories in First Impressions. *Clothing and Textiles Research Journal*, 12(1), 9–15.

Chattaraman, V. & Rudd, N.A., 2006. Preferences for Aesthetic Attributes in Clothing as a Function of Body Image, Body Cathexis and Body Size. *Clothing and Textiles Research Journal*, 24(1), 46–61.

Dalkey, N. & Helmer, O., 1962. *An experimental application of the delphi method to the use of experts*, Santa Monica, CA: The RAND Corporation.

Elliot, A.J., Greitemeyer, T. & Pazda, A.D., 2013. Women's use of red clothing as a sexual signal in intersexual interaction. *Journal of Experimental Social Psychology*, 49(3), 599–602.

Fan, J., 2009. Psychological comfort of fabrics and garments. In J. Fan & L. Hunter, eds. *Engineering Apparel Fabrics and Garments*. Woodhead Publishing Limited, 201–250.

Flusser, V., 1999. *Shape of Things: A Philosophy of Design*, London: Reaktion Books.

Fourt, L. & Hollies, N.R.S., 1969. The comfort and function of clothing. , (June).

Green, W.S. (William S.. & Jordan, P.W., 2002. *Pleasure with products: beyond usability*, Taylor & Francis.

Hekkert, P. & Leder, H., 2008. Product Aesthetics. In *Product Experience*. 259–285.

Hekkert, P., Snelders, D. & Wieringen, P.C.W., 2003. 'Most advanced, yet acceptable': Typicality and novelty as joint predictors of aesthetic preference in industrial design. *British Journal of Psychology*, 94(1), 111–124.

Hermann, N., 2005. *Ética e Estética: a relação quase esquecida*, Porto Alegre: EDIPUCRS.

Hosseini Ravandi, S.A. & Valizadeh, M., 2011. Properties of fibers and fabrics that contribute to human comfort. In *Improving Comfort in Clothing*. 61–78.

Jordan, P.. & Green, W.S., 2005. *Pleasure with Products. Beyond Usability*, London and New York: Taylor & Francis.

Kamalha, E. et al., 2013. The Comfort Dimension; a Review of Perception in Clothing. *Journal of Sensory Studies*, 28(6), 423–444.

Kuckartz, U., 2014. *Qualitative Text Analysis: a Guide to Methods, Practice and Using Software.*, SAGE Publications.

Kupfer, J., 1994. Clothing and Aesthetic Experience. In M. R. DeLong & A. M. Fiore, eds. *Aesthetics of Textiles and Clothing: advancing multi-disciplinary perspectives*. Monument, CO: International Textiles and Apparel Association, 97–104.

Levinson, J., 2009. Philosophical Aesthetics: An Overview. In J. Levinson, ed. *The Oxford Handbook of Aesthetics*. New York: Oxford University Press.

Li, Y., 2001. the Science of Clothing Comfort. *Textile Progress*, 31(1–2), 1–135.

Linden, J.C. de S. van der, 2005. O conceito de conforto a partir da opinião de especialistas. , 1–5.

Linstone, H.A. & Turoff, M., 2002. The Delphi Method – Techniques and Applications. *The delphi method - Techniques and applications*, 1–616.

Macdonald, A.S., 2002. The Scenario of SensoryEncounter: Cultural Factorsin Sensory-AestheticExperience. *Pleasure with products: Beyond usability*, 113–123.

Maffesoli, M., 2012. O tempo retorna: formas elementares da pós-modernidade. *Rio de Janeiro: Forense Universitária*.

Norman, D.A., 2013. *The Design of Everyday Things*,

Roy Choudhury, A.K., Majumdar, P.K. & Datta, C., 2011. Factors affecting comfort: human physiology and the role of clothing. In *Improving Comfort in Clothing*. 3–60.

Rudd, N.A. & Lennon, S.J., 1994. Aesthetics of the Body and Social identity. *Aesthetics of textiles and clothing: Advancing multidisciplinary perspectives*, 163–175.

Sackman, H., 1974. Delphi Assessment: Expert Opinion, Forecasting and Group Process. *United States Air Force Project RAND*, 1, 130.

Sanches, M.C. de F., Ortuño, B.H. & Martins, S.R.M., 2015. Fashion design: the project of the intangible. *Procedia Manufacturing*, 3, 2311–2317.

Schifferstein, H.N.J. & Spence, C., 2008. Multisensory product experience. In *Product Experience*. Elsevier, 133–161.

Slater, K., 1977. Comfort Properties of Textiles. *Textile Progress*, 9(4), 1–70.

Slater, K., 1997. Subjective Textile Testing. *Journal of the Textile Institute*, 88(January), 79–91.

Slater, K., 1986. The Assessment of Comfort. *Journal of the Textile Institute*, 77(3), 37–41.

Veryzer, R.W. & Hutchinson, J.. W., 1998. The Influence of Unity and Prototypicality. *Journal of Consumer Research*, 24(4), 374–385.

Wong, A.S.W. & Li, Y., 2006. Prediction of clothing sensory comfort. *Clothing Biosensory Engineering*, 178.

Yu, W., 2011. Achieving comfort in intimate apparel. In *Improving Comfort in Clothing*. 427–448.

Textiles, Identity and Innovation: Design the Future – Montagna & Carvalho (Eds)
© 2019 Taylor & Francis Group, London, ISBN 978-1-138-29611-4

Sublimation, color and emotion

F.M. Marques
CIAUD, Lisbon School of Architecture, Universidade de Lisboa, Portugal

ABSTRACT: In textiles the sublimation only allows to pass the color to fabrics of synthetic base, normally the most applied fabric is the polyester, or that has a minimum base of polyester, this technique enables a kind of ink fusion (without going through the liquid state called sublimation), the process is simple, from a digital or scanned, print-quality drawing to a printer that is fed by a proper paper to withstand high temperature that does not absorb the printing ink. In this article we will examine the advantages of sublimation over other more traditional printing processes.

1 SUBLIMATION, COLOR AND EMOTION

1.1 *Introduction*

The sublimation textile printing is a color-coding technique for a textile or rigid material produced for this purpose, usually the ceramic surfaces produced for that purpose which are covered with an absorbent resin, which supports high temperatures near 200°C. In textiles, sublimation only allows color to be transferred to synthetic-based fabrics, usually the most applied fabric is polyester, or having a minimum base of polyster normally the most commonly used has 90%, this technique enables a kind of ink fusing (without going through the liquid state called sublimation), the process is simple, from a digital or scanned, print-quality drawing to a printer that is fed by its own paper to withstand the high temperature that does not absorb ink printing, called "transfer paper". It is printed with 4 base cartridges, magenta, cyan, yellow, and black in a CMYK process, where white is the substrate base that presumably receives the color. In the process of sublimation by transfer there is no white ink cartridge, however Epson has another direct sublimation printing process that can be coupled to a white cartridge, but this process will not be addressed in this paper and only allows the printing of flexible substrates, in this article we will analyze other substrates.

1.2 *Sublimation ecological and economic advantage*

Sublimation is a process that has been gaining commercial ground in relation to other more traditional textile printing processes, either the screen-printing process or the offset process because it is faster, of lower economic cost and especially less polluting because it is a dry process and without diluents. In the traditional offset for textile printing requires photolith for sensitization of the sheet metal of offset, and each sheet only prints one color, so, to reach the four-color process the machines are very large, consume a lot of energy, and require long print runs to be compensatory.

The alternative process for textile printing of small quantities is the serigraph process, which requires a large skilled workforce, a lot of space for the application of the process, and a procedure of execution of photoliths, delayed and of great technical exigency, especially for four-color printing, where a silk screen sensitized by each applied color is required. It is true that there are many direct colors and chromatic effects that sublimation does not understand, namely all metallic and white colors, but one process does not exclude the other. In the serigraphic process one of the big questions is the correct color, when there is polychrome, so that for each direct color must be open a screen, and the process becomes more difficult as more colors exist, because it requires the screen to be adjusted with the drawing established so that moiré does not exist. The fact that working with liquid and solvent colors is a wasteful process, which requires a lot of washing, and extremely polluting, and is not an eco-logical process even when using paints based of acrylic solvents, since even this type of paint has a heavy ecological footprint both upstream and downstream, that is to say both in its manufacture and in its use, with the greatest impact on the washing of utensils after use.

The great advantage of the screen-printing process is that it is possible to apply to all materials, whether textiles or other materials, including glass, in terms of color is an almost unlimited process, allowing applications in sectors of innovation, with the incorporation of nano-technology, able to respond to the sectors of higher efficiency, from the incorporation of electric conductive materials, photo-luminescent or others, with the great advantage of being executed in cold without using complicated processes of preparation and can be applied to more or less absorbent materials without restriction. The use of suitable paints is only necessary, while sublimation requires preparation of the materials in advance and demands ink absorption and requires materials that have resistance to high temperatures.

As regards textile materials, the silk-screen process is possible on all fibers, whether natural or artificial, whether fabrics are woven or non-woven

The reliability of the serigraphic process is enormous and its duration proven over time, the process had been chosen by the militaries to identify all their equipment, whether clothing or weapons. Even in the identification of pumps or plates, the technique used is serigraphic printing. According to Enric Vidal , he is vice-president of Graphispag 2003, a trade show of products for printing, y ex-president of Serigraph 2001, a trade show for the presentation of techniques for silkscreen printing. As is president of "Sectoral Commission of Suppliers for silkscreen Printing" from Spain.

> *"La serigrafía es un sistema de impresión en constante crecimiento gracias a la continua aparición de nuevos materiales".*
>
> *"Silkscreen printing is a printing system in constant growth thanks to the continuous appearance of new materials." (own translation)*

Serigraph (Silkscreen) printing has always been closely linked to the textile industry, due to the economic factor and the possibility of small print runs is its greatest advantage, according to Josep Tobellas, he is a silkscreen printing technician with a long experience who began his career in 1971, and is president of FESPA, headquartered in Spain, is an international association of the sector of silkscreen printing, digital printing and textile printing, organizes events and exhibitions and conferences with knowledge of this sector of activity.

> *"La serigrafía es un sector, que en cierta medida ha estado considerada como la hermana pequeña del mundo de la impresión y es un sistema de impresión al que no se le ha prestado demasiada atención, dentro de la industria gráfica, fuera del mercado textil, hasta hace unos pocos años. En la actualidad hay un fuerte empuje en este sector gracias a la impresión funcional en la que está cogiendo gran protagonismo."*
>
> *"Silkscreen(serigraph) printing is a sector, which to a certain extent has been considered as the small sister of the printing world and is a printing system that has not been given much attention, within the graphic industry, outside the textile market, Until a few years ago. Currently there is a strong push in this sector thanks to the functional impression in which it is taking great prominence." (own translation)*

The process of sublimation printing, is a recent technology and as such in constant evolution. Its development is due to the advantage of being an entirely digital process with an inexpensive workforce, accessible to the general public, and without the need for great technical knowledge of the operator, other great advantages is the price, it costs the same unitarily to execute a print or a thousand The great limitation of this technology is that until now the process is exclusively for synthetic textiles, there is no quality in textiles of organic origin, with the values to go up when you want to print in cotton, there the trensfer paper should be special, the most used is the "Forever Sublidark 201", over other materials are limited to the ceramics pre prepared to receive this type of impression.

To be possible the sublimation printing is necessary upstream a computer that sends signal of 300dpi for a plotter or printer that prints using 4 CMYK cartridges as previously mentioned, for a transfer paper, which only absorbs a minimum amount of ink, everything that has ink will be printed the absence will be white or the color of the substrate support, the transfer should have the least possible contact with the human hand because the fat on the skin can cause stains in the final heat process.

In a more industrial process the thermal (thermal) machine is a thermal roller calender with a width sufficient to transfer in a continuous process to the width of the fabric. In a more artisanal process the machine is more accessible in both dimensional and economic terms, it may be the same machine that makes the traditional transfer application, as long as it reaches temperatures in the order of 200°C, the dimensions vary according to the manufacturers, but in most cases the most usual will be a machine that allows the printing of a sweater.

When the sublimation transfer paper is placed in the machine over the fabric, by thermal action and pressure, the printing will go from solid state to gaseous without passing through the liquid state (called the sublimation process), it is this chemical reaction that will enable the ink to be absorbed at the level of nano particles by the carrier, being impregnated in the textile fiber without any transformation of its texture or touch. In the ceramic this needs to be coated with a proprietary resin which will then be sewn into a furnace. There is glass also prepared to receive sublimation impression.

The great ecological advantage of this process is that there are no washes, solvents, it is quick, and only the necessary amount of ink is used to pass the information to the support, it allows the unit or multiple production without adding economic value.

The reading quality is far superior to that of serigraph printing or other traditional printing processes, allowing a large definition depending on the texture of the support.

Figure 1. Illustration of the designer João Bettencourt Bacelar, on a transfer paper for sublimation, the printing is done in a mirror, when to go to the substrate to be read from left to right. Source: Own elaboration.

Figure 2. Detail of the illustration of the designer João Bettencourt Bacelar, after the thermal transfer from the paper to the lycra fabric, all the details of the drawing are visible in the lycra. **Source: Own elaboration**.

Available print colors are all that can be printed with digital printers, the quality may be photographic, depending on the brightness of the support that receives the print. It is not up to know to print special colors such as metallized or gold, when this type of effect is intended we should turn to other processes.

2 THE APPLICATION OF SUBLIMATION ON SUBSTRATES OF ANOTHER NATURE

It is the object of this study, still in the initial process, the application of sublimation on several surfaces not prepared with the sublimation resin, since the resin can alter the perception of the material.

This study was only possible by the close collaboration with the design studio of pressure screens, "Oficina das Formas", who made the prints and supplied the textile material (mostly lycra, and also fabric for sublimation, this one of more yellowish base) for tests.

Several materials were tested, some of them because of their characteristics, did not withstand the necessary pressure which together with the required thermal factor damaged the material or in materials where the result was not satisfactory and for these reasons will not be addressed in this article.

2.1 *Sublimation and its definition*

The textile definition of sublimation is superior to other printing processes, the main reasons being due to the very fabric of the ink receiving substrate to be printed and also by the process itself which does not pass through the liquid state, so that there is no trace of the ink by the textile absorption.

As can be seen in the photograph above the definition of the line or the volumetry is of great care, both in the fine line that makes up the petal of the flower as in the volumetry of its stem or in the gradient of the leaves.

Printing would not be more or less defined if printed on a textile or paper substrate.

When the substrate is not white or when it has already been printed with a certain color, a color impression on its surface will change our perception of color, because the colors will be changed due to its overlap.

In the image of figure 4 we have an overlap of cyan blue on orange the result will be close to a greenish gray with brownish lavas, a color without great definition, unlike when it is printed on white background.

3 SUBLIMATION IN SUBSTRATE OF WOOD AND METAL

Wood as a substrate can be very interesting for sublimation due to the characteristics that it acquires on the wood grain.

We can print on a printed wood without being equipped, this impression allows the precession of the grain and the wood hair , being possible to have the precession of the printed drawing.

In Figure 5 we can have the precession of typography and color on wood, in this case plywood without any type of finishing, and already used, which may have fats that prevent the good impregnation of the

Figure 3. Example of sublimation printing on fabric. The vivid colors of the print and the fine traces of the illustration drawing are passed to the fabric, without great texture with the same quality as if it were printed on a paper. Source: Own elaboration.

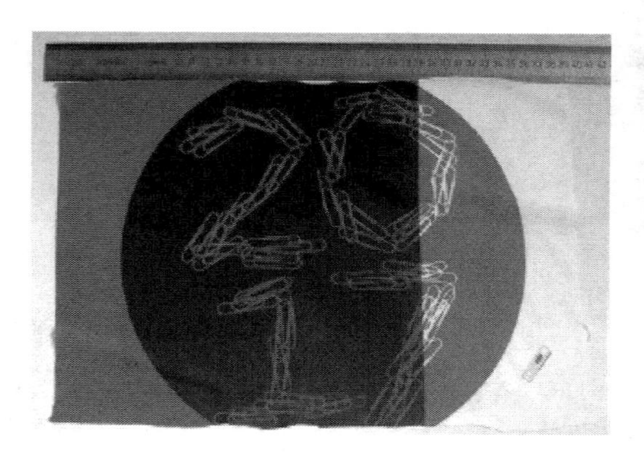

Figure 4. Sublimation print of an illustration in blue on orange (pre-printed) and white background. Source: Own elaboration.

pigment in the wood fiber, even so the definition of the print is remarkable.

Sublimation printing is also possible on metal, the results may be more difficult and resins similar to those used on ceramics may be required, which requires a baking step for a period of time, and at a temperature which does not all metal composites can withstand.

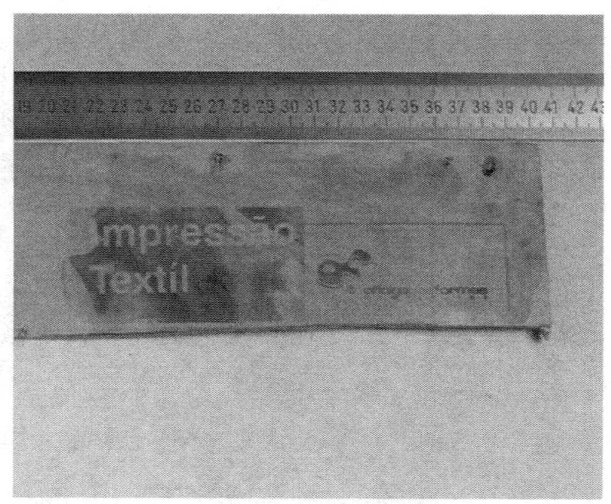

Figure 5. Example of sublimation printing on woodwithout finishing. The color printing was printed on a wood not equipped, not sanded, with texture, being visible however the printed typography except in the place in which the wood has a difference in its surface. Source: Own elaboration.

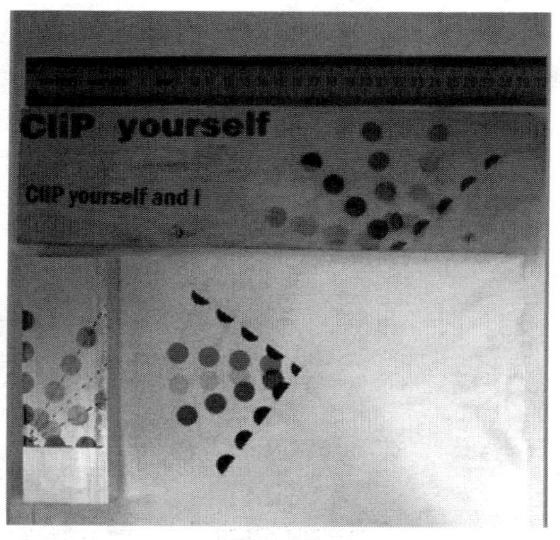

Figure 6. Sublimation printing on different substrates Wood without finishing Aluminum Lycra textile. Source: Own elaboration.

If printing is direct without recourse to varnishes on metals or sublimatic resins the impression remains as a weave on the wood as can be seen in the image that we see the CYMK impression on fabric, wood and metal to be able to see the differences between substrates.

In the previous photograph (figure 6) we have the CMYK primary colors printed on different substrates, wood, aluminum, and textiles, the latter with advantage in the definition of the print, for being a material with less texture than wood and for absorbing better while the aluminum does not absorb and the wood has its fiber hairs.

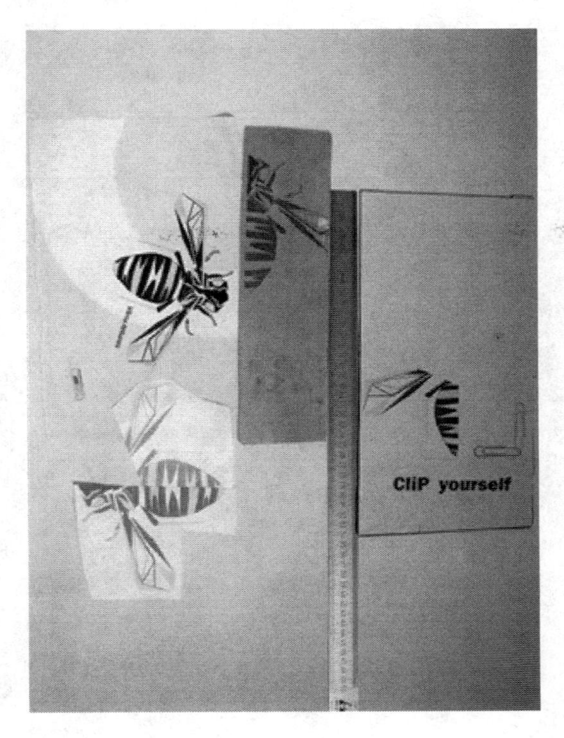

Figure 7. CMYK printing on various substrates Plywood without finishing White lacquered aluminum (alubond) Lycra textile. Source: Own elaboration.

4 CONCLUSION

When printing typographies or illustrations in sublimation printing on materials other than the conventional substrates for this type of printing, it must be borne in mind that the drawing will lose some definition because of the non-opacity of the printing ink, however having a definition of great detail and can be taken advantage of a subtle transparency of the design or printing press.

This transparency of the printing can be advantageous in some situations where the overlapping of layers or textures can be taken out, as in the case of wood, in which its design of the shaft is visible even if it has a color on top.

ACKNOWLEDGMENTS

The authors of this paper wish to thanks the Centre for Research in Architecture, Urbanism and Design (CIAUD) of the Lisbon School of Architecture of the University of Lisbon and FCT for founding this project.

REFERENCES

AEDES. (2017). Noticias del sector de la impresion. Entrevista a josep tobella maestro y tecnico serigrafo. [Electronic Version]. Retrieved 13-06-2017 from www.fespa.com/es/noticias-del-sector-de-la-impresion/funciones/entrevista-a-josep-tobella-maestro-y-tecnico-serigrafo.html

Arrausi, J. J. (Ed.). (2008). *Disenho e Impresión de Tipografia. Barcelona*: ediciones CPG.

Barbero, S., & Cozzo, B. (2009). Ecodesign. Königswinter: H.F.Ulmann, Tandem Verlag GmbH.

Beresniak, D. (2000). *O fantástico mundo das cores* (2 ed.). Lisboa: Editora Pregaminho.

Berger, W. (2009). *Glimmer how design can transform your business, your life, and maybe even the world.* London: Random House Business Bookes.

Bertagna, G. (2013). *Tessiture – Textures – Il fascino delle tessiture apparenti* [Electronic Version]. Retrieved 15-12-2013 from http://giuliobertagna.webnode.it/texture/.

Beylerian, G. M., & Dent, A. (2005). *Material Connexion; The global resource of new and innovative materials for architects, artists and designers* (1 ed.). London: Thames & Hudson.

Birkeland, J. (2005). *Design for sustainability, A sourcebook of integrated eco-logical solutions*. London: Earthscan Publicatios Ltd.

Creativecopias. (2017). Sublimação [Electronic Version]. Retrieved 18-06-2017 from http://blog.creativecopias.com.br/o-que-e-sublimacao-ou-transfer/.

FESPA, https://www.fespa.com/es/noticiasdel-sector-de-la-impresion/funciones/entrevista-a-josep-tobella-maestro-y-tecnico-serigrafo.html [visitado 03/01/2017]

FEIGRAF. Federación Empresarial de Industrias Gráficas de España Enlace: www.feigraf.es

Freepaint. (2017). *Como funciona a sublimação?* [Electronic Version]. Retrieved 18-06-2017 from https://www.freepaint.pt/o-que-e-a-sublimacao-e-como-funciona/.

Lefteri, C. (Ed.). (2007*). Making It, manufacturing tecniques for prouct design.* London: Laurence King Publishing Ltd.

Marques, F. M. (2011). *Cristal valências do desperdício para o design*. Unpublished press, Universidade Técnica de Lisboa, Lisboa.

Norman, D. A. (2004). *Emotional design: Why we love (or hate) everyday things*. New York: Basic books.

O2:, & design, a. i. n. f. s. (2007, 20 February 2007). Guidelines for Ecodesign Retrieved 25-05-07, 2007, from http://www.o2.org/www.pre.nl

Peneda, C., & Frazão, R. (1995, fevereiro 1995*). Ecodesign, no desenvolvimento dos produtos*. Cadernos do INETI, 1, 3–75.

Ryn, S. V. d., & Cowan, S. (1996). *Ecological Design*. Washington, D.C.: Island Press.

Stattmann, N. (2003). *Ultra Light – Super Strong, A new generation of design materials.* Basel, Boston, Berlin: Birkäuser – Verlag für architektur.

Thompsom, R., & Thompsom, M. (2013). *The manufacturing guides – Sustainable Materials*, Process and Production. London: Thames & Hudson.

Textiles, Identity and Innovation: Design the Future – Montagna & Carvalho (Eds)
© 2019 Taylor & Francis Group, London, ISBN 978-1-138-29611-4

Digital imaging: Textile surfaces and the virtual environment

L. Santos
Lisbon School of Architecture, Universidade de Lisboa, Portugal

G. Montagna
CIAUD, Lisbon School of Architecture, Universidade de Lisboa, Portugal

ABSTRACT: The present paper is an essay about the impact of digital image in the contemporary society, with a special focus to the representation of virtual textiles. The emergence of digital images in the XX century is greatly limited by the "crisis of the representations", generated by the doubt felt regarding the reference of the images. There is a necessity to make informed, unprejudiced studies that try to understand the true capacities of digital images.

In order to understand these questions, a literature review was made, consulting the main authors in the area of image and digital studies. As a result of this research we were able to understand the actual limitations and potentialities of virtual textile representations, and to analyze some of the connections actual researchers are discovering between the virtual and the physical.

The digital image is a natural evolution of our society that is silently changing the way we live and deal with ourselves and each others. The study of these subjects is essential to trace the cultural paths we are making as global community of gradually disconnected individuals.

1 INTRODUCTION

Nowadays, our life is filled with images. As Joly (2012) and Rancière (2011) claim, we live in civilization of images, where everything is an image.

With the diffusion of the digital paradigm, our world experience depends on "screens", the dominant medium where present images manifest themselves (Belting, 2014). We are gradually estranged from the "world of the directly visible objects", to be transported on a daily basis to the immediate "world of the observable objects" (Amsterdamski apud Massironi, 2015). The physicality of textiles is being borrowed to their digital representations, that seek to evoke the same sensations and emotional responses.

The "hyper-connection" to the new forms of access and production of images generates a constant pressure from the *now* and from the *here*, absorbing all the moments of our daily life (Lipovetsky, 2016).

It is important to understand how these new "mediums" of image transmission are generating new human forms of interaction with textile surfaces.

2 IMAGE

2.1 *Crisis of the representations*

The concept of image is linked to the notion of "resemblance" and of "analogy", because its function is to evoke, through representation, something other than itself, making visible what is not present physically: something from the "real" world.

This resemblance produces similar feelings of those produced by the "real" objects, making the observer to construct an untrue "equivalence" between image and object (Joly, 2012). This "equivalence" leads to a gradual replacement of the "real" objects by their image surrogates, alienating the observer from the "real" world.

The "crisis of the representations" is generated by the doubt felt regarding the reference of the images. The "real" object is progressively lost, being replaced by images that tend to be more and more disconnected from the sensible world, only referring to themselves, and denying any analogy to past images and with the world itself (Belting, 2014).

We stop trusting in images because of these extremist notions of "resemblance": if on the one hand the excess of resemblance provokes a confusion between the image and the represented object (for example in the actual cinematographic productions (Fig. 1), where the Computer Generated Images equate to real images); on the other hand, the lack of resemblance causes a disturbing illegibility (Fig. 2) (this dissemblance, is the basis of what we presently call "art" (Rancière, 2011)).

It is through this scenario that we stop trusting in images, especially in those created or invented in a way that distances themselves from what we interpret as reproductions (Belting, 2014).

Figure 1. Textile surfaces simulations produced recurring to Computer Generated Graphics in The Amazing Spider-Man 2 movie (2014). https://www.youtube.com/watch?v=cG2qDR5ka5w [03 July 2017].

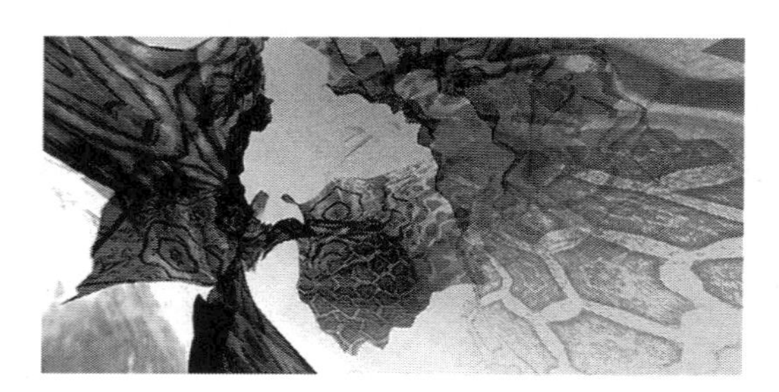

Figure 2. Mathematically generated, never repeating, fractal textile fabric by Daniel Brown in 2012. http://hoxtron.com/ [03 July 2017].

2.2 *Image and culture*

Our comprehension of the world is increasingly made through images. It is also through those images that we express our relationship with the world and with ourselves.

This inter-connection man-image-world is related to the three acts that are related to the image existence: the act of "fabrication", the act of "perception" and the act of "interpretation". If on the one hand the image is the product of one "medium" (like photography, painting or video) rooted in the act of human creation and imagination (the act of "fabrication"), on the other hand, the image also accrues from a process of recognition and interpretation by the individual (the acts of "perception" and "interpretation"). All of the three image acts are firmly rooted in the culture, and consequently, in each other. This means that the creation of images, both physical or mental, depends on the present culture, and that the mental processes of perception and interpretation are rooted to our cultural background (Belting, 2014 e Joly, 2012).

One example of this fact is the influence of science fiction in our concept of future fashion. This notion is born from the connection between the

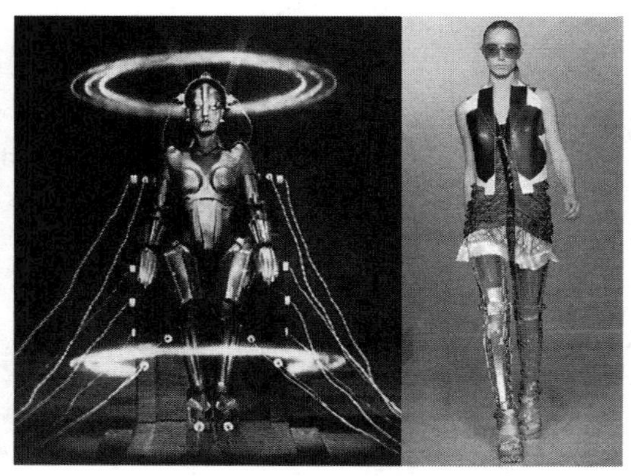

Figure 3. The cultural influence between the *Metropolis* movie from 1927 and the Balenciaga Spring-Summer 2007 Fashion Show. http://www.vogue.com/article/sci-fi-fashion-trends-barbarella-balenciaga [03 July 2017].

"physical image construction" (movies, photographs, drawings, etc.), the "mental image construction" (imagination, dreams and memories), and the "image perception/interpretation" (the way that clothes' form, fit and surface are seen and interpreted). These three acts form a net of mutual validations, where the collective production of sci-fi images standardizes the individual notion of future fashion, on both physical (the fashion creations) and mental (what we categorize as futuristic) clothing constructions (Fig. 3).

In other words, the perception that we have of the world (our culture) transforms the way we see things (Jenny, 2014).

If the culture acts as a catalyzer in the creation of personal and cultural images, cultural changes propitiate the renovation of languages and ways of seeing. Even the most naïve act of selection, analysis or appreciation will act as a cultural filter in the adoption of images (Barthes, 2015), fueling the constant development of the visual representations.

Images are manifestations of the present, misfit for future generations. That is why, once fulfilled its purpose, each image induces another (Belting, 2014).

3 DIGITAL IMAGE

3.1 *The digital medium*

To talk about image as a mechanism of visual communication, two interdependent concepts arise: the concept of transmitted "visual information" (or "content") and the concept of "carrier" (or "medium") that enables the materialization of that same information for further reading by the spectator (Munari, 2009). Only through the existence of these two concepts, can an image be perceived and interpreted by an individual.

The "medium" affects how images are interiorized because, depending on the present technical conditions, it transports the images in a certain way.

Figure 4. Balenciaga Spring Summer 2017 Campaign, where the "artistic concept" overlays the communication of the characteristics of textile surfaces. http://www.sleek-mag.com/2017/01/11/ss17-fashion-campaigns [03 July 2017].

However, despite influencing our perception and interpretation, that "medium" is distinct from the images it carries. A photograph is not understood as a piece of printed paper, but as a piece of reality that, detached from its carrier "medium", comes to us as a pure image, ready to be mentally perceived (Belting, 2014).

Also, the digital image has its "medium", accrued from the concept of frame, established by the painting discipline, where the illusion of dominium over the world is created in the spectator (Manovich apud Belting. 2014). But while in paintings the images are fixed to the "medium", in the digital domain the images are volatile because they are not physically linked to any "medium", only to exist in a "hyper-medium", abstract and immaterial.

This abstraction creates a new connection between image and "medium", creating infinite results in the equation "image = content + medium", through the variation of the parameter "medium". This way, the content connects to all the "mediums" where it appears, not to be captured by any of them, and the image gradually depends less on the "medium" that carries it. That is why the digital image is "inter-medial": it uses the different "mediums" to be carried (like photography, cinema, television and video), never to compromise with any of them.

Nowadays, the same fashion image can be seen in a tablet, mobile phone or any other digital "medium", never to be fully captured by any of them. Many of the physical characteristics of the represented objects are absent or manipulated, and the true essence of the garment is lost (Fig. 4).

This new "inter-medial" existence of images requires from the spectator another type of visual understanding, where image and "medium" are separated, aggravating the "crisis of the representations", where the image is disconnected to any physical existence. The image, separated from physical reality becomes auto-referenced, representing nothing more than itself (Fig. 4).

This immaterial perception of images is not casual because it reflects the immaterialization of the human being, that constantly seeks an escape from the physical body, pursuing a digital existence, manipulable and disconnected from the raw reality. The dialogue between body and image is not lost and continues to be, like in past eras, the result of a connection between image "mediums" and corporal perception (Belting, 2014).

The abstraction of images makes the digital "mediums" independent from the contents they convey, transforming these "mediums" in visual prosthesis of a new human existence, a door to a new world where all images and our post-modern being live.

3.2 *The digital image*

The digital image is, like its predecessors, a different visual representation. Because of its imateriality and inter-mediality, the digital image is able to generate new mental productions in the user, of a virtual richness that is natural to our body (Alliez apud Belting, 2014). This new kind of images has the potential to establish a profound imaginative dialogue with the individual, an unexplored field by its predecessors.

The "inter-medial" and "hyper-real" existence that turns the digital image into an "immaterial equivalence" (in other words, a resemblance with reality, that only occurs in digital media, therefore totally apart from reality), leads to the creation of alternative versions of reality, through the falsification of any image apparently "real" or the creation of purely synthetic images (Fig. 5).

In any of the situations, the digital image have the power to mimic the "real" (a virtual image is produced by the extension of the light rays - an image in a mirror, for example; virtuality is the illusion of being in the presence of reality, even when we're not (Joly, 2012)) with so much perfection that they can become "virtual" and construct the illusion of reality, even if they're not real.

But even the virtuality needs the be supported by reality, says Belting (2014). Even if disconnected from their mediality, digital images maintain their relationship of interdependence with reality and with the human being that, as seen before, produces, percepts and interprets them.

The evolution of "mediums" and representational techniques are directly connected to the visual content produced in a certain era. The computer, one of the most iconic "medium" of our time, has an immense creative potential but is no more than an instrument (Munari, 2009). It can perform some "intelligent tricks", like to simulate ambients and objects, yet it is still unable to dialogue in an intelligent way with the user, during the creative act (Lawson, 2005).

Figure 5. *Nobody Wants to Be Here and Nobody Wants to Leave*, a digital art creation from Troy Ford. https://creators.vice.com/en_us/article/8qvenb/futuristic-3D-love-virtual-art-museum-digital-age [03 July 2017].

All the digital "mediums" still are an extension of the human being that created them.

4 THE TEXTILE SURFACE

The human is a sensorial being, connecting with the surroundings with the help of all the five senses. To touch and to be touched is an essential sensation to the human existence. The textile surface connects with us in this sensorial way, touching and evolving us from birth to death. These sensorial experiences are then crafted into memories, letting us gain an understanding of materials, by attaching to them an emotional meaning (Auch, 2016) through sensations, which can be mechanical, thermic or visual.

According to Quinn (2010):

> "a textile's surface is more than just a façade — it's a curious layer of aesthetics and identities, and a contentious site of exploration and resistance. (. . .) Surfaces are also manifestations of social norms and moral codes, (. . .) [and] have perceptual qualities, their mutability enabling the body to take on chameleon-like characteristics. Because surfaces are interpreted by sight and touch, physical contact with the surface initiates a complex multisensory, emotional and cognitive experience that provides a uniquely individual interpretation of the world."

4.1 *Virtual textile surfaces*

A textile surface is a vessel of meaning and emotion, imbued by those multi-sensorial memories generated by previous contacts with textiles.

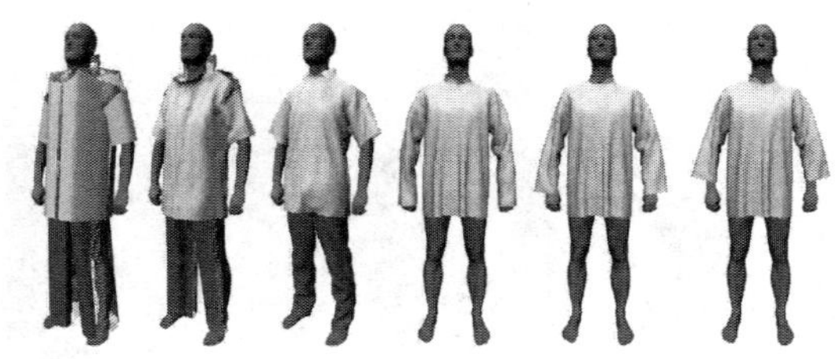

Figure 6. Simulation of shirt and trouser (left). Interactive changing of sleeve length. (right). In Divivier (2004).

The perception of a textile depends of an interweaved dialogue between sight and touch. When we see a representation of a textile in a virtual form (a photograph, an image) we are unable to access some of the essential sensorial characteristics related to the original physical object, limiting the emotional response through the observed textile surface.

Japanese Asahi 3D CAD displayed full digital representations of textile surfaces, for the first time, during IMB 1997. These images questioned the way artificial representations of textiles can evoke the same emotional responses than the real material (StichWorld, 2011).

3D simulation softwares present themselves as a fast and economical way to represent a garment during design, production, communication and sales stages of the product lifecycle (Fig. 6). Such softwares are able to assist the designer in ways the physical prototype cant, like presenting garment tightness to the virtual body, allowing easy pattern adjustment. Also, virtual bodies can have various anatomies (customized manually or automatically by 3d scanner technologies) allowing the production of garments in a more personalized way (StichWorld, 2011).

It is important that those digital representations of textiles have the right "look and feel" of the selected cloth, capturing the real textile visual characteristics: weight, thickness, sheerness, drape and stretch (Spence, 2000).

For Divivier (2004), the most critical part of the virtual representation of cloth is its reflectance (the object's percentage of reflected light). In order to address to the right sensorial feelings, the virtual cloth's reflectance needs to be concordant with the real material characteristics like color and texture.

4.1.1 *The virtual environment*

Technology represents a portal to the world of virtual images, allowing the user to interact with digitalized images.

There are many experiences regarding the way we interact with this virtual world, trying to create effective ways of emotional and sensorial connection.

Like seen before, mobile technologies like smartphones, tablets and personal computers are an easy way to access this virtual world, recurring to digital platforms created to emulate the real feel of handling

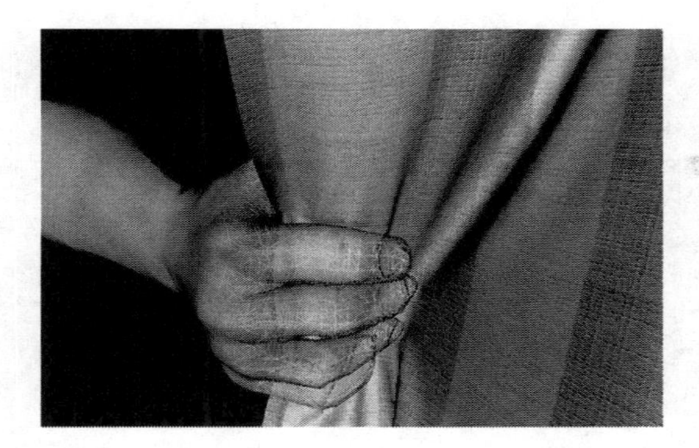

Figure 7. The HAPTEX system (2007), that produces a realistic 3D representation of a virtual textile. http://haptex.miralab.unige.ch/ [02 July 2017].

clothing. Many clothing brands invested in virtual try-ons that allow the user to interact with textiles generated digitally.

Another type of digital interaction is through virtual mirrors which provide a life-size display of virtual human representations (avatars), dressed in the chosen pieces of clothing, even allowing manual adjustment of the garments on the body (Divivier, 2004).

The human–machine interaction is also being enhanced through *haptics*, a discipline that studies touch as interaction. The human is a haptic being, sensing and manipulating through tactile and kinesthetic sensations (of movement or strain in muscles, tendons, and joints). By crating systems that work with computer haptics, it is explored the bidirectional relationship between avatars and virtual objects. The user, through the avatar is able to interact with virtual objects. That virtual contact then generates kinesthetic stimuli on the user, through the use of mechanical devices (Saddik, 2011). One example of these devices is the HAPTEX system (Fig. 7), that produces a realistic 3D representation of a virtual textile, animated in real time through a haptic/tactile interface that allows the users to "feel" the displayed virtual textile (HAPTEX, 2007).

4.1.2 *Textile surfaces beyond reality*

Like seen before, the digital image has the ability to evoke reality, but also to go beyond it.

To obtain the real "look and feel" of textiles we need to address their real physical characteristics, in order to link these representations to the mental images we have stored in our memories. For that to happen, the representations need to have a high degree of approximation with the real textile (Spence, 2000).

But the virtual world, free from the real world's physical constraints can generate new forms of imagetic discourses (Fig. 5). These new forms of creation can open doors to new representations that instead of emulating reality can be explored with the purpose of understanding the digital materiality. The understanding of the uniqueness of simulated (virtual) and physical (actual) properties of materials will open paths to informed explorations of future textile materials (Smitheram, 2016).

Even though, like seen before, digital images will never be totally disconnected from the cultural reality of the human being, that will always establish an emotional connection with the visual forms that he creates.

5 CONCLUSIONS

Through human history many types of images emerged, all carried by a contemporary "medium" that allows their perception. But this physical connection between image and "medium" is always momentary, because the mental perception of an image always requires a detachment from the carrier "medium". We are able to conclude that any image exists independent from its "medium".

The digital image, reflection of a virtualized society, declares the independence of images, which presently live in a "hyper-medium", ready to be summoned by any digital "medium" with the simplicity of a "click".

The digital representation of fabrics suffers from the legacy of the "crisis of the representations", where the spectator mistrusts the veracity of images. The overly realistic visual feel of virtual textiles (Fig. 1) and the creation of virtual realities apart from the real world (Fig. 5) generate confusing emotional reactions in the spectator, because these new realities overly resemble the real or present alternative digital realities that play with the cultural background of the viewer.

Once we can understand the unique potentialities of these new visual representations, the mutability inherent to digital images will open a richer, open dialogue with the viewer and with the creator of images. The language of digital textiles will be expanded by its own rules, questioning limitations imposed by the physical reality. Once freed from preconceptions imposed by past image representations, the virtual textile will be able to craft a self-identity and influence in its own way the human being in his physical and digital existence.

On the other hand, in production and consumption areas, studies try to counterbalance the lack of materiality of virtual textiles by creating interfaces that connect both virtual and material worlds in a natural way. Investigations in haptics and digital "mediums" are diluting the boundaries between virtual and real, connecting these two sides of the contemporary individual.

ACKNOWLEDGMENTS

The authors of this paper wish to thanks the Centre for Research in Architecture, Urbanism and Design (CIAUD) of the Lisbon School of Architecture of the University of Lisbon and FCT for founding this project.

REFERENCES

Auch, M. 2016. The Intelligence of the Hand. In Kane, F., Nimkulrat, N., Walton, K. (eds) *Crafting Textiles in the Digital Age*: 171–188. London/New York: Bloomsbury Academic.

Barthes, R. 2015. *A Câmara Clara*. Lisboa: Edições 70.

Belting, H. 2014. *Antropologia da Imagem: Para uma ciência da imagem*. Lisboa: KKYM+EAUM.

Divivier, A. et al. 2004. Virtual Try-On: Topics in Realistic, Individualized Dressing in Virtual Reality. In *Virtual and Augmented Reality Status Conference 2004*

HAPTEX. 2007. "HAPTEX: HAPtic sensing of virtual TEXtiles" http://haptex.miralab.unige.ch/ [2 of July of 2017]

Jenny, P. 2014. *Um olhar criativo*. São Paulo: Gustavo Gili

Joly, M. 2012. *Introdução à Análise da Imagem*. Lisboa: EDIÇÕES 70.

Lawson, B. 2005. *How Designers Think: The design process demystified*. New York: Architectural Press.

Lipovetsky, G. 2016. *Da Leveza: Para uma Civilização do Ligeiro*. Lisboa: Edições 70.

Massironi, M. 2015. *Ver pelo Desenho: aspectos técnicos, cognitivos, comunicativos*. Lisboa: EDIÇÕES 70.

Munari, B. 2009. *Design e Comunicação Visual*. Lisboa: Edições 70.

Quinn. B. 2010. *Textile Futures: Fashion, Design and Technology*. New York: Berg.

Rancière, J. 2011. *O Destino das Imagens*. Lisboa: Orfeu Negro.

Saddik A. et al. 2011. Haptics Technologies. Springer-Verlag Berlin Heidelberg.

Smitheram, M. 2016. Haptic acts of making: A surface imaging design practice using digital and virtual tools. In. . ..

Spence, A. 2000. Sensing the fabric: To simulate sensation through sensory evaluation and in response to standard acceptable properties of specific materials when viewed as a digital image. In *Lecture Notes in Computer Science*.

StitchWorld 2011. 3D Garmet Simulation. In *StitchWorld*: 28–30.

Textiles, Identity and Innovation: Design the Future – Montagna & Carvalho (Eds)
© 2019 Taylor & Francis Group, London, ISBN 978-1-138-29611-4

Costume and characters' construction in cinematographic fiction and drama

C.M. Figueiredo & A. Cabral
CIAUD, Lisbon School of Architecture, University of Lisbon, Portugal

ABSTRACT: In the Feature Films Vertigo, La Strada, or The Philadelphia Story, both the quality of costumes as the judicious choice of them for each scene goes with the direction and production design, as they assume precise rolls in the composition, narrative, script and meaning of the characters – making clear that without them the stories of their supposed authentic life would not be told in their fullness.

An actor's performance, driven by a posture and attitude inherent in the character he plays, is reinforced in the projection of his role, because it is framed by a dialectic in which the costumes, in its broadest sense – including make-up and accessories – constitutes a relevant item in the film narrative, in the fictional world, and in the shot composition. Emotions triggered simultaneously converge with the illusion of realism intended in the script and storytelling intended by the director.

1 THE THEMATIC COSTUMES IN THE ART OF CINEMA

"Between the culture of touch and the culture of thought is the culture of symbolic connection" (Garcia 2011: 101). This reflection refers to the gesture of dress and use, which are tactile elements and social meaning linked to fashion, which demonstrate the fragile border between what fashion design creates for everyday life and what the design of costumes produces for the fiction.

The same consideration also reminds us of what we expect from a movie as a spectator, that is, of being involved in the drama presented to us, in such a way that we become convinced that we are part of it, raising an amount of awareness. The recreation of experiences and emotions, when viewing the scenes – with their representations, narratives and meanings – is achieved through a meticulous communication project, in which the visual and fictional world created by the production design is able of unleashing on the viewer a state of immersion, representing for this a phenomenon of "metamorphosis [which] is the motto for the experiential renewal of each human" (Vasco 2009: 258).

Without us realizing it, fashion plays a key role that is more than simply dressing a character at a period that fits the script and film narrative. This article aims to analyze the use of everyday clothing in fictional everyday life, in order to create specific speeches around the construction of characters in *Vertigo* (Alfred Hitchcock 1958), *La Strada* (Federico Fellini 1954) or in *The Philadelphia Story* (*Cukor* 1940).

Fashion has a far more significant role than the simple dressing of a costume. It contributes to the characterization of the fictional world where the action unfolds, meeting the intentions of the realization in a whole, corresponding methodically to a discourse in which the signifiers of fashion converge with all the syntactic and significant devices of the cinematic narrative.

2 PERCEPTION IN FILM SPEECH: SIGNIFICANCE OF CONFLUENCES OF UNIVERSES IN FASHION AND CINEMA

In *Vertigo* (Hitchcock 1958), fashion is present in every moment of the film hinge. Hitchcock introduces to us, at the beginning, a fashion designer who develops a brassiere with an innovative constructive technology. This detail, apparently unreasonable, evidences the precision with which fashion elements are shown throughout the narrative.

Let's see: that detail more illuminating than a shoe foot in some delicate "scarpins" to be dragged by a staircase up to demonstrate the fragility and the dead weight of a driven character to an event on the brink of disaster?

However, it is the framing of the scene, occurring on an old wooden staircase – itself unsecured and giving access to the top of a bell tower – as the facts of the shot be in low-key, the set of low angle and high angle extreme shots, that gives to it much of its symbolic content. They are in fact women's feet, but the skin, the condition of the vulnerable body, stands out in the near darkness.

In the film, in general, the details stand out from the background not only by specific framing and lighting, but also by color and its contrasts. It is a color palette of the era, rich in pastel and neutral tones, highlighted by saturation control details in need cases.

The figure of *Judy*, appearing in the direction of *John*, is, for these reasons, unavoidable to the viewer's gaze. But this is achieved in order to meet the (re)created fictional world, because the costume belongs to that place and we feel that the character is just one person among the many who walk in that street of S. Francisco. Consistency comes from seeing it with a point-of-view from *John*.

It turns out in fact that "the film piece (. . .) conveys a message, aimed at a target audience and its effectiveness depends on an intensive analysis of the script and intentional use of visual meanings" (Carpinteira 2011: i).

Let us now think about the film *La Strada* (Felini, 1954). The contextualization of the social condition of the characters is clearly demarcated by the costumes. The characters *Gelsomina* and *Zampano* are clearly poor; this is reinforced by a peculiar aspect of *Zampano*'s costume, when he intermittently bears a chalk line, in contrast to the usual look – careless, consisting of a bomber jacket, a coarse sweater, and a pair of looking trousers dirty, already worn. We understand the character in his rude character, information that is completed by the expression and gestures of the actor, in tune with his house: a motorcycle with trailer, dark and messy (figure 1).

The film clearly falls within the professional situation of the characters, leaving no doubt about who the street circus performers portraying, strengthening neo-realistic nature of it, that is, showing a post-war in a poor country where everything begins again, and where the circus would be a food of the soul. The designation *La Strada* suggests the scenic space and defines both the path and the destination of the characters. With the surrounding space – the street: often the space of artistic performance – being annulled, it remains the characters, depending on which the film lives mostly.

In *The Philadelphia Story* (Cukor 1940), the costumes essentially reinforce aspects of personality that the main character (*Tracy*) reveals, but is necessarily linked to the family and social environment in which the plot unfolds. There is a wardrobe for each specific situation; for example, swimming in the pool requires your own attire, a bathrobe, in addition to the simple bathing suit, because swimming is a ritual, just like riding.

The costume delimits the personal space of the character – and even the professional one in the cases of some characters with whom he / she works – but not only, it also denotes an insistence on revealing a lifestyle appealing to that period, full of sophistication and opulence, of which is an example the family home, privileged space of action. This is where the wedding

Figure 1. Zampano on his motorcycle; *La Strada* (Fellini 1954); women dressed according the period in *The Philadelphia Story* (Cukor 1940).

and the engagement party of *Tracy* take place. This time, "it is worth emphasizing the importance given to the lifestyle, considered as a way of living private, expressed in activities, interests, opinions . . . [which] delineates a whole pattern of action and interaction of people" (Garcia & Miranda 2010:68).

This is evident in this party, when several women of similar physiognomy are filmed at the bar counter, from the front, and we see a "frieze" of looks of the period, from hair to clothes, from jewels to their poses (figure 1).

3 CREDIBILITY OR MANIPULATION: THE INNER LIFE OF THE CHARACTERS WHO INHABIT THE FICTIONAL WORLD GETTING STARTED

The costumes in the films under study play a narrative role, although differently in each one of the films analyzed. The clothing has a close connection with the psychological characterization of the characters, their connection and hierarchy in relation to each other, their contextualization in the space of action, in which the scenes unfold, defining it. The hierarchical relation-ships between characters are evident

Figure 2. Gelsomina ignored; La Strada (Felini,1954); Judy subdued; Vertigo (Hitchcock, 1958).

Figure 3. Gelsomina in her role; La Strada (1954),Federico Fellini; Tracy's sister acting; The Philadelphia Story (1940), George Cukor.

through the use of clothing, by comparing the types of clothes, props, characterization, gestures, attitudes and scenarios, the latter contributing greatly to the characterization of the characters because "they offer or suggest the appropriate background for action" (Carpinteira, 2011:16).

In *Vertigo* (Hitchcock, 1958), the costume is used with such subtlety that it becomes evident only in the eyes of the viewer when the director wants fashion to bring specific connotations to the storyline. This is seen in detail as a proof of shoes – again, the women's favorite accessory stands out – when Judy, upset, puts on those that *Madeleine* would wear.

We see her figure only from the waist down, covered with a skirt that swings with the floor, in great prominence and to the center, the two gentlemen on each side sitting and contemplating it with overjoy (figure 2).

Here there is no doubt that the hierarchy is established with the contribution of the costume, in the conjuncture of the scene, as it happened in the scene previously described of *La Strada* (Felini 1954). But the control does not belong to the personage emphasized to the center, because the scenic composition that frames it between two masculine figures, makes it subordinate to that event.

Let's see also the inner transformation of characters in others that themselves interpret, with the collaboration of the costumes. It is given as in fashion consumption, the "investment in values in a novelty of the moment manifested by the look, [in which] the consumer becomes another, discovering himself different from the one in which his routine had converted him" (Garcia & Miranda 2010: 36). Let us see, then, Judy transformed into Madeleine, in *Vertigo* (Hitchcock 1958). This character lived a routine that his social condition would never allow. The opulence of an evening dress worn at a dinner party in the fashionable restaurant, or a classic suit inspired by a Christian Dior H-line suit, would be high society luxuries.

All the details of the clothing and the fictional environment give full credibility to the moment lived and felt by her. The need to feel different from what one is, which is expressed by clothing, is also evident in *La Strada* (Felini 1954).

We identify an apparent shift in *Zampano* as he emerges on a superman's cape to raise his feeble circus number (break chains with pectoral muscles), convincing himself he's a hero.

In his case, he clearly does so as a person who "acts influenced by an understanding of a particular situation" (ibid., 2010: 68). In *Gelsomina*'s case, change is fleeting, but genuine, for it reveals the fulfillment of a desire.

His vulgar poncho becomes the central piece of clothing when he contrasts with *Zampano* as a bird.

It is not clown painting and scratched blouse that characterize it as such, but the cover of every day. It is with her that she finds her place to represent, fluttering, convincing the audience and surprising herself (Figure 3).

In these two situations, as in artistic performance, it takes place a series of intimate gestures that last a few minutes, "but that could last for hours" (Goldberg, 2011: 8), which reveal desires for expression and certain sensibilities framed in particular discourses.

Let's not forget *The Philadelphia Story* (Cukor 1940), which is relevant in another aspect of this topic: to deceive the remaining characters by the inconsistency of the context, rendering their performance unreliable. Tracy's sister emerges from high-heeled slippers in a lopsided ballet, striking an exaggeratedly large diamond necklace (Figure 3); she also plays piano and speaks French.

Such attitudes intrigue and deconcentrate the infiltrated journalists, who seek clues that discredit the family, thus facing a supposedly undefiled life. *Tracy* emerges from the action, dressed eccentrically, leading them into the sitting area and sitting in front of them on a center table, in a total relaxed mood, that further highlights her haughty figure. The illumination and close-up of his malicious smile, framed by a collar with a simple bow, make him look innocent, even courteous, but in the end, she is a bit cynical.

Figure 4. Tracy in a kiss at moonlight; The Philadelphia Story (1940); Zampano in sorrow; La Strada (1954), Federico.

4 EMOTION: STIMULI AND STATES OF MIND OF THE BEHOLDER

Regarding the costume, the spectator feels the emotions without realizing the details that are in front of his eyes, because the accessories, typologies and details of the clothing provide an intuitive reading. As objects of daily life, they are, in the form of links to events already experienced by the spectator, belong to the objects of their experience or at most of their imaginary, and attract him because "our gaze (...) tends to look for familiar elements, ones of affectation" (Santos, 2008:57).

The director and production designer, aware of this fact, use this feature to snatch and captivate, when they decide to do so. The costumes can help to surprise the viewer with the perception that he had of a certain character, and can be used to create suspense, as well as series of emotions inherent to social conditions or affective ties between the characters or from those with the viewer.

In *Vertigo* (*Hitchcock* 1958), the wardrobe circumscribes the personality of John, meticulous detective, impeccably dressed in a classic way. Simultaneously, it is also the costume that can deceive him, and its use controls the whole development of the drama. Involved in scenarios and props, the costume contributes to delineate their actions, thoughts and daydreams – including the spectator in their expectations, anxieties, doubts and revolts.

John falls in love with the fragility and sophistication of Madeleine, visible in the solemnity with which she dresses to envelop him in enigmatic situations, whether in a museum to contemplate a portrait of a deceased or in a mysterious village of his obsession.

Judy's slip pendant, a symbolic accessory of the weave, appears more as a revealing garment of it than as a fashion accessory appropriate to the dress and the season (a pearl necklace would be more expected). In this case, the costume is not only to confirm the suspicion of the viewer, it also serves to surprise the characters of the storyline itself – this is how John discovers that he was the target of a scam.

In *La Strada* (*Felini* 1954), a personal pendant is provided to form a parting seemingly harm-less, but contains a coated assigned gesture. Let us see in *The Philadelphia Story* (*Cukor* 1940): what more exuberant element to frame a kiss taken by the moment, than a hug in a sleeve embroidered with crystals that glitter in the moonlight? (figure 4).

If we make a careful analysis, we conclude that fashion, from the most sophisticated to the most mundane element, feeds the films in an improved and even sentimental way. *Zampano* ends by redeeming himself for his insignificance in the fact that he exalted him most;

Figure 5. Profile of Judy and Madeleine; Vertigo (1958) Alfred Hitchcock.

is dressed in scratched suit that feels all the hurt of having lost.

It is also at a moment of revelation that Tracy, in *The Philadelphia Story* (1940), eventually finds herself. Her engagement dress denotes a valorization of the event in question: a symbol of the epiphany of the monotony created around her social life, conservative and of which she is trying to liberate herself.

However, the overall costumed reading is done not only through the framing of each scene, but also through the anachronistic perception of scenes. The viewer's acquired knowledge of the character is constructed from meanings that he is reading through the symbols he observes, but the composition of *mise-en-scène*, the musical background and the repetition of the character's gestures contribute to the understanding of the overall, as well as for the prediction of future actions that she latter may undertake. Consider the case of *Vertigo* (1958) in which the images of the faces of *Madeleine* and *Judy* are filmed in the same perspective, lateral to their face, and observed by the same character, *John* (figure 5).

Despite the differences in terms of hairstyles or make-up, there is an implicit suggestion that fuels a suspicion in the viewer's mind: will they ultimately be the same person?

When *Judy* emerges from *Madeleine*, in the end, her silhouette is circumscribed in green, as if it were a ghost, to confirm that for *John*, her beloved, whom she thought dead, was resurrected. In fact, green is the color that unites the three characters, guiding the viewer: it is in John's pole (which he wears after saving *Madeleine* from the fall to the river), in *Madeleine*'s scarf sitting at the table in the restaurant, in *Judy* dress when she is recognized by *John* on the street.

Even in *La Strada* (Fellini 1954), there are repetitions that suggest the spectator: we see that the predictability of *Zampano* stems from his vanity, and years after having obtained the suit and *Gelsomina* has died, he reappears on the scene with the same, indicating that his wandering life did not change and corroborating the wandering personality that the spectator had created in his mind.

For her part, in *The Philadelphia Story* (*Cukor* 1940), *Tracy* accustoms us to a visual that stands out due to the details of refinement, showing that he is a demanding person. The beginning of the film foretells its end, analyzing the costumes carefully. The opening scene shows her to separate from her husband, driving him out of the house in his nightshirt, completely exposed; is with the same rectitude that he remarries to him at the end, dressed in a simple dress, which reveals her as she is, freed from the impositions of others, who see her on a pedestal. She says, "And you know how I feel? As a person, as a human being."

5 CONCLUSION

The global reading of the costume is characterized by a series of formal and cultural premises of the cinematographic discourse, not forgetting the body's gesture and its connection to fashion and its idiosyncrasies, especially those that manage to involve the viewer in the narrative, by the fact of being familiar and equal to situations like those of their daily life and cultural context.

The costume related to these situations is selected with subtle criteria and details, so that the performance of the actor is natural and spontaneous. The cultural context to which the costumes belong is equally evident in the attention given to décor, contributing to convey emotions and narratives convergent with the unfolding of the plot and with the representation of the fictional universe in which it takes place. The costumes are therefore a primordial element of the kinematic narrative, being a fundamental element for the construction of the characters and drama of the cinematographic fiction.

ACKNOWLEDGMENTS

This research was only possible with the support of CIAUD – Research Centre in Architecture, Urbanism and Design, Lisbon School of Architecture of the University of Lisbon, and FCT – Foundation for Science and Technology, Portugal.

REFERENCES

Carpinteira, Y. (2011). Design de Produção: a intervenção do Designer de Produção na Peça Cinematográfica. Mestrado. Lisboa. Faculdade de Arquitectura da Universidade Técnica de Lisboa.

Garcia, C. (2011). Imagens Errantes: Ambiguidade, Resistência e Cultura de Moda. São Paulo, Estação das Letras e Cores.

Garcia, C. & Miranda A.P. (2010). Moda é Comunicação, 2ªed. São Paulo, Editora Anhembi Morumbi.

Goldberg, R. (2011). Performance Art, 3rd ed. London and New York, Thames & Hudson.

Vasco, N.M. (2009:258). Arte: Comunicação ou Não Comunicação? Da Objectividade Elementar à Subjectividade Artística. Doutoramento. Aveiro. Departamento de Comunicação e Arte da Universidade de Aveiro.

Santos, D.R. (2008). Anything Goes? Uma Discussão sobre a Necessidade de uma Orientação Ética na Arte Contemporânea. Mestrado. Lisboa. Faculdade de Arquitectura da Universidade Técnica de Lisboa.

Films:

La Strada (1954), Dir. Federico Fellini, Itália.

The Philadelphia Story (1940), Dir. George Cukor, USA.

Vertigo (1958), Dir. Alfred Hitchcock, USA.

1.3. Textile products

Textiles, Identity and Innovation: Design the Future – Montagna & Carvalho (Eds)
© 2019 Taylor & Francis Group, London, ISBN 978-1-138-29611-4

The fabric of belonging: Place-based textile community engagement

Anne Marr
TFRC, Central Saint Martins, University of the Arts London, UK

ABSTRACT: This article discusses how textile making can enhance the mental health and wellbeing of homeless and vulnerably housed residents as well as stimulating greater community exchange. The author uses the London-based 'Home and Belonging' arts programme as an action research case study to explore new methods for place-based textile engagement activities and how they can be utilised to enhance the social capital of urban neighbourhoods. As part of the programme a group of homeless and vulnerably housed residents mapped their personal journeys, explored their local neighbourhood and created a collection of upcycled chairs as well as bespoke textile designs. The positive effect of communal making activities becomes evident through visual data collection which demonstrates the increased confidence and self-esteem of the participants leading to reduced isolation as well as increased community engagement. The paper concludes to recommend the importance of community visibility in order to facilitate holistic urban fabric exchange.

1 INTRODUCTION

'Home and Belonging' is a bespoke arts-based programme delivered in partnership with Crisis Skylight Brent (CSB) – part of Crisis, one of the biggest charities addressing homelessness in the UK – and Central Saint Martins (CSM), University of the Arts London. The project was funded through the Brent Council Voluntary Sector Initiative Fund and will be running until September 2018. This paper portrays the findings of the first project phase from October 2016 to July 2017.

The aim of the Home and Belonging project is to improve the lives and wellbeing of homeless and vulnerably-housed residents in the London Borough of Brent, building their resilience whilst supporting them to participate in the wider community. The project encouraged Crisis service users in expressing narratives of their own experience of community belonging, whilst nurturing a variety of soft, transferable skills as well as providing opportunities to creatively participate in local regeneration activities. Main project objectives are:

- to improve the mental health of homeless and vulnerably housed residents by reducing their isolation, increasing their confidence and self-esteem through textiles as a vehicle for purposeful learning and self-expression
- encouraging local community engagement to give homeless and vulnerably housed residents a voice to respond to the changing nature of their neighbourhood, bringing greater arts provision into the borough and promote community cohesion and integration through public events.

The Home and Belonging team set up new textile classes tailored to the needs of homeless and vulnerably-housed residents in Harlesden, Brent. 12 different workshop activities were offered, ranging from college visits to story telling, print-making, photography and tapestry. During the first year, the main programme engaged participants in making textile artwork, adopting environmentally-friendly ideas and showcasing their work in order to test out different place-based neighbourhood activities. About 25 homeless and vulnerably housed beneficiaries and 500 members of the wider community engaged with public exhibitions and community events during phase 1.

The following paper will portray the context of Home and Belonging, outline new methods of textile engagement, and analyse its findings in order to recommend transferable tools and successful approaches to community enhancement through urban fabric interventions. The author aims to question the traditional understanding of 'urban fabric' by extending its boundaries to include more textile-based approaches to respond to societal needs and create more empathetic urban communities.

2 BACKGROUND

An outline of the different Home and Belonging project partners as well as the current context of homelessness in the London Borough of Brent.

2.1 *Crisis skylight brent*

Crisis is an UK wide charity organisation which offers education and training about housing, employment and health for people who are currently homeless, have experienced homelessness in the past three years or are in immediate danger of losing their homes and don't have a partner or a family member to support them. Crisis Skylight Brent (CSB) was funded in April 2016, when it evolved out of a previously existing local self-help group called LIFT. CSB had no experience in delivering art and design classes and the project coordinator had to set up new workshop facilities. However CSB already had an active local network with earlier LIFT service users, which was utilised for the Home and Belonging project recruitment.

2.2 *Central saint martins*

Central Saint Martins (CSM) is a multi-disciplinary art college located in Kings Cross in central London and is part of the University of the Arts London (UAL). The author of this paper and initiator of the Home and Belonging project is the Course Leader for BA Textile Design as well as a senior researcher based at the Textile Futures Research Community (TFRC) based at CSM. The TFRC Urban Fabric research area explores sustainable and innovative textile design to enhance urban environments and their positive impact on daily life.

2.3 *Local context*

The Home and Belonging project took place in Harlesden in the London Borough of Brent, in one of the top 5% of the most deprived areas in the UK.

The employment rate in Harlesden is 59.3% (UK Census 2011), which is significantly lower than the London and National levels which is 70%. The proportion of those claiming job seekers allowance is particularly high at 18.3% compared to 3.6% nationally in 2014.

Homelessness and housing insecurity are key issues facing many Harlesden residents. Brent is in the national top 10 local authorities for the number of rough sleepers and has the second highest rate in London. There are currently 3,000 people in temporary accommodation in Brent, and from 2010/15 there was a large growth in homelessness acceptances (amongst the highest levels in the country) with an increase from just over 300 acceptances in 2010/11 to around 850 in 2014/15. The London based CHAIN database shows that London rough sleeping levels have risen by 104% to 8,096 since 2010 (CHAIN annual report 2015/16). This surge is primarily due to a decrease of tenancies in the private rented sector, and Brent's recent submission to the House of Lords CLG Committee signalled that it expects 'this pattern [of increasing homelessness] to worsen significantly over the next few years as Local Housing Allowance is frozen for 4 years whilst market rents will almost certainly increase.' Homelessness. Communities and Local Government Committee (2016).

Like many London neighbourhoods, Harlesden is experiencing constant urban development changes – resulting in increased property prices and shifting demographics. The area has currently a population of 17,500, an increase of 40.43% since 2001 (UK Census 2011). There is a local concern about the effects of these changes, particularly on those most vulnerable in the area and the Home and Belonging project aims to test a place-based neighbourhood approach to help address these issues. Whilst Harlesden has many environmental challenges, at the same time it has a lively community, which manifests itself in the diverse and vibrant economic and cultural folklore of its High Street. Local residents have recently initiated the Harlesden Neighbourhood Forum to enable residents and business owners in Harlesden to begin to play a more active role in their community to tackle common issues such as fly tipping.

Furthermore the project responds to a gap in local arts and cultural provision: Harlesden has no community arts facilities or programmes, and an alternative arts programmes at St Mungo's residential home had to close in 2016 due to lack of funding.

2.4 *Health and wellbeing*

Health is a gauge of the physical and social fabric of our communities and of our individual journeys through life directly mirroring the nurturing opportunities as well as deep challenges along the way (Harkins 2016). Research indicates a strong link between homelessness and ill health as well as substance abuse. Poor mental and physical health is both a cause and consequence of homelessness. People who have experienced homelessness are over twice as likely to report a physical health issue then the general public. 44% of homeless people have a mental health diagnoses, in comparison with 23% of the general population (Health Needs Audit, 2016). Homelessness is often a very stressful experience (I was all on my own, 2015), leading to loneliness, lack of self-esteem and raised suicide rates, which are nine times higher than in the general population (Homelessness Kills, 2012). Additionally 41% use drugs and alcohol to cope with mental health issues (The unhealthy state of homelessness, 2014). In 2015/16, 45% of the people who accessed Crisis services identified themselves as having or having had mental health issues, a figure echoed at a local level at Crisis Brent.

3 URBAN FABRIC

Urban fabric is commonly used as a term to describe the physical aspect of urban design and streetscapes, often, however, excluding economic and sociocultural qualities. Following on from the idea of place attachment by recalling the past (Lynch 1972; Altman and Low 1992) Antonovosky identifies one's place identity as influencing the meaning and significance attributed to place (1987). This project attempts to extend the traditional perception of urban fabric in order to include a wider range of social, cultural and environmental attributes to holistically interweave hard tangible and softer intangible strands of spaces. Settings rich in place icons such as churches or parks, may evoke a stronger local relatedness and sense of community (McMillan and Chavis, 1986). Currently 85% of people in the UK agree that the quality of the built environment influences the way they feel. (Creative Health Report 2017). This creates a fine balance between a healthy community, a healthy environment and personal wellbeing. Additionally helping to create an environment creates closer community attachment. People will therefore manage and maintain it better, reducing the likelihood of vandalism, neglect and the subsequent need for costly replacement (Wates, 2014).

3.1 *Social capital*

Inner city areas like Harlesden constantly change both physically and culturally, which makes it difficult for residents to anchor themselves in a local community. Lynch explores the relationship of time and place and how residents benefit from a deeper understanding of their urban surroundings: "Instances of environmental transformations are common. On one hand, people must endure them, and we see their efforts to preserve, create, or destroy the past, to make sense out of rapid transition, or to build a secure sense of the future. On the other hand, the initiators and regulators of change (....) struggle with these transformations in another way, straining to comprehend and control them." (Lynch, 1972:3). A positive local association as well as the weaving-together of people in communities develops social capital as "a resource based on trust and shared values" (Gauntlett, 2011:133). Creating a collective place identity nurtures social capital and community identity. Social capital enables a community to resolve problems more easily because members know that they can rely on each other: "It fosters awareness of the ways in which our fates are interlinked, and encourages us to be more tolerant, less cynical, and more empathetic." (Gauntlett, 2011:147).

3.2 *Community engagement*

Art acitivities have achieved positive outcomes for homeless people including reducing isolation, promoting social networks, self-esteem and communication skills, as well as therapeutic benefits: they can specifically benefit mental health issues by addressing isolation and supporting relaxation and daily routines.

Making workshops have been commonly used as a 'process-led' praxis in textile design. However, in the context of urban fabric research, communal making becomes a communication tool to establish new synergies between people and places, and enhance the social capital of a community. Wates (2014) identifies a number of benefits of community engagement such as empowerment and community feedback as neccessary to inform holistic urban design solutions. At the same time "an impressive and growing research suggests that civic connections help make us happy, wealthy, and wise. Living without social contact is not easy." (Putnam, 2000:287).

Working alongside a group of makers automatically creates an open forum for exchange: sharing the process of learning and creating a textile establishes an immediate rapport within the group. Gauntlett describes the personal value of group making "... people often spend time creating things because they want to feel alive in the world, as participants rather than viewers, and to be active and recognized within a community of interesting people" (2011:222). Textile making is experienced as a positive act as it enables the maker to express their identity through ideas, colours, material and techniques. Makers benefit as they "internalise their learning, making it part of themselves, and relate it to their real life." (Brockbank and McGill, 2007:42). The Home and Belonging project aim is to enhance both making skills as well as personal wellbeing through textile workshops and to present and communicate emerging design products and personal narratives in a wider community context.

4 RESEARCH METHODOLGY

This project was set up as an action research case study in order to explore the participatory and experimental nature of the project and generate an open dialogue amongst participants and facilitators, based on the theoretical research methods framework from Reason and Bradbury (2001). Action research is appropriate for this study because of its focus on social interaction and experimental approach. Utilising qualitative measures, data for this study were collected in the following ways:

- written and photographic diary
- interviews with project participants
- visual check in/out data during workshops
- questionnaires and enrollment data
- review of contextual information and literature.

5 PROJECT TIMELINE

5.1 *Scoping and recruitment*

The main objective for Crisis Brent was to set up a new integrated arts programme in Harlesden, whilst the author was particularly interested in exploring new methods of textile engagement within inner city communities.

Over a period of three month both institutions set up the general logisitics and visited similar existing art workshops at Crisis headquarters in Central London. It was highly beneficial to learn from the already existing good practice and the tutors gave valuable tips on the shared parameters of working with Crisis members. However it emerged that the external set up was also quite different. Crisis Brent had only a 15 sqm office room available to host art classes and other meetings, in comparsion to the 100 sqm custom-built art studio space at Crisis headquarters. Whilst

Figure 1. Mapping of personal belonging: hostel bed layout.

Figure 2. Community Pop Up shop in Harlesden.

participants at the established art classes in central London come from many different London Boroughs, Crisis Brent had strong local connections, with all service users living in Brent, offering unique community insights.

In order to promote the new programme, and recruit more Brent participants, the arts co-ordinator set up a number of small outreach acitivities in local hostels and community hubs, as well as distributing leaflets. This led to a first taster workshop with seven participants to test out preferred textile activities and scope areas of special interest. Participants were invited to try out 20 minute rotations of weaving, wood block printing and badge making. Non of the textile acitivities required any further knowledge and all materials were provided. The feedback from the initial sessions then led to the selections of the main workshop activities as well as the most suitable timings for sessions to be accessible a large number of people.

5.2 *Workshop programme*

Over a period of four months CSM academics and applied-arts practitioners delivered 12 two hour long workshops on Wednesday afternoons, as well as two CSM site visits and additional drop in sessions on Tuesdays. In the first 6 months 25 crisis members participated in the programme: eleven were female and fourteen were male; four were Asian/ Asian British background, ten Black / Black British, nine White/ White British as well as 2 from other ethnic backgrounds. The participants were between 25 and 65 years of age.

Session 1: Creative explorations
General introduction and creative mapping of personal journeys as well as experience of local neighbourhood.

Acitivities: brainstorming and collaging, exploring the colours, sounds and smells of Harlesden. Secondary and primary research generation for later sessions as well as show and tell group work.

Session 2: Field trip to central saint martins
Visit to the CSM museum and textile workshop tour.

Activities: Show and Tell, object handling session, student tours and demonstration in the weave and knit workshops.

Sessions 4–5: Tapestry weaving
Transformation of fly-tipped furniture into 3D tapestry objects and personalised chairs.

Activities: Selection of individual colour schemes, large and small scale tapestry weaving, embellishment, painting and decorating.

Sessions 6–7: Digital design
Digitial image making responding to Harlesden's vibrant High Street folklore.

Activities: research, photography, graphic design and heat transfer printing.

Sessions 8–12: Analogue design
Analogue image making and T-Shirt slogan development inspired by local observations. Certificate award ceremony

Activities: research, stencilling, typography, drawing and painting as well as heat transfer and screen printing processes.

Session 13: CSM degree show visit
The programme ended with a visit to CSM to look around the degree show in order to reflect and relate the textiles outcomes of the Home and Belonging project to a wider art and design context.

Actvities: Project presentations by CSM students.

Additional drop in sessions
Due to irregular attendances, overwhelmingly enthusiastic responses as well as time-consumingly intricate work, the Crisis arts co-ordinator arranged additional optional drop in sessions on Tuesdays to extend the workshop access times. These were also vital in order to churn out a 'collection' of refined designs for the pop up shop and exhibition.

5.3 *Project exhibition and dissemination*

At the end of the project the organisers arranged a pop-up shop showcasing the Home and Belonging designs as part of a local community event bringing different local stakeholders togther such as the Harlesden Town Garden team as well as the Neighbourhood forum. The event was attended by about 150 members of the community as well as the Mayor of Brent, and the public were invited to leave feedback about local urban planning initiatives.

Additionally, CSM researchers curated an exhibition based around Urban Fabric showcasing two London-based academic research projects as well as the Crisis project outcomes. For a month Crisis T-Shirts and embellished tapestry chairs were exhibited in the CSM Window galleries, attracting a diverse audience of students, and academics as well as at least 500 members of the general public.

6 FINDINGS

Throughout the workshop programme qualitative and quantitative data has been collected via 25 participant 'enrolment' forms. Additionally the project has been monitored through feedback sessions, photographic diaries, emoji check-ins and course evaluation forms, which were completed by seven participants at the end of the programme.

6.1 *Strong attendance and retention*

Key to the project's success was the locally based arts co-ordinator whose continued presence and positive support acted as a proactive agent to gel the group. This led to high group retention rates with eight members attending 85% of workshop sessions, demonstrating their strong commitment to the project. An additional group of nine participants attended three or more sessions whilst six participants only came once. Orginally the tutors were planning to brief all members at the beginning of each session, however, due to differing personal circumstances as well as mentoring appointments people popped in and out during the afternoon. At the same time participants got very absorbed in the making process and worked well beyond the two hour slot to finish ambitious projects. In response, the organisers extended the course hours, added a drop in session on Tuesdays and changed group briefings to the middle of the workshop. There was certainly no lack of enthusiasm: some participants took voluntary 'homework' with them or completed work for others.

The workshop took place in a small office space with all participants working alongside each other on a communal table. Whilst this constrained the making of large scale work, at the same time it encouraged lively conversations and a natural integration of new members into the group. It was noted how much more sociable and convivial the workshop in Brent was in comparision to the spacious but more individualistic set up at the Crisis headquarters in central London. The Home and Belonging project benefitted from the shared local background of the participants as well as the communal table layout in the workshop space as it nurtured common ground amongst the participants.

6.1.1 *Universal textiles communication*

In respose to their weekly observations as well as participants' feedback on the day, tutors adapted course content throughout the duration of the programme. Whilst none of the participants had any previous knowledge of textiles, it was observed that all of them expanded their colour, composition and technical skills. The intricate hand co-ordination of tapestry weaving and the complexity of digital design applications were experienced as most challenging by some members. Whereas heat-transfer press printing, drawing and painting appealed to newcomers and advanced participants because of their immediate accessibility and 'instant' results.

At CSM, textiles generally attract a predominately female audience and it was promising to see an even gender balance in the group. Textiles are particularly accessible because of their universal presence in daily lives both as clothing and home interiors. During the workshops it was observed how individual material and colour selection can be utilised as a powerful design tool to express cultural identity. For example, two participants worked almost exclusively in red-green-yellow Jamaican colour schemes whilst another participant focussed on 'minimal' black and white arabic typography.

The group invented a specific collaging technique by recycling small colourful cut-off shapes of heat transfer paper and printing them onto T-Shirts and hard surfaces. This design was later called 'Harlesden Camouflage' as it resembled traditional camouflage patterns but with an Urban colour scheme. These camouflage designs were also very well received at the public pop up shop events and sold out quickly. In addition to their growing textile design skills, participants have also gained transferable knowledge such as time management, team work and presentation skills.

6.2 *Self-esteem and the creation of a safe space*

Participants have overwhelmingly reported an increase in confidence, self-esteem and in their social connections since taking part in the Home and Belonging sessions. This is a direct result of the supportive and open atmosphere at Crisis Brent, which created a safe space to build strong group identity and ownership. Clifford and Herrmann observe: "unless your group already know each other prior to the project, time must be spent fostering a sense of 'the group' whereby barriers are broken down, they share common ideas, concerns and goals and individuals gain a sense of belonging." (1999). Even though the group consisted of a diverse mix of generations, cultural backgrounds as well as gender it was observed that textiles were a material that all could relate to. Staff had been impressed by participants' enthusiasm throughout the programme and how they built effective working relationships with one another. In response to the course evaluation question 'What have you enjoyed most?' participants stated:

- Gaining new art skills
- Expressing myself through art
- Getting to know people
- Looking at things I hadn't noticed before
- Looking at art
- Everything!

100% of the participants who responded to the feedback form stated that:

- Their relationships with other people and their trust in others have improved.
- Their confidence and self-esteem have improved
- Their time management / motivation and skills with working in a group have improved.

Furthermore, all participants apart from one have given their full consent to using their image / artwork for future promotional purposes. This was not the case at the beginning of the project, and demonstrates an increase in self-esteem and trust.

6.2.1 *Visual feedback on personal well-being*

It was noticed that some participants had difficulties expressing themselves through written English and at the same time the group voiced a general questionnaire and paperwork fatigue. In response, the author developed a new visual evaluation method to elicit deeper feedback on participants' experience of each session. Participants were encouraged to check in and out of each workshop by choosing a printed emoji sign (happy, sad, frustrated etc.) that best described their mood at that time and put it up on a board. This proved a very well received communication tool, and the emojis became a comfortable and accessible expression of the mental state of each participant. The majority of the group progressed from sadness, cautiousness or frustration to happy emoji signs by the end of the class. Participants also invented additional personal emoji signs as well as adding words like 'unity' to the board. This method has been positively received by other partner organisations, which have also implemented it to inform their evaluation reports.

Figure 3. Emoji check in and check out.

6.3 *Outreach and community exchange*

Course feedback forms show that 100% of participants feel more engaged in their community. Existing Crisis members who also access partner organisations quickly promoted the textile workshops to other clients. This self-promotion is how the majority of new clients from external addiction support schemes came to join the group. Additionally four participants have signed up to be community researchers and other participants volunteered to take part in promoting the project to other homeless and vulnerable people. Four participants are still involved in Crisis Brent's neighbourhood planning work and have creatively contributed to the design and delivery of neighbourhood forum events.

The pop up shops in Harlesden and at Central Saint Martins evidenced the importance of bringing different stakeholders together. In particular as they provided a relaxed set up to connect participants to a diverse audience, invite exchange and promote their designs. It was significant for the participants to have a physical presence through their textiles and find an open-minded audience for their stories. Many of the printed textiles displayed personal messages such as: "Sorry not my Problem", "Love", "Messy by Design", "Be" and "Do it with Passion". People who bought the products openly related to a specific narrative by wearing and displaying an authentic Home and Belonging voice. This correlates to Couldry's findings on the importance of a voice as a form of agency and the act of taking responsibility for the stories one tells: "For a voice to be meaningful, people need opportunities and tools for self-expression but they also need to be heard." (2010).

7 CONCLUSION AND RECOMMENDATIONS

7.1 *Textile transformations*

The author acknowledges the limited scale and particular environment in which the Home and Belonging research was conducted. At the same time the findings of this project correspond strongly with other recent reports and the unique value of textile engagement is clearly transferrable to related sectors. An all-party inquiry into health and wellbeing strongly evidences the benefits of creative activities and concludes with a recommendation to establish a more strategically aligned approach: "We recommend that leaders from within the arts, health and social care sectors, together with service users and academics, establish a strategic centre, at national level, to support the advance of good practice, promote collaboration, coordinate and disseminate research and inform policy and delivery..." (Creative Health: the Arts for Health and Wellbeing Report 2016). Thorpe and Gamman, who have been working with inmates on a number of design projects in prisons, have also identified multi-layered advantages for individual citizens as well as society: ". . . the use of creative activities and tools is a means for the accumulation and transmission of social knowledge. It influences the nature, not only of external behavior, but also of mental functioning of individuals." (2015). The specific findings of the Home and Belonging project echo these recommendations as well as highlighting the benefits of locally embedded textile interventions to enhance the social fabric of an urban environment. This has led to the following recommendations for place-based textile community activities:

Textile workshops

- Set up early outreach / recruitment activities to reach participants from different feeder institutions and offer taster sessions to gauge interest
- Create a 'safe space' to enable open exchange and conversations
- Encourage a responsive and flexible course set up for reciprocal learning (including tutors)
- Include a range of materials and colours to reflect the culturally diverse aesthetic preferences of all participants
- Design 'instant' making activities that are accessible for all abilities as well as study visits to introduce a wider context
- Invite critical reflection and the expression of personal local narratives through new forms of textile communication

Community engagement

- Increase local opportunities and spaces for cohesion and integration amongst homeless and vulnerably housed residents and other community members
- Expand interdisciplinary approaches to bring different partner institutions together
- Create more opportunities for co-design: to share feedback, listen to diverse voices and develop a holistic local urban fabric design approaches.
- Integrated provision of services to homeless and vulnerably housed residents with the innovative use of applied arts for community wellbeing

Textiles have much to offer in order to deliver community engagement as a transactional and mutual process and bring all stakeholders to the table. This is especially valuable for fluctuating communities to co-design a place identity reflecting multiple cultural narratives from past and present experiences. There is much scope for place-based textiles to respond to societal or urban needs, enhance wellbeing and create more empathetic neighbourhoods.

ACKNOWLEDGEMENTS

The Home and Belonging project was only possible with funding support from Brent Council, Crisis and the University of the Arts London. A very special thanks to all collaborators: Atara Fridler, Sumathi Pathmanaban, Gabriel Parfitt, Chryssi Tzanetou, Peter Close, and last but not least to all participants for their invaluable enthusiasm, inspiration and feedback.

REFERENCES

Altman, I. and Low, S.M. (eds) 1992. *Place Attachment.* Human behavior and environment, Vol.12. New York: Plenum Press

Antonovsky, A. 1987. *Unravelling the Mystery of Health: How People Manage Stress and Stay Well.* San Franciso: Jossey: Bass

Brockbank, A and McGill, I. 2007. *Facilitating reflective learning in higher education* (2nd ed.). Maidenhead: SRHE and the Open University Press

Clifford, S., Herrmann A. 1999. *Making a Leap - Theatre of Empowerment.* London: Jessica Kingsley Publishers

CHAIN Annual Report 2015/16. Weblink https://data.london.gov.uk/dataset/chain-reports [20.7.2017]

Creative Health: the Arts for Health and Wellbeing Report. 2016. Short Report. Weblink http://www.artshealthandwellbeing.org.uk/appg-inquiry/ [20.7.2017]

Couldry, N. 2010. *Why Voice Matters, Culture and Politics after Neoliberalism.* London: Sage Publications

Gauntlett, D. 2011. *Making is Connecting.* Cambridge: Polity Press

Harden A., Sheridan K., McKeown A., Dan-Ogosi I., Bagnall A.M. 2015. *Evidence Review of Barriers to, and Facilitators of, Community Engagement Approaches and Practices in the UK.* London: Institute for Health and Human Development, University of East London

Harkins, C. 2014 *Creative Health Report*, Glasgow: Centre for Population Health

Health Needs Audit. 2015. Weblink http://www.homeless.org.uk/facts/homelessness-in-numbers/health-needs-audit-explore-data [15.07.2017]

Homelessness, Third Report. 2016. Communities and Local Government Committee. Weblink. https://publications.parliament.uk/pa/cm201617/cmselect/cmcomloc/40/40.pdf [17.7.2017]

Homelessness Kills. 2012. Crisis Report. Weblink https://www.crisis.org.uk/ending-homelessness/homelessness-knowledge-hub/health-and-wellbeing/homelessness-kills-2012/ [24.07.2017]

I was all on my own. 2015. Weblink https://www.crisis.org.uk/ending-homelessness/ homelessness-knowledge-hub/health-and-wellbeing/i-was-all-on-my-own-2015/ [17.7.2017]

Lynch, K. 1972. *What Time Is This Place?* Cambridge: MIT Press

MacCallum, D., Moulaert, F., Hillier, J. and Vicari Haddock, S. 2009. *Social Innovation and 'Territorial Development.* Farnham: Ashgate Publishing

McMillan, D.W. and Chavis, D.M. 1986. *Sense of Community: a definition and theory.* Journal of Community Psychology Volume14:6-23. Hoboken: Wiley

Putnam, R.D. 2000. *Bowling Alone, the collapse and revival of American community*. New York: Simon and Schuster paperbacks

Reason, P., Bradbury, H. (eds) 2001. *Handbook of Action Research: Participative Inquiry and Practice*. London: Sage Publications

The unhealthy state of homelessness. 2014. Weblink http://www.homeless.org.uk/sites/default/files/site-attachments/The%20unhealthy%20state%20of%20homelessness%20FINAL.pdf [20.7.2017]

Thorpe, A., Gamman L. 2015. Could design help to promote and build empathetic processes in prison. In Jonas, W., Zerwas, S. von Anshelm, K. (eds), *Transformation Design, Perspectives on a new Design Attitude.* 83-100. Basel: Birkhäuser

*UK Census 2011.*2011. https://www.ons.gov.uk/census/2011census [17.7.2017]

Wates, N., 2014. *The Community Planning Handbook* Oxon: Routledge

Photo Credits

Figures 3 by Gabriel Parfitt; Figures 1 & 2 by Anne Marr

Textiles, Identity and Innovation: Design the Future – Montagna & Carvalho (Eds)
© 2019 Taylor & Francis Group, London, ISBN 978-1-138-29611-4

Telaio 31: A hybridization between creativity, experimental and industrial process

P. Ranzo & P. Maddaluno
University of Campania "Luigi Vanvitelli", Naples, Italy

G. Gentile
IPCB-CNR (Institute for Polymers, Composites and Biomaterials – national Research Council of Italy), Pozzuoli (Na), Italy

ABSTRACT: *Telaio 31* is an innovative textile project that is collocated in between textile and fashion design. Moving from an open planning, *Telaio 31* suggests "rereading" the traditional jacquard machine through an innovative process of "total manufacturing". The aim is to create a clothes item directly from weaving. The links between project, process and product have been rethought, while applying a functionalizing property (hydro/oleophobic properties) to the fabric being created. A methodology in line with the *European Textile Technology Platform* programme has been experimented, also in line with the *The Ted – Zero Waste Dress strategy*. The establishment of a creative workshop was a fundamental step, a laboratory for advanced research inspired by the models of *community of practise* and of *knowledge sharing*. In line with the process of *design thinking*, the project also set up a multidisciplinary network involved in the ideation and realization of an only project.

1 INTRODUCTION

The strategic lines drawn up in 2006 by the Agenda of *European Technology Platform* have seen, in ten years, a marked transformation of textile production, both at large enterprises and SME.

The support of research, mainly university campus, to the world of production and the consideration of major social and consumer changes have made it possible to achieve forecasts in 2006.

Quality, experimentation of new technologies, consideration of environmental and of new consumers as the main actors of all processes have produced transformations for small and large revolutions and for the IV industrial revolution of the industry.

A revolution characterized by increased capacity by small and medium-sized businesses to generate continuous innovation focused on quality and specialization rather than increasingly cost-effective and environmentally unsustainable products.

A very interesting fact is that innovation is not only tied to technology, but above all to the inclusion in the production cycle of qualified human resources and research, able to interact closely with production in a short time and to form real teams, throughout the product testing cycle (research, design, prototyping, process validation, and product).

The flexibility of the European SME in the textile sector is fundamental.

The European community's forecasts to the textile industry underline the strategic importance that will be the ability to develop collaborative and integrated solutions: "By 2025, the textile and clothing industry (...) will operate according to a globalized and Efficient circular economic model that maximizes the use of local resources, exploits advanced manufacturing techniques and engages in cross-sectorial collaborations and strategic clusters". (European Commission – Joint Research Center Industrial Landscape Vision Study 2025, January 2016).

In the creative and cultural fashion industry, design and its ability to create value through the creation of large design contexts strongly linked to contemporary transformations have often anticipated new research and production lines that can be considered as reference paradigms.

The research that we present, *Telaio 31*, has matured in a team based on sharing a context of innovation shared by an Italian company, a creative team, an interdisciplinary research group.

The initial goal: to combine more creative, research and production specificities in order to introduce multi-level, project, product, functionalization and consumer innovations in a sustainability context. The name, *Telaio 31*, was born from the company's decision to devote a chassis number 31 to researching and experimenting with a team led by a PHD student.

1.1 *"Telaio 31"*

Telaio 31 is a project that is collocated in between textile and fashion design.

Languages considered by contemporary critics to be both near and distant. Similar areas, often interwoven, but characterized by studies and researches that have highlighted the diversity.

A possible initial "merger" has been theorized by Fiorani in an essay titled *Corpi di stoffa* (Bodies of fabric).

It is a definition that takes inspiration from the analysis carried out by Mario Perniola on the concept of the"fourth body". "It is a suggestive vision, that opens up to [...] a new approach to the worlds of fabric, interior design and fashion, that allows to grasp the connections between the corporeal dimension and the sensorial and feticist dimensions and their hybridizations with the visual and technological arts"(Fiorani 2009).

Telaio 31 takes inspiration from this modern way of intending this relationship, with the objective of investigating the role played by the organic-planner in the process of fashion innovation. A designer who is no longer a director or a fashion designer-movie director, but a sensible inventor of the horizontality of the creative processes. A figure that gives life to "soft architectures" (Mendini 1985), where the return to handicraft is merged with a fascination for artificiality. A figure who questions the use of new technologies and is influenced by the codes of contemporaneity.

This scientific research revolves around two points: "product planning", characterized by the dimension of research into the culture of planning, with attention to environmental issues; and "process planning", based on the connections between different knowledges and experimenting the shift of specific knowledges from one sector to another.

The aim is to put forward a methodological and productive perspective that is multidisciplinary, and implement a planning process that is continuous and open.

This choice is motivated by the structural change in contemporary fashion: "Fashion design, like all other design oriented sectors, is being affected by these new operational modalities. [...] The evolution of the process of multidimensional innovation-planning-production and the study of its dynamics [...] cannot be addressed by one discipline alone" (Conti 2007).

What is first of all needed, in order to obtain a competitive and innovative product, resulting from a significant cultural awareness, is transversality of knowledge.

Given this framework, the study identifies a disciplinary area that we could define as "fiber design". A category that looks at things in their original phase.

It analyses the primary source of the project-clothes item: it studies its characteristics, to enhance and also to rethink them. The starting point is the essence – the fiber – that is modelled into unexpected shapes, into possible geometries. Also to bestow a new identity, and surprising qualities.

Poised between tradition and innovation, *Telaio 31* tries to "reread" the potential of the jacquard machine.

In particular, the project questions the technical dimension. Everything is technique, Braudel used to say, and its function is primordial, so sooner or later everything ends up depending on its intervention. Its breadth is that of history. It is material culture and 'human science', as André G. Haudricourt already defined it.

What is examined is the transformation of the chosen yarn, the procedures of manufacturing and the modes of employment. And, at the same time, the project questions the infinite possibilities of weaving patterns that the jacquard loom offers. "The functioning of a loom lends itself well to researches on primary shapes. Shape is not merely added to fabric, it is enacted directly with colour, through the woven material. Clearly, a thorough knowledge of the material and of the textile structures is necessary, a sensitive attention to their potential, to the receptiveness to light, the structures of the interweaving, the colour. The fabric is thus returned to its corporeity and seen as an 'organism'. Its secrets can be explored and we can hear the 'sound of the colour' in the material, grasp the colour of pure silk, the coldness of artificial silk, the roughness of hemp and wool. [...] Invention is not only a technical solution of a problem: it is an event in which a complex process precipitates, that merges scientific ideas with technical experiences: it is a flow of models, objects, methods. Behind the objects that technology brings into being [...] is the creation of a new language, of a new way of seeing the world and of thinking about the body and oneself inside the world" (Fiorani 2012).

The processual phase is central in this research: to rethink modes and modalities that characterize the planning of a clothes item.

The study thus aims to remodel the instruments of multidimensionality and the codes of multidisciplinarity.

With multidimensionality we intend the range of instruments that are necessary to define and carry out a project.

With multidisciplinarity we intend the space in which the codes and different cultures come together, to be joined and to be "blended".

Since the identification of the concept, and then during the briefing, the aim has been to make the time of the planning phase coincide with the time of production, in order to obtain a short cycle.

Telaio 31 is founded on the implementation of the *Zero Waste Dress* methodology: no packaging; reduction of waste produced in the textile sector in the pre and post consumer phase. And advanced research workshop.

1.2 *Applied methodology*

It was established an advanced research workshop that draws from the models of *community of practise* and *knowledge sharing*.

In line with the process of *design thinking*, a multidisciplinary network has been set up, involved in the ideation and realization of one project. The aim of this research was to propose a "rereading" of the traditional jacquard machine through the innovative process of "total manufacturing". The aim is to create a clothes item directly from weaving, so as to make the planning time coincide with the production time. The links between project, process and product have been rethought, while applying a functionalizing property to the cloth being created. The result that is a *seamless*-jacquard.

A methodology in line with the *European Textile Technology Platform* programme has been experimented, also in line with the *The Ted – Zero Waste Dress* strategy, based on the absence of packaging and on the reduction of waste production in the textile sector in the pre and post consumer phase. The establishment of a creative workshop was a fundamental step, a laboratory for advanced research inspired by the models of *community of practise* and of *knowledge sharing*. In line with the process of *design thinking*, the project also set up a multidisciplinary network involved in the ideation and realization of an only project.

The laboratory has been coordinated by an "organic planner" and has involved a scientific network and a productive network. An experimental and open production chain divided into two macro areas: ideational (dedicated to research) and planning (dedicated to the development).

The "actors" involved are different.

A company in the textile-manufacturing sector located in Lurate Caccivo (Co), founded in 1985, the Bernasconi Tessuti e/o BiSeta (for the tertiary production) that is leader at an international level in the use of jacquard looms. The company is an interpreter of a consolidated tradition of the made in Italy, and was chosen after an accurate study of the districts and production chains present on the Italian territory (Fig. 1).

The intervention of the company has been indispensable to access a specific *contextual-knowledge* that has allowed the process to evolve.

A second actor is the IPCB-CNR (Institute for Polymers, Composites and Biomaterials – national Research Council of Italy) located in Pozzuoli (NA), whose scientific research has been dedicated for many years to the experimental study of textile material; finally the Department DICDEA of the Campania University of Luigi Vanvitelli.

The project is articulated in the following phases: the ideational/project phase; process development 1; process development 2; experimental prototype for testing; textile functionalization.

1.3 *Ideational/project phase*

In this phase the network for experimentation and briefing of the project was defined. In particular, the moodboard was identified. Characterized by some key words: open and sustainable planning; enhancement and safeguard of the made in Italy. The concept can be described in two points: "total manufacturing" and textile functionalization.

Figure 1. Bernasconi Tessuti e/o BiSeta, manufacturing textile company: internal view of the company. On the right: *Telaio 31*.

1.4 *Process development 1*

The process that lead to the realization of *Telaio 31* had to overcome some delicate criticalities: from the definition of the design of the module to the resolution of weaving techniques; from the study of the fibre to the reinterpretation of the same fibre.

The first phase consists in the creation of the "element" and of the textile structure.

A young Iranian artist was involved in this phase, Soheil Saeed Naderi. From the start a graphic expression founded on the line was elaborated. A choice able to highlight the *seamless* finishing and enhance the design of the contour with regard to the portability of the clothes item.

After some meetings, Naderi elaborated 20×20 samples in white handmade cotton paper (boiled in water at a high temperature and left to dry in the sun) on which he engraved an abstract composition of patterns with a dynamic rhythm.

This paper was chosen because visually close to the jacquard weave.

For the first weaving trials, after having addressed some technological aspects together with the manufacturing company, it was decided to weave using three threads (1 warp + 2 weft) and a "plural" range of textures, with different densities (stex) and a rich palette of colours (white, ecru, lurex, beige, sand).

A technical and functional choice. A way to test the consistency of the fabric, its grammar, the softness, the chromatic play of the weave of the module. The initial samples of weavings were realized with mixed yarns.

Polyester + nylon: to analyze the three dimensionality of the jacquard.

Polyester + scoured silk + cotton: to test the lustre. Polyester + raw silk + cotton: to observe the opacity. The outcomes were not convincing with regards both to aesthetic and tactile quality.

Raw silk, which in part retains sericin, is rough and opaque; less compatible with wearability.

Scoured silk, on the other hand, is lucid, shiny and slippery to the touch, but also very thin. It is compatible with wearability.

For this reason the "planar communication of the surface", to quote Branzi (2010), was rethought.

The line was reassembled with a tangent rhythm in one point and extended with an all over ratio. Attention then was focused on the design of the clothes item.

The creation of the models was carried out following not the traditional modality of tailoring but with an unstructured approach (in line with the *seamless* modality).

After studying the various aspects it was decided to operate with a mixed composition: 90% silk + 10% polyester. A choice which allows to confer fluidity to the fabric with a view to wearability (Se) and preservation of the structure (Pl).

Figure 2. Clothes item-cloth realized.

1.5 *Process development 2*

The first difficulties encountered during the manufacturing process were: the creation of a double cloth using the jacquard loom (to allow immediate wearability); and the "engraving" of the contour of the clothes item on the cloth (its perimeter is composed of both sewn parts and parts to be unsewn).

Naturally, this type of machine produces single cloths and iconographic or "abstract" patterns on the whole surface. Unlike a circular loom that uses single threads for warp and weft, the jacquard loom uses an only thread for the warp but cuts the weft thread. For this reason, it cannot be used to create tubular fabric.

After a series of attempts, aided by the staff of the manufacturing company, it was possible to achieve the first innovation in this process. A "jacquard tubular" was created and various weaves depending on the cut and on the necessary stitchings. It was then possible to identify the yarns, balancing the dtex and defining the colour range.

Starting from a three thread weave (1 warp + 2 weft; 90% Se + 10% Pl) the chromatic palette was divided into four section: cognac (warp) + carrot and honey (weft); white (warp) + pale 1 and scoured 1 (weft); beaver cloth (warp) + noisette and mastic (weft); cement (warp) + steel gray and smoke (weft).

Having defined the "grammar of the project", it was possible to further the study of the warping and then of the weaving.

Another objective was to collocate this project in the framework of the *Zero Waste Dress*. In order to adhere to this strategy, a study on the dimensions of the cloth was carried out. The width of an electronic jacquard loom can be standard (1,40 m) and the length variable (Fig. 2).

In this sense, it was decided to use a cloth measuring 110 × 110 cm, in order to have a minimum amount of offcuts, from which two clothes items can be cut out (Fig. 3). The overall dimension of these elements corresponds to the sizes 38-40-42-44 (slim line) (Fig. 4).

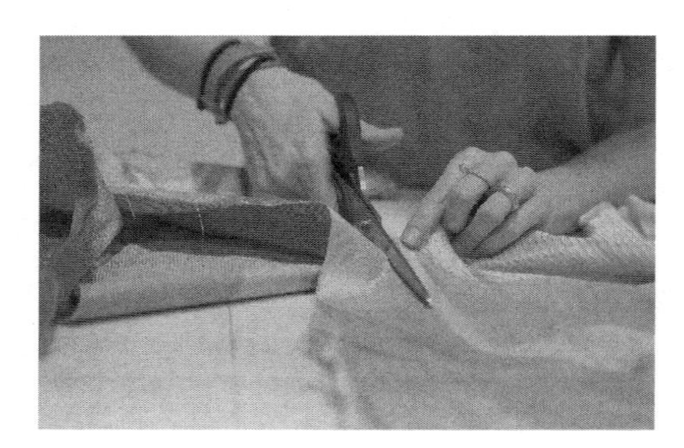

Figure 3. Cutting phase.

This way only small strips of cloth at the sides remain not used. The height and weight of the finished cloth remain unvaried with respect to the total height and weight generated in the weaving phase, because the fabric does not undergo RAM. The percentage of the waste material thus obtained is 2% for the 110–110 height and 399 Gr/Mtl.

1.6 *Experimental prototype for testing*

Once the weaving operation is over, the cloth is cut and wrapped around a very resistant cardboard cylinder 15 cm in diametre.

A member of the manufacturing company's staff transfers the tube to the quality control department for the fabric inspection machine. The cloth is laid on an oblique counter, and analyzed in every single part with a cold light (so as not to harm the texture), both inside and outside.

The criticalities which may emerge are: knots, fraying, scratches.

With some of these problems it is possible to intervene manually thanks to the support of seamstresses.

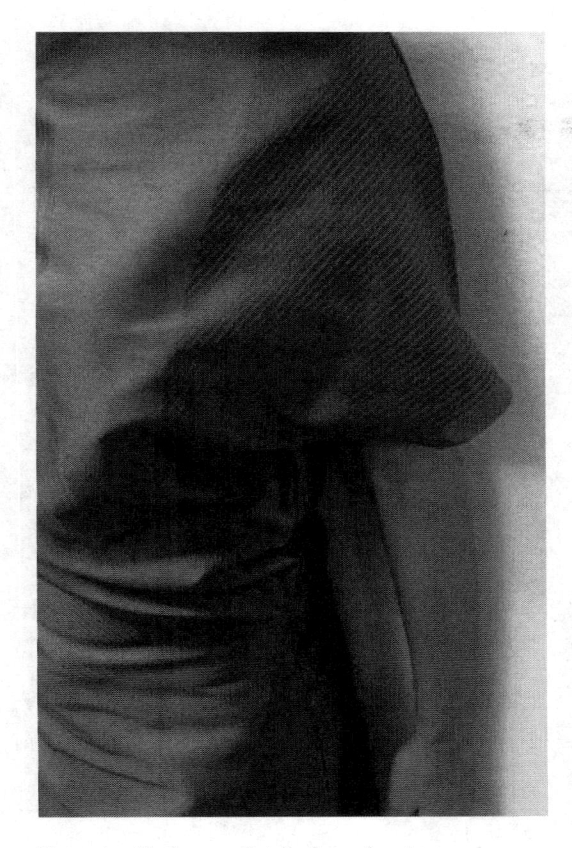

Figure 4. Cloth worn. Detail of *seamless-jacquard*.

Once the quality check is completed, the cloth is transferred to a large table for the cutting phase. It is a very delicate action, that requires attention.

Also this aspect has been the object of an accurate study.

Initially it was decided to adopt a laser cutting machine able to burn the fibre and avoid fraying of the cloth and/or of the salvedge.

Because of the strong colour, however, this type of cut blackened and altered the colour with the high temperature.

For this reason, hand cutting was preferred; parts of the edges are distinct from the weave. Continuous tightly woven marks indicate areas that will fit the body (head, arms, legs).

The discontinuous loosely woven marks correspond to where the clothes item will be closed.

Once the cut has been made, the clothes item-cloth can be worn.

1.7 *Textile functionalization*

Along with the activities in the manufacturing company the first functionalizations treatments of textile samples were carried out, with the aim of creating a stain-resistant coating, hydrophobic and oleophobic, without altering the handle, the aesthetic quality and the breathability of the fabric.

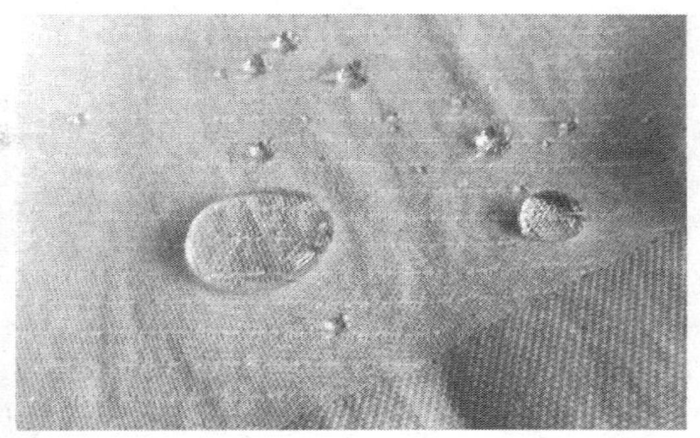

Figure 5. Hydro/oleophobic properties on the fabric.

With this aim, the activities carried out revolved around the identification of a treatment based on the application of a nanocoating based on polytetrafluoroethylene (PTFE), nanoparticles in water dispersion.

To guarantee a good durability of the treatment, the application of PTFE nanoparticles was combined with the application of a polymeric phase that, acting as continuous matrix, was able to partially incorporate the nanoparticles and anchor them to the surface of the fibres.

As the polymeric matrix, a water dispersed polyurethane was used.

The developed treatment allows to avoid the drawbacks that are typical of the hydrophobization treatments presently adopted in the textile industry, based on the chemical treatment of clothes with perfluoroalkylated substances.

The employment of these substances is in fact strongly criticized by campaigns promoted by international organizations (Detox Campaign by Greenpeace), that call for the commitment on the part of the textile industry to stop using toxic or polluting components in industrial processes.

Part of the clothes were then treated applying the developed polyurethane/PTFE nanocoating by a single-step dip-coating process, obtaining a final prototype characterized by excellent hydro/oleophobic properties (Fig. 5).

2 CONCLUSIONS

- Textile product characterized by a jacquard weaving. Inside two *seamless* dresses to be cut and worn.
- This research proposes a rereading of the traditional jacquard machine through the innovative process of total manufacturing: the creation of a finished clothes item directly from the jacquard weaving, in line with the *Zero Waste Dress* code (no packaging; reduction of the waste produced in the pre and post consumer phase; simplification with regard to the duration of employment and of manufacturing) to achieve a result that is a *seamless-jacquard*. Finally,

to impart a functionalization to the fabric without altering its handle and the aesthetic aspect by means of a process with a low environmental impact.

- Textile product characterized of a weave using three threads (1 warp + 2 weft) and a mixed composition: 90% silk + 10% polyester. A choice which allows to confer fluidity to the fabric with a view to wearability (Se) and preservation of the structure (Pl).
- Textile product characterized by the construction of a "tubolar jacquard" frame on which the two dresses were made. The parts of the edges are distinct from the weave. Continuous tightly woven marks indicate areas that will fit the body (head, arms, legs). The discontinuous loosely woven marks correspond to where the clothes item will be closed.
- Another objective was to collocate this project in the framework of the *Zero Waste Dress*. In order to adhere to this strategy, a study on the dimensions of the cloth was carried out. The width of an electronic jacquard loom can be standard (1,40 m) and the length variable. In this sense, it was decided to use a cloth measuring 110 × 110 cm, in order to have a minimum amount of offcuts, from which two clothes items can be cut out. The overall dimension of these elements corresponds to the sizes 38-40-42-44 (slim line).
- The clothes item-cloth has been cut by hand-tailored scissor; once the cut has been made, the clothes item-cloth can be worn.
- In the future the cloth made by jacquard weaving can be industrially developed and having a serial production. The *seamless-jacquard* can be applied on a wide range of clothing to wear. For example: outerwear, trousers, jerseys. And various accessories: scarves, gloves, bags, hats.

Telaio 31 is the result of the joint work of the different parts of the network of experimentation. Each "actor" involved in this process questioned his or her own specificity. To give life to an open project, characterized by sustainability and innovation.

REFERENCES

Bertola, P. & Conti, G. 2007 (eds). *La moda e il design. Il trasferimento di conoscenza a servizio dell'innovazione*. Milan: Edizioni Poli.Design.

Bertola, P. & Manzini, E. 2006 (eds). *Design multiverso. Appunti di fenomenologia del design*. Milan: Edizioni Poli.Design.

Braddock Clarke, S. E. & O'Mahony, M. 2002. *Techno Textiles. Tessuti rivoluzionari per la moda e il design*. Milan: Ascontex Editoriale.

Branzi, A. 2010. *Scritti presocratici. Andrea Branzi: visioni del progetto di design 1972/2009*. Milan: FrancoAngeli.

Canonico, P. 2014. Una strategia europea per il 2020. Innovazione tecnologica, forme di aggregazione multidisciplinari, nuovi modelli di business per lo sviluppo del Tessile Abbigliamento. *Tex Innovation 18* (June): 7–13.

Caves, R. E. 2000. *Creative Industries: Contracts between Art and Commerce*. Harvard University: Press.

Conti, M. G. 2007. Considerazioni finali. In Bertola, P. & Conti, G. (eds), *La moda e il design. Il trasferimento di conoscenza a servizio dell'innovazione*. Milan: Edizioni Poli.Design.

Fiorani, E. 2009. Corpi di stoffa. In Soldati M. G. & Conti M. G. (eds). *Fashion e textile design. Percorsi paralleli ed evoluzioni storiche*. Milan: Lupetti.

Fiorani, E. 2012. Tessere idee e mondi di stoffa. In Soldati M. G. & Sabbioni A. (ed). *Tecnologia e innovazione per l'industria tessile*. Milan: Lupetti.

Li, S., Huang, J., Chen, Z., Chen, G. & Lai Y. 2017. A review on special wettability textiles: theoretical models, fabrication technologies and multifunctional applications. *J. Mater. Chem.* A(5): 31–55.

Mendini, A. 2004. *Alessandro Mendini. Scritti*. Milan: Skira.

Mora, E. 2009. *Fare moda. esperienze di produzione e consumo*. Milan-Turin: Bruno Mondadori.

Quaid, M. Mc 2005. *Extreme Textiles: Designing for High Performance*. New York: Princeton Architectural Press.

Quinn, B. 2013. *Textile Futures. Fashion, Design and Tecnology*. Paris: Berg International.

Quinn, B. 2013. *Textile Visionaries: Innovation and Sustainability*. London: Laurence King.

Yetisen, A.K. Qu, H., Manbachi, A., Butt, H. & Dokmeci, M.R. 2016. Nanotechnology in Textiles. *ACS Nano*, 22: 3042–3068.

Textiles, Identity and Innovation: Design the Future – Montagna & Carvalho (Eds)
© 2019 Taylor & Francis Group, London, ISBN 978-1-138-29611-4

A new way for the bobbin lace tradition

I. Bieger, C. Carvalho & G. Montagna
CIAUD, Lisbon School of Architecture, Universidade de Lisboa, Portugal

ABSTRACT: This article is part of the PhD in Design research. The object of study here featured is the Peniche bobbin lace, and its decline over the years. Because there have been fewer and fewer people interested in this traditional craft and also because there has evidently been little progress of the same, we have conducted a research on the possibilities of making a change to the design, in order to develop a new avenue for this tradition, regarded as a heritage in the city of Peniche – Portugal. With an interventionist methodology, we suggest replacing the traditional cotton thread with a yarn made of carbon fibre and another made of fiberglass. Materials with a certain plasticity, may be woven using the bobbin, without changing the local technique or the traditional stitches. Through the results, we suggest the creation of new items with the inclusion of bobbin lace, replacing the doily, which was until then, lacemakers first choice item.

1 INTRODUCTION

The bobbin lace is object of study and research of the PhD in design by the Faculty of Architecture of the University of Lisbon. The bobbin lace has sparked our interest due to it being a declining textile craft. There are increasingly fewer people interested in buying it. Peniche, located in Midwest Portugal, is a small peninsula with about 10 kilometres in perimeter where bobbin lace is regarded as part of its culture. The strong relationship between the lace knots and those of fishing nets, accounts for the proximity of bobbin lace to fishing communities.

Lace knots are almost always associated with the knots made on fishing nets, most of which made by fishermen's wives. In Peniche, bobbin lace was the way of life for most of the town's women and, for many families, it was a resource over wintertime. In Peniche there are very few women who do not have any historical connection to bobbin lace. Over the years, identity relationships have been built towards this popular know-how (Calado 2003).

At a very tender age, women began crafting bobbin lace, over the course of years, this practice and the hours dedicated to this craft, have led to its perfection, revealing a fine workmanship following the many working hours. Mastering the activity shows a given level of maturity (Sharpe 2010).

The learning mode of most of Peniche's lacemakers was still during childhood, through family members. Many of them learnt from their mothers or by watching someone work on it. There were also the so-called "little popular schools". These so-called little popular schools not only taught the lace craft, but also taught the children how read, write and pray, among other learning activities, deemed important for the child to learn. At these little popular schools, children were introduced to bobbin lace from a very tender age, from around the age of four (Calado, 2003).

An Industrial School was later created which, among other things, taught older children the art of bobbin lace. To this effect, the *Casa de Trabalho das Filhas de Pescadores* ("House of Work for Fishermen's Daughters") was created. In addition, the Bobbin Lace Workshop was later created, which operated on the premises of the Stella Maris Club. We can also speak of the creation of Rendimar, a worker's production cooperative, "the aim of which was the exercise of activities primarily related to the manufacturing of lace, but also for producing other women's crafts" (Calado, 2003, p. 206). As evidenced, there were several initiatives to try and maintain the tradition of bobbin lace weaving alive. Nowadays, there is the *Escola de Rendas* ("School of Lace"), operating nearby the tourist office. The School of Lace is maintained by the Peniche Municipal Council and is the meeting venue for the city's lacemakers. The school is open every day and has a full-time teacher, who provides assistance to lacemakers.

Each town that makes bobbin lace has its own trait, such as the way in which the bobbins are held. In Peniche, the tradition is to hold the bobbins with the hand facing upwards, whilst in other places, it's held facing down. Likewise, we can talk about the materials used for weaving lace, specific materials, which are in turn necessary for its execution.

Among the necessary materials are the bolster cushion, the bench to hold the cushion in place, the pricking card, threads, sewing pins and the bobbin, the instrument that gives the technique its name. The bobbin is a thin lightweight cylinder, usually made of wood, to where the thread is attached, the end of which is shaped as a ball or a pear. Bobbins can be made from several types of wood, as well as ivory and other

Figure 1. Weaving of bobbin lace. Source: the author. Photograph taken on 4 March 2015.

materials. Bobbins can be quite different, depending on the places where the lace is made and also on the financial status of each lacemaker.

Author Pamela Nottingham argues that bobbins should be easy to handle, without any sharp edges or deformities, so that the thread does not get stuck. "Bobbins are made of bone, wood or plastic; some are highly decorated, others are plain and inexpensive. It is essential that the slim North Bucks bobbins are weighted with a rings of beads" (Nottingham, 1995, p. 2).

There is no defined size insofar as the cushion to be used. however, according to Bridget Cook (1987), the essential bolster cushion features are that it be sufficiently strong and sturdy to withstand the sewing pins, as well as the lace work that is woven on it. The author shows the many cushion shapes available in the market and their easy acquisition.

In Peniche, the card is normally handmade in a sequence of steps. "The pricking card (or pattern, design, mould) is the basis on which the lace is woven" (Calado, 2003, p. 120). Several sheets of waxed paper, glued to each other using a flour glue, are required. On both the first and last sheet, a thicker sheet of foolscap paper is glued. Following the entire drying process, the papers are painted orange, using aniline dye, a colour which is achieved using saffron. This painting process assists the lacemaker's sight, because the orange contrasts with the thread, normally of a white or beige colour (Calado, 2003).

After the pricking card is ready, it is necessary to transfer the design and prick it using a special tool, called the pricking pin. The result will be small holes where the pins will be pushed into. Mincoff and Marriage (1907) state that these small perforations serve as a guide, so that the lacemaker knows the exact location in which to place the pin. For Kellogg (1920) it is very important that the lacemakers have a very good understanding and know how to interpret the pricking card, which is the pattern for each design to be woven. According to the author, this is regarded as the most difficult part of the bobbin lace weaving process. But only a good reading of the pricking card shall enable a good quality lace.

"a very important step in the process of the formation of lacemakers for it allows the domain of the product allowing the autonomy for the creation of new products and a bigger domain the process, which refers to the reading of patterns and the initial positioning of the bobbins" (Saldanhar & Almeida, 2012, p.688).

As for the threads used in bobbin lace weaving, in Peniche, it was predominantly cotton thread, with a few occasional uses of linen thread, always in white or beige (Calado, 2003). The thread used is a subject for debate. Traditionally, lacemakers use cotton or linen threads.

2 ROAD TO INNOVATION

Comparing what is done today with back then, bobbin lace continues to be woven in the same manner. In the opinion of some of Peniche's lacemakers, only the quality of the lace has improved. However, the preferred materials for its weaving remain the same, cotton or linen thread. There are very few lacemakers who seek other threads for their work.

Considering Poirier when he states: "everything changes in a permanent and irreversible manner" (Poirier, 1999, p. 25), this leads us to reflect upon the lack of innovation that bobbin lace has undergone. In order for lace not to become a mere memory of past generations, strategies must be implemented for its development and a new meaning must be given to its use.

The poor demand for new materials and the non-use of new technologies linked to a traditional handcrafted product, does not allow room for its evolution. The same results are always being achieved. However, we reinforce our concept with a quote from Lalkaka, who says "a torrent of technology-based goods hits the market every week, improving the quality of lives in some ways while also creating complexity and dislocation" (Lalkaka, 2001, p. 2).

We recognise the need for innovation to expand the Market and its demand, thus reaching younger generations with more appealing products. As evidenced by Samli "Therefore radical innovations that would make a major impact on the economy and on quality of life are lacking" (Samli, 2011, p. 5). However, if we want a greater scope and development for the product, we need to rethink its current ways.

Products made from bobbin lace have not changed over the years. The doily continues to be lacemakers chosen item for weaving. However, this is not a product that is on the shopping lists of younger generations. Indeed, one must think of bobbin lace as a commodity, instead of it being regarded as a simple hobby. It's a question of survival of the lace and a stimulus for the self-esteem of lacemakers, also improving their quality of life.

"Innovation is about knowledge – creating new possibilities through combining different knowledge sets. These can be in the form of knowledge about what is technically possible or what particular configuration of this would meet an articulated or latent need. Such knowledge may already exist in our experience, based on something we have seen or done before. Or it could result from a process of search – research into technologies, markets, competitor actions, etc. And it could be in explicit form, codified in such a way that others can access it, discuss it, transfer it, etc. – or it can be in tacit form, known about but not actually put into words or formulae" (Tidd, Bessant, & Pavitt, 2005, p. 15). For Rainey (2005), there is a high success rate in the transformation of created products. For this, we propose weaving bobbin lace with materials other than cotton or linen thread. Resulting in redefined products with added value and which bring greater visibility. According to the author, turning bobbin lace into a modern and contemporary style is a way for repositioning lace on the market. This way, bobbin lace gains a new value and a more embracing meaning.

For Rainey (2005), one of the aims of innovation is to add value to all elements involved, the product, the organisation, the customers. This way, providing benefits and positive results. Moreover, the author argues that the most obvious sources of new product opportunities are existing product lines. Solving problems, repositioning products, meeting new market segments, new needs and desires, and providing additional benefits and performance are sources of opportunity for new products.

"And it is change that always provides the opportunity for the new and different. Systematic innovation therefore consists in the purposeful and organized search for changes, and in the systematic analysis of the opportunities such changes might offer for economic or social innovation" (Drucker, 2002, p. 34–35).

When we think of new materials for weaving bobbin lace, a series of connections between new and traditional materials, between typical and modern objects, come to mind. According to Brownell, we are in a fertile period for innovation, in particular innovation in materials. These affect the society in which we live, and likewise, affect our conduct. According to the author, new materials bring advantages in the various uses, in the products we use, thus breaking down obsolete standards (Brownell, 2006).

For Braraston, the profound knowledge of techniques and taking into consideration previous experiences is of paramount importance towards the new steps into new researches. It provides clarity as to what can be done and the difficulties to be overcome. The new findings can be enriching, providing new and unique opportunities for product development. "The discovery of how a material might respond to processes more commonly associated with other disciplines or materials is intriguing and can provide opportunities for expanding previously uncharted frontiers" (Bramston, 2009, p. 20).

By integrating new materials into lace weaving will result in us having new products. However, new and potential markets may be identified. Douglas Montgomery (Montgomery, 2001) highlights the need for a series of tests that will provide experimental results, these being liable to change the processes of use of materials and shall, consequently, alter the system variables.

3 NEW MATERIALS

In search of innovation, we have looked-up materials that could be woven using a bobbin. That is, materials with a given plasticity and flexibility whilst, at the same time, non-slippery, so that the thread would not get caught in the bobbin. We know that, in the bobbin lace technique, the thread is wound around the bobbin in order to be woven. With these specific features, we intend to maintain the traditional techniques of bobbin lace and search for innovation through new materials. This way, we propose to weave lace using fiberglass and carbon fibre yarns.

We have chosen these two yarns because we consider it unlikely that they have already been used in this technique. Both yarns can withstand high temperatures. The concern with this feature relates to the objects that can be created from the tests undertaken.

4 RESULTS

4.1 *Carbon fibre*

As regards test analysis, the yarn allows for all stitches to be executed well. Stitches capable of being woven: basket, braids, plait, open rows, filled rows and cloth stitch. Using the braid stitch, the result is more consistent and the lace presents the desired texture, given the characteristics of the Peniche bobbin lace. The yarn has the required flexibility for executing all stitches. Its texture does not allow for the presentation to be perfect, as the yarn shows some shredded threads.

4.2 *Fibreglass*

The yarn has the required flexibility for executing all stitches. The stitches tested were braids, basket, open rows, filled rows, cloth stitch and plait.

5 WORK PROPOSALS AND DIFFICULTIES ENCOUNTERED

The difficulties encountered are related to the care required in handling these materials. It is necessary to

Figure 2. Bobbin lace test using carbon fibre yarn. Source: the author. Photograph taken on 13 December 2015.

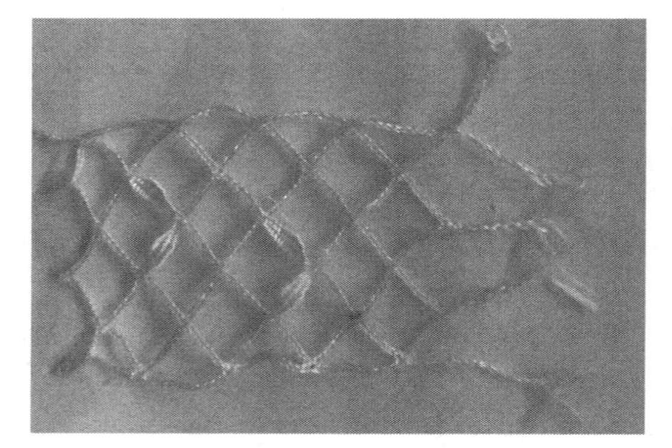

Figure 3. Bobbin lace test using fibreglass yarn. Source: the author. Photograph taken on 13 December 2015.

wear protection gloves and masks, given that the materials release microscopic particles adverse to health. Another difficulty encountered was related to the lack of colours available on the market for both materials tested, thus restricting the choice.

From the tests undertaken, and the knowledge regarding the materials to be woven, we intend to carry out other interventions using the tests presented.

As both yarns, carbon fibre and fiberglass, are resistant to high temperatures, we propose the creation of objects by incorporating other materials, such as glass. We can create the most diverse objects by fusing these materials. The objects may be decorative or artistic, such as bowls, vases, chandeliers. The merger of glass and lace, submitted to high temperatures, may originate some more appealing objects.

A traditional product made using a technological product and fused with glass may result in creative and innovative objects. This could lead to overcoming the market crisis, which bobbin lace is currently in.

6 CONCLUSION

Bobbin lace is regarded as a cultural heritage for the city of Peniche and it has its value in view of the community. It is necessary to find alternatives so that bobbin lace does not become a memory of the past. Turning it into something more attractive and innovative should be a local political pursuit and a road towards preventing its decline.

We conclude that it is possible to introduce new materials into bobbin lace weaving. Should bobbin lace, which is regarded as a traditional handcraft product, be woven using these new technological materials presented herein, it can maintain its features and its traditionally woven stitches. The technique and stitches shall, therefore, remain, allowing for the swift identification of the technique and of the bobbin lace. This result becomes an increasingly more modern and contemporary product, when linked to glass. Innovations that can give greater visibility to this product, which will gradually fall into decline if not rescued by new interventions.

ACKNOWLEDGEMENTS

We would like to thank Capes Brasil for funding Isabel Bieger's PhD in Design at the University of Lisbon.

REFERENCES

Bramston, D. (2009). *Basics Product Design – Material Thoughts.* Lausanne: AVA.

Brownell, B. E. (2006). *Transmaterial* New York: Princeton Architectural Press.

Calado, M. (2003). *História da Renda de Bilros de Peniche* (ed. Author). Peniche: Gráfica Torriana.

Cook, B. M. (1987). *Practical Skills in Bobbin Lace.* London: Batsford.

Drucker, P. F. (2002). *Innovation and Entrepreneurship – Practice and Principles.* New York: HarperCollins Publishers.

Kellogg, C. (1920). *Bobbins of Belgium.* London: Funk & Wagnalls Company.

Lalkaka, R. (2001). Fostering Technological Entrepreneurship and Innovation. *Millennium Book Chapter.*

Mincoff, E., & Marriage, M. S. (1907). *Pillow lace: A Practical Hand-Book.* New York: E. P. Dutton and Company.

Montgomery, D. C. (2001). *Design s and Analysis Experiments* (5 ed.). New York – Arizona State University: John Wiley & Sons.

Nottingham, P. (1995). *The Technique of Bobbin Lace.* London: Dover Publications.

Poirier, J. (1999). *História dos Costumes – O Homem e o Objeto* (Vol. 3). Lisbon: Editorial Estampa.

Rainey, D. L. (2005). Product Innovation-Leading Change through Integrated Product Development. *Cambridge: Cambridge University Press.*

Saldanhar, M. C., & Almeida, . D. (2012). Situated modelling in the drawing workshop for bobbin lace. *IOS Press*, 683-689.

Samli, A. C. (2011). From Imagination to Innovation – New Product Development for Quality of life. *Jacksonville: Springer.*

Sharpe, P. (2010). Lace and Place: women's business in occupational communities in England. *Women's History Review*, 283-306.

Tidd, J., Bessant, J., & Pavitt, K. (2005). *Managing Innovation – Integrating Technological, Market and Organizational Change* (Third Edition ed.). Chichester: John Wiley.

Textiles, Identity and Innovation: Design the Future – Montagna & Carvalho (Eds)
© 2019 Taylor & Francis Group, London, ISBN 978-1-138-29611-4

Comparative study of the main physical properties of denim fabrics used in jeans manufacture

L.N. Souza
School of Arts Sciences and Humanities, University of Sao Paulo, Sao Paulo, Brazil

R.A. Sanches
School of Arts Sciences and Humanities, University of Sao Paulo, Sao Paulo, Brazil,
CIAUD – Research Centre for Architecture, Urbanism and Design, Lisbon School of Architecture, Universidade de Lisboa, Portugal

H.A. Gomes & C.R.G. Vicentini
School of Arts Sciences and Humanities, University of Sao Paulo, Sao Paulo, Brazil

F.M. Moreira da Silva
Lisbon School of Architecture, Universidade de Lisboa, Portugal

ABSTRACT: According to the Ministry of Industry, Foreign Trade and Services (MDIC), Brazil had an increase of 12,17% in denim exports throughout 2016, a significant change if compared to the period of 2012–2014, when there was a decrease in sales. The strategy selected by companies to follow this level of expansion is to ensure that their denim product also includes functional properties. The purpose of this research is to perform a comparative study of the properties of denim fabrics used in jeans manufacture. Four different fabrics used in the manufacture of jeans trousers were selected: fabric 1 – composition: 68% cotton/30% polyester/2% elastane, fabric 2 – composition: 56% cotton/42% polyamide/2% elastane, fabric 3 – composition: 90% cotton/9% polyester/1% elastane e fabric 4 – composition: 97% cotton/3% elastane. Physical test methods were performed to establish, among the raw material studied, which of them adds more functionality to the final product.

1 INTRODUCTION

In Brazil, there is a strong market trend emerging in the world of fashion: denim fabrics known as "stretch", obtained by mixing cotton with elastane. Many companies are also in search of fabrics that resemble in comfort those employed in sport apparel, often knitted fabrics, but with a new aesthetic approach, prioritizing patterns and technology used in woven fabric production.

In order to meet market needs, such technology can be found in the composition of blended yarns made of cotton and synthetic fibers (polyester and polyamide) and elastane used in the production of denim fabric.

The purpose of this research is to draw a comparative study of the characteristics of denim fabrics used in jeans manufacture.

2 LITERATURE REVIEW

Textile fiber properties aid us in understanding how textile articles behave and in explaining their respective performances.

The final product must meet customer needs and add functionality. According to Kadolph and Langford (2006), clothing apparel must offer: attractiveness (hand, drape, appearance), protection (heat, water, cold), ease of maintenance, comfort and durability.

Aesthetics is related to appearance and hand of the final product. Therefore, this property is evaluated by important characteristics such as luster, resulting from reflected light on the fabric, classified as bright, lustrous or opaque; drape, concerning how the fabric adjusts over a three dimensional form, such as a body or table; pattern, which describes the nature of fabric surface and hand, the sensation made by fabric on skin (SANCHES, 2011).

Fabric durability can be defined as guarantee of the conditions of use for a product during a period of time. Durability may be tested in laboratory, but experiments may not reliably reproduce actual conditions of use (KADOLPH; LANGFORD, 2006).

Fabric must be comfortable when worn or during its use. Comfort varies according to individual and personal preferences, different weather conditions and degrees of physical activity (BROEGA; SILVA, 2008).

Ease of maintenance can be understood as the guarantee of conservation of the textile article during its use and after washing procedures (SANCHES, 2011).

2.1 *Comfort*

Comfort is a fundamental attribute for human beings. Weather consciously or not, mankind seeks to improve its state of physical or psychological comfort.

According to Soutinho (2006), comfort is defined by four aspects with regards to clothing:

a) Thermophysiological (thermal) comfort: must protect people from cold and heat alike and must simultaneously allow moisture transfer between layers of clothing.
b) Psychological comfort is achieved through the combination of factors such as: durability, aesthetics, fashion, social and cultural environment, predominantly related to fashion trends followed by society.
c) Physical comfort (sensorial): is caused by mechanical and thermal contact between fabric and skin, thus related to contact induced sensations by the fabric, this contact may be static or dynamic.

   Pressure during static contact depends on clothing construction, size and weight, also fabric flexibility and compressibility.

   The mechanical effects are more complex in dynamic contact during movement caused by the user. The cutting of fabrics used in clothing manufacture and their elastic properties have their importance heightened. Friction between threads and weight of the final product cause mechanical energy loss, which lessens the physical sensation of comfort (NEVES et al., 2008).

   In order to provide this kind of comfort, clothing must have characteristics such as softness and flexibility during use and should not scratch, irritate or strain the body.
d) Ergonomical comfort is related to the shape of the clothing. The most influential factors over this type of comfort are cut, seam, pattern shape and anthropometrical tables. Factors associated with the ability to perform corporal movements related to type and structure of material used are also important.

Comfort characteristics sought by customers in apparel can be viewed as their functional and aesthetical specifications, the latter provides psychological comfort while functional specification guarantees physiological and sensorial comfort (BROEGA; SILVA, 2008).

3 MATERIALS AND METHODS

3.1 *Materials*

For this study, four denim fabrics were selected from the market:

Table 1. Pressure gauge according to the type of fabric.

Type of Material	Example	Pressure (cN/cm2)
Thin	Blankets, knitted fabrics	0,35–35
Medium	Woven fabrics, carpets	1,40–144
Thick	Felts, canvas	7–700

Source: ASTM D 1777–06.

Fabric 1: Composition: 68% cotton/30% polyester/2% elastane
Fabric 2: Composition: 56% cotton/42% polyamide/2% elastane
Fabric 3: Composition: 90% cotton/9% polyester/1% elastane
Fabric 1: Composition: 97% cotton/3% elastane

3.2 *Methods*

The following test methods for physical properties were performed: determination of linear density (ASTM D 3776–96), determination of thickness (ASTM D 1777–06), abrasion resistance (ASM D 4966), determination of elongation (JIS L 1018–02), determination of dimensional change (NBR 10320–02) and moisture management (AATCC Test Method 195–01).

3.2.1 *Determination of linear density*

In accordance with standard test method ASTM D 3776–96, five specimens with 100 cm^2 area must be prepared using a template. Samples must be weighed by employing an analytical scale so that the linear density of the article in grams per square meter can be calculated from the obtained measurements.

3.2.2 *Determination of thickness*

Test method ASTM D 1777–06 determines that pressure applied by the equipment must be gauged according to table 1. The test consists of applying pressure on the specimen for five seconds, cautiously, in order not to alter the original sample state and then read the thickness value directly from the equipment. This standard recommends taking five measurements of each fabric sample.

3.2.3 *Abrasion resistance*

In accordance with standard ASM D 4966, three specimens with 140 mm diameter of the abrasive cloth must be provided and attached to the base of the Marthindale apparatus for this test. Three specimens measuring 38 mm in diameter must also be prepared, fixed to polyurethane foam and fitted to the apparatus. For fabrics weighing more than 500 g/m2, foam should not be used. Apply pressure of 9±0.2 kPa for light fabrics and 12±0.3 kPa for heavy fabrics.

Use one of the following criteria:

a) For fabrics, determine the number of cycles until two or more threads are ruptured.

b) Determine the number of cycles until colour or appearance reaches greyscale grade 3.
c) Determine mass loss in percentage after a number of cycles.

Note: For the analysis of this study, 12,000 cycles were performed on each fabric sample. The samples measured 38 mm in diameter and were weighed before and after the tests to determine the percentage of mass loss.

3.2.4 *Determination of elongation*

In accordance with JIS L 1018-02 (adapted procedure for Fabric Extensometer equipment), five specimens cut in the direction of the warp (woven fabrics) must be prepared for this test using specified templates. The extending apparatus should be adjusted with a load of 300 gf/cm (2.94 N/cm) and the distance between jaws, marked at the beginning of the experiment at 7.5 cm (distance L0). The test consists of applying the load weight for a period of one minute in the specimen, then measuring the distance between the marks after its elongation (distance L1). The specimen is removed from the apparatus following the predetermined period of time and the distance between the initial marks made at the beginning of the experiment (distance L1) is measured again.

The percentage values of elongation (A%) are obtained after the experiment through equation 1:

$$A(\%) = \frac{L_1 - L_0}{L_0} x100 \tag{1}$$

3.2.5 *Determination of dimensional change*

Standard NBR 10320-02 for dimensional stability tests, specifies that three measurements of 25cm should be performed and marked on specimens of 38 × 38 cm, both longitudinally and transversely on the woven fabric analyzed. Measurements of the initial distances (L0) were carried out before the article was washed with detergent in an automatic machine. Measurements of the initially marked distances (L1) were then taken after washing, drying and conditioning of the specimen.

The percentage values of dimensional stability (ED%) can be calculated through equation 2:

$$ED(\%) = \frac{L_0 - L_1}{L_0} x100 \tag{2}$$

3.2.6 *Moisture management – MMT (Moisture Management Tester)*

According to AATCC Test Method 195-01, five specimens measuring 8.0 × 8.0 cm cut in the transverse section of the fabric are required for the test. First, the samples are washed and dried on a smooth surface. A predetermined amount of distilled water and sodium chloride solution (synthetic sour) is placed in the equipment, then the specimen is placed on the lower sensor with the reverse side of the fabric facing upwards. The test duration is 120 seconds.

According to Borelli (2013), the equipment provides the following indices to characterize the moisture management capacity in the tested sample:

- Wt (s): Wetting Time: WTt (top surface) and WTb (bottom surface) indices show the amount of time (seconds) required for the top and bottom surfaces of the fabric to become wet after the test is started. This measurement can be compared with the drop absorption time in the test specified by standard AATCC 79.
- AR (%/s): Absorption Rate: ARt (top surface) and ARb (bottom surface) rates indicate the average moisture absorption capacity (%/s) of fabric surfaces during the pumping of the test solution.
- MWR (mm): Maximum Wetted Radius: The MWRt (top surface) and MWRb (bottom surface) represent the maximum radii of diffusion of the liquid in mm and are determined by a set of concentric rings on the top and bottom surfaces when the slope of the total water content becomes greater than tan (15°).
- SS (mm/s): Spreading speed: The SSt (top surface) and SSb (bottom surface) measurements represent the spreading speed of the accumulated moisture in both surfaces of the sample in mm/s, from the center to the maximum wetted radius.
- R (%): Accumulative one-way transport capability: the difference in moisture accumulated between the two surfaces of the fabric, considering a single direction of transport (from top to bottom), expressed in percentage and determined according to equation 3:

$$R = Ub - Ut \tag{3}$$

Where:

Ut: percentage of absorbed liquid on the top surface during the total time of the test.
Ub: percentage of absorbed liquid on the bottom surface during the total time of the test.

- OMMC: Overall moisture management capability: an index that indicates the fabric capability to manage moisture, ranging from 0 to 1, where 0 represents the worst performance and 1 the best. The index considers three aspects:
- Moisture absorption rate on the bottom surface: ARb
- One way liquid transport capability: R
- Moisture spreading speed on the bottom surface: SSb
- Overall moisture management capability (OMMC) is defined by equation 4:

$$OMMC = 0{,}25xARb + 0{,}50xR + 0{,}25xSSb \tag{4}$$

3.3 *Analysis of results*

Classification and selection of the most suitable denim fabrics for clothing were made through analysis of box plot graphs.

3.3.1 *Box plot*

A box plot is a graph that provides information on location, dispersion, symmetry, whisker length, and aids in finding outliers in a data set. A box plot displays

Table 2. Linear density experimental values of tested fabrics (g/m2).

Tests	Fabric 1	Fabric 2	Fabric 3	Fabric 4
CP 1	311,3	341,1	298,8	293,3
CP 2	310,7	342,5	298,4	289,6
CP 3	314,1	339,2	301,5	287,2
CP 4	320,0	347,9	298,3	292,1
CP 5	321,3	350,0	297,5	287,8
Mean	315,5	344,1	298,9	290,0
Standard Deviation	4,91	4,60	1,53	2,65

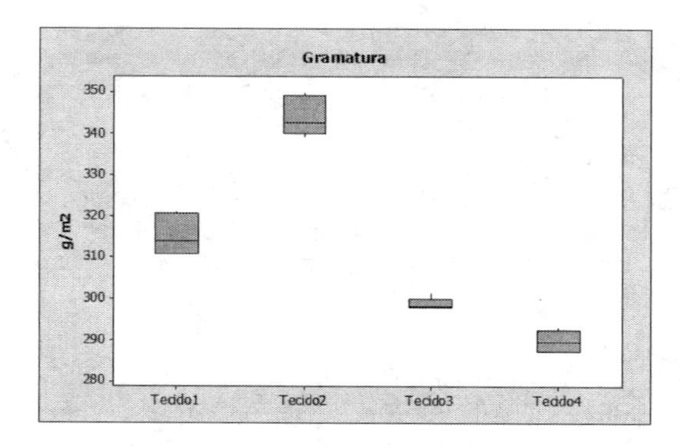

Figure 1. Box plot – Linear density.

in a simple way values that can be used as cut-off points in a survey, since atypical values can adversely affect decisions made based on the analysis of values that are not properly disregarded.

According to Triola (2005), a box plot is a rectangle which indicates the confidence interval of the mean, while its horizontal segments delimit the values that are homogeneous among them, namely those that follow the normal curve. The rectangle is aligned vertically with two half-lines, one on each of the opposite sides of the rectangle. The height of the rectangle is defined by quartiles Q1 (25% of the data) and Q3 (75% of the data). A line divides the rectangle by the median Q2 (50% of the data). Outliers are usually represented by asterisks, circles, or squares.

4 RESULTS AND DISCUSSION

Four denim samples (fabrics 1, 2, 3 and 4) were purchased. Analysis of experimental results was executed through Minitab software using box plot graphs.

4.1 *Determination of linear density (ASTM D 3776–96)*

Table 2 shows individual values, means and standard deviations obtained in the linear density tests.

Figure 1 shows the box plot graph comparing values of linear density (weight) of the tested fabrics.

The box plot shows that the fabrics have different grams per square meter mean values. From the graph it is possible to classify these fabrics as a function of weight per unit area: fabric 4 is the lightest, then follows fabric 3, fabric 1 and finally fabric 2 which is the heaviest.

4.2 *Determination of thickness (ASTM D 1777–06)*

Table 3 shows individual values, means and standard deviations obtained in the fabric thickness tests.

Figure 2 shows the box plot graph comparing thickness values of tested fabrics.

The box plot shows that the fabrics have different thickness mean values. From the graph it is possible to classify these fabrics according to their thickness: fabrics 4 and 1 are lighter, followed by fabric 3 and lastly fabric 2, the thickest.

Table 3. Thickness experimental values of tested fabrics (mm).

Tests	Fabric 1	Fabric 2	Fabric 3	Fabric 4
CP 1	0,52	0,66	0,57	0,50
CP 2	0,54	0,65	0,58	0,48
CP 3	0,53	0,64	0,60	0,48
CP 4	0,50	0,64	0,60	0,49
CP 5	0,49	0,63	0,58	0,48
Mean	0,52	0,64	0,59	0,49
Standard Deviation	0,02	0,01	0,01	0,01

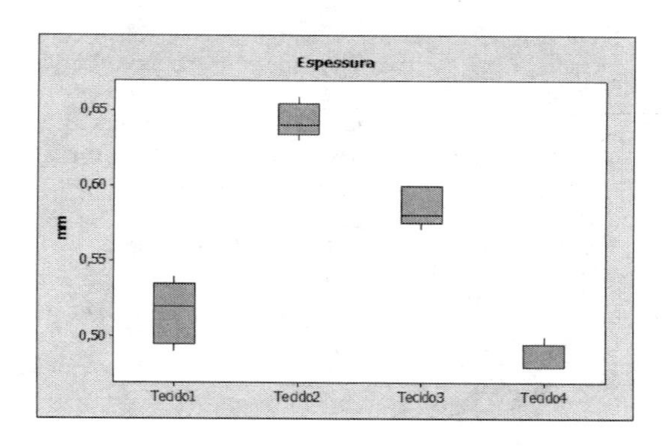

Figure 2. Box plot – Thickness.

4.3 *Abrasion resistance (ASM D 4966)*

Table 4 shows individual values, means and standard deviations obtained in the fabric abrasion resistance tests (%).

Figure 3 shows the box plot graph comparing values of abrasion resistance of the fabrics.

Table 4. Experimental values of abrasion resistance of the tested fabrics (%).

Tests	Fabric 1	Fabric 2	Fabric 3	Fabric 4
CP 1	1,21	0,87	0,98	1,29
CP 2	1,23	1,30	1,01	0,81
CP 3	1,47	1,08	1,50	1,34
Mean	1,30	1,08	1,16	1,15
Standard Deviation	0,14	0,22	0,29	0,29

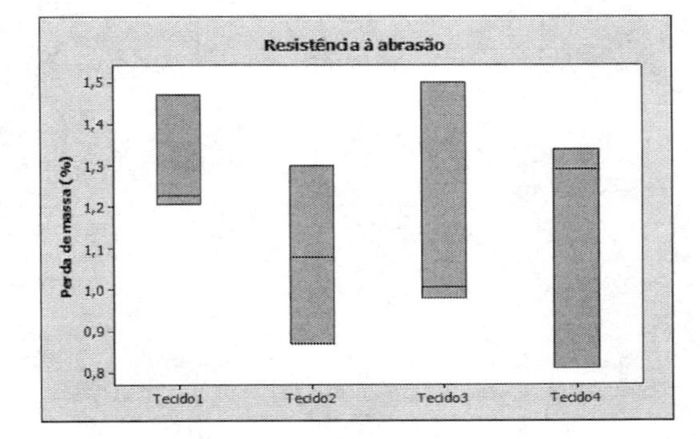

Figure 3. Box plot – Abrasion resistance.

Table 5. Experimental values of elongation & (%).

Tests	Fabric 1	Fabric 2	Fabric 3	Fabric 4
CP 1	9,33	25,33	18,67	36,00
CP 2	9,33	26,67	17,33	36,00
CP 3	9,33	24,00	17,33	37,33
CP 4	10,67	25,33	16,00	37,33
CP 5	10,67	26,67	16,00	36,00
Mean	9,87	25,6	17,07	36,53
Standard Deviation	0,73	1,12	1,12	0,73

The box plot shows that the fabrics have statistically equal mean values of abrasion resistance.

4.4 *Determination of elongation (JIS L 1018–02)*

Table 5 shows the individual values, means and standard deviations obtained in the elongation tests in the transverse section.

Figure 4 shows the box plot graph comparing elongation values in the transverse section of the fabrics.

The box plot graph shows that the fabrics have different mean values of elongation percentage. From the results, the fabrics can be classified according to their percentage of elongation: fabric 4 has a greater elongation percentage, followed by fabric 2, then fabric 3 and finally fabric 1 with the lowest elongation.

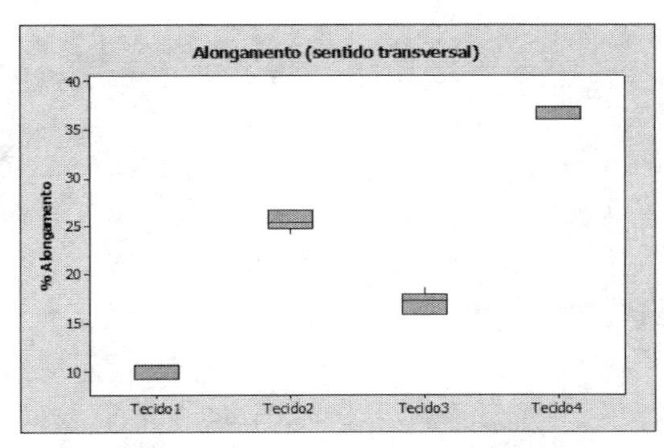

Figure 4. Box plot – Elongation of tested fabrics.

Table 6. Experimental values of dimensional change in the longitudinal section (%).

Tests	Fabric 1	Fabric 2	Fabric 3	Fabric 4
CP 1	1,19	0,00	1,19	0,00
CP 2	0,00	0,00	1,20	0,40
CP 3	1,19	0,00	1,59	0,40
Mean	0,79	0,00	1,33	0,27
Standard Deviation	0,69	0,00	0,23	0,23

Table 7. Experimental values of dimensional change in the transversal section (%).

Tests	Fabric 1	Fabric 2	Fabric 3	Fabric 4
CP 1	0,40	4,70	4,98	4,58
CP 2	0,79	4,38	4,77	4,32
CP 3	0,40	4,40	4,22	4,98
Mean	0,53	4,49	4,66	4,63
Standard Deviation	0,23	0,18	0,39	0,33

4.5 *Determination of dimensional change (NBR 10320–02)*

Tables 6 and 7 show, respectively, the individual values, means and standard deviations obtained in the tests for determination of dimensional change in the longitudinal and transverse sections of the fabrics.

Figures 5 and 6 show, respectively, the box plot graphs comparing dimensional change values of the fabrics in the longitudinal and transverse sections.

The box plot shows that the fabrics present different mean values of percentage of dimensional change. From the graph, it can be concluded that fabric 1 has the mean dimensional stability percentage (longitudinal section) statistically equal to fabrics 2, 3 and 4, and that fabric 2 has dimensional stability percentage statistically equal to fabric 4.

The box plot shows that fabric 1 presents lower dimensional change percentage mean values than fabrics 2, 3 and 4.

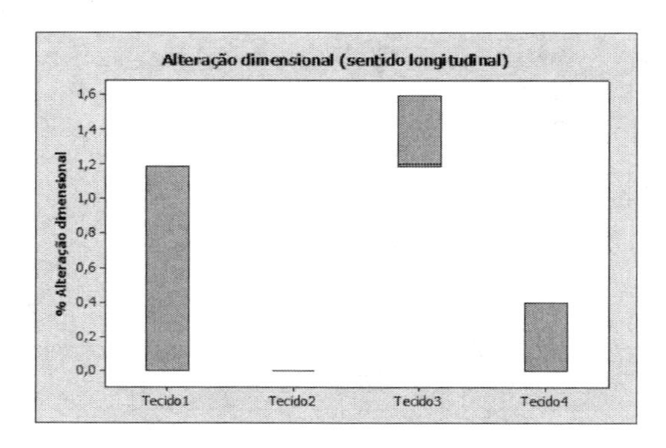

Figure 5. Box plot – Dimensional change in the longitudinal section.

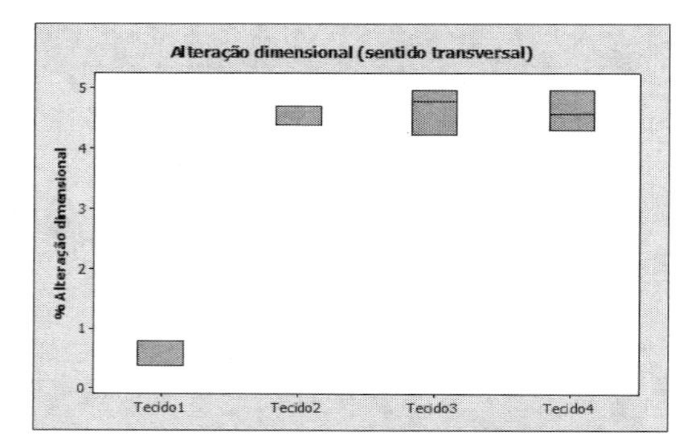

Figure 6. Box plot – Dimensional change in the transverse section

Table 8. Fabric experimental values (index) of the moisture management test.

Test	Fabric 1	Fabric 2	Fabric 3	Fabric 4
CP 1	0,1239	0,0778	0,1200	0,0904
CP 2	0,1209	0,0810	0,0838	0,0892
CP 3	0,1184	0,0762	0,1067	0,0859
CP 4	0,1181	0,0730	0,1010	0,0852
CP 5	0,1200	0,0778	0,0862	0,0919
Mean	0,1203	0,0772	0,0995	0,0885
Standard Deviation	0,0023	0,0029	0,0150	0,0029

4.6 *Moisture management (AATCC Test Method 195–01)*

Table 8 displays individual values, means and standard deviations obtained in the laboratory tests using the MMT (Moisture Management Tester) equipment, to assess moisture management & of fabrics.

Figure 7 shows the box plot graph with the comparison between (comparing) OMMC values of fabrics.

The box plot graph shows that fabric 1 has the highest moisture transport speed, fabrics 3 and 4 have

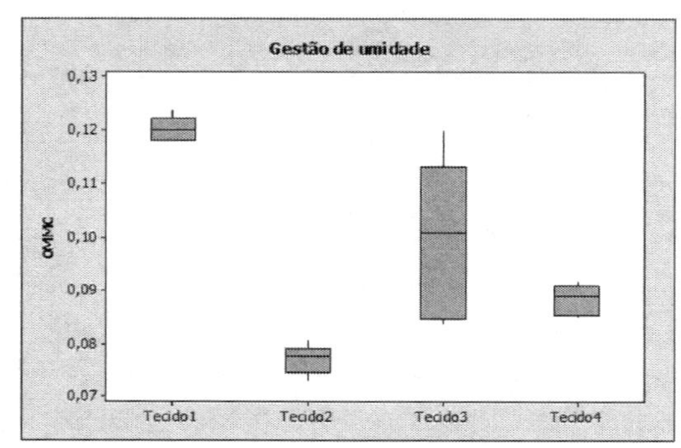

Figure 7. Box plot – Moisture management of fabrics.

similar speeds, while fabric 2 has the lowest moisture transport speed.

5 CONCLUSION

Industries are continuously improving product development processes, perfecting them with the introduction of new materials, procedures, equipment, etc., in order to meet needs of consumers, who today seek not only quality, comfort, aesthetics but also usability in fashion products.

Studying fabric properties is key when ascribing clothing functionality. Some types of fabric incorporate science and technology within their fibers, thus offering better functional properties.

In this research, comfort was assessed in an objective way through the following tests: moisture management, determination of linear density and thickness of fabrics. Fabric durability can be defined as guarantee of the conditions of use for a product during a period of time. This property was tested with elongation and abrasion resistance laboratorial methods. Ease of maintenance is understood as the guarantee of the textile product conservation during its use and after washing, tested through determination of dimensional change (SOUZA, 2016).

Results obtained in this experiment show that when compared to the other fabrics that were studied, fabric 1, in all analyzed characteristics – composition: 68% cotton/30% polyester/2% elastane, has an intermediate weight per unit area, lower thickness, abrasion resistance statistically equal to the other tested fabrics, good dimensional stability in longitudinal and transverse sections, higher moisture transport speed and lower elongation percentage in transverse section. Therefore, fabric 1 has greater potential to meeting consumer requirements.

ACKNOWLEDGEMENTS

This research was only possible with the support of CIAUD – Research Centre in Architecture, Urbanism and Design, Lisbon School of Architecture of the University of Lisbon, and FCT – Foundation for Science and Technology, Portugal, and USP for their financial and institutional support.

REFERENCES

AMERICAN ASSOCIATION FOR TESTING AND MATERIALS. ASTM D 1777–06; 1.1.1. Standard Test Method for Thickness of Textile Materials. West Conshohocken, 2006.

AMERICAN ASSOCIATION OF TEXTILE CHEMISTS AND COLORISTS. AATCC TM195. Liquid Moisture Management Properties of Textile Fabrics. Research Triangle Park (NC), 2012. 12p.

AMERICAN SOCIETY FOR TESTING AND MATERIALS, ASTM D 3776–96; Standard test method for mass per unit area (weight) of fabric. West Conshohocken, 2006. 5p.

AMERICAN SOCIETY FOR TESTING AND MATERIALS, ASTM D 4966–98; Standard test method for abrasion resistance of textile fabrics (Martindale Abrasion Tester Method). West Conshohocken, 2006. 4p.

ASSOCIAÇÃO BRASILEIRA DE NORMAS TÉCNICAS, NBR 10320; Materiais têxteis – Determinação das alterações dimensionais de tecidos planos e malhas – Lavagem em máquinas domésticas automáticas. Rio de Janeiro, 1988. 6p.

BORELLI, C.; Comparativo das propriedades de transporte de umidade, capilaridade, permeabilidade ao vapor e permeabilidade ao ar em tecidos planos de poliéster. Campinas, Faculdade de Engenharia Química, Universidade Estadual de Campinas, 2013. PhD thesis.

BROEGA, A.C.; SILVA, M.E.C. O conforto como ferramenta do design têxtil. 3° Encuentro Latinoamericano de Deseño. Buenos Aires (Argentina), July 2008.

JAPAN INDUSTRIAL STANDARDS, JIS L 1018-02; Testing methods for knitted fabrics. Tokyo, 2006. 55p.

KADOLPH, S. J.; LANGFORD, A.L. Textiles. Ed. Prentice Hall. New Jersey, 2006.

MDIC – Ministério da indústria, comércio exterior e serviços citado em GBLjeans. Available at: http://gbljeans.com.br/noticias_view.php?cod_noticia=7400. Acessed in 24 March 2017.

NEVES, M.M. et al, Projecto de vestuário termicamente confortável. 5° Congresso Luso-Moçambicano de Engenharia e 2° Congresso de Engenharia de Moçambique. Maputo, September, 2008.

SANCHES, R.A.; Estudo comparativo das características das malhas produzidas com fibras sustentáveis para fabricação de vestuário. São Paulo, Escola de Artes, Ciências e Humanidades, Universidade de São Paulo, 2011. Habilitation thesis.

SOUTINHO, H. F. C., Design funcional de vestuário interior. Braga (Portugal): Escola de Engenharia, Universidade do Minho, 2006. MSc dissertation.

SOUZA, L. N. de. Proposta de metodologia para adaptação de vestuário para pessoas com deficiência física (cadeirante). São Paulo, Escola de Artes, Ciências e Humanidades, Universidade de São Paulo, 2016. MSc dissertation.

TRIOLA, M. F. Introdução à estatística. 9. Ed. Rio de Janeiro: LTC, 2005.

Textiles, Identity and Innovation: Design the Future – Montagna & Carvalho (Eds)
 ISBN 978-1-138-29611-4

Textile elements in a design project of children's furniture

C. Salvador
CIAUD, Lisbon School of Architecture, Universidade de Lisboa, Portugal

ABSTRACT: In the final phase of a product design research about the adaptation of children's furniture to the child and sustainability, namely high chairs, a literary review based research about textile fibres and its properties and utilities, seemed necessary to begin a decision-making process in terms of materials for the textile elements on a high chair project, being the comfort cushion's textile cover for babies or toddlers, the main target. Given the specific problems inherent to a child's sensitive skin, as well as the characteristics of the most common fibres of each fibre group, it was possible to conclude that natural fibres from vegetable source such as bamboo or man-made fibres with cellulose such as lyocell, stand out. Due to its hygroscopic, allergenic, resistance, thermal and tactile properties combined with sustainable and eco-friendly origins, a combination of bamboo raw material with lyocell manufacturing process seems to be the best option for the project.

1 INTRODUCTION

Focusing on the child's wellbeing, a Product Design research about the adaptation to the child and sustainability of children's furniture, namely high chairs, is being conducted. Besides its theoretical essence, the knowledge obtained is put to practice on a children's furniture design project of a high chair, more suitable for children from 6 months up to 7 years of age, both physically and psychologically. But also providing solutions, which may enable extended product life cycles, contributing to sustainable development.

Following a definition of the briefing for this project and starting by designing the structure of the high chair, it seemed necessary to lead a research about textile materials in order to know more about the existing possibilities to produce the textile elements of the high chair's prototype.

After the identification of a main element (Fig. 1) – comfort cushion for children from 6 months up to 2–3 years of age (including safety belt system), a literary review based research about textile fibers and its utilities, began, focusing on the importance of the cushion's textile, given its contact with the baby or toddler's skin.

The sensitive and prone to change nature of the child, makes designing any product for this target, quite difficult. Some studies of ergonomic nature have been previously made (Salvador, Vicente & Martins 2014) (Salvador 2015), in order to better understand children's behavior towards this kind of equipment, but its textile elements were not analyzed in detail.

With the aim of seeking for information on the diversity of the types of fibres, its properties and utilities, this paper describes an important step in this children's furniture design project, as the cushion's textile can be decisive on the adaptation of the high chair

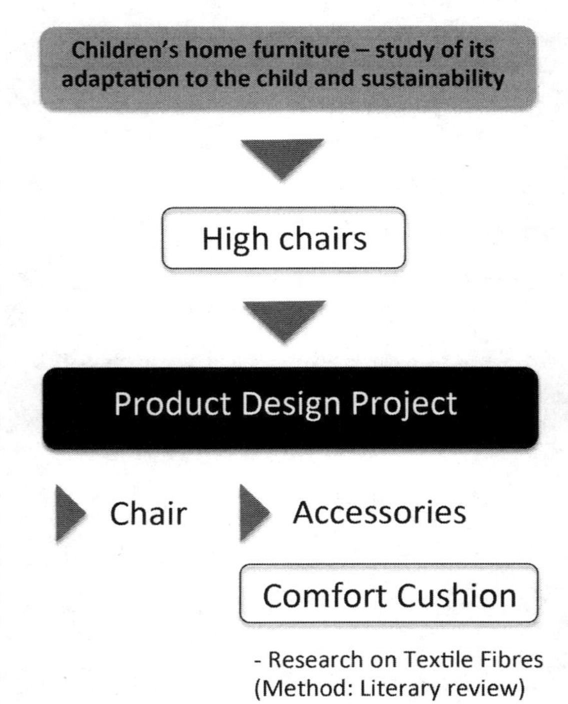

Figure 1. Schematic representation of the research process. Digital illustration: Cristina Salvador, Portugal.

to the child's need for comfort as a baby or toddler. Essential issues to deal with are the perspiration problems, thermal comfort, adaptation to sensitive skin, texture and tactile comfort, maintenance, etc.

2 THE CHILD'S COMFORT

Babies and toddlers are a specific population with developing systems of thermoregulation and specific size, shape and temperature profile (Parsons 2002). With a relatively large head, large surface area to mass ratio, high blood flow and limited sweating capacity, babies are prone to heat loss and gain. Measuring thermal discomfort may be done through behavioral measures such as crying or moving away, including mood changes, distraction and irritability.

Human skin is a big breathing organ and when it's normal, it is colonized by a variety of bacteria, typically not harmful. But in case of Atopic dermatitis, (AD) which is a common, chronic fluctuating skin disease (Hipler and Elsner 2006) with prevalence in children of between 7 and 17%, commonly within the first 2 years of life (in approximately two-thirds of cases will persist into adulthood), there's an aggravated risk of intolerance to bacteria, especially *S. aureus*. A high proportion of infants with AD has a positive family history of AD, asthma or allergic rhinitis and will go on to develop further atopic complications later in childhood. AD is a multifactorial disease in which both hereditary and environmental factors play a role (Baker 2006).

A textile cover for a children's chair cushion is an element, which will be in direct contact with the baby or toddler's skin, maybe for long periods of time as it is frequent the child also falls asleep after meals while seated. Being its target, children from 6 months of age (average age the child can sit autonomously) up to 3 years of age (Fig. 2).

It needs to be a protective shield to the child, covering a strange element to the baby's skin – a piece of furniture. But it also has to allow the skin to breathe and perform heat loss and gain through perspiration, fighting harmful bacteria effect and preventing irritation, allergy, rash and unpleasant odor. It also has to be resistant to wear, friction and washing, with a smooth touch so it doesn't irritate the baby's sensitive young skin. There also must be a balance between hygroscopicity and the need to repeal dropped liquid to prevent stains. The textile must also be certified to have no harmful chemicals and substances.

The eco label OEKO-TEX® Standard 100 was introduced in 1992 (OEKO-TEX 2017), aiming to achieve uniform standards for the use of chemicals in textile production and to set up environmental criteria for potentially hazardous substances in textiles. The OEKO-TEX® tests and criteria are divided into different classes of products depending on the intensity of skin contact. Products Class I are textiles and toys for babies and small children, up to 3 years of age (underwear, rompers, bedding, soft toys etc.) and products Class II are textiles with a large part of their surface in direct contact with the skin, such as underwear and linen. The comfort cushion for this high chair project would be part of Class I products group.

The tests seek for possible threatening substances in textiles such as formaldehyde, several extractable and heavy metals like mercury or lead, pesticides, chlorophenols, phthalates, carcinogens and allergens in dyes, polycyclic aromatic hydrocarbons, flame retardants, solvents, detergents, polyfluorinated substances and emissions of volatile substances, among others. Woven and knitted fabrics, raw materials and manufactured goods can be tested in general, but there are studies and tests focusing on children's seats, more specifically car seats for babies and toddlers and other textiles, in Denmark (Kjølholt et al. 2015).

3 TEXTILE FIBRES

3.1 *Natural from vegetable source*

Plant fibres are the largest group of natural fibres and can be extracted from leafs, bast, seeds, grasses and reeds, fruits and wood.

Cellulose $(C_6H_{10}O_5)_n$ is the main component of plant fibres (Rowe 2009). Its molecules comprise a succession of glucose rings (Carr 1995) (Fig. 3).

The cellulose that is most free from encrusting substances is found in cotton and ramie fibres (83 to 99%). The higher the cellulose content and the higher its

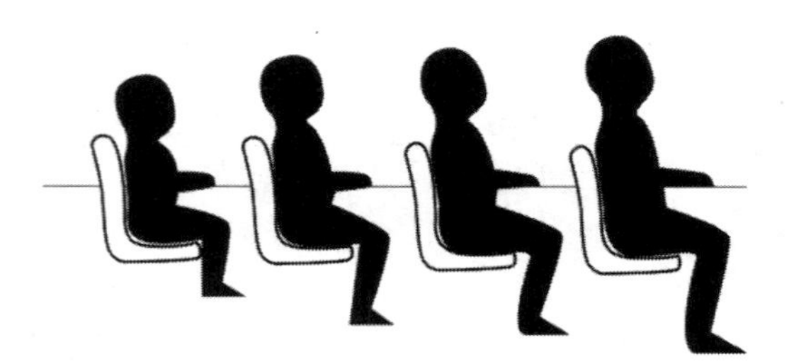

Figure 2. Graphics with age target for the high chair's cushion. Digital illustration: Cristina Salvador, Portugal.

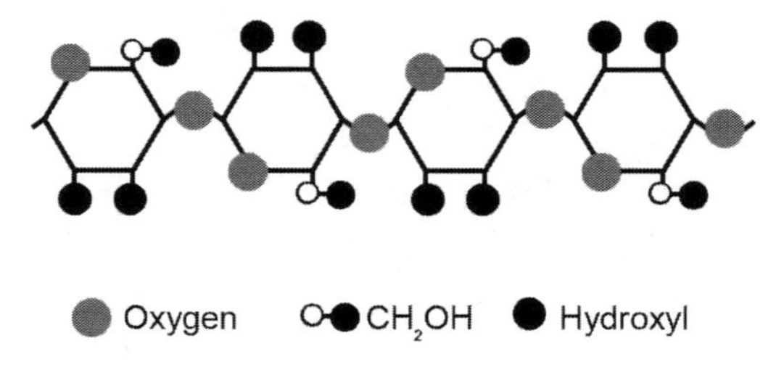

Figure 3. Graphics representing cellulose molecules based on published data (Carr 1995). Digital illustration: Cristina Salvador, Portugal.

purity, the better and more valuable the fibre is. Apart from cellulose, plant fibres contain other substances that accumulate in middle plates and fibre walls, such as lignins, pectins and hemicellulose.

Bast fibres include flax (linen), which is one of the oldest textile fibres in the world and can be grown in warm or cold climates. Its characteristics make it very comfortable in clothing, especially in warm climates, with good air permeability, high hygroscopicity, coolness, softness, low susceptibility to electrostatic charges on its surface and ultraviolet protection efficiency, having a positive impact on human physiology.

Another bast fibre is ramie, extracted from the stem of a perennial plant belonging to the Utricaceae family, grown in Eastern Asia. An intense lustre, good resistance to atmospheric conditions and quite high resistance to bacterial action characterize ramie fibre. Ramie can be used for light clothing fabrics, technical textiles and upholstery and waste fibres are used for paper production.

Cotton is the most common seed fibre that grows around the seeds of the cotton plant (Gossypium), native to tropical and subtropical regions around the world. Soft and staple, it has natural abundance and good inherent properties (Hipler and Elsner 2006) such as good folding endurance, better conduction of heat, easy dye ability and excellent moisture absorption. The waxes and fats present on the surface of the fibre, make it smooth and flexible. Cotton is widely used for clothing and also used for interior products, pure or in blends with other natural and man-made fibres. It can be cultivated organically without using pesticides and chemical additives in fertilizers.

Bamboo, which is a grasses and reeds fibre extracted from a perennial plant that grows in monsoon climates, has been a target of increasing interest due to its moisture absorbing properties, antibiosis, bacteriostatic and deodorization functions and protection against ultraviolet radiation. Bamboo has been used for centuries in applications such as construction, carpentry, weaving and plaiting (Rowe 2009).

With a rapid growth, less need for water and almost no need for pesticides, bamboo seems to be a sustainable (Liese and Koln 2015) and easily renewable resource, capable of producing a light, strong fabric with naturally round and smooth fibres suitable even for sensitive skins, requiring less dye than cotton. A lot of Bamboo fabrics are made through a viscose manufacturing process, as it involves not so eco-friendly solvents. Although quite a sustainable harvest, in fabric form doesn't seem so sustainable.

3.2 *Natural from animal source*

The main representatives of natural fibres from animal origin are several kinds of wool (animal's hair), such as sheep wool, alpaca or angora, leather, natural silk and spider silk.

The largest producers of sheep wool are Australia and New Zealand, Argentina and South Africa. Merino is one of the best-known breeds of sheep and its wool accounts for about 30% of the total world production of wool fabric (Rowe 2009). The fibre has many valuable properties, such as high hygroscopicity, absorbance of sweat due to its porosity, absorbance of UV radiation, resistance to external moisture and high abrasion resistance. Wool fibres have a natural lustre, depending on the type of wool and the climatic conditions. A characteristic feature of wool that is different from plant fibres and silk is felting, which determines the thermal properties of wool products. However these hair fibres can be irritating to sensitive skin, due to sharp spurs.

Natural silk (Bombyx Mori) is considered to be the most noble, natural fibre of animal origin and is obtained from the cocoons of mulberry or oak (tasar) silkworms. Silk is the only natural fibre, which exists as a continuous filament. Each Bombyx mori cocoon can yield up to 1600 metres of filament. The silk fiber's triangular cross-section gives it excellent light reflection capability. China is the greatest producer of silk in the world, although silkworms are bred in Europe (Greece, Spain, Italy and Poland). Natural silk can be used for luxury textile products and carpets. Silk fabrics are characterized by their lustre, are smooth and thin, and very pleasant to touch. Silk was used in the past, as the basis for paintings, especially in oriental art. It has been used for stockings, dental floss and cosmetics, and was used for parachutes (up to World War II). Tasar silk is soft and smooth, with a beautiful lustre. It is used for clothing production, fabrics such as shantung and pongee, and also for paper. Although, not as resistant as animal hair fibres, silk can absorb up to 30% of its own weight in moisture without creating a damp feeling. When moisture is absorbed, it generates 'wetting- heat', which helps to explain why silk is comfortable to wear next to the skin.

3.3 *Natural from mineral source*

Natural mineral fibres such as glass fibres, ceramic fibres and others can be utilized in cases requiring improved fire resistance (Rowe 2009). Some of these can resist temperature of greater than 1200°C (for example, basalt and alumina fibres). However, its comfort properties preclude their widespread use in interiors.

3.4 *Man-made fibers from natural source*

These regenerated fibers from cellulose and its derivatives are made through a physical rearrangement of the molecules originally present in a natural source, such as wood pulp (Carr 1995).

Viscose rayon was originally produced in 1891, attempting to create artificial silk. It usually has a high luster quality giving it a bright shine. The distinguishing property of regular rayon is its low wet strength. As a result, it becomes unstable and may stretch or shrink when wet (Mass 2016). Many of its properties are similar to those of cotton or other natural cellulosic fibers. Rayon is more moisture absorbent than cotton, soft, comfortable to wear, drapes well, and is easily dyed in a wide range of colors. It does not build up static electricity and does not insulate body heat making it ideal for use in hot and humid climates. Rayon has moderate dry strength and abrasion resistance. Like other cellulosic fibers, it is not resilient, which means that it will wrinkle. Rayon withstands ironing temperatures slightly less than those of cotton. It will mildew and may be attacked by silverfish and termites, but generally resists insect damage.

Modal is regenerated cellulosic fiber and a variation of rayon. Lenzing Modal® is made from sustainably harvested beech trees in PEFC (Programme for the Endorsement of Forest Certification schemes – the world's largest forest certification organization). Modal fibers with a high wet modulus were originally developed in Japan in 1951. Lenzing (a fibre producer company from Austria) started selling modal fibres in 1964. In 1977, Lenzing started using an environmentally friendly bleaching method for pulp for their cellulosic fibers.

Modal's distinguishing characteristics are its high wet strength and its extra softness. In addition to its use in general apparel, its softness makes it especially ideal for body contact clothing such as lingerie and under garments. Due to its high wet strength, modal can be machine-washed and tumble dried. Modal fibers are dimensionally stable and do not shrink or get pulled out of shape when wet like many rayons. They are also wear resistant and strong while maintaining a soft, silky feel. Modal is about 50% more hygroscopic, or water-absorbent, per unit volume than cotton. It's designed to dye just like cotton and is colorfast when washed in warm water. Even after repeated washing, modal remains absorbent, soft and supple.

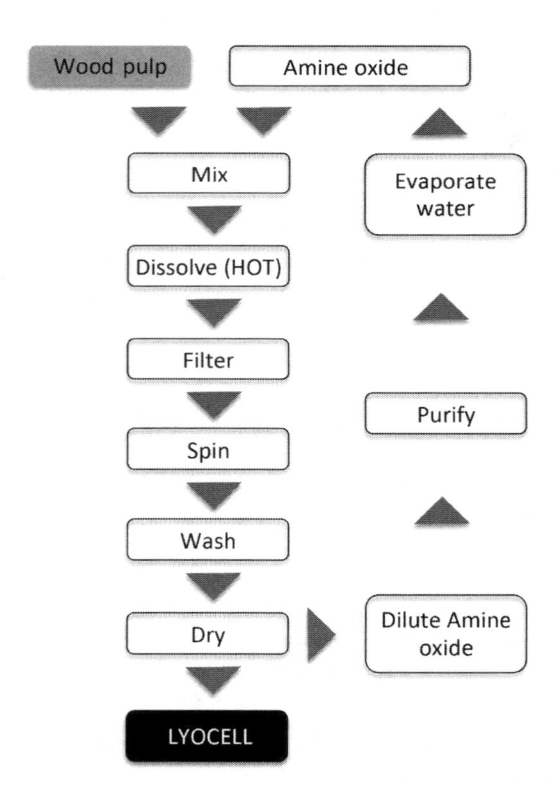

Figure 4. Graphics representing lyocell manufacturing process based on published data (Woodward 2017). Digital illustration: Cristina Salvador, Portugal.

Lyocell is another regenerated fibre, in production since 1990. Tencel® is Lenzing's brand name for lyocell (Tencel 2017) and its advantages include environmental friendliness of the chemical processing combined with its softness, drape, resistance to growth of bacteria which create odors, and other properties.

Lyocell is the first cellulose fibre to use nano technology. The nanofibrils are hydrophilic (a strong attraction to absorb water) and optimize absorption of moisture with excellent cooling properties by releasing moisture to the air. Lyocell controls and regularly absorbs moisture, 50% more than cotton and even more than wool. It supports the natural ability of the skin to act as a protective shell to regulate body temperature and maintain water balance. A subjective feeling of wellbeing depends considerably on moisture absorption and on surface structure of the fibres. The microscopic surfaces of lyocell fibres, due to the nanofibrils, are smoother than the surfaces of modal, cotton and wool.

Lyocell prevents the growth of bacteria, which cause odors, naturally without the addition of chemical treatment which may cause allergic reaction and are environmentally unfriendly, often used on synthetics and cotton. Bacterial growth is prevented through the moisture management of the fibre. When moisture is produced it is directly absorbed from the skin and transported to the inside of the fibre. Thus no water film is produced on the skin where bacteria could grow. Clothes remain odor free for multiple wearing, much longer than cotton. This also means fewer washings and saving on water and energy as well as on the wear and tear that occurs on any fabric from the washing and drying processes. Lyocell itself is hypoallergenic, meaning that it is not likely to cause an allergic reaction in sensitive individuals. This is why it is used for clothing and home furnishings, by individuals with Multiple Chemical Sensitivities (MCS) and those with allergy sensitivities, psoriasis, and neurodermatitis. It is also anti-static and doesn't cling. Lyocell's manufacturing process (Fig. 4), with not harmful and recyclable solvent is eco-friendly and can also be made with cellulose from birch, oak, eucalyptus or bamboo, being the last, a quite sustainable source.

3.5 *Synthetic polymer fibres*

These fibres are made with artificial polymer molecules (Carr 1995) from sources like petroleum. The most popular kinds of synthetic fibres used for interior textile manufacturing are polyamide, polyester, polypropylene, polyethylene, acrylic and fibres of the meta-aramid group. Manufacturing of interior textiles also requires non-flammable synthetic fibres, such as meta-aramids and polyamides (Rowe 2009). The chemical structure and properties of these fibres do not differ from analogous fibres used for clothing or other technical applications.

Polyamides (PA) are some of the best-known thermoplastic polymers. In 1935, W. H. Carothers (DuPont, USA) produced the first polyamide (nylon 6.6). The first intended application was as a synthetic replacement for natural silk. Polyamides are light polymers with a density of only 1.14 g/cm3; they are abrasion-resistant and, due to this property, are very popular in carpet manufacturing. Although not commonly used for other interior textiles, polyamides are the synthetic fibres most used in carpet manufacture, and blended polyamide and wool yarns are often used for this application too. Even small quantities of polyamide (up to 20% of the blend) increase the dynamic properties of woolen yarns and thereby the durability of carpets made from them. Polyamides have a low resistance to sunlight and are also not highly resistant to concentrated minerals and organic acids, though they are resistant to alkalis. Since polyamides are colorless and also easily dyed, their other application is for decorative goods. Polyamides can be dyed topically or in a molten state (solution dying); they can be printed easily and have excellent wear characteristics. However, it is important to note that, due to dye sites on the fibre, polyamides tend to stain very easily, and these dye sites need to be filled to increase polyamide stain resistance.

Polyester fibres (PES) are made from polyethylene-terephthalate (PET). PET was invented as a plastic in W. H. Carothers' laboratory (USA) in the 1930s but after the discovery of nylon, work on polyester slowed down. In 1939, the British scientists J. R. Whinfield, J. T. Dickson, W. K. Birtwhistle and C. G. Ritchie resumed W. H. Carothers' investigations and, in 1941, created the first polyester fibres, which they called Terylene. In 1946 the DuPont company (USA) bought all author rights and developed polyester fibres, which they called Dacron. In 1951 polyester was introduced to the textile and clothing market. The market for polyester fibres grew until the end of the 1960s but sales drastically fell in the 1970s due to discomfort caused by the fibres, especially in clothing. A new age of polyester use started in the 1990s after the manufacture of microfilament fibres began in 1986. Due to their versatility and easy care, polyester fibres are now one of the most widely used synthetic fibres for clothing and interior textiles. It is resistant to acids and alkalis, and is also resistant to sunlight. It is important to note that, due to their thermoplastic nature, the field of their application is sometimes restricted since, like all thermoplastic fibres, flame resistant polyester starts to melt at high temperature.

Acrylic fibres (polyacrylonitrile, PAN) were first made in 1941 by DuPont (USA) and trademarked Orlon; just one year later, acrylic fibres were also made in Germany. Its commercial manufacture however, started only in 1950. Acrylic fibres are produced from acrylonitrile (first made in Germany in 1893). A very important property of acrylic fibres is their wool-like handle: they are very soft and warm, and, furthermore, they have a good resistance to light and chemicals. Consequently, acrylic is used for drapes, fur imitation, carpets and other decorative interior textiles. Some modified acrylic fibres have high anti-pilling or flame resistant properties; as is the case with flame resistant polyester, the field of application of these modified acrylic fibres is sometimes restricted due to their thermoplastic nature.

4 CONCLUSIONS

Textile fibres seem indeed to be a vast subject, which would require further and more profound research, specially, in terms of chemical structure and its consequences on the properties of each fibre. However, the knowledge obtained with this research was very important for the children's furniture design project, as it helped unravel the possible choices for a textile fibre, suitable for the high chair's comfort cushion's cover for babies or toddlers.

Being sustainability one of the main issues to deal with in this project, the natural fibres from vegetable source and the man-made fibers from regenerated cellulose seem to be the most suitable groups of fibres, mainly because of its origin of sustainable vegetable source combined with its hygroscopic, allergenic, thermal, resistance and tactile properties.

Both bamboo and lyocell show improved properties facing the traditional cotton or flax (linen), being very resistant to bacteria, flexible and smooth, although bamboo is a sustainable harvest, after the manufacturing process to obtain the fibre, lyocell is more eco-friendly. A bamboo lyocell would be an option, although might be more expensive than most eco fabric alternatives (Simplifi Fabric 2017), but more sustainable in terms of raw material and with overall good performance in terms of manufacturing process and transformation, also facing cotton (Jesus 2011).

These vegetable source fibres seem to be the most suitable (available in the market) fibres for the child's sensitive skin, which needs to be protected in the daily contact with furniture.

ACKNOWLEDGMENTS

This research was only possible with the support of CIAUD – Research Centre in Architecture, Urbanism and Design, Lisbon School of Architecture of the University of Lisbon, and FCT – Foundation for Science and Technology, Portugal.

REFERENCES

Baker B. S. 2006. The role of microorganisms in atopic dermatitis. *Clinical and Experimental Immunology.* April, Vol. 144(1): 1:9.

Carr, C. M. (ed.) 1995. *Chemistry of the Textiles Industry*. Glasgow: Blackie Academic and Professional.

Hipler U.-C. & Elsner, P. (eds.) 2006. *Biofunctional Textiles and the Skin . Current Problems in Dermatology. Burg, G. (ed.), Vol. 33*. Basel: Karger.

Jesus, S. 2011. Novas Bases Têxteis para Novas Exigências Sociais – a sustentabilidade das fibras sintéticas. Master Thesis, Faculdade de Arquitectura da U.T.L. online: http://www.repository.utl.pt/bitstream/10400.5/4710/1/DISSERTA%C3%87%C3%83O%20FINAL.pdf. available: 30th July 2017.

Kjølholt, J. et al. 2015. Chemical substances in car safety seats and other textile products for children, Survey of chemical substances in consumer products No. 135, 2015. Copenhagen: The Danish Environmental Protection Agency.

Liese, W. & Köln, M. (eds.) 2015. *Bamboo – The Plant and its Uses*. Hamburg: Springer.

Mass, E. 2016. *Rayon, Modal and Tencel – Environmental Friends or Foes*. Yes, it's Organic. online: http://www.yesitsorganic.com/rayon-modal-tencel-environmental-friends-or-foes.html#axzz4gnHsj2sb. Available: 10th May 2017.

OEKO-TEX®, 2017. *OEKO-TEX® Standard 100*. online: https://www.oeko-tex.com/en/business/certifications_and_services/ots_100/ots_100_start.xhtml. Available: 10th May 2017.

Parsons, K. 2002. *Human Thermal Environments – The Effects of Hot, Moderate and Cold Environments on Human Health, Comfort, and Performance*. Boca Raton: CRC Press/Taylor and Francis.

Rowe, T. (ed.) 2009. *Interior Textiles – design and developments*. Cambridge: Woodhead Publishing Limited.

Salvador, C. 2015. Contributions of Ergonomics to an Affective Sustainability in Children's Furniture. *6th International Conference on Applied Human Factors and Ergonomics (AHFE 2015) and the Affiliated Conferences*. Elsevier Procedia, Science Direct.

Salvador, C., Vicente, J. & Martins J. P. 2014 Ergonomics in children's furniture – emotional attachment. In: *Proceedings of the 5th Conference on Applied Human Factors and Ergonomics (AHFE 2014)*, Krakow. Abraham, T., Karwowski, W., Marek, T. (eds.). CRC/Taylor and Francis.

Simplifi Fabric 2017. Lyocell Fabric. online: https://www.simplififabric.com/pages/tencel-lyocell. available: 30th July 2017.

Tencel, 2017. Tencel® by Lenzing. online: http://www.lenzingfibers.com/en/tencel/the-fiber/. available: 10th May 2017.

Woodward, A. 2017. Lyocell. Made How, Vol. 5. online: http://www.madehow.com/Volume-5/Lyocell.html. available: 30th July 2017.

Textiles, Identity and Innovation: Design the Future – Montagna & Carvalho (Eds)
© 2019 Taylor & Francis Group, London, ISBN 978-1-138-29611-4

Ergonomic clothing design for care-dependent elderly women

A. Caldas
Federal University of Piauí, Brazil
DET, School of Engineering, University of Minho, Portugal

M. Carvalho
DET, School of Engineering, University of Minho, Portugal

H. Lopes
Institute for Interdisciplinary Research, University of Coimbra, Portugal

M. Souza
National Service for Industrial Training, Brazil

ABSTRACT: This paper describes the development of ergonomic clothing suitable for the body characteristics of the care-dependent elderly women at four institutions: two located in the city of Guimaraes (Portugal) and two in the city of Teresina (Brazil). The prototypes were developed based on the needs of the elderly according to some functional factors as well as their physical conditions, aiming an easier handling by the caregiver and comfort for the user. The functional properties of the applied materials and accessories, the pattern design process, and the garment construction are described.

1 INTRODUCTION

The integration between design and ergonomics improves product development. This relation is stated by Dul & Weerdmeester (2004) considering ergonomics as an applied discipline to product design, aiming to improve safety, health, and comfort. For Mozota (2002), design is an area that contributes to the resolution of problems from the creation of coordinated and systematic activities, together with management. According to Burdek (2006), product design should answer to specific problems of a target population. This is the case of this study, focused on elderly women. From this perspective, it is necessary to highlight some particularities regarding the body of the elderly person, in particular their body changes due to aging.

From the this relationship between ergonomics and design, Gomes Filho (2003) refers that it is made possible knowing the basic ergonomic factors of garment design, such as the requirements for comfort, body posture, and safety, as well as visual, tactile and synesthetic actions. The author (Gomes Filho, 2003) also reinforces that the designer must consider the use of clothing, maintenance and cleaning. The comfort for the user, in this case the clothing for the elderly person, is related to the suitability to the body and other requirements, such as practicality, quality, and functionality, as highlighted by Araujo (2009).

The integration between design and ergonomics is one of the intents of this effort, an undergoing PhD study in Textile Engineering at the University of Minho (Portugal). It aims to develop clothing with greater comfort to care-dependent elderly women, besides facilitating handling by caregivers during the daily tasks of dressing and undressing. Therefore, the ergonomic, sensorial and psychological aspects of comfort for the user assume the central core of this research.

2 THEORETICAL FRAMEWORK

For the users' satisfaction with clothing, when considering ergonomic functions, the garment should be designed according to the variance of height and body shapes of the target population, considering their anthropometric data, as well as their needs in terms of movement and interaction with others. In the development of the proposed clothing, in addition to the contribution of ergonomics and anthropometrics, it is necessary to observe the complex movements of the users, in particular those performed by the caregivers, as they are the main responsible for the tasks that involve their handling.

All designers, at the time of development of a new product, have the responsibility to guarantee an adequate interaction of the product with the user. It is important to use new assessments, yet in the first design phases, to re-evaluate the design proposed, with the goal of reaching the best fit with the human dimensions (Panero & Zelnik 2002). This is of major relevance when a product is going to be in direct, and in long, contact with the user's skin, as it is the case of clothing.

According to Iida (2005), ergonomics can contribute at different moments, as also identified by Wisner (1987): during design and after design (correction), to improve awareness or to promote participation. In the design phase, the contribution of ergonomics occurs during the product design. In correction ergonomics, it occurs when it is applied to existing situations, aiming to correct current problems. In the awareness ergonomics, the actions are focused on problems that were not solved during design or correction phase, as well alternative, training programs are promoted. Finally, in participatory ergonomics, the user is involved, aiming to use his practical knowledge when seeking the solution for the problem.

When the proposal is the development of a product focused on the elderly population, the design phase requires a suitability of the product to the user but also an acceptance by their caregivers. This is why a toile (garment prototyping) is so relevant since caregivers are responsible for handling the elders for dressing, undressing, health caring, and cleaning.

Iida (2005) mentions two authors (Mcclelland & Brigham, 1990), who share the same vision regarding the relevance of ergonomics during product development and also the knowledge of the user's profile, product utility, usability and the interaction with the user, it is recommended to clarify the users' profile, their needs, values, and desires. To reinforce this, Dul & Weerdmeester (2004) show that ergonomics includes aspects such as posture and body movements (sitting, standing, pushing and pulling), environmental conditions (noises, vibrations, lighting and climate), and tasks (appropriate and interesting). The integration of all these requirements allows the development of safe, healthy, comfortable and efficient products for everyday life.

The ergonomics contribution usually occurs in the initial phase of the product design, in the work environment or in specific situations that already exist, aiming to solve problems related to safety, fatigue, quality improvement, etc. Its contribution is being widened as it has been used to improve the interaction between humans and artifacts. This involves both physical and organizational aspects, covering planning and designing activities. Ergonomics aim to promote health, safety, satisfaction and well-being for the individuals.

According to Iida (2005) there are two ways of performing ergonomic experiments: one is under artificially constructed and controlled conditions, like laboratory research, the other occurs in the field studies, from the observation of the phenomenon under real conditions.

Furthermore, to reach this study goal, it is of paramount importance to apply anthropometric data. Anthropometry is the scientific area related to the study of the human body, volumes, shapes, their movements and joints (Petroski, 2007). Anthropometry is also useful for the nutritional analysis of the elderly, as it is a simple and good prognosis method for studying possible diseases, functional disability and mortality, and can be used for both diagnosis and disease monitoring (Brazilian Health Ministry, 2011).

3 METHODOLOGICAL STAGES

3.1 *Process for the design of clothing*

The proposed design was based on the main body measurements, collected from a group of 101 participants, aged over 65 years for the evaluation of the physical and psychological conditions regarding the selection of the profile of the probable participants of the anthropometric study, resulting in the collection of 79 measurements of the body. This data allowed the development of an experiment to define values of more specific measurements for the development of the basic pattern block supporting the design of basic clothing prototypes for the dependent elderly. The basic construction pattern block was designed in medium size (M), in order to reach a wider range of users. For Jones (2005) and Heinrich (2005), the measurements applied in the basic design process are defined by companies, designers or by the standards for clothing sizes developed by the countries that already have performed anthropometric studies. The medium size makes easier designing larger and smaller sizes and can be used during the grading process of the construction patterns.

By using these clothing design techniques, it is possible to structure the product development, keeping the focus on the proper understanding of the design. According to Fischer (2010), this requires a good observation of the design proportions. This way, body shape should be observed (considering the most adequate ease values for each part of the construction pattern block) as well as the notches insertion (location of clothing' parts in pockets, openings, hems, among others) and other sewing parts (like pockets, collars, and others).

Still, for designing ergonomic clothing, it is imperative to consider the social, physical and psychological needs of the users, and their caregivers' tasks, through the analysis of the collected data from the registered elderly women and the work practices of their caregivers through direct observation and photographic registration. The method used involved the observation of two caregivers, with different years of experience in their function, while performing their daily activities. This way, it was possible to understand that with experience in their function, practicing the same tasks on a daily basis, the handling of the elderly women is performed in a mechanized way, meaning that the task becomes more simplified. Ferreira, Leal and Guimaraes (2004) report that observation constitutes a subjective interaction between observed and observer, which are embedded in a single reality from which the findings occur.

Through the use of the basic patterns block, it was possible to structure the development of the product, keeping the focus on the adequate understanding of the model regarding the body modification that the person acquires in the course of aging. According to Fischer (2010), this understanding requires a good observation about proportions.

The garments were designed according to the specific needs and local context of the elderly participants of the cities of Guimaraes (Portugal) and Teresina (Brazil).

3.2 *Elderly from Guimaraes, Portugal*

Five models were designed to be used during the pleasant temperature days, close to Summer, the hottest European season. The values of the average monthly temperature change regularly during the year, reaching the highest values in Summer (>27°C) during the months of July and August. The lowest monthly average temperature occurs from December to February, reaching below zero temperatures in the highest altitude areas of Spain and Portugal (Climatological Atlas of the Iberian Peninsula, 2011).

Besides taking into consideration the environmental characteristics, it was also necessary to know some behavioral aspects of the elderly and to analyze information provided in the caregivers' survey. The garments selected for prototyping are composed of dresses, trousers, skirts, blouses, considering the complementary details of the models, such as raglan sleeves (diagonal cut lengthening to the neckline) and kimono sleeves, soft scoop necklines and narrow collars, easy openings with zippers and buttons. 3/4 sleeves models were designed both in raglan and kimono formats. Two of these models, do not require the definition of the shoulder slope. The 3/4 sleeves were used as an option for the long sleeves, in consideration that this is preferred due to the extended cold period of the region.

3.3 *Elderly from Teresina, Brazil*

In Teresina the garment was developed especially for the Summer, considering that in this region the hot days last almost all year. Teresina's monthly average temperature varies from 26.9°C to 37.1°C, reaching extreme temperatures (above 40°C). Also, the relative air humidity reaches 75 to 83% in the urban area (Medeiros, 2014). In addition to data collected from the caregivers' survey, other observations such as physical and financial conditions, beyond the climatic conditions, were important for the clothing design, and wider fits were proposed.

Puccini & Wolf (2015) carried out a study focused on the development of garments for the elderly population. The authors realized that dresses are considered the most comfortable garment by the individuals. This statement is in accordance with the proposed design by this study, emphasizing the dresses with wider and short raglan sleeves, wider kimono sleeves with lappers, boat and V-shape necklines, seam finishes with the same fabric of satin taping, and easy openings (with buttons and zippers). The design was focused on necklines, specifically the place of openings that provide inserts such as various types of collars. Models with larger round necks, boat neckline, and V-shape were developed. In some models, narrow collars and half collars (a type of collar that ends at the shoulder) were inserted with external finishes to avoid discomfort in the neck region. Regarding the sleeves (most of them are designed to be fixed in the armhole), the arm movements were taken into consideration. The body movements were observed, avoiding overlapping seams and cuts located in places with more skin contacts due to body movements.

One of the peculiarities in most of the designed models with sleeves is the raglan sleeve. This type of sleeve allows for better body heat release when compared to the sleeve with the armhole at the shoulder. This is because it shifts the armhole (commonly located where the arm is joined to the torso) for the neckline, in a diagonal line; that is, the shoulder slope, providing a greater area in the direct contact with the arm. Another type of sleeve, the kimono sleeve, was also used by moving the seam from the shoulder cavity to the lower triceps branchial (upper arm muscles). These sleeves do not have direct contact with the arm joint, this to avoid friction during upper limb movements. As an example, in the study by Schiehll, Silva, and Simoes (2014), with groups of women with some difficulty in dressing and undressing, the long-sleeved blouse with a defined and adjusted shoulder height was the most difficult for the participants because of the reduction in the amplitude of the armhole's encounter with the sleeve, limiting the arm extension when combined with the forearm flexion movement.

4 MATERIALS, METHODS AND RESULTS

Flat fabrics of plain textile structure were used requiring enough openings to fit the user's body, as this textile structure does not provide enough elasticity (except when composed of elastane). In this study, cotton fibers were the most used, due to their characteristics, allowing a greater versatility. This natural fiber has several advantages, such as a soft touch, low tendency to cause allergies, absorbability, good tensile strength, wash resistance, easy dyeing, and breathability. When combining cotton with other fibers, the fabrics present excellent comfort levels and maintenance properties. Table 1 represents the main characteristics of the materials used in the clothing of the elderly in both cities.

The first prototypes were produced using 100% cotton fabrics (Figure 1), to analyze the ease values and the garments construction methods. The second prototypes used 50% cotton and 50% polyester fabric (Figure 2), the type of fabric that has the most similar texture when compared to the 100% cotton fabrics.

Table 1. Main characteristics of the materials used.

Origin	Portugal	Brazil			
Reference	TC64	5150	5170	T-2133	T-1318
Description	TC64	Techno Polo	Techno Fit	Ferrari Tinto	Fio
Supplier	Somelos	Santista Tavex	Santista Tavex	MN A. Gerais	Marango
Structure	Plain	Plain	Plain	Plain	Plain
Composition	96CO 4EA	100CO	62CO 35PES 3EA	100CO	100CO
Density (y/cm)					
Weave	66	49	58	59	60
Weft	38	27	29	30	34
Weight/m^2 (g/m^2)	142	168	181	106	119

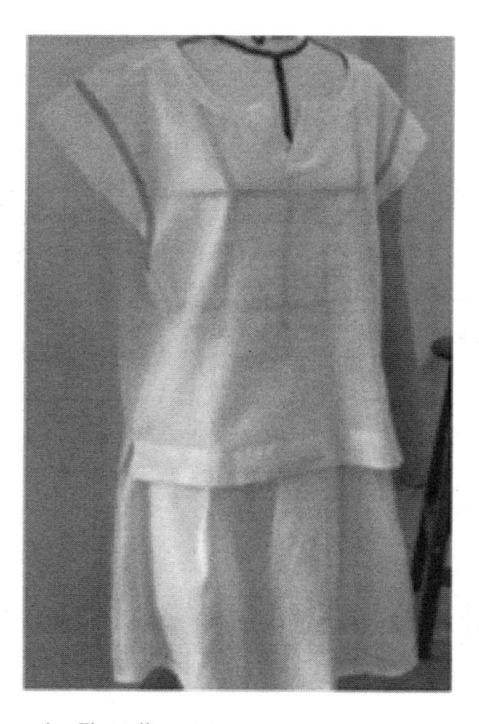

Figure 1. First toile prototype.

The designed models were evaluated on a manikin to detect the defects regarding the positioning of openings, necklines, collars and sleeves, and also to define the best seam finishes for the garment construction.

In order to obtain more comfortable garments, cuts requiring seams were avoided, as any seam can cause skin irritation and increase friction in some body parts. The proposed garment was designed with appropriate openings and closures, repositioning seams, taking into account the characteristics of the elderly skin, which becomes more sensitive with aging.

Figure 2. Second toile prototypes.

The garment construction process took into consideration the required sewing machines according to the type of material used, model, and seam types, the appropriate accessories for each model and the required quality.

The essential qualities offered by a product suggested by Iida (2005) are technical-constructive qualities (specifically in the case of clothing, the pattern design and the construction), ergonomic qualities (comfort and safety), and aesthetic qualities (visually pleasing). These are quality requirements that enable the design of artifacts with characteristics that meet human needs. Sensory comfort was also taken in consideration when choosing an appropriated type of garment construction and seam finishes, in this case, lapped seam and seam finishes with satin taping, hiding the external seams. In openings with zippers, using a zipper fly with double lining (Figure 3). These finishes provide less friction when in contact with the skin of the user and a better aesthetics for the garment.

In the prototypes' finishes, clasping accessories such as zippers and buttons were used. The polyamide zippers were chosen due to their easy handling and smoother (less rough) touch. Polyamide zippers were chosen due to their easy handling and less harsh touch, allowing better sensorial comfort.

For the production of the final prototypes, plain fabrics were used, according to Table 1, all available in the Portuguese and Brazilian markets, with ideal characteristics and attributes of comfort, addressing the identified characteristics and needs of the users. Although texture perceptions are subjective in nature, the user's sensitivity and preference can be related to the materials structural properties and are more commonly found in flat fabrics.

Regarding the buttons, none of the available options in the market offered features of flexibility, elasticity and touch softness, which could avoid the users' discomfort and facilitate the caregivers' tasks. To meet

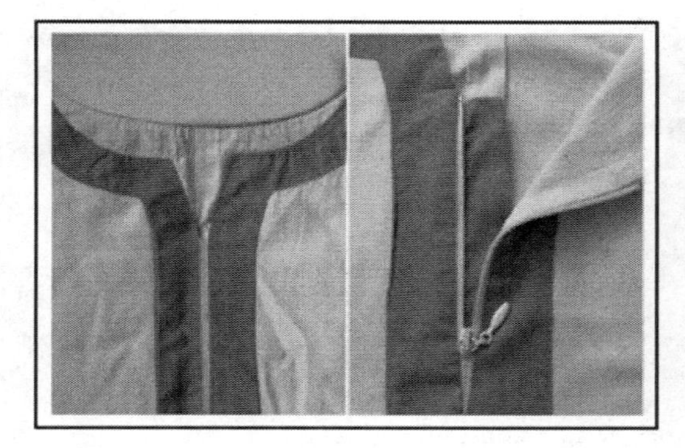

Figure 3. Zipper fly with double lining.

this specific need, flexible buttons were developed (Figure 4) through a partnership with researchers from the Mechanical Engineering Department of University of Minho. Prototypes of a new button were developed using a 3D printer - *Prusa Model 13*. Different types of flexible filaments (named thermoplastic elastomer – TPE) were tested. In a second phase, for technical reasons, such as calibration and productivity of the available 3D printer at the University, as well as due to the high production costs, a company was contacted, allowing the button to be obtained in the colors and formats required. Löbach (2001) reinforces the conclusion that the design allows the materialization of an idea in the form of designs and models, resulting in a possible innovation to be industrially produced.

The button produced is composed by a plastic component (polyvinyl chloride), commercially known as *Plastisol L/100 Bianco Ral 901*. It is a material that has plasticizing additives such as phthalate, with viscous appearance, that must be protected from freezing and stored at temperatures ranging from + 5°C to + 35°C. Also, the four-hole button was chosen because it can be better fixed to the garment when compared with the two-hole button. The later, with daily manipulation can be easily loosened and subsequently removed. The four-hole button assures higher safety levels for users, like in the example reported by a caregiver of an elderly woman, who used to remove the buttons from her clothing and swallow them. The button developed is characterized by its flexibility, being foldable, and presenting a good resistance with greater comfort to touch.

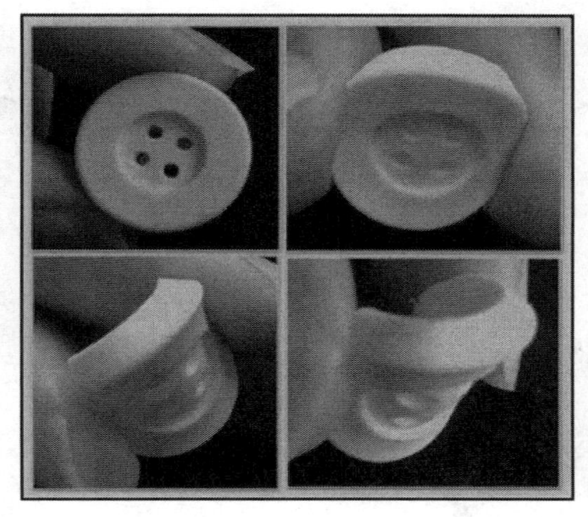

Figure 4. Flexible buttons.

5 CONCLUSIONS AND FUTURE WORK

In this research, it was possible to identify the main needs of the care-dependent elderly women, understand their limitations regarding the use of clothing, as well as the impact that this has on their caregivers during dressing/undressing for health treatments and cleaning tasks. The anthropometric characteristics and the ergonomic needs were considered during the development of the clothing pattern design blocks for these specific individuals, considering the aging effects on their bodies. The first developed toile prototypes took into account this information, as well as the suggestions of the caregivers. Different materials and accessories were tested in the prototypes that will be evaluated by the elderly of the two cities. The final prototypes were developed using flat plain fabrics, available in the Portuguese and Brazilian markets, which present ideal characteristics and attributes of comfort. In the last research stage, the validation process (usability, adjustability, handling, and functionality) of the prototypes was defined, considering three basic comfort requirements (ergonomic, sensorial and psychological) for the elderly comfort, and weighted according to the caregivers' point of view: ergonomic comfort (60%); sensorial comfort (30%); psychological comfort (20%). For this procedure, user and caregiver surveys were applied, with analysis and discussion of the results for validation of the proposed clothing. The result of this last undergoing phase of the study will be presented in a future paper.

ACKNOWLEDGMENTS

We would like to acknowledge UFPI-Federal University of Piauí and 2C2T-Science Center for Textile Technology from University of Minho.

This work is financed by FEDER funds through the Competitive Factors Operational Program (COMPETE) POCI-01-0145-FEDER-007136 and by national funds through FCT-Portuguese Foundation for Science and Technology, under the project UID/CTM/000264 and CSF/CAPEs.

REFERENCES

Araújo, M. dos S. de 2009. *Design de vestuário para desportistas deficientes motores*. Tese de mestrado em Design e Marketing – vestuário. Universidade do Minho.

Atlas Climatológico da Península Ibérica. 2011. Available in: <http://www.ipma.pt/resources.www/docs/publicacoes.site/atlas_clima_iberico.pdf>. Accessed in: 5th May 2017.

Burdek, B.E. 2006. *Design: história, teoria e prática do design de produtos*. São Paulo: Edgard Blücher.

Dul, J. & Weerdmeester, B. 2004. *Ergonomia prática*. São Paulo: Edgard Blücher.

Ferreira, C. J. R., Leal, M. T. B. e Guimaraes, M. G. N. 2004. "Desvelando o método etnográfico: que contributos para a ciência de enfermagem?", *Revista Sinais Vitais. – ISSN 0872-0844. – n° 56*, pp. 52–55.

Fischer, A. 2010. *Fundamentos de design de moda: construção de vestuário*. Porto Alegre: Bookman.

Gomes Filho, J. 2003. *Ergonomia do objeto: sistema técnico de leitura ergonômica do objeto*. São Paulo: Escrituras.

Heinrich, D.P. 2005. *Modelagem e técnicas de interpretação para confecção industrial*. Nova Hamburgo: Feevale.

Iida, I. 2005. *Ergonomia: projeto e produção*. São Paulo: Blucher.

Jones, S.J. 2005. *Fashion design: manual do estilista*. São Paulo: Cosac Naif.

Löbach, B. 2001. *Desenho Industrial: bases para configuração dos produtos industriais*. São Paulo: Edgard Blücher.

McCletland, I.L. & Brigham, F.R. 1990. *Marketing ergonomics – how should ergonomics be packaged?* Journal Taylor Francis Online, vol. 33. 519–526. Published online: 16 Dec 2010.

Medeiros, R.M. de. 2014. *Caracterização de mudanças climáticas por meio de séries meteorológicas para o município de Teresina/Piauí.* Revista Pernambucana de tecnologia. Vol. 2, no. 2, abr. 2014. ISSN 2319–0949.

Ministério da Saúde do Brasil. 2011. *Orientações para a coleta e análise de dados antropométricos em serviços de saúde: Norma Técnica do Sistema de Vigilância Alimentar e Nutricional – SISVAN.* Brasília: Ministério da Saúde.

Mozota, B.B. de. 2002. *Design management*. Paris: Éditions d'Organization.

Panero, J. & Zelnik, M. 2002. *Dimensionamento humano para espaços interiores*. Barcelona: G. Gili.

Petroski, E.L. 2007. *Antropometria: técnicas e padronizações.* Porto Alegre: Palotti.

Puccini, C. & Wolff, F. 2015. *Desenvolvimento de coleção ergonômica para mulheres acima dos setenta anos*. Revista *Icônica,* vol. 1, No. 1, ago. 65–77.

Schiehll, L.O. et al. 2014. *Design de vestuário inclusivo para mulheres com limitações funcionais: projetando autonomia.* In: Italiano, Isabel Cristina et al. *Pesquisas em design, gestão e tecnologia de Têxtil e Moda*. São Paulo: Escola de Artes, Ciências e Humanidades.

Wisner, A. 1987. *Por dentro do trabalho: ergonomia: método e técnica*. São Paulo: FTD.

Textiles, Identity and Innovation: Design the Future – Montagna & Carvalho (Eds)
© 2019 Taylor & Francis Group, London, ISBN 978-1-138-29611-4

Textile design experiments for automotive innovation

Maria Antonietta Sbordone
Università degli Studi della Campania "Luigi Vanvitelli", Caserta, Italy

Gennaro Gentile & Giacomo Cesaro
ICTP-CNR, Napoli, Italy

ABSTRACT: New artificial fibers evolve through cutting-edge technologies, they represent the intersection among technology, materials and project in the automotive sector. The composites reinforced with natural fibers (FN), called eco-composites, are materials that affect many scientific research projects and the economic exploitation opens new ways for applications in new business areas. The key element for the development of these materials is the compatibility of processing with the industrial technologies used for the production of traditional composites. In this area, an important growth is planned in terms of productivity and industrial viability in the coming years. The growing scientific and industrial interest in composites reinforced with natural fibers is due to their high performances in terms of mechanical properties, good processing properties, chemical resistance and the low ratio between cost and density. The main limit to the diffusion of eco-composites reinforced with natural fibers is not so much the mechanical performance, which in some cases is comparable if not superior to the composites currently used, but due to the costs of the materials, and in particular to the high cost of polymer matrices obtained from renewable and/or biodegradable sources. The design idea for the experimental research activity has been directed to the production of a multi-layered, eco-sustainable material, flexible as a fabric and characterized by properties suitable for the possible applications in replacement of bodywork elements. The material has been designed to have excellent thermal-acoustic insulation properties, essential for high-value added applications in the automotive industry. On the basis of the results obtained, a preparatory method was developed for the production of some samples of multilayer prototypes.

1 INTRODUCTION

1.1 *Introduction*

The project described in this paper is about new artificial fibers evolve through cutting edge technology, they represent the intersection among technology, materials and project in the automotive area. The composites reinforced with natural fibers (FN), called eco-composites, are materials that affect many scientific research projects and the economic exploitation opens new ways for applications in new business areas.

Composite reinforced with natural fibers (FN), also named eco-composites, are materials that concern many scientific research projects and the economic exploitation opens new ways for applications in new business areas.

2 ECO-SUSTAINABLE MATERIALS

2.1 *Natural fibers*

One of the greatest environmental problems we face today is the problem of plastic waste produced on a global scale. Production growth, increased use of plastics, and waste disposal, are important phenomena; scientific research point to eco-suitable materials that are easily biodegradable as a solution. The term eco-composite is usually used to indicate a composite material with environmental and ecological benefits over conventional composites.

By definition, an eco-composite can contain natural fibers (NF) and polymers obtained from renewable sources. Using a narrower definition eco-composites can be considered as the product of the combination of biodegradable polymeric matrices with natural fibers (NF).

2.2 *NF as reinforcement for eco-composites*

The growing scientific and industrial interest in composites reinforced with natural fibers is due to their high performances in terms of mechanical properties, good processing properties, chemical resistance and the low ratio between cost and density.

In the long term environmental interest could lead to the replacement of traditional reinforcement materials (inorganic fillers and fibers) with organic reinforcements from renewable sources.

Natural fibers, in fact, are an important ecological alternative to conventional reinforcement fibers (glass, carbon, kevlar). The benefits of natural fibers compared to traditional fibers are the low cost, the high-tenacity, the low density, the specific strength properties, the reduced wear of processing plants, the lower energy use, the positive balance of CO2 emission, and the biodegradability.

Due to their hollow and cellular nature, natural fibers can also be used as acoustic and thermal insulators and show an apparent reduced density.

3 ECO COMPOSITES. INDUSTRIAL RESEARCH AND DEVELOPMENT

3.1 *Industrial research and development*

The key element for the development of these materials is the compatibility of processing with the industrial technologies used for the production of traditional composites. In this area, an important growth is planned in terms of productivity and industrial viability in the coming years.

Generally, the processes for the production of composite materials reinforced with natural fibers, thanks to some adaptations, are similar to those for the production of composite materials reinforced with other fibers, such as glass fibers.

During the processing, the temperature should not get through 200°C, and the material retention time at high temperatures should not be too long, to avoid the degradation of the fibers. Many technologies are common for composite materials reinforced with natural fibers or glass fibers, for example: injection molding and compression molding.

For making composites, natural fibers are used in the form of fabrics, non-wovens and mats, through technologies of bulk molding compound and sheet-molding compound, and sometimes as unidirectional reinforcing fibers (UD).

Below there are some examples of projects that have led to eco-composites commercial development during the last decade.

Daimler Chrysler has developed the so-called EXPRESS procedure, for the processing of composites reinforced with thermoplastic matrix fibers, in which a natural fiber mat is compressed. Fused thermoplastic matrices are introduced into the mold directly from the extruder, while natural fibers mats are loaded into the mold discontinuously before each polymer load, along with the additives.

The German Agricultural Engineering Institute in Potsdam has presented a simple technology, suitable for the processing of canvas, linen, coconut fibers, for the production of composite panels for thermal and acoustic insulation for the design, the construction area and for the automotive industry, announcing that it has reached a processing capacity of raw materials up to 3 t/h.

The US company Pheonix™ LLC Biocomposites, uses wheat straw in its Biofiber ™ composites, and sunflower discards in its Dakota Burl composites, both for use in design applications.

Important research on eco-composites was done at the German Aerospace Center (DLR). Recently, Riedel and Nickel have reported the state of the art on the future prospects of eco-composites, mainly based on unidirectional laminates of flax and canvas fiber in thermoplastics (PHB-Biopol, PCL-Capa, based on SCONACELL A starch).

Other research groups have oriented development activities towards eco-composites reinforced with vegetable fibers based on soy protein.

Chaba and Netravali have worked on the characterization of green composite materials based on yarn of flax and cross-linked soy.

It should be noted, however, that most of the efforts aimed at the marketing of products reinforced with natural fibers have been consider the use of traditional polymer matrices. Currently, PP is the most used material and is used in a large number of recyclable eco-composites.

Visteon and Technilin have developed a polypropylene matrix material reinforced with flax fiber, R-Flax®. Considering the very high specifications imposed by Opel, which include critical safety requirements, R-Flax® can be used for interior elements (car door panels), where aesthetic qualities can be an element of choice for the user.

NFC Wageningen has submitted a patent for the commercial production of granular of composites "PP/natural fibers granulated", GreenGran NF30, 50, 70, which can be easily processed by extrusion. Due to their high content of natural fibers, GreenGran NF30, 50, 70 granules have better insulating properties (thermal, acoustic), better flame resistance properties, improved dimensional stability at high temperatures with no sharp edges, after a car accident (safety requirements for internal parts of a car vehicle).

In the area of the ECO-PCCM project, promoted by the EU FP6-INCO program, a number of PLA, PHBV and PP reinforced with different natural fibers were created, with properties suitable for making panels for internal elements of eco-houses.

Many LCA studies, have already proved that composites reinforced with natural fibers are potentially better than composites reinforced with synthetic fibers for many applications, including those for the transporting area.

Current socio-economic trends are pushing more and more to replace traditional polymers (such as polyolefins) with biopolymers from renewable and/or biodegradable sources, even for eco-composites, the industrial interest will tend towards green materials completely, in which both the matrix and the reinforcement will consist of environmentally sustainable materials. For this reason, future opportunities for further and wider applications of eco-composites based on biopolymers will depend on two key factors. The first element is related to the increasingly competitive marketing of polymer matrices characterized by thermomechanical properties comparable to those of polyolefins. The second one will depend on the development of effective and economic methods of modifying natural fibers or the development of effective compatibilizers to improve the properties of eco-composites and make them more suitable for a larger number of applications.

4 ECO-COMPOSITES FOR THE AUTOMOTIVE AREA

4.1 *Performances and limits*

As mentioned above, the main limit to the diffusion of eco-composites reinforced with natural fibers is not so much the mechanical performance, which in some cases is comparable if not superior to the composites currently used, but the costs of the materials, and in particular the high cost of polymer matrices obtained from renewable and/or biodegradable sources.

This is also true for composites used in the automotive industry.

However, there is a strong interest from car manufacturers in the constant development of eco-composites for car efficiency and marketing reasons. For example, eco-composites can contribute, with the same mechanical performance, to lighten the overall weight of the vehicles, while improving both driving performance and energy consumption of the car.

4.2 *Renewable sources of materials*

In recent years the use of materials from renewable sources in the automotive sector has grown steadily. Composite materials reinforced with natural fibers have been used in the internal and external of a vehicle.

In particular, with regard to the internal parts, examples of sustainable materials applications from various car companies are used for the construction of dashboards, of interior panels of the doors, of hoods, of seats, of backrests and interior trim linings for vehicles.

For examples of details, in 1996, Mercedes-Benz used an epoxy matrix reinforced with jute fibers in the door panels in its E-class vehicles.

In 2000, Audi then launched a mid-range car, the A2, in which the door panels were made of polyurethane reinforced with mixed fiber of sisal/linen (flax).

During the last years, Toyota has relaunched the use of bioplastics instead of traditional plastics for many interior components, trying to position itself as a leading brand in the adoption of environmentally friendly materials.

The use of composites reinforced with natural fibers was started in Toyota already in 2003 (RAUM model) for the spare wheel cover system, made of PLA reinforced with kenaf fibers.

5 EXAMPLES

5.1 *Concept realization of fabrics/polymeric foams*

Other examples of eco-composites applications are some internal components in PBS reinforced with bamboo fibers and mats made of mixed blends of PLA/polyamide. Toyota then used seatfoams with soy materials in Matrix and RAV4 models.

More complicated is the problem of using composites reinforced with natural fibers for the external components of vehicles. The main problem is the capability for materials to handle extreme conditions, temperature, changes in humidity, exposure to weather, and mechanical shocks.

The problem of absorption/desorption of water by the fibers, resulting in dimensional variation, is still an obstacle to overcome for the creation of composites characterized by high durability, a key element for the automotive area. Despite this, there are various examples of eco-composite applications for external components.

In 2000, for the Travego bus, Mercedes fitted the engine compartment with a polyester composite flax reinforced to improve sound insulation. Although this component is installed in the engine compartment and the classification as an external component has been questioned, this is considered the first case for the use of natural fibers for components installed outside the passenger compartment of a vehicle and so is a milestone in the application of these materials. Other applications are more related to development projects than to applications on vehicles in production. In the concept car ECO Elise (2008), Lotus has replaced fiberglass composites reinforced with canvas fibers for the design of bodywork and spoiler panels.

6 RESEARCH ACTIVITY ON NEW MATERIALS

6.1 *Realization of fabrics polymeric foams. Introduction*

In the first part of this document, the state of the art of eco-composites reinforced with natural fibers has been reported, with particular attention to developments in scientific research and commercial programs aimed at the realization of new materials and products with a specific focus on automotive applications. The analysis made, has shown that most of the composites reinforced with fibers are designed for application as internal components to the vehicle.

However, recent trends show a strong interest in the automotive industry for the external use of environmentally-friendly composite materials, possibly in replacement of bodywork elements.

In this area, the projects developed by some automotive companies for the replacement of bodywork with textile items are very interesting. In particular, the first studies in this direction were conducted by BMW, which in the prototype GINA (2008), instead of the traditional metal bodywork, installed a traditional elastic fabric that fits perfectly with the structure (Fig. 1).

Most recently, the German engineering Company Edag (Fig. 2), has presented at the Geneva Motor Show (2015), a concept car Light Cocoon, in which the chassis is clearly transparent, because a polyester

Figure 1. Carboard by BMW-Gina 2008.

Figure 2. Cardboard Eco-Elise by Lotus 2008.

multilayer fabric, which is elastic, waterproof is installed in place of the bodywork (154 g/m^2).

6.2 *Realization of fabrics polymerics foams. Concept*

The design idea for the experimental research activity has been directed to the production of a multi-layered, eco-sustainable material, flexible as a fabric and characterized by properties suitable for the possible applications in replacement of bodywork elements. The material has been designed to have excellent thermal-acoustic insulation properties, essential for high-value added applications in the automotive industry.

The designed multilayer consists of two external skins made of natural fabrics. Flax and cotton fabrics have been identified as important materials for these layers. For the internal layer there are two possible solutions, one based on commercially available material and the other one on an eco-friendly material created through an innovative but not directly transferable process on an industrial level without a right adjustment phase of the methodology of realization. In both cases, the core is made up of a polymeric foam (Fig. 3).

For the production of samples, allow modulus polyurethane foam has been selected in the case of the commercial product, with properties that ensure good adhesion to the textile skins without the need for adhesives. In the case of the eco-sustainable material, a foam based on polyvinylalcohol (PVOH) was

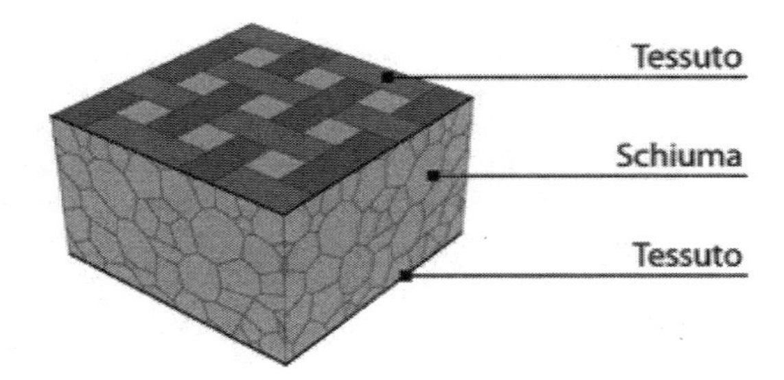

Figure 3. Multilayer scheme fabric/foam/fabric.

produced and reinforced with varying amounts of cellulose fibers, capable of modulating the properties.The added value of the material is in the properties of the core, which is able to trap air thanks to its microstructure so it can increase the insulation and sound absorbing and shock protection characteristics, which are basic features for paneling of the automotive industry, with environmental impacts limited, compared to other competing materials.

During the design stage, the attention was on the optimization of three parameters: i) structure of the expanded material; Ii) adhesion to the expanded interface textile/matrix; Iii) optimization of the finishing thickness multi-layer.

7 EXPERIMENT

7.1 *Materials*

The utilized materials are listed below.

- Cotton, cloth fabric, weight 197.7 g/m^2, linear density of warp and weft 47.7 tex, obtained from EMPA (Switzerland).
- Flax, cloth fabric, weight 271.6 g/m^2, linear density of warp and weft 82.8 tex, obtained from EMPA (Switzerland).
- Polyurethane foam Helix®, obtained from Uniflex S.p.A. (Italy).
- Polyvinyl alcohol (PVOH, 87–90% idrolizzato, Mw 30000–770000, obtained from Aldrich Chemical Co. (Italy).
- Glycerol (G, pureness >99%), obtained from Aldrich Chemical Co. (Italia).
- Cellulose fibers, commercial name Arbocel BC1000, average fiber lenght 700 μm, fiber diameter 18–22 μm, obtained from JRS Pharma GmbH (Germany).

7.2 *Sample preparation*

Multilayer fabric samples containing an intermediate layer in polyurethane foam were made according to the following protocol:

- Flax and cotton linen fabrics (textile) have been cut following the desired length (with a maximum surface area of 100 cm^2);

- The textiles were wet with distilled water to increase adhesion properties with polyurethane foam;
- The textiles have been scratched off with polyurethane foam to obtain a better interface between the core and the lower and upper skins;
- The polyurethane has been applied and leveled to obtain a constant thickness on one of the two sides of the textile. Excess foam has been eliminated with the aid of a spatula. The upper layer of fabric (textile) was then applied;
- After 48 hours, and after the complete drying of the sample, it was pressed to obtain samples of different thickness and apparent density; pressing was done at room temperature using a P200 Dr. Collin Gmbh press (Germany).
- The last stage provided for sample shaping.

Multilayer fabric samples containing an intermediate layer of PVA/cellulose foam were made according to the following protocol:

- The sample core was made from PVA plasticized with glycerine and reinforced with cellulose fibers Arbocel BC1000. Flax and cotton fabrics were used for the outer skins of the multilayer systems.

Samples were made according to the preparatory procedure described below:

- Evaporation of PVA in distilled water at a concentration of 0.08 g/ml, addition of the appropriate amounts of glycerol cellulose fibers (refer to the material sheets for weight ratio between components) under agitation at room temperature;
- The foaming of the solution was done by mechanical stirring with a Bimby Vorwerk Thermomix™ 3 l for 2 minutes at a stirring rate of 3000 rpm;
- The foam obtained from the previous stage was poured into molds (made in different sizes and thicknesses, from $4 \times 4\,cm^2$ up to $8 \times 8\,cm^2$, with thicknesses between 7 and 14 mm), closed to the top with textiles described before, previously wet with distilled water to improve adhesion between expanded textile/material;
- The molds thus filled were also quickly closed at the top with another fabric (previously wetted with distilled water); excess foam has been eliminated with the aid of a spatula;
- Samples were quickly frozen in liquid nitrogen and maintained for 24 h at $-18°C$. The samples were finally lyophilized to P $5 \times 10\text{-}2$ torr in an Edwards Modulyo freezedryer.

7.3 *Caracterization of samples*

The multilayer fabric samples obtained were characterized according to the following experimental protocol:

1) Measure of linear weight and density;
2) Fourier transformed IR spectroscopy analysis (FTIR) in ATR mode, performed on a Perkin Erlmer ONE spectrophotometer equipped with a modulus for diffuse attenuation measurements;

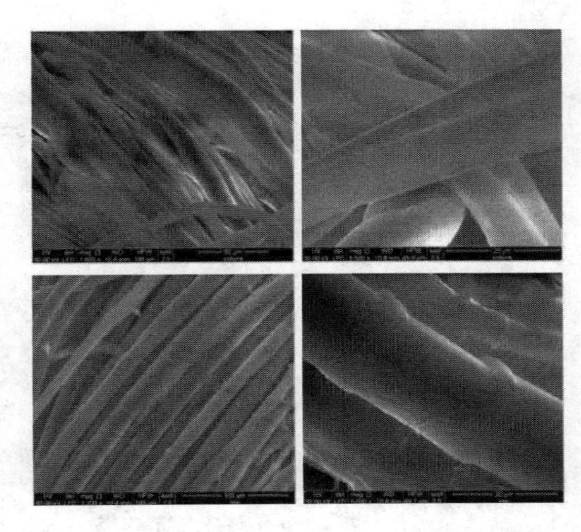

Figure 4. SEM of: a) skin of cotton of a multilayer cotton/foam of PVOH/flax fabric; b) skin of linen of a multilayer linen/foam of PU/flax linen.

3) Low-power scanning electron microscopy by electronic scanning microscope (SEM) FEI Quanta 200 FEG, 30 kV acceleration voltage, secondary electron mode, low vacuum operating conditions (0.75 torr);
4) Materials thermal insulation is being tested.

7.4 *Results and material samples*

Both systems have shown a good degree of adhesion between the fabrics and the expanded material. The prototypes are characterized by a high lightness and a good rigidity, as well as a good "hand" easily declinable according to different styles and designs depending on the choice of different natural fabrics.

The first stage of analysis was to evaluate the skins, to ensure that during foaming, part of the applied expanded material did not pass through the fabric and reach the outer surface of the multilayer. Clear evidences of transfer of the expanded material to the external surface has been pointed out for the coupling of PVOH/flax fabric. This problem was essentially attributed to the weaker texture of the selected fabric. In contrast to PVOH/cotton fabric and PU/linen fabric combinations, morphological analysis by SEM (Fig. 4) and FTIR spectroscopic analysis, did not reveal material transfer phenomena. In both cases, the fibers are not contaminated with traces of polymeric material due to the passage of foam under application.

Flax and cotton spectra have no characteristic features of the presence of polyvinylalcohol or polyurethane, whose spectra are reported by comparison. For this reason, cotton/PVOH/cotton and flax/PU/linen systems were selected for prototype preparation.

Below there are some samples produced (Fig. 5, Fig. 6, Fig. 7, Fig. 8).

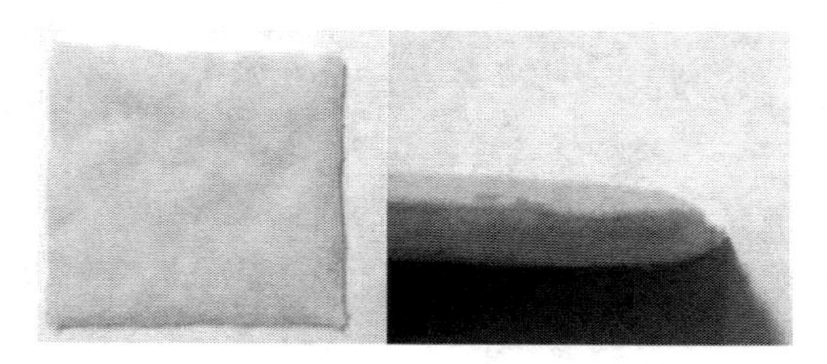

Figure 5. Sample N. 05 Multilayer in cotton (Skins) PVOH, Glycerol, Cellulose (core).

Figure 6. Sample N. 07 Multilayer in cotton (Skins), PVOH, Glycerol, Cellulose (core).

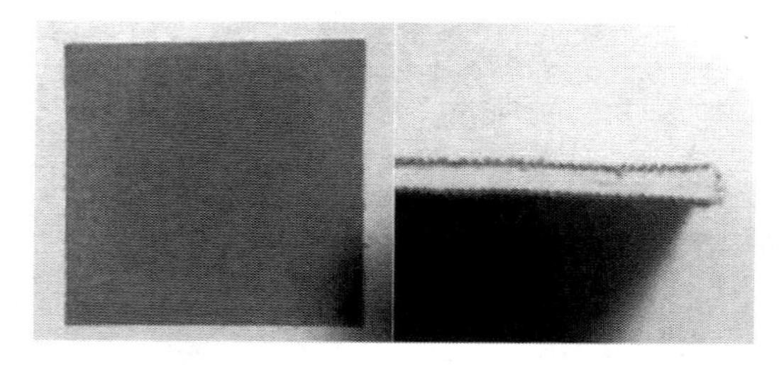

Figure 7. Sample N. 07 Multilayer in linen (Skins), Polyurethane foam (core).

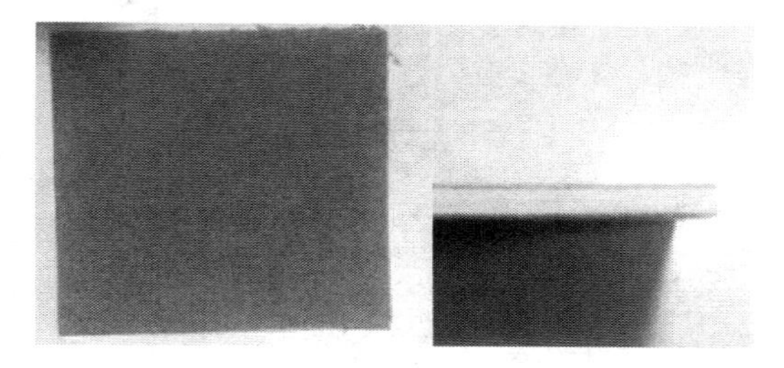

Figure 8. Sample N. 09 Multilayer in linen (Skins), Polyurethane foam (core).

8 CONCLUSIONS

On the basis of the results obtained, a preparatory method was developed for the production of cotton/PVOH/cotton/ and flax/PU/flax multilayer systems, with defined thicknesses, densities and weights, combining, when required, the foaming and coupling stages with the fabrics to the following pressing step.

The characterization of the multilayer systems was carried out in terms of:

1. Analysis of technological properties (apparent density, weight);
2. Morphological analysis by electronic scanning microscopy of some samples.

The following activities are still in progress:

1. Analysis of thermal insulation properties;
2. Completing morphological analysis to evaluate textile/foam adhesion;
3. Realization of prototype multiplayer characterized by complex shapes.

REFERENCES

Avella, M. Bogoeva-Gaceva, G. Bužarovska, A. Errico, M.E. Gentile, G. Grozdanov, A. 2007. Polymer Engineering and Science, 47(5), 745–749.

Avella, M. Bogoeva-Gaceva, G. Bužarovska, A. Errico, M.E. Gentile, G. Grozdanov, A. 2008. Australian Journal of Crop Science 1(2):37–42.

Avolio, R. Graziano, V. Pereira, Y.D.F. Cocca, M. Gentile, G. Errico, M.E. Ambrogi, V. Avella, M. 2006. Agronomy for Sustainable Development, 26, 251–255 .

Avolio, R. Graziano, V. Pereira, Y.D.F. Cocca, M. Gentile, G. Errico, M.E. Ambrogi, V. Avella, M. 2015. Carbohydrate Polymers 133:408–420.

Cocca, M. Avolio, R. Gentile, G. Di Pace, E. Errico, M.E. Avella, M. 2015. Carbohydrate Polymers 118:170–182.

Davies, G. 2003. Materials for automobile bodies. Oxford: Replika Press Pvt. Ltd.

Henning, F. Ernst, H. Brussel, R. 2003. "Innovative Process Technology LFT-D-NF Offers New Possibilities for Emission Reduced Long-Natural Fiber Reinforced Thermoplastic Components," in 3rd Annual SPE Automotive Composite Conference, Troy, MI, USA, September 9–10.

Ichiara, Y. Takagi, H. 2002 in International Workshop on Green Composites, Tokushima, Japan, November 26 (2002).

Joseph, K. Mattoso, L.H.C. Toledo, R.D. Thomas, S. Carvalho, L.H. Pothen, L. Kala, S. James, B. 2000."Natural Fiber. Reinforced Composites," in Natural Polymers and Agrofibers Composites, E. Frallini, A.L. Leao, and L.H.C. Mattoso, editors, San Carlos, Brazil, Embrapa.

Korolis, G. Silva, A. Fontul, M. 2013. "Green composites: A review of adeguate materials for automotive applications", Composites Part B: Engineering 44:120–127.

Mundera, F. 2003. "Advanced Technology for Processing of NFP for Industrial Applications," in 7th International Conference on Wood Plastic Composites, Madison, WI, May 19–20.

Rowell, R.M. 1998. "Property Enhanced Natural Fiber Composite Materials Based on Chemical Modification," in Science and Technology of Polymers and Advanced Materials, P.N. Prasad, J.E. Mark, S.H. Kendil, Z.H. Kafafi, editors, Plenum Press, New York. http://www.compositesworld.com/news/cwweekly/2006/marchnews.

Textiles, Identity and Innovation: Design the Future – Montagna & Carvalho (Eds)
 ISBN 978-1-138-29611-4

Comparative study of the manufacturing technologies of Sports Bras aiming at adjusting the productive processes of the Brazilian apparel enterprises to the model Industry 4.0

F.M.P. Silva
SENAI Technology College Antoine Skaf and School of Arts Sciences and Humanities, University of Sao Paulo, Sao Paulo, Brazil

A.Y.S. Duarte
Nossa Senhora do Patrocínio University, CEUNSP, Salto/São Paulo, Brazil

W.C. Ming
SENAI Technology College Antoine Skaf and School of Arts Sciences and Humanities, University of Sao Paulo, Sao Paulo, Brazil

R.A. Sanches
School of Arts Sciences and Humanities, University of Sao Paulo, Sao Paulo, Brazil and CIAUD – Research University of Lisbon, Lisbon School of Architecture, Universidade de Lisboa, Portugal

F.M. Silva
CIAUD – Research Centre for Architecture, Urbanism and Design, Lisbon School of Architecture, Universidade de Lisboa, Portugal

ABSTRACT: The objective of this paper is to present a preliminary study of the technologies adopted on the manufacturing of Sports Bras in Brazil, aiming at presenting different perspectives of how the technology is adopted by the Textile and Apparel Industry to achieve the Industry 4.0. Three technologies were analyzed: traditional manufacturing, *seamless* knitting (circular machine of average diameter) and *seamless* knitting with control of feeding connected to the circular machine of average diameter. It was noticed that, in comparison to the traditional production, seamless technology presents a reduction in manufacturing time, labor cost, energy and raw material consumption, and less use of the physical spaces. Regarding the final product, the articles produced with the seamless technology present advantages such as ability to meet different biotypes, bigger sustenance, compression, comfort and facility of maintenance. The seamless knitwear production with an equipment that controls the feeding system has all the advantages of the seamless technology manufacturing, and can be seeing as a mini factory, a production model compatible with the Industry 4.0.

1 INTRODUCTION

The term Industry 4.0 is considered the Fourth Industrial Revolution, and was firstly used in Hannover Messe, in 2011. In the North America, similar ideas are arising with the name of Industrial Internet (Drath & Horch, 2014). At the moment, in the United States, the concept gained the terminology Advanced Manufacturing (cf. nsf, 2015c). In spite of the growing use in enterprises, consultancies, centers of inquiry and universities, the term Industry 4.0 still doesn't have a common definition. Hermann et al (2015) define six elements associated to this term:

1) Interoperability;
2) Virtualization;
3) Decentralization;
4) Real-time capacity;
5) Support services;
6) Modularity.

As stated by Azmeh and Nadvi (2014), the model of government of the Global Value Chains is not static. The apparel industry is highly competitive and, with the development of the fast-fashion, was pushed by the increase of the quality, diversity of articles, fashion contents, costs and prices reduction. The fast fashion, so, designates the possibility of offering products that incorporate some element of style with short life cycles (Ciarniene & Vienazindiene, 2014). The fast fashion can be understood also as a strategy that makes products with high frequency of collections, which tries to attend the demand of consumption in contrast with the relatively low prices.

The objective of this paper is to present a preliminary study of the technologies adopted on the manufacturing of Sports Bras in Brazil, aiming at presenting different perspectives of how the new technologies are being incorporated by the Textile and Apparel Industry in order to achieve the Industry 4.0 production model.

2 LITERATURE REVIEW

2.1 *Textile and apparel chain in Brazil*

The origin of the textile production is closely related to the origin of humanity. Clothing, as well as food and shelter, is considered a basic human need. In addition to this, it is a form of self-expression and sense of belonging that becomes even more important for individuals and social groups (Ha-Brookshire & Labat, 2015).

The use of natural fibers may be one of the reasons why man has organized small communities and combined different tools for the handmade production. Findings from 30000 years ago include spun fibers and natural dyes, baskets for carrying food, water, small amounts of materials and even infants and children (Lundborg, 2014).

Since the pre-industrial period – where handmade textile products were produced by artisans that had full control of the production system and ownership of their tools – until nowadays – with the mass production and decentralized system – the Textile and Apparel Chain played and still plays an important role in the history of industrialization in the world.

In Brazil, this activity originated in the colonial period, developed in the twentieth century and reached maturity in the 1940s (Kon & Coan, 2009). In its recent history, the Brazilian textile industry has experienced several economic crises due to external competition and the decreasing production volume.

The textile and apparel industry presents a wide range of operations that covers from the raw material extraction until the post-use, with a diversity of materials, industrial processes and the heterogeneity of the manufactured products (SANCHES, 2011).

The lack of investments in the 1980s, due to the economic stagnation, directly affected the Brazilian industrial park, and especially the textile industry. The industrial park of this sector became obsolete that created some serious difficulties for the implementation of new technologies and processes, leading in the 1990s to the closure of many manufacturing units, mainly in the artificial and synthetic fabrics' production (Kon & Coan, 2009).

During the 1980s and 1990s, the Total Quality Management (TQM) was the focus of the industrial production. Then, new methods were developed based on the principles of human resources management.

Brazil is the fifth largest textile producer in the world, the fourth in apparel and the fifth in cotton production. However, this country participates with less than 0.4% of the market share. About 50% of the world's textile production is concentrated in Asia, especially in China. Asian countries lead all the statistics of the sector: export, production, employment, cotton production, investments and number of companies (ABIT, 2014).

The Brazilian's Textile and Apparel Chain employs 1.7 million people; the Fashion industry is the second largest employer in the manufacturing industrial park (ABIT, 2014). Due to the successive economic crises, this chain has been losing its competitiveness. In 2000, the importation accounted 2% of all textile inputs consumed, while in the year 2014 the value reached 17.5% (Alves & Conceição, 2015).

There are many textile and apparel manufacturers' poles in Brazil. The South Region, which includes the states of Paraná, Santa Catarina and Rio Grande do Sul, concentrates the production of knitting and bed/table/bath articles; in terms of technology, this pole is one of the most advanced poles in the country. The local consumer market and economic incentives have made the South region the second largest textile center in Brazil. Also in this region, there is a "Silk Valley", in which a great number of companies are producing silk cocoons. The Northeast Region pole produces denim, cotton fabrics and polyester yarns, mainly in the States of Ceará and Pernambuco (Bezerra, 2014, Obiettivobrasil, 2016). The Southeast Region is the largest textile production center in Brazil with a large variety of textile articles. The State of Rio de Janeiro has its textile pole in the mountainous region that produces underwear, beachwear and sports outfits. Minas Gerais presents a large knitting volume production and industrial districts of spinning and cotton articles. The State of São Paulo concentrates the largest apparel manufacturing industries and retail trade centers in the country.

2.1.1 *Apparel manufacturing in Brazil*

The apparel manufacturing is responsible for the manufacturing of the yarns, fabrics, knits and improvement textile.

The fibers and filaments sector produces the raw material for the textile industry. The raw material can be classified according to its origin. The natural fibers can be from animal, vegetable or mineral; the chemical fibers can be manufactured from natural or synthetic polymers.

The confection turns the fabric and knit into finished pieces for the final consumer. The stockings and *seamless* segments turn the yarns into final product.

In number of formal manufacturing unities, the textile chain and production industry is composed by approximately 32 thousand enterprises, being that the end of the chain is composed by nearly 29,3 thousand productions (ABIT, 2014).

As for the size of the enterprises, in the beginning of the productive chain the manufacturers are little and intensive in capital, the end of the chain is composed by a great heterogeneity of small and medium enterprises (DEPEC, 2017). The confection segment is composed by 21 different sectors, including articles of bed, table and bath, intimate apparel, outfits of any type and accessories (ABRAVEST, 2015). According to the IEMI (2016), in 2015, the confection sector was responsible for 69,7% of the turnover of the textile chain.

2.2 *Industry 4.0*

The Fourth Industrial Revolution, namely Industry 4.0, is an integrative cyber-physical system based on modern control systems, embedded software systems and Internet addresses (Anderl, 2015).

According to the German Academy of Science and Engineering, the Industry 4.0 corresponds to the next industrial revolution. The Industry 4.0 is not only a technical challenge but it will also change the Industries' organizational structure.

Unlike the three previous Industrial Revolutions, the Fourth Industrial Revolution is assessed *a priori*, and not *ex-post*; for the first time in history, this industrial revolution is characterized as a prediction of what is to happen, and not an assessment of what has passed (Hermann et al, 2015).

The main goal of Industry 4.0 is to improve the value chains among the product's lifecycle. In this context, the improvement of industrial competitiveness can be achieved by organizing and controlling value, new business models and networks (Anderl, 2015).

The result of the Fourth Industrial revolution will be the Smart Factory, where Cyber Physical Systems (CPS), Internet of Things (IoT) and Big Data are the key technologies for achieving the production goals.

The Cyber Physical Systems (CPS) are technical systems containing both virtual (cyber) and real (physical) systems. By "cyberizing the physical" and "physicalizing the cyber" (Lee, 2010) it is possible to specify physical subsystems with software-controlled behavior (Anderl, 2015).

Kagermann et al (2013) describe the vision of Industry 4.0 as a global network that incorporates new machinery, storage and production systems in the form of CPS, which include intelligent machines, storage systems and production facilities able to autonomously exchange information, triggering actions and controlling each other independently.

The Internet of Things (IoT) is an approach to equip real systems with embedded systems so that they become interconnected in the so-called "smart systems" (Anderl, 2015).

Through the IoT, physical objects and real processes have virtual representations that allow the interaction between them without limiting factors of the physical environment, such as geographic position and time, making the collaborative process faster and more effective. The IoT is when physical elements have virtual identities and personalities and operate in spaces through intelligent interfaces to connect and communicate (Sabo, 2015).

Data exchange is the foundation for the Industry 4.0; thus, the amount of data generated must be efficiently integrated during the product development process among the stakeholders. Big Data include information from a multitude of sources, and emphasize a change in data quality and not only quantity (Chandler, 2015). Big Data has seven dimensions:

- Volume: the amount of data generated and collected;
- Speed: the speed of data analysis;
- Variety: the diversity of the types of data collected;
- Viscosity: the data flow resistance;
- Variability: the flow rate and data types;
- Veracity: quantifies noise and reliability of the data set,
- Volatility: the validity of the data and for how long they should be stored (Desouza and Smith, 2004).

Besides the Smart Factory, there will be changes in the product's configuration and consumption. The Smart Product contains information from the production process and communicates with all production chain, which facilitates the failure control e monitories the real time production conditions. The Smart Manufacturing is related to smart machines that are able to self-optimize their own configuration in order to anticipate errors.

Regarding the consumption, the consumer will play an active role; in this new scenario, the purchase decision will be based on the initiatives of companies. Concepts such as sustainability, local production and direct communication channels must be integrated in the product development process.

2.3 *Industry 4.0 in the textile and apparel production in Brazil*

The use of digital technologies in the Brazilian industry is poorly spread. According to inquiry carried out by the Brazilian National Confederation of the Industry (*Confederação Nacional da Indústria*), 58% of the respondents know the importance of these technologies for the industry competitiveness, but less than the half uses them. The focus of the Brazilian enterprises is to improve the production processes aiming at the increasing of the productivity (CNI, 2016).

In another inquiry, 48% of the respondents adopts, at least, one of these digital technologies: digital automation without sensors, quick prototyping or 3D printing, cloud storage associated and the incorporation of digital services in the product (Journal O Povo, 2016).

In a complimentary inquiry, Schiewig (2016) pointed that 47% of the respondents uses digital tools in the product development process and 33% in new products and new business models (SCHIEWIG, 2016).

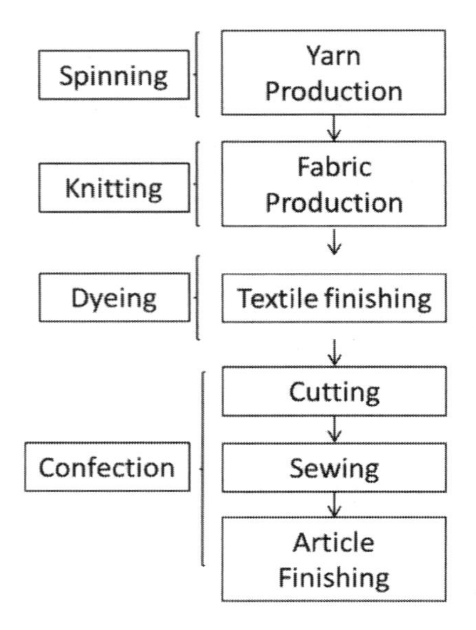

Figure 1. Main stages for the knitwear manufacturing in traditional production (Authors).

3 RESEARCH METHODOLOGY

The methodological procedure adopted in the present study has predominantly exploratory character and it is based on a comparative study between the three productive systems used in Brazil for the apparel manufacture.

The delimitation of this research is the Sports Bras manufacturing. Polyamide yarns and elastane, which are produced by chemical industries from the manufacture of the synthetic fibers, are the raw material.

Three production flows for the manufacture of the Sports Bras were analyzed: traditional method (Article 1), hosiery *seamless* machine for the manufacture of the article without lateral sewing (Article 2) and hosiery *seamless* machine with preparation of the raw material for the manufacture of the article without sewing (Article 3).

Article 1 was produced in a traditional confection, following the productive process: the knit fabric was produced in large-diameter circular hosiery knitting machine, benefited and finished in a confection. Figure 1 summarizes this flow.

The terminology *seamless* designates the concept of clothes without sewing in which the lateral sewing is removed or drastically reduced. The final product presented the following production process: the knits without sewing were produced in circular mill of middle diameter, assembled and finished in the same enterprise. Figure 2 presents the production's flow of Article 2.

Article 3 was produced was produced in a circular mill of middle diameter and finished in the same enterprise. Figure 3 illustrates all industrial activities.

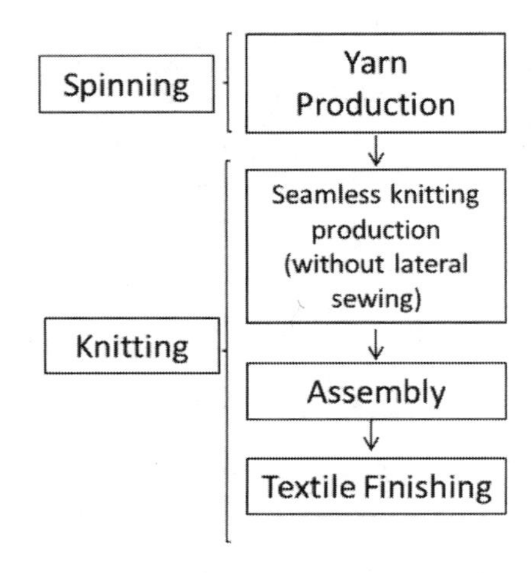

Figure 2. Main stages for the articles manufacturing in a seamless knitwear (without lateral sewing) (Authors).

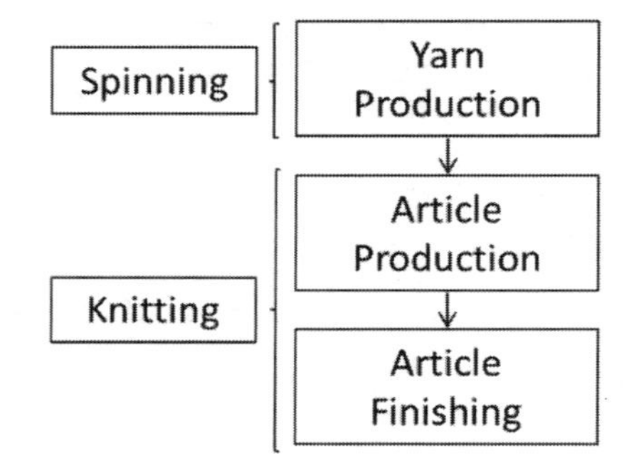

Figure 3. Main stages for the articles manufacture in a seamless knitwear (without sewing) (Authors).

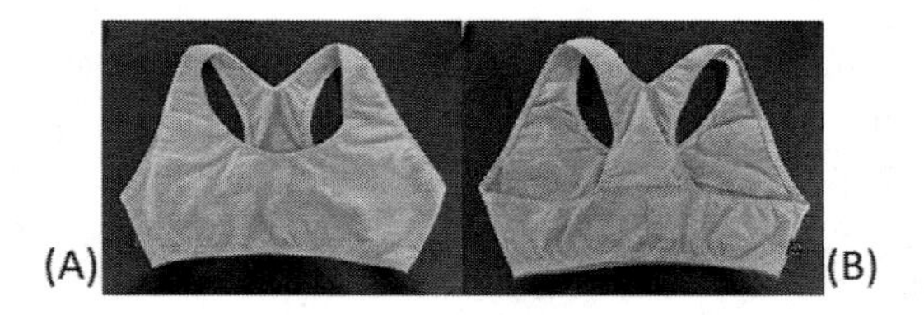

Figure 4. Article 1 – Sport Bra manufactured in a traditional confection – front (A) and back (B) views (Authors).

4 PRELIMINARY RESULTS

Article 1 was produced in large-diameter circular machine and finished in dry-cleaner. The final product was developed in a traditional production, following the next stages: creation, modeling and cut; assembling or sewing and finishing. Figure 4 presents an example of final product manufactured with this technology.

In large-sized and in some middle-sized garment confections the development, assembly and part of the finish are supported by automatic and semi-automatic

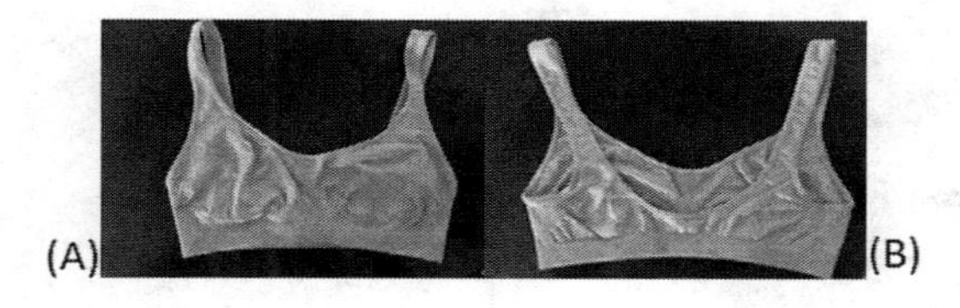

Figure 5. Article 2 – Sport Bra manufactured in a *seamless* knitwear – front (A) and back (B) views (Authors).

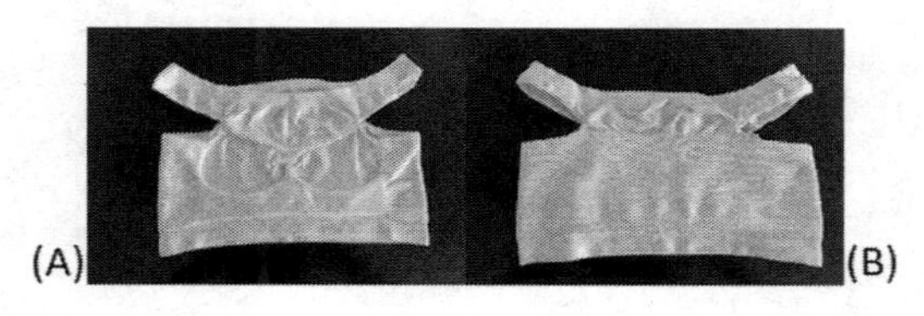

Figure 6. Article 3 – Sport Bra manufactured in *seamless* knitwear with feed control – front (A) and back (B) views (Authors).

machines with software's support. However, in small-sized and some middle-size companies, these activities are made manually.

Article 2 was produced in a *seamless* knitwear with the next manufacture flow: texture of the pieces without lateral sewing, assembly, dyeing and fixation (figure 5).

Article 3 was produced in knitwear, with the next manufacture flow: texture of the pieces without sewing, dyeing and fixation. An equipment that controls the feeding speed and tension was connected to the knitting machine. Figure 6 illustrates the production flow.

The apparel without lateral sewing has some advantages on the traditional, such as:

- Reduction of the quantity of sewing in the clothes, since the clothes are manufactured in the form of a tube;
- A great number of varieties in contextures, modeling, colors, yarns, and drawings, besides the possibility to produce anatomical pieces;
- Less raw material and time waste;
- Reduced space, number of processes and energy cost.

In the *seamless* knitwear, the manufacturing process is discontinued. The texture of the tube is provided by the automatic machines, the assembly of the article is done in manual machine and the improvement of the final article in automatic machine.

The manufacturing of an article without sewing in mill *seamless* with equipment that controls the feeding has the advantage of being fully automatic. This production system is closely related to the guidelines from Industry 4.0. Article 3 can be an example of the attempt to transform the industrial park into a fully automatic system; in order to achieve the Industry 4.0 model, it is necessary to join the production with the consumer online demand, as a mini factory.

Additional challenges rely on digitalization, modularity and customization of the production and reduction of the time to market.

5 CONCLUSIONS

The Brazilian enterprises focus on the improvement of production aiming to increase the productivity, but neglect the opportunities to implement new product development methods new business models. The advanced manufacturing still is little spread in the Brazilian industrial sector. The familiar management makes it difficult to adopt modern techniques of administration, manufacture and productivity control.

Besides, even with the enhancement of the manufacturing park, it is possible to notice that the apparel manufacturing presents craft characteristics, in which the product quality is associated to the skill of a worker.

In the knitting production, the clothes without lateral sewing are still considered a relatively new technology, especially in Brazil. As for the productive process, in comparison to the traditional production, *seamless* technology presents a reduction in manufacturing time, labor cost, energy and raw material consumption, and less use of the physical spaces, among other advantages.

Regarding the final product, the articles produced with the *seamless* technology present advantages such as ability to meet different biotypes, bigger sustenance, compression, comfort and facility of maintenance.

The *seamless* knitwear production with an equipment that controls the feeding system has all the advantages of the *seamless* technology manufacturing, and can be seeing as a mini factory, since the final product is produced without the necessity of subsequent sewing.

ACKNOWLEDGEMENT

The authors thank CIAUD, SENAI, UNICAMP and USP for the financial support.

REFERENCES

ABIT – Associação Brasileira da Indústria Têxtil e de Confecção. Agenda de Prioridades Têxtil e Confecção – 2015/2018. São Paulo, 19p., 2014.

Alves, Cristina; Conceição, Claudio. The need to modernize Brazilian industry. The Brazilian Economy, p.8–16, nov., 2015.

Anderl, R., 2015. Industrie 4.0 – technological approaches, use cases, and implementation. at-Automatisierungstech., v. 63, n. 10, p. 753–765.

Azmeh, S. and Nadvi, K. Asian firms and the restructuring of global value chains. LSE Research Online Documents on Economics from London School of Economics and Political Science, LSE Library. Disponível em: http://econpapers.repec.org/paper/ehllserod/56666.htm, 2014.

Bezerra, Francisco Diniz. Análise retrospectiva e prospectiva do setor têxtil no Brasil e no Nordeste. Informe Técnico de Estudos Econômicos do Nordeste, Fortaleza, 2014.

Chandler, D. A world without causation: big data and the coming of age of posthumanism. Millenn-J. of Int. Stud., v. 43, n. 3, p. 833–851, 2015.

Chi, T. Business Contingency, Strategy Formation, and Firm Performance: An Empirical Study of Chinese Apparel SMEs. Administrative Sciences, n.5, p. 27–45, 2015.

Ciarniene, R. and Vienazindiene, M. (2014) Management of Contemporary Fashion Industry: Characteristics and Challenges. Procedia—Social and Behavioral Sciences, 156, 63–68, 2014.

CNI – Confederação Nacional das Indústrias. Indústria 4.0: novo desafio para a indústria brasileira. Indicadores CNI, Ano 17, Número 2, Abril 2016.

DEPEC – Departamento de pesquisas e estudos econômicos. Têxtil e confecções (março/2017). Disponível em: https://www.economiaemdia.com.br/EconomiaEmDia/pdf/infset_textil_e_confeccoes.pdf. Acesso em 20/04/2017.

Desouza, Kevin C.; Smith, Kendra L. Big Data for Social Innovation. Stanford Social Innovation Review, p. 39–43, 2014.

Drath, R.; Horch, A. Industrie 4.0: Hit or Hype? IEEE Industrial Electronics Magazine, n.8, p.56–58, 2014.

Ha-Brookshire, Jung; LaBat, Karen. 2015. Envisioning textile and apparel research and education for the 21st century. International Textile and Apparel Association – ITAA Monography #11, 38p.

Hermann, M., Pentek, T., Otto, B., 2015. Design principles for Industrie 4.0 Scenarios: a literature review. Working Paper n.01/2015, Technische Universität Dortmund, 15p.

IEMI – Instituto de Estudos Mrketing Industrial. Relatório setorial da indústria têxtil brasileira. IEMI, São Paulo, 2016.

Jornal O Povo. Estudo aponta cenário da indústria 4.0 no Brasil. Disnonível em: http://www.opovo.com.br/noticias/economia/2016/05/estudo-aponta-cenario-da-industria-4-0-no-brasil.html. Acesso em: 15/03/2017.

Kagermann, H., Wahlster, W., Helbig, J., 2013. Recommendations for implementing the strategic initiative Industrie 4.0: final report of the Industrie 4.0 Working Group. 82p

Kon, Anita; Coan, Durval Calegari. Transformações da indústria têxtil brasileira: a transição para a modernização. Revista de Economia Mackenzie, v. 3, n. 3, 2009.

Lee, E.A., 2010. CPS foundations. In: Proceedings of the 47th Design Automation Conference. ACM, p. 737–742.

Lundborg, Göran. The Hand and the Brain: From Lucy's Thumb to the Thought-Controlled Robotic Hand. ISBN 978.1-4471-5333-7, Springer Science & Business Media, London, 2014.

Manyika et al. Manufacturing the future: the next era of global growth and innovation. McKinsey&Company, 2012. Disponível em:

Obiettivobrasil. O setor têxtil no Brasil. Disponível em: http://www.obiettivobrasil.com.br/pt-BR/news/pesquisas/O%20setor%20t%C3%AAxtil%20no%20Brasil. Acesso em 16/01/2017.

Sabo, Filip. 2015. Industry 4.0 – a comparison of the status in Europe and the USA. Austrian Maschall Plan Foundation, 33p.

Sanches, R.A.; Estudo comparativo das características das malhas produzidas com fibras sustentáveis para fabricação de vestuário. São Paulo, Escola de Artes, Ciências e Humanidades, Universidade São Paulo, 2011. Tese de Livre-docência.

Schiewig, I.; Indústria 4.0: desafios e oportunidades. Disponível em: http://cio.com.br/tecnologia/2016/03/21/industria-4-0-desafios-e-oportunidades/. Acesso em 30/04/2017.

2. Identity

2.1. Textile art

Textiles, Identity and Innovation: Design the Future – Montagna & Carvalho (Eds)
© 2019 Taylor & Francis Group, London, ISBN 978-1-138-29611-4

Our past and future relationship with textiles

C. Gale
School of Fashion and Textiles, Birmingham City University, UK

ABSTRACT: Our existing and past personal relationship with textiles is considered and parallels are drawn between the experience of textiles and the making of textiles. The future is then considered in the context of the arrival of new technologies that combine robotics and artificial intelligence. The role of computing, androids and robotic hands is discussed and the potential impact they may have on the origination and making of textiles and textile goods. Also raised is the question of the potential challenge to the status of humans as the sole originators and source of material culture.

1 OUR RELATIONSHIP WITH TEXTILES

On bright summer evenings, when I was a little boy, perhaps about 6, I would lie in my bed looking at the light coming through my bedroom curtains and trace, with my eyes or finger, the shape of a cowboy's lasso. The curtains featured Roy Rogers and a rearing horse called Trigger. The imagery and the experience has stayed with me all my life, the curtains seemed a coarse open weave, I remember an effect like that of hessian, the hues were the pale stone colour of some desert, say New Mexico or Arizona, punctuated with green cacti. At the time America, cowboys and deserts seemed exotic and distant from life in the UK and American culture had yet to truly infuse into the Western world. I looked, as a child, at the exotic and distant, the fabled and mythological and entered a dream world. Preparing this text made me think that perhaps I could find the curtain design on the internet 54 years later. The experience was like looking through some astronomical telescope trying to find a distant galaxy. I think I found my curtain design, a tiny thumbnail (Fig. 1), a remnant from a deleted webpage. There were things I had forgotten, the lasso spelling his name, the cowboys at a campfire and perhaps as I had grown older there were other things and associations I had layered onto my memory – Navajo Native Americans, Pueblos and Zapotec rugs. A life of watching movies and a period of teaching and studying textiles had merged with a powerful emotional memory of comfort, fantasy and the exotic. For me, like many, experiencing textiles has been like being a tourist and, at their most powerful, textiles leave a complex memory that glows in recall but also accretes other associations, meanings and feelings.

Figure 1. Roy Rogers and Trigger fabric.

Once in the early 2000s when I was writing a book about textiles I felt I needed to find an example of that meditative dreamspace but not so much from the point of view of audience but instead from the point of view of the practitioners, the creatives. I had been teaching craftspeople and textile artists and textile designers for about a decade and so was kind of familiar with the intensity of working and, often, the quiet of it. I came across something written by the French novelist Colette about her daughter: "If she draws, or colours pictures, a semi-articulate song issues from her, unceasing as the hum of bees around privet. It is the same as the buzzing of flies as they work, the slow waltz of the house painter, the refrain of the spinner at the wheel. But Bel-Gazou is silent when she sews, silent for hours on end, with her mouth firmly closed, concealing her large, new-cut incisors that bite into the moist heart of a fruit like saw-edged blades. She is silent, and she – why not write down that word that frightens me – she is thinking." (Colette, 1974)

This description, although obviously sentimental and emotional, summed up for me the calling of textiles, how and why people were drawn to it. In the 1990s I had been interested in theories about artistic knowing, intentionality and action such as the various philosophical nuances of *jouissance* and Heidegger's *'ready to hand',* explanations of why or how we just *do.* My art background had perhaps made me opposed to linguistic and symbolic explanations of the artist and their processes. Instead I preferred the instinctive world of actions and how those instincts can relate to the material world.

Some years later my friend Jasbir told me about 'flow' and the work of Mihaly Csikszentmihalyi, his theory says that people are happy when they are in a state of *flow*, a state where they are so absorbed in their activity they forget everything else and become exceedingly satisfied and fulfilled (Csikszentmihalyi, 1990). The explanation seemed to fit many practitioners, particularly those involved in textile art and craft. They tended to become highly absorbed in play and exploration and a kind of addictive selflessness and sometimes even, a kind of disembodiment. These theories seemed to me the explanation, or at least a plausible description, of the craftsperson and the textile or fiber artist, perhaps too their audience but perhaps, not always the textile designer. In my world in UK Higher Education in the mid 1990's and early 2000's the artists and craftspeople seemed to start to disappear, replaced by aspirational designer-makers and entrepreneurs. Fine art seemed to appropriate the discourses of textile art, students on fine art courses using textiles found themselves challenged to be 'more fine art' while textile students had to choose between complex philosophical explanations or simpler commercial ones. Perhaps also it was diminishing studio space or the increasing cost of being a student or the marketing of fine art culture or, just one of those trends, but the textile artists and craftspeople seemed gradually to disappear and became a minority. It has always kept going as a practice but some 20 years later I have found it is doing well in the education world in China. Spurred by economic growth, spacious accommodation, building projects, a lack of art snobbery and a different cultural and philosophical tradition, the textile artists and the craftspeople once again make installations and meandering articulations of playing with awe, the dream, materials, colour and form (Fig. 2). It would seem that there is a kind of imminence in our relation with textiles, under particular sets of circumstances, our engagement with textiles will follow a particular path, whether its dreamily staring at a set of curtains or teasing out a complex manifestation of thread and fabric or suddenly deciding we must make scarves to sell.

Figure 2. *The Season of Fluorescence*. Chen Yanlin.

The psychological and emotional pull of textiles, why we are drawn to it, has then some ontological premise to it, as does they way we interact with it. The social political and human aspects of that engagement are evidenced in a 50,000 year history or longer. For most of us the first thing we feel after the human skin is fabric and for most of us we will die in or on cloth. This epic history and highly personal relationship tends to make us think we know what textiles are, that we understand fashion, trend and style, we understand making cloth, we understand how we might use it and what individually or collectively we can expect of it. As with the case of the Chinese textile artists or with Colette's daughter, we have a social anticipation about our behavior and our opportunity with textiles. Within our psychological and anthropological understandings we see the potential of self-expression and ritualization. Within our economic understanding we know what value textiles have and how they have it. However since the turn of the millennium, I am less sure that the next 100 years will be the same or similar to the last 50,000. The reasons are many and complicated but their potential impact is of science fiction proportions. The forthcoming changes that will affect all our societies may also alter our cultural appreciation of textiles and the human relationship with textiles. The changes themselves are to do with the means of production and the forms of consumption, which will affect us as creative authors, producers and as consumers. In this paper I will just discuss one novel future means of conceiving and making textiles and textile goods.

2 THE FUTURE OF PRODUCTION

When we consider the means of production of textiles there are various historical landmarks – powered machinery, dyeing techniques, the development of synthetic yarns, coating. As well as yarn technologies we also have had fabric technologies such as pleating machinery, automated pattern cutting and so on. None of these scientific and technological innovations has

Figure 3. An exhibition in 1979, *Drawings*, at SFMOMA showed a robot creating drawings. Collection of the Computer History Museum, 102627449.

Figure 4. *Sophia* a robot made by Hanson Robotics appears with Jimmy Fallon on *The Tonight Show*. Credit Andrew Lipovsky/NBC.

fundamentally interrupted the way we creatively play with textiles or confronted our social satisfaction with our own forms of creative expression. Also we tend to have in our minds an idea of what scientific and technological change is and an idea of what we consider is its pace and how societally we absorb those changes. Within these models of change we assume some things are too tricky for a technological solution or things like our aesthetic satisfaction can only be stimulated by the expression of some culturally attuned peer. As an example we still have sweatshops with seamsters and sewing machines, the fashion industry has made no significant advance on a model of production (with some exceptions) since the early 19th century. We also saw, particularly in the late 20th century attempts to get computers to be artistic in a human way, whether through the programming of permutations or this in combination with some stochastic and 'interfering' environment. Some of these were sustained attempts at the pursuit of automated creativity such as the work of Harold Cohen and his AARON system (Fig. 3). However public reception of the outcomes generally did not match the accompanying AI hyperbole. As a result as much as we are aware of technological progress we are also aware of technological failures or, at the least, we do not expect there might be radical developments in the offing, that a technological development might throw us off balance. Let alone the possibility of a technological and industrial revolution that might overturn the human role in that 50,000 year textile tradition and give it a run for its money.

We could be forgiven for thinking that technological progress arises within an industry but more often than not it seems that it arises from outside an industry, that industries are usurped by new arrivers, thus it was with the electronics industry, the companies that made valves more often than not didn't make it to manufacturing microchips or the stages in between. The worlds of textiles and fashion equally do not exist in industrial isolation and they, like so many others, are likely to be affected by the coming of a new industry, the robots or perhaps, more correctly, androids and these in turn will find synergy with the steady creep of artificial intelligence and big data.

So what's so special about an android and why should it make such a big difference? An android is a humanoid robot designed to imitate a human, there has always been something of a question in robotics as to why you would want to make an android but the fact is it matches with mythological and science fiction aspirations and these tend to be a big driver in technology innovation, for example what was the logic in wanting to fly? Also the further society pursues the android the more it sees applications and income streams. I can remember a time when media sociologists didn't think the games industry was of much importance but the revenue base proved to be enormous and the image processing demands of computer games became a key driver of the development of computer systems. The fact is, as crude as it sounds, we want to see those androids, we keep making movies about them and how they are going to be and what they can do for us. There was a time when the android was simply nowhere near being a reality but the convergence of technologies, the bits and pieces of androids, sometimes made for different purposes recently start to make them a possibility and more likely a probability (Fig. 4).

It is perhaps the fact that the potential uses of androids are so broad that they seem likely to be made and also consequently that they will and must find a role in relation to textiles and fashion. This last point is about an industrial revolution in the sector potentially more transformative than any we have ever seen.

"The development and introduction of a robot seamstress looks more viable than it was a decade or so ago. Algorithms relating to fabric performance are utilized for everything from CAD visualisations to film animation (Choi and Ko, 2005), sensor technology and the processing of sensory feedback are highly advanced (Sonar and Paik, 2016), the development of robotic hands and arms moves apace (Li, Qiao et al, 2016). So too the ability to process highly complex data very quickly (Chen, Hu et al, 2016). The cognitive processes of proposition, rhetoric and speculation seem within the grasp of new models of AI (Gao and Zhou, 2016).

All the parts and faculties of a robot seamstress seem present or coming very soon." (Gale, 2016)

If you take an android (or even aspects of it) as factor X and insert into the world of textiles you start to play a game of what can't it do? It could search the internet for textile imagery, techniques and processes and remember more than any human, it may not be able to create as we might understand it but it could replicate and derive permutations in a way which is unimaginable to us. We are already used to the idea of embroidery equipment and sewing machines such as the Janome 7318 with its 18 'built-in decorative stitches', the convergence of computing and textiles equipment has already taken place, but the logical extension of such technologies would be the ability to connect to the internet and begin a process of searching for and interpreting available stitch forms. The further extension of technology into the arena of domestic robotics and androids could be the basis of a revival or a democratizing of couture techniques and products. Androids could be home tailors and spark a huge wave of custom and bespoke production. Androids could equally sustain or appropriate cultural heritage and textile history. In associated industries in fashion and interiors the location and nature of the supply chain could become radically altered and long-standing ideas about flexible manufacturing become a reality. Suddenly within this context what is the role of the human being, not just the factory worker but also the textile artist, craftsperson or designer? There is a suggestion that we will in some way be collaborators with robots but there is implicitly an issue of competition around the areas of pre-existing skill and knowledge. There are some analogies to be drawn with the commercially successful practice of copying in manufacturing or where the model of consumerism is inverted and the consumers become the authors as with dub music. Whichever these can be seen as a radical alteration of the origination of cultural content, in the same way social media subverts the hierarchy of the press, so a new technological model of making might subvert the hierarchy of cultural agency in textiles.

There is an awkward question for any practitioner, which is if their audience is removed from them or is no longer interested in their style of work, who are they making for? The world of robotics will challenge the identities of makers and the meaning and purpose of their lives in a way no technology has before. At the same time it may augur a more complex textile experience for all. In my life I have both been able to appreciate the textiles of others and to enjoy my own opportunities for creative expression. I have also realized that personal creativity can shift between creating, editing, managing and dissemination and that these activities are satisfying in different ways. As much as an android might challenge the model of the creative maker that we have come to valorize it may also prove to be the seed of a creative culture of commissioning, I might be happy to commission and see again any of the thousands of textile artefacts I have been privileged to see. Commissioning will require imagination

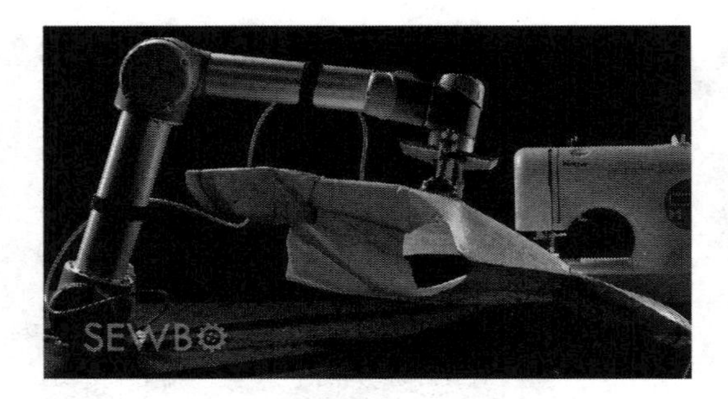

Figure 5. A fabric handling robot by Sewbo Inc.

and interpretation, the ability to identify projects that could range from the architectural to the intimate. It is here perhaps that our ability to visualize or articulate the mythological, utopias and dystopias will leave us (at least for a while) some edge over the machine. It is still a little hard to see when more advanced robotics technology becomes tasked to work in say the garment manufacturing sector or in embroidery production but a survey of the internet is enough to clarify that is a matter of when not if (Fig. 5). If we add to this the potential autonomy and mobility of the android then we augur a new concept of the relationship we have with equipment and the means and locus of textile production.

The history of patents is enough to tell us that technology doesn't always evolve in the way we might imagine. In the present day anyone who has been or is a creative person would consider as ridiculous a proposal that humans might cease to be the primary source of cultural development. It is the old 'robot as artist' game and all evidence was that computing technology lacked the grammar of cultural production. Somehow also it seems morally offensive, an attack on the preserve of what is human, what I am, what we are. However patterns and signs of the erosion of primary authorship are already visible in the creative sectors. As an example these days many of the students where I work think they are being creative by trawling the internet for pictures and putting them on Pinterest or creating a mood board or blog with them. The very presence of the internet has cultivated an editorial style of practice that was not present in the practice of textile design twenty years ago. This is not necessarily a better or worse kind of creativity, it is just different. Recently, if you visit trade shows like Première Vision, you find a big trend is vintage but, unlike a trend of period retro as you might have found in the past, you can pick from pretty much the entire 20th century design cycle and more. My sense of this was that designing or selecting textiles was becoming a bit like Pinterest, a culture of editing, perhaps too a revisiting of the methodology of the readymade.

Daily human experience is not as it was when I was born in the 1950s, computer technology has changed the very nature of 'normal' and it will continue to do so (Kelly, 2016), it will undoubtedly change the way we work and live. Faced by a technological unknown,

Figure 6. *The Shadow Dexterous Hand* by the Shadow Robot Company.

some prophets talk of catastrophes, in essence the loss of the need and socioeconomic purpose for people. Of course we cannot know the future but we can at least acknowledge the likely arrival of a new catalyst within the field of textiles, very different to previous technological and scientific revolutions and one that might just undermine aspects of our cultural purpose. For most who work in textiles they know it is not simply a visual culture, it is a culture of touch. Sometimes I think simply looking at textiles stimulates some synaesthesia, I sense the feel of textiles simply by looking at it. Perhaps that was my childhood experience looking at those Roy Rogers curtains. The relationship between our bodies and material culture, between us and textiles supports every kind of description ranging from mysticism to cognitive science but whichever explanation we choose, it is simply there, it is an inviolable form of embodied cognition (Wilson & Foglia, 2011). Our relationship with textiles as consumer or audience seems therefore permanently secure however the 50,000 year role of the human hand and mind in the creation of textiles and textile goods looks like it will soon be challenged by our pursuit of biomimesis and the productive, encyclopedic, android hand (Fig. 6).

The *Dextrous Hand* comes with a hefty price tag but it is of course just part of the beginning of what promises to be the evolving technology of replicating the human form. That final envelope, the android, will become the gift to machine intelligence. In the early days of AI the desire had been to pursue sentience in computers, an intelligence like our own. In the event, some half a century later, we have created a form of intelligence different to ours. Now we wish to recreate our bodies in machine form and will no doubt combine it with the intelligence we have made. The skeptic might well say '... what will that be to do with creativity and textiles?' the answer of course is '... having invented such a remarkable thing why would we not expect it to be involved with something that has been so fundamental to our culture and history and has such personal value for all of us?'. The biggest question we will have to ask is what will our relationship be with the stranger in the room and how will our creative path change in response to their presence?

REFERENCES

Chen, D., Hu, Y., Cai, C. and Zeng, K. (2016), 'Brain big data processing with massively parallel computing technology: challenges and opportunities'. *Software: Practice and Experience*, Oxford: Wiley Online Library.

Choi, K and Ko, H. 2005. 'Research problems in clothing simulation'. *Computer Aided Design*, 37(6), Amsterdam: Elsevier.

Colette 1974. 'Bel Gazou', in *Colette: Earthly Paradise*, R.Phelps (ed.), Harmondsworth:Penguin, 225–41.

Csikszentmihalyi, M. 1990. *Flow: The Psychology of Optimal Experience*, Harper and Row.

Gale, C. 2016. 'The Robot Seamstress' in *Advanced Research in Textile Engineering 1(1)*, New York: Austin Publishing.

Gao, J. and Zhou, C. 2016. 'A system for modeling robotic intellectual belief'. *Mechatronics and Automation (ICMA)*, New York: IEEE.

Kelly, K. 2016. *The Inevitable*, New York: Viking.

Li, X., Qiao, H., Ma, C., Li, R. and Zeng, K. 2016. 'A dynamical compliant grasping strategy for dexterous robotic hands with cushioning mechanism'. *Proceedings Intelligent Control and Automation (WCICA) 12th World Congress*, New York: IEEE.

Sonar, H. and Paik, J. 2016. 'Wearable PZT sensors for distributed soft contact sensing'. *Design and Signal Conditioning Manual*, rrl.epfl.ch.

Wilson, R. & Foglia, L. 2011. Edward N. Zalta, (ed.) "Embodied Cognition". *The Stanford Encyclopedia of Philosophy (Fall 2011 Edition)*.

Textiles, Identity and Innovation: Design the Future – Montagna & Carvalho (Eds)
 ISBN 978-1-138-29611-4

Focus on textiles – performance studies

Ângela Orbay & Anabela Mendes
FLUL, Faculty of Letters, University of Lisbon, Portugal

ABSTRACT: The first module of the PhD seminar on Topics in Theatre Studies designated as Focus on Textiles at the Faculty of Letters and which spanned over several weeks, unveiled and inquired about diverse performing arts works. Our work focused on the capacity to create a discourse and desire for different aspects of the textile element given its performance. This allowed us to appreciate and expatiate on what contained for us, simultaneously, something new and old.

1 INTRODUCTION

Textile is a narrator, a communicator fluent in several languages. "Narrative discourse indeed relies so heavily on textile metaphors that we could conclude that narrative is actually a textile concept" (Goett 2016).

By looking at the scene of various performance projects, we can easily identify the communicative power of fabrics in the (re)interpretation of many issues and in the involvement of many causes.

Its performance is intimately linked to the human condition, portraying it or often times being itself that human condition. The diverse physical and material functions, dematerialize themselves with the ingenious and elusive processes of creating the message, often propelled by the power of its interpreters. Choreographers, dancers, actors, musicians and playwrights manipulate textile materials, integrating them selectively into their visual or sound inventions in order to suggest cognitive or sensory dimensions or even in the development of other areas. The sequence of the discourse presented in some performing arts works have made them a subject of interest and allows us to think about textiles and thus place a *Focus on Textiles*.

2 CHOREOLOGICAL TEXTILES/TEXTILES AS A MEANS IN CHOREOGRAPHIC CREATION AND COMPOSITION

Choreology according to Laban is the "theory of the laws of dance events manifested in a synthesis of spatial temporal experience [which] deals with logic and the order of dance balance... It is the grammar and syntax of choreography, inseparable of emotion and dramaturgy" (Laban 1966).

Figure 1. *Case-study Illustrations.*

Choreological textiles are an honorary title that my eyes confirmed. an attentive gaze of a textile designer with a great love for dance that understands this *form* as a prolific agent of vocabulary/syntax/dramaturgy in dance.

Let us distance ourselves from the figurine, the composed figure, in order to focus on the textile as a pure state. Let us deliberate on the textile as an aesthetic entity, an actor/narrator, a material of opportunities where we can find a significant value to submit to the attitudes of intuition. The fabric, which is the most common term, has been included by several choreographers in their creations, taking *advantage of their physical properties* for the production and creation

Figure 2. *Lamentation photo:* Barbara Morgan, 1935.

Figure 3. Héla Fanttoumi Photo: Laurent Philippe, 2011.

of movement. It is not only mechanical/physical movement, but also movement capable of expressing the body-mind/body-emotion/body-reason/body-non-body into choreographic quality.

This practice is developed through the challenge between two forces, the dancer's body which is the driving force that generates action and textile body which is the actor that reacts. The choreography is the result of this experience between the two systems that proliferates in sequences of movement of a strong dramatic expression.

The proposal for this reflection is to look at this type of narrative and open our perception and conscience to this materiality. It is one that reinvents the body and dares to invade the scenic space, fulfilling its function of giving meaning to the choreographer's intention in its different and multiple interpretations.

2.1 *Material – Composition – Behaviour – Symbolism*

Martha Graham (1991) comments: "I use a long tube of material to indicate the tragedy that obsesses the body, the ability to stretch inside its own skin." In Lamentation (Fig. 2) 1930, the choreographer explores the characteristics and elastic properties of the tubular mesh in the deformation and dilation of the body boundaries, to symbolize the second skin, and the solitary desire to leave the body condition.

The elasticity and resilience of the tubular mesh are related to human elasticity and resilience. The dynamics of "Effort" combine body-mesh-skin in which the dancer distributes the body weight inside the spatial limits of the fabric which in turn supports the contortions, impulses of tension, amplification and changes in direction.

The dramatic intensity is amplified by the game of motricity, by the eloquent interaction with the textile form, resulting from an ingenuous physical unit that describes the conflict, the trajectory of pain, guiding us into that habitat which is so particular and specific to that of solitude.

"... About this play, in which the dancer wears a kind of sack Martha Graham said: The fabric is the skin. Now the space of the body is not more than that, a second skin..." (Gil 2002).

The repertoire of fabric in movement in Graham's work is generously documented in Barbara Morgan's photographic archive. The series of images that compose it gives us examples that confirm a grandiose involvement with an ample range of fabric and a great variety of properties applied to different discourses. We can add Isadora Duncan and Loie Fuller to Marta Graham with their veils and flying silks as compelling examples that have brought new vocabulary to dance and moulded the modern era.

2.2 *Dance questions the veil*

Manta is a play about the Ijab a women's clothing in the Islamic world composed of cotton, polyester and/or other fibres that completely cover the woman's body, revealing only the eye area (Fig. 3). The controversial point is the obligation of the Ijab, a veil that constitutes an identity barrier. Integral veil/Covers integrally/Integrity.

The mantle dehumanizes the body and distances it from its physical form to emphasize an emotional state.

From this underlying idea Héla Fattoumi builds an interpretive tool through the game of fabric on the body and in space. The actions of folding, spreading, shaking or abandoning the fabric intensify the psychological process and establish themselves in a coherent way with the issue at hand.

"I wanted to make this experience of being inside that fabric, that symbol of clothing, and seeing how physically, I could move, ... see what could be born on the stage from this experience" (Fattoumi 2016).

The gestures allow for the construction of their own story. It is about the female body that is suffocating, escaping and rejecting the frock. This triggers strong images with frenetic movements and leads to moments of trance. The interaction with the fabrics through movement should provoke a deep reading. Dance is not only movement, just as materials are not only extensions of the body that is moved. The fabric and the body are devices that Fanttoumi articulates to raise questions of a cultural, political and social nature.

Figure 4. First Born photo: Cynthia Quinn.

Figure 5. Killing time Orchestra photo: Daniel Lopera.

2.3 *Aerial – Ethereal Textiles*

Aerial textiles are a group of light fabrics with specific characteristics such as: lightness, transparency and fluidity from where several types of fabrics proliferate including: gauze, tulle, voile, muslin, organza, sash, among others.

The particularities of these fabrics have to flow, move or hover, the way they move in space or around a body are present during the production of the gesture, sense and meaning and applied in the creation of symbolism or the representation of elements.

The tulle reveals a dynamic quality that is distinct with particular period of gravity.

The fine mesh is slow it hovers between the body and space completing the silence. The bodily weights of the dancer/fabric instinctively form a fluid material with a skilful and mystical behaviour with a natural grace that dramatizes a dream, expressing the states of the soul.

The tulle used in First Born (Fig. 4) take perceptible forms, with a metaphorical force that plays the role of signifying a place of immateriality, which can also reveal an inner life.

The fabric becomes an element, a secret that covers bodies in a journey between heaven and earth. The delicate essence of the material transmutes to something pure and divine. The traces blown by the wind, by the gentleness of the movement of bodies, express an eternal time of great plastic beauty.

3 PROCESSES, LABOUR AND MACHINERY ON STAGE

The processes and instruments of manufacture occupy the space of acting and the laborious moment is the active space of the message. The intention is conveyed by the simple gestures of sewing, knitting, or embroidering. The show presents new instincts and new environments to the practices and traditional techniques that weave readings for the public that discover other functionalities attributed to a knitting machine, a weaving machine, a spinning wheel or needles that operate visual and sound effects into a new environment to reveal other realities.

3.1 *Acoustic sources*

We can identify some textile processes that transform fibres in thread into fabric or meshes. The traditional techniques and primary functions of their objects are maintained at the same time as they explore the sound environment.

Some *craftivists* are using the practice of fabrication of "Meshes in Loco" as a mechanism for political and social message. *Knitting Nation* by Liz Collins, reinvented the dynamics of the flat knitting machines. The quintet is composed by an army of knitters in repetitive actions that reproduce sound and fabric. Everything was placed in order to be seen and transformed into a concert-manifest-performance-installation that is intensely fed by a thread, by the movement of needles and audio processing. The militarist attitude that appeals to the social conscience of the workers is allied to electronic music and the art of noise, culminating in a *Craftivist* staging that is paced by a "family of noises" stemming from the mechanical movement of the machines that remind us of the principles of *Russolo*'s "futurist orchestras" (Roselee 2007).

Jobina Tinnemans is a contemporary composer and the author of "Killing Time" Orchestra (Fig. 5). The composition process of this piece includes the capture and manipulation of sound from a group of knitting *musicians*. The idea of timelessness is marked by subtle advances of each row of knitting that grows stitch by stitch in meditative gestures, thus forming an orchestral zone of ancestral female labour that deal with the particular and remote. The tick tack of the needles marks a breathing period in the music with draw backs and sensitive advances in the historical landscape that the hands of those women convey. ". . . Every concert the performing knitters make additions to the knitted "score" which are a living document of the history of the piece, visualising the passage of time through its trace in fibre" (Tinnemans 2014).

The implications of the materials and techniques in dramaturgy progress in the construction of complex forms, in the creation of scenic spaces, in themes that are concerned with questions of identity, with collective and private cultural issues. The timeless archetype of textile is exposed to the world to a large extent that it builds or incorporates it.

4 STUDENTS' WORK

"We are born into stories as we are into textiles" (Goett, 2016).

4.1 *Knitting my Way*

"Knitting my way or the phd process transformed into textures" by Cátia Faísco is a meaningful work not only for the proper implementation of the technical aspects, but mainly due to the form in which it was created and the response it generated.

For instance, the image of books piled side by side with the character "*PhD student*" knitting recreates the environment of her thoughts.

"Play or Walk Away: equally came to be the start-up theme for a knitting experience ... I promised to myself that I would knit a single piece before I finished the PhD. . ." (Faísco 2017).

The routine is a laboratory of interaction of knitting looms that construct a mesh of information woven with words and rows of stitches and give form to the thesis "*Play or Walk Away: Sexual Desire in Contemporary British Dramaturgy*".

Thinking and doing, unweaving and weaving again until a possibility appears.

To the character identified with work clothes and needles in hand, we add threads, serigraphed words, images of text in loop to describe the challenge that she faces in her research, "where to place limits to such a great variety of materials?"

Knitting my Way shows the student's perception, her carefulness in searching stitch by stitch in the exploration of a new concept where the words and needles are still necessary to supply another dimension.

4.2 *Sehrazade*

Was the work presented by the student Sajad Yari. It is the result of an imaginary that originated in the theme of his thesis. –"Violence of Technology, Postdrama Theater and Sexuality Narrative"

The narration of sexuality and violence relied on various artefacts to describe the atmosphere of life and death, vengeance and desire.

The short story tells that Sherazade resorts to a trunk of fabrics in an attempt to escape with life and put to an end to the king's vengeful project of assassinating all of the sultan's virgins. The lacework, voiles and other aspects of white, red and black silk give potential to the context of sexual lust. The usage of *malicious fabrics* normally used for women's intimate apparel cover and discover the bodies in between veils and flying fabrics. The music, Persian poetry, the voice with a cursing intonation, over backgrounds with illustrated images, collages, juxtapositions of fabrics in video recordings compose the frame of a thousand and one nights in *Lace & Lust*. The fabrics make a direct correspondence to the instincts of violence, sexuality and female sensuality, ideas and intentions of the mind.

The proposed assignment led students to construct a teaser for their PhD theses and was carried out according to the following steps:

1. Translation of words into images; creating a panel with key images, key words, key symbols, and the visual tidiness of the stimuli.
2. Construction of analogies and meaning; selection of materials bearing in mind their functionalities, the associations that they entail, the way they can be understood and the messages they may generate.
3. Practical exercise, the construction of a narrative that resorts to diverse techniques such as: serigraphy, stamping, video, sound recording, knitting, sewing, collage, text and voice.

The student works: Knitting my way by Cátia Faísco and Sherazade by Sajad Yari fulfilled the proposed challenges of the Focus on Textiles syllabus in an original way.

5 CONCLUSION

The variety of materials, their physical properties and versatility in term of processes, have added a symbolic utility to textile media. The performance projects under analysis reveal an attentive recurrence to the lexicon of the textile industry, but the construction of the show changed their destiny. the creations proliferate into orchestras of sewing machines, in the use of stage processes and technologies, such as in knitting machines for the production of sound and visual effects. the accessories such as scissors and needles have an extended role that come to mean danger and fear.

The function of the tools, objects and technologies, is reformulated and provides a service to major and small questions. the cotton or the mesh transcends from raw material to landscape. the lexicon associated to the textile becomes iconography the metaphorical material that constructs a narrative, just as the actions of sewing, embroidery, ripping, cutting, and knitting have already won multiple meanings. the elastic properties of a tubular mesh amplify the boundaries of the body and skin. the circular skirts in the dervish dance reach brightness and velocity given the weight and structure of the bleached white cotton fabric. The implications of the materials in dramaturgy progress in the construction of composed forms, in the creation of a scenic space rooted on characteristics such as: the lightness and transparency of aerial fabrics, such as tulle, gauze, organza ,the weight and opacity of structured fabrics in the dispersion or elasticity of threads and meshes and through the versatility of fibres and interactive materials.

The focus on textile sessions covered several performing arts works, focusing on the various aspects of the textile, its performance, and on intelligent usage options in multidisciplinary projects. The proposals under observation were **accessorized** with audiovisual means and with demonstrations/contact with materials and techniques that were prepared for action.

REFERENCES

Fattoumi, H. 2016 http://viadanse.com/spectacle/manta/ [31 of July of 2017]

Gil, J. 2002. *Movimento Total, O Corpo e a Dança*, São Paulo: Iluminuras.

Goldberg, R. 2007. *A Arte da Performance do Futurismo ao Presente,* Lisboa: Orfeu negro.

Graham, M. 1991. *Blood Memory*, New York: Doubleday.

Jefferies, J.; Wood C.; Diana, C.; Hazel 2016. *Hand Book of Textile Culture,* London: Bloomsbury Academic.

Laban, R. 1966. *Choreutics*, UK: Macdonald & Evans

Lenira, R. 2003. *Dicionário Laban*, São Paulo: Annablume.

Tinnemans ,J. 2014. https://jobinatinnemans.com/portfolio/killing-time/ [31 of July of 2017]

Textiles, Identity and Innovation: Design the Future – Montagna & Carvalho (Eds)
© 2019 Taylor & Francis Group, London, ISBN 978-1-138-29611-4

Brazil's social representation of Dom João VI in the literary work of Oliveira Lima: Fashion, habits, customs and urban aspects

M. Araújo, M.S. Barros de Held & J.P. Pereira Marcicano
Escola de Artes, Ciências e Humanidades, University of São Paulo, Brazil

ABSTRACT: The main challenge of historian is to find a point between reality and fiction. The Brazilian historian and diplomat Oliveira Lima wrote the narrative *Dom João VI no Brazil* in 1906, which purposes are to portray the arrival of the Portuguese court in 1808 and its stay until 1821, by possession of historical documents, reports of travellers and iconographies. He also provides the reader an overview of Rio de Janeiro's society in the early decades of the nineteenth century, emerging features of inhabitants, habits, customs, urban aspects and fashions. In the novel style, the author gives life to his narrative, using the rigor of the plot, with all literary resources. This work aims to contrast part of the literary work of the historian with two important historical sources cited in *Dom João VI in Brazil*, which are the reports of Spix and Martius and the iconography of Jean-Baptiste Debret.

1 INTRODUCTION

The year of 1808 was a decisive moment for the Brazilian people. In this year, Dom João VI and his court arrived in Brazil from Portugal. One of the most important acts of Dom João was to open the Brazilian ports to all friendly trading nations. This measure transformed Brazil colony, which was soon to assimilate the customs and institutions, ideas and methods of France and England, the great European metropolis. Efforts were made to establish the prince regent, his family and the Portuguese Court in the country until then colony of Portugal

For three centuries, Portugal had carefully closed its American possession to foreign commerce, kept it with impeccable zeal. The transfer of the Portuguese court to Brazil was the great transformation for a people renegade to colonialism, whose developments resulted in territorial enlargement and moral value, since the country assumed the position of the sovereign nation, on the way to economic independence. At that time, Dom João founded Brazilian nationality, whose developments culminated in its independence in 1822, as Calmon (2002) stated.

The thirteen years of the Portuguese court in Rio de Janeiro led to deep changes in the Brazilian economy, politics and society, especially in the city of Rio de Janeiro, a living testimony to the events surrounding the life of Dom João VI, his family and his government. Because of the new position of the country, in 1815, Brazil was raised to United Kingdom of Portugal. Acquainted with Brazil, Dom João insisted on staying in Rio de Janeiro, as this city gave him peace and plenty, returning only in 1821 to Portugal, due to the revolutions that erupted there. Upon returning to Lisbon, he handed Brazil the Regency of his son and heritor Dom Pedro I (Lima 1945).

The Portuguese prince was aware of his arrival changes and benefits to the country and he knew these changes would be important and essential elements for the construction of Brazil history. Thus, the monarch had the sensitivity to record the episodes involving his life, his court, his family, his government, his new country and the city of Rio de Janeiro, his new home. For this, He hired a commission of great and famous European artists, called *Missão Artística Francesa* (French Artistic Mission). One of the purposes of these artists was to portrait the Royal Family and an overview of Brazilian society, especially the Rio de Janeiro population, with their customs, manners and fashions. However, not only did the artist's iconography made this record. Even travelers, explorers and other foreigners who visited Brazil at that time corroborate to build this chapter of history of Brazilian society in the nineteenth century, whether for scientific purposes or by attracting wealth and gains in this new land, or even by a simple curiosity to see the Court of Dom João VI.

Therefore, based on these historical documents and eyewitness, the Brazilian diplomat and historian Oliveira Lima (1945) wrote *Dom João VI in Brazil*. The author's work represents one of the largest historical narratives that portray the arrival and the residence of the Royal family in Brazil, a fact of great relevance for the Brazilian history. Brazil was considered the unique American territory owning a King and a European Court installed on its land. As time passed Brazil loose European influences and gradually adapted to colors and tropical climates, taking the aspects of the Brazilian nation. Dom João VI was a popular king, a sensitive man, diplomat and easygoing, sometimes described as burlesque by many historians, and the real founder of Brazilian nationality. According to Lima (1945), the Portuguese monarch felt in love at first sight with the lands that offered him tranquility whereas Portugal took him the peace, setting Portugal aside in favor of Brazil. But in April 1821, the king returned to Lisbon in due to the threats of invasions and revolutions that occurred in Portugal and the Portuguese people requests.

Though the main points of the author's narrative were the political acts of Dom João, the diplomatic issues related to the arrival of the Royal family in Brazil and the problems surrounding the Dom João's government, a critical look at the particular details was reserved. Relevant aspects concerning Rio population in the nineteenth century as habits, customs, manners and fashions, and the daily life of the city was highlighted, according to historical documents. Thus, Oliveira Lima outlined Dom João's life, his family, his new society and its problems; a narrative composed in a novel style, with rich stylistic resources and prominent presence of comedy, by approaching the historiographical narrative to the novel style based on historical records. With his work, the author brought the historical truth to his reader regarding the noble and heroic acts, the success and the failures of the government of the king.

In this way, the work of Oliveira Lima was conducted along the same way of the realistic novel, with a detailed description of the facts. The writer tried to reproduce the spirit of that society, taking as premise the reports of national and foreign documents of historians, travelers as Martius and Spix, the paintings of important artists, noting here the iconography of Jean-Baptiste Debret, who, according to Lima (1945), is one of the artists who offers the most complete and interesting artistic documentation of Dom João's stay in Brazil. In other words, a fairly accurate idea of what Rio de Janeiro was in the early nineteenth century.

Thus, the descriptions contained in Oliveira Lima's narrative over Rio of the Monarchy, the Portuguese court and the city habits can be contrasted with the iconography of Debret, as well as with the travel reports of Martius and Spix. It is important to emphasize that the narrative Dom João VI in Brazil was composed within the criteria of historical research established by Ranke (2010). This criterion is the love of historical research, the documentary research, the deep analysis of the documents, and the interest in the foundation of the causal nexus. The chapters chosen for presentation were Chapter II, "A ilusão da chegada: a nova corte", chapter XXV, "O espetáculo das ruas"; and chapter XXVI, "As solenidades da Court". These chapters offer subsidies to contrasts between the historical documents, paintings or historical travelers reports, and the author's historiographical narrative.

To understand the city of Rio de Janeiro in economic terms and the way of living between different social classes in a country just transformed from colony to Metropolis, Oliveira Lima in *Dom João VI no Brazil* highlights the inhabitant's composition of the Rio de Janeiro city. As was said by the author, before court's arrival, the African slaves formed the majority of the Brazilian population, followed by a half-breed, Indians and a minority of white. When Portuguese court arrived, the population was about 50,000 inhabitants and 110,000 in 1817. According to the statistics of Spix and Martius (1981), nearly 24,000 Portuguese arrived in Brazil increased the population of whites. The sudden arrival of thousands of Portuguese represented a substantial increase in Rio's population and had an immediate impact on the city's residents.

In 1820, the population was 150,000 inhabitants in its total. With this population growth, the city faced all sorts of problems, from the urban to public health issues. The diseases were common, typical of the tropics, derived from the lack of personal and public hygiene. Oliveira Lima explained that "the cleaning of the city was all trust to the buzzards [. . .] and some travelers described it as one of the dirtiest human settlements existing in the world, fatally destined to a pest nursery" (1945). However, Spix and Martius (1981) stated those diseases as common as in other parts of the world: bladder and syphilis. The existing hospitals were the Mercy Hospital and the Royal Military Hospital.

The author narrates that, with the increasing number of people in the city, in addition to the problems of hygiene and public health, the city was exposed to all kinds of violence. The thefts were frequent and increasing day by day, as well as the knife fighting, as the number of the disorderly rabble increased. "The city became anarchized by releasing the capoeira groups, which were armed with razors, spreading terror at parties and social meetings" (Lima 1945) (translated by the author). Despite highlighting the violence of the city at work, the historian states that the occurrences of robbery were not higher than in other capitals.

In the landscape aspect, the writer describes that when the Royal family arrived in Rio, the city was suffocated by the forest, a place invaded by the natural landscape, with swamps and virgin forests, meaning little distraction for the inhabitants of the new metropolis. Brazilian civilization was surrounded by "fantastic and poetic natural beauties". The inhabitants did not have much fun, only the "*Passeio Público*" and the "*Aqueduto da Carioca*", which was "frequented by slaves, gypsies and sordid beggars" (Lima 1945) (translated by the author).

The entertainment of the wealthy was the backgammon game at night, the meetings at Manoel Luís Theatre, "an old, dirty, hot and airless place, too little lit by olive oil a wooden chandelier and sconces of leaves of Flanders, with an orchestra and shows of rude realism" (Lima 1945) (translated by the author), which had the worst and rebel actors already commenting for the national revolution. However, despite the city offers little attraction, its humor was compatible with Brazilian tropical characteristics.

Rio de Janeiro gradually got a better aspect, especially with the arrival of foreigners attracted by the ambition of places and gains, or by mere curiosity and challenge. The city gained new streets and the old ones were cleaned. The construction of new buildings, gardens and landscape gardening gave the remarkable appearance to Rio de Janeiro. The new cultural incentives provided social actions for the entire metropolis. Also the residence of foreigners contributed to the love of arts, Sciences and industries, especially the pleasure of comfort, luxury and social life. Besides. All these facts influenced positively on the acquisition of new habits for Brazilian social life. (Lima 1945).

In the educational aspect, said the author that there was almost no public education, especially after the expulsion of the Jesuits from Brazil. In Rio de Janeiro, free education was provided by the religious schools of *São José* and *São Joaquim*. In the first one, the studies included Greek, French, English, rhetoric, geography, mathematics, philosophy and theology; in the second one, the ecclesiastical subject. According to scientist Spix and Martius (1981), there was a strong influence of French culture in Rio society. Thus, the scientists were surprised the native language was not eliminated from upper stratum of society. Learning the French language was allowed only to male gender and the language was very common in the upper class.

On December 04, 1810, the Military Academy was founded at Largo de São Francisco de Paula. The studies at Academy included arithmetic, algebra, geometric analysis, rectilinear trigonometry and drawing, among others. At this time, intellectually, the open circulation of *Correio Braziliense* newspaper was allowed, and its loyal reader was Dom João VI. The usual reading was Voltaire and Rousseau, both banned in Portugal, and translations of Shakespeare, Poppe, Nelly and Klopstcok (Spix et al. 1981).

Martius and Spix (1981) narrate with surprise the influence of French fashion not only in fashion articles but also in language and literature, mainly in Voltaire and Rousseau, which draws the attention of the two travelers. According to them, the German language and Poetics were practically unknown by Brazilians. The only literary journal was the Patriot and the newspaper was *A Gazeta* do Rio de Janeiro.

There was an attempt to transform the city of Rio de Janeiro into a city of European aspect. However, Oliveira Lima (1945) points out that, although there were efforts to provide the city a pleasant European appearance, welfare and comfort of the Royal family, they were just efforts. In the opinion of the writer, the city had no charm, and no elegance could be given to the city due to its urban and social disorders. In contrast, the vision of the German researchers Spix and Martius, Rio was a place where the language, the customs, the architecture and the inflow of products from all over the world gave the city a European aspect, although they felt themselves in a strange continent because in everywhere they went, they face: "The variety legion of black and mulattos and the working class, as soon as I put my feet on the ground. This aspect was more of a surprise than pleasure. The rude and inferior nature of these importunate and naked men hurts European sensitivity, which just left the delicate customs and refined manners of its land" (Spix et al. 1981) (translated by the author).

They also describe that "The buildings of Rio are in general miserly and similar to the old part of Lisbon [. . .]. The presence of the Court has influenced favorably in architecture. The Casa da Moeda and several private houses in Catete and Mata-porcos are examples of this new architecture [. . .] The water distribution is made by black slaves poorly through black unorganized, in open bottles and exposed to the Sun, which should draw the attention of the public (Spix et al. 1981) (translated by the author).

In addition, in the narrative of the Germans it is stated that Rio de Janeiro was an expensive place to live in. According to the reports of foreigners who lived there, the Brazilian products as sugar and coffee cost almost the same price of these products in Lisbon.

However, the city was in progress. The members of the court were interested in establishing the pattern of life they had overseas, especially the manners. In chapter XXVI, "Solenidades da Corte", Lima describes the main events of the Court of Dom João VI in Brazil. The first great royal event was held in celebration of Queen Dona Maria's birthday, Dom João's mother. In 1810, people celebrated the princess Maria Teresa's marriage. In 1817, the royal wedding of Prince Dom Pedro I. But the most important celebrated event was the coronation of Dom João VI in 1818. At this time, Rio de Janeiro received the title of capital of the monarchy. The fashion of Napoleonic decorations, which was in vogue, was required for this event in Rio. French artists were responsible for the decoration of the *Largo do Paço*, in Greek style for the temple, in Roman Style for the arch and in Egyptian style for the obelisk (Lima 1945).

The ceremony of the "beija-mão" was also considered an important event, and it used to hold in *São Cristóvão* palace, from Monday to Saturday. In this occasion people could get close to royalty and the high class had the opportunity to show their formal clothing: black jacket, white vest, white knee-length breeches, stockings, and black hat (Lima 1945).

In Chapter XXV, "Espetáculo das ruas", Lima (1945) offers the reader a delight about the manners and fashions of the city, all in accordance with Debret's iconography (1989) at the time. Rio de Janeiro was a mixture of people in a city that did not have adequate resources to attend the population demand. In a colonial way, all types of sellers were on the streets, prayers of souls, beggars for saints wearing green, Scarlet and blue mantle; and sellers of rue. There was some African ritual such as the funeral of little children and marriage among slaves of wealthy families.

According to Lima (1945), the city of Rio had pronounced African culture, reason why the funeral events, parties and parades were frequent. But as the

Figure 1. Une dame allant à la messe dan as chaise à porteur Debret, *Picturesque and historical travels in Brazil*, 1989.

black population decreased, these events were over. German travelers Spix and Martius (1981) clarify that the decrease occurred from 1808 as the white population consisted of British, French, Dutch, Germans, Germans and Italians arrived in Brazil. They established themselves as traders after opening the ports. Prior to that, the number of blacks and mulattos was greater than the white number. (Spix et al. 1981).

However, the historian states that the black population still played a valuable role in the Brazilian economy. This population was entered in sales, by black slaves or by freed blacks, and in constructions. They were also wandering barbers, basket makers, grocers, animal dealers, sellers of hearts of palm, milk, grass forage, corn, coal, onions and garlic, thatch for mattresses, and coffee, and oxen car conductor.

Concerning transportation, in *Dom João VI no Brazil* the writer narrates the transportation used by the Court. It was the horse-drawn carriages conducted by white livery lackeys, contrasting to the old carriages pulled by mules and dirty black that there was before the arrival of the Portuguese court. Also the palanquins and sedan chairs came and went in the streets, as shown by Debret in Figure 1. The painter explains about the sedan chair: "The bather, imported from Lisbon, is used in Brazil as the litter in France. The sedan chair in Rio is recognized by its coverage, always adorned with more or less golden adornments – conducted by livery slave [. . .] can oppose it the luxury of some mulatto concubines that take advantage of the holidays to display in the Church all the ridiculous of their bad taste, generally inappropriate and exaggerated, which call attention even in the streets to the luxury their sedan chairs with very delicate ornaments, and covered in a profusion of varied colors [. . .]. The honest woman, as it should be, unlike, keeps her curtains closed, reserving, however, the possibility to open it, showing herself. (Debret 1989) (translated by the author).

Lima (1945), according to Debret's iconography describes that "Apart from the moments that demand ceremony, officials, and even the nobles, were seen at home in a comfortable way, with a beard grown, messy hair, their shirt with the sleeves rolled up and the often loose over shorts, bare legs and clogs on the feet" (translated by the author).

By contrasting to Oliveira Lima, Debret, in his paintings describes the rest moment in this way: "Here is the portrayal of the moment in which Brazilians, after their works, rest in a comfortable way, with light clothes in a hot day in Rio" (Debret 1989) (translated by the author).

In *Dom João VI in Brazil*, women had not been set aside. The writer observed a very important and recurring theme in the novels of the nineteenth century: the condition of women in Brazil. They suffered the effects of seclusion and constant dealing with slaves, relationship that "characterized on the one hand by the arrogance and the other by the degradation of the spirit, in the same way that the lack of exercises damaged their slender shapes, since the sedan chair represented the almost their unique way of locomotion. [. . .]. In female gender, ignorance was naturally the most extensive and marked feature, almost limiting the instruction of the most distinguished ladies enough to pray, store accounts, knowing the language of flowers to match with the Saints and with their lovers." (Lima 1945) (translated by the author).

Women were also present on Spix e Martius' reports: "Although the beautiful sex is included in the general Court changes and is more present on the streets and in the theater, it still retains more or less the same social position that Barrow (1806) quoted in his description in the year 1792 (Spix et al. 1981) (translated by the author).

The women of the Joanine period, in their reclusion, had as companies, the children and the slaves. The only occupation was manual work with needles. So, the historian describes their day by day: "[. . .] cross-legged on a mat, the housewife, surrounded by domestic slaves, sews, weaves lace, makes silk and paper flowers, make delicious cakes" (Lima 1945) (translated by the author).

Here there is notoriously the appeal of the narrator, when he interferes with the action "makes delicious cakes", and also in conveying information clearly to the reader.

However, Debret goes beyond to their explanations in his painting, which were represented in Figure 2. Below, the description of the female routine by Debret: "The system of European governors, in the Portuguese colonies, constantly serves to leave the Brazilian population deprived of education and isolated in the slavery of their routine habits. This led the education of ladies to the simple care of her domestic work. Thus, since our arrival in Rio de Janeiro, shyness, a result of lack of education, reduced the ladies at meetings more or less numerous and, even more, barred from all manner of communicating with foreigners. [. . .] Here, I tried to portray a mother of small family in her home where we found her seated, as usual, on her marquise [. . .] place that serves, by day, as fresh and comfortable couch in a hot country, to rest all day, sitting on her legs in an Asian way. Immediately at her side and well within reach is the gongá (wicker basket) designed to contain

Figure 2. A lady in her house with her daily occupations Debret, *Picturesque and historical travels in Brazil*, 1989.

the sewing works [...] The domestic slave works in the tea, at the foot [...] (Debret 1989) (translated by the author).

These families used to set the table with porcelain plates full of food and British crystals with wine and a lot of sweets, when at meetings or celebrating a birthday. However "They didn't have good manners; they vulgar customs like cleaning the knife in the tablecloth and eating with their mouth inside the plate". (Lima 1945) (translated by the author).

Still in chapter "*O espetáculo nas ruas*", Oliveira Lima, for the delight of the reader, provides a broad description of the customs of the society at that time: leisure, processions, food, fun and, especially clothing. The parties have become much more fun and interesting from the arrival of the Royal family, with fireworks to celebrate some important birthday or event, as well as the presentation of high society with elegance, distinction and luxury. Processions of all kinds took place daily in the city of Rio de Janeiro: "The processions passed in the crowded streets to the sound of religious songs and the rockets through the crowd which followed the processions by devotion and pleasure. In these days, there was a trade of sweets, black grocers selling cakes, and great trade of candies, which the most respected shop was at "Ajuda" Street." (Lima 1945) (translated by the author).

Fashion is a valuable source of information for historians regarding habits and customs of a particular society (Barnard 2003). Since there is an understanding of fashion as a factor of significant relevance to the industry and market, being linked to the customs, art and economy, it appears that the studies of the history of fashion are relevant to historians, economists and sociologists. As a cultural phenomenon, fashion contributes as much to the history of humanity as to the history of the economic sector. In this way, fashion and its accessories were common themes in narratives and reports of travelers, being present in the Oliveira Lima's work. Debret, in some paintings highlighted the clothing of the lower class: slaves and ordinary citizens, as well as the upper classes fashion: officials, nobles and Court.

When the Court moved to Brazil, the clothing was one of the most important sources of social stratification. Through the wardrobe people showed their purchasing power. Clothes were considered possessions so valuable that could be part of the individual's inventory. The clothes left clear the position of the individual towards society, in which he lived, revealed the genre, religion, regional origin and profession. Besides, clothing identified the individual through the uniforms, whether school uniform, hospital, military etc. Fashion in the nineteenth century was divided into Empire, Victorian, Romanticism and the *La Belle Époque*.

Rio de Janeiro was the center of the Portuguese monarchy and home to an independent empire. Brazil now had noble titles and, with a diplomatic corps, sought to mirror itself in Parisian society for luxury and glamor. This new condition required services for its level, especially in the garment sector and luxury items. Thus, there was a need to enlarge the staff involved in the garment sector, as tailors, dressmaker, shoemakers, hatters, florists and weavers. Gradually, the British and French dominated the Brazilian market and improved Brazilian's choices and styles. The fabrics and fashion products that arrived here from France and England or somewhere else were very expensive. The use of jewels enhanced the luxury women dresses. The Brazilian ladies were not intimidated to show their luxury in a city with very little comfort, with poorly urbanized streets and substandard housing.

Fashion in Brazil, in the mid-800, depended on the labor of the Polish and the French dressmaker, whose were established in Rio. According to Garcia (2003), the first group of French dressmakers arrived in Brazil on 26 March 1816, probably to attend the requirements of the Portuguese court. They settled in Ouvidor Street. Before that, Ouvidor Street, the main fashion street in Rio at that time, was a street of bad repute. This passage was observed in the narrative of *Dom João VI no Brasil,* and the author described that "In the Ouvidor Street, which had pretensions of elegant of fashion, black barbers abounded until shortly after the arrival of the Royal family, when the hairdresser of the court, Monsieur Castilhos, with his lotions and perfume, established in the street, as also Mme Josephine, the best fashion dressmaker, opened her store" (Lima 1945).

In the early years of the nineteen century, the French Empire style was in vogue, inspired by the Greco-Roman style. The dress had a fitted bodice ending just below the bust, loose sweater-like waist below the bust, giving a high-waisted appearance, a long and loosely skirt, supported by petticoats in light fabrics, preferably white color, muslin or linen and transparent and, for the night, always low-cut. From 1810 to 1820, dresses changed little. The waistline remained high sleeves were fuller at the shoulder, skirts with the conical silhouette, and heavy ornamentation around the dress near the hem. As ornaments, women wore

Figure 3. Formal ladies costumes. Debret, *Picturesque and historical travels in Brazil*, 1989.

pearl earrings, stones, cameos, feathers on the head, fans and gloves. The men's clothing has the British style, more sober, coat, Knee-length breeches, stockings, high collars, scarves tied around the neck, and the permanent use of boots. (Braga 2005).

For the processions, which occurred daily, the people wore their best clothing. In important processions, such as the one of the Blessed Sacrament, it was custom to adorn the streets for the feasts and processions with damask bedspread and fabrics of crimson damask from India and silk from China. From de windows and on damasks bedspreads and gold brocade, princesses and ladies of the court show themselves adorned with gauze turbans, shiny and large diadems, feathers on the head enjoy the processions. "The women were dressed in low-cut in the daylight, silk, and heavy jewels," according to figure 3 (Lima, 1945) (translated by the author).

It was one of the few opportunities women had to show their elegant and luxurious costumes or dress court. The spectacle was so great that the historian compared the processions with *Long Champs* (hippodrome of Paris, located in the *Bois de Boulogne*) in Paris, where women used to show their new dresses and new evening costumes. In the figure below, the Court garments (Lima 1945).

In some passages, Oliveira Lima describes fashion in all social classes, according to Debret's paintings, "elegance joined the order of the day, and the dedication in the costume and attachment to the ceremony reached the point that customs officers were in the uniformed service all the time, wearing bicorn hats, buckles and rapier sword" (Lima 1945) (translated by the author). Figure 4 displays the representation of Debret.

Oliveira Lima (1945) describes the clothing of a middle-class family, a public servant, his daughters, wives and slaves, in hierarchical order. He emphasizes the men's clothing: "Black coat, embroidered vest, big buckles of fake diamonds tightening on the knees, knee-length breeches, and stockings".

Regarding slaves, Lima describes the African wore turban, white raw cotton skirt, crossed with a colored

Figure 4. Customs officer walking on the street. Debret *Picturesque and historical travels in Brazil*, 1989.

Figure 5. Black cook from Recôncavo Baiano. Debret, Viagem Pitoresca e Histórica ao Brasil, 1989.

stripe fabric and dropped shoulder shirt. The beauty of African slaves was described by the author: "Black women, with their embroidered chambray shirts, their skirts of raw cotton, turbans on their heads, are very original, very sexy and very pleasant, in a picturesque town and nice neighborhood. A detail for the French, who had no words to glorify the Recôncavo Baiano". African clothing described by the writer can be observed in Debret, as figure 5 (Lima, 1945) (translated by the author).

By contrast the author's work with important historical documents as reports of travelers and Debret's iconographies, it's possible to recognize Oliveira Lima's style and his great work. As a historian, Lima used the resources of narrative fiction as a vehicle in the historiography to entice his reader. His narrative contains ordered and coherent presentation of sequence events. Thus, establishing an important connection between history, art and literature, the historian sought to follow the pattern that is characteristic of a literary work, by the interconnection of history and literature. This process can confuse the position of a historian with the novelist, as the historian has a dual task in creating your art work of fiction. The first one is to build a coherent and meaningful narrative; the second one is to present the image of the facts as they really are. The exact adaptation of the poetic art and its rhetoric to its way of looking at the past makes a historical narrative an admirable novel.

REFERENCES

Barnard, M. 2003. *Moda e Comunicação*. Tradução de Lúcia Olinto. Rio de Janeiro: Rocco

Barrow, J. 1806. *Rio de Janeiro in A voyage to Cochinchina in the years 1792–1793*. London: Printed by T.Cadell and W.Davies, p. 95

Braga, J. 2005. *História da Moda, uma narrativa*. São Paulo: Anhembi Morumbi

Calmon, P. *História Social do Brasil: Espírito da Sociedade Imperial*.1 ed. São Paulo: Martins Fontes. 2002, p. 3

Debret, J. 1989. *Viagem Pitoresca e Histórica ao Brasil*. São Paulo: Edusp

Garcia, C.; Miranda, A. 2005. *Moda é comunicação: Experiências memórias e vínculos*. São Paulo: Anhembi Morumbi

Lima, O. 1945. *Dom João VI no Brasil*. 2 ed. Rio de Janeiro: José Olympio Editora

Ranke, L. 2010. *Sobre o caráter da ciência histórica in A História Pensada: Teoria e Método na Historiografia Europeia do Século XIX*. 1 ed. São Paulo: Contexto, pp. 202–214

Spix, J.; Martius, Karl F. 1981. *Viagem pelo Brasil 1817–1820*. Vol 1. São Paulo: EDUSP. P. 52

Textiles, Identity and Innovation: Design the Future – Montagna & Carvalho (Eds)
© 2019 Taylor & Francis Group, London, ISBN 978-1-138-29611-4

Textile design in lace: Creative practice based on collections study

V. Felippi
PGDesign, Faculty of Architecture and Engineering School, University of Rio Grande do Sul, Brazil

E. Rütschilling
Faculty of Art Institut and Design Graduate Program, University of Rio Grande do Sul, Brazil

G. Perry
Design Department, University of Rio Grande do Sul, Brazil

ABSTRACT: This study aims to discuss the interaction between textile design and museum lace collections in view of the creative practice using traditional techniques. It is a documentary, bibliographic research following technical procedures of action research. It is considered that by adding the ancient cultural value of lace and design it was possible to obtain an innovative result in the traditional technique of needle lace. We also find that lace collections are inexhaustible sources of references and possibilities for promoting products, uniting areas of knowledge, valuing, preserving, continuing and renewing handicraft and cultural traditions.

1 INTRODUCTION

Lace has always held a prominent position among the fabrics and, regarding its study within the museum scope, it is in the international scene that the subject stands out. This is because institutions in countries such as England, France, Italy and Australia conduct investigations and interactions between lace collections and researchers, academics and professionals of several areas for theoretical and practical research. In the Brazilian scenario, this kind of interaction is practically non-existent, perceiving a gap to be filled and that should be stimulated.

Publications such as "Lace, Here, Now" (Briggs-Goode & Dean 2013) and "The metaphorical value of lace in contemporary art: the transformative process of led inquiry" (Joy Buttres 2013) are references which represent the state-of-the-art in the study and practical and theoretical investigations on the theme of laces. The book "Lace, Here, Now" intends to show that lace goes far beyond just a limited and stereotyped perception and illustrates how such issues can be flanked by creative ideas from academics and practitioners. The book also points out how the Lace Archive at Nottingham Trent University's School of Art and Design is active on the historical and practical lines, promoting interactions and experiences that result in theses, exhibitions, activities linked to museums with the aim of building and strengthening practices with the art and the design.

Joy Buttress, who defended her thesis at the School of Art and Design, also at Nottingham Trent University (2013), discusses the theme of lace in a theoretical and practical setting. The author describes lace as a multi-faceted fabric such as: delicate and decorative associated with femininity and playing an important role in the ornamentation of bodies. Additionally, it can convey signs of status, wealth, power and eroticism. It is a fabric that is inserted in religious rituals in the form of the veil of the bride, the veil of the widow in mourning, the dress of the priest and the child in the ritual of baptism. Taking into account all these aspects and her experience with the collection of lace that belongs to the same institution in which she defended her thesis, Buttress created works and developed research that brings up all these questions.

From these observations, the objective of this study is to discuss the interaction of textile design with lace belonging to museum collections in view of the creative practice of contemporary object production using traditional handicraft lace weaving techniques.

To support the discussion, the practical result of the interaction between the mentioned areas is presented in this study. From the visual, formal and technical study of an item of lace belonging to the Fashion and Textile Museum of the Federal University of Rio Grande do Sul, Brazil, a garment was created using a traditional needle lace technique.

This activity contributes to the demonstration that research in the field of textile design employing traditional techniques presents several ways of research and that, by adding the knowledge contained in the collections, the possibilities increase and doors are opened to unite areas of knowledge, contribute to the preservation and valuing of traditional knowledge, as well as giving continuity and renewal to such practices.

2 MUSEUM COLLECTIONS AND CREATIVE PRACTICES: LACES IN CONTEXT

According to Meneses (2011), museums are privileged spaces because they are platforms that link scientific, cultural and educational functions. Faced with such possibilities, in this study we focus on the potential of museum collections and their relation to creative practices. For such, we must consider Ramos' (2004) comments that visits to museums should begin in the classroom with game activities that use daily materials. According to the author, if we learn to read words, we must also exercise the act of reading objects, of observing the history that exists in the materiality of things. Any object can be treated as a source of reflection and stimulate inter-disciplinary dialogues, and studying museum objects is not only about visiting the past, but about stimulating studies about past times in relation to what we live in the present (Ramos 2004).

Matos (2014) points out that although studies in the area of museology have developed in recent years, museums are being underemployed in their educational activity. Another factor is the use of the museum only to verify something previously taught in the classroom, rather than being used as a partner in the construction of knowledge.

And, within this context, laces are included. Beautiful, complex, delicate and intricate objects that offer possibilities for studying and researching not only their visual, but their historical, cultural and technical aspects as well. In the international context, they are widely studied, as in publications by Earnshaw (1982, 1983, 1991, 2000), Browne (2004), Kraatz (2006) and Etcheverry (2013), as well as those by the specific museums and research institutions.

Such publications and knowledge contribute to the study of laces and their reinterpretations, clearly showing that they are increasingly evident in current creative practices (Briggs-Good & Dean 2013). In the contemporary context, there is a tendency to value the actions and knowledge of this ancient craft. This valorization is due both to the interventions of museological and educational institutions, uniting their collections with students and professionals.

In Brazil the interaction with lace collections is little explored and divulged. The interactions that occur are associated with the use of textile museum object images that are taken into academic spaces by teachers, that is, the secondary sources stand out. Not that this is negative, but interacting with primary sources contributes towards minimizing the impacts of third-party interpretations that the secondary sources carry. However, as Muller (2006) points out, it is interesting that sources be interwoven and confronted for a broader understanding.

It is understood that there is a long way to go in the activity of joining lace collections and design practices in the Brazilian scenario. This activity should be stimulated because it offers competitive advantages and allows researchers and scholars to know about and disseminate knowledge regarding the cultural, historical and aesthetic richness contained in museum objects.

The integration of research activity and museum collections of whatever nature does not necessarily require any mediation, but, in some cases, expert knowledge is recommended. In this sense, as in the experiences cited by Briggs-Goode (2013), for the interaction of lace collections, scholars select their sources, while mediators provide stories about the objects, their construction process (lace design), and the value of collections for future projects and collaborations. Interacting with museum textile objects can be transformed into a class on history, design and the relationship of the matter with society (producer and user).

3 TEXTILE DESIGN

Textile design is the activity of fabric conception (Neves 2000) in both industrial and artisanal processes at a small scale or to support mass production lines. Among the applications for the fabrics we can mention: clothing (fashion, sports, safety), decoration, automotive industry, medicine, construction and geotextiles (drainage systems, erosion systems, filters, among others).

Fabric, in general, is the result of interlocking, in either orderly or disordered forms, of textile fibers or yarns. In the industrial process the typologies are divided into: 1) woven fabrics: they are characterized by the perpendicular interweaving of the warp yarns and weft; 2) knitted fabrics: consists in either vertical or horizontal yarn loops, and 3) non-woven fabrics: a flat and flexible structure constructed of fibers or filaments (Maluf, Kolbe 2003). It is important to note that in the manual process some techniques fall into the above typologies and others, such as laces, employ a wide range of supports to provide their own characteristics.

Regarding laces, the focus of the study, we understand them as a typology of fabric that can be woven in both manual and industrial processes. It has as its characteristic an independent structure, which does not need support to exist and has open and closed spaces, which give it lightness and transparency, contributing to highlight the motifs and the bases of connections between them. Etcheverry (2013) points out not to confuse the ornamentation of laces with only the most enclosed spaces, since we can lose sight of the areas with open points that are also fundamental. Lace designs show a compositional order that may or may not be repeated. When they are repeated it is considered that there is a rapport in which are contained all the formal elements of the drawing.

The manual process of lace weaving consists basically of three large groups: the bobbin lace; the needle lace; and the lace formed by knots. In each of the groups there is a diversity of stitches, which may be used in the same lace to provide it with visual and technical richness.

Manual laces were the ones that drove and influenced industrial production. Until the seventeenth century production was basically homely (Etcheverry 2013), but with the excessive and abundant use of laces, the evolution towards industrial processes was inevitable. The first industrial machine equipment appeared in 1809 (Middleton 1938) in England, and mechanical and technological developments advanced in the following years. This was because, initially, some looms were not able to attach details to the lace, such as outline yarns, but over the course of the century improvements were implemented and looms were capable of producing lace with complex designs, on a larger scale and at a lower price.

In the industrial process lace can be woven basically by embroidery (for example, in chemical lace or embroidery on a net) and by weaving, in which a series of specific looms are used, as well as warp knitted.

Earnshaw (1991) believes that machines hold the key to the future of fashionable lace and, with increasing demand, designs become more complex and beautiful, and more economical to produce. While handmade lacing is perpetuated as art, preserving for posterity ingenious techniques of old lace, machines are ideal for spreading them into a society currently occupied.

In both manual and industrial processes, various raw materials (textile yarns of different compositions) are being used, resulting in distinct laces in tactile and visual effects and also in applications.

Professionals who work as textile designers, whether of lace or other types of fabric, necessarily needs to have in-depth knowledge of fibers, yarns, textile bindings (thread interlacing system), rapports, colors, equipment, beneficiation processes, textile finishing and product applications. Besides that, their performance is based on the knowledge of the project methodology and its production process. This is because the fabric to be produced must meet not only the desired aesthetic parameters, but must be designed taking into account the materials that best respond to its function, the best production method and the level of quality to be guaranteed (Neves 2000). According to Neves (2000), when designing a fabric, textile designers must be in tune with the market for the product, the cost of the product, the conceptual quality and the productive quality, to lastly stick to the aesthetic and structural quality.

It is important to mention that textile designer performance is linked to both industrial production and artisanal production interaction. The latter is very significant in the Brazilian scenario and features mostly a character of collaboration with artisan groups. The professionals who work at the craft level can receive the same training as those working in companies, but they need to know the characteristic aspects involving the context. This is because it is necessary to know the values and identity of each culture, in addition to their traditions, since such elements take a primary standing in their differentiation and anchor the valorization, preservation, continuity, recognition and renewal of those traditions, if applicable.

In the action with the handicraft, textile designers aim to develop new products or assist in the pursuing references for improvements in existing products. In this situation, it is the professionals who are inserted as a support tool to reinforce product attributes and preserve the cultural identity of the location and they must be aware of the abundant raw materials of the region, the fauna and flora, the practices and the behaviors, and contribute to a product proposal that assists the dialogue of the community with the consumer public. This all adds up to the knowledge of all issues involving production processes, technologies, materials, applications, and so on.

It is believed that combining the knowledge of traditional craft practices with design is one of the paths for innovation and continuity of craft traditions. This is the path taken in this study.

4 METHODOLOGY

To reflect upon the interaction of textile design using traditional craft techniques with museum collections, a bibliographic and documentary research was initially done. Following that, the lace from the collection at the Fashion and Textile Museum of the Federal University of Rio Grande do Sul was analyzed and, after selecting one lace, the action-research methodology was applied to create a textile object for the garment.

The study made use of the action-research method that, according to Swann (2002), is applied to the design as a non-linear method allowing for back and forth possibilities to analyze the issue and adjust the solutions. In this way, it is characterized as an iterative method.

According to the author, the design process can only be effective by reexamining, revisiting and reformulating the issue. Thus, the most important feature of action-research is that the results help the researcher develop new theories or expand existing scientific theories. The method involves planning, acting, observing and describing, reflecting and evaluating. Throughout the research process, those involved were able to develop a grounded justification for their work. Theories and solutions produced through this method should be disclosed to stakeholders, be they participants, scholars, professionals or community members, thus arousing interest in the activity (Swann, 2002).

5 RESEARCH AND DEVELOPMENT

To link theoretical research with practice, the first step was the selection of lace from the collection. The collection to which the lace belongs was studied by the

Figure 1. Lace belonging to the collection of Fashion and textile Museum-UFRGS. The circle highlights the point studied and put into practice in the creative process.

author for her master's dissertation and is also part of her doctoral research.

It was the handling during the research that aroused the desire to explore elements of the lace collection and combine with her professional activity as a textile designer. The combination resulted in the creation of a garment with a contemporary design employing the ancient practice of making needle laces. During the stage of reflection and evaluation, the author takes on the position of researcher, fashion designer, textile designer, lace maker and artisan, since she was responsible for all stages of the process.

After visual analysis of all the laces in the collection, the detail of one of them stood out. It is a lace woven in an industrial process with the technique of embroidery on a net decorated with floral motifs and organic forms. The motifs show an outline yarn that contributes to highlight them and form drawings as well, whose highlight is the volume that is in the lower left base, according to the circle in Figure 1. That is, the outline yarn either occupies the place of the element that highlights the floral motifs of the lace and or occupies the place of the motif due to the shape (spiral) into which it was embroidered.

The choice of that lace item was due to the visual effect of the motif and to the fact that it is a lace made in an industrial process. Thus, we have a lace produced in an industrial process that inspired the creation of a lace produced in a manual process.

As a creative practice, the product to be created and produced was the upper part of a dress (a kind of blouse), both for the versatility and for the expected time of execution.

The technique used was needle lace, having a construction process similar to those used in Irish and renaissance laces, that is, they are constructed with ribbons or strings that are tacked on a support and later connected with stitches to weave the lace.

The differences between the lace presented here and the Irish and renaissance laces are:

1. The ribbon that defines the drawing is made with the crochet technique, characterized by being a thread with relevant elasticity, typical of this technique;
2. In most creations, the drawing is not made onto the support (which may be any type of fabric), resulting in laces of a similar visual language, but with different drawings. When the lace requires symmetry, some guidelines are drawn on the support;
3. Unlike renaissance and Irish laces, it is malleable and flexible, that is, it does not restrain movements and fits easily into diverse models.

Cotton yarn was used as the raw material for the production of the lace, which is easily found in the retail and wholesale trade. The thread in question has been used by the author in her work with hand weaving for more than 15 years in the knowledge that it is of great quality regarding comfort (pleasant to touch) and color fixation (fade proof).

The steps of the construction of the lace and the garment are described below:

1. Definition of the model and the support to be used. The support is made of polyester and elastane fabric, cut to the intended dimensions and of a color that contrasts with the yarn used in the lace;
2. Construction of approximately 100 m of ribbon (made in current point) using crochet needle from 5 100% cotton thread cables, title Ne 8/2, in blue;
3. The ribbon is then threaded onto the model. During basting, the design is built. To facilitate the symmetry of the drawing some lines were drawn onto the support;
4. Once whole ribbon is stitched together on the model, the process of "joining" the ribbons begins with a manual sewing needle and a cotton thread. This joining is made with a cross-stitch and its function is to be the basis of bonding between the laces. During the making of this seam, the sequins (decorative element), which are part of the fabric structure, are applied;
5. The next step consists in untacking the lace from the support (model).

Figure 2 illustrates the process of basting the ribbon to the model.

From the support, it is possible to notice that the drawing was not previously marked onto the model, but only some guidelines to delimit the areas aiming at a symmetry of the parallel lines that are located on the sides. The parallel lines not only contribute aesthetically, but also value the spiral motifs that are arranged side by side and also overlap, contributing towards the visual effect of contrast between volumes and empty spaces. Even with the drawing not having been scratched, there is no impediment for the object to be repeated, since the support is not destroyed in the process.

Figure 2. Process of making the lace: basting the ribbon onto the support and sewing the lace.

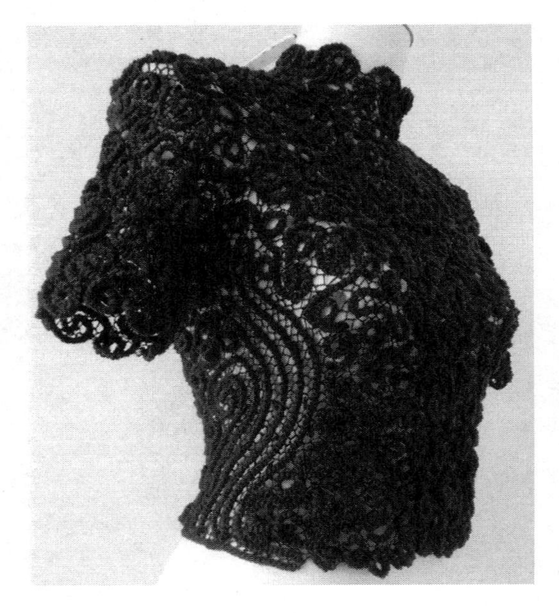

Figure 3. Finished lace.

Figure 3 shows the finished lace.

The piece execution time was of approximately 40 hours of work. It is important to note that both the execution time and the visual effect may differ if it is made with different thicknesses of the ribbon or the yarn that joins it. One of the characteristics of this piece is the lack of stitching on the sides that is usually present in clothing items. Furthermore, it offers versatility, as it can be used in either formal or casual events, depending on how it is combined with other garments and accessories.

6 RESULTS

The knowledge acquired during the research, combined with the practice and experience in the area, was relevant to realize that new possibilities for creation are possible, taking inspiration from technical or visual elements of lace collections. By investigating and incorporating the elements (forms of construction, motifs and relations between empty and full spaces) the possibilities for new results are broadened, even when one process is transposed to another. But to do so, it is necessary to carefully observe the constituent elements, the mastery of the technique to be executed and the raw material to be used in order to exploit the resources that may be relevant.

The construction activity of the presented object is almost a ritual of repetitions, albeit not automatic, since the nuances of the surface and applicability, e.g. volumes, empty and full spaces, visual effects of the drawing, position of the motifs, symmetry, usability, timelessness, etc are taken into account. Furthermore, some characteristics can be altered at the moment of form construction.

Ganem (2013) comments that in the interaction between designers and artisans, oftentimes one does not know the particularities of the other, and when this happens, great difficulty arises in developing a piece. And, in this study it was realized that when the designer performs both roles there is no border or dividing line between the areas and the difficulties are practically non-existent.

According to the United Nations Conference on Trade and Development's Creative Economy Report (UNCTAD, 2010), design is a viable tool for functional creations, as it is based on knowledge, producing goods or services with creative content, a cultural and economic value, and a market purpose. In this way, design is a strong ally to connect with handicrafts and the potential of museum objects, as they are able to bring forth the intrinsic and extrinsic cultural references of objects from collections by placing them under the spotlight, perpetuating and strengthening all aspects of these areas.

In addition to this, museum objects can be revived and brought to the present. Their historical, technical and cultural references are revisited and re-evaluated. The resulting products will be linked to the history and memory of the object of the collection and, depending on the lace, the resulting product will be timeless.

7 CONCLUSION

This study verified, through the presented practice, that laces are sources rich in historical and technical references, enabling creative practices due to the innumerable and complex relations that can be established by the interpretations.

Besides that, it demonstrates that it is feasible to combine textile design with the traditional craft practices of lace making by looking at contemporary results using museum objects as references. In this sense the practice reflected an original result, thanks to the chosen source, the experience and the mastery of the lace construction technique contributing to perpetuate traditional activities of fabric weaving loaded with historical and cultural references.

The project methodology used was approved and is intended to encourage further research and practice. The object of the project has a clear identity with craftsmanship, where manual skill, knowledge and dexterity are of relevant importance. These points confer added value, the possibility of producing pieces with exclusive designs and a high degree of difficulty for imitation through industrial processes.

Regarding the raw material used, cotton, its choice considered characteristics and intrinsic properties, such as being durable under suitable conditions, renewable, biodegradable and recyclable.

It is also observed that in contemporaneity artisans do not necessarily possess only an empirical and family formation, it is possible that they also have a scholarly formation.

The cultural and identity aspects that are part of the artisan context and impel the creations cannot be left aside when there is the performance of the designer, because it is these references that will give due importance and value to the object. The author's performance as a designer, as a weaver/artisan and researcher allowed for this verification and contributed to the application of creative possibilities and sensitivity to visual, technical and historical stimuli.

Discussing the boundary between craftsmanship and design consists in employing the design methodology, which allows the object to be repeated. The design process takes into account requirements such as originality, sustainability, clear functions, image, versatility, marketing and management. Combining the points cited with the retrieval of visual, technical, memory and cultural elements from the museum collections to develop objects/products may be deemed a valid and promising strategy.

The integration relations between museum objects, especially within the theme of lace, and the design is still little diffused in the Brazilian scenario. The possible difficulties of interaction are understood, but it is necessary to give visibility to these activities so that the intrinsic knowledge of the object may surface under a new look. And this new look contributes to valuing cultures, stimulating research, preserving, renewing and providing continuity to handicraft traditions, especially in the field of lace. This is because studies point out that many techniques are doomed to disappear due to a lack of interest on the part of weavers as well as of those who acquire/use them. It is believed that in some cases it is the link of the design that is missing to unite the two points. As Buttress comments (2013), lace is versatile and can fit into areas such as literature, art, design and fashion. Thus, one can perceive broad performance options for researchers, scholars, designers, artisans and weavers.

REFERENCES

Andrade, E.G. 2011. Entre o tradicional e o efêmero: renda irlandesa e novas instruções de uso. In: Anais *Seminário de estudos culturais, Identidade e Relações Interétnicas*. Aracaju: Universidade Federal de Sergipe. Available at: http:// 200.17.141.110/pos/antropologia/seciri_anais_eletronicos/ down/GT_02_2011/ANDRADE_E.pdf[Acessed 26.08.13].

Andrade, R. 2014. Historicizar indumentária (e moda) a partir do estudo de artefatos: reflexões acerca da disseminação de práticas de pesquisa e ensino no Brasil. *ModaPalavra e-Periódico 7(14):* 72–82. Available at: http://www.revistas.udesc.br/index.php/modapalavra/article/view/5099 [Acessed 24.11.15].

Andrade, R. & Paula, T.C. 2009. Estudar e pesquisar roupas e tecidos no Brasil. *II Seminário Nacional de Pesquisa em Cultura Visual*. Faculdade de Artes Visuais/UFG, Goiânia/GO. Available at: https://seminarioculturavisual.fav.ufg.br/up/778/o/2009.GT3a_Rita_Andrade_e_Tereza_Cristina.pdf [Acessed 1.03.16].

Barros, K.S. 2009. *Análise antropotecnológica do desenvolvimento de novos produtos na produção artesanal: caso das rendeiras de bilro da Vila de Ponta Negra em Natal*. Dissertação, Natal: Centro de Tecnológia, Universidade Federal do Rio Grande do Norte.

Briggs-Goode, A. & Buttress, J. 2009. A Taxonomy of Pattern Through the Analysis of Notthingham Lace. In: *Association of Fashion And Textile Courses*, Inglaterra: University of Liverpool.

Briggs-Goode, A. & Dean, D. 2013. *Lace, Here, Now*. London: Black Dog Publishing, 2013.

Browne, C. 2004. *Lace from Victoria & Albert Museum*. London: V&A Publications.

Bruno, M.O. 2006. Tecidos e Museologia: perspectivas para a formação profissional. In: Tereza C.Toledo de Paula (ed.), *Tecidos e sua conservação no Brasil: Museus e coleções*. São Paulo: Museu Paulista da USP: 131–133.

Buttress, J. 2013. *The Metaphorical value of lace in contemporary Art: The Transformative Process of a Practice-Led Inquiry*. PHD, Thesis, Nottingham: Nottingham Trent University.

Costa, E.L. 2008. *O invento de Jacquard e os computadores: alguns aspectos das origens da programação no século XIX*. Dissertação, São Paulo: Universidade Católica de São Paulo.

Earnshaw, P. 1983. *Bobbin & Needle lace – Identifications and care*. London: Batsford Craft Ltd.

Earnshaw, P. 1988. *A dictionary of lace*. London: Shire Publications Ltd.

Earnshaw, P. 1991. *Lace in Fashion-from the sixteenth to the twentieth centuries*. Guildford: Gorse Publications.

Earnshaw, P. 2000. *The Identification of Lace*. 3th. ed. London: Shire Publications Ltd.

Etcheverry, D. 2013. *Esncajes: história e identificación*. Argentina: Fundación Museo Del Traje.

Ganem, M. 2013. *Design dialógico: uma estratégia para gestão criativa de tradições*. Dissertação, Bahia: Universidade Federal da Bahia.

Kraatz, A. 1989. *Lace-History and Fashion*. Reino Unido: Edit Rizzoli.

Maluf, E; Kolbe, W. 2003. *Dados Técnicos para a indústria têxtil*. São Paulo: IPT-Instituto de Pesquisas Tecnológicas do Estado de São Paulo: ABIT-Associação Brasileira da Indústria Têxtil e de Confecção.

Matos, I. Educação museal: o caráter pedagógico do museu na construção do conhecimento. *Brazilian Geographical Journal: Geosciences and Humanities research medium*,

Ituiutaba, v. 5, n. 1, p. 93-104, jan./jun. 2014. Available at: http://www.seer.ufu.br/index.php/braziliangeojournal/article/view/23630/13811 [Acessed 10.02.17].

Meneses, U.B. 1998. Memória e cultura material: documentos pessoais no espaço público. *Revista: Estudos históricos*, 21. Available at: http://bibliotecadigital.fgv.br/ojs/index.php/reh/article/view%20File/2067/1206 [Acessed 12.03.16].

Meneses, U.B. 2011. A comunicação/informação no museu: uma revisão de premissas. In: Gabriel M. Forell Bevilacqua[ed.], 1 *Seminário Serviços de Informação em Museus*, São Paulo, 25–26 nov. 2011. São Paulo: Pinacoteca do Estado. Available at: http://biblioteca.pinacoteca.org.br:9090/publicacoes/index.php/sim/article/viewFile/6/5 [Acessed 12.04.16].

Middleton, G (1939). Imitation of Hand-made Lace by Machinery Part II. *Bulletin Of The Needle And Bobbin Club Volume*. V. 23, N. 1 – New York. Available at: <http://www.cs.arizona.edu/patterns/weaving/periodicals/nb_39_1.pdf>. [Acessed: 6 set. 2012].

Müller, L. 2006. Fontes para o estudo da história da indumentária, 1750-1900. In: Tereza C.T.de Paula (ed.) *Tecidos e sua conservação no Brasil: museus e coleções*. São Paulo: Museu Paulista da USP, 127–130.

Neves, M. 2000. *Desenho Têxtil – Tecidos*. Portugal: TecMinho, 2000. Vol. I.

Nottingham Trent University. Disponível em: http://www.ntu.ac.uk/research/groups_centres/art/108389.html#. [Acessed 25.01.17].

Ramos, F.L. 2004. *A danação do objeto: o museu no ensino de História*. Chapecó: Argos. Available at: http://www.pead.faced.ufrgs.br/sites/publico/eixo4/estudos_sociais/a_danacao_do_objeto.pdf [Acessed 19.02.16].

Relatório de economia criativa 2010: economia criativa uma, opção de desenvolvimento. Brasília: Secretaria da Economia Criativa/Minc; São Paulo: Itaú Cultural, 424 p. Available at: http://unctad.org/pt/docs/ditctab20103_pt.pdf [Acessed 27.04.17].

Swann, C. 2002. Action research and the practice of Design, *Design Issues* 18 (2), Massachusetts Institute of Technology.

Textiles, Identity and Innovation: Design the Future – Montagna & Carvalho (Eds)
© 2019 Taylor & Francis Group, London, ISBN 978-1-138-29611-4

Fashion: Spokesman of a generation that will not be silent

V. Szabo & M. Mendes dos Santos
USP, EACH, University of Brazil, Brazil

P. Maria da Silva Costa
UNASP, Master Degree in Health Promotion, Adventist University Center of Sao Paulo, Brazil

M. Sílvia Barros de Held & João Paulo Pereira Marcicano
USP, EACH, University of Brazil, Brazil

ABSTRACT: Fashion is expression. Clothes carry symbols of values, status and the belonging of each individual in society. In regards of gender, we will show that over the history, not always the clothing meant distinction. The unisex concept already represented the equality of rights, the equality of power, sometimes just style or practicality. This unisex proposal, goes back and forth in order to express a cause. Talking about the future, the values and behavior that are being cultivated within the children and teenagers are seen, they are the Z Generation or iGen, and the world where they are immersed in, is a digital world, fast and interconnected. This article invites you to take a look and understand this generation, instead of opposing to the innovations that are revolutionary and promising in all sections, from the industry to products and services. And what about the fashion? It is more likely that the unisex proposal will be absorbed by many groups supporting their causes, in which the clothing is the pacific sign of this expression.

1 INTRODUCTION

This current research's goal is to translate the models of clothes as a way of expression and peaceful struggle for equality and freedom rights, that aims the demystification of sexual choices embedded in choosing the costume.

Considering the design, fashion is the review of the world through the dressing art, what is called clothing, in reality are objects featuring symbols, and how one is going to use it will turn into the silent voice, consciously or unconsciously, of his or her judgement about the world and beliefs that guides them. Notice that the individual is a product of a society who changes itself concerning the comprehension and assimilation of what is feminine and masculine.

Today, the generation Z is found as a protagonist of this scenery, and carries in its ideals different understandings of gender, identity and the body in which resides.

Coming from the "politically correct" inside the social role that each one has to develop, full of this guesswork, the Z's demystify old fashioned conclusions from common sense that there are obligations directed to a respective gender. The metamorphosed law, would be the adaptation of this roles to every person who wants/needs to develop, without being labeled.

However, there are some aspects in the generation Z, which brings an ancient idea of the meaning of the clothing to the being, this would be a sociological term of belonging to a certain group. Once the individual is sociable, there are measures that are taken consciously or unconsciously, to choose which social circle would fit to his or her precepts, being no different, of course, from the representation of this values from the clothes and the body that grows into a showcase of this process.

What's new? The article will show the potential of the theme in the cycle that surrounds the generation Z, which with an amazing dimension and speed will hit the global scale, since this generation is equipped with technologies that didn't exist in previous generations.

The communication networks will have a special place in this article, after all they represent the voice of a generation that uses the internet to rise, behind the screens, a flag of equality towards the judgement war which bullies, hurts and brings bitter consequences to the society through ages, oppressing its weakest group.

There will be a review from another angle of what is unisex, where genderless clothes will translate the equality in roles and social responsibility, showing rights attained to make people see the necessity of the being in "being" what he or she really wanted to be and not only just showing the "politically correct" to be accepted.

Figure 1. Adam and Eva in "as vestes que Deus me vestiu". Source: Palavra Viva.

Figure 2. Babylonian costumes. Source: Vestuario da antiguidade.

2 THE GENDER AND THE CLOTHES IN ITS BEGINNING

The bible has in its pages, one of the first registries about gender, "male and female he created them" from Genesis 1:27, where is contrasted with the creation of two genders, female and male, another report is about the clothing, after eating the forbidden fruit and noticing their own nudity, the first couple uses, as described in Genesis 3:7: leaves to make themselves loincloths as shown in the Figure 1.

Right after giving the judgement, God prepares to the couple more resistant clothes, made from animal skins according to the book, in Genesis 3:21.

In the evolution scenery developed by Darwin, and as Ramos and Lencastre says (2013, p.2):

Of course, to Darwin, the sexual selection consisted in the choice, made by each partner, of singular morphological characteristics and behaviors considered useful in the reproductive point of view, in which, with time, they would mix creating male and female characteristics of their own..

Considering this way, only aspects that would enable the reproduction of the specie, regardless the necessary gender to fecundation. Furthermore, beyond the influences of the environment, the type of gender to be defined would be gradually built.

2.1 *Unisex clothes in the eastern civilizations*

Few things are known about the way of dressing of old civilizations, by looking to engravings of that time which resisted through time and made possible the study of ancient societies, their clothes and their life

The mesopotamians, one of the biggest civilizations of their time, brought in their concepts the unisex fashion, their clothes were equal to men and women, the changes only occur to differ status, as Silva (2009) says.

In either Assyria or in Babylon, the typical costume was a kind of tunic with short and tight sleeves that looked like the egyptian Kalasiris. In the lower social classes, this were the clothes for men and women, just changing the use of a belt. Even in the prosperous periods, the noble's slaves still used this costume.

> The man from the higher rankings wore the same outfit with short sleeves, but longer, reaching the feet. Almost all wore decorated belts according to their own status, their clothes were also more or less decorated and embroidered. (Silva, 2009, p.6).

This people are from completely different times compared to ours, yet they carry in the history of their cultures the genderless factor in which were fabricated and used their clothes. Without any exaggerated distinction between men's and women's costumes, that didn't stop them from belonging to a specific gender just because they were using similar clothes. The contrast promoted by the society were only for the status, the noble costume were more complex than the regular one.

For living in a hot zone, as the babylonians, the egyptian used simpler clothes, which had the power to distinguish the individuals socially, since the poor people, apart the gender, didn't had any clothes. Remaining the focus in the status brought by the costume.

> The typical egyptian costume is the Chanti, a kind of male loin-cloth, and the Kalasiris, a long tunic used by men and women (Silva, 2009, p.6).

2.2 *Unisex clothes in classical civilizations, Greek and Rome.*

The version of genderless clothes in the west remains with pretty similar bases of the east, where just a few differences were found between the feminine and masculine costumes, returning to the eastern concept that its meaning were focused in differ status, not gender, having a slightly alteration when used in special occasions. For example, the ones used to go to temples, these had to reflect the purity of the being, and

Figure 3. The egyptian and their clothes. Source: Guilherme.

Figure 4. Greek/roman clothes. Source: Pinterest.

required different colors from the common wore daily, showing the connection that fashion will have with spiritual concepts, bringing the representativeness of the devotion of the individual towards the gods.

The greek and roman civilizations, will have in their clothes, besides the esthetics valorization: the status, bringing to the surface the representation of the age and the roles of each person.

In rome, like Silva (2009) says, people in political offices as senators were obligated to wear, under punishment threat, a white robe, so that could show his position in society.

In general the clothes worn by greeks and romans were basically rectangles tied with brooches and belts, the rich men used not only the rectangle, but also a draped cloak. Because to them, the bigger the volume in the costume, the bigger the power and the influence.

The clothes mentioned by Silva (2009), didn't include clothes with eroticism. Focusing in esthetic and therefore reflecting, like the senators, the voice of the society which he represented, in other words, the costume meant much more than the personal tastes, but his or her behavior, in a silent way, translating the being.

Figure 5. Matrimonio Arnolfini. Source: Espaço da arte.

2.3 *Gender with its specifics clothing, the history of the Gothic Europe/middle ages scenery*

The clothes carry within them the meaning of ages, related to the representation of masculine and feminine gender (Zambrini, 2016).

The differences in the costumes start to get bigger in the Gothic Europe, where man begin to wear pants and pantaloons.

In the middle ages as Zambrini (2016) says, the differences of clothes between the genders are more vast since the Catholic church starts to condemn the woman as an existence that induces to sins.

The clothes wouldn't stop reflecting the time, politics and beliefs, and for this reason women begin to wear costumes that could hide their bodies.

In the modern age, a change in the men's wardrobe is noticed, where they start to use embroideries, laces and colorful cloth, common in the feminine universe, like Zambrini says (2016).

Notice that in this point the term genderless in the clothes starts to resurge, this time with men using materials typically feminine.

Parallel to this changes, the market sees in this a rising source, starting to invest in the production of clothes, systematizing the way of dressing, so then, consequently the bourgeoisie could attain a considerably rise in the amount. And one more time the clothes would differ the status and role of each individual in society, but with an extra, now the limitation of the territory by gender, between what would be feminine clothes and masculine clothes.

3 MARIE ANTOINETTE AND HER CLOTHING, PROTESTS TRANSCRIPTED IN CLOTHES

Marie Antoinette, austrian, married with king Louis XVI, lived in Versailles in the Baroque period where the rococo style prevailed in the wooden art in general, was seen also in her clothes.

About the clothing, the queen wore in the beginning at a young age as a hobby, in Versailles, going from

Figure 6. Middle Ages clothes. Source: Bitsfashionblog.

Figure 7. Le Boudoir de Marie- Antoinette. Source: Sana.

Figure 8. Fashion history. Source: Dernière.

Figure 9. Fashion history. Source: Dernière.

two to three times a week to the parisian capital, in order to be known by her subjects, who felt extremely proud for having the queen in their city, squares and stores.

However what was a hobby in the beginning, started to influence the subjects of the queen, showing a never seen proximity from the court towards commoners, since her clothes were different from the traditional court clothes and similar to the bourgeois women clothes, becoming a reference to all of them, her clothing were remarked with touches of extravagances, escaping from the traditional, exhibiting competence, strong opinion and capability against the court.

Antoinette compromises herself in transmitting this ideals in the clothes, like the reinterpretation of the redingote coats, which were typically for men and used in riding, were used a lot by the queen.

Marie Jeanne Rose Bertin was her favorite stylish, but the other ones that helped choosing the models, weren't exclusive professionals from the court, deals were made with them so that after the queen wore the clothes, one week later they could sell those to the rest of the population.

Marie Antoinette used the clothes to show respect and credibility (Crowston, 2014). The fashion to the queen were an instrument so that she could use it to demonstrate her power, coming from her feminine power, which influenced subtly and indirectly, all the sovereignty of her country, emphasizing that "to have credit with an empire, it is necessary that you look like you already have it", consequently, Antoinette focused her fashion in a way that she could always be the center of attention wherever she was.

Skillful in the creation of new styles, dominating the branches of fashion and using it to show an image, Marie Antoinette, through her clothes, presented to society a hidden claim to obtain prestige by the court, roving that even though she was a woman, was powerful and influent.

According to the american researcher Caroline Weber, author of queen of fashion: what marie antoinette wore to the revolution: caroline weber queen of fashion: what marie antoinette wore to the revolution "she used fashion as a political instrument, in a way that she could increase or sustain her authority in moments when she seems to be at risk". It was by the appearance, how Antoinette showed herself as a queen: beyond any other woman in france. Unlike Cleopatra or Elizabeth I, Antoinette didn't dressed herself to intimidate; her idea was to astonish; in the costume balls always searched for the most gorgeous dress, that could make her remarkable in the middle of the crowd. This way, everyone's eyes would be upon the queen, and in a few days later, her style would be copied by every women in France, noble or bourgeois. Never before, a french queen were that alluring. They used to be discreet. Antoinette dared to go against the court by her fashion style, and for a long time was admired and imitated, like a celebrity nowadays or Lady Diane in the United Kingdom. Becoming the maximum reference in fashion: it was her that dictated the tendency about dresses, hairstyles and makeup. She was copied by nobles in Versailles and the rich bourgeoises. The sequence marks the moment of transformation of the character: from a fragile teenager to the Marie Antoinette who caused all that commotion. She starts to use her position as a queen to live a dream's life. Her biggest pleasure becomes the construction, day by day, of her own glamorous image (Sant'Anna; Homsi, 2011, p.11).

But Gomes (2016) says that, before her tragic death, Antoinette left behind all the royal fancy clothing to wear simple cotton dresses, the glamorous hats didn't fit anymore in the image that she wanted to show, and instead she chooses the straw hat, to transmit the idea of crises that the court was being through.

Lately, with the king's death, Antoinette wore only black, representing grief and a silent claim against her own accusers.

Afterwards, being condemned to the guillotine, her choice was a white dress which pleaded in favor of her innocence, showing peace and plenitude, emphasizing her courage and steadiness.

> To Marie Antoinette, Fashion were not only about fancy clothes and tendencies, but, a mean of communication, a way of express sensations, feelings, indignation, and mainly, protest against what she judged wrong. (Gomez, 2016, p.1).

4 FASHION AND THEIR INTERPRETATIONS IN THE TWENTIETH CENTURY

After comings and goings the genderless idea earns voice and expression with a bigger reverberation in a moment that fashion becomes the silent claims showcase, the XX century, filled with changing, when the fashion had a strong participation between the decades.

The 20's will be a scenery of social changing, where the woman starts to fight for her rights in the job market, looking for other roles besides the mother and wife ones and the same acceptance as the men as a capable professional. These new women choose clothes with a straight model, leaving behind the use of the corset and other kinds of dressing that would emphasize their silhouette.

In the following decades, as Steverson (2012) says, the genderless visual gains strength because of the second world war, the women start to obtain more space and their clothes start to show more and more their fights and achievements. Pants, boots and mannish knits turn into popular items among them, considering the subliminal message about their busy days, giving more comfort and practicality.

Highlighting the 60's, when the genderless fashion is labeled as unisex, fed by lots of important people who helped in the dissemination of the idea, like the very short hair of the famous model Twiggy, the revolution of the woman's wardrobe using "stolen clothes from a man's wardrobe", by Yves Saint Laurent exemplified by "Le Smoking", and Coco chanel with the cardigan and the skirts above the knees (Baker, 2015).

In the musical world, Annie Lennox with tie and suit, the singers Boy George and David Bowie with his pantaloon pants.

The genderless fashion with their comings and goings, little by little demystify and rebuild social perceptions and show the importance of the clothes in culture, spreading not only innovations in the social scope but a showcase of their behavior and tendencies, which will bring a new point of view to the future generations so they can glare and change the world around them.

What used to be insane, woman wearing pants, today is completely normal, because the clothes let the individual performs his or her right to be free, free from the corsets that by mere esthetic oppressed the daily life of a lot of women and free from the judgements.

The woman is not less feminine for not wearing high heels or for wearing pants and a blazer, to looks like a man, trying to show credibility and trust in her clothes, otherwise is judged.

The past differences still remain in the present, unfortunately the current moment is of an unequal society, in this context the genderless appear giving to the individual, without judging, oppress or label the chance of standing against the world, without the determinism forced by the social construction around the gender.

5 UNIFORM: WITHOUT GENDER FOR EQUALITY

Uniform (from Latin uniformis) is a pattern of clothes worn by members of an organization. If used as a noun, the word uniform is synonym of garb, attire, a costume worn by elements of a same group. Regardless the gender, they differ only to contrast the status between the elements of the group.

In places and organizations like the army, police academy, surgery centers, civil aviation, and many others, to men and women the clothes don't have any difference, reflecting the acknowledgement for equality between gender, and that they can achieve high rankings by their own efforts.

Figure 10. Candidates from 2017 participate in their first graduation. Source: Viaeduca.

Figure 11. Jaden Smith and his girlfriend. Source: Pinterest.

6 FASHION IN THE GENERATION Z – IGEN

Known as "digital natives", they are the ones born after 1995, teenagers nowadays, their subjectivity is built and expressed through "online", leaving the intimacy behind and becoming public because of the internet.

They rise from a technological culture marked by the easy access to information, with a small tolerance to the limitations of space-time.

In their process of subjectivation is notable the influence of the pairs (Caniato, 2008) and the formation of groups (Santos; Lisbon, 2014).

The childhood of this generation occurred alongside the constant technological evolution, making them more globalized than the previous generations, due to the daily contact with technologies like cellphones and computers since birth, as Ceretta and Froemming (2011) say.

The virtual universe, provided by technology influences their social interactions. The Z's are not limited by their gender, biological gender, sexual orientation, they share the same preferences about their clothes or ideals (Ceretta; Froemmin, 2011).

They are interactive individuals, updated, realist, social activists, recognize equal rights in society despite the gender, "they demand equality between people from different races and genders. Want to change the world and support the local communities", according to Dan Schawbel, creator of "Workplace-Trends.com" and author of the best seller "Me 2.0: Build a Powerful Brand To Achieve".

The generation Z illustrates the theory of the canadian sociologist Marshall McLuhan, who says that we live in a global village. The internet changed and keeps changing the world. Any cause spread in the social networks or in the internet, echoes, gains followers, breaks physical and hierarchical boundaries, comes from a regular anonymous person to reach the world audience affecting even decisions and positions of world leaders. It is already in human history the Arab Spring, developed because all these factors. Not forgetting, that the spreading of cyber bullying, terrorism acts, xenophobia are evil to society, but the point of this article is to focus in the positive side of the innovations.

To Gilles Lipovetsky (Siqueira,2013), in "Hypermodern times", the rule is innovation, "the model that is repeated is not having a model to repeat", fashion is what seduces to buy and makes the consumption an essential part of the composition of the hypermodern being, the social consciousness is the urge to recognize the other as a different being. It is the necessity of living in a world where everybody is equal because of their own differences and the acknowledgement of these differences as something in their nature is everyone's right.

The fight for equality of rights and roles in society, for the definition of one's sexuality, for the feeling of belonging or just for style, the genderless theme or neutral gender is one of the topics of discussions of generation Z, diffused in digital medias, highlighting people, behavior and products.

In brasilian portuguese, following the generation Z genderless concepts, they replace the letter "a" and "o", which represent female and male gender, for the letter "e". Giving a genderless meaning to the word. Like the word "amigo/amiga" (male friend/ female friend) changes to "amigue" (neutral) (Allegretti, 2016).

Through the fashion they express, through "selfies" they sell and through likes they validate themselves. Instagram, Facebook, Youtube, Vine, Snapchat, Skype are all portals that transport them from the real to virtual world in an obsessive feedback, because their relation with time cares about the present and the immediate answer, they test everything, nothing is irrational. What doesn't fit, will be easily replaced by the new things.

This generation will bring their own sexuality as something in construction. And homosexuality, bisexuality and pansexuality are themes that are more and more discussed in classes and in the daily life of teenagers, their behavior, vocabulary, the way they dress and the color of the painted hair or nails, reveal

Figure 12. Jaden Smith – Louis Vuitton Campaign. Source: Dailymail.

Figure 13. Shiloh Pitt. Source: Inquisitr.

their choices. The unisex is rewritten as the genderless fashion.

7 CONCLUSION

As a result of most of the members from generation Z still being children and teenagers, many of their adults characteristics are yet to be examined, being watched and analysed through the next years.

The current information available about this generation shows that they are getting more self-conscious, independent, innovative and focused on the goal, through immediacy and practicality. They didn't live in an era before the social medias, they live in the online world, which resulted in deep implications in their own relationships, time usage, learning and consumption. They arrived in a world where gay marriage is allowed by laws, teenagers take part in suicide attacks, while others turn into billionaires creating applications for cellphones or becoming famous because of Youtube.

Question what exists and this creates big challenges and opportunity to all kinds of organizations and industries. Understanding them, instead of resisting, will allow a study of data that will drive the directions of a business to the success, as well as helping recording another step into the evolution of the human being. Fashion is a peaceful manifest at a hand in the past, in the present and forever.

REFERENCES

Allegretti, F. Amigues para sempre. Revista Veja, fevereiro 2016, edição 2465, p 62–69.

Bitsfashionblog. Middle Ages clothes. https://bitsfashionblog.wordpress.com. Access may 03, 2017.

Boucher, François. História do Vestuário no Ocidente. São Paulo: Cosac Naify, 2010.

Caniato, A. M. P. A subjetividade na contemporaneidade: da estandartização dos indivíduos ao personalismo narcísico. Silveira, A. F. et al., org. Cidadania e participação social [online]. Rio de Janeiro: Centro Edelstein de Pesquisas Sociais, 2008. P 5–22.

Ceretta, S. B.; Froemming, L. M. Geração Z: Compreendendo os hábitos de consumo da geração emergente. RAUnP, Natal, ano 3, n. 2, p. 15–24, abr./set. 2011. Disponível em: <http://repositorio.unp.br/index.php/raunp/article/view/70>. Acesso em 28 de julho de 2017.

Conti, Flavio. Como reconhecer a arte Rococó. Lisboa, Edições 70, 1978.

Crítica: Com imagens que encantam, 'Azul é a Cor Mais Quente' é maior destaque do ano. Disponível em: http://www1.folha.uol.com.br/ilustrada/2013/12/1381274-critica-com-imagens-que-encantam-azul-e-a-cor-mais-quente-e-maior-destaque-do-ano.shtml. Acesso em 04 de maio de 2017.

Crowsron, C. Maria Antonieta Foi a Primeira Mártir da Moda. Disponível em: https://www.vice.com/pt_br/article/z4bq4y/maria-antonieta-foi-a-primeira-martir-da-moda-v21n2. Acesso 28 de julho de 2017.

Dailymail. Jaden Smith – Louis Vuitton Campaign. Avariable in: http://www.dailymail.co.uk/femail/article-3384070/Jaden-Smith-17-revealed-new-face-Louis-Vuitton-WOMENSWEAR. Access may 04, 2017.

Debord, G. A sociedade do espetáculo. Rio de Janeiro: Contraponto, 2000.

Derniere. Le Boudoir de Marie- Antoinette. Avariable in. http://maria-antonia.justgoo.com/t127-marie-antoinette-et-les-corsets-un-rapport-conflictuel. Acesso em 28 de julho de 2017.

DICIONÁRIO INFORMAL. Aldeia Global. Disponível em: http://www.dicionarioinformal.com.br/significado/aldeia%20global/11077/ – Aldeia Global. Acesso em 04 de maio de 2017.

Espaço da arte. Matrimonio Arnolfini. Avariable in: https://sekcastillohistoriadelarte.wordpress.com/orna. Access on may 3, 2017.

Ferro, Marc. Cinema e História. Rio de Janeiro: Paz e Terra, 1992.

Fraser, A. Maria Antonieta. São Paulo: Record, 2007.

Gomez. P. Maria Antonieta, a expressividade da ex-Rainha da França através da moda. Disponível em: http://universoretro.com.br/maria-antonieta-a-expressividade-da-ex-rainha-da-franca-atraves-da-moda/. Acesso 28 de julho de 2017.

Guilherme. The egyptian and their clothes. Avariable in: http://guilhermeribino.blogspot.com.br/2011/05/roupas-egipcias. Access on may 3, 2017.
HISTÓRIA DA DEFINIÇÃO DOS GÊNEROS NA MODA. Disponível em: http://xicogoncalves.com.br/historia-da-definicao-dos-generos-na-moda. Acesso em: 03 de maio de 2017.
Hobsbawm, E. A Era das Revoluções. Rio de Janeiro: Paz e Terra, 1977.
Inquisitr. Shiloh Pitt. Avariable in: http://www.inquisitr.com/3486465/angelina-jolie-and-brad-pitt-sh. Access may 4, 2017.
KPMG. Geração Z: Apostas, Valores, Visão e Empresas. Disponível em: http://noticias.universia.pt/destaque/noticia/2017/04/12/1151572/geracao-z-aposta-valores-visao-empresas- (KPMG). Acesso em 04 de maio de 2017.
Lever, E. Marie Antoinette: The last Queen of France. London: Portrait, 2006.
Morin, E. Cultura de massas no século XX – O Espírito do Tempo, 1: Neurose. Rio de Janeiro: Editora Forense Universitária, 2002.
O ESCANDÁLO DA RAINHA: O MARTIRIO DE MARIA ANTONIETA – CONCLUSÃO. Disponível em: https://rainhastragicas.com/2012/06/07/o-escandalo-da-rainha-o-martirio-de-maria-antonieta-conclusao-4/. Acesso 28 de julho de 2017.
Palavra Viva. Avariable in: http://miss-joelmagoncalves.blogspot.com.br. Access on may 3, 2017.
Pinterest. Greek/roman clothes. Avariable in: https://br.pinterest.com/miicdm/hstoria-da-moda-grecia/?lp=true. Access on may 3, 2017.
Pinterest. Jaden Smith and his girlfriend. https://br.pinterest.com/explore/jaden-smith-novia. Access may 04, 2017.
Racinet, A. The complete costume history. Taschen: New York, 2007.
Ramos, M. C.; Lencrastre, M. P. A. Feminino e o masculino na etologia, sociobiologia e psicologia evolutiva: Revisão de alguns conceitos. Disponível em: http://www.scielo.mec.pt/scielo.php?script=sci_arttext&pid=S0874-20492013000200002. Acesso em: 03 de maio de 2017
Roche, D. A Cultura das aparências: Uma História Da Indumentária (século XVIIXVIII). São Paulo: Editora Senac São Paulo, 2007.
Sana. História da Moda. Avariable in: http://modahistorica.blogspot.com.br/2013/05/a-moda-na-era- rococo.html. Acesso em 28 de julho de 2017.
Sant'Ana, P; Homsi, L. Maria Antonieta: conexões entre moda, cinema e negócios. Disponível em: http://www.usp.br/anagrama/Expressao_MariaAntonieta. Acesso 28 de julho de 2017.
Santos, W. P.; Lisboa, W. T. Características psicossociais e práticas de consumo dos "nativos digitais": implicações, permanência e tendências na comunicação organizacional. Comunicação & Mercado/UNIGRAN, Dourados – MS, vol 03, n. 06, p. 98–110, jan–jun. 2014. Disponível em: http://www.unigran.br/mercado/paginas/arquivos/edicoes/6/7.pdf. Acesso em 28 de julho de 2017.
Silva, U. C. História da Indumentária. Disponível em: https://webcache.googleusercontent.com/search?q=cache:vE3Pcwl w9wJ:https://wiki.ifsc.edu.br/mediawiki/images/e/e2/Hist%25C3%25B3ria_da_Indument%25C3%25A1ria_vers%25C3%25A3o_02.pdf+&cd=3&hl=pt-BR&ct=clnk&gl=br. Acesso em: 03 de maio de 2017.
Silveira, A. F., et al, org. Cidadania e participação social [online] Rio de Janeiro: Centro Edelstein de Pesquisas Sociais, 2008. p. 5–22. Disponível em: http://books.scielo.org/id/hn3q6/pdf/silveira-9788599662885-02.pdf. Acesso em 28 de julho de 2017.
Siqueira V. Os tempos hipermodernos de Gilles Lipovetsky-Resenha.2013. Disponível em: http://colunastortas.com.br/2013/09/05/os-tempos-hipermodernos-de-gilles-lipovetsky-resenha/. Acesso em 04 de maio de 2017.
Sousa, G. *Como as Redes Sociais provocaram a Primavera Árabe*. Disponível em: http://www.estrategiadigital.pt/como-as-redes-sociais-provocaram-a-primavera-arabe/. Acesso em 04 de maio de 2017
Stevenson, N.J. Cronologia da moda: de Maria Antonieta a Alexander McQueen. Tradução Maria Luiza X. de A. Borges. – Rio de Janeiro: Zahar, 2012.
Vanoye, F.; Goliot-Leté, A. Ensaio sobre a análise fílmica. 2ª edição. São Paulo: Papirus Editora, 2002.
Verdú, D. Geração Z mudará o mundo. Disponível em: http://brasil.elpais.com/brasil/2015/05/02/sociedad/1430576024_684493.html. Acesso em 04 de maio de 2017.
Vestuario da antiguidade. Babylonian costumes. Avariable in: http://vestuariodaantiguidade.blogspot.com.br/2011/02. Access on may 3, 2017.
Viaeduca. Candidatos do ano de 2017 participam de sua primeira formatura. Avariable in: http://www.viaeduca.com.br/candidatos-do-ano-de-2017-participam-de-sua-primeira-formatura/. Acesso em: 30 de julho de 2017.
Weber, C. Rainha da Moda: Como Maria Antonieta se vestiu para a Revolução. Rio de Janeiro: Jorge Zahar Editor, 2008.
Zambrini, L. Olhares sobre moda e design a partir de uma perspectiva de gênero. Disponível em: https://dobras.emnuvens.com.br/dobras/article/view/452/409. Acesso em: 03 de maio de 2017.

Textiles, Identity and Innovation: Design the Future – Montagna & Carvalho (Eds)
© 2019 Taylor & Francis Group, London, ISBN 978-1-138-29611-4

Before textile • after wall

J. Afonso
CIAUD, Lisbon School of Architecture, University of Lisbon, Portugal

ABSTRACT: The architectural proposal for a new Textile Museum in Portugal was developed in parallel with the research for a light textile wall in relation to Architecture and Interiors furniture Design.
What better place to test and experience the applicability of textile in architecture than the building where, for years, textile has been produced, and where we can now re-discover the textile manufacture from the spinning to the weaving machines and see the old machines working again.

1 IDEA

Love, passion, and attention to detail are the ingredients of every textile design, same as in architecture. Picture 01 was taken on my first photo expedition to the "CORGA" Factory, it was the genesis of my concept for the textile Wall: the mix of the place and the pastel colors were the key elements I am usually chasing at every location.

Quite often we forget that one of the most important things in life is our happiness, and we, as architects, can make a big difference in the happiness and in the lives of the people that are affected by our architecture.

In this project, with important influences of the past, marked by many generations who worked in this factory, our objective was to highlight all the work and bring back an environment in which the main figure is the textiles.

We combined all these important ingredients that resulted in a wall that joins the textile with the architecture. What better way to put a smile on your face than being surprised by this wonderful wall with superficial textile and inside led's light (LIGHT WALL) *"... for cloth, like body, is a mediating surface through which we encounter the world..."* Boontje (2007), So in his experience of working with happiness, Boontje emphasis the importance of textiles amid ourselves, and of the application of textiles in architecture.

Figure 1. Corga factory.

In this first phase of creation and not only in this phase, would be essential speak of an inspiring and influential source of critical awareness – namely Art. Art, therefore, deserves a place in the heart of our society. It brings together visionary writings by authors with diverse social and cultural backgrounds, writings that can inspire reflection and action.

Art can play a role in this process through its capacity to create unprecedented situations, present parallel worlds, and turn invisible ideas into visible, in these ways Art can spark individual awareness of ethical, social, or political issues and speak to the motivations, convictions, or emotions underlying rules or laws.

We think the texture of a tissue used, has a great importance in the evolution of: *"... We want to play with the sensory capacity of textile to give architectural spaces a different touch and feel ..."* Moor (2014).

2 MUSEUM CONCEPT

We feel that we always have a responsibility to our community, to our country...

The methodology of conservation-revitalization design for industrial architecture heritage facing demolition has to take this into consideration.

In the project of the Textile Museum, in Castelo Branco/ Retaxo, Cebolais, that we are rehabilitating, as well as restoring the old wool manufacturing machines. The historical industrial space was reborn by keeping its historical value and a new function.

If we can "re-design" these buildings respecting their historical value, we can preserve and re-use them.

This redesign exercise allows dividing the all in a few essential elements. We can then hold these

Figure 2. Textile machines.

Figure 3. Textile workers.

elements in the imagination, being taken to dissolve a series of static fragments in a dynamic chain: moments of collective memory, labor moments between man and machine! ... The thoughtful creativity of a new museum is expressed in a platform of emotional experiences, finding authenticity in its diversity and its own identities.

We must rebuild for people, we must keep in mind all the different generations who have, in any way, inhabited these industrial buildings, constructing them, working there, walking in them, whose lives have crossed these spaces and who are usually forgotten. We should integrate all these realities as a building's latest piece of maturity, according to a dialectical-materialist interpretation of history.

Re-thinking the idea and concept of the museum was a challenging task, partially because all the generations who had the knowledge of textile manufacture wool, have now disappeared and there was the risk of losing all the skills and memories of this rich cultural heritage. On the other hand, it was possible to gather a reasonable collection of machines and textile equipment to be displayed in the museum. The machines, particularly the looms, were in need of deep repairs, but it was feasible to provide proper conservation very quickly so that the equipment could be shown operating in its original function maintaining the authenticity of the wool manufacturing industrial site.

We look for comfort in a space that is both a living museum with the old machines and a retrospective anthropological, economic, social, and cultural history of the region Beira Baixa, where the textiles were produced. The textile machinery working, processing the woolen yarn, may well be the singularity of this Textile Museum. . .

> *" surface becomes a structure, and structure becomes geometry it goes beyond a pure transmutation of textile techniques into design techniques: it is, a completely textile way of thinking in architecture. . . " Ludovica (2006).*

The purpose of architecture is people, therefore we need to think a lot more about how we can recycle buildings and spaces reintegrating them in peoples lives.

The more diverse questions the architect asks, in the act of pursuing the solution, the more likely it is to trigger interesting proposals and bring an interdisciplinary flourish with the development of the project.

This new Museum inside the small community of Cebolais de Cima / Retaxo promotes a lifestyle where you can be eco-friendly and carefree aiming to re-connect the native inhabitants to their roots.

We realized the potential to create something unique in this old industrial building, and attention to detail and nuanced natural light design are the real game changers.

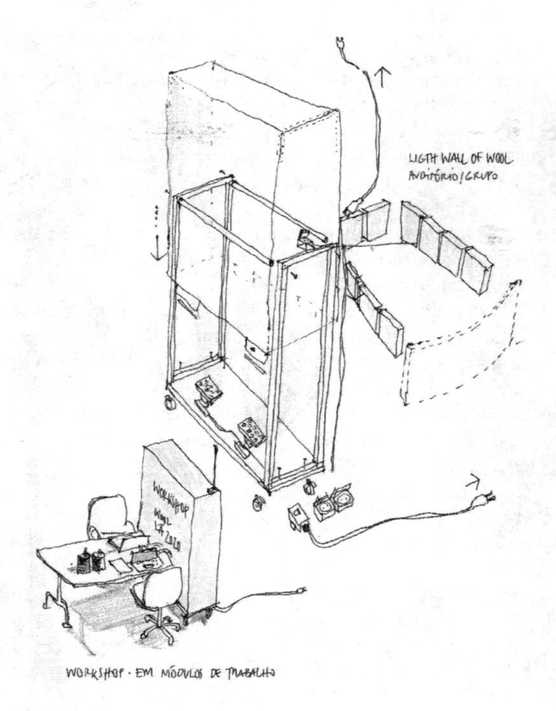

Figure 4. Textile Wall/Sketch.

Figure 5. Light Wall / Inside the Pavillion.

An alternative natural light through skylights has emerged in the ceiling (Experimental Pavilion).

This space is seeking smarter answers to improve museum mobility in a broader sense. The necessity of a versatile and smart Textile Museum, not only in terms of energy efficient but also flexible in its uses, pleasant and welcoming for all potential occasions, including: art meetings, café gatherings, a multipurpose space where different generations can interweave and different activities can thrive. In this way, finding the right balance between the space that values the cultural heritage of the region and a space that is fun, stimulating and attractive to the younger generations. In this way achieving a new and holistic concept for modern museums.

An example of the application of this concept is the Musicological Pavilion's integration of old and new – in this case, the application of "non-aggressive" technology elements to an almost industrial form: the isolation leads to a focus, we want to share this inspirational environment with future visions.

Progress is often shaped by an economic logic. However, the most important motivation is to improve one's well-being and "untouched" industrial spaces with old machines and looms, and textiles "*. . . provide protection from heat and cold, absorb noise and give the control the amount of light entering the view. Their manifold characteristics and application potentials make textiles a highly interesting architectural material. Textiles further possess very special, sensually tangible often poetic aesthetics. . .*" Krugër (2009).

3 TEXTILE WALL

We think the main challenge and singularity of this project was to find a driving force in the act of preserving the industrial heritage of an industry which has virtually disappeared leaving behind deactivated and deteriorated factory buildings.

In this way, there was a research work of the element that would re-create a space with a past with so many memories and a huge impact on all the population living in Cebolais de Cima.

After a long and exciting process of creation, our big focus was to join the architecture world with the textiles directly and still give a fundamental paper to the light. The architecture is done by opaque walls, and we think this concept can change doing walls with interlacing threads of the fabric. Like this, we got a clear selection between environments but creating something slight and nice.

The light is another fundamental element that gives life to the wall and to the building. As an element translucent, it achieves that the light happens between the threads of fabric and creates a connection between environments. Another function is to be removable, can change of place and can be used in various different situations. All this makes sense by the situation, the place and by the circumstances that it is done this piece.

"WALL-LIGHTING" collaborative work was the forefront of everything we do. The renovation is not an end, but a starting point, bringing the partners and our team together to respond to contemporary demands as it occurred with this dynamic textile wall.

Each element adds up elaborate layers to an unpretentious design. The sculptural "WALL-LIGHTING" combines warm pieces of metal frames, wood and white textile with a smooth design and a discreet cable management that makes it possible to attach, for example, a workstation... in forthcoming workshops in the Experimental Pavilion.

Our intention was to merge light and shadow in a single piece (a textile wall) achieving a sense of natural light and perceiving the shadow at the same time.

We reassess the "textile wall" appearance in the quest for this effect and when we think about a special furniture now we think about texture, touch, size... there as to be a physical and emotional connection.

4 CONCLUSION

Making this architectural project took about six months in a filtering process and the construction one year.

It was a process that cannot be counted on the days that it was done but by the effort and dedication. It was a project that has met all the goals that compromise. All the promises of a return to revitalize that manufacture, and to put the machines operating again were fulfilled. It was a long journey, not only by the architectural project but also it was a long learning process. I went to the village, I met the way of being and living of the people that live in that city and especially the people who worked in the old textile manufacture. These people were fundamental because they have made possible the reconstruction and repair of old machines that were forgotten by many years. In this way, we could all together and with a particular worth of the old employees, return to give life to something that was a reference to that village, and for all the generations that worked and spent their lives there.

It is a slow process, but it is the only way we know how to do it. To build in a context... it was more of a labor of love.

With knowledge and perception, man can face the challenges, providing he has the strength needed to cope positively with change. Design, as a tool for human achievement, can be a mechanism to receive changes in a good way. The balance between what we know, our existence, and what the future provides, will become the basis for responsive Design... *" art helps us accomplish a task that is of central importance in our lives: to hold on to things we love when they are gone..."* Botton (2013).

ACKNOWLEDGMENTS

The authors of this paper wish to thanks the Centre for Research in Architecture, Urbanism and Design (CIAUD) of the Lisbon School of Architecture of the University of Lisbon and FCT for founding this project.

REFERENCES

Boontje, Tord, Martina Margrets, New York, Rizzoli 2007, p.195

Botton, Alain, John Armstrong, Art as Therapy, London: Phaidon 2013, p.8

Ludovica, Maria, Textile tectonics, an interview with Lars Spuybroek, London: Architectural Design Architextiles vol 76, 2006, p.28

Moor, Tina, An Aestic Approach to the use of Textiles in Architecture, LUASA / Stoffwechsel: www.dns2014.org.

Krüger, Sylvie, Textile Architektur, Berlin: Jolvis 2009, p.6

Textiles, Identity and Innovation: Design the Future – Montagna & Carvalho (Eds)
 ISBN 978-1-138-29611-4

A woven language: Tais as an expression of Timorese culture identity

Maria José Sacchetti
CIAUD, Lisbon School of Architecture, University of Lisbon, Portugal

ABSTRACT: This paper focus on tais, the name given to the Timorese traditional woven textiles. The study of these textiles essentially woven by women, are relevant because they offer important insights into the history, culture identity and customs of the Timorese people. These cloths are an important part of the cultural heritage of this people, which is very fragmented into different ethnics groups and religions. These textiles distinguished themselves not only by their stylistic and technical weaving terms but also by their cultural meaning. These tais carry in them the signs of the tribes, their values, their beliefs, traditions, rites of passages and so on. This first investigation took place in site on the island of Timor by the author and it should be looked as a starting point for its further examination, especially in the case of the tais of East Timor, where previous wars and political instability did not make these kind of study possible.

1 INTRODUCTION

Timor Island is situated in the Indonesian, part of the Malay Archipelago, the largest in the world. Situated under 500 kilometres from neighbouring Australia, it is a mountainous island with a narrow shape that has been compared to the one of a crocodile. The island is divided into two separate parts: the west half, which is ruled by the Indonesian government, is predominantly Islamic but still showing signs of the previous Dutch control; the relatively newly independent east side is predominantly Portuguese-influenced and mainly Catholic.

This paper focuses on tais, the name given to the Timorese traditional woven textiles.

As in many other societies within Indonesian, Timorese textiles are an essential piece of the island's cultural heritage. Since the island different ethnic groups produce them, these textiles distinguished themselves not only by their stylistic and technical terms but also by their cultural meaning, where they play a relevant role in community rituals. Traditions on the island are rarely static, and new ideas and techniques – some of them imported from other distant islands – have been absorbed and reinterpreted over the centuries to accommodate contemporary social and economic demands.

In ethnographic terms, the Timorese are divided in two main groups: the *Atoni* from Melanesia and the *Tetum* from Southern Belu, originally believed to have come from Malaca.

Timor textiles are not only prominent for their quality and their colours but also for the range of the decorative techniques used to produced them. *Warp-faced Ikat* is a technique used throughout the island. Along with it we may find *Sotis*, a warp-faced floating weave, which give it a reversible appearance and last *Buna*, a discontinuous supplementary weft that resembles embroidery.

Women weave these textiles during the dry season. Since the government, mainly in the western side, has encouraged cotton growing, hand spinning is still wide spread, especially for distinctive textiles. Nowadays, commercially sold cotton, pre-dyed thread and chemical dyes can be easily found in regional street markets, where the towns of Los Palos and Desa Rasa, in East Timor, are both known for textiles produced using commercial thread and chemical dyes. Nevertheless, natural colours are widely used throughout the island. More so than any other island in the archipelago, the dominant colour – for unexplained reasons – is red. Although some experts suggest that the bougainvillea that blooms during the dry season inspires the colour, many Timorese communities traditionally associate it to life, blood and bravery, which I believe is a more plausible explanation.

While western dress is commonly worn on a daily basis on both sides of the island, local textiles continue to play a relevant role in the rituals that celebrate changes which come at different phases of social life and in social, spiritual rituals and others related to agriculture. During ceremonies, the men wear regular rectangular cloths called *selimut* that consist of two or three panels sewn together and wrapped around the waist, while the women wear similar *sarongs* around the waist or crossed over the chest. Small *selendang* or scarves are popular as gifts or for use in barter, as are belts, bags for *shiri* or betel (natural stimulants that are chewed) and items of headgear. In general, all these pieces are decorated with beads, *sotis* or *buna* rather than *ikat*. Contemporary *ikat*, especially from the Kupang area, is used to produce bags, hats, purses and other items of clothing such as waistcoats, trousers, skirts and jackets.

The decorative motifs used throughout the island continue to be of traditional origin: anthropomorphic figures with outstretched arms and hands are common, as are zoomorphic image or birds, cockerels, crocodiles, horses, fish and water bugs, and plants, trees and leaves. Geometrical designs similar to hooks and lozenges, locally called *kaif*, are generally interpretations of the Dong-Son culture.

Timor is divided into several units called 'regencies' that are subdivided into districts. These can be used to observe similarities and differences in the textiles produced. Thus, Kupang includes Amarasi, Amfoang and Fatuleu; the Centre-South whose capital is Soe, includes Amanatun, Amanuban and Molo; the Centre-North whose capital is Kefamenanu, includes Biboki, Insana and Miomafo; Belu, whose capital is Atambua, includes Lamaknen, Malaka and Tasifeto. Ambenu, whose capital is Oecusi, is the part of East Timor that includes the administrative capital Dili and is divided into thirteen districts.

Contemporary weaving in Timor is fundamentally traditional in terms of style and commerce. Many of the textiles can be found in an art centre in the outskirts of Kupang or in a large number of specialist shops in the same city. The art centre also has demonstrations of how to weave *sotis*. The weavers accept commissions to work names and logos into their textiles. The centre also focuses on a training program for weavers from other islands that wish to study the local craft. Some of these activities are subsidised by government programs, others by agents interested in promoting and developing local arts and crafts. This concept of subsidising the exchange of weaving skills between islands offers interesting potential for the future of textile production, as these programs clearly foster and stimulate the exchange of ideas, designs, colours and concepts.

2 METHODOLOGY

The Weaving Co-operative in Kupang

PT Timor Agung Flobamor is a craft studio where textiles are produced using hand-tied and dyed threads that are then woven on back strap looms (held by a band that goes around the weaver's back) called *cedogan* or semi-automatic ATBM operated by men. It takes around three weeks to weave a 75-meter roll. In this case, the weavers use commercially spun cotton and chemical dyes, but the time consuming process of producing the pattern by tying the threads and dying the pattern into the threads uses the same methods employed in the remotest villages. As in other islands where ATBM looms are used, the *ikat* is broader and longer than that produced on back-strap looms. However it is still *ikat* and not an imitation, as it may seem.

2.1 *Wrap ikat in amarasi*

The leading textile producing regions are located around Kupang, Amarasi, Amfoang and Fateleu. The Amarasi weaving centre is in Baun, whose fabric is characterised by a white central panel flanked by wrap *ikat* in reddish-brown with supplementary wrap floats of *sotis*. The *sarongs* are made up of two sections of wrap *ikat* decorated with *sotis*. In general Amfoang textiles have a white centre with lateral panels in *buna*, whereas those from Fataleu also have a white centre with *buna* motifs in the centre and on the lateral panels. In the outskirts of town in the Amarasi region, they produce *selimut*, *sarongs* and seledang with wrap *ikat* that uses ancient designs and techniques. In a practically unaltered weaving process each of the approximately twenty-three communities satisfy local tastes, producing traditional weaving whose designs are passed from mother to daughter.

The image of a good dyer as a medieval alchemist applies in Amarasi. Every weaver in the area has her own secret recipe to achieve the shades of brown, blue, green, yellow or pink that she wants. The dying process can take anything between two or three days to several years, depending on the complexity of the colour and the number of colours that have to be mixed.

The thread used for *ikat* is either hand-spun or purchased in Kupang. The threads that are to be dyed a specific colour and that require mordant have to be soaked in candlenut or tamarind seed oil for approximately one week. The tied threads are then placed in a bath of the appropriate dye. Indigo, locally called *nila*, produces blue, while morindin (Indian mulberry), known as *uru*, is the basis for red, as is a group of other local leaves, barks and roots from indigenous trees and plants. Mud is used to produce a black dye. Furthermore, yellows, pinks, browns and greens are also used in cloths either as plain wrap bands or *ikat*. However, the most used colour in *ikat* is the distinctive reddish-brown that contrasts with the natural colour of the threads. Contemporary *ikat* textiles from the Baun region also include panels with *ikat* in pale blue and black. The threads used for the *ikat* parts are tied together in three or six sections, just the enough number of wrap threads required to produce a complete cloth or *sarong*.

After the threads have been dyed but before they are woven, they are treated with a solution of tapioca and water to stiffen them, which makes it easier to create the pattern, which should be tightly woven and clear. Once the wrap threads have been woven with a single weft thread in one colour, the cloth is repeatedly washed with cold water to soften it and dissolve the solution. The dyes are prepared with such care and skill that there is practically no loss of colour. The end result are soft and subtle, almost subdued tons with *ikat* motifs that appear to be the negative of the threads' natural colour. However, nowadays traditional motifs in Amarasi style can be found made exclusively using chemical dyed *ikat*, with the motif in black on a vibrant red, orange or yellow background.

The weavers weave the textiles on a body-tension loom. These women live in local communities where along with their families are responsible for the entire process: preparing the threads, tying the threads to create the pattern, dying the threads and ultimately weaving the cloth. The production process often includes combining the *ikat* technique with *sotis*. The *seledang* takes between eight hours and a day to weave, while a section of *ikat* for a *sarong* takes between one or two weeks.

The patterns and motifs are of great significance both for the Timorese who weave the textiles and for those that wear them. At first sight, many designs linked to the hook and the lozenge forms seem very similar. Many designs represent elements of daily life in the weavers' surrounding environment: plants, trees (bau neki) and leaves (kres kou), birds (esi ana) or snakes, as well as abstract motifs such as boxes (pan bua ana) and belts to hold coins and weapons (pasu kena). These motifs are usually produced using *ikat*, although they can also be made with *sotis*.

As in other areas of Timor, the *selimut* – normally consisting of three pieces sewn together – is produced for local man. The two outer panels are decorated with *ikat* and *sotis* patterns, while the central panel is left undecorated in white. The women wear *sarongs*, sewn so that they fit snugly around the body with just one fold to allow them to walk. The *sarongs* are worn as skirts and combined with blouses. In most cases, a smaller cloth is added wrapped around the waist and hips. The *sarong* itself is made up of three narrow panels and can include strips of a single colour with a natural and chemical dye and bands with *sotis* produced with threads that are also dyed with natural or chemical colours. The *selimut* and *sarongs* are worn for ceremonies, religious rites and celebrations, and are also highly esteemed presents among members of the community.

The Amarasi weavers produced *ikat* and *sotis* textiles more for themselves than for sale. The few such items available are mainly purchased by tourists and visitors to the island, rather than inhabitants from the region. The communities that do receive visitors on a more regular basis, such as Baun, produced other items that are sold as souvenirs rather than offering the more elaborated traditional pieces. The weavers also sell individual parts from larger cloths or *sarongs* with sections in *ikat* to be used as handkerchiefs or shawls, and even smaller items to be used as belts, mats, table settings, some of which can be ordered these days.

2.2 *Wrap ikat in Amanuban, Amanatum and Molo*

The main textile-producing regions in Amanuban are Niki-Niki, Boti and Painite, while Soe is the main regional market. The characteristic features of textiles from Amanuban are its three-panel centres in wrap *ikat* whose borders are decorated with red stripes and textiles made up of two panels decorated entirely with *ikat* stripes. Their traditional *sarongs* have a similar form to the *selimut*, consisting of three panels with *ikat* in the central panel. Characteristic of the cloths of Niki-Niki are gentle shades of indigo and red and images of human figures. Amanuban is considered the most important centre of weaving in the region using *ikat*.

There are many notable areas in Amanatun, including Ayotupas and Oinlasi, which is also one of the region's main markets. Textiles from Amanatun feature *ikat* centres with hook and lozenge patterns, a white central panel with *buna* or a black centre, while the two lateral patterns are striped *sotis*. *Selimut* with a black central panel can be found as they are in Insana, but others with two entirely *ikat* panels, sometimes interspersed with strips are also popular in the region. Ayotupas is known for its traditional *sarongs* in four panels decorated with *buna*.

In contrast, the centre of textiles from Molo is either white or has *ikat* decoration with crocodile patterns.

2.3 *Wrap ikat in Biboki, Insana and Miomafo*

Highly coloured textiles in shades of reddish pink and brown *ikat* are typical of Biboki, but can also be found in the markets of Atambua and Oelolok. They normally consist of six *ikat* bands on a two-panel cloth (three bands in each panel) alternating with narrow single colour bands in red or green. The Biboki region is close to Belu, and some of the locally produced textiles have adopted the latter's style, which consists of two main and other narrower *ikat* bands. Some of the *selimut* produced in this area are also similar to those from Amanuban, with narrow red bands separating the *ikat* patterns. Traditional *sarongs* from Biboki mix *sotis* with indigo *ikats*. While many older pieces were produced using hand-spun and dyed cotton, others (mainly modern copies) do not include *ikat*. The region's central market is located in Kefamenanu, which is an outlet for many textiles produced in the area.

The weavers from Biboki region has been encouraged to continue practicing their art and have been provided with a sales point via the non-profit-making Tafaen Pah Yayasan foundation, set up in 1989 in Kefamenanu.

Many of the textiles produced in the Insana region can be found Oelolok, itself also a major weaving centre that is famed for its *buna* work, although *ikat* is still relevant, especially when produced in indigo. A *Selimut* from this area has three panels with central bands in black or white and a variety of decorative techniques on the lateral panels, or two panels with several *ikat* bands in indigo. The most common pattern shows a cockerel, usually portrayed in shades of dark brown, blue or black.

2.4 *Wrap ikat in Belu, Lamaknen, Malaka and Tasifeto*

These areas are all near the East Timor border. This proximity is reflected in the work, as the narrow *ikat* bands are more characteristic of East Timor than of the western half of the island. Modern pieces from this area often have a red or orange and black background with many narrow stripes in bright colours, a design that is also popular in East Timor. The *selimut* and *sarongs* from this area use *ikat* and the *buna* and *sotis* decorative techniques.

2.5 *Wrap ikat in Ambenu and East Timor*

Ambenu, previously known as Oecusi, is a small enclave that belongs to East Timor but situated within the west part of the island. Its traditional cloth is the *selimut* with a white centre decorated with *buna* and *sotis*. Contemporary pieces consist of small *selendang* with black and yellow *ikat* and the *selimut* with *ikat* in broad stripes with large motifs inspired by European and religious themes. These are, to some extent, similar to those found in the Sikka region on the island of Flores. Here the cloths formed by two panels seem to be symmetrical, but in fact one is narrower than the other.

East Timor has greater regional variety than the West in terms of weaving. However, due to the political instability on the East side of the island and that as a consequence it has never been an attractive tourist destination, the decorative art did not develop significantly, nor had there been an in-depth study of the craft, as it happened in West Timor.

In the context of the textile production, the techniques are similar on both sides of the island. The seeds are removed using crude wooden cotton-gins (ledú), followed by the spinning of the thread. The colours are mainly achieved using natural dyes obtained from the leaves of the *Táo* tree, roots of the *Menuc* tree and *Mivem* leaves. These are trodden, then mixed with lime and water and finally boiled in clay pots. The threads are then kept in this liquid for several days until they absorb the desired colour. In Balide, a town half an hour's drives from Dili, pre-spun thread and chemical dyes are used along side hand-spun cotton and natural dyes. The production of *ikat* is less common than in West Timor, but *sotis* and *bunda* are still the dominant techniques. The background colours used for *ikat* are red, orange and yellow, which contrast with black. European-style floral motifs and religious imagery are the most common designs – most likely as a result of the Portuguese influence, which left some relevant marks on life on this half of the island – substituting the hook and lozenge motifs found in West Timor.

3 CONCLUSIONS

The study of the art of weaving in the island of Timor is relevant to the study of the history and customs of the people and the culture on the island. It is a rich decorative art not only due to the variety of techniques that are used in the production of the local textiles – from natural dying processes to weaving techniques – but also because of the identity and cultural values signaled by the finished cloths. The recipes and some inherent secrets to their success are transmitted through generations and so they risk being lost if not documented properly. These tais carry in them the signs of the tribes, their values, their beliefs, traditions, rites of passages and so on. The textiles are intertwined with the peoples' lives, where the craft is part of their daily customs. These cloths are offered to each other on important occasions or for especial reasons which embedded them with emotional and personal values; they are passed down from generation to generation.

Particularly in the case of East Timor, the study of its tais is pressing. As mentioned before, a well-documented study does not exist with the exception of a sample of tais spread throughout a few ethnological museums. This first investigation took place in site and it should be looked as a starting point for its further examination. The war and political instability that East Timor lived in the past years gave place to a new country that needs simultaneously to reflect on its history and look into its future. Without a doubt the tais play a role in both processes.

ACKNOWLEDGMENTS

This research was only possible with the support of CIAUD – Research Centre in Architecture, Urbanism and Design, Lisbon School of Architecture of the University of Lisbon, and FCT – Foundation for Science and Technology, Portugal.

GLOSSARY

ATBM – *Alat Tenun Bukan Mesin*, a semi mechanical loom used throughout Indonesian.
Betel – Leaves of the climbing *betel* plant and other ingredients and other ingredients are mixed and chewed as a stimulant.
Buna – A decorative technique of discontinuous supplementary weft from Timor, similar in appearance to embroidery.
Ikat – Decorative technique where the threads of the wrap or weft are tied with strips of dried plants to form a pattern before the threads are dyed and woven.
Indigo – Natural blue plant dye.
Morinda – Indigenous tree whose roots produce a natural red dye.
Nila – Another name used for the natural blue dye.
Sarong / Tais Feton (in Tetun) – Cloth for female clothing, normally sewn into a tubular shape to fit close the body and worn from the waist down.
Selendang – Rectangular cloth smaller than the *selimut,* that is usually worn over the shoulders.
Selimut / Tais Mane (in Tetun) – Cloth for male clothing, larger than the *selendang* (approximately 2 x 1.30 m) and worn from the waist down.
Shiri – Stimulant chewed throughout Southeast Asia, made of betel leaves, lime and other ingredients.
Sotis – A supplementary wrap float technique found on Timor that appears reversible.
Tais – Timorese traditional cloth.
Uru – The name used in Amarasi for the natural dye *morindin*.
Wrap-faced ikat – When the *ikat* technique (tying before dying) is only used for the wrap threads before they are woven.

REFERENCES

Arte Textil en Indonesia, 1989, Barcelona: Fundación Folch.

Cinatti, Ruy, 1987, *Arquitectura Timorense*. Lisboa: Museu Etnologia.

Exposição de Timor: Catálogo, 1931. Lisboa: Sociedade de Geografia de Lisboa.

Forshee, Jill, 2001, *Between the Folds*. Honolulu: University of Hawaii Press.

Gillow, John, 1992, Traditional Indonesian Textiles. *New York: Thames & Hudson.*

Gillow, John and Brian Sentence, 2000, Textiles: Le Tour du Monde Illustré des Techniques Traditionelles. *Paris: Éditions Alternatives.*

Gittinger, Mattiebelle, 1985, Splendid Symbols: Textiles and Tradition in Indonesia. *New York: Oxford University Press.*

Gunn, Geoffrey C., 1999, Timor Loro Sae: 500 anos. *Lisboa: Livros do Oriente.*

*Hamilton, Roy and Barrkman, Joanna (ed*s), 2015, Textiles of Timor: Island in the Woven Sea. *Los Angeles: Fowler Museum at UCLA.*

Hitchcock, Michael, 1997, Indonesian Textiles. *Hong Kong: Periplus Editions.*

Holmgren, Robert J. and Anita E. Spertus, 1989, Early Indonesian Textiles. *New York: The Metropolitan Museum of Art.*

Linguagens Tecidas: Têxteis Ikat Indonésios da Colecção de Peter ten Hoopen, 2014. Lisboa: Museu do Oriente.

Ruy Cinatti: fotografias e textos inéditos, 1997. Ler: Livros e Leitores. Lisboa. N.° 39.

Sacchetti, António Emílio, 2000, *Timor em formação*. In *Segurança e Defesa*. Lisboa: Comissão Cultural da Marinha, pp.167–175.

Seiler-Baldinger, Anne Marie, 1994, Textiles: A Classification of Techniques. *Washington, D.C.: Smithsonian Institute Press.*

Warming, Wanda and M. Gaworski, 1981, The World of Indonesian Textiles. *New York: Kodansha International.*

Textiles, Identity and Innovation: Design the Future – Montagna & Carvalho (Eds)
© 2019 Taylor & Francis Group, London, ISBN 978-1-138-29611-4

Textiles as media for visual identity

M. Neves
CIAUD, Lisbon School of Architecture, Universidade de Lisboa, Portugal

ABSTRACT: Visual identity has been used as a broad designation for the visual presence of brands. A set of elements represents brands values and main ideas, allowing for intended audiences to clearly identify certain brands. Nevertheless, identity becomes mistaken with recognition and brands tend to part ways with user experience, by not allowing a connection between their visual messages and personal habits or behaviours. In this context, objects that use textile materials promote proximity and allow identification of users with such objects, which in turn, communicate a brand. Thus, increasing relationship between both. This position paper discusses such relationship, between constitution of graphic marks and the notion of visual identity with the possibility of textile clothing user experience. It argues textile based objects may claim a better connection between brands and their audiences. With the need to ensure a regular presence among its users, trademarks and their graphic design may benefit from materials that ensure greater closeness to people.

1 GRAPHIC BRANDS AND THE CONCEPT OF IDENTITY

Graphic brands are a phenomenon based and practiced within the concept of visual identity and are usually constituted in a monolithic (Mollerup 1997) or dynamic (Van Nes, 2012) way.

By visual identity we may understand visual components of brand communication. This is supported by elements of graphic origin and whose purpose is to obtain a composition which can be identified and recognized by people, making it different from the rest (Olins, 2003).

Although the area under study and its professional practice is often treated as a purely two-dimensional execution, referred to as "symbol" or "logo", it is possible to observe its growth, in terms of importance for companies, institutions and people, and in terms of diversity and complexity in its design and presentation.

However, this criterion of identity has been restricted to visual recognition, where users tend to make sense of a brand graphic component. This is especially noticeable in projects of visual and corporate identity that make use of a unique construction, present in all media which communicates the brand. Combination of elements of visual expression (drawing, symbols, typography, colours, textures, alignments, etc.) is based on maintenance and preservation of compositional constraints that ensure the exact replication of a graphic mark in various media.

1.1 *Monolithic brands*

In visual identity projects for companies and institutions, it is generally used as a rule, repetition of a graphical visual composition, for it to be identified through memory. When it appears to users, this composition replaces any representatives of these entities (Mollerup, 1997) and, consequently, the same visual signature is always used.

One of the first examples of consistency in this kind of projects is Allgemeine Elektricitäts Gesellschaft (AEG) visual identity, developed by Peter Behrens, appointed architect and designer of the company in 1907, and then followed by other companies equally known:

> *"First known as 'house style' – which was a series of rules that gave consistent treatments to all elements within an organization – this was developed in the 1930's by Olivetti in Italy and the Container Corporation of America as Corporate Identity. The central feature was the trademark, a kind of emblem with or without the company or brand name, which was always in the same style of lettering"* (Hollis, 1997).

This characteristic has become mandatory in all works of such genre. Behrens' work *"represents one of the very first – if not the first – attempts at an integrated identity for a corporation"* (Frascara, 2004) and this capacity for integration has come to be confused with the design of any visual identity. This strategy has been used and developed ever since.

1.2 *Dynamic brands*

Monolithic brands can impart a certain security and stability of work and communication. However, design practice has changed due to a set of contemporary factors (Davis, 2008). Accentuated use of new digital technologies and their ability to introduce new channels of communication, but also their constancy and habituation (Tapscott, 2009), introduces user experience as a transversal concern (Neves, 2016; Walker, 2017), to which brands are not unrelated. As a matter of fact, their contribution may not be based on repeating messages, but developing approaches where users may notice common elements (Shilum, 2011). Within this context, dynamic or mutable marks exploit modes of operation at experience level which move away from most common regularity (Hsu, 2013). Van Nes (2012) presents the best collection so far of such cases, making it clear there is far more study to be made.

It is of interest in this regard to mention NAi (Netherlands Architecture Institute) visual identity project, shown in figure 1, designed by Bruce Mau Design, since it originates from a suggestion of identity recognizable in people (Mau, 2000). That is, a single individual transform over time and does not remain visually alike during his or her life. All visual records we may produce throughout our lives prove the possibility of recognizing ourselves and simultaneously detecting a set of changes that do not call into question such recognition. NAi brand is based on this idea of diversity that transposes to printed matter and to the field of graphic design. Several objects produced for this purpose are presented in figure 2.

Figure 1. Several possibilities to present NAi's visual identity, Bruce Mau Design.

Figure 2. All objects display a different version of NAi's visual identity. Bruce Mau Design.

The advantage of this visual identity is not its graphic form, because it will always be different, neither will it be other elements of graphic composition, but rather this potential relation with users, through their attention, facing the unexpected and by providing a divergent identification from any expectation.

Dynamic brands are based on a fixed use of at least one visual element, releasing all others to a variation, which does not, however, interfere with recognizing the brand itself (Van Nes, 2012).

This mutability of communication objects, understood and handled by their users, reveals the importance of studying and constituting visual identity projects through concepts dissociated from mere graphic composition, perception or enjoyment.

Objects now integrated in a communication group of a certain brand must assume a relationship with users and allow for their participation, in moments when brand values and ideas are experienced.

2 RANDS AND THEIR (DES)CONNECTION WITH USERS

Most of brands and therefore, visual identity projects, extend to a greater or lesser number of media, objects and communication means, guaranteeing visibility and presence in front of people. Ensuring in this way, brands are likely to be known, and probably more important, recognised.

Kaplan (2016) argues that brands in their broader sense, can already be defined as experience. They are indeed a set of several experiences users can obtain by contact with all different materials, messages and people, representing brands. A brand is the resulting perception we get by add many of those experiences.

However, not all media operate in a way which ensures a close relationship with users, nor does it ensure user identification with the brand itself. Possible recognition, so sought after in graphic design projects does not imply proximity and identification of users with the core concepts of a brand.

One of the difficulties in graphic design and especially in the design of graphic brands and processes of visual identity is a relative omission of users as central characters. It is common in these contexts to think of consumer or client, overlapping the notion of brand to that of communication or the presence of a material object in which a brand is conveyed. This set of produced media is understood as declination of graphic marks and rarely, as an occasional contribution to construction of a total brand understanding.

Nevertheless, various objects used by brands should have as their purpose, to distinguish themselves from their competitors, but also to take advantage of contact with people to allow an affective connection (Wheeler, 2009). Because there is an intuitive phenomenon between people and objects which may lead to a better and more connected use (Norman, 2004). Or even because, the perception of designer work nowadays goes directly to elaboration of an experience and no longer just for the object itself (Margolin, 1997).

3 TEXTILES AND THEIR CONTRIBUTION

One material which incites this relationship and experience is precisely textile, by its capacity for personal interference. Textile means of reproduction ensure a close relationship with users.

Each person, when wearing a garment or accessory, places himself as a contributor to brand communication (Ozipek, Tanyas and Mahmutoglu Dinc, 2012). As Baudrillard (1998) suggested, one would appropriate objects insofar as he or she would communicate values inherent in objects themselves. They are material culture which integrate a personal project and each person will intend to be identified with what the object conveys. When using a garment with a graphic mark reproduced, user becomes an agent of communication and dissemination of such brand. This may not always be a voluntary action. However, a person consciously establishes a process of identification with the brand, sharing its concepts and values, using one of its media and affirming the brand in its fullness.

Textiles have become a strong tool for individual identity, much in the same way as brands. They create an emotional connection and are main features in establishing modern self.

This has been obtained through mental constitution of an image that carries with it certain values, which can easily be induced and not confirmed. Branding is always a formalization based on some trickery.

In fact, fashion adequately uses imagination and creativity to convince people into buying clothes they probably would not need (Tungate, 2005).

Clothing brands occupy a relevant place in this matter, because they cross both concerns. Constitution of their graphic marks is based on the same logic as the rest, and therefore on criteria of memory and recognition. They undertake the same material effort, even if it is partly obtained by commercialized products themselves. But on the other hand, they understand graphical constitution of clothing as something important in the relationship users create with the brand itself.

The use of textiles in branding efforts becomes increasingly important, if we attend the needs of companies addressing several stakeholders. Some companies need clothing and accessories for their staff, for customers, for sponsors and partners. For regular business, for special events, for promotions or new products. Depending on these very different scenarios, all these textiles to be produced will fulfil distinct functions. Employee clothing for instance will have to signal someone as belonging to the communicated brand, while sponsor and partnership activities will be designed to promote and cause a certain impact.

As examples, we present 2 simple cases where textiles were used as part of a visual identity project. Both cases were graphic marks designed for two different events, with limited time frame.

3.1 *Case 1*

The first one was Accuracy 2006, an international symposium on spatial accuracy assessment. Textiles

Figure 3. Textiles produced for Accuracy 2006, Author.

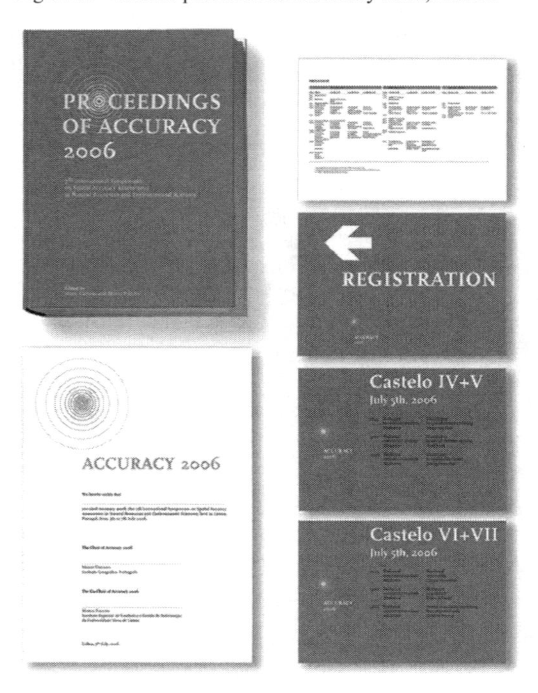

Figure 4. Print materials produced for Accuracy 2006, Author.

presented in figure 3 were produced to be distributed to participants, so they could be used during the symposium and create two levels of identification: with the event and its research contents, thus creating a sense of group between peers; and to allow others to identify participants of this specific event. In figure 4 we can see several other project materials produced.

3.2 *Case 2*

The second one was AgIM, an educational project on agriculture precision management. Textiles (in figure 5) were conceived to identity students in research field work and group meetings to discuss results

A very close identification with the rest of communication materials was necessary, for outside people to immediately recognize such individuals.

In both cases, textiles played a key role in defining brand values and origin, helping to communicate effectively.

It is known brands play a vital part in clothing consumers choices (Ewing, Mostert and Puth, 1999), but we are yet to understand how much clothes contribute to brand engagement, especially those brands which do not have clothes in their core activities or products. It would be of good value to further explore user experience when it comes to such materials.

Since the notion of user experience is precisely the moment of contact of a person with a certain object ready to be used (Garrett, 2011) and the preparation of the exact same object to meet the needs felt by users (Battarbee, 2004).

4 CONCLUSION

Graphic marks have been contained in a narrow perspective of identity. Those which use constant repetition of a visual composition opt for security of recognition, but may not reap benefits of users for lack of interpellation. On the other hand, graphic marks which use a dynamic or mutation of their visual presence, interfere in the relationship that users develop with such brands. These examples focus on modifying, albeit partial, the composition of various graphic elements, allowing different and quite diverse reproductions. This regular transformation of the way a brand presents itself to its public requires not only brand's memory (a desirable and sought-after factor in any brand-building work), but a close, familiar, temporal relationship with it. Such an approach to brands, while increasing, is still small, at least in what concerns the impact of use. Something verifiable in several means of communication.

Within the group of materials usually used by brands to present themselves, those who select textile presuppose a level of proximity to users, not comparable to others. In a moment when communication and experience strategies are developed for brands that significantly involve users, objects that use textile, namely clothing, may be preferred, due to their contribution in making this relationship better.

It may be beneficial for brands, which gradually but intensely develop user experience research, to enter markets or simply to consolidate themselves, to address the advantages of textiles for reproduction of

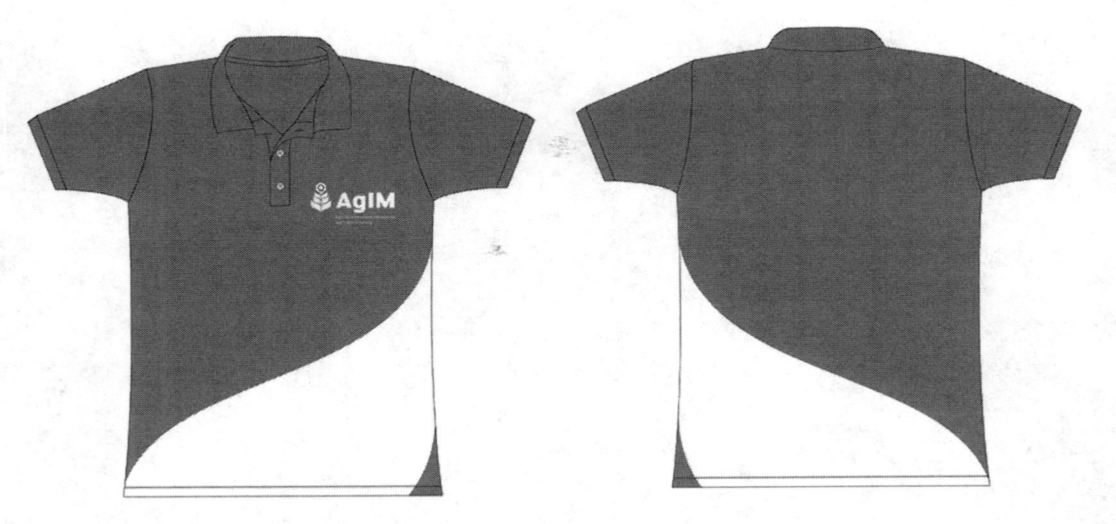

Figure 5. Textiles designed for AgIM, Author.

their graphic brands. Increasing along the way, levels of identification with people through use of these materials.

ACKNOWLEDGMENTS

The author would like to thank the funding support by the Foundation for Science and Technology of the Ministry of Science, Technology and Higher Education of Portugal under the project UID/EAT/04008/2013 (CIAUD).

REFERENCES

Battarbee, K. 2004. *Co-Experience: Understanding co-experiences in social interaction*. Helsinki, University of Art and Design Helsinki.

Baudrillard, J. 1998. *Le système des objets*, Gallimard. Paris.

Davis, M. 2008. Why Do We Need Doctoral Study in Design? *International Journal of Design,* 2(3), pp. 71–79.

Ewing, M., Mostert, P., Puth, G. 1999. Consumer perceptions of mentioned product and brand attributes in magazine advertising. *J. Prod. Brand Manag*. 8(1), 8–49.

Frascara, J. 2004. *Communication design: principles, methods and practice*. New York, Allworth Press.

Garrett, J. J. 2011. *The Elements of User Experience: User-Centered Design for the Web and Beyond*. Berkeley, New Riders.

Hollis, R. 1997. *Graphic Design: a Concise History*. London, Thames and Hudson.

Hsu, M. 2013. Annotation of Dynamic Identities in Interactive Aesthetics. *Advances in Journalism and Communication*. 1(4), 41–49.

Kaplan, K. 2016. *Brand Is Experience in the Digital Age*. In https://www.nngroup.com/articles/brand-experience-ux/

Margolin, V. 1997. Getting to know the user. *Design Studies,* 18, 227–236.

Mau, B. 2000. *Life style*. London, Phaidon Press.

Mollerup, P. 1997. *Marks of Excellence*. London, Phaidon Press Limited.

Neves, M. 2016. Enhancing User Experience in Graphic Design: a Study in (unusual) Interaction. in Rebelo, F., Soares, M. (Eds.) *Advances in Ergonomics in Design*. Advances in Intelligent Systems and Computing, 485, 801–810.

Norman, D. 2004. *Emotional design: why we love (or hate) everyday things*. New York, Basic Books.

Olins, W. 2003. *On Brand*. London, Thames & Hudson.

Ozipek, B. Tanyas, M. Mahmutoglu Dinc, N. 2012. Factors Affecting Branding with Special Reference to Clothing Industry. *RMUTP International Conference: Textiles & Fashion*, Bangkok.

Shillum, M. 2011. Branding is about creating patterns not repeating messages. *Co-Design*. Access <https://www.fastcodesign.com/1664145/branding-is-about-creating-patterns-not-repeating-messages>

Tapscott, D. 2009. *Grown up digital: how the net generation is changing your world*. New York, McGraw Hill.

Tungate, M. 2005. *Fashion Brands: Branding Style from Armani to Zara*, London and Philadelphia, Kogan Page.

Van Nes, I. 2012. *Dynamic Identities: How to create a living brand*. Amsterdam, BIS Publishers.

Walker, R. 2017. How Logos Became the Most Important Quarter-Inch in Business. *Fortune*. Access <http://fortune.com/2017/06/16/business-logos-evolution-importance/>

Wheeler, A. 2009. *Designing Brand Identity*. New Jersey, John Wiley & Sons.

2.2. Culture and craft

Textiles, Identity and Innovation: Design the Future – Montagna & Carvalho (Eds)
© 2019 Taylor & Francis Group, London, ISBN 978-1-138-29611-4

Textile semantics: Perception and memory

S. Andrew
School of Art, Design and Architecture, University of Huddersfield, UK

ABSTRACT: This paper focuses on a semiotic exploration of textiles, examining the relationship between the designer's authorial intention as the maker and the interpretation of their work by the viewer, particularly how visual images within textiles are read, and if the designer shares a common 'visual grammar' with their audience. The first stage of the paper discusses cultural perceptions of the tactile qualities of textiles and how the visual image in textiles has been used to communicate with the viewer. The role of context as an influencing factor in the meaning generated from image and cloth/fabric, and the dominant cultural codes by which textiles are read, are considered. The second stage examines viewer perceptions of a printed textile installation designed to communicate a family history narrative. Viewers' interpretations of the narrative and the installation sites are discussed, highlighting examples where personal memory and shared cultural memory influenced their responses.

1 INTRODUCTION

Individual memory is built from many facets, mediated through different communication channels it incorporates perceptions transmitted from those with whom we have formative and on-going relationships, and the shared experiences (real, virtual or hyper-real) that form our collective cultural memory. Erll (2008) notes that the material (artefacts, media), the social (people, social relations, institutions), and the cognitive (culturally defined ways of thinking) contribute to the formation of cultural memory. Viewer interpretations of textiles and the context in which they are used or viewed, are informed by both individual memory and shared cultural memory, which act as triggers for the generation of shared perceptions that in semiotic terms reveal dominant cultural codes of meaning.

1.1 *Tactility and perception*

Although we may think that we respond individually to how a textile looks and feels, a survey of viewer perceptions of textile samples suggests that our fundamental responses to many textiles are culturally learned (Andrew, 2008), creating a hierarchy of meanings associated with each fabric. When viewers were asked to match descriptions such as expensive, cheap, luxurious or hardwearing to unnamed fabric samples that they could touch, most respondents recorded similar attributes for Pahjah raw silk as for jute. When these fabrics were shown as named samples in a duplicate survey, most viewers gave different attributes to these two fabrics. The word 'silk', rather than the experience of the fabric itself, elicited a learned cultural response from viewers that silk was more refined, expensive and luxurious (Andrew, 2008).

The meanings we derive from textiles are learnt from an early age and their meanings become culturally embedded. Whilst some fabrics are associated with luxury, others are associated with work wear. Barnard (1996) acknowledges the communication potential of the perceived qualities of textiles and how this can be a highly effective tool in advertising. He notes that the connotations of textiles such as linen and flannel have been used in advertising fragrances, whilst silk is often associated with hair products and cosmetics.

A brand of chocolate is well known in Britain for employing the 'why have cotton when you can have silk' strapline in their television advertising campaigns (www.campaignlive.co.uk), successfully associating chocolate with qualities attributed to a textile by asserting the luxurious connotation of silk over the functional, everyday connotation of cotton. Pairing this with an up-market, low-lit apartment and the musical introduction to Rhapsody in Blue in their television advertising in the late 1980's, a sense of anticipation and indulgence was communicated. Even the lamp in the apartment, a serpent design, functioned as a signifier of the Garden of Eden and temptation. By 1992 the company updated the television advert and added a 'why have cotton when you can have silk' voice-over to clearly articulate the associations between their product and the qualities of silk, using a chocolate coloured Dior dress to denote silk and provide the added connotations of taste and wealth. Recent adverts for the chocolate still include the textile comparison strapline.

1.2 *Context and perception*

Dominant cultural readings of textiles are pervasive. Responses to textiles are developed through our first childhood encounters with the tactile qualities of woven cloth and knitted fabric, but cultural influences later inform our responses to textiles. These are then subject to perceptions of the product they are made into, and the context in which they are viewed. Davis (1992) and Kaiser (2002) examine the context dependence of clothing messages, where garment and context impose their own layer of meaning on our understanding of the textile. Davis exemplifies this through the different meanings generated by black lace in a funeral veil as opposed to a nightgown. Black lace is a (redundant) signifier of lingerie, as lingerie is the dominant reading of the textile in many western cultures.

When we see an image, or form a mental picture of it, we rely on a mix of dominant cultural codes and memory to decode it. Dominant codes create shared understanding by conveying the preferred (or majority) values and cultural perceptions of a society. Dominant codes are those we expect. They are known as 'broadcast codes' (Fiske, 2010) as they operate on the basis of commonality and rely on cultural understanding that the majority of people share. We become aware of 'broadcast codes' informally through day-to-day experience from childhood through to adulthood. However, codes are informed by many different facets of cultural knowledge, ideology and personal experience, and the expected meaning we try to communicate through what we perceive as the shared dominant codes of a particular group might not be received in the way we intend. In some cultures black lace may signify 'funeral veil' but it is likely to be a subcultural reading of the textile, not the dominant reading. This reading of black lace could be categorised as aberrant decoding (Eco, 1977; Fiske, 2010). 'Funeral veil' may have been the dominant cultural code associated with black lace in western cultures in the past, but it is in marked contrast to the dominant cultural perception of the textile in western cultures now. Context can change this; if a person were standing in a church and asked to think of black lace, a dominant cultural reading of 'funeral veil' would not be surprising. Denim also elicits a strong learned cultural response concerning the meaning of the cloth and the contexts in which it is seen, making a denim wedding dress an unlikely choice due to the connotations of the cloth and the shared cultural memory associated with it.

1.3 *Dominant cultural codes in textile practice*

Artists and designers can mix and subvert the cultural codes that are commonly use to understand images and textiles, but only if they have a good understanding of what the dominant cultural codes and sub-codes are that apply to the images and textiles in any given context. Dolce and Gabbana incorporate lace and successfully reference both lingerie and

Figure 1. Left: 'Spreckley 2' (chest detail). Right: 'Lumley' (front and back view). Both by Paddy Hartley for the exhibition Project Façade (2007–2008). Images: courtesy of Paddy Hartley.

funeral attire in garments described as their 'Sicilian widow look', whilst Yinka Shonibare's works such 'Dressing Down Textiles in a Victorian Philanthropist's Parlour' (1996–7) and 'Scramble for Africa' (2003) incorporate 'African' printed textiles into Victorian costume. This change of context forces the viewer to consider the contrast in meanings communicated by the printed cloth and the garment shape, challenging their perceptions of cultural identity and colonialism and actively subverting dominant cultural codes. Jefferies (2008: 51) notes:

> The history of fabric used, the ambiguous materials and motifs of west African textiles, seem to symbolize the rich complexity of post-colonial cultures in that, while the patterns and colours are thought to be authentically African, they actually originate from Indonesian Batik work, a technique which was industrialised by Dutch traders historically active in Africa.

Shonibare's re-contextualisation of the printed textiles in Victorian costume is effective because the dominant cultural reading of the patterns and printing style is that they are typically 'African'.

Paddy Hartley's work 'Project Façade' (2007–2008) exemplifies communication through the relationship between the visual content in the work, the object itself, and the context it is viewed in. His pieces communicate the stories of WW1 service men that had pioneering facial reconstruction surgery. The pattern cutting and stitching back together on the facial masks echo the process of the surgery, with image and text on the soldiers' experiences combined into modified army uniforms (Figure 1). Exhibited at the Imperial War Museum, the relationship between content, object and context is unified. Hartley works within the parameters of dominant cultural codes to enable the viewer to deal with difficult subject matter.

Nina Edge particularly utilises clothing as a form of political activism. Her exhibition 'Terra' (2005) included camouflage hijabs to emphasise differences in ideology between cultures. Floral textiles were embroidered with fighter jets and camouflage fabrics embroidered with flowers to communicate and contrast the impact of the military on the domestic and vice versa.

In Shelley Goldsmith's work printed images of natural disasters were used to communicate human vulnerability, such as the piece 'Baptism' (2003). The use of this type of printed image within infant dresses adds an additional level of meaning to viewers' perceptions. The viewer reads meaning into the image and the garment simultaneously, making connections between the disaster and the child's dress.

1.4 *Dominant cultural codes and gender*

Some of the most pervasive dominant cultural codes apply to the use of colour and image in printed textiles for children's products that reinforce gender perceptions. A British high street retailer was highlighted by the campaign group 'Let Clothes Be Clothes' in January 2015 for manufacturing and marketing Natural History Museum clothing with dinosaur images only for boys (Hill, 2015). In an open letter to the company the campaign group provided examples of their t-shirts with printed images of kittens for girls and 'bug expert' for boys, noting that the textiles in coats for girls were described as 'cosy and dry' and 'pretty and practical', whilst the textiles in boys coats were described as 'made with a host of clever fabric technologies to ensure maximum durability and long wear' (https://letclothesbeclothes.uk). Highlighting the 2011 Bailey review (www.gov.uk), the campaign group raised issues such as age appropriate fit and coverage of clothing, and the lack of gender-neutral clothing, across a range of retailers.

As one of the highest crowd funded children's clothing projects in Kickstarter history to date, the Princess Awesome brand have rejected typical images for girls and pink colour palettes on their clothing, in a small step against the dominant cultural codes associated with gender. The aim of the brand is to produce clothing for girls that features prints such as dinosaurs, robots and transport, images that are more usually associated with boyswear. The response from consumers has been positive. Creators Eva Saint Claire and Rebecca Melsky state 'It shows that parents and children alike are rejecting the gendered straightjacket that mass consumer culture forces them into' (St Claire, 2015, quoted in Hill, 2015: 10). The images on the clothing communicate a different message to the viewer about the child wearing it and to the child themselves. St Claire states:

> This matters for so many reasons. When people meet a little girl in a pink, sparkly dress, they talk to her about how sweet and pretty she looks. If she's wearing a dress with robots or trains on, they'll talk to her about them instead – that is, about technology and engineering. That opens up her mind, her world and her future. (St Claire, 2015, quoted in Hill, 2015: 10).

1.5 *Printed textiles as communication*

The idea of using printed textiles to communicate with both user and viewer is not a recent phenomenon.

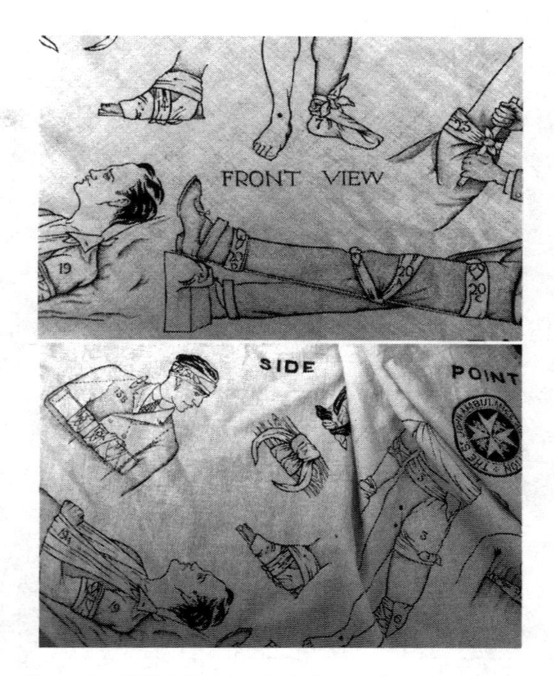

Figure 2. WW1 St John Ambulance sling (two details). Images courtesy of Janet Haigh.

Historically 'conversational prints' were designed with figurative images (stylised or realistic) to convey a simple narrative or message. In World War One St John's Ambulance slings were printed with instructions for tying bandages and tourniquets (Figure 2) and several British fabrics produced in the 1940's functioned as both conversational prints and propaganda. The dress fabric 'Coupons' (1941–42) by the Calico Printer's Association notes how many coupons are needed to purchase each garment shown in the print, whilst 'Victory V' (1941) aimed to boost morale and demonstrate the patriotism of the wearer (http://collections.vam.ac.uk).

Toile de Jouy is a particular example of textiles as communication where the visual narrative tradition can be traced through to contemporary practice. The printed cloth was originally produced in Jouy en Josas, France, from 1760 to 1843 (Bredif, 1989) for interior furnishings and fashion fabrics. The first prints incorporated idealised scenes of country life but designs were also created to commemorate specific events. The tradition continued, with 1930's art deco examples by Ruth Reeves such as 'Manhattan' showing idealised vignettes of industry and city life within the repeat composition. But 'Glasgow Toile' (2004) by Scottish print company Timorous Beasties demonstrated a more contemporary interpretation of the toile de Jouy tradition. This interior fabric moves away from idealised scenes to a more subversive and realistic version of life in a modern city. Kent Henricksen's treatment of toile de Jouy takes this a step further using print and embroidery to add ominous masked figures within ornate repeat designs such as 'Absence of Myth'

(2007), exploring themes of violence, submission and slavery.

Subversive prints are more common within fashion, particularly where a designer external to the brand alters a company logo to convey a new message. Wayne Hemmingway modified the Shell oil company logo to read Hell, and changed the Hoover domestic products logo to the less controversial Groover for a range of Red or Dead t-shirts in the 1990's. More recently design company Liquid Sky adapted the adidas logo for their 'abducted' t-shirt and Judy Blame changed the 'Keep Britain Tidy' campaign image into 'Keep the World Tidy' with the campaign figure discarding a swastika symbol.

1.6 *Paradigmatic and syntagmatic choices*

In the construction of their work, all artists and designers make 'paradigmatic' and 'syntagmatic' choices. These terms originally derive from linguistics (Saussure, 1983) and apply to the choices we all make when we communicate, choices that are often subliminal and instantaneous. Fiske (2010: 62) notes that 'paradigms are composed of units with an overall similarity, but with distinctive features that distinguish them from one another'. In semiotic terms these units are called 'signs'. A sign is made up of the signifier (such as the object, word, image, sound, taste or smell) and it's signified meaning to the viewer (Barthes, 1967). A syntagm is the message into which chosen signs are combined. All messages involve selection (from a paradigm) and combination (into a syntagm), whether they are visual, spoken or written (Fiske, 2010). Consider religious icons; instead of selecting one signifier and its signified meaning from the paradigm of 'religious icons' to create a printed textile design, Wayne Hemmingway of Red or Dead chose many religious icons and combined these in a visual syntagm in the form of a print entitled 'Guru' (1997). This had to be withdrawn after protests from a range of religious communities who did not want their religious icons, and what each of them signified, associated with the meanings of others, or with the term 'Guru' (Kingswell, 1998). Artists and designers also undertake what Roland Barthes (1967) described as 'commutation tests'. This test is primarily one of substitution where alternative signifiers are considered and evaluated in terms of their effect on the meaning generated. Commutation tests function as conscious or subconscious, instinctive actions within the ongoing practice of design. The designer carries out successive cycles of selection and elimination as they analyse, for example, the substitution of one image for another within a composition and its impact on the compositional relationship and meaning of the work.

2 PERCEPTIONS OF A TEXTILE INSTALLATION

Many textile practitioners are concerned with narrative in one form or another, as stories play a significant role in the continuity of cultural knowledge and traditions. Millar (2007) notes that visual narratives are particularly context dependent, that they can be renegotiated as social situations change and are open to subjective editing by the viewer. We set out to visually communicate, but is shared meaning between the maker and their audience generated in the way the practitioner intends?

Perception, memory and subjectivity in the representation and interpretation of visual narratives formed the basis of a research project to test how meanings encoded in printed textiles by the maker were decoded by viewers. A visual narrative was constructed based on the experiences of John Edgar Bell, a Quaker and conscientious objector in WW1, who was imprisoned for refusing to fight. Two installations, each consisting of three textile panels were produced. The panels were designed to communicate the family at the start of the war and the social exclusion they faced, the imprisonment of the conscientious objector, and the hostility towards the objector's family that continued into WW2 and beyond. The panels were located in galleries, museums, churches and corporate spaces without any title or description to collect over 450 viewer interpretations of the textiles via 1:1 interviews and self-completion questionnaires. Colour images of the textile installations can be found at:

> http://www.ahrc.ac.uk/research/readwatchlisten/imagegallery/2014galleries/atextilenarrativeofjohnedgarbellconscientiousobjector/

Initially, relationships between photographic images and other visual signifiers such as colour and mark were explored to create micro visual syntagms and consider the meanings they might convey. This was supplemented by investigation into the experiences of pacifists and their families in WW1, contacting English Heritage to arrange access to Richmond Castle to photograph the cells where objectors were imprisoned.

The textile development progressed with paper, digital, and cloth experiments. Specific images were incorporated to develop the visual narrative, such as crossed out medals to connote lack of bravery, white feathers as symbols to signify cowardice, and stamps to denote the era (and suggest 'for king and country') (Figure 3). Due to John Edgar Bell's conscientious objection, the family experienced abuse from their community, with graffiti and crosses scrawled on the door of their home. This was translated into the textile work through printed and stitched crosses and brush marks over and around key signifiers in the panels. Images such as small white crosses, to serve a dual communication function as a Christian symbol and a grave marker, and crossed out medals, were included in the first installation panels.

These images were expected to operate via broadcast codes and create shared interpretations from viewers. Other printed elements, such as crossed keys, operated via 'narrowcast codes' (Fiske, 2010), in that only a few individuals might know their correct

Figure 3. Stamp detail, panel one, installation one.

meaning as a Christian symbol relating to Saint Peter. In the first installation shadows of prison bars were imposed over the image of the objector and a uniform pattern placed under his image (uniforms were thrown into imprisoned objector's cells on a daily basis, wearing the uniform was agreement to conscription).

2.1 *Installation one*

The predominant reading of the first installation was that the panels communicated a war narrative, with themes linking war, family and suffering emerging in viewer responses. Aberrant decoding was evident in some responses, such as the installation communicating 'nature versus the developed world' from one participant. Only a small number of viewers stated that the panels related to conscientious objection.

The crossed out medals were incorporated as an anti-war/anti-heroism signifier, to indicate the objector's beliefs, his refusal to fight and his perceived lack of bravery by society. It was expected that they would be very redundant signifiers and easily understood, but the majority of viewers did not perceive the crossed out medals an as anti-war statement, or connoting non-participation in the war effort. Fighting in the war, bravery, heroism, war dead and military were the predominant readings. The medals were removed in the development of the second installation panels but images to specifically denote WW1 replaced them, such as guns, bombs and warships.

Responses to the first installation revealed that most viewers' associated the small white crosses with graves rather than Christianity; so larger decorative examples were incorporated into the development work for the second installation. Most viewers interpreted the feather correctly but others suggested it represented 'pigeon carriers at the front line', 'elements of nature', 'a symbol of peace and love' and 'a writing implement'. Very few readings of the first installation related to imprisonment, and the prison bar shadows as a signifier were too subtle and were not noted in any responses. The second installation therefore incorporated images with greater semiotic redundancy (barbed wire, tally marks and lock) to generate the imprisonment reading. A key was later added to reinforce this, but also suggest potential for release.

2.2 *Installation two*

Interpretations of pacifist meanings increased in responses to the second installation and shared interpretations of the panels included war and wartime, imprisonment, suffering, and family relationships (some in relation to war). Aberrant decoding was still evident though, with one viewer interpreting the visual narrative as:

> The struggle of non-British man travelling to Britain on work related journey. He is out to earn to maintain his family, however, encounters problems and is left with choice to go without much or stay and prosper with new family. He ends up doing the latter, which involved breaking his marital oath and religious law, and separating him from previous family. Ultimately his new offspring inherit British status.

The image of the ship (Figure 4) predominantly signified war and defence to viewers, but was also interpreted as migration. One viewer described the warship as a steamer signifying emigration to America or Australia. Their overall reading of the installation was 'Family history, imprisonment, emigration, disposal, memory'. The praying, tied and bandaged hands across the second installation panels created meanings of 'Hope for the future but trapped in a fast changing world', and 'Ties to enforced ideas'. One viewer read the hands in each panel as a micro narrative of 'Praying? Then shackled, then bandaged, track history of the man, was his objection from religious causes – sacrifice of human life?' Aberrant decoding of the hands included 'swearing in the oath of allegiance to the king' and marital breakup. Readings of imprisonment increased in responses to the second installation, with prisoner of war, POW camp and internment noted. One viewer described the 'face of man behind barbed wire' as 'a prisoner of war or deserter condemned to firing squad', another described the figure as 'priest, persecution for religion'. The larger decorative crosses incorporated in the second installation generated the intended interpretations of religion. One viewer noted 'As seen on the pulpit/altar, must be something to do with pacifism and church, being against war'.

The decorative nature of the crosses, combined with the increase in scale, generated a shift in the signification of the imagery towards religious meanings in the second installation. The image of the family and white feather was intended to communicate the on-going ostracism they faced in the years after WW1. One viewer particularly interpreted this image as 'carrying disgrace of white feather, suffering through loss and convictions of others, betrayal of country and church

Figure 4. Ship detail, panel one, installation two.

Figure 5. Woman and barbed wire detail, panel two, installation two.

to support them'; others suggested 'defence of family (pressures to go to war)'. Photographic portraits across the panels generated readings related to family and the impact of war, particularly loss, suffering, love, absence and separation, such as 'the fragility of an individual family in a wartime situation' and 'innocent people caught up in wartime events not of their own making'.

The photographs led some viewers to reflect on memories of friends and family histories, one viewer responding that the woman and barbed wire (Figure 5) reminded them of 'someone I knew who was in an internment camp as a child in World War Two', another stated 'memories evoked of my own family photos', whilst one viewer described the 'fear in woman's eyes' as meaning 'ambiguity of serving England'.

2.3 *Impact of location*

The influence of context on interpretations of the installation clearly emerged at some sites, with viewers associating the images with the history of the area, and some to their own family histories, rather than the intended conscientious objection narrative.

Saltaire United Reformed Church was chosen as a site because the religious context had a direct relationship to the content of the work. But the geographical area also had a history of textile manufacturing, with Salts textile mill located opposite the church, and both buildings being part of Saltaire village, a world heritage site. John Edgar Bell and his family had lived in Saltaire. This formed a link between the content of the work and the context of the church and surrounding area. Interpretations of the installation at the church included:

> The social and economic history of Saltaire presented by work, duty, family, church and environment. The link with textile industry and Saltaire.
>
> Context in Saltaire made me think of industrial heritage, also family photos. Northern industrial towns, factory life and culture – the material itself reflects these qualities.
>
> Firstly the church, the way it sits in the church. It makes you think more about it; also your senses are being stimulated, because of where it is.

Others viewers stated the meaning of the narrative as:

> Family life connected to the church. The involvement of mills and woollen products during the war.
>
> It reminds me of my family (grandmother, aunt, uncle, cousins) who worked in Salts Mill.
>
> Maybe some sort of link with textiles but that might just be because of where we are.

Walford Mill Gallery and Lloyds TSB banking headquarters provided 'neutral' contexts, as the work had no relationship to the function of the buildings, or the history of the surrounding area. Lloyds also offered a corporate context for the installation that was not a feature of other sites in the study. But viewers at this site were still influenced by context in their reading of the narrative:

> The key, is that perhaps some reference to the, you know, the lock works? I just thought we were looking at Birmingham and the bigger picture, Birmingham and the immediate surrounds.
>
> Is it all to do with local history? Well, I mean, most of it comes from Willinghall, which is like a big lock making, or used to be a big lock making area, so that was the one thing I thought of, but I think you'd only pick up on that if you lived in that sort of area.

Sort of got history to it I guess, it's probably locally relevant some how. It certainly looks like it's probably something round here but I'm not from Birmingham so I'm not sure.

Manchester Museum of Science and Industry was included in the study as it provided a historical context related to textiles and the geographical area also had a history of textile manufacturing. The work was shown in the textile gallery of the museum that had displays of working textile machinery and exhibits relating to Manchester's historic role in the cotton trade. Several responses about the meaning of the installation at this site reflected this heritage:

The panels are representative of the development of the textiles industry interspersed with historical artefacts mostly relating to imperialism."

Conveyed the convergence of people to Manchester to work in the textile mills.

The true cost of our textiles. The struggles of early manufacturing and price paid by workers with their heath and lives.

Are the plants depicted cotton? If so it could show the link with Manchester.

I assume by the fact that it's here it is to do with possibly the production of cotton or some sort of other industry.

They're obviously very old images aren't they; I would guess that they are representing something of the heritage of the industry of this area.

From what I can see it looks like something that relates to this museum.

Viewers across the sites also responded to the way the work was made, with comments such as:

Feeling of past wars through aged textiles and photographs. Lack of support from country and church. We would not have spent as much time interpreting the works in other mediums.

Textiles have an added meaning to me – the history of cloth – everyday use and the fading/rotting qualities that evoke memories of their own.

Materiality of textiles always speaks of history to me.

The textile panelling gives a feeling of both antiquity and reality which I feel wouldn't exist in another medium.

Textiles 'carry' history more intrinsically than other media – they are quite literally amplified with memory of life lived. The texture and worn nature of these pieces vividly convey the aspect of a personal memorial.

3 SUMMARY

Olick (2010:156) notes that groups share 'publicly articulated images of collective pasts' but questions whether individual memory and collective memory are really separate things as collective memory refers to 'a wide variety of mnemonic products and practices, often quite different from one another' (Olick, 2010: 158) and mnemonic practices are simultaneously both individual and social.

The development of the textiles as a memorial to a family history was a form of mnemonic practice. Although communication intentions for the textile panels were specific, each image and composition was open to multiple interpretations. No single signifier in the work elicited a consistent interpretation from all viewers. The readings of visual syntagms within the textiles were complex, as some viewers formed unexpected relationships between images across the panels and made different paradigmatic choices of meaning to those of the maker, thereby constructing alternate readings of the installations. Images incorporated into the work with an expectation of shared readings were informed by viewers' personal and cultural experiences, their own memories and histories, and the boundaries of their wider social and historical knowledge, often creating divergence from the intended meaning of the narrative. When a substantial number of viewers drew the meaning of an image from a different paradigm to that of the maker, the dominant cultural paradigm that the image belonged to was revealed.

The influence of context on viewers' readings of the installation was evident in a small number of responses at each site in the study. Viewers drew on their knowledge of the social history of the area and the function of the site (e.g. a museum or a church) to inform their reading of the textile installations. When the textile panels were exhibited in Saltaire, local viewers' particularly drew on oral histories passed on through families to inform their responses to the narrative. At the Lloyds site in Birmingham the images of the crossed keys and single key (installation one) had greater impact on the reading of the narrative than at other sites due to the history of lock and key making in the area (which the maker was unaware of). This has wider implications for the exhibition of communicative textile artefacts, as each site brings its own history, associations and expectations which impact on the reading of the work, particularly with local viewers.

However, some common readings of the narrative and individual signifiers were evident across a substantial number of responses to the installations across the sites. These provided evidence that the textiles functioned as a mnemonic product with viewers at a group level (in addition to an individual level). Collective cultural memory informed a shared visual grammar that had greater influence than location on viewers' readings of the narrative in most cases.

REFERENCES

Andrew, S. 2008. Textile Semantics: Considering a Communication Based Reading of Textiles. In *Textile: the Journal of Cloth and Culture*, March (issue 6.1): 32–65.

Barnard, M. 1996. *Fashion as Communication*. London: Routledge.
Barthes, R. 1967. *Elements of Semiology*. London: Johnathan Cape. (Originally published in French by Editions du Seuil, Paris).
Bredif, J. 1989. *Toile de Jouy*. Paris: Adam Biro.
Davis, F. 1992. *Fashion, Culture and Identity*. Chicago: University of Chicago Press.
Eco, U. 1977. *A Theory of Semiotics*. London: Macmillan Press.
Erll, A., Nunning, A. & Young, S.B. (eds.) 2008. *Cultural Memory Studies: An International and InterDisciplinary Handbook*. Berlin: Walter de Gruyter GmbH & Co. KG.
Fiske, J. 2010. *Introduction To Communication Studies (Studies in Culture and Communication)*. London: Routledge.
Hill, A. 2015. Dancing ninja dresses join battle against pink hordes. In *The Guardian,* (February 21st): 10.
Jefferies, J. 2008. Contemporary Textiles: the art fabric. In N. Monem (ed.) *Contemporary Textiles, the fabric of fine art*. London: Black Dog Publishing: 51.
Kaiser, S. 2002. *The Social Psychology of Clothing: Symbolic Appearances in Context*. New York: Fairchild Publications.
Kingswell, T. 1998. *Red or Dead: the good, the bad and the ugly*. London: Thames and Hudson.
Millar, L. 2007. *Cloth & Culture Now.* Espom: University College for the Creative Arts.
Olick, J.K. 2010 From Collective Memory to the Sociology of Mnemonic Practices and Products' in Erll, A. and Nunning, A. (eds.) *A Companion to Cultural Memory Studies*. Berlin: Walter de Gruyter GmbH & Co. KG.
Saussure, F. 1983. *Course in General Linguistics*. London: Duckworth. (Translated by R. Harris from the original version in 1916).
http://www.campaignlive.co.uk/article/galaxy-revive-silk-strapline-11-years/1112282 [accessed 20/04/2016]
http://collections.vam.ac.uk/item/O110717/victory-v-dress-fabric-calico-printers-association/ [accessed 20/04/2016]
https://www.gov.uk/government/collections/bailey-review [accessed 01/07/17]
https://letclothesbeclothes.uk/category/dinosaurs-for-all-campaign/ [accessed 01/07/2017]
http://princess-awesome.com/collections/girls [accessed 20/04/2016]

Further information on the research study is available in the following publications:

Andrew. S. 2014. Image and Interpretation: Encoding and Decoding a Narrative Textile Installation. In *The Journal of Textile Design Research and Practice*. Vol.2: 2 (November 2014): 153–186.
Andrew. S. 2013. The Medium Carries the Message? Perspectives on Making and Viewing Textiles. In *The Journal of Visual Arts Practice*. Vol. 12: 2 (August 2013): 195–221.

Textiles, Identity and Innovation: Design the Future – Montagna & Carvalho (Eds)
© 2019 Taylor & Francis Group, London, ISBN 978-1-138-29611-4

Multiculturalism in fashion design: A case study of three emergent designers in Macau

Ana Cardoso
University of Saint Joseph, Macau

João Cordeiro
London School of Film, Media and Design – University of West London
CITAR, Research Centre for Science and Technology of the Arts – Catholic University of Portugal

ABSTRACT: The aim of this paper is to gain a better understanding about the Fashion Industry in Macau, focusing on how local designers are influenced by foreigner culture and how they balance market, creativity, aesthetics and multiculturalism in their professional activity. For the purpose of the study, we have followed a mixed methods approach, including a questionnaire survey followed by a case study research (with in-depth interviews and visual analysis of portfolios), resulting in quantitative and qualitative data collection. We can conclude that the local Fashion Industry is expanding, following a different route of the Textile and Garment Industry of the 1980s. Nowadays, the local Fashion Industry is more focused on the design and creativity, rather than industrial production. The three local designer participating in our case study include multicultural inspiration in their designs and follow similar creative and market strategies, such as designing two different lines (tailor vs ready-made) and using on-line platforms for marketing and sales.

1 INTRODUCTION

1.1 *Macau as synonym of multiculturalism*

Macau is a special administrative region of China, situated in the western side of the Pearl River Delta, 70 Km from Hong Kong. Having been greatly shaped by the Portuguese presence since 1557 (as noted by Mendes (2013), several authors mention different years for the definitive settlement date of the Portuguese in Macau, within the period of 1553 and 1557), and later having been under Portuguese Administration until 1999, Macau has shown to be a place for sociocultural exchanges between Asian and Western cultures. From that time, until our days, Macau represents a peculiar fusion of Oriental and Western influences (Loureiro, 1999).

The several fluxes of immigrants of various ethnic origins made Macau a place of cultural diversity. Since the beginning of the Portuguese outpost established in Asia, the arrival in Macau of Malays, Indians, Africans, Japanese, Thais, Indonesians, Filipinos and Europeans, reveals a multicultural presence that is still reflected nowadays in the city.

"Multiculturalism is not new to Macau. Macau was first colonized by the Portuguese in the 16th century, the first modern foreign territory within China. Because of its early entrepreneurial background Macau developed into a trading and manufacturing hub for specific products and tourism." (Woodside, Megehee and Ogle, 2009, p. 7).

These ethnic and multicultural exchanges have produced a diverse society, present in the habits of local people and their ways of living. With an industry, largely dependent on the casinos, tourism and leisure activities transform Macau into a core of economic attraction and major tourist destination.

1.2 *The modern textile industry in Macau: From 1890 to 1999*

Looking into Macau's industry side, it can be said that it is divided into three stages "early industrialization (1890's–1950's), industrial expansion, based on the production of light goods for exports markets (1960's–1980's) and industrial relocation to China (1980's–1999)" (Sousa, 2009).

During the early 1900's, Macau has established small cottage industries, including fireworks, matches and incense, but it was the gambling, first established in 1847, that maintained the economic prosperity of the colony.

Macau's prominent industries during the second stage were the textile industry, footwear, toys, incenses, machinery, enamel, firecrackers, wooden furniture, Chinese wines and electronic goods. The production and distribution of textiles and garments for global markets has formed the core of Macau's export-based industrialization over five decades. On the third stage, Macau's textiles and garments industry was composed of a small number of small-scale manufactures and a large number of export firms, exporting mostly to China. But unfortunately, facing the rising production costs in a small territory of limited industrial resources, Macau had to transfer most of its textile and garment production bases to the province of Guangdong.

1.3 *The current perspective over Macau fashion promotion, events and educational institutions*

With the relocation of a significant part of the manufacturing industries to Guangdong, Macau's local industries had to change their focus to other areas, such as gambling, entertainment and tourism. Since 2003 the local government has put an emphasis on local creative industries, including the promotion and development of "Made in Macau" label, and sponsoring of local fashion events and initiatives such as the Macau Fashion Festival (MFF), Macau Fashion Parade and the Macau Fashion Link (MFL).

On the other hand, the Cultural Affairs Bureau provides different subsidies to help local designers produce and promote their creations. One of the platforms is "The Subsidy Programme for Fashion Design on Sample Making", launched in 2013, which is an annual competition that gives opportunities to local designers to showcase their design collections and provides an exchange platform between participants and the jury, where local fashion designers receive professional advice in order to elevate their creativity standards as well as their marketing plans. The main objective of the programme is to popularize and promote local creations. Additionally, the local government has been working together with local institutions to promote local fashion designers. One of these institutions is the Macau Productivity and Technology Transfer Centre (CPTTM), founded in 1996. It is a non-profit organization jointly establish by the government and the private sector, which aims to support the local design industry, offering Fashion, Design, Textile and Marketing professional training courses. CPTTM organizes every year, since 2010, the Macau Fashion Festival, featuring fashion shows (that include clothing, jewellery and other accessories), seminars and showcasing exhibitions. It includes local designers and designers from Taiwan and Mainland China. This event is incorporated as a part of the Macau International Trade and Investment Fair (MIF) held in the Venetian Macau, every year in October. According to Mr. Shuen Ka Hung, the Director of CPTTM, "the festival is intended to boost Macau's fashion industry, providing an opportunity for designers to showcase their collections and to build a network which will allow them to take their brands overseas" (Pinto, 2014).

On the educational side, the University of Saint Joseph (USJ) and the Macau Productivity and Technology Transfer Centre (CPTTM) have joint efforts in 2016 to open a Bachelor Degree in Fashion Design, the only one currently operating in the territory. Their common resources include "the latest garment-making software, Pantone, colour and trend forecasting, as well as a wide range of reference books to provide students access to global information" (Macau Idea, 2015).

"The Fashion Design Program was designed considering a diagnostic of requirements and critical facets within: the current socio-cultural developments derived from the convergence of east and west; the most recent technological innovations; the advent of new design materials; the new trends in aesthetics; and the paradigm shifts in the social and productive reality of this industry" (Fashion design Bachelor, 2017).

In this paper, we focus on characterizing the Local Fashion design industry from the perspective of the designers and how the multicultural essence and economic context of Macau have shaped their design and production process.

2 MATERIALS AND METHODS

In order to better understand the emergent fashion design scene in Macau, we have adopted a Mixed Methods approach, based on the collection of both qualitative and quantitative data. In particular, we have followed the Sequential Explanatory Research Design Strategy, proposed by Creswell (2014), which includes a first stage of quantitative data collection, followed by qualitative data collection. Both stages are explained next in detail.

2.1 *Stage one: A survey with fashion designers in Macau*

This first stage of the research was dedicated to get an overview of the current Fashion Design Industry in Macau, from the perspective of the creators. In order to do that, we have defined a questionnaire with open and close-end questions.

According to several sources (among them, the Creative Macau database (CCI, 2005)) the population of Macau Fashion Designers is around thirty-five designers. After contacting them, ten have replied positively and agreed to answer the questionnaire. The ten respondents actually represent a valid sample, as they cover a wide variety of designer's background, such as different origin countries, different culture, education levels and brand ownership.

The survey was conducted online using ©Google Forms service and was sent to the designers by email and ©Facebook messages. The survey was conducted during April 2015.

The questionnaire focused on how local designers structure and conduct their design activity and was divided in five segments: 1) designer's background, 2) brand, 3) production and selling, 4) Government and local Industry and 5) designing process.

2.2 *Stage 2: Case study with three emergent fashion designers in Macau*

The second stage of the study was based on an Exploratory Case Study approach, dedicated to get deeper insights into the phenomenon of the Fashion

Design Market in Macau, consisting of 1) face-to-face in-depth interviews (some of the interviewees were sent additional questions by email after the interview) and 2) visual analysis of designers' portfolios (on websites and social media sites).

These three designers have been selected and interviewed based on the following criteria:

- The interviewees have established and worked for their own brand for more than 3 years.
- Have studied and graduated in Fashion design outside Macau.
- Their age ranges are between 28 to 31 years old (targeting emergent/young designers).
- In their brand, they produce tailor-made and ready-made fashion collections.
- They sell for different markets, such as Macau, Hong-Kong and online market.

For the analysis of the in-depth interviews we used content analysis techniques, coding and comparing the responses for each question across the three respondents. The responses were then divided in nine categories, which defined the subchapters of the results' section (3.2).

The designers' portfolios combine images and textual information and were compiled from the following sources:

- Macau Fashion Festival promotion catalogue.
- Designers portfolio that where sent in PDF format before the interview.
- Web portfolio, such as brand websites, Facebook, Instagram.
- "Macau Fashion Parade", catalogue from Kong-Kong Convention and Exhibition Centre.
- Local institution and online websites, such as CPTTM, Creative Macau and Macau Ideas.

The analysis of the portfolios was mostly based on image content analysis of the collections and their textual descriptions, defining the main trends, aesthetics of the designs and individual creative processes, thus providing a visual narrative and explanation on how their collections have evolved and developed.

"(...) for many researchers, images carry different kinds of information from the written word. (...) are often more evocative of the sensory, as well as richer in information, than interview talk or written text can be, (...)" (Rose, 2016, p. 330).

3 RESULTS

In this section, the results of the collected data are divided in two main groups: Survey and Case Study.

3.1 *Survey Results*

The survey results were divided in five sub-sections, based on the five segments of the survey questionnaire. Due to pages' constraints, we present only the most relevant data.

All the respondents are from Macau and have studied fashion-related courses (30% have postgraduate studies). 50% studied in Macau, 40% in UK and 10% in Hong Kong.

80% have their own brand and 20% work for other brands and during their career have already created around 5 to 10 collections. 50% of our sample reply's that have been involved in fashion career between 2 to 5 years, 20% 6 to 10 years, 20% over 10 years and 10% under 1 year. 90% of the sample subjects design for women's wear. 60% design outfits that are custom-made while 40% produce in large quantities. Local designers show their collections in fashion shows in Macau and Hong Kong, websites and in catalogues.

70% buy raw materials in China and Hong Kong and 30% buys in Macau. The selling places are Macau, Hong Kong and China. 40% sells online, 30% in their own store, 20% in their own showroom, 10% in shared showrooms. Common clients are 70% local people, 30% Others citizens, such as people from Europe, Hong-Kong and mainland China. On the Government support, 40% reply's local government support financially on local producers and designers, but also 40% reply's that government support is insufficient, and 20% support with promotional activities.

About the fashion technologies, 90% thinks that technology is the future of fashion and 10% think that technology is appeal to emotions and will persuade clients to buy. Will they use new technology on their designs, 80% reply that will use technology in their design, while 20% reply's no.

60% characterize Macau's fashion industry as a mixed cultural market. They think that Macau is a mix cultural place, and that fashion is also a mix market.

3.2 *Case study results*

The case study results were originally divided in nine categories, based on the in-depth interviews structure. Each category combines data from the interview and portfolios' analysis. In this paper, we have decided to present the information organized by interviewee.

The first interviewee, Isabella, was born in Macau and did a BA in Fashion Design in London, Kingston University (Figure 1). In 2010, after graduation, with a sponsorship from a buyer, starts by selling shoes, that were worn by several Hong Kong artists. In 2011, she developed her clothing line and established her brand Nega C., participating in different shows in Hong Kong and Macau. In the beginning her style was more "crazy", maybe due to her relation to the showbiz industry. After opening her first store in 2012, her collections' lines were more focused on market demands: a blend of street fashion and fast fashion, with a mix and match aesthetics, from Asia to Europe. "I don't intentionally mix different aesthetics, [I do it] maybe because of the demand of the global market style." She has a tailor-made and ready-made line. Most of her collections are produced in China/Zhuhai, and details or small productions may be produced in Macau. Most of their clients are local people (from Macau), but also has clients coming from Japan, and

Figure 1. Nega C. Graduation collection from Kingston University, 2010.

Figure 2. Cocoberryeight, Summer beachwear collection, 2015.

other Asian countries (both tourists and non-tourists). Since it sells more to the Asian Market, she adapts her designs to the Asian body measurement (ex. Shorter sleeves). Sizes are mostly S and M, and some items are free size. Isabella has two stores in Macau (exclusive for her own brand). Additionally, she sells in C-shop (in order to further promote the "Made in Macau" label, the Cultural Affairs Bureau opened the C-Shop, offering a wide range of services, to let local visitors to enjoy the design and creativity of the local producers) in Macau and online platforms (her own website and other e-commerce platforms). In Hong Kong, sells in Volusvous shop. Sales volume per year is around 1200 pieces. Age range of her clients is between 20 to 40 years, and the price range (in MOP) for piece is around 350 to 1000 on the ready-made collection and the tailor-made collection is up to 2000, depending on the client's needs. She never used government support but believes that local government should give more support on selling places, instead of giving just money.

The second interviewee, Barbara, is Portuguese (mother from Portugal and father with Chinese ancestors) and has graduated from the London Fashion College. Spontaneously, she started her own brand Cocoberryeight, "a young, modern, urban brand with a hint of bohemian style, hippie, sometimes hemp" ('Macau Fashion Festival Promotion Catalogue', 2015). Initially she focused more on her clients' demands (tailor-made) but currently she developed a beachwear collection (ready-made), which follows a more market-conscious approach (Figure 2). The tailor-made collections were produced by herself in Macau, while the beachwear collection is being produced in Hong-Kong. She works for an international Market (most of the clients are Portuguese, Australian, Asian, Romanian, Croatian), not adapting to a specific market (Asia or Europe). Sizes range from XS to XL. Barbara mostly sells in stores in Hong-Kong and in her online platform. Sales volume per year is around 5000 pieces. Age range is between 18 to 40 years, and the price range is around 500 to 1300 on the beachwear collection, the tailor-made depends. She already asked for government support but her idea is that "it's a complicated process". "Government helps, but normally not enough, they help but in a wrong way, we need more space to sell with low rent on stores".

The third interviewee, Lalaismi Wai is from Macau and her parents are from China. She graduated in 2012, in ESMOD (École Supérieure des Arts et Techniques de la Mode), Beijing.

Before establishing her own brand, she has worked at Macau Fashion Concept Store (C-Shop), and in 2012 established her own brand Pourquoi. Her brand style in the beginning was influenced by the Japanese's style, but recently, with the emergence of a global market, her designs have a mix of different fusion elements, such as Lolita style, mixed with a more comfortable wear and lifestyles with a more Asian culture design (Figure 3). Initially, Lalaismi focused on a more creative market (tailor-made) and gradually shifted to a more commercial line (ready-made). For both collections (tailor-made and ready-made), she gets inspired by western designers. In the past, the brand was produced in Macau but nowadays, due to the low cost and production efficiency, they are produced in Mainland China. Her major clients are "young costumers, that have a strong personality and want to be different." Sizes range from S to L. Pourquoi sells on pop-up Stores (such as C-shop) in Macau, as well as on the brand's online shop and other e-commerce platforms. Sales volume per year is around 60 pieces. Tailor-made collection is over one thousand and ready-made collection is below one thousand dollars. Lalaismi participated and got the government subsidy to produce a

Figure 3. Porquoi "Time Traveller" fall-winter collection, 2016.

collection and also has asked for exhibition funding. She thinks that the local government support has been great to more vigorously promote the culture, in recent years but "if the government can take a more proactive approach to promoting the relevant activities, would be better."

4 DISCUSSIONS

On the survey, we can observe that the main production place of their collections is in Macau, China and Hong-Kong. In the case study, the three designers produce for small scale products in Macau and two designers produce their collections in Mainland China (due to the low production cost) and one produces in Hong-Kong (due to a language barrier, since she only speaks Portuguese and English). They all create two types of collection lines, one being more commercial and price conscious, the "ready-made collection" and the other, a "tailor-made line" that is more expensive and where they introduce their own aesthetics, creativity and express their imaginary world. On the commercial-oriented line, their products are more standardized while in the tailor-made line products are more personalized. In the ready-made collection, they adapt the brand style to the fashion trends and mainly to the current market position, while the tailor-made line is dependent on the demands of the client and the events where the client will use the outfit.

Local designers are influence by the multicultural reality of Macau, by using different aesthesis and inspiration trends in their collection. Additionally, we can also observe multiculturalism on their portfolio, by the use of diverse materials and fabrics in their collections, as well as different cuttings, silhouettes, pattern designs and colours. As an example, Nega C. uses on her designs European influences (lace, godet, pleats, transparencies, asymmetric lines, ruffle and pastel colour) and Asian influences (silks, golds and silver colours; Asian patterns and flowers embroideries; a conservative silhouette, such as high necklines). While Cocoberryeigh, is a more European style (crochet, minimalist line, bohemian style on patterns) but also with an Asia touch (sequins, gold and silver colour). Pourquoi, on the other hand, is influenced more by Japanese and Asian aesthetics (Japanese fabric patterns, silks, gold and red colour, sakura flower details on embroideries and geisha make-up as we can observe on the (Figure 3), but also, on some collections, by a European style (godet, pleats, polka-dots, laces, transparency and white colour). White colour in the Chinese community is used during the time of mourning, death and during ghost festivals, therefore Chinese people will wear white during a funeral or while summoning ghosts. On the other hand, white colour on the Western community represents or signifies purity, innocence, and light, normally used on festivities, such as weddings. Gold, red and silver colour is a favourite colour used in the Chinese community, it means wealth, good luck, celebration, happiness, joy, vitality and long life.

Two of these three designers adapt to the Asian market body measurement (maybe because of the production place), while one of them (since she produces beachwear) offers more standard sizes. Another important observation on the local fashion industry is the adaptation to a global market, with more and more online platforms developing in Macau; local designers, besides selling their products in their own shops, ateliers, showrooms and pop-up stores, they also sell on e-commerce platforms.

5 CONCLUSIONS

Macau as a former Portuguese colony, and currently as a Chinese Special Administrative Region, lives in the frontier of these two different cultures. Therefore, it is the perfect starting point for reflection on how the translation of cultures can be relevant to the deepening of the relationship between the two countries. It has a monolithic economy, which Macau's government is making an effort to diversify, exploring other industries outside of the gaming sector. For that, the government is investing in the emergent Creative Industries, boosting the local "Made in Macau" brand, which can help Macau to create a brand identity and improve on the tourism business. Throughout the 21st century cross-cultural and historical influences had a profound impact upon fashion design. The multicultural reality is also influence on the designer's creations and this cultural background can be an advantage inspiring their creative process. The findings of the study indicate that local designers adapt to foreign and local markets by designing a vast and varied product selection and by introducing different aesthetics and trends' concepts in their products.

By analysing the data collected by the interview and their portfolio collection we conclude that the three designers were influenced by multiculturalism.

Due to their foreign educational background, they have received foreign educational methodologies and have contacted with people from diverse cultural backgrounds and this fact has a direct influence on their designs' aesthetic and creative process. Local designers used different aesthetics in their brand, due to the demand of the global market. For fashion designers, create and show a collection is not just creating something with their intuition. Their cultural background also influence on their creative process, such on the research trends and inspirations. The way they read and interpret the culture tendency is the main focus. With the advance of technology development, it let us to access to new fabrics, new integrations of electronics, new resource and with that, designers can have more possibilities on their creations. As we can observe on the case study results, the local designers have been influence by their individual costumers. Subsequently, many of them, designs for local artist or musical bands, this culture group as no influence on their brand style. Local fashion designers implement on their clothing a unique arrangement of colours, textures, lines and shapes. The way they combined and arranged this specific details or shapes, is what is gives characterization and distinctiveness on their creative product. Since their all produce two type of collection line, one drives by the trends and market, and other drives by the private costumer. Moreover, all this aspect is influence by the body proportions, skin colour, as well as, physical aspects and its symbolic relation to the culture. Fashion for many creators is drive by the culture issue and nowadays, street wear, social media and fast information are also the demand.

We hope this study can be provide a significant contribution to characterize local Fashion Industry and assist those interested in exploring how the types of mixed retail brands could be expanded in Macau. The local fashion industry has a strong base and is expanding, and Macau can become a starting point to showcase and promote local fashion designers' work.

ACKNOWLEDGMENTS

I would like to thank all the designers that have replied the survey, and a special thanks to Barbara Ian, Isabella Choi and Lalaismi Wai for giving me the opportunity to take them as my subjects for my case study. They have been very helpful throughout my work.

I would also like to thank my tutors for guiding me. A special thanks to Mariana P. L. Pereira for her help in improving the manuscript.

REFERENCES

BBINK J. O. G. A. (n.d.). Culture of Macau. Retrieved May 28, 2016, from http://www.everyculture.com/Ja-Ma/Macau.html

Boxer, C. R. (1963). *Racer Relations in the Portuguese colonial empire (1415–1825).* Oxford University Press, London.

Boyce, C., & Neale, P. (2006). *Conducting In-Depth Interviews: A Guide for Designing and Conducting In-Depth Interviews for Evaluation Input* (Vol. 2). USA: Pathfinder International.

CCI. (2005). Creative Macau, Macau Fashion designers. Available at: http://www.creativemacau.org.mo (Accessed: 20 February 2015).

Choi, I. (2010). Nega C. Graduation Collection. Retrieved June 20, 2017, from https://www.facebook.com/NegaC.Fashion/photos/a.1077830438897323.1073741827.262416443772064/1077832562230444/?type=3&theater

Creswell, J. W. (2014). *Research Design: Qualitative, Quantitative and Mixed Methods Approached.* 4th Ed. London: SAGE.

Domingues, C. (2014, October 6). Macau Apoia Promoção da Moda Portuguesa na Ásia. Plataforma Macau. Retrieved from http://www.plataformamacau.com/macau/macau-apoia-promocao-da-moda-portuguesa-na-asia/

Fashion design Bachelor. (2017). University of Saint joseph. Available at: http://www.usj.edu.mo/en/courses/fashion-design/ (Accessed: 15 July 2017).

Ian, B. (2017). Cocoberryeight Facebook. Retrieved June 20, 2017, from https://www.facebook.com/Cocoberryeight/?fref=ts

Institute, M. T. & I. P. (2011). Macao ideas. Retrieved April 15, 2017, from http://macaoideas.ipim.gov.mo/en/index.do

Leong, Y. (n.d.). Made In Macau, Outlook for Fashion Industry, José Tang: Growing an Enterprise by Expanding Sales. Retrieved April 3, 2017, from http://www.c2magazine.mo/en/reports-en/chi-chat-en/8122/

Loureiro, R. M. (1999). *Guia de história de Macau: 1500–1900.* Comissão Territorial de Macau para as Comemorações dos Descobrimentos Portugueses – Macau – CTMCDO.

'*Macau Fashion Festival Promotion Catalogue'. (*2015). Macau: CPTTM.

Macau Fashion Gallery. (2017). Retrieved April 3, 2017, from http://macaofashiongallery.com

'*Macau Fashion Parade'.* (2016).Hong Kong: Kong-Kong Convention and Exhibition Centre.

Macau Idea. (2015). 'Learning in Style', Macao Ideas, 27 July. Available at: http://macaoideas.ipim.gov.mo/macaors/html/article/20150727/MI_172.pdf (Accessed: 19 April 2017).

Mendes, C. (2013). *Portugal, China and the Macau Negotiations, 1986–1999.* Hong Kong: Royal Asiatic Society.

Pinto, C. (2014). *Macao Fashion Festival | Catwalk showcases local and overseas designer's creations*, Macau Daily Times. Macau. Available at: http://macaudailytimes.com.mo/macao-fashion-festival-catwalk-showcases-local-overseas-designers-creations.html (Accessed: 10 April 2017).

Rose, G. (2016). *Visual Methodologies: An Introduction to Researching with Visual Materials.* 4th Ed. London: SAGE.

Sousa, M. (2009). *Regional integration and differentiation in a globalizing China: the blending of government and business in post-colonial Macau.* University of Amsterdam.

Wai, L. (2017). Pourquoi. Retrieved March 12, 2017, from http://www.lalaismi.com/home.html

Woodside, A. G., et al. (2009). *Perspectives on Cross-Cultural, Ethnographic, Brand Image, Storytelling, Unconscious Needs, and Hospitality Guest Research, Advance in Culture, Tourism and Hospitality Research.* UK: Emerald.

Textiles, Identity and Innovation: Design the Future – Montagna & Carvalho (Eds)
© 2019 Taylor & Francis Group, London, ISBN 978-1-138-29611-4

Textiles as a tool for social cohesion and active aging in social projects: Cases exploring the creative dimension of participants

A. Souza & R. Almendra
CIAUD, Lisbon School of Architecture, Universidade de Lisboa, Portugal

R. Porto
Lisbon School of Architecture, Universidade de Lisboa, Portugal

A. Vasconcelos
Faculty of Fine Arts, University of Lisbon, Portugal

ABSTRACT: This study presents a diagnosis about the use of creative techniques in social projects adopting textiles as a tool for social cohesion and active aging. Aging process among populations and functional, psych-cognitive, and social-economic difficulties triggered by it highlight to the need for new Design practices which can contribute with satisfactory solutions to the elderly people. The dimension of creativity linked to textile activities developed aiming at mitigating social limitations of the elderly is exposed through two social projects conceived on a local scale, being one in Lisbon, Portugal, and the other in Uberlândia, Brazil. Semi-structured open interviews, and documents review are used for constructing the case studies. This study's findings lead to the identification of the challenges involved in performing such nature of initiatives as well as the procedural description about creative and co-participation activities.

1 INTRODUCTION

In the overview of the current world's population aging process it is imperative to develop, discuss, and think about strategies addressing the elderly social integration, assistance, and active development. In this context, projects oriented to physical, psychomotor, and cognitive needs including the elderly in participative activities aiming at the maintenance of their social relationships are emerging in the field of Design. Especially in participative-character projects it is possible to register a wide range of cases involving textile materials that allow preserving and using traditional knowledge, valuing labor capabilities, and offering the participants an active way of working.

Therefore, there is a lack of documents in these cases to describe the creative activities arranged, tools used, methods chosen, and benefits achieved during the process. Thus, the starting point of the present study lies on the following question: what are the opportunities and challenges involved in social projects for elderly social assistance using textiles and creative techniques as cohesion tool?

The goal of this study is describing general features of creative process in order to produce useful information to designers, creative intervenient participants, and researchers interested in social projects involving textiles and the elderly public from the perspective of participative activities.

The research method chosen is the analysis of two cases involving individuals in the same age range, but in different countries. In both cases the following goal of study was established: the craftsmanship; the preservation of the "know how" status of a traditional or popular constructive technique; and the proposition of a change towards elderly wellbeing.

2 THE CONTEXT OF THE SURVEY

2.1 *Design for the elderly population*

In general, world's population is going through a demographic transition process in which elder populations are being structured. The rate of older people, at 60 years or older, must be doubled from 2007 through 2050, while the rate of those older than 80 years must quadruplicate and reach almost 400 million of people (ONUBR 2017). From such estimates there is a common sense that elderly population progressively plays a fundamental role in the constitution of contemporary society, bringing deep changes to the current public health and social assistance policies, among others.

The changes occurring on population's age structure in Brazil are evident. Data presented by the Brazilian Institute of Geography and Statistics show that elderly population, estimated as 10% from the total Brazilian population as of 2010 along 40 years, that is until 2050, will represent 29.3% from the total population (SDH n.d.). In 2008 the ratio of elderly individuals to each group of 100 children aged from 0 to 14 years old was 24.7. It is estimated that in 2050 the ratio will be 172.7 (IBGE 2008).

In 2015 aging index reached 143.9% in Portugal (INE & PORDATA 2017). In order to comprehend the meaning of such mark properly one must notice that the aging index represents the number of individuals older than 65 per each 100 individuals younger than 15 years old. When these values are high they mean a demographic transition in advanced stage. Unlike when the value is smaller than 100 it means there are less elderly than younger individuals.

During aging process it is usual to accomplish significant changes on individuals functional, psycho-cognitive, and social-economic aspects (Mazo et al. 2004). In the functional aspect, changes on the physical appearance and diverse functional systems of bodily physical structure are visible. In the psycho-cognitive aspect there are changes on cognitive perception, such as memory records, as well as learning process and response timing. In general, there is a reduction of psychomotor abilities, which in turn causes difficulties to motor and sensorial capabilities. In social-economic terms, when an individual retires for age causes, the distance from social/professional circle and the loss of purchase power cause the individual to become isolated.

Design seeks to offer alternatives generating wellbeing, comfort, and safety to the elderly. Some initiatives concentrate on the development of products that can answer to the elderly functional requirements, by adding ergonomics and universal Design knowledge. The "cane collection", designed Francesca Lanzavecchia and Hunn Wai, is an example of that.

Other Design initiatives pursue generating answers to the elderly psycho-cognitive limitations. An example of cognitive limitation that takes place, usually from the 65 years on, reflecting on several functions such as memory, concentration, and language is sporadic Alzheimer disease. In the (ongoing) Doctorate project called *Design for Alzheimer's Disease,* by Portuguese Designer Rita Maldonado Branco, tools are being created, for example a card game, a table cloth, and a family book to the diseased individuals. Such practical tools are Design trials that stimulate the handicapped individual independence, causing the family interaction to become easier. In the context of initiatives aiming at social-economic changes among the elderly, projects proposing solutions by means of Design services are seen, and those usually depend on the active engagement of the elderly. The *Action for Age* project, in forth simultaneously in Lisbon and London, challenges students to sketch Design services that can bring improvements to orderly's life quality, based on the establishment of intergenerational relationships. Even though such services are developed at conceptual level, it is important to notice that the proposals involve participative experiences offering the coexistence among several social and age groups, as well the efficient use of public spaces, sharing crafting and traditional knowledge, and boosting social networks.

A common point to both projects *Design for Alzheimer's Disease* and *Action for Age* is the inclusion of creative experience. According to Sanders (2016), creativity is manifested in four levels of complexity, from the most basic known as doing, in which the individual feels himself as a creative one when performing daily tasks, such as producing a handmade product like a meal; the second level, adapting, involves adapting some existing thing, personalization of an artifact, for example customizing clothes; the third level, making, requires developing hand and brain abilities in making something brand new, for example constructing a household artifact; and the highest level, creating, covering the self-expression of the one leading to innovation, such as generating a new service.

A recent example of social action stimulating elderlys creativity is the *Project Lata 65* initiative which offers *doing* type activities – from the simplest level. This project's main idea is teaching urban art to people older than 65 years, more specifically *graffiti* skills, while creative techniques such as visual reference collection, drawings, and stencil templates for later trial of spraying walls. Under the advisory of Portuguese architect Rodrigues, the project has received financial support from the Municipal Council of Lisbon for its implementation.

The examples described demonstrate Design answers to the struggles of the elderly as to social relationships while revealing the potential for developing creative activities as a co-creation for maintaining relationships and wellbeing among this population. Therefore co-creational activities involve the generation of social, monetary, and experience value (Sanders & Simons 2009).

Co-creational activities in the Design process that focus on producing social value are based on human aspirations and pursue sustainable ways of life. The authors Sanders & Simons (2009) indicate some prerequisites for practicing co-creation which bring social transformation: i) the most suitable phase for its stimulation in the group is the *front end* phase; ii) to believe that everyone is creative and able to participate as long as proper conditions are available; iii) the formation of multidisciplinary teams enclosing individuals maintenance of constant dialogue and offer of learning workshops; e v) to keep focused on the final product and service but also on the experience brought to the participants and stakeholders.

A common point among the examples of co-creative activities addressed to the elderly that use textiles is the use of a popular hand technique or a traditional constructive technique. Exchanging experiences about such knowledge helps in the cooperation and involvement of all participants in the activities. Besides that, the application of popular and traditional techniques contributes to make visible all social and cultural elements in the artifacts, providing help to the valuation of both product and its manufacturer (Krucken 2009).

2.2 *Textiles as a tool for social improvement*

Most scientific studies in textile field relate to manufacturing, materials, performance, innovations, and clothing. Under the Design perspective, textiles are assessed from their process of designing creation and structures specific to them, as the production of patterns for multiple surfaces applications. It is notable that there is an effort to investigate the physical aspects of textiles, but on the other hand, recent studies take social aspects into account, such as wellbeing feeling and the comfort brought by the physical contact with the textile product.

As to the social perspective it is right to say that textiles and the material used for producing clothing both affect the physical and psychological aspects of elderly daily life.

In this state, initiatives like the European international project *EASYTEXT – Aesthetical, Adjustable, Serviceable and mainstay Textiles for Disabled and Elderly* rise. This is a project that assumes the offer of clothing and special textiles to the elderly must be improved, and in which a data bank is created to supply manufacturers with clothing and textiles information, and a system of industrial clothing production is created considering the individual adaptation using the automated technique for body scanning and shifting patterns according to individual dimensions (Meinander & Varheenmaa, 2002).

An example of social recovery by textiles manipulation takes place in the *Estabelecimento Prisional de Tires (EPT)*, in Portugal, designed exclusively to the secluded women. At first, labor workshops were developed by the Order of Religious Nuns of Bom Pastoral, in 1954, in which the nuns created drawings from the churches tiles and made graph paper models. It is a model of work profession in extreme episodes of social seclusion, where women are delivered technical knowledge – they learn how to stitch in *Arraiolo* design (tapestry) – and interact with exterior community through the market of the items produced from such knowledge (1996). In this case, the amount acquired with the sales is shared among the secluded women and the prison installation.

The number of social projects involving textile handmade practical activities and unemployed women in the Portuguese context is increasing. In Porto city there is a project called *Vintage for a Cause (VfC)* which focus on women older than 50 years with no work activity and proposes the customization of old used clothes into *vintage* style clothing. The project aims at the promotion of active aging, personal development, interpersonal relationship, and social inclusion by being busy, with two perspectives present: interpersonal perspective, related to the conviviality among the participants; and technical/operational perspective, related to sewing knowledge. The customized pieces are unique ones, conceptualized with the cooperation of Portuguese stylists who help in the process of stimulating creativity and valuation of the pieces produced. An interview performed by the authors with the current coordinator of the project reports the major techniques used in the creative process are visual references collected from magazines and brochures. Also, performing social-cultural activities such as workshops, allows for the acquisition of new knowledge, the interaction between the participants, and the enthusiasm for social life. Since the project focus on social values, total income gathered by the sales of products integrally revert into the project, aiming at its self-sufficiency and sustainability.

In Lisbon, the City Council finances the *Tecidos de Autonomia (TdA)* project, designed to women suffering economical deprivation associated to unemployment, disease, poor jobs, lack of social and family support, lower qualification levels, emotional instability, and support dependence and social help conditions. Creative techniques between collective and individual tasks are disregarded in this initiative. The project is about creating a formation atelier for developing technical, personal, and social capabilities; building up a social business for selling products made out from the reuse of textile wastes; and a contest among porches decorated with textile-made ornaments.

In the Brazilian background two cases are highlighted for they translate the idea of reuse, income generation, and enterprising for low revenue groups by means of working with textiles. The first one, the *Cardume de Mães (CdM)* project, is a production group assembling 10 women living in peripheral regions who produce items using wastes from advertisement banners and their transformation into fashion items and appliances. In a partnership with the *Rede Design Possível*, which invests on social and environmental initiatives bringing positive changes to society, the *CdM* project offers training, advisory, new products development, and strengthening of the business identity. The second case, the sewing workshop *Pano Pra Manga (PPM),* works with people living in an outermost community of São Paulo, being composed by women who were outside the labor market and learnt a new occupation from being in contact with textiles. Since 2008, when qualification workshops were launched, the project technically formed over 100 people from the neighborhood.

So far it was noticed that besides female elderly population the textiles are used as a tool in social projects addressed especially to women under social vulnerability conditions. Most of the cases identified thus far have a participative approach, which means they look forward involving the participants of the project in the activities so they can contribute to the maintenance of the social project as listed in Table 1. In turn, economical sustainability of these social projects

Table 1. Design approach and economical sustainability of social projects.

Initiative	Design	Sustainability
EASYTEXT	focused on the user	subsidy
EPT	participative	self-sustaining
VfC	participative	subsidy
TdA	participative	subsidy
CdM	participative	subsidy
PPM	participative	subsidy

is a weak point, being mostly necessary to claim for private, public, or non-profit institutions subsidies in order to maintain the activities going.

3 STUDY CASES IN BRAZIL AND PORTUGAL

3.1 *Construction of study cases*

The present study develops from the analyses of two study cases. The criteria adopted for selecting cases are: i) projects aiming at promoting better social conditions to elderly population; ii) projects adopting creative techniques grounded on participative process; and iii) ongoing initiatives.

Data were collected from documents offered by City Councils of each of the cities, from several media publications, and semi structured interviews with the leaders of the project.

The elements used for analyzing the cases are: i) the category corresponding to the field of research in Design; ii) the phase of inclusion of co-creative activities; iii) the project focus (on product, service, or experience; or a mix); iv) the level of complexity of the creative activities; v) the creative techniques adopted; vi) the use of popular knowledge and valuation of local qualities; vii) the typology of working teams.

3.2 *The A Avó Veio Trabalhar Case, Portugal*

The initiative *A Avó Veio Trabalhar (AVT)* is mainly sponsored by the City Council of Lisbon and aims at using Design to promote the active aging by creating experiences where co-participation methods are used. The space is currently maintained by Foundation Calouste Gulbenkian on consignment basis, and presents as a creative laboratory where people can use materials and machines for sewing, confection, and embroidery, besides learning with the most experienced participants. The elderly share their knowledge among the community during the workshops, teaching the participants how to embroider, knit, sew, or screen-printing. The space reserved to the activities works simultaneously both as school and point of sale. The creation of collections with the items manufactured is envisaged in the activities with the elderly. Each collection is developed in three-month periods and defined as co-creative process, with new ideas, use of traditional techniques, study of trends, and the participation of different artists. The *AVT* project counts the participation of 70 elderly participants since its foundation up to now.

3.3 *The Fios do Cerrado Case, Brazil*

The *Fios do Cerrado (FdC)* initiative aims at promoting the wellbeing to elderly under deprivation conditions from the learning and execution of crafts activities. The proposal is centered in activities that promote the active aging through sharing experiences and practical activities. Free professionalizing courses are offered to that population twice a year. In the courses, the participants learn all processes involved in hand weaving, from spinning, through cloth patterning, weaving, to finishing. After learning period there is a possibility to make part of the team of participants so that the participant keeps improving, sharing knowledge and working profitably.

The project launched in 1993 is subsidized by the City Secretary for Social Development and Work and maintained by the Cultural and Care Foundation Filadélfia, a non-profitable entity that supports initiatives aiming at preserving traditional construction techniques. It has its own space larger than 1,200 m^2 for developing its activities, besides its own shop for selling products. The project directly incorporates 23 fixed artisans who work in town and assembles approximately 15 students in the courses twice a year. Among the products developed by the artisans the collection of rugs, curtains, blankets, and table cloth are highlighted. The production of the pieces takes place in steps: cleaning cotton, carding, spinning, warping (determining length, width, color, and cloth pattern), weaving, finishing, and sales. Pieces are sold at token prices that cover part of the space costs.

4 DISCUSSION

4.1 *The nature of projects in terms of type of research*

In the context of Design knowledge field, according to the structure proposed by Cross (2007), the two cases studied correspond to the perspective of Design praxeology, probably due to the fact that there is scarce theory about the issue in discussion. Opposites, participation practice brings more information about the elderly population in terms of behavior, which is crucial to Design interventions in the search for improvements in the elderly population wellbeing. Therefore, our suggestion is that more research must be performed about Design processes and practices on projects involving textiles and elderly population. New tools, techniques, and methods must be developed to address specific needs of elderly population.

Table 2. Analysis of creative activities

Case	Level of complexity			
	Doing	Adapting	Making	Creating
AVT		x	x	
FdC		x	x	

4.2 *Creative activities*

Regarding to the inclusion of creative experience in the activities developed in each case, it is followed the analysis proposed by Sanders and Simons (2009) as shown in Table 2.

Occupations that refer to the creation of artifacts developed by the elderly participants in the *AVT* project configure as activities to adapt existing textile items, related to the adapting level, as well as creating collections under the advisory of Design and Fashion professionals, related to the making level. In the *FdC* case, creative activities are guided by more experienced participants, corresponding to adapting level. The first line of patterning was conceived by a local artist and weaver, Edimar de Almeida. Suggestions of changes that do not interfere with the visual quality of patterns to the cost of constructive viability have been proposed. In a more advanced way, some activities involve the third level described by the authors, the making one, since the participants explore hand abilities in order to construct new objects.

In the context of creative techniques adopted in the process of artifacts creation, the techniques adopted in the *AVT* case envision the use of visual references from magazine clippings and the internet, while in the *FdC* case the means for testing composition, reading, and matching of existing patterns and colors are explored.

With regards to the step of inclusion of co-creative activities into Design process, both in the *AVT* and in the *FdC* projects, creative activities are expected to take place during conception and planning of the project, from the front end phase, reaching the preconditions indicated by Sanders & Simons (2009).

4.3 *Product, service, or experience*

The *AVT* project generates: a) products, since the elderly makes artisanal artifacts in a participative process; b) services, when offering learning and knowledge exchange spaces among different generations of participants; and c) experience, when stimulating wellbeing among elderly population and thus facing negative aspects of aging process such as isolation, depression, and low self-esteem. In the *FdC* project weaving products are traditionally explored by low wages population. When the project was proposed the main idea was preserving material culture and patrimony produced by a local activity while supporting active aging. Besides the experience of sharing the "learn to do" thing and living in a community, the project provides a service of social assistance since, further than focusing on elderly individuals, it cares the most vulnerable ones.

4.4 *Work teams*

In both cases the work teams are multidisciplinary ones, corresponding to Sanders & Simons (2009) ideas as to the work teams' preconditions to the co-creation practices that imply social changes. The leaders of the Portuguese project aré trained in different professional areas and reveal previous experience in social featured projects. As to the Brazilian project, the tutors are usually young people who learnt the technique in the same Center and volunteer to the initiative. Often, in such nature of projects, participants come from different areas of knowledge, having no restrictions as to coworkers' backgrounds. It is also seen, as indicated by Sanders & Simons (2009), that dialogue is sustained between leaders and participants, as well as the offer of learning workshops to them with the presence of experts and artists in residence.

4.5 *The use of popular knowledge and the valuation of local qualities*

In the activities of *AVT* project the elderly use constructive hand techniques that have emotional and symbolic values (Krucken 2009). Either the appropriation of local identity textiles or even the insertion of a local culture in the artifacts created are not evident. The confection under reduced scale is prioritized, in an exclusive-piece manufacturing model, which justify the high cost of the masterpiece. The *FdC* case uses weaving popular knowledge, a constructive technique that used to be massively explored in Uberlândia. From the statement that this such constructive technique would disappear, the *FdC* project is launched. This initiative makes use of local material culture preservation in order to show up their knowledge, thus enhancing symbolic, cultural, and social values (Krucken 2009).

5 FINAL CONSIDERATIONS

From the present study it is noticed that some adjacent issues still have to be further approached in future studies, such as: the idea that gender-related engagement is limited; the stimulation of social relationships among different generations, overcoming the binomial elderly and youngster people; and the ways to support and maintain the activities, that is, the sustainability of social business.

It is noticeable that most of the social projects addressed to elderly population involving textile manipulation and co-creative activities aim at female gender. It is important to investigate in future studies along the projects' leaders by survey interviews, the reasons why male gender is not taken into account in activities of such nature.

Only one of the cases indicates actions on intergenerational exchange of knowledge and experiences, and it still can be granted for limited to the offer of achievements conducing to the recognition of elderly value in our society.

Finally, reporting to social projects sustainability, the cases analyzed get financial support for their implementation, most of which from public management programs. After implementation, and the public management supporting time is over, it is necessary to search for partnerships with private institutions in order to maintain the activities due to the short term financings. It is also shown that when teams are not multidisciplinary ones it is harder to get findings, as well as assuring the sustainability of the social business.

ACKNOWLEDGMENTS

This research was only possible with the support of CIAUD – Research Centre in Architecture, Urbanism and Design, Lisbon School of Architecture of the University of Lisbon, and FCT – Foundation for Science and Technology, Portugal.

REFERENCES

Bouthoul, G. 1949. *Traité de sociologie.* Paris: Payot.

Braddock, S. & O'Mahony, M. 2005. *Techno textiles: revolutionary fabrics for fashion and design.* London: Thames & Hudson.

Cross, N. 2007. From a Design Science to a Design Discipline: Understanding Designerly Ways of Knowing and Thinking. In Michel, R. (ed.), *Design Research Now*: 41–54. Birkhäuser Basel.

IBGE. 2008, December 3. *IBGE: População Brasileira envelhece em ritmo acelerado.* Retrieved from http://port.pravda.ru

INE & PORDATA. 2017. *Número de indivíduos em idade activa por idoso.* Retrieved from http://www.pordata.pt/

Kawamura, Y. 2005. *Fashion-ology: An introduction to fashion studies.* Oxford: Berg.

Krucken, L. 2009. *Design e valorização de territórios: valorização de identidades e produtos locais.* São Paulo: Studio Nobel.

Mazo, G., et al. 2004. *Atividade física e o idoso: concepção gerontológica.* Porto Alegre: Sulina.

Meinander, H. & Varheenmaa, M. 2002. *Clothing and textiles for disabled and elderly people.* Vuorimiehentie: VTT Technical Research Centre of Finland.

ONUBR. 2017. *A ONU e as pessoas idosas.* Retrieved from https://nacoesunidas.org/

Provedoria de Justiça. 1996. *Relatório sobre o sistema prisional.* Lisboa: Gráfica Telles da Silva.

Sanders, E. & Simons, G. 2009. A Social Vision for Value Co-creation in Design. *The Open Source Business Resource*, 27–34.

SDH. n.d. *Pessoa Idosa.* Retrieved from http://www.sdh.gov.br/

Textiles, Identity and Innovation: Design the Future – Montagna & Carvalho (Eds)
© 2019 Taylor & Francis Group, London, ISBN 978-1-138-29611-4

Cultural theory and textile manufacture

L. Tigre & M. Araujo
School of Arts Sciences and Humanities, University of Sao Paulo, Sao Paulo, Brazil

ABSTRACT: This article aims to make a brief discussion about cultural identity, using different concepts and thoughts about culture and how it is transmitted.

Using mainly the thoughts of Stuart Hall and Ruth Benedict, the article uses the concepts of anthropology to start a discussion that is recurrent in the field of social theory, identity.

Textile production is used as an example, since it is a common activity for different peoples and serves to demonstrate different cultures.

1 INTRODUCTION

Cultural identity has been a subject of broad anthropological debates and its interest is mainly due to the subjective character of identity and its various forms of manifestation, generating criteria for analysis.

Use as a reference the author Ruth Benedict that discusses in her research of cultural diversity, and the brief text of Sahlins "The Brief Cultural History of Culture", in which the author describes a great culture that would involve the whole world. For Benedict, culture is formed by Customs and these are passed down from generation by generation, being taught by, and an act not innate. Sahlins says that the traditions will be replaced by a great culture, a consequence of globalization.

This research aimed to make a discussion on the issue of cultural identity and address the issue that, according to Stuart Hall, "is being extensions while discussed in social theory". The anthropological vision was used to help establish what kind of concept it will use in the research and how to choose the people who would be studied.

The Moche and Chimu's cultures were chosen because they are the ones best know today and are the areas where the techniques appear to be most numerous. All the people of this region were remarkable weavers (D'HARCOURT, 1962). Manual textile production had a strong influence in the formation of the identity of peoples.

2 METHODOLOGY

The methodology used was bibliographical and documentary, with exploratory review of documents, articles and books. The library of the University of Sao Paulo, Dedalus, was used to select the material with keywords such as: culture, identity and anthropology.

The main aspects considered during the reading were the relevance of the author to the areas of textile technology, anthropology and archeology and the relation that they made of the techniques with the developed cultures and technologies. The analysis was of content, in which they were interpreted and formed a judgment about the authority of the authors and the value of their work and ideas.

The weaving techniques are rooted in the culture in which the development is carried out from generation to generation in an oral way, bringing much of the people in the manufacture of this type of textile, which makes it difficult to research the methods and meanings of each motif (D'HARCOURT, 1962).

In digital media were also found articles and sites about anthropology. Selected authors are references in the field of anthropology and treat mainly on topics related to culture. Stuart Hall is the main researcher on the issue of identity today, which made his texts indispensable for this article.

For this research were used the theories concerning the cultural identity, to relate with weaving manufacture and perform the analysis were taken into account theories that if immediate habits and cultural production, disregarding behaviors, another aspect studied in cultural theory.

3 LITERATURE REVIEW

3.1 *The culture*

The question whether or not there is an identity is one of the most recurrent topics in discussions of social theory (HALL, 1992) studied by anthropologists, ethnologists and sociologists.

When looking for the definition of what would be an anthropologist, we came across definitions as given by Bernardi in "Introduction to studies ethno-anthropology" (1992). For him, the anthropologist has always to explain his own identity and illustrate the difference between anthropology and ethnology. Although "legitimate" definitions according to Bernardi, they are limiting and can lead to confusion. With this differentiation of the terms, companies and scientific branches specific to each area were formed and ethnology came to be applied to the study of cultural or spiritual and social aspects of human activity (BERNARDI, 1992).

The two branches eventually join and the Division passed to physical anthropology, which study the shapes and structures of the human body, and cultural anthropology, which observes and asks the meaning and the structures of life as production and expression of mental activity (BERNARDI, 1992).

There are several meanings for culture, but in order not to exclude the peoples that were not in the most dominant countries, several formulations of anthropological vision were made (BERNARDI, 1992).For Tylor (1871) "culture or civilization is complex whole which includes knowledge, belief, art, morals, law, custom, and any other skills and habits acquired by man as a member of society". Despite being an evolutionary definition, this definition of Tylor helps include more primitive people, in a cultural study (BERNARDI, 1992).

On the translation made by Bernardi, Tylor refers to a unit complex and it brings up the tradition, a value that is passed from father to son and so on generating a tradition amongst the members of that community. This transmission of knowledge is mentioned by Benedict (s.d.) and starts to direct us to the question of identity.

Boas (1930) also set out to make a definition of culture and for him "culture encompasses all manifestations of a community's social habits, the reactions of the individual as affected by the habits of the group in which they live and the products of human activities while Determined by these habits."

In contrast to the two, Sapir (2012) does not give a definition to the term, because for him "indefinite areas of thought, that move, are restricted or amplified according to the point of view of the one who uses them." For him culture can have three senses: the technical sense, used by ethnologists; the sense led by common sense, often known as cultural snobbery; and the third that "is the most difficult to define and illustrate satisfactorily, perhaps because rarely those who use them are able to offer a perfectly clear notion of what they themselves mean by culture "shares similarities with the previous two and" the cultural conception we are now trying to grasp is intended to include in a single term those general attitudes, visions of life and specific manifestations of civilization that assure a specific people their distinct place in the world" (SAPIR, 2012).

The third design exposed by Sapir is the conversation with the subject of this article, which purports to discuss the question of identity, because this topic is too with a similar question, nationality: "In an attempt to find embodied in the character and civilization of a people some peculiar excellence, some distinctive force that is remarkably his. Culture, in this sense, becomes almost synonymous with the "spirit" or "genius" of a people" (SAPIR, 2012).

3.2 *Identity- A question*

For a long time, the question of identity stabilized the social world. But with greater interaction between people and the arrival of postmodernity, there was a decline of old concepts, predefined identities and stereotypes, and the appearance of a modern individual and identified with different cultures (HALL, 1992). This fragmentation made was to what anthropologists are calling an "identity crisis". This new condition is seen as part of a wider process of change increases, which is displacing the structures and processes of modern societies and changing models of stereotypes people, transforming the social situation in the world (HALL, 1992).

Stuart separates the identity in three very different conceptions of the others. He explains that there is the subject of the enlightenment, totally unified gifted capabilities-centered reason, conscience and action, an individualist conception, the sociological subject that reflected growing complexity of the modern world and dribbled to postmodern (HALL, 1992). The subject brings a sociological concept interesting for the discussion of identity, according to Hall. In this case, the identity is between the interior and exterior, "designed to" ourselves "in these cultural identities and to same time we internalize their meanings and values, making them "part of us"" (HALL, 1992).

The subject today is becoming fragmented, i.e. is composed not of an identity, but of many, and sometimes contradictory or not resolved. According to Hall, the identity is defined historically, i.e. being built over time and as well as Sapir (2012), says that culture can change, different identities can form at different times and it doesn't have to be unified.

As we get more experience and are exposed to more stimuli, new signs emerge. In this way, the system of meaning and cultural representation multiplies and is presented with the multiplicity of possible identities and we can connect with each other at least temporally (HALL, 1992).

In counterpoint to the multiplicity of Hall, Sahlins says in "the Brief cultural history of culture" (1994) that this fragmentation is a phenomenon of the 20th century and that assumes that with the globalization, the so-called peripheral countries want to integrate, so allow themselves to undergo a culturalization. The author says that it would be possible to reach a world culture. So instead of having several identities and connect depending on the stimulus to which we are exposed, to Sahlins everything becomes one.

3.3 *Ancient peoples and postmodernism*

Western civilization expanded much more than other peoples and spread their beliefs and behaviors. This meant that people that had been discovered or conquered often had more awareness of the role of culture, because they saw their customs be replaced and massacred by Europeans (BENEDICT, s. d.), as Persian rugs that have lost quality and religious and cultural significance to be exported by Europeans (CAMPANA, 1991).

The concept presented by Stuart Hall is for post-modern individuals. As he says, before there were individuals whose identities were more indefinite. The study of so-called primitive societies is important for providing material for the study of forms and cultural processes. Local cultures allow the study of these processes (BENEDICT, s.d.).

3.4 *Culture as usual*

When we think of the term "meaning" used by Stuart, we can establish a dialogue cultural theory discussed by Geertz in "A dense description" (2008), in which he uses semiotics and thinks about the concept of culture as something essentially semiotician.

Geertz discusses the culture as something public and brings an interesting thought to the discussion of cultural identity. "If the culture is a standardized conduct or State of mind or even both together, somehow loses the sense" (GEERTZ, 2008).

Thinking of a standardized culture, it is possible to include in this discussion the thoughts by Ruth Benedict in "Patterns of culture" (s.d.) and applying her analysis about phonemes and linguistics from different people with different weaving methods. According to Barty Philips (1994), there is no culture that does not have its own weaving technique and few that do not include tapestry between your skills. Although the basic process remains the same (weft and warp intertwined) every people has a "default" to produce a different type of fabric and a different node to do a design.

Ruth (w. d.) speaks of culture as customs and these customs are passed down from generation by generation, forming a tradition and something that is part of the "arc" of that particular civilization. The author believes that anthropology studies humans as products of life in society and this produces conventions and values. These points distinguish one community from all others that belong to a different tradition (BENEDICT, s. d.).

Benedict says that everyone sees the world with a vision conditioned by a defined set of customs, institutions and ways of thinking, being impossible to separate their conceptions of stereotypes (BENEDICT, s.d.). When we think of the identities of people who don't exist anymore, so we can observe their customs we must take into account the Customs and values targeted today are loaded with the conceptions of who studies.

3.5 *Identity in the context of national culture*

Currently not national cultures are something so well defined, but still define ourselves by saying we are

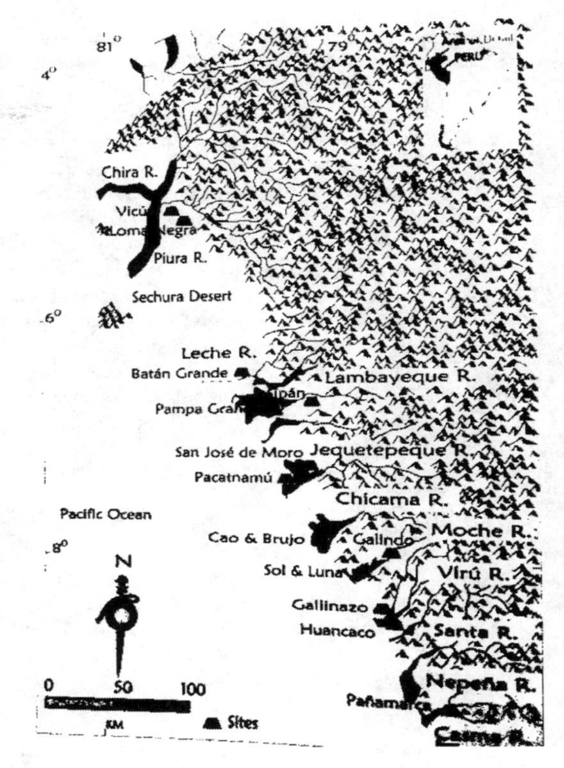

Figure 1. Map of Peru (Source: QUILTER. The Moche of ancient Peru, 2010).

Brazilians or Peruvians, and somehow we still have a national cultural identification (HALL, 1992).

This need to define is crafted by the philosopher Roger Scruton (apud. HALL, 1986), which helps us to understand that although the man acting as a as he identifies as something wider.

National identities are innate, formed and transformed by representation. They are the set of signs and meanings of the place where we live (HALL, 1992) that form webs that weave the culture (GEERTZ, 2008).

Second Arlene Dávila (1997) people don't know the culture, making it a means of Government use of stereotypes created by common sense to make nationalist policies, especially in countries that are under the influence of any economic power.

3.6 *Culture Mochica or Moche and Chimus*

Peru's north coast is a barren strip of desert fronting one of the world's richest maritime habitats (Figure 1.), a bountiful fishery created by the cold waters of the Humboldt, or Peru, current. Inland the desert gives way to mountains and plateaus of the Andes, dissected by rivers. From remote antiquity until today, well-engineered irrigation systems drawing these rivers have nurtured vast green fields of maize, beans, and squash in the valley bottoms (QUILTER, 2010).

The name "Moche" comes from the Moche Valley, the location of two large, prehistoric adobe structures translated as "the Temples of Moche". It was there that the first archaeological excavations of Moche artifacts took place in the late nineteenth century (QUILTER, 2010).

Priests and warriors occupied the highest ranks of Moche society, ruling over members of the lower classes, who tilled the fields and produced the elaborate paraphernalia for rituals and war. Next to the largest temples stood cities with streets and avenues along which lay residential compounds. Within them and at specialized workshops, artisans produced some of most beautiful metal objects found in the New World. The Moche' s representational art style, depicting gods, priests, warriors, animals, plants, and seemingly the full spectrum of everyday life, appeals strongly to modern tastes and sensibilities (QUILTER, 2010).

For some art styles, written texts can provide these sorts of external references. We have no such references for Moche, which, like the other ancient South American cultures, had no writing (QUILTER, 2010).

Although we cannot establish the beginning of the Andean culture, some excavations show that pre-Inca cultures already made textiles with differentiations that may indicate that the drawings were more than just aesthetic adornments.

The archaeologist Heinrich Ubberlohde-Doering in 1938 made excavations in an archaeological site of Pacatnamú. His works focused on a middling mound that was disfigured by the activity of looters. The odds of finding intact tombs of some importance were remote, but the archaeologist managed to find three. They began to appear tombs of different types of what is believed to be the principle of the Mochica culture until the end of the Chimú tradition (BUTTERS, 1999).

The three tombs found were well preserved and the E1 call was the one with the most items. Even with textiles in singular conditions, due to the contact with the bodies (BUTTERS, 1999).

These fabrics are the ones that best represent the Mochica textile art with known provenance. There are other suitably preserved textiles such as those excavated by Christopher Donnan and Guillermo Cock at this same archaeological site (BUTTERS, 1999).

Some of the artifacts recovered were large Mochica gears, such as the Kelim rugs found covering the main individual. They highlight human characters with mystical attributes (BUTTERS, 1999).

3.7 *Textile manufacture*

As was described previously, all civilizations have some sort of weaving (PHILIPS, 1994). Traces were found in Eastern Europe that proves the presence of weaving in the Paleolithic period (up to the year 10,000 b. c.). In this period the main reason the use of textiles was cover the body to control the cold (PEZZOLO, 2007).

Figure 2. Cotton fabric that has the representation of a sacred condor with a snake curled up in your womb.

The first civilizations (Assyrians, Babylonians and Egyptians) the clothing was already used to differentiate the social classes. In these civilizations the natural fibers were used to weave (wool and linen). The raw material and technology available determined the type and style of clothing was developed (PEZZOLO, 2007).

Despite being an activity as old, each civilization has a different method of interlacing wires (PHILIPS, 1994). In the case of the Andes, for example, there are different endings for each style. These methods are not repeated in other regions of the world (HOCES, 2006).

The aborigines had at their disposal as primary materials the subdivided agave fiber; two kind of cotton, natural brown and white; and the glossy wool of the llama and domesticated alpaca, as well as the finer and silkier wool of the wild vicuña. D'Harcourt (2010) says that in rare instances they added human hair and an old chronicle mentions that the Incas sometimes incorporated gold and silver threads in their fabrics, but he has never seen a textile with these.

Although Europeans and Americans export their own versions of the most characteristic models (Persian, Turkish and Andean Orientals), these productions are still characterized as typical of a particular location, as with manual production in Brazil, which are characteristics of a given region, are widely sold in other regions and acquire status typical.

3.8 *Manual technology*

The appearance or origin of Peruvian textiles comes around (2.500 to 4.000 BC) with the manufacture of the first "tissues" prepared with techniques "pre-looms", that is, only with the skill of the hands (CHOQQUE ARCE, 2009).

Of these techniques, the most important were the interlacing, "curly" and tied, with which raised purses, carpets and rags. In this period, called "Huaca Prieta" were found more than 3000 inter-connected fragments, most cotton, noting especially the emergence of the first textile design corresponding to the representation of a condor with outstretched wings worked structurally (Figure 2). This indicates that the drawings were achieved through the interweaving of warp wires with weft yarns (CHOQQUE ARCE, 2009).

Figure 3. Reproduction of a scene painted on a vase from Trujillo.

The yarn, twisted by hand, without the aid of a spindle rotated or governed by a flywheel, was usually even tightly twisted, often excessively so. Yarns were used in single form or in several joined strands, in conformity with the purpose for which they were required. Double or two-ply yarns predominated; of equal thickness of the single smoother and stronger than single yarns (D'HARCOURT, 1962).

The looms were quite rudimentary. They consisted basically of two parallel bars, between which the yarns of the warp were stretched (Figure 3). The bars were sometimes kept separated by four stakes driven into the ground, or, for more delicate pieces, were attached to a small frame, but usually one of the bars was attached by means of a cord to the branch of tree or to a beam or rafter, while the other was held firm by means of a strap passing around the lower back of the weaver, who, by a single bodily movement, could increase or diminish the tension of the warp (D'HARCOURT).

A type of fabric found of these civilizations was the fabric Karwa. The fabric was a high-demand activity, whose growth was caused due to the use of fur of camelids and the discovery of the loom, which allowed a greater complexity in the textile and weaving techniques (CHOQQUE ARCE, 2009).

On Karwa, 8 km south of the cemetery of Paracas Ica department, the tissues look like painted with a recurring representation of anthropomorphic beings holding walking sticks that end in heads of snakes, eccentric eyes and thick lips with fangs, which show a relationship with the Chavin people, especially in the drawings of cats and birds Harpies (CHOQQUE ARCE , 2009).

The main structural techniques of the Moche fabrics are Kelim rugs or carved, eccentric (Figure 4), interleaving, flat screens with a discontinuous supplementary weft to form fringes, modular technique that

Figure 4. 3 fragments embroidered with motifs of birds and lizards. (Source: Textiles Prehispanics)

allows Zoomorphic or anthropomorphic representations of geometric character with blocks on top of each other (CHOQQUE ARCE, 2009).

The Chimu culture developed textiles strongly influenced by architectural drawings and metalworkers. In many fabrics are geometrized birds views, similar to those recorded on the walls of the Citadel of Chan Chan. The predominant figures are shown frontally with the arms extended and holding walking sticks, sometimes sitting under covered structures or staggered platforms (MAPLE CHOQQUE, 2009).

The Chimu relies on reason with coastal flora and fauna. There is a wide variety of animals, especially cats, snakes, birds and there are also geometric and non-figurative themes. A frequent motif in Chimu were the animals with cephalic appendices, Crouching and body wavy tail, which is represented in profile or sitting (CHOQQUE ARCE, 2009).

The Chimu culture worked with techniques of tapestries, gauze, painted fabrics, embroidery, feathers and use that to their tissues are the large canvases to decorate the walls of their palaces for clothing such as skirts, shirts, turbans, thongs, among others (CHOQQUE ARCE, 2009). Each craftsman or Weaver that he weaved these pieces had a great importance to the culture to which he belonged. It was a true art that represented what was around that people.

3.9 *Andean textile art today*

To understand the situation of the historical textile in today's society it is necessary to make a connection to what happened in Europe when the academies have suffered a decline in your prestige, and the situation started to turn handicrafts (SIMIONI, 2008).

During the Art Nouveaux movements and, especially, by the movement Arts Crafts led by & William Morris, there was a revaluation of textiles (SIMIONI, 2008).

Several changes, the transformation of the craftsman in artist, the creation of aesthetics in your modern sense and the separation of the arts with the religion, contributed to the textile arts were ignored. In the modern age there was an attempt to resume the textile art as a noble art, but after the Industrial Revolution and

Figure 5. Fragment of painted fabric. Technique: Eccentric Carpet (Source: Site.

weaving the tapestries were already linked to manual labor and automatic, losing the characteristics that made them be regarded as an art. (PEVSNER, 2002).

"William Morris proposed the resumption of traditional methods and crafts, because in them the worker participated in all stages of production. In this context the textile production resumed and valued within the field of "Haute Couture"." (SIMIONI, 2008).

This is where the rugs are large-scale production and Andean rugs enter the system that Lipovetsky and Serroy (2015) call capitalism artist. Andean rugs are no longer pieces made by those peoples and their descendants and for their own use and have been aesthetic products which refer to those people and are sold to those who can buy. The weavers before manufactured textiles with care and time could be included, in which Sennett (2009), are called craftsmen, to those who are dedicated to art for art's sake.

The artificer had your value recognized in the classical era and in various civilizations continued to be recognized. (SENNETT, 2009).

Since most primitive weaving was an activity reserved for women, which gave them respectability in public life. Crafts like weaving were practices that contributed to civilize the tribes of hunter-gatherers (SENNETT, 2009).

These productions performed by women were a kind handmade, with productions in local markets and isolated, and were not yet included in the system of modern capitalism (LIPOVETSKY; SERROY, 2015).

As occurred with Persian rugs and Turks, which were manufactured by Europeans during the Industrial Revolution (CAMPANA, 1991), Andean rugs are sold industrially and even sold as souvenirs in museums on the pre-Columbian cultures, such as the Andean Museum in Chile (source: Andean Museum Website).

One of the greatest Peruvian artists of our time, Maximo Laura, works on elements shown in "typical" articles in manufactured tapestries (Figure 5). A way to keep the tradition and bring a bit of modernity. The tapestries made by him are sold on a website and displayed in galleries around the world. Despite designing and painting the tapestry project, they are

Figure 6. Maximo Laura Tapestry (Source: Site Maximo Laura).

not made by Maximo, but by a selected group of Master Weavers who have been trained by Mr. Laura on his own techniques. To keep the Peruvian tradition of past techniques from generation to generation, Maximo trains all his weavers with his own technique and process (source: Maximo Laura website).

4 DISCUSSION

Anthropology has different concepts for culture, but most of them are related to activities and productions of civilizations. The very definition of what is the work of anthropologist brings ambiguity regarding the term identity. Cultural anthropology studies the mental production and activities. As we saw with Geertz, each civilization has your culture generated by different signs and meanings.

Thinking of cultural production as something collective, it is possible to understand that each civilization produces a certain culture. The difficulty of establishing a division is due to the fact that the anthropologist is already inserted in his own culture when analyzing others, so his vision before another is already loaded with the stereotypes and prejudices to which he has been exposed since his birth.

The discussion of identities before globalization occurs not because of the mix that exists in society today, but because the characterization for the definition of a cultural identity has been made.

If we consider traditional activities, passed from generation to generation, we can establish that this production is characteristic of local finishing, although we must also consider the time realized, because different times bring different meanings.

Thinking today and adding to the example of traditional textile manufacture, the fragmentation defined by Hall is evident, since one can have the methods used by the Andes, for example, made in other regions of the world.

These productions are commonly included in the identity explored by governments as a form of nationalistic politics, to attract tourists exploring stereotypes, created on the view that all culture in a country is similar. However, some ways of exploiting this identity can help maintain traditions that would otherwise be consumed by modernity and globalization.

5 CONCLUSION

Stuart Hall says the identity today is somewhat fragmented and we can conclude through this article that, even if the focus of the study is the oldest civilizations, our interpretation of the cultural production of a people will be influenced by our own culture and are created stereotypes that don't necessarily represent.

The discussion on identity is far from over and in this brief article we can only conclude that the identity depends largely on the interpretation of cultural productions, until random cultures that lived before Postmodernism are today defined with an interpretation influenced by signs to which we are exposed today.

Traditional activities help establish that a manufacture is characteristic of certain place, but it is also the need to consider the time performed, because different seasons bring different meanings and influence on current cultural production from this same location.

The traditional productions established in an ancient conception of identity are commonly included in incentive cultural policies. Are explored by Governments to attract a tourist, which cultivates stereotypes and caricatures and stresses in other countries a vision that they already have about a false identity.

This highlights the need to expand the studies of cultural identity, so that clearly describe, that local issues influencing decisively the perpetuation of traditional activities that differentiate between people. These studies may indicate what information and how they should be organized and disseminated, to give you an appreciation of local issues, cultural identity. In the case of Brazil, we can move much in the specification of various aspects of these issues, which featured Brazilian cultural diversity.

In this context, the textile manufacture provides a wide range of information for identity analysis. We can use your fragmented and long process, as the object of research, because differences between the processes, materials and the specific characteristics of the local issues have a strong influence in the formation of cultural identity.

REFERENCES

Benedict, Ruth. Patterns of culture. Lisbon: Edition Books to Brazil, s.d.

Bernardi, Bernardo. Introduction to ethno-anthropological studies. Lisbon: Edition 70, 1992

Boas, Franz. Anthropology. Enc. Of Social Science, V. 02.1930

Campana, Michele. Oriental carpets. São Paulo: Martins Fontes, 1991

D'Harcourt, Raoul. Textiles of ancient Peru and their techniques. Seattle. University Of Washington Press, 1962

Geertz, Clifford. "Dense description: by an interpretative theory of culture". In: The interpretation of cultures. Rio de Janeiro: LTC, 2008, pp. 2–24

Hall, Stuart. Cultural identity in postmodernity. Politic Press/ Open University Press, 1992

Images of Carpets. Available at: <http://www.tiwanakuarcheo.net/13_handicrafts/textiles.html> Accessed on 28. Apr. 2017

Laura, Maximo. Creation Process. Available at: <http://maximolaura.com/creation-process/> Accessed on Apr 25, 2017

Lima, Paula Garcia. "Fashion, necessity and consumption". In: Design Minutes. Palermo, 2010. Available at: <http://fido.palermo.edu/servicios_dyc/publicacionesdc/archivos/148_libro.pdf> Accessed on February 11, 2017

Pezzolo, Dinah Well. Fabrics – History, workmanship, types and uses. São Paulo: Editora Senac São Paulo, 2007

Philips, Barty. Tapestry. London, 1994

Quilter, Jeffrey. The Moche of ancient Peru. Cambridge, Massachusetts: Peabody Museum Press, 2010SAPIR, E. Cultura: autentica e espúria [tradução de José Reginaldo Gonçalves]. In: Sociology & Anthropology. V.02, N° 04, 2012, pp. 35–60

Sahlins, M. A brief cultural History of Culture. Paper prepared for the world Commission on Culture and Development. UNESCO, 1994

Tylor, E. B. Primitive culture. Boston, 1871

Textiles, Identity and Innovation: Design the Future – Montagna & Carvalho (Eds)
© 2019 Taylor & Francis Group, London, ISBN 978-1-138-29611-4

IN-*Between*. Designing within complexity

F. Vacca & C. Cavanna
Design Department, Politecnico di Milano, Milan, Italy

ABSTRACT: The paper aims to investigate on specific areas of the textile industry, experimenting knowledge transfer processes and hybridization of techniques between craftsmanship and advanced manufacturing in order to revitalize peculiar Italian productions with a strong market potential. In particular, it aims to demonstrate the potentiality of an IN-Between design and craft approach through the integration of advanced manufacturing processes into typical production of craftsmanship, to generate a new typology of objects that became distinguished both for their innovative meaning and connection with advanced manufacturing processes. Therefore, the paper focuses on the cultural background connected to identity, craftsmanship and know-how on the textile industry and present in detail IN-between, an experimental project for the development of new dyeing and printing techniques in order to create unique garments in an industrial way. The inspirational research is inscribed in the contemporary complexity related to the textile sector and it aims to experiment a new approach in design to stimulate its current and future economical and cultural growth, thanks to the transmission of local identity interpreted by the new technologies, without losing the value of handcrafted item. The result is a textile research based on a new aesthetic that combines craftsmanship with the serial industrial production expressed with the inkjet printing.

1 DESIGN & CRAFT: BETWEEN MATERIAL CULTURE, TERRITORIAL IDENTITIES AND NEW TECHNOLOGIES

Whereas, until recently, the model of craft production was considered as an expression of hostility towards progress and the innovation of languages, we are now witnessing a renewed interest in the distinctive practices and knowledge of *authentic* productions (Castells, 2004, Kapferer and Bastien, 2009, Pine & Gilmore, 2007). This is done through the rediscovery of traditional processes and the enhancement of the *cultural capital* of a specific social community, which has led to understand the relation *global vs local* in a very different way.

Territorial identities lose their specificities and uniqueness to gain a pluralistic and hybrid character, built on convergence and fade paradigms of relationship flows which reinvigorate the interaction between places, cultures, communities, people, customs, rituals and iconographies (Manzini, 2015). While the 'local' becomes *cosmopolitan specialties* (Fiorani, 2006) to attract more and more markets by leveraging the theme of exoticness and difference, the 'global' undergoes a process of *indigenization* (Appadurai, 2001) of new cultural forms, thus acquiring new formal values. This process of *cultural transition* (Appadurai, 2001) across the *cultural capital,* engages with the heritage to connect signs and social meanings through the innovative reinterpretation of routes that combine the richness of ancient and distinctive techniques of a given culture with the new languages of contemporaneity to develop products and services in which differentiation and customization are the main features of a renewed design culture. The recoding of craft-based knowledge (Sennet, 2008) feeds narratives that become *cultural biographies of things* (Kopytoff, 1986). In this regard the deposits of material culture and craft practices typical of a specific community and its surroundings become real *cultural capital* and not only the heritage of memory (Throsby, 1999). I.e. they are an aggregate of tangible and intangible factors that are awarded a cultural value which is continuously reinvested in the development of new artifacts, goods and services as well as new values and meanings. Furthermore, in a complex, globalized and *deterritorialized* age as the one we are living in, it is important to keep in mind the flows of cross-cultural migration which, being in constant evolution, strongly clash with the erroneously-established idea that identity is one and immutable and that belonging is local or localizable (Appadurai, 2001). The different panoramas of interaction between cultures and their fruition processes are no longer recognizable or definable in a "spatial" sense but must be explored in a "temporal" way, in their evolution and in their complicated and continuous interactions. In this context, design plays an important role of mediation, contributing to the processes of enhancement and actualization of the *cultural capital*, a resource from which design draws within projectual processes.

This "territorial wealth" has been fuelled by an equally rich "human wealth" which has been able to preserve and value knowledge, techniques, processes, in a continuous process of hybridisation between Design and Craftsmanship, tradition and innovation, triggering a "migration" and "cultural" translation process of meanings and techniques from one geographic-cultural context to another and experimenting increasingly advanced and innovative processes and methodologies.

In an era of increasingly accessible technological resources (Micelli, 2011), the Italian manufacturing industry is therefore faced with a singular correlation and integration between traditional crafts and the most advanced digital technologies. This paradigm shift not only involves the absorption of new technologies, but rather it highlights a system where there are no pre-defined value hierarchies between crafts, industry and technology: a coexistence and mutual exchange between yesterday's, today's and tomorrow's knowledge, in a game of analogies and contrasts, of typical and atypical.

As we wish to demonstrate with the present paper, there are great opportunities in the textile industry to test the new design and production ideas, to create business networks, and to complement forefront technologies with more craftsmanlike skills, providing the basis for rethinking not only the production processes, but also the design of artefacts.

2 HYBRIDISATION OF LANGUAGES

The contemporary designer has to operate in an economic, political, and social context which is a complex system that deals with volatility, plurality of expressions and hybridisation (Manzini, 2015); it is therefore indispensable to be able to identify a potential design space where complexity may become an added value, the creator of new expressive languages. The project aims to begin from the origins of craftsmanship know-how, the fundamental value of Italian history and culture, and to combine it with design culture, promoting a profitable synergy between evolved and traditional production processes. This interaction can become virtuous, generating value by transferring knowledge between actors and reconfiguring traditional codes and languages, outlining a process which can be repeated over time, able to penetrate and compete on international and globalised markets (Vacca, 2013).

In this context, the project is supported by the fact that the Italian fashion system is permeated by a constantly evolving design culture, rooted in history but fully flexible and open to accommodating changes. The Italian fashion system can boast great potential as well as production and manufacturing flexibility thanks to a system of small and medium enterprises which have always invested in research and innovation" (Becattini, 2001, Bertola & Colombi, 2014, Micelli, 2011). And this is exactly where we need to restart, i.e. from innovation, which is never random, but the result of an important investment and research process which must permeate all the corporate culture. In the field of textile design, integrating industrial production and craftsmanship know-how is one of the fundamental requirements for developing new products.

Today, more than ever, it has widely been demonstrated that fabric plays an important and central role in the overall concept of clothing; in fact, the choice of a specific material is essential for the characterisation, success and performance of garments. Many designers in fact base their collections on the fabric; often the fabric itself becomes the protagonist, exalted by simple structures that leave room for the communicative power of the material with which they are made. The textile surface design is also based on merging two aspects; in fact, by distilling the analysis path, two types of operating models, one structural and one emotional, can be identified. The first consists of manipulating matter, the second of colour, linked to a more superficial sphere. The manipulation can be reinterpreted in the textile field as curling, pleating and folding, while colour is expressed through print or dye. By analysing the various manual practices with which it is possible to work on textile fabrics to change the aesthetic features and increase their value, it emerged that the technique which best embodies this feature is Shibori, an ancient Japanese textile dyeing technique, where the initial fabric is intrinsically modified by folds, curls and ligatures before being dyed. During the research and experimentation, the various techniques of Shibori were studied, each of which has a different name and aesthetic outcome:

- KANOKO: tie-dye by binding which consists in designing a drawing and "pinching" with the fingers the parts of fabric which will be tied with a string and will then be the area affected by the tie-dye.
- MIURA: ring binding tie-dye where a small hook is used to hold the fabric lifted which will then be bound. In this case, the binding occurs by winding the string twice around the fabric, without any knot, to then move on to binding the next part of fabric.
- KUMO: folding and binding tie-dye where portions of fabric are folded tightly and evenly to then be bound from bottom to back and vice versa to then pass on to the next cone. The result of this technique is a unique and special texture resembling the shape of a web.
- NUI: binding and topstitching tie-dye which uses topstitches to create tie-dye areas on the fabric.
- SUJI: pleating tie-dye which is usually carried on the entire garment by vertically pleating the kimono along its length. The folded head is tied with a thread to keep it in shape in order to only expose the peaks of each crease to the dye, thus obtaining a vertical strip pattern.
- ARASHI: wrapping tie-dye which consists in wrapping the fabric diagonally around a pole, then securely tightening it by fastening it with a string.
- ITAJIME: tie-dye with wooden blocks which involves folding the fabric sandwich-like between

two pieces of wood which are then held together firmly by means of a string tied around them.

As Shibori is a craftsman technique, it has a strong casual component, as the aesthetic result of the textures obtained through tie-dyes is not entirely controllable. Obviously, practice and experience, combined with design, make it easier to get a rough idea of the aesthetic outcome which can be obtained, although there are external factors that cannot be avoided, which are not predictable and controllable, but which produce unexpected and random effects. And it is precisely the impossibility of controlling the outcome of the process that makes these craftsmanship techniques so interesting and contemporary. The need and the challenge of linking such a complex craftsmanship technique to an industrial one was born from this reflection, with the ambition of transferring the charm of randomness to something made in series, still maintaining an across-the-board "in-between" attitude.

3 BETWEEN: DESIGNING RANDOMNESS

Shibori thus becomes the starting point for experimental testing applied to fabrics, the resulting textile manipulations represent the "structural level" of the project, while printing replaces dye on an "emotional level".

In the field of research and experimentation, traditional Japanese fabric manipulation techniques have been synthesised into two macro operating levels:

- *folding level:* WRAP (wrap and re-twist) vs. PLEAT (fold and pleat) offering a different level of randomness in the final output
- *binding and fastening level:* BOUND (bind, topstitch, curl) vs. CLAMP (blocking by using external objects)

Once the "structural level" was defined, it was necessary to analyse the possible printing techniques most suitable to the "emotional level" project objectives. Block printing is still based too much on manual skills, therefore it is not suitable for the purpose as the project would be limited to an artistic craftsmanship level. Silk screen printing may be a good compromise, but it is not supported by a sufficiently flexible technology to adapt to every type of design and print substrate. Inkjet printing, on the other hand, provides more creative possibilities, it is supported by a highly flexible technology, and takes full advantage of the technological developments in the digital era in which we live.

By relating inkjet technology with textile manipulation techniques (Fig. 1), two fundamental and critical points, closely linked to each other, emerged:

- *Colour absorption methods.* When it comes to dyeing, colour penetrates into the fibres and is absorbed by the fabric in a way that is compatible with the resistances to which it has been subjected: folding, curling, ligature. In printing, however, colour is distributed over the fabric's surface and this is why we cannot talk of colour absorption. If the samples are folded or wrapped over and over again, colour will only deposit on the outer surface directly in contact with the print head.
- *Sample thickness and printing surface.* The thicker a fabric or sample is once it has been handled, the more difficult it is to print it as it cannot pass under the head of the printer. In fact, the thickness and the printing surface are linked to the fabric dragging in the machine, which is carried out by rollers driven by an electric motor. For this reason, when entering the printing machine fabrics should not have creases and/or curls which could rub against the print head, causing stains. In fact, the head cannot be lifted too much and the maximum height it reaches is about 12 mm. This implies a narrow choice of materials on which the print may be carried out: not-too thick and smooth surface fabrics.

In the specific case of a shibori technique, the fabric necessarily has irregularities related to the supports on which it is folded or curled, and, additionally, the thickness, which is generated as a result of manipulation, is too high to allow it to pass through the printer. For this reason, we began to consider the origins of inkjet technology in order to look for a solution in a field that is not the one of textile printing, which only works on flat surfaces.

Among those available we found flatbed printing, designed to print on rigid supports, which solves the problem of the fabric jamming the machine, as in the case of textiles. Not all flatbed printers, though, can print on rather thick and irregular materials. One of the few machines which has these features is the Mimaki JF1610, a UV inkjet printer, capable of printing on large format rigid supports, all with high resolution and competitive prices. Mimaki is an American company that introduced its first UV inkjet printing machine in 2004 for printing non-treated three-dimensional materials and since then, Mimaki UV systems have gained a good reputation in on-demand printing of many small size industrial products. Recently, the market has shown the need to print on large format rigid materials, and various types of UV plotter have been suggested. Mimaki therefore decided to invest in a real flatbed, exclusively dedicated to UV printing. This type of flatbed guarantees very precise printing on heavy and large media, or on multiple small-sized objects for serial productions.

The product offers a 1200 × 1200 dpi high resolution with photographic quality, as well as the ability to six-colour print with a white covering ink, to create vibrant images even on transparent materials.

UV technology allows to fasten and dry the colours directly on the printer, this is why the prints can be instantly handled and are ready for use. The two UV lamps, placed on the sides of the print head, polymerise the colour as soon as it leaves the nozzles and fix it firmly. In addition to having a very large surface (160 × 310 cm), the printing bed has a powerful

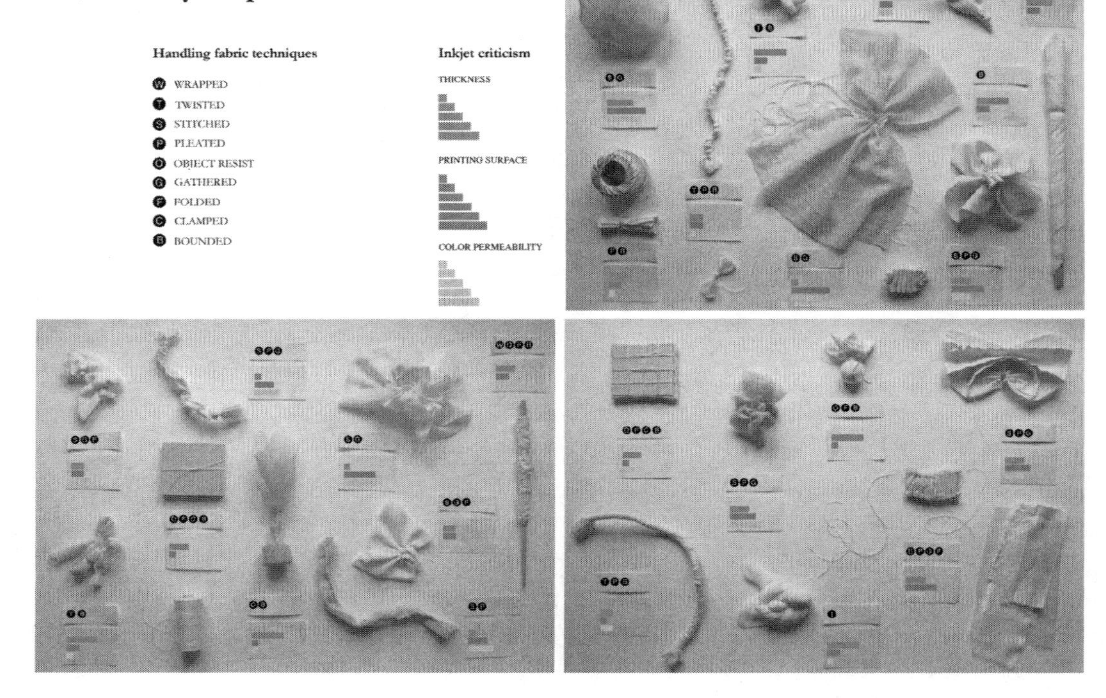

Figure 1. From dye to print. Relationship between inkjet technology with shibori manipulation techniques. Textile manipulations developed by Chiara Cavanna.

suction system that keeps the material perfectly still and secure so that it can be printed with high precision on both heavy media and on materials which easily misalign. Its other great potential is the print thickness which reaches up to 50 mm.

Mimaki JF1610 is also equipped with an electronic mechanism which allows the operator to easily adjust the height of the print head according to the thickness of the materials; there are also sensors which prevent the print head and the printed media from colliding, these, in fact, keep track of the thickness of the material both during its positioning and during all the printing process. The inks used are UV Flexible with transfer eco solvent. These have a very low level of VOC (volatile organic components) and therefore do not release organic solvents into the air. Overall, the printer is eco-friendly, both because of the type of inks used and the fact that it does not produce ozone and short-wave ultraviolet rays.

Based on these considerations, we therefore decided to exclusively identify and select the textile manipulation techniques most compatible with the issues highlighted by the technological limits and consequently to identify the folding and fastening techniques, thus defining precise aesthetic outputs: curls, double fabric stitching, regular topstitching.

As occurs in the Shibori techniques, where fabric manipulation is functional to defining the resulting texture, the chosen fabric manipulation techniques result in the image breaking up (Fig. 2).

The result is fascinating as it combines the randomness typical of craft dyeing processes with the modernity of new printing technologies. The image, imprinted on the irregular and three-dimensional textile support, breaks down when the support is opened and freed from the resistances. We obtain a two-dimensional surface, but with the pattern of a previous "structure" printed on it. The process remains imprinted on the fabric as a story and, at the same time, this very casual and unexpected resulting effect becomes reproducible as a mass-production.

4 IN CONCLUSION

The *In-Between* textile sperimentation aims to have a scientific and technological impact as it works both on a process level and on a product level, using technology developed in industries other than textile and exploiting their potential applications in the field of traditional manufacturing. What makes this project interesting is the role that design plays in the process of hybridization with craft and advanced manufacturing. In fact, the design approach investigates the possible implementations for both design thinking and design methods in order to transform the projectual approach

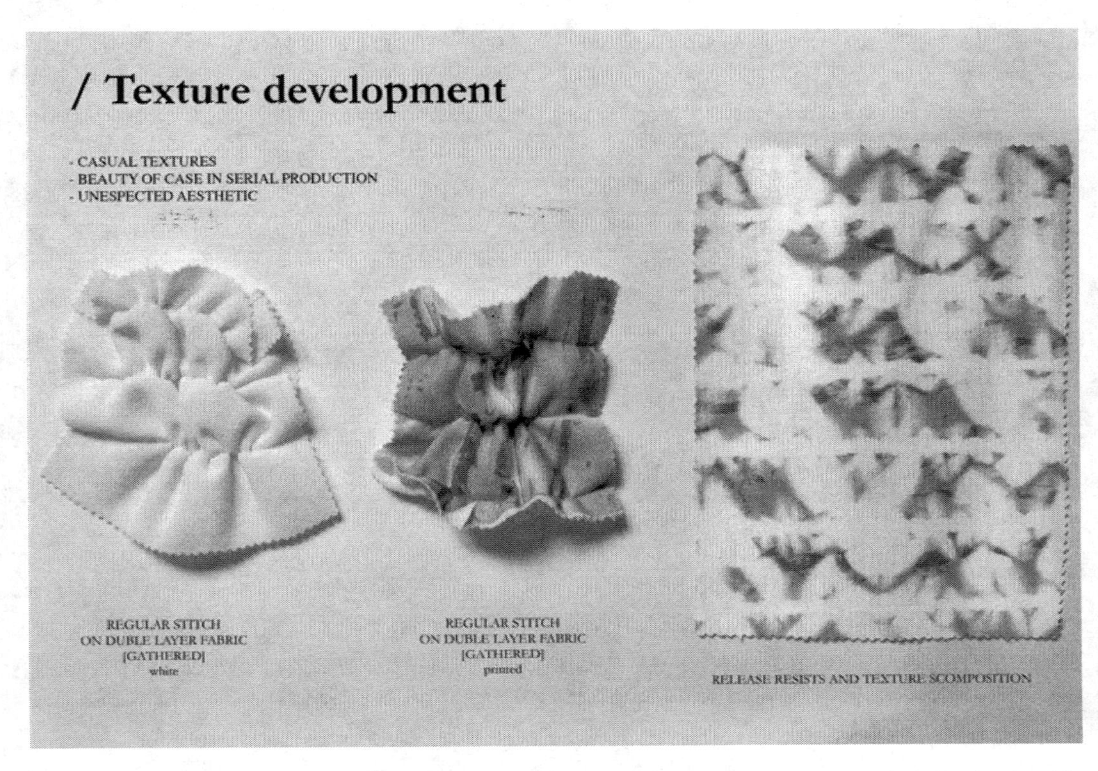

Figure 2. Texture development in-between inkjet technology and shibori techniques. Textile manipulation developed by Chiara Cavanna.

which is not focused exclusively on the final product but on the chance to experience advanced production processes that can increase the value of the project generating value for all stakeholders and not just for customers.

In those specific fields of practice, Design shapes future artefacts rooted in the reinterpretation of the past. We rediscover the value of craftsmanship and combine it with the potential of new technologies: the uniqueness of manual work may be transferred to mass production.

The project infact identifies a way in which design can become a promoter of processes of continuous innovation in favour of traditional and artisan activities without distorting the meanings or neglecting their identity and culture of which they are the representation. The ancient fabric binding and folding techniques in the Japanese dyeing process of Shibori merge with the newest digital printing technologies.

The outcome of this experimental encounter consists of not-entirely predictable textures, where the printed image randomly breaks up on the three-dimensional surface of the fabric. Each print will be unique and unrepeatable, just like a work of art or craftsmanship, but at the same time this process can be carried out industrially and in series. The case therefore becomes an indispensable component of the evolutionary process and a new key to interpreting the design path.

Given this perspective, the Design & Craft approach has demonstrated to be able to establish a system of universally recognized values emphasizing the transience of local-provenance and identities.

AUTHORS' NOTES

The paper is the result of common research and findings, nevertheless, sections 1 and 4 was edited by Federica Vacca and section 2 and 3 was edited by Chiara Cavanna. The methodology presented is the result of research studies and projects undertaken in recent years by Fashion in Process, multidisciplinary research collective within the Design Department at Politecnico di Milano University. [www.fashioninprocess.com].

REFERENCES

Appadurai, A. 2001. Modernità in polvere, Rome: Meltemi.

Becattini, G. 2001. *Dal distretto industriale allo sviluppo locale. Svolgimento e difesa di un'idea,* Torino:Bollati Boringhieri.

Bertola, P. & Colombi, C. 2014. "Re-Branding Made in Italy: a Design Driven Reading". In P. Bertola & C. Colombi, (eds.), *Fashion Practice. Fashion Made In Italy. Special Issue*. (6)11, 175–200. London: Bloomsbury.

Castells, M. 2004. *The Power of Identity*. West Sussex: Blackwell Publishing.

Fiorani, E. 2006. *Moda, corpo, immaginario*. Milan: POLI.design.

Kapferer, J. & Bastien, V. 2009. "The specificity of luxury management: Turning marketing upside down" in *Journal of Brand Management*, 16(5–6), 311–322.

Kopytoff I. 1986. "The Cultural Biography of Things: Commoditization as process". In A. Appadurai (eds.), *The Social Life of Things: Commodities in Cultural Perspective*. pp. 64–94. Cambridge: Cambridge University Press.

Manzini, E. 2015. *Design When Everybody Designs*. Boston: MIT Press.

Micelli, S. 2011. *Futuro Artigiano*. Venezia, Italy: Marsilio.

Pine, B.J. & Gilmore J.H., 2007. *Authenticity*, Harvard Business School Press.

Sennet R. 2008. *The Craftmen*. Milano, Italy: Feltrinelli.

Throsby, D. 1999. "Cultural capital". In *Journal of cultural economics*. 23.1–2. pp. 3–12.

Vacca, F. 2013. *Design sul filo della tradizione*. Bologna, Italy: Pitagora Editrice.

Textiles, Identity and Innovation: Design the Future – Montagna & Carvalho (Eds)
 ISBN 978-1-138-29611-4

Cultural territories: Textile constructions in the work of the stylist André Lima

O. Maneschy
FAV / UFPA, Faculty of Visual Arts, Federal University of Pará, Brazil

Y. Maia
Center of Humanities and Social Sciences, CCHS, University of the Amazon, Brazil

ABSTRACT: Among fabrics, embroidery and stamping, the designer André Lima transits between diverse cultural territories, in an anthropophagic movement, resulting in collections that feature a profusion of colors, textures and models, all of which were researched and created by him. This article aims to reflect on the concepts of culture, issues of cultural identity and complex processes in the construction of own fabrics of the stylist's work.

1 INTRODUCTION

The stylist André Lima, born in Belém, Pará, Amazonia, Brazil, developed a fruitful creative process, establishing creative paths in which issues of identity, material culture and textile research gained unique space and projection in the brazilian scenario, where the most varied techniques of composition and textile printing played a fundamental role. Rare artist, his first collection in a national fashion week took place in 1999, in the Casa de Criadores (Casa de Criadores and a fashion week held in the capital São Paulo dedicated to show the work of young stylist; in 1999, André Lima parades his own brand for the first time in this event. In 2007, the event celebrated 10 years of history and the stylist honored with a retrospective of his work was André Lima) project, São Paulo. Since this "first" collection, located in the "center" of brazilian fashion, the stylist focused all the creative development using fabrics brought with him to São Paulo. His family had an intimate relationship with textiles, his father's fabric trade in the Amazonian interior of fine stores like Paris N'América (built at the height of the Belle Époque with materials from Europe inspired by the Galeries Lafayette in Paris; its outer shell with stones from Portugal, steel structure from Scotland, and tiles from Germany), as well as embroidery and brocades for various uses, and even for decoration.

Figure 1. Untitled, editorial made for André Lima at the Museum of Zoology of the University of São Paulo, photo: Orlando Maneschy, 1999.

This universe, along with embroidered quilts by his grandmother, among other rarities of the time, were brought with him to São Paulo, as references which ended up becoming the exhibted collection there by the designer. Wild, the characters of Lima, whose faces were covered by hair, first appeared with a mixed soundtrack with spoken words and music by the brazilian singer Maria Betânia, and with the tinkling of gypsy gold bracelets worn on the feet. They wore unique clothes, with embroidery, applications, etc. His creations had begun with hybrid cultural references, soon transformed into objects of desire and receiving a good critique.

For that matter, this article points to the relationships present in the work of the stylist between material culture, issues of cultural identity and complex processes in the construction of his own fabrics for his collections, as well as exclusive textile printings. Images of the history of art, ethnic photographs, divas of Brazilian popular music and aesthetic references of several decades feed the universe of André Lima. In addition, the stylist always, when traveling, acquires ethnic fabrics, as well as clothing from thrift shops for research and studies, observation of details, which also feeds his development of surface design in his unique constructions.

2 ANDRÉ LIMA: TEXTILES AND CULTURAL IDENTITIES

On one hand, clothing is linked to the information possibilities that the object carries with it, to its aesthetic principles that guide the stylist's choice, its various uses, its materials, the confection techniques, its social relations, its history, and, on the other hand, they are visual trails that carry the moment of creation and the clues to the interconnections of this information with the sociocultural interactions that involve the creator.

According to Santos (2000), "cultural identities are neither rigid nor even immutable. They are always transient and fleeting results of identification processes. (...) Identities are, therefore, identifications in progress." From this point of view, we can understand that a fashion collection carries both records storing the creator's time memory as it is traversed by other discourses and by a network of memories and identities constituted by the encounters that the stylist establishes in order to create new meanings for its projects.

"If identity is a social construction and not a given, if it is within the scope of representation, this does not mean that it is an illusion that would depend on the subjectivity of the social agents. The construction of identity is done within social contexts that determine the position of agents and therefore guide their representations and their choices." (Cuche, 2002).

Thus, André Lima (2012) tells about the social and family context in which he was inserted. He says that he was a very curious child in relation to time, other worlds, other times. He would go to the library to research and discover, from a young age, cultural connections from one subject to another. He fiddled with his aunts' wardrobe. He watched his grandmother sew. He had an extreme curiosity since childhood.

"There [at his house] had at the same time [the successful records of Brazilian Popular Music] the *Gal Tropical*, by Gal; Alibi, by Maria Bethânia; and *Alerta Geral*, Alcione. And we heard these songs. So obviously, listening to these women, who were the biggest expression of MPB, and then Fafa de Belém among so many there, as a soundtrack of course was already an edited mixtape. (...) I found Gal a solar person, and I as a child identified myself a lot. This record is what she is on the cover with flowers on her head, one red and one yellow, two roses. It's something that came in and then I used it in a parade, I used it at my first *Fashion Week*, then at the gypsy parade, then we used a flower made out of straw. This thing of the flower in the hair, surely comes from the Gal, and from the *Gal Tropical*. The record was the great moment of her career, in which she consecrated herself as a diva. She had costumes made by Guilherme Guimarães [successful brazilian stylist between the 1960s and 1980s in Rio de Janeiro]. She was a star. I find this fundamental in my language. Exactly this clipping. Obviously parallel to this is the cut of Bethania that is less image and more density and that was stronger, even because the immersing and identification were much deeper. The things she said, the things she believed. From a mystical relationship with things, a relationship with nature, half animal, half wild, which I then boarded" (Lima, 2012).

We understand that identities are not capable of totalization or centralization in a single individual. It is in circulation with social assemblages and is assumed by individuals in their particular existences. Therefore, this type of cut, where cultural references are activated, materializes in the partnerships developed with special weaving, in which diverse materials are used in the weave of the fabrics. Distinct threads, a combination of materials for the textile elaboration articulated with his own desire, developed exclusively for Lima, based on research directed by the creator. Thus, unique fabrics appeared in his creations. As much in textile printing as in the applications, embroidery, knitting, etc., cultural references that are mixed as if their characters were sometimes beings of a lost tribe, and at others time a princess of a mystical world.

We can present, as an example, a summer collection of 2002/2003 (Figure 2), in which the designer worked with crochet weaves, fringes, metals, lace, feathers, mandalas, starting from reference images that punctuated the creation of the collection, in a visual relation between images and what is part of his identity construction, in this case, based on the reference of the singer Maria Bethânia and her relation with nature and mysticism (Figure 3, image of the book of the stylist with an illustration of Maria Be tânia), as well as images of the group Dry and Wet (lysergic band from the 1970s). Wings present in an engraving in which the singer is portrayed, result in fringes that guarantee the symbolic atsmophere evidenced by the styling developed for the fashion.

Therefore, a significant part of the images that compose its collection, reveals the exuberant and hybrid nature of the Amazonian visuals, mixing cultures of this vast region. We can see these relationships in the 2006/07 Collection in which André Lima sought references in the Kuna Indians (Figure 4) – living in Colombia and Panama – with resigned geometric prints and construction of silhouettes that refer to the typical costumes of this culture.

Figure 2. Untitled, Summer Collection 2002/2003 (fashion show photo), André Lima.

Figure 4. Untitled, Summer Collection 2006/2007 (fashion show photo), André Lima.

Figure 3. Untitled, page 46, from the book Coleção Moda Brasileira André Lima, 2008.

This transit through different cultures leads us to realize that André Lima, in his creative process, works with one of the most important concepts to think brazilian culture: Anthropophagy, as the curator Paulo Herkenhoff points out in the catalog text of the XXIV Bienal de São Paulo, of 1988, recognized, internationally, as one of the most important occurred in the country.

"'Scripts. Scripts. Scripts. Scripts. Scripts. Scripts. Scripts.' Seven times the word 'Roteiros' is repeated in Oswald de Andrade's 'Cannibalist Manifesto'. It is found between the paragraph 'Against the reversible world and the objectified ideas' and 'The carahiba instinct'. 'Scripts…' is a presence between the mechanics of 'deadened' ideas and cannibalism, etymologically originated in Carahiba. Africa, Latin America, Asia, Canada and the United States, Europe, Oceania and the Middle East are our 'Scripts…', defined without a single criterion, as a continent, economic bloc or regions. It is not an expanded species of allegories of the four continents, developed by European art of the seventeenth century. The plural noun 'Scripts' connotes multiple points of view." (Herkenhoff, 1998).

In figure 4 we can see how Lima establishes this mechanics between the ideas of his visual researches, in anthropophagic process cannibalizing references, swallowing cultures; Transforms them, resignifying them, into fabrics and modelings that transit through different repertoires, as in the example presented below, in which part of the Kuna universe is activated in their creations.

For the paraense stylist, there is no creation if it is not through desire, since looking at the world and separating what it likes is already a personal edition that starts from the creator filter. For him what matters is to put together things, mix cultural references, edit and feel the clipping that is there, but that mixes several references in the same collection.

Part of this "memory" stored by Lima is made of clothes, which for him are important for the mixture of times and places, and this materializes in clothing and images that there, in his studio, compose a fragment of fashion classics of the 20th century, as well as urban landscapes, everyday details, femininity, and his multiverse Amazon, in other words, affective cuts and their cultural identities guide the development of his projects.

It is important to understand the role of memory in the construction of other realities. According to Halbwachs (2006), individual memory is the result of a complexity of combinations, images, thoughts that come from the experience of various scenarios and that lead to a new particular order, in that sense the individual memory carries a collective memory.

In the research on these recolletions of André Lima, we can begin to understand his passion for different cultures, strong women, for the profusion of references and colors, for the sensual female body.

In the 2006 winter collection, his research brings together Nepal, India, China, Africa and Brazil, in a profusion of references collected, translated into flowing silhouettes, such as princesses from distant times, who can inhabit the Forbidden City (Beijing). In addition to the images in the fabrics and textile printings, hair intertwines in delicate work, as we can see in the figure, in which braids are mixed with earrings that fall over the clothes.

Therefore, to develop a fashion collection is, above all, a manifestation of the desire to express itself through clothing, but that begins by looking at these open, interstitial and cultural spaces and discover new possibilities, and the individual view depends on the cultural identities that mould the creator, since they guide his visual choices.

"The culture of a given social group is never its essence. It is a self-creation, a negotiation of meanings that occurs in the world system and which, as such, is not understandable without the analysis of the historical trajectory and the position of this group in the world system" (Santos 2000).

Santos (2000) helps us to understand fashion creations and dressing act as spaces of exchange, where clothing assume a mediating role in the construction of cultural identities between society, the stylist and those directly involved in the fashion industry, by interacting with cultures and each other through clothing.

André Lima and his collections go through an identity universe in metamorphosis, constituted within

Figure 5. Untitled, winter 2006 (backstage photo), André Lima.

the space-time displacements of the stylist. Through his life narrative and his reflections on multiple cultures, we note details of these existential territories: his predilection for ethnic references, the 1970s and 1980s; a predilection for fashion classics, for Brazilian Popular Music divas, concern for the perfection of the textile finishing, experimentation in fabric, and extremes.

3 CONCLUSION

In 2014, with the massive arrival of international brands, which impaired the production of fabrics of various weavings, and the production costs of an item with rich details, the stylist decided to diversify his production, expanding his operation area, consulting and developing specific collections for other brands. In this context, Lima, based on invitations from renowned institutions, made a plan to distribute his acquis in several important collections, which are present in four brazilian institutionsAnhembi Morumbi University (UAM), Armando Alvares Penteado Foundation (FAAP), Museum of Art of Rio (MAR) and Federal University of Pará (UFPA).

However, most of his collection of clothing and accessories of fashion shows, in addition to the documents of the creation process, clippings, photographs of fashion shows, etc., were integrated into the Amazonian Collection of Art of UFPA (the Amazonian Art Collection is the result of the research, articulation and perception of the curator Orlando Maneschy, of this delicate relationship between Amazon and Brazil, the Amazon and current art, the Amazon and the world; the acquis is deposited together with the contemporary collection of the UFPA Museum and brings together works by artists who traverse the region at different times, the Technical Reserve of Museology of UFPA brings together these works as well) in a section called Fashion. This collection is now in the Technical Reserve of Museology of UFPA. In the collection donated by André Lima, 775 objects have already been enrolled (enrollment "is the act by which all objects belonging to the museum are counted, creating a numbered list to the control and general identification of the museum collection" (Padilha, 2014)): 237 pieces of clothing, 184 accessories, 220 paper documents, 78 drawings and 56 samples of fabric and textile printings.

When entering a museological space, other places and other meanings came to those objects and images: from the continuous and non-linear flow of a contemporary fashion studio, whose relation is from object-consumption to a signification in objects-documents belonging, now in a museogical acquis.

Thus, we understand that this network formed by the imaginary, wearable and textile territories of the stylist are accomplices of his ways of being and doing; And they point to his processes of subjectivation, since they affected the creator's look and perception and traverse more or less his creation projects and clothing. Textile objects become objects to reflect and to signify, turning into a record of the collective and personal memory of the stylist.

We understand that "identities" are not capable of totalizing or centralizing the individual, since they engender, at each moment, new subjective processes. They circulate in social assemblages and are assumed by individuals in their particular existences. André Lima is a stylist of the superlatives, free imaginary transit, far from seeking fixed and shallow territories, the stylist seeks contrasts, investigations and experimentation, being taken by a wild force in his creative processes.

REFERENCES

Cuche, Denys. 2002. A noção de cultura nas ciências sociais. Bauru: EDUSC.

Halbwachs, Maurice. 2006.A memória coletiva. Tradução de Beatriz Sidou. São Paulo: Centauro.

Logullo, Eduardo. Vale das bonecas. In: Andre Lima: Coleção Moda Brasileira. São Paulo: Cosac Naify, 2008.

Maia, Yorrana. 2013. Cartografia de Si: territórios particulares e compartilhados do processo de criação do estilista paraense André Lima. Dissertação (Mestrado em Comunicação, Linguagem e Cultura). Belém: Universidade da Amazônia.

Santos, Boaventura de Souza. 2000. Pela mão de Alice: o social e o político na pós modernidade. São Paulo: Cortez.

Textiles, Identity and Innovation: Design the Future – Montagna & Carvalho (Eds)
© 2019 Taylor & Francis Group, London, ISBN 978-1-138-29611-4

Ornamentation of textile surfaces: Proposal of classification and application in fashion events in Minas Gerais, Brazil

W.G. Amorim & M.R.A.C. Dias
Universidade do Estado de Minas Gerais, UEMG, Brazil

ABSTRACT: The ornament as a textile surface design feature is the focus of this article. Initially it was proposed the revision of the techniques of ornamentation and the elaboration of a classification capable of organizing its understanding, especially by the designers and the teaching of these techniques. In addition to the techniques, ornamentation is also seen as a way of organizing the ornaments on the projected surface. Finally, we presented examples of ornaments identified in the last two fashion shows of the Minas Trend 2016 of winter and summer, in the city of Belo Horizonte, MG, Brazil.

1 INTRODUCTION

Throughout the civilizing process, man instituted communicational patterns through ornaments (Gombrich, 2012). The artifacts and its ornaments are derived from the material culture of a territory while also characterizing it. Material culture reflects the impacts of artifacts on human life, contributing to the formation of cultural identity (Cardoso, 1998). One of the earliest ways of creating communication patterns was the ornamentation of the human body (Gombrich, 2012). It modified the body through paintings, tattoos, jewelry, accessories and clothing in a variety of combinations. Influenced by the nature that surrounds him, man invested in the creation of artifacts and in abstract and subjective characteristics for himself. Among primitive peoples, tattoos on the face and body were a mechanism of expression. Through tattoos, they sought to impress terror on their enemies, or new parameters of beauty (Jones, 2010). The search for power and authority led man to create clothing and ornaments that would identify him with his heroes, animals and gods. This gave him the belief that he was able to control magical powers, nature, and also his own fate (Ruthschilling, 2008).

Man, from the primitive to the present, seeks effective methods of interaction, communication and individual expression through his own body. The desire for adornment is a strong instinct that accompanies the process of development of human culture (Jones, 2010). There is not even a culture in which there is no tradition of body ornamentation throughout the history of mankind. Body and culture reflect and mirror each other, changing according to social regiments (Pires, 2005). The ornaments also translate relations of belonging into images. Social and cultural regiments change over time, as does the aesthetic and symbolic perception of ornaments in body and clothing.

In this sense, ornaments are especially used to refer individuals and objects, to distinguish zones and planes, to establish rhythms and sequences, to associate, oppose, distribute and hierarchize. Any surface, whether natural or fabricated, is always supported by signs of classification and distinction (Pastoureau, 1993).

2 SURFACE ORNAMENTATION

Ornamentation is part of the design process. Few objects have their forms solely directed by the practical function. The ornament has functions that extrapolate the structure of the object and presents itself through figurative, representational, abstract and symbolic elements (Pedroni, 2013). Such elements potentiate semiotic and psychological factors of emotional appeal (Norman, 2008).

The textile surface design is a technical and creative activity that aims at the elaboration of two-dimensional or three-dimensional images reproduced through the objects, providing expressive perceptual characteristics to the appearance of the surfaces of the products (Ruthschilling, 2008). With the design of colors, graphics and texture effects are ones of the most effective methods for renewing the look of a clothing (Jones, 2005). It has the capacity to create instruments of differentiation through the design of structures, prints, embroidery and unique textile manipulations; besides the systematization of the use of new techniques and technological resources applied to the tissues.

Images created on textile surfaces have elements that combined can transmit messages. These elements are part of its visual language, such as: the point, line, shape, direction, tone, color, texture, dimension, scale, and movement of the visual and tactile elements. The clothing surface gathers all this information in terms of combinations and options. The interaction that these elements establish between them expresses meanings, usually oriented by messages preestablished in the conceptual design of a product.

For a purpose of communication within a collection of fashion products and driven by specific ideas and goals, designers design textile surfaces to create images that drive consumers to make choices. In this way, the textile surfaces become very important for the elaboration of the products in a fashion collection, since the elements materialized in the visuality of the surfaces of the fabrics can provide identification to their consumers, promoting sensory-cognitive interactions between subjects and clothes. The surfaces have a communicative character, constituting a space of active experimentation with the observer/subject, who acts and reacts to it (Schwartz, 2008). The design must be appropriate to the wishes and needs of the brands, the consumer, and the materials and processes available.

Thus, textile surfaces and their basic compositional elements act as an integral part of the language of fashion, expressing discourses that can be "read as a text". Fashion can be understood as expression within the human possibilities created for communication. As it is in the expression grounds that fashion constructs its meaning, its content is contained in its textual manifestation, that is, in its plastic configuration (Castilho, 2009). The manipulation of materials and visual compositions are part of this fashion plastic, which expresses through the resources related to the designs of its products.

Figure 1. Proposed classification of the ornaments divided into structural, chemical, applied and manipulated, examples and graphic icons adopted.

3 ORNAMENTATION TECHNIQUES

Several techniques, technologies and processes of ornamentation are applicable to the textile surfaces of clothing. The technique of making a fabric, in some cases, determines the ornamentation of its surface. For example, a pattern may be woven using fibers of different colors, pigmented through natural or artificial dyes. Fabrics can be stamped by hand or machine after the elaboration of its structure. Fabrics can also be ornamented with embroidery and adding textures or embellishments. Textiles are flexible, their fibrous nature enables transformations through the seam. In this way each technique used has specific characteristics in relation to the materials and processes used.

Once the processes of transformation of fibers into yarns, yarns to the structural construction of fabrics, its articulations with the garment industry and the methodologies for ornamentation, it followa to the description and classification of the different techniques used in the processes of ornamentation of fashion clothing This classification aims to group similar resources and facilitate an understanding of its applications. A distinction of the techniques was made from four design possibilities: structural, chemical, applied and manipulated. In all technical classes it is possible to find variations produced in small, medium or large scale, manually designed through handcrafted or mechanized (with the aid of technological equipment) means. In order to exemplify as different design possibilities, brief descriptions of some groups of techniques were made for each class of ornamental features (Fig. 1).

3.1 *Structural ornaments*

Structural resources are those designed from interwoven fabric yarns. They feature textures, graphics elements, and patterns according to the interleaving of different color yarns or materials. Flat fabrics, knits and lace can provide for an ornamental configuration in the structure once designed for this purpose.

- Tufted fabrics: they have complex patterns of interweaving, both in flat fabrics and in knitwear (Jersey). The technique was created by the French mechanic Joseph Marie Jacquard, in the seventeenth century. It is operated from a system that, in creek looms, selects thread by thread through the storage of information in punched cards.

- Machining: machines that produce fabrics from this technique raise or lower the warp threads (those that run the length of the fabric), to create patterns such as stripes, plaid and small designs. Machining fabrics have less complex textures than jacquard.
- Knitting: knits that can be made manually or with the aid of machines, from a set of two or more needles. The technique involves the interlacing of the yarns in an organized way, being possible to interlace yarns of different colors and textures, constructing patterns and graphics.
- Lace: fabrics in open and transparent mesh that form three-dimensional drawings from the structural configuration of the yarns. The number and variety of intersection points between the yarns determines the sharpness of the designs, which can be made by hand or machine. In Brazil several traditional techniques of making handmade lace are practiced, such as Crochet, Labyrinth, Renaissance, Filé, Bilro, Nhanduti, Irish, Frivolité, among others.
- 3D printing: 3D printing technologies are constantly evolving and have allowed designers to experiment with different shapes and materials to create surfaces that could not be produced using traditional methods. The most popular process in this feature group is layer deposited printing of polymer filaments that physically build a surface designed in a digital environment.

3.2 *Chemical ornaments*

The ornamentation through chemical processes happens in the treatment of textile structures already conceived. With the aid of dyes, pigments or reagents, drawings or textures are printed. Pigments and dyes are substances that when applied to a material give it color. The main difference between pigments and dyes is that, when applied, the pigments are insoluble and the dyes are soluble. Another difference between the two products is coverage: when the pigment is used in an ink it simultaneously promotes coverage, opacity, dyeing and color; the dye only promotes dyeing without providing coverage. In this way, the dye maintains the transparency of the object; as the pigment gives color and takes the transparency.

Reagents can promote other types of chemical interactions with textile materials other than coloring, such as degrading a specific type of fiber, if applied in a particular area. This class of techniques can happen through machinery or through artisanal processes of painting; are considered stamping techniques.

- Direct and immersion painting: it occurs from artisanal techniques that use materials and tools of low complexity like brushes, blocks of wood, small matrices, ropes, waxes or even the hands. They can be done cold or raising the temperature of fabrics, dyes and pigments; applying controlled amounts of paint to the surface or dipping a fabric with pre-insulated areas in a container with colored liquid for dyeing. Examples of direct or immersion painting techniques: Watercolor, Block Print, Stencil, Gradient, Marmoring, Batik, Tye Die, Shibori, etc.
- Cylinder: it is characterized by the transfer of the color through a cylinder that is mechanically rolled on the surface of the fabric. It is used metal cylinders of nickel, chromium or copper which have microperforations where the ink comes into contact with the fabric. The special dye is inserted into the cylinder and, by means of a ruler, the ink is transferred to the tissue from the inside out. In this process it is necessary to fix the dye to the fabric with a vapor drying.
- Silkscreen printing: is a printing process that requires the preparation of a matrix. The matrix is a frame with an open weave fabric, generally of silk, nylon or polyester, which is coated with a photosensitive emulsion to be revealed in a light table. This revelation creates spaces programmed for the printing of drawings. With the matrix positioned on the surface to be printed, the ink is deposited and guided by the extension of the matrix with the aid of a squeegee, performing the transfer of the desired graphics to the fabric.
- Direct digital stamping: printing happens by means of ink jets directly on the surface of the fabric, which generates a transition between art production and printing instantly. For the printing and fixing of the stamps, most technologies covering this process require the previous treatment of the tissue with chemical reagents and afterwards with the washing of the same ones
- Indirect digital stamping: known as transfer or sublimation (physical process of transition from solid state to gaseous), is considered an indirect digital technique because it requires the transfer of a previously printed design to the surface of the fabric. In the first step of the process an image or stamp is printed on a special paper (transfer) with polymer-rich pigments suitable for sublimation. In a second moment the printed paper is put in contact with the fabric under pressure and high temperature, in presses or thermal calanders. The pigment of the paper adheres the textile fibers giving rise to a dyeing. This technique is compatible only with synthetic fiber fabrics.

3.3 *Applied ornaments*

This classification includes the embroideries and applications. They consist of adding material through threads and needles or through textile adhesives. Tissues with interventions of these techniques acquire three-dimensional aspects through relief.

- Embroidery with threads: this technique uses only threads of greater thickness and density, and can be divided into two groups: (i) free and (ii) counted threads. (i) Free embroidery happens on scratches, transferred to the fabric and filling these designs can combine stitches in a random manner.

(ii) Embroidery on Counted Threads does not require any scratches, as it is worked by the yarn count of the fabric itself, each stitch being worked on an exact number of threads, generating patterns and textures. The materials most used to perform this technique are needles, threads and racks to keep the fabric tensioned.

- Embroidery with stones: it consists in fixing, with the help of threads and needles, stones and beads with drill holes (glass beads, beads, natural stones, etc.). The technique requires the creation and transfer of a design to the fabric and the filling of that design by sewing the materials.
- Textile adhesives: also known as thermofilm, are thermoplastic adhesives; it has the quality of softening when exposed to a certain temperature. Under these conditions, the molecular structure of its fibers is broken, being possible to join pieces of different composition. The technique allows to couple graphics of different textures with the aid of thermal presses.
- Applications: as with the embroidery with stones, it consists on attaching decorative elements to the surface of the fabric with the help of threads and needles, but in this case, there is no limit regarding the materials. The most common applications are cutouts fabrics and laces different from the textile base to be applied.

3.4 *Manipulated ornaments*

This class includes the most different types of textile manipulations for clothing, including thermoforming processes, restructurings, folds, wrinkles, pleats, frills, gores, ribbing, and other differentiated seams. The constructive process of textile manipulations emphasizes specific parts of the body and clothing, influencing and helping to generate volumes and supports. Associated with drapingou moulage, which are three-dimensional modeling techniques of clothing, textile manipulations always consider the volumetric effect that is intended to be achieved in a certain region of the body. Textile manipulations promote the restructuring or reconstruction of fabrics by means of cuttings, seams or heat. By the sculptural action on the support for the materialization of the idea, these techniques are concretized as the making process advances and are the result of an appropriation of the elements of the three-dimensional language like threads, flat, volumes, proportions, textures among others which are present in the product configuration setting.

- Termomolding: process of recording structural reliefs on a textile surface with the aid of matrices or supports. The technique modifies the structure of the tissues by subjecting it to heat. The two most recurrent processes in this technique are pleating and embossed. The pleated fabrics have an accordion effect acquired from a folding system subjected to heat. As in the embossed requires the elaboration of matrices that have furrows and are made in pairs that fit (male and female). The conformation of the tissue occurs when it is positioned between these matrices and is subjected to heat and pressure.
- Pences, pleats and draperies: These techniques also have the idea of "folds", but are always supported by a seam. They are used to give three-dimensionality to the clothing, creating volumes that consider the curves of the body.
- Geometric maps: diagrams are created to predict the stitches and to do desired volumes. These techniques consist of marking the reverse of the fabric with stitches programmed to receive tacks. Geometric maps generate patterns of textures with repetitive layout or occupy specific areas of a modeling through localized layout.
- Reconstruction: wear and tear, jointing of cutouts or the creative insertion of stitches allows restructuring fabrics with already preconceived structures. This is a free technique and with variations determined by the creation process of the designer.
- Laser cutting: consists on the removal of material through cuts designed digitally and performed with specific machinery. Laser is short for Light Amplification by Stimulated Emission of Radiation. This powerful light, when programmed, creates planned cuts on the textile surface.

4 ORNAMENTS LAYOUT

The distribution of ornamental elements on the textile surface of clothes may occur in a random manner, without a precise or organized order, in a clear and definite order. It is up to this study to investigate the intentional (organized) processes of ornamentation. These processes require methodologies and techniques that allow the structuring of perceptual information. Such processes presuppose the planning of the organization of ornamental elements and are pertinent to the field of Surface Design.

The spatial organization of visual information on a surface can be projected from a module (Rubim, 2004; Rüthschilling, 2008; Schwartz, 2008; Rinaldi, 2009; Freitas, 2011; Mol, 2014). The module is the set of patterns (graphics, textures and colors) arranged in an area of predetermined size (Freitas, 2011). It is the measure adopted to regulate the proportions, being able to be diagrammed in a specific area or designed to cover all the modeling of clothing. A module represents the smallest common measure containing the different elements that enter into the visual and tactile makeup of the clothes ornaments. There are two basic possibilities of ornamentation in clothing considering different types of modules: repetitive layout and localized layout (Fig.2).

4.1 *Repetitive layout*

In the repetitive layout (Fig. 2) the modules are juxtaposed or overlapping; it has forms programmed for

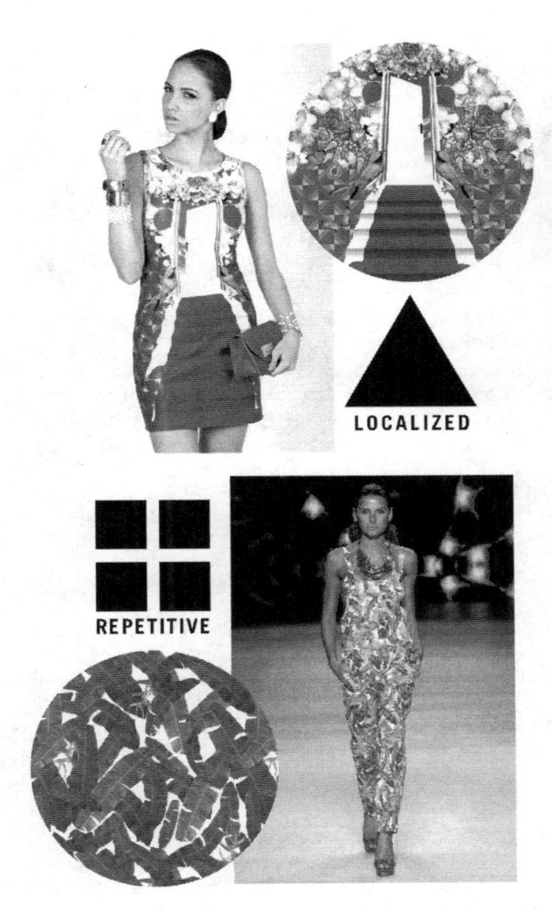

Figure 2. Examples of repetitive, localized diagramming and respective graphic icons adopted.

repetition, and do not consider the limits of clothing modeling. The articulation between the modules generates a pattern according to a pre-established geometry. Programming for repeating modules is also known as rapport (Ruthschilling, 2008).

4.2 *Localized layout*

In the localized layout one or a set of modules are arranged with a specific position relative to the flat mold of a clothing and may occupy it partially or totally. The shape and layout of the modules should consider the visual aspect that is intended to be achieved when coating a human body.

5 TEXTILE ORNAMENTATION IN MINAS GERAIS FASHION

Regarding the fashion made in Minas Gerais, the use of ornamentation resources is recurring. In order to investigate the incidence of these resources, it was chosen as research source the main product launching platform for the sector in Minas Gerais: Minas Trend. It is an

Figure 3. Examples of ornament features identified in the Minas Trend 2016 Winter fashion show.

event organized by FIEMG (Federation of Industries of the State of Minas Gerais), which gathers prominent companies in the clothing industry, for five days, twice times a year.

In this event there is a business hall, where buyers and exhibitors interact, and individual and collective fashion shows, where the conceptual launches of the new products are carried out. The collective shows take place at the opening of the event and contemplate the work of different local brands. As a cutout to be investigated, the collective shows held at MinasTrend from 2013 to 2016 were analyzed, which totals six fashion shows. From the examination of images provided by FIEMG's Public Relations, Ceremonial and Events Integrated Management, produced by Fotosite Agency, we selected images considered relevant to exemplify the recurrence of resources in each edition researched.

The different techniques of the Textile Surface Design of the companies and designers were analyzed, and classified according to parameters established in the presented classification.

The images are accompanied by a "zoom" that allows to clearly visualize the identified ornamentation feature, identified by the corresponding icon. In the figures are shown some models of the collective opening parades of Minas Trend 2016 Winter (Fig. 3) and Summer 2016 (Fig. 4).

6 FINAL CONSIDERATIONS

In the definition of Aurelio (2010) dictionary, ornament is "an accessory or adjunct", and can be "sophisticated or decorative". It is something used to "beautify", or that, naturally, does this: a decoration, "embellishment". It is the quality of expression that can

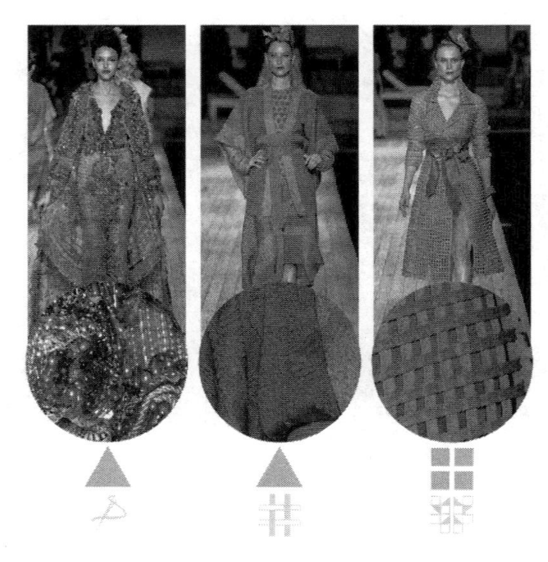

Figure 4. Examples of ornament features identified in the Minas Trend 2016 Summer fashion show.

give "chandelier or glory" to everything that adorns or embellishes. Ornamentation is an elemental and universal mechanism common to all aesthetic manifestations, as a phenomenon of apprehension (Pantaleão, Pinheiro, 2011). It is the "art in addition to art" that "makes people happy; it represents all that makes life worth living" (Trilling, 2003, p. 89). The etymology of the word ornament derives from the Latin ornare, which means "adorn" or "equip". In this sense, ornament is not limited to the addition of apparently superfluous attributes; it represents an increase of quality, an improvement (Gola, 2008).

The creative design of textile surfaces requires the designer to use an intelligible language so that all stages of production enable the transposition of the idea to materialize the final product adding to it new qualities. The stages of this process precede the actual production of the product. From a generation of alternatives, the designs that will be tested and possibly entered into the portfolio of offers from a fashion company are selected. For the execution of the proposals, technical drawings that allow to simulate colors and the desired layout of the ornaments are made, besides the discrimination of the application of the materials to be used. In the design of clothing surfaces, generally, samples, prototypes and textile models are made, which are called tests or pilot pieces. At the beginning of the development process these pilots can be elaborated in small scale or in small cuts for evaluation and validation of the initial idea. The adjustments are recurrent in projects of textile ornaments; they can redefine colors, shapes, materials, scale or even discard a proposal that is not adequate.

Therefore, the recognition of the possible resources of textile ornamentation, pertinent to the field of Textile Surface Design, is of extreme importance for the performance of designers. The classification of resources, according to parameters established in this article, can contribute to the projective action of designers. Textile Surface Design can be applied to products from different industry segments such as furniture, decorative items, home textiles (bed, table and bath), among others. The techniques used in the preparation of textile surfaces can add value, provide exclusivity and create stimuli.

In this way, the Surface Design is perceived as the technical and creative activity, which aims at the elaboration of two or three dimensional images reproduced through the fabrics, providing expressive perceptual characteristics to the appearance of the artifacts.

REFERENCES

Cardoso, R. 2010. *Uma introdução à história do design*. 3. ed. São Paulo: Blucher.

Cardoso, R. 2012. *Design para um mundo complexo*. São Paulo: Cosac Naify.

Castilho, K. 2009. *Moda e linguagem*. São Paulo: Anhembi Morumbi.

Comunicação, Universidade do Estadual Paulista, Bauru, 2009.

Freitas, R. O. T. 2011. *Design de superfície*: ações comunicacionais táteis nos processos de criação. São Paulo: Blucher.

Gola, E. 2008. *A Joia: história e design*. São Paulo: Editora Senac.

Gombrich, E. H. 2012. *O sentido de ordem*: Um estudo sobre a psicologia da arte decorativa. Porto Alegre: Bookman.

Jones, O. 2010. *A gramática do ornamento*: uma coleção de mais de 2.350 padrões clássicos. São Paulo: Senac.

MOL, I. A. 2014. *Superfícies de um lugar:* proposição de método de ensino para design de superfície a partir de valores culturais brasileiros. Dissertação (Mestrado em Design) – Escola de Design, Universidade do Estado de Minas Gerais, Belo Horizonte.

Norman, D. A. 2008. *Design emocional: por que adoramos (ou detestamos) os objetos do dia a dia*. Rio de Janeiro: Rocco.

Pastoureau, M. 1993. O pano do diabo: uma história das listras e dos tecidos listrados. Rio de Janeiro: Jorge Zahar.

Pedroni, F. 2013. Por uma definição do ornamento. In: *Anais do IX Encontro de História da Arte* – UNICAMP, 2013, Campinas. Anais do IX Encontro de História da Arte – Circulação e trânsito de imagens e ideias na História da Arte. Campinas, SP: Unicamp. pp. 80-85.

Pires, B. F. 2005. *O corpo como suporte da arte*. São Paulo: Senac.

Rinaldi, R. M. 2009. *A contribuição da comunicação visual para o design de superfície*. Dissertação (Mestrado em Desenho Industrial) – Faculdade de Arquitetura Artes e Comunicação, Universidade do Estadual Paulista, Unesp, Bauru.

Rubim, R. 2004. *Desenhando a superfície*. São Paulo: Rosari.

Rüthschilling, E. A. 2008. *Design de Superfície*. Porto Alegre: UFRGS.

Schwartz, A. 2008. Design de superfície: por uma visão geométrica e tridimensional. Dissertação (Mestrado em Design) – Programa de Pós-Graduação em Design, Unesp, Bauru.

Textiles, Identity and Innovation: Design the Future – Montagna & Carvalho (Eds)
 ISBN 978-1-138-29611-4

The charm of nonchalant elegance. Stories of Sicilian tailoring for men

Giovanni Maria Conti
Politecnico di Milano, Milan, Italy

ABSTRACT: When we talk about fashion and design, we take for granted the concept of contemporaneousness. Starting out by studying traditions before producing new objects could come across as a paradox and, very probably, to a certain extent, it is indeed in this paradox that the very charm of studying tradition lies. This study has investigated how the men's tailoring industry has found in the south of Italy, and more, precisely in Sicily, its own characteristic expression and well-defined identity, fruit of the meeting between the particular historical and social fabric of the location and the uniform of the middleclass man, codified by the England of the 1800s. In this way, differentiating itself from the other Italian tailoring traditions, the tradition of the south generated a characteristic product and a way of wearing it that was just as distinctive, but always in line with the context in which it was introduced. With a view to analyzing the local traditions, the study has also identified the use of hand-sewing techniques as another typical element, which today persists in Sicily. At the end of this study, we will understand the approach of a design that, based on the strong points of the local traditional heritage, no longer aspires to create artisanal products, but products of design, unique objects capable of telling a story, promoting the past and the spirit of the period.

1 THE SUIT AS THE UNIFORM OF THE MIDDLE-CLASS MAN

If we think about how the western man dresses, we immediately think about a uniform that is always the same and has undergone few substantial variations through time. And the expression "uniform" is even more fitting if we think that the men's suit as we know it today dates back to when the middle class of the nineteenth century decided to clearly distinguish the differences in terms of social class and therefore distance themselves from the fanciful, bizarre fashions of the Ancien Régime. The roots of this phenomenon lie in the ascent of the middle class, which, from the 1700s on, became the modern social class and the mirror of the changing times; gentlemen whose wealth often came from a profession and not from an inherited noble title.

The suit composed of tails, long sleeved waistcoat, shirt and trousers, remained unchanged and was made in plain coloured fabrics, usually black, leaving aside velvet and silk, which were symbols of another social class, and favouring wool and cotton with no decorations or embroidery. It was a conscious form of sobriety chosen by those who wished to distinguish themselves from the lavish spending of the aristocrats and courtiers, countering their idleness with a productive and cultural commitment that modelled itself on that of landowning tradition and ecclesiastic intellectuals (Morini, 2010). In the same period, a middle-class aesthetic model of clothing was also invented for women, consistent with the men's design, and in stark contrast to that of the cynical and amoral courtiers. However, as we will see, the middle-class women's model, which also paid tribute to the concepts of moderation, virtue and comfort, would change as fashions changed, whereas the men's version would become an institution.

In his "Treatise on elegant living", Balzac not only tackles the theme of the middle-class suit, but most of all, that of the concept of elegance, *"the great prestige that most people assign to education, purity of language, good manners, [...] the perfection of what derives from the person"*. The problem that the "new" man has to resolve in a changing society is therefore one of "moral superiority". In the same text, Lord Brummel, the highest authority on elegance during that period, conveyed man's elegant thought in his material expressions, such as the way he spoke, his gait, his manners; as these were elements directly descended from the interior essence of man, they were therefore subject to elegance.

Therefore, there was a need to focus the attention of social living to a wider concept of "elegance" that included the values of simplicity, sobriety, correctness and naturalness.

If it is true that clothes speak (Squicciarino, 1992) and that clothing is the expression of society, the way in which people dressed was not long in following suit. The suit made the social man, in the sense that on the one hand, it communicated his position and role, and on the other, it dictated his behaviour, adapting it to suit the occasion. The choice of a suit that would be appropriate for the occasion on which it was to be used was not yet an expression of elegance, but it was already a form of good manners.

The right clothes, however, matches the right fabric; if it is true that the dress describes man, the fabric describes what type of dress will be what that man will wear. Fabrics are a key component of material culture, of trade, of the technological development related to their dyeing and production as they become items of clothing or furniture. It is known that the wool industry, along with construction, was the most important productive activity of the medieval cities and the matrix of the revolution that leads to the modern.

Fabrics, however, are more than this: they are the skin of culture, its bodies and furnishings, they are the bearers of symbolic messages and a text regarding the construction and hybridization of identities, while they are also objects of design and technological innovation. They are, in fact, closely linked to the extraordinary and varied world of techniques and technologies.

2 THE SUIT OF THE SOUTH OF ITALY: A SOCIETY, A SOCIAL CLASS

The diffusion of the middle class man's uniform can therefore be associated with the rise of the middle class in Europe, and is the fruit of a particular ideological tendency rather than an aesthetic one. By no coincidence, the combination of English tailoring and the social and cultural fabric of the south of Italy generated a unique interpretation of the way men dressed and of traditional tailoring, which still exists today and which formed the personality of the men's clothing style now identified as Italian. Unlike the events that were unfolding in most of Europe, in the Bourbon kingdom of southern Italy, society had remained tied to a medieval structure. The economy of the kingdom was mainly founded on agriculture and was based on a feudal system of large estates, often used as pastureland or cultivated with large crops. These were tended to by a few families with noble titles and privileges over the population, but they did so with very little heed for innovation. The middle classes, wedged between a small noble class and a large class of ordinary people, would inherit and adopt the behaviour and ways of doing things that were typical of the small nobility, showing just how weak their own social and cultural identity was, as their highest aspiration was to acquire the noble titles and lifestyles which had belonged to the old land-owning aristocracy. The middles classes in the south, which had grown up in the shadow of the feudal system, had inherited the ancient barons' legacy without any dramatic struggles, transforming slowly and contradictorily, without being able to express their own real hegemony over the rest of society. In this social context, the way people dressed and the concept of elegance have different consequences, in some cases displaying elements of unique eccentricity and shamelessness; indeed, the story goes that Baron Francesco Agnello of Agrigento, who owned large estates in the province, habitually wore shirts with diamond buttons to show off his immense wealth to all and sundry.

In a context like this we have to think about the importance of the materials and their appearance: for example, the shirts in popeline was one of the favourite materials for its fit and especially for its "touch" in contact with the skin. Or the typical jacket, single-breasted button, silk satin revers, made in twill or wool or in Vicuña.

Unlike in many other European contexts, in which the English style of dressing was adopted as the result of the compliance with the ideological model it represented, another unique element of the diffusion of men's tailoring in the south was the type of contact that it had with its location and society. The modern version of the English "middle class" suit reached the south due to the effective presence of English people in the area. As is well-known, from the 18th century, the South of Italy was one of the most important stopovers on the Grand Tour, the educational journey that English and then European noblemen embarked upon in search of their historical and cultural origins; indeed, after Rome, the young Englishmen arrived in Naples, where they not only came into contact with the city, which at that time only came second to London in terms of size and population, but also with the ancient ruins of Pompeii and Herculaneum and, in the Phlegraean Fields, with the volcanic phenomena of Vesuvius. Another extremely important stopover was Sicily, with the island's volcanoes and Greek and Baroque treasures, about which Friedrich Maximilian Hessemer wrote in his "Letters from Sicily" at the beginning of the 1800s: *"Sicily is the dot on the i of Italy, [. . .] to me the rest of Italy simply resembles a stalk, placed there to support this wonderful flower"*. Another important channel of contact with England was that opened by trade. The Mediterranean and Southern Italy were an important market for the products of the English economy, which was beginning to introduce large-scale production thanks to the launch of the Industrial Revolution. As early as the end of the 18th century, a community of English merchants and businessmen, which revolved around the wine and Marsala industries, had formed in Sicily. The Sicilians established profitable trade relations with the English communities that had settled in Sicily and also began to adopt their social traits and lifestyles: the aristocrats of Naples and Palermo, two of the largest centres that hosted the English in that period, easily welcomed their English friends into their drawing rooms, often learning to speak their language. For example, the book-sellers of Palermo would sell English texts in the original editions to as many Sicilians as Englishmen and, in general, the hosting communities warmly welcomed the English way of life, to such an extent that they also looked favourably on mixed marriages between members of the different communities.

The English thus exported their lifestyle, and with this, not only their ideas and culture, but also their consumer goods. One of these was the English fashion reserved for men only. At first, this gave rise to the opening of branches of English tailoring companies such as Tayler, Gutteridge, and Lennon and Murray in Naples, while later local tailors went to train as apprentices in Savile Row: this was the case for the famous La Parola of Palermo, whose descendants finished their training at Henry Poole and Stovel & Mason. Having exported the English culture of dressing to the largest centres of Southern Italy, the uniform of the "European middle-class man" was interpreted based on new principles, which were sometimes so different that they came into stark contrast with those that had generated the modern men's suit.

Firstly, the meaning of the term middle-class had to be significantly put into perspective, as the aristocratic classes, together with the "aristocratised middle classes", were the first to submit to the charm of sober elegance in their way of dressing. This required a significant economic commitment on the part of those who wanted clothes that were notably different to those produced by the local tailors, and they often resorted to having their clothes made directly by the tailor shops in London.

When the Sicilians adopted the modern way of dressing, this also substantially changed the perception of the message that the model in question had originally been designed to convey. Indeed, the men's uniform had been the fruit of principles of democracy, hard work and discretion, but this ideological aspect was completely neglected by the nobles who wore it; in fact, those principles were not in the slightest contemplated by the noble wearers. The interpretation of the men's uniform conveyed by the nobles from the south of Italy was based first and foremost on the communication of a privilege, as the new fashion was the indicator of an opening towards foreign culture, and the concept of modernity was interpreted as promoting practicality, even in the way one dressed. Noblemen, such as, for instance, Prince Pietro Lanza of Scalea, wore their tails with silk satin lapels both in the city and in the countryside, perhaps during a stroll within their feud. Hard work was not one of the values upheld by the nobles, and, unlike in the Calvinist culture, neither was it particularly highlighted as an element useful for saving Catholic souls, or at least not in such an exteriorised, extremist form. The principle of discretion was interpreted in two ways: while, on one hand, it ran openly in contrast with the luxury and splendour heralded by the Ancien Régime silks, on the other it was worn with casual indifference, thanks to a conduct that can be likened to the dandyism that was immediately embraced by the Sicilian and Parthenopean gentlemen. It was particularly the ideological de-contextualisation and the above-mentioned attitude that allowed the English model to be criticised, reworked and recreated in line with the local values. However, it is important not to forget another two aspects, which are more strictly technical but equally as important in the development of a typically Southern Italian tailoring style. The first aspect is the weather, as the mild climate undoubtedly created the need to adapt garments that had been conceived and created for other geographical environments and climatic conditions. The other aspect is the notable difference in the anatomy of an inhabitant of northern Europe and the smaller proportions and different shapes of a Mediterranean man, who would have been significantly shorter and stockier.

3 THE FIGURE OF THE TAILOR AND HIS TRAINING IN THE SOUTH OF ITALY

The reports that tell of trade brotherhoods, such as the Confraternita dell'Arte dei Giubbonari e dei Cositori (Brotherhood of jacket makers and tailors) founded in Naples in 1351, let us understand how firmly rooted the culture of artisanal tailoring was in southern Italy and just how highly qualified a craftsman the tailor was considered, with the presence of a complex system of organisations and associations of craftsmen, at least in the urban areas. Indeed the etymology of "custureri", or tailor in Sicilian, which dates back to the Norman reign (from "couturier"), confirms that this form of craftsmanship had already been well-structured over three centuries previously. In the rural areas, although the tailor played an essential role within the country community, he was often forced to also work in other fields as a side-line, for example, as a barber. When the wealthy classes, who spent short periods of the year in the province, were on holiday in the feuds, they used foreign tailors or tailors from the city. The artisanal industry was often supported by the local nature of elements upstream of the fabric supply chain, such as for example in the case of silk, which was produced in centres in San Leucio, Messina, Palermo and Catanzaro. Instead, the European fashions arrived thanks to the circulation of foreign magazines, or as a result of direct contact with travellers and foreign merchants. Despite the dynamic nature of the clothing industry and the obvious flow of ideas and new tastes, the tailoring sector was still founded on a solid artisanal tradition: the know how involved therefore remained unspoken, almost secret, and children were sent as apprentices to learn under "o mastru" in the tailor's workshops as early as when they were ten years old. The learning process was gradual and, although it did not follow an official code, was structured in such a way as to increasingly refine the manual skills of the apprentice by having him perform parts of the tailoring process, or complementary activities, such as making the trousers or the necks of the jackets. Although it had been invented in the mid-1800s, the sewing machine never officially made its entrance in the sector of men's tailoring; at most, it might have been used to provisionally tack the fabrics when the pieces to be assembled were being prepared. As the apprentices would learn during their long period of study, the 8000 stitches needed to make a jacket were all to be done by hand, one by one, because, according to the master tailors, sewing a garment by hand enabled it to better follow the outlines of the body and did not stress the fabrics. As he was the keeper of such a specialised and exclusive know-how, the master tailor managed to establish a dialectic relationship with his customer, and interpret his requests exactly as a freelancer would have done, rather than a mere executor. Therefore, it is not surprising that it was indeed in these kinds of contexts that profitable relationships with customers were established, and these not only helped to encourage the customer to buy regularly, but also generated excellent opportunities for creating innovative projects as customers and tailors exchanged requirements and proposed solutions. Although it was based strongly on the individual tailor, the tailoring industry generated forms of association that were more modern than the old brotherhoods. This is the case of the Circolo Mediterraneo dei Sarti and the Centro Internazionale Moda Sarti (CIMIS) established in Naples, one in the 50s and the other in the following decade, by the group of master cutters that operated in the city at that time.

Although Sicilian tailoring comes from the same origins as its Neapolitan counterpart, which is better known and characterised by a distinctive style and product identity, it was not as widespread or even well known outside the island itself. Compared with Neapolitan tailoring, which was set in an important cultural context such as the city of Naples, and so had a strong identity, Sicily spread its way of working over three of the more important cities, Palermo, Messina and Catania. As these cities were different and their inhabitants historically tended more towards mistrust than alliance, the Sicilian tailoring sector also struggled to identify itself on a regional level as a local "tailoring tradition". In Sicily, the English were truly at home: between Bronte and Maniace in the province of Catania, in 1799 Ferdinand I of Bourbon, King of the Two Sicilies, gave the English admiral Horatio Nelson the abbey of Santa Maria in Maniace to thank him for having beaten back the so-called Parthenopean Republic, thus saving his life and kingdom.

This extended permanence of the English had a precise effect on Sicilian tradition: it formed a style that was extremely relaxed and comfortable, but considerably more balanced and less exuberant, and also very much more in line with the local spirit of the region, which was completely different from the Neapolitan one. Indeed, the Sicilian public loved to take care of themselves and bathe in luxury and fine tailoring, but compared with the Neapolitan clientele, they preferred to do so in a more elegant way, with more nonchalance.

4 THE PAST AS THE REDISCOVERY OF FUTURE TRADITIONS

If we consider what has already been said, such a society and tailoring industry, which were so different from their English counterparts, could not help but produce a product that was significantly different from the uniform of the middle class Englishman. Although the type of garments produced were always men's suits and other items of the traditional wardrobe, the perspective in which these designs are located changed completely, as did the requirements and design parameters they had to meet.

On the contrary to the sustained, modelled jacket of Savile Row, the jacket made by the southern Italian tailoring industry is extremely soft, almost completely disinterested in conveying the importance and austerity of the English gentleman, but instead determined to be a more flexible, comfortable, casual garment, which can be worn with ease, and with the disdainful nonchalance of princes, barons and middle class land-owners. The jacket has folds and vertical pleats in several points, elements that English tailoring would have looked upon as anti-aesthetic errors, but that are clearly present here for the purpose of comfort. The key points of the jacket are the shoulder, the sleeve and the front. The shoulder is natural, that is, not corrected or extended beyond its true dimension by epaulettes or shoulder pads. It rises quite high towards the collar and then descends, leaning entirely on the body wearing it, so as to distribute its weight evenly over the wearer. The shoulder is joined to a very tight-fitting sleeve, which allows the arm to move comfortably without the jacket moving. A large sleeve is introduced into an almost circular slot, skilfully reduced to a minimum with needle and thread, and the seam turns back internally on the shoulder, making it more flexible thanks to some small folds known as "a mappina". The cran, that is, the triangular gap between the collar and the lapel, is located very high up, to give the figure a tall, upward feel. The lapel is folded back with a modelled curve that gradually recedes. Often its fold is positioned so that the three-buttoned jacket in fact only has two useful buttons, with the highest folded down inside the lapel, which can be turned over and closed. The waistline is reached gradually; this detail has the effect of apparently lengthening the body; an extremely useful trick, considering that the stature of a man from the South of Italy was smaller than that of a Northern European.

Our research and a look at contemporary history unfortunately shows us that the South has slipped considerably: Sicily has not managed to look beyond its history and all that remains of that particular handcrafted tailoring is a memory stamped in the photos of the nobles and middle class of long ago. What is the challenge faced today by those operating in the design of tailored garments? In contemporary design, tailoring once again occupies a place of honour in the fashion world and in the opinion of consumers, aided by the rediscovery of craftsmanship and the recent attempts to promote that which is slow and the value of collective memory. Faced with a crisis in terms of the cultural and economic standards by which we set store, over the last few years we have been inundated with blogs, magazines, books and publications which tell us that tailoring is cool. Unlike in the past, when the Italian industry was based on different types of tailoring in Sicily, Naples and Abruzzo, not to mention Mantua and Lombardy, today, tailoring offers a framework for experimenting, innovating and rediscovering. Thus, history takes on a different meaning, giving us a wonderfully vast archive providing new generations with a study of lines and customs to be analysed and to draw on for the purposes of innovation. The meaning of the suit changes, as does the concept of gender, with the industry sometimes preferring to design for the no-gender model rather than for a well-defined body. But one thing that doesn't change is the concept of elegance, especially that of garments "made in Italy". This is intentional; part of our quest to prolong and conserve over time the taste and "savoir-faire" of a population that operates in such a rich and jagged context as Italy.

5 FINAL CONSIDERATIONS

However, nothing could have been done in Italy if, Italy itself, was not one of the largest manufacturing countries in Europe.

The textile-clothing represents a very important sector in the Italian manufacturing field and, at the same time, one of the leading sectors of made in Italy that has contributed over the years to "the definition of the concept of good Italian taste and quality of life, producing positive effects on the image of Italian products in the world." (Sistema Moda Italia, 2003) (Maggioni, 2009). In Italy the textile sector is very big and vast and it has generated two of the largest and extensive productive districts in Europe. For example, the industrial district of Biella and the manufacturing district in the area of Carpi-Modena that represents the Italian excellence worldwide in terms of quality of the material, the semi-finished product and the final product.

Historically, in Italy, scientific and technological research on both natural as well as artificial fibers will mark out a new course for textile which, at first, will be blended with wool, silk, cotton and rayon and lanital, transforming itself into the sector which would see many innovations from the early years of the twentieth century. "In Italy the production of artificial fibers took its first steps in 1904 in a factory belonging to Cines of Padua, which was owned by the Count of Chardonnet. The product, however, is expensive and fairly dangerous and is unable to establish itself in the textile industry. It is during the First World War that the production of artificial fibers in Italy develops: 1918 saw the founding of the Châtillon Silk Company which adopts the viscose process at its plant in Val d'Aosta." (Garofoli, 1991)

In Italy, the textile industry is perhaps the oldest and most complete, where there had been a strong tradition of cultivating fibers, or raising livestock, in order to obtain materials that could be used in the textile and fashion sectors. *"Know-how"* attitudes that have characterized whole areas of Italy until they become the territories in which the artificial-fibers industry also took root in "a growth that astonished the world for its contemporary progressive improvement of quality, for the rapid approval that came from the consumer, for the movement in terms of the proportions of textile fibers employed". (Garofoli, 1991)

To speak of a textile product, then, we cannot disregard its soul, as represented by thread and the different qualities of yarn that exist; it is a product that lives in the contemporary world and, since technology has permeated this particular field as well, undergoes various treatments to increase its performance characteristics. The finishing treatments can be divided into two main groups: there are those designed to improve technical qualities such as antistatic, crease-resistant, non-shrink, anti-infiltrating, impermeability, waterproofing, fireproofing and mothproofing treatments. Or else there are those treatments that affect the appearance of the fabrics and sometimes even disguise the original material such as ironing, dyeing, scouring, felting, flocking, coating, embroidery, counter gluing and lacing.

Especially in the period of Italian autarchy, the development of new fibers and yarns will ensure that thoughts will turn to dressing in a different way; it will no longer be necessary, because of their total or partial absence, to wear garments made with the better-known yarns like cotton, wool or silk, but gradually innovative fabrics will be placed in the market, including those stemming from previously unknown production processes. The invention, for example, of the spinning centrifuge in 1900 provided a major contribution to the development of the processing of viscose, destined to become the world's most popular man-made fiber. Another interesting example was SNIA Viscosa, the producer of the first Italian casein fiber, which was even celebrated by Filippo Tommaso Marinetti in "Poema del vestito di latte". Or, 1937 will see the introduction of Cisalfa, "the new fiber for wool producers which presents a number of economic solutions for the department stores that would allow a still agricultural Italy to dress to impress". (Giordani, 2012)

The fascination of fabrics structures the imaginary and aesthetic sensibility of the various eras in which experiments on fabrics plays a central role and also determines a new approach to design and new scenarios for the combination of fashion-industry and technology. We find this in all cultures and eras, but to stick only to fashion in the modern sense, we only have to think of the fabrics of Bauhaus, those designed by the Futurists and Constructivists in which the materials become bearers of new ideas and configurations, Chanel's use of the jersey, the employment of materials like cloth or canvas for the art of Capucci, the experiments of Paco Rabanne, to the mental clothes that Miyake designs in connection with textile designers, to the new ultra-strong, flexible, lightweight, bewitching technological materials, made available by research on smart materials in which are outlined new products and a new ontology of the artificial.

REFERENCES

Baroncini, D., 2010. La moda nella letteratura contemporanea. Torino: Bruno Mondadori.

Buzzacarini (De), V. 1992. L'eleganza dello stile. Duecentanni di vestir maschile. Milano: Lupetti.

Buzzacarini (De), V. 1994. Giacche da uomo. Modena: Zanfi Editore.

Calanca, D., 1992. Storia sociale della moda. Torino: Bruno Mondadori.

Correnti, S., 2003. Storia di Sicilia. Come storia del popolo siciliano. San Giovanni la Punta (CT): Brancato Editore.

Garofoli, M. 1991. Le fibre intelligenti. Un secolo di storia e cinquant'anni di moda, Electa, Milan.

Gigli Merchetti, A., 1995. Dalla crinolina alla minigonna. Bologna: Clueb.

Giordani, B. in AA.VV. 2012. Maglifico! 50 anni di straordinaria maglieria Made in Italy, Skira, Milan.

Iuliano, L. A., 2010. Lungo il filo di Aracne. Fili, trame e tinte Calabria mediterranea. Soveria Mannelli (CZ): Rubbettino Industrie Grafiche ed Editoriali.

Maggioni, M.A., 2009. Il distretto tessile biellese; l'eccellenza sfida la crisi ("Quaderni Fondazione Fiera Milano"), Scheiwiller, Milano.

Maing, J., 2011. Sleevehead's guide to sicilian tailors. Ebook.

Morini, E., 2006. Storia della moda. XVIII–XX secolo. Milano: Skira.

Naldini, M, 2005. Uomini e moda. Mezzo secolo di abbigliamento maschile nel racconto di Giuliano Angeli. Milano: Baldini Castoldi Dalai editori.

Pollan, Michael. 2006. The Omnivore's Dilemma: A Natural History of Four Meals. New York: Penguin.

Taglialatela, M. A. (edited). 2010. La creatività sartoriale campana. Abbigliamento maschile e moda mare. Napoli: Arte.

Vacca, F., 2013. Design sul filo della tradizione. Bologna: Pitagora Editrice.

Villarosa, R., Angeli, G., 1990. The Elegant Man. Milano: Idealibri.

Zumbino, S., 2014. Trattato di confezione moderna. Moda Maschile. Tradizione dei su misura italiano. Milano: Blue Nozca.

Textiles, Identity and Innovation: Design the Future – Montagna & Carvalho (Eds)
© 2019 Taylor & Francis Group, London, ISBN 978-1-138-29611-4

Warps and textures in symbology

Alexandra Ai Quintas
CIAUD, Lisbon School of Architecture, University of Lisbon, Lisbon, Portugal

ABSTRACT: In the 19th century, there was an architect and professor whose work had a great influence on the formation of many architects of the Modern Movement, Gottfried Semper. In his writings especially in The Four Elements of Architecture, he established the importance of the relation between the walls and partitions in the spatial organization of a dwelling. He also explored the relevance of the use of tapestries, especially in the ancestral cultures settled in mild climates in Antiquity. Centuries later, tapestries, carpets or rugs, manufactured either the Middle East or In Europe, were used in the residences of the Noble or the upper Clergy. In what concerns the issues of the ornamental motifs depicted in the decoration of either the oriental or the western origin tapestries, both use inspiration of biomorphic subjects. Nevertheless, while the first ones tend to use stylized designs of plants and animals, the second ones tend to follow a far more naturalistic kind of representation. Symbols pertain to both traditions but are definitely more hermetic and geometrized in the oriental pieces.

1 INTRODUCTION

There was a relevant architect in the decade of the 40's of the 19th century, Gottfried Semper. He developed an architectural theory that went against the Vitruvian triad and derived from a system where there were considered the gravity laws, as well as the materials. The written work of Semper, *The Four Elements of Architecture*, is a kind of study of theoretical nature, in the sequence of his intense practical activity in the cities of Hamburg and Dresden.

In a conference that he led in 1848, Semper referred to the existence of primordial forms *Unformen* and develops two ideas that generated the first shelters: the closure (*Umfriedung*) and the roof. In another conference of the same period, he adds the hearth to those other elements and the involving wall *(Einfassungsmauer)*, as the primary element of ancient architecture of people from the south and the primordial seed *(Urkein)* for the shelter, the temple or the city.

He put into evidence the textile character, and therefore non load-bearing of the partitions of nomadic cultures and the hegemony of textile production, as well as the connection of Man to the built form, upon which resides the art of tectonics.

According to Semper, the structural symbolical essence of tectonics would find itself closer to the cosmological and ritualistic arts of music and dance, better that the figurative arts of painting and sculpture.

This distinction came to influence Loos in the discrimination between the tasks related to the activity of building and the role of architecture as monumental art, capable of erecting, both a tomb or a monument. Both attributes of structure and ornament can be inscribed in the Greek expression *Kosmos,* synthetizing the ideas of "universe" and "decoration". Traditionally, the symbolical function of architecture would be an attribute of its monumentality, meaning by its classical language, a certain definitive universality.

According to Semper, the concept of "enclosure", defined by the partitions, acquires its architectural value, defining a new spatiality or an interior world, confined or protected from the exterior and also organized around the hearth, which works as the social and spiritual center of the shelter, dwelling. In the 19th century, the column, the beam and the wall lost their metaphorical meaning, serving intentions of expressive character.

The importance of the wall as architectural element in all Semper's theory establishes a hierarchy in the definition of architectural space, as well as it tends to influence some concepts of the Modern Movement of the 20th century.

Being the wall one of the defining elements of space and having been traced is genealogy since nomadic civilizations, with the tents and fences made of intertwined branches, more easily one can understand the tradition of architecture which, in the present study, is the base of the concepts of spatial transfiguration.

Semper felt forced to go back to the primeval conditions *(Urzustände)* of human society to reach what he proposed to establish, the more synthetized possible way. The first alliances were formed around the fireplace along with the first primitive religious cults. The sacred fire was the focus around which form order would take place.

Architecture was thus defined into two different types of dwelling that of the cloister, with dominant walls from the south, and the dominant roof architecture from the north.

Almost after the presentation of his theory of the *The Four Elements,* Semper insists in closure and begins to draw what later on will be his thought about the idea of cladding *(Bekleidung).* The original idea was based in the action of creating fences by the action of intertwining flexible vegetal fibers. Later on, there were the hung tapestries or carpets woven or in wicker. These elements in wicker would serve as a separation of private property. Rugs and carpets would prevent the air drafts and protect against too much heat or cold, being at the same time a substitute of masonry in climate favored regions. This latter required another type of building techniques. Thus, wicker used as original spatial division may be considered the first essence of the wall.

According to Semper, hung tapestries were the true walls, the visible frontiers of space. The walls were necessary for reasons of security and load bearing. The interior structure would hide behind the legitimate colorful woven representatives of the walls. The *stucco* will evolve from the development of the wall as a division element for the system of oriental polichromy, deriving from the looms and dyeing tanks of the Assyrians. Obviously, the need of structural, thermically resistant walls, would have made tapestries to be substituted by mortars, wood, metallic plates, adobe blocks, alabaster or granite panels.

Figure 1. Iconographic source: Greek vase of the VI century AC. Greek knight with a garment that exhibits the characteristics of the fabric very close to the decoration of *Kilim* carpets.

2 THE ORIGINS OF TAPESTRY

The origin of tapestry goes back to high Antiquity, when tapestries and fabrics were manufactured according to the technique in Greece and in Rome. Moreover, the then used technique of weaving conferred to the Coptic fabrics, many of which with rich with figurative and ornamental compositions, the feature of authentic tapestries.

The conquest of Egypt by the Arabs encouraged the fabrication of those fabrics by Muslim weavers, thus developing the traditions of the Coptic artists. When the hegemony of the Arab civilization spread out into the West, they tended to generalize to the Iberian Peninsula that branch of sumptuary arts.

Illuminated manuscripts did contribute to inspire and serve as models for the *cartoni* from the middle ages. Nevertheless, one cannot consider that there were in fact real *cartoni* dating from that historical period.

The "Tapestry from Bayeux" is no more than a piece or strip cut out of rough linen cloth that was embroidered.

The "cartoni" dating from the 14th century were extremely summary in their essence, presenting some notes about colors or tonal values that were depicted in "grisaille".

Some tapestries (a whole series from the 15th century) that were manufactured for the Église de Sainte Madeleine de Troyes were made out of drawings that were executed on some bed sheets that were sewn together by a seamstress.

In fact, the first known "cartoni" date from the early 15th century, having been painted on paper with tempera technique, according to drawings by Raphael and pertaining to the collection of Elizabeth II from England.

In what concerns the areas of production that we can establish as "oriental", Iran, or Persia has been one of the most relevant ones, other regions being Turkey, Caucasus, Turkestan, India and China. Most of the patterns used in rugs or carpets from Persia are florals but many of these manufactured artifacts especially those that have their origin in nomadic tribes (in the Caucasus) obey to the geometry typical of this region. Heriz and Shiraz are two of the more prominent cities that produce geometric patterns.

The famous *Bokhara* is from Turkestan. For the correct identification of the rug it is essential the knowledge of the knot used in weaving: the Turkish knot (or "Ghiordes") from the city of the weavers in Turkey or the Persian knot (or "Sehna"). The most used raw materials are wool on oriental rugs, therefore mostly of "angora goat" – which is used in Turkoman rugs – while silk is used in Chinese carpets or rugs.

Before 1900, there were mainly used vegetal dyes, but around the end of 19th century, the demand in Europe was so remarkable that synthetic dyes were beginning to be used, so that from 1920 on, the vegetable dyes based practically disappeared.

3 THE DYES AND THE COLORS

Very often these dyes were fabricated from insects, – the *Kermes*, – which were gathered during the Summer time and then killed and put into vinegar. Maybe it was the richest and permanent of natural dyes. The *Coccinelle* is an insect from some *Coccus cacti*, from Mexico that was introduced into the Orient relatively in more recent times that was used for dyeing in the the same color. The third type of red was obtained from a root of "madder wood", *(Rubia tinctorum)*, put into boiling water and afterwards drained. To obtain a violet color it had to be mixed with milk and yeasted grape juice.

The Indigo blue (*Indigofera tinctoria*), brought from India to Europe in the 16th century was presented in dye cakes, which were not water solubles and could be reactivated, by the means of a catalyzer, with the reagents.

Other dyes were obtained from other species such as weld yellow *(Reseda luteola)* and madder red Archil violet *(Roccella tinctoria)*. Brasil wood (Caesalpinia) came from the Indias.

On the topic of colors: in the Middle Ages, and even as late as the Renaissance, people knew and used relatively few namely blue, red, yellow, green, brown, white, black, and ash-grey. By the fourteenth century a total range of 20 to 24 colors and shades was quite common.

Figure 2. The Bayeux Tapestry, 11th century. Musée de la Tapisserie de Bayeux, Bayeux.

Figure 3. Details series *Dame à la Licorne*, c. 1500, Musée de Cluny.

4 EARLY EUROPEAN TAPESTRIES

In Europe the craftsmanship industry of tapestry became more developed from the end of the 14th century on. One can detect a considerable gap between the "Apocalypse of Angers" and some well-known European fragments such as the "Cloth of St Gereon" (Cologne, 11th century) or the "Badishol" (Norway. 12th–13th centuries), all of them being pieces of a considerable dimension, whose aspects seem developed as a response to certain needs at the end of the Middle Ages.

The uncomfortable castles of those times required tapestries that could turn them into more habitable residences. Those light tapestries concealed walls made of rock, and adjusted to their rough surfaces. As pieces of luxury, tapestries were considerable forms of ostentation, thus proclaiming some sort of status. Mobility, ease of transportation and of the storing the tapestries was considerable. So, literally, one can consider that design and color covered the nude walls.

English wool, that had been considered superior since the times of the ancient Roman Empire, came through the Guilds to the production centers. Flanders was gaining from the commerce of the product (raw material). Flanders was gaining from the commerce of the product that arrived to Bruges from London.

In the 16th century, there was a break of the production centers in Flanders. There was a migration phenomenon of habilitated craftsmen to England and then there was the establishment of the supremacy of England in textile industry.

Superior competences and powerful patrons protected the craftsmen against the ordinary workers employed in the textile manufactures. The first center of its industry was Paris. Ten were integrated in the "corporation of tapestry weavers" as "tapissiers de la haute lisse" – using high-warp or vertical looms. There are registers of the rapid evolution of that industry in evolution.

Paris was replaced by Arras that gained by the ruin of the industry of Paris. The fame of its looms was so much spread that a tapestry in England came to be designated as an "arras" or an "arazzo". While Arras flourished, there were also looms in Oudenaarde, Bruges, Brussels and in Tournai. The city belonged to France but was surrounded by Burgundy.

The representation of human beings and animals is very common in the medieval tapestries under the form of a young man "Damoiseau" or a young woman, "Damoiselle", normally with a symbolical animal like Lion, (meaning strength) or a Basilic (meaning Jealousy) and a vegetalist background, the "entrelacs". The colors used in the background are symbolical. Very often, they tend to the use of green and blue.

Figure 4. Details series *Dame à la Licorne*, c. 1500, Musée de Cluny.

Figure 5. Harrier dog, series Dame à la Licorne, c. 1500, Musée de Cluny.

The young lady sometimes is depicted between two Fabulous/monstruous animals with the head of a Basilic with Peacock feathers, symbolizing for instance carnal desire and heat in combat.

Other subjects, which are very frequently depicted and related to privacy, were a pair of lovers *"Les Amoureux jouant aux échecs"* or *"jouant aux cartes"*, under a tent, or inside a *hortus conclusus*.

Other animals often depicted such as the unicorn symbolize chastity, as the deer the desire for God. Those tapestries showed a chromatic harmony – the balance of reds, greens and blues – is completed by a bright saffron yellow that accompanies the text of the banners or streamers: "It is God who has created all" (Unicorn); "That may be praised His Trinity" (Deer).

The series of the *Dame à la Licorne* and the *The Unicorn Hunt*, one must understand that there is a complex symbology related to that mythical animal.

Figure 6. "Hands on hips" female figures.

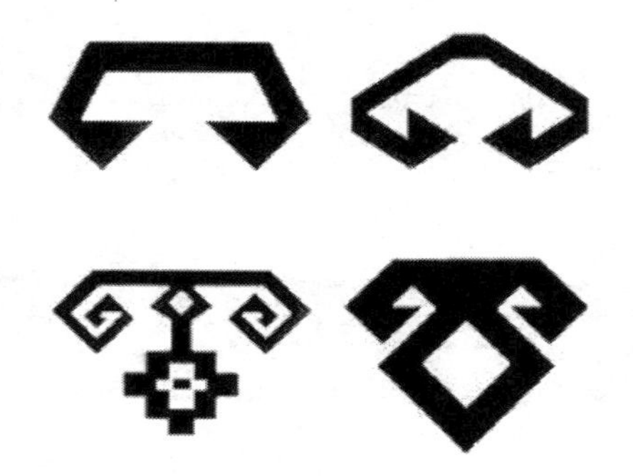

Figure 7. Ram's horns.

We have recently dedicated a full article in 2017 to the first Series and the issue of the Five senses and the Sixth one.

5 ORIENTAL RUGS AND CARPETS

Even though tapestries are truly connected to family patterns, there are not two which are similar. Biomorphic representations are very complex and also the symbols used have a huge variety. In the nomadic tribes and communities there are stylized animals or biomorphic shapes as in the European tradition. Symbols tend to be, in these compositions, extremely rich and with huge variations.

Even though the design is based in a stylized human female figure, the pattern received different designations in Anatolia (such as *gelin kiz, cocuklu kiz, aman kiz, karadoseme, seleser, kahkullu kiz, cengel, sarmal, cakmakli, eger kasi, turna katari*). It stands for the symbol of motherhood and fertility. To observe the evolution of the motif we may start with a mother goddess statuette dating from 3000 BC found in Ahlatlibel, near Ankara.

Known as *boynuzlu yanis, boynuzlu, koclu yanis, gozlu koc basi*, the ram's horns is a symbol of fertility, heroism, power and masculinity.

The double combination of hands on hips and ram's horns indicate the male and the female symbols. The eye motif in the center of the composition keeps the intention of protecting the family against the evil eye.

This is a symbol of dualism, inherited from the Far-East and imported to Anatolia, suggesting harmony between masculine and feminine principles.

A girl using this motif is trying to inform her family that she wants to get married.

The human eye is the source of an evil glance. Its harms may be prevented against by a human eye.

Figure 8. Bronze statuette 3000 near Ankara.

Figure 9. Fertility results from the combination of “hands on hips” and “ram’s horns.

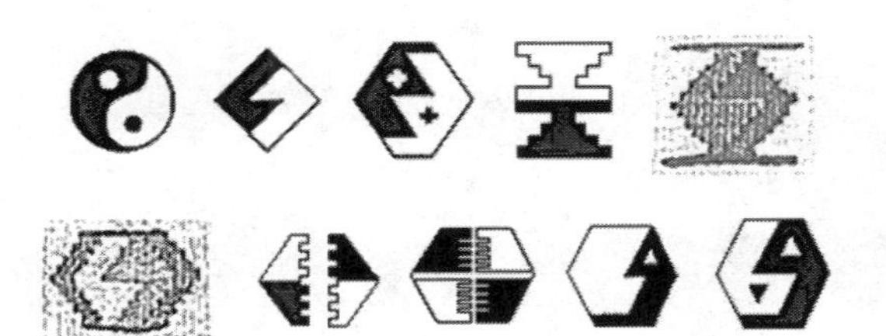

Figure 10. Duality: Love and unison.

Therefore a shape of the eye is based on a diamond divided into four. The triangle can be the the stylized form of the eye.

The belief in Anatolia against the evil eye is based on the cross, obtained by a horizontal and a vertical line, dividing its power into four. One can find crosses on the walls of Çatal Hüyuk.

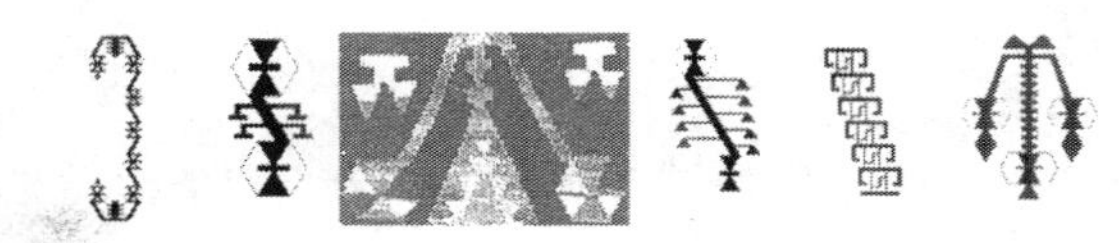

Figure 11. Earrings: the girl communicates she wants to get married.

Figure 12. The eye.

Figure 13. The cross.

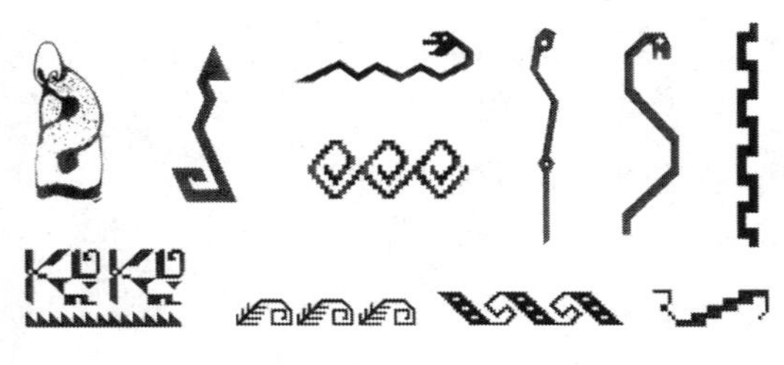

Figure 14. The black snake.

Figure 15. The scorpio.

Figure 16. The star motif.

The history of mankind is pervaded with myths with snakes. The Black snake is the symbol of happiness and fertility.

The Dragon is a mythological winged creature with paws like the lion, a tail like a snake’s. He is the master of air and water. Furthermore, the fight of the dragon and the Phoenix is believed to bring fertile rains of spring.

Believed to be a great serpent, is the guard of treasures and secret objects as well as the tree of life. People believed that it was the cause of lunar eclipse.

The star motif on an Anatolian weaving generally means happiness The six pointed star – known as Solomon's Seal – is being used in Anatolia since the time of the Phrygians who lived long before the time of Solomon. Taking into consideration the mother goddess statues where the star symbolizes the womb, it could be said that the motif stands for fertility.

This article tried to put into evidence the huge variety of biomorphic representations either in eastern or western traditions of tapestry design.

ACKNOWLEDGMENTS

This research was only possible with the support of CIAUD – Research Centre in Architecture, Urbanism and Design, Lisbon School of Architecture of the University of Lisbon, and FCT – Foundation for Science and Technology, Portugal.

REFERENCES

Ates, Mehme. Turkish carpets, the Language of Motifs and Symbols.Istambul.

Bennett, Anna Gray, 1992. Five Centuries of tapestry from the Fine Arts Museums of San Francisco. Fine Arts Museums of San Francisco, CA. USA, Chronicle Books.

Bennett, Ian, 1972. Book of Oriental Carpets and Rugs, Hamlyn, London, NY, Sidney, Toronto.

Reynolds, Graham, 1966. The Raphael cartoons. Victoria and Albert Museum, London.

Sevensma, W.S., Tapestries, 1965. The Merlin Press Ltd, London.

Schlosser, Ignaz,1963. European and Oriental rugs and carpets. B. T. Batsford, Ltd, London.

Webgraphy

https://www.kilim.com/kilim-wiki/kilim-motifs. acceded 18-01-2018.

https://historywriterblog.wordpress.com/2014/11/11/medieval-art-and-literature-the-bayeux-tapestry/. acceded 18-01-2018.

2.3. Marketing and consumption

Textiles, Identity and Innovation: Design the Future – Montagna & Carvalho (Eds)
© 2019 Taylor & Francis Group, London, ISBN 978-1-138-29611-4

Design Camp LOGO UFSC: A proposition of immersive experience in the textile market for design students

N.C. Salvi, I.L. Guedes, F.P. Pasqualotto, L.S.R. Gomez & M.M. Gonçalves
Universidade Federal de Santa Catarina, Florianópolis, Brazil

ABSTRACT: Based on the recognition of the need to promote Design innovation aspects in the textile sector, LOGO UFSC Design Camp presents itself as an event that proposes an interaction between different actors in the area on the improvement of methods of generating creative, innovative and effective proposals to the dilemmas of the sector. These actors are Design students in their different modalities and levels, professors, employees, and companies. This article presents a Design Camp LOGO UFSC model, computing the results of its first edition. The work configures itself as applied and descriptive, of a qualitative nature. The first edition of the event involved a total of 46 students, 19 companies, and 17 educational institutions. In all, 24 challenges were proposed and resulted in 96 projects of creative solutions. The results are intended to contribute to the improvement of co-operative and collaborative processes and promote a better understanding of innovation and design in both the academy and the textile market in Santa Catarina, Brazil.

1 INTRODUCTION

Santa Catarina is one of the Brazilian states recognized for its manufacturing capacity in the textile, clothing and footwear sectors and has been working for years to position itself as a creative center in these sectors. The state has achieved satisfactory results in this regard, in part thanks to the work of educational institutions, which work to encourage students and the market to seek the development of the creative process in areas related to fashion rather than the habit of copying products – widely diffused in the sector; But also as a result of the efforts of industries to differentiate themselves from their competitors.

In this context, this Design Camp model assumes a fundamental role in the development of this thinking. This model gives the students of different areas of design a proximity to the labor market, through an immersive and cocreative/collaborative experimentation platform. Students and Professionals with the technical and scientific support of professors and collaborators are involved in solving problems faced by the industry of Santa Catarina, sharing knowledge, experiences and bringing results to the development of design value. The first edition of LOGO UFSC Design Camp was held in partnership with *Santa Catarina Moda e Cultura* (*SCMC*) (Santa Catarina Fashion and Culture) which shares the same objectives. *SCMC* is an association of companies and educational entities from the state of Santa Catarina that since 2005 has been working to promote the state's creative identity through the interaction between the various stakeholders in the Design and Fashion sector. The study stated in this article is part of the Master's Dissertation "Design Camp: a model of creative immersion" (SALVI, 2017), presented to the post graduate program in Design (UFSC).

Design education has developed throughout the country and, according to the report *Design no Brazil* (SEBRAE, 2014), Santa Catarina has 17 graduation courses in Design, 35 Bachelor's degrees in Design, two specialization courses and two postgraduate courses (master's and doctorate's) covering industrial design, product design, fashion design, graphics, interiors, games and others. This growth in the education sector is due to the process of replacing unskilled professionals from the market – a common situation due to the recent presence of training courses for the profession of designer.

Based on these considerations, this work intends to describe the Design Camp project, addressing its methodological steps and the results achieved. These objectives classify the present work as descriptive since it seeks to "describe the characteristics" (SILVA and MENEZES, 2005) of the Design Camp event, showing the processes, their results and the impressions of those involved in their application. The project for the development of the event, whose research is of a qualitative nature, can be classified as applied since its objective was to generate knowledge for a practical application and solution of a specific problem. (SILVA and MENEZES, 2005). The project was divided into three distinct phases: the process of preparation and selection of students throughout 2015, through the participation of students in lectures and workshops; The event itself, where the projects were developed by the students for the participating companies in the form of a competition; And finally the evaluation of the results of the projects and the respective awards.

2 THE EVENT

LOGO UFSC Design Camp was developed by the Laboratory of Orientation of Organizational Genesis (*LOGO – Laboratório de Orientação da Gênese Organizacional*) of the Federal University of Santa Catarina (UFSC) and was conducted under the guidance and coordination of Professor Dr. Luiz Salomão Ribas Gomez, Ph.D., with the help of his collaborators – professors and students LOGO UFSC.

The Design Camp consists on the integration of students or professionals from the spheres of Design with companies from various branches of the creative industry of Santa Catarina, immersed in the same environment over a week. With a methodology build on Problem Based Learning which proposes the development of projects based on the presented problem adapting the most appropriate methodologies according to the desired goal. The event was developed with different activities promoting integration, creativity, and innovation in solving the problems presented by companies.

The model acts in the generation of knowledge, through the Design, being an event model and basically, in the accomplishment of Design activities, guided by a goal and realized by a determined group of people (SALVI, 2017. P.33).

The Design Camp was divided into two stages, the first being the preparation stage, named Conecta, and the final event, the Design Camp itself. The first stage consisted of workshops and lectures in order to guide and enable students to the experience during the final event, as well as analyze and select them according to their skills and performance. This first stage resulted in monthly meetings, which took place over six months during the year 2015, in different cities of the state of Santa Catarina. These meetings involved the students of participating institutions and newly trained professionals in the areas of design. The Conecta stage also had a virtual platform in the social network format, which helped the communication, exploration, and sharing of knowledge, as well as for the evaluation of those enrolled to participate in the second stage. Both activities were intended to generate knowledge, skills, and familiarity among those involved with the design contents, making them able to participate in the final event entitled Design Camp. In this way, both the activities developed in the preparatory events and the frequent manifestations in the social network gave the participants a score, which served as a qualifying indicator for the selection of the subscribers to the Design Camp.

2.1 *Design Camp LOGO UFSC – Pilot Event*

The first edition of LOGO UFSC Camp Design took place in partnership with *Santa Catarina Moda e Cultura (SCMC)* (Santa Catarina Fashion and Culture). Altogether forty-eight subscribers entitled Campers were approved for participating in the Design Camp. Divided into four teams, each with a name and a representative color, they developed a total of 96 projects, being 24 per team, solving the challenges proposed by the 19 participating companies.

The division of the teams followed criteria of interdisciplinarity, different educational institution, different aptitudes, age difference, level or training course. These criteria were used in order to generate greater diversity among the members of each team, prioritizing the balance of competencies between groups.

Each team elected a general leader and assigned members the responsibility to lead each of the challenges. The overall leader was responsible for managing the activities, the deliveries of the team as a whole and ensuring the good internal relationship. The leaders of the challenges were concerned with regulating the progress of activities and ensuring that deadlines were met. It was also centralized to the leader of each task to maintain communication with the representatives of each company, who were present at the event and available for consultation, exchange of information and follow-up of the work. Each company was represented by one or more employees who remained full-time in their respective stands to support all teams. It should be noted that there were no members of the companies present within the teams.

To encourage and certify that all participants had contact with all companies, the stamp exchange method was established. All the visits made by the Campers to the companies' stands were registered by giving the team a sticker, proving the passage.

One of the main intentions was to promote the generation of solutions with limited resources, as this could promote more creative results. Robinson (2012) argues that many of the creative works emerge from constraints imposed as guidelines. In this objective, the time and the scarce financial resources, aimed at the generation of alternatives of low cost, and the exploration of more creative solutions. Despite this, some conditions were extended aiming at the quality of the prototypes, as well as the presentations of them.

2.1.1 *Preparation*

As a traditional practice of the *SCMC* program, some workshops were held during the year, in order to train and engage students and professionals interested in promoting and improving Design in Santa Catarina. In all, six training events were held, with different themes, promoting the involvement and preparation of participants for the final event.

The workshops took place monthly, in different cities of the state of Santa Catarina, in order to facilitate and involve the largest number of educational institutions. The workshops took place between May and November of 2015 at the educational institutions SENAI Unit of Blumenau, UNIVALI campus Balneário Camboriú, Institute Orbitato in Pomerode, *Pontifícia Universidade Católica de Jaraguá do Sul*, UNIVILLE in the city of Joinville, as well as the Sapiens Park complex, which simultaneously broadcasted one of the events with the purpose of increasing the attending audience. The themes were as follows: Focus on Passion, Focus on Ideas, Focus on Multidisciplinarity, Focus on Courage, Focus on Innovation and Focus on Autonomy. All of them were represented by a hashtag, which was used as the name of the event and assisted the identification of the volume of Campers' participation in the event's social network.

The first workshop was given by Luiz Wachelske and Andreia Schmidt Passos, entrepreneurs of the textile sector. The second event was held in partnership with the TEDx Floripa event, in Florianópolis (SC). At this workshop in particular the students witnessed the presentation of 14 lecturers, from different areas of activity, in the TED talks which consist of quick and objective conversations about their respective cases or domain subject. At the time, participants were able to enjoy the technological and innovative structure of Sapiens Park Technology Center, with simultaneous transmission of the lectures that took place in another part of the city with insufficient capacity to attend LOGO UFSC Design Camp participants. The third event was hosted by lecturer Dr. Maurício Manhães, Ph.D., a professor at the Savannah College of Art and Design. The fourth workshop was guided by the cases of Ronald Heinrichs and Celaine Refosco. The fifth preparatory event was held under the guidance of Dr. Luiz Salomão Ribas Gomez, Ph.D. The sixth event closed the series of workshops and was attended by Professor Dr. Nelson Pinheiro Gomes, a specialist in trends studies.

All workshops brought a similar structure, interspersing theoretical content with practical activities in order to develop participants' creative and experiential skills. In addition, the oral presentation of the activities to the rest of the teams favored familiarity and the development of the ability to present and defend one's own ideas.

Intercutting the occurrence of the workshops, the *SCMC Cotidiano* program was included, which consisted in the visitation to some companies participating in the project. In this platform, the interested parties were able to know the structure, the products and the daily routine of important companies such as Cia Hering, Dudalina, Estudio Elaiá and Audaces.

2.1.2 *Daily routine at the design camp*

The immersion began on November 23, 2015, Monday, with the reception of the selected participants and the presentation of the structure of the event. Rules and information of conduct were also passed on, which are important for the maintenance of a good environment among the Campers. Following, a first welcome conversation presented the representatives of the companies and reference people in the organization of the event. Then the opening was given with the division of the teams and their respective professors of support. The first challenge was also launched, which was organized by the company Lancaster.

The teams organized themselves internally, according to their goals and deadlines, distributing the activities among the members and prioritizing those with greater urgency or complexity. The meetings of validation or orientation with the companies were followed in a natural way to the emergence of doubts or solutions.

Although it was a very heterogeneous group in several aspects such as training, skills, age, level of education or training, it was possible to observe significant professionalism among Campers. Most of the time there was a good interaction between them, healthy discussions in the search for better results, suggestions for improvement of the presented ideas, or disapproval of the resulting solutions.

The opportunity to encourage less conventional solutions has contributed to the satisfaction and profitability of companies and their representatives. The companies have brought real problems to the challenges of Design Camp LOGO UFSC, which can return in solutions more useful and more adapted to to reality. This factor also contributed to the Campers having a direct contact with the reality lived in the market, having experiences in various branches of the creative industry, living with their problems and difficulties.

Regarding deadlines achievement, the most agitated days were the third and fifth, due to the large number of validations, and the proximity of final de liveries, respectively. In these two moments, as well as on the fourth day – although with more tranquility, the Campers continued working through the night, aiming at better results to the challenges.

Parallel to the work of the teams in the solutions to the challenges, there were activities open to the community. *Moulage* workshops as a creative resource for design, taught by Roger Arend; Fashion technical drawing with the software Audaces Idea (Software developed by the company Audaces, exclusively to meet the fashion market), made by Naiane Cristina Salvi; Creativity Workshop, with Professor Rose Zanchett; *Jeansvendando*, with Priscila Locatelli, about the segment of Jeanswear; Fashion Marketing with Luana Pacheco; And the workshop Handmade Sewing Stitches, given by Eliana Gonçalves and Lucas da Rosa. There were also workshops to exercise the body and make the Campers release the pressure, such as the Acrobatics Circus workshop, offered by Monica Costa.

There were also talks presented by important industry names like Rui Hess de Souza, talking about the 10 years of the *SCMC*, and by important design names such as Meredith Levine, with the lecture "Future-Proofing For A Digital Age: Using Social Media To Build A Brand", and Renato Cunha, who presented the theme "Sustainability and wearable technology: the two strands of the future of fashion".

In addition to the lectures, there were some talk-shows, with companies participating in the event, such as Automatisa, which provided laser cutting technology for the use of Campers; the company Nanovetores, also in the technology sector, which has a series of projects and products intended to the textile market; and Bortolini, the company responsible for the construction of the workstations occupied by the participants. There were also round tables with representatives of the companies involved in the project, guided by the themes "Design reaches the chain" and "Our label has the force".

To break the intense routine of work and to promote the relaxation of the Campers, there was a game of treasure hunt. Requested by the team of employees, the first team that gathered the amount of stamps from the participating companies won the race and received as the prize a box of energy drinks, to renew the forces and to keep working.

Deliveries came to an end on Saturday, November 28, 2015, around 2:00 p.m. All the projects and prototypes were put for presentation in niches, for later evaluation by the companies, the professors and the team of collaborators.

The observed points were different for each group of evaluators. The representatives of the companies based their evaluation on the execution and the solution given to the challenges, guided by the achievement of the proposed objectives and by the viability, creativity and innovation presented in the results. The companies returned an individual evaluation of each Camper, choosing one in particular, that had better discernment in the development of the activities. The professors evaluated the teams in general, towards technique and finishing presented. The collaborators, in turn, considered the individual aspects of behavior, conduct during the event and the autonomy of the Campers in the search for aid or exploration of possibilities.

Figure 1. Facilities – accommodation and food areas.

Figure 2. Workstations.

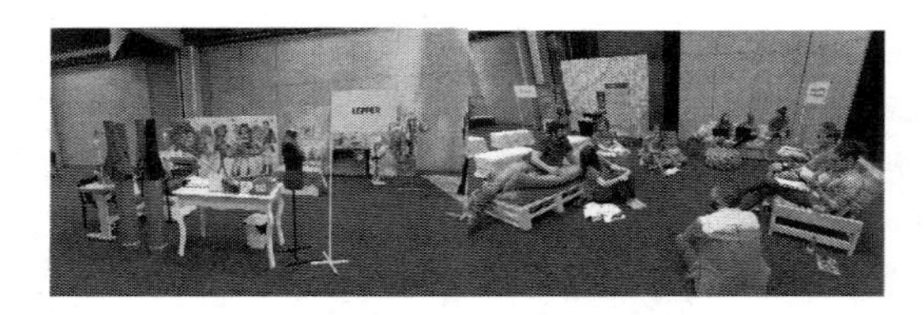

Figure 3. Installation of company and area of coexistence.

3 APPLICATION SCENARIO

To be creative, a place need not be crazy, eccentric and located in Northern California. The prerequisite is a social and spatial environment where people know they can experiment, take risks, and exploit all their skills. (BROWN, 2010, p.30).

The convivial environments designed for the Design Camp mainly took into account the need for creative exploration of the participants. In this sense, from the familiarization and intimacy between the members of the teams to the construction of the spaces, they were designed to highlight these needs (SALVI, 2017).

The Campers were accomodated at *the Governador Luiz Henrique da Silveira* Convention Center in Florianópolis-SC, space provided by the Government of the State of Santa Catarina. Located in the Sapiens Park complex, it was inaugurated during the event Design Camp LOGO UFSC, this being the first official event in the new convention center of the capital. The site has more than 2000^2 m divided into two floors.

Design Camp LOGO UFSC participants received all necessary support regarding accommodation and food (Figure 1) during the six days they were camped there. Each Camper received a kit for accommodation consisting of a tent, a sleeping bag, a pillow, a blanket, a bath towel, a face towel and a bathrobe. Other gifts provided by partner companies also accompanied the set of items.

The tents were concentrated in a space protected from daylight, near the bathrooms and the dining room. For the hygiene of the participants, a container with showers and sinks was installed. The dining area was composed of tables and chairs, and some armchairs for resting. The event offered seven meals a day. In addition to the main meals – breakfast, lunch and dinner, the participants could count on intermediate meals, as well as a table with coffee and biscuits available day and night.

The work environment was designed by the company Bortolini Corporate Furniture using planned furniture, thinking about the best use of space, and making it pleasant and facilitating the creativity. In a wide area, the structures (Figure 2) were installed for the Campers on one side, and on the opposite side were placed the tables for occupancy of the representatives of each participating company.

Around the workstations were arranged modular armchairs, for relaxation or meetings between the teams. Facilities developed by the companies with spaces and structure that could facilitate and complement the activities such as drawing materials and other utensils also composed the environment.

In the case of a majority of textile and fashion companies, a sector was set up with sewing machines, designed for the production of prototypes and rapid

Figure 4. Lancaster challenge results.

tests. There were also available for use of the Campers printers, computers, office supplies, a laser cutting machine and wireless internet connection covering the Event Center.

Figure 5. Result for challenges of the companies Loa, Altenburg and Audaces.

4 RESULTS

According to the proposed objective, different classifications were assigned to the challenges presented by the companies. The projects graded as initial returned descriptive solutions and basic studies to support the feasibility of the proposal. To the projects classified as intermediates, fashion collections were presented in illustrated drawings and contemplating the basic items of a collection or a product, such as indication of materials, color charts, sketches and semantic and conceptual orientation of the collection. The projects classified as advanced resulted in complete projects with a theoretical support of the researches, sketches, and prototypes of the final product.

Several proposals presented by Campers as solutions to the challenges of the companies were considered innovative and with market value. It is worth mentioning the results presented by the Campers for the companies Lancaster (figure 4), Loa, Altenburg and Audaces (figure 5).

The company Lancaster operates in the textile printing industry and asked the Campers to develop prints with themes related to Central America. Below are some of the proposals developed by the students (figure 4).

The result presented below involved the challenges of the companies Loa, which operates in the manufacture of socks and underwear, and the company Altenburg, which is inserted in the market of bed and bath. Both brought to the Campers challenges related to the companies' textile waste. These are basically leftovers from the process of confection of the products, large volumes of small tangled threads and small pieces of cloth. The developer team was able to combine the challenge of the company Audaces (from the technology area of software development for the garment industry) that sought creative solutions to the need for social projects. In this way, the team developed the products that could use the waste as a filler, could be produced by poor communities and the final product would sent to public schools and daycares. Below is one of the proposed products (figure 5).

On the last day of the event the four teams delivered a total of 96 projects to solve the 24 challenges presented. The projects were evaluated by three of the four groups of actors involved: representatives of companies, professors, and collaborators. Each group evaluated the projects under a certain perspective contemplating technical, economic and scientific criteria. The employees were also entrusted with evaluating the performance and behavior of the Campers. The companies were also able to indicate the Campers of better prominence and resourcefulness in the execution of the projects.

After the results were determined, the winner projects, the winner students, as well as the team with the best score were announced.

The prizes awarded by the partner companies, such as bicycles, watches, sunglasses and sewing machines, were given to the most well-ranked students. The team with the best score received the privilege of being the cover of a Trendbook, which will bring together the best works presented by teams at LOGO UFSC Design Camp to be developed by FIESC (*Federação das Indústrias do Estado de Santa Catarina* - Federation of Industries of the State of Santa Catarina) and will be nation-wide available. After the disclosure of the results and the awardings, the Campers were able to celebrate the achievements and overcome a week of intense work.

All those involved in the various contexts of the event, from students, professors, businesspeople, LOGO UFSC team and *SCMC* team, were able to collect from this experience significant professional and personal growth, with knowledge and maturity of great value, making them professionals better qualified and prepared to promote Design in Santa Catarina as a differentiating and quality value.

5 FINAL CONSIDERATIONS

Considering the pioneering role of LOGO UFSC Camp Design in promoting a unique context for generating creative and innovative solutions to the real problems of textile companies, many were the considerations regarding the format, execution and results of the model.

Regarding the preparatory process (*SCMC Conecta*) it was pointed out by students and professors involved the need for a greater approach of technical skills to solve the challenges proposed during the event. During the development of the projects, the students' different abilities were evident within their specific backgrounds, which due to the several limiting factors of the process did not always translate into sharing and leveling of knowledge among group members or between groups. For the improvement of the proposed model it is worth the reflection on the content of the workshops in relation to the results expected by the representatives of the companies.

Regarding the format of the final event, it is important to highlight the positive impact of formatting the physical space for the achievement of the project objectives. The provision of furniture, the constitution of spaces with specific purposes (work space × living space), the availability of physical resources (machines, tools, work tools) and other aspects of this kind were crucial in the good development of the works, deserving specific attention and remaining as a suggestion for the development of future works.

Still in the development of the activities proposed in the final event it is necessary to denote the crucial role of the group of collaborators (LOGO-UFSC professors and students) both in the technical support of the students and in the motivational aspect as a whole. Considering the limitations imposed by the format – total immersion, team definition, deadlines, expectations regarding deliveries, the whole follow up by this group of volunteers was of utmost importance in solving students' doubts, in the technical improvement of the projects and in maintaining the positive atmosphere favorable to the creative process.

Another elementary point in the execution of the proposed creative process was the involvement of the participating companies in the projects developed. In the evaluation of the results obtained by entrepreneurs who were actively present in the process and of those who did not follow the development of the tasks, it is possible to identify a notorious difference both in the quality of the solutions presented and in the relevance of these results in the generation of insights for the development of innovation in the organizational culture of their companies.

It was possible to observe that the immersive environment contributed to the maintenance of the focus and concentration of Campers. The accommodation at the event facilitated the dynamism and distribution of time according to the interest and strategy of each team, having the freedom to extend working or resting times. The ease of access to the group of company representatives, as well as to the professors, facilitated more aligned and coherent results to the objectives and facilitated the co-creation between the members of each team, as well as the other groups of actors involved.

In the improvement of the application of co-creation and in the development of other methodologies focused on innovation, it is proposed to carefully evaluate each stage of the experimental process proposed by Design Camp LOGO UFSC. For future work, it is suggested to deepen the questions about how academia and the market can benefit from this type of experiential practice in the improvement of its designs.

REFERENCES

Águas, S. 2012. *Do design ao co-design: uma oportunidade de design participativo na transformação do espaço público*. On the W@terfront, n. 22, p. 57–70.

Brown, Tim. 2010. *Design Thinking: uma metodologia poderosa para decretar o fim das velhas ideias*. Elsevier, Rio de Janeiro.

Dziobczenskia, P. R. N.; Lacerda, A. P.; Porto, R. G.; Seferin, M. T.; Batista, V. J. 2011. *Inovação Através do Design: Princípios Sistêmicos do Pensamento Projetual*. Revista Design e Tecnologia, v.2, n. 03, Universidade Federal do Rio Grande do Sul, Porto Alegre. 54–63p.

Gonçalves, Eliana. 2015. *Santa Catarina Moda e Cultura: parceria entre UDESC e empresas*. Florianópolis: Dissertação de Mestrado – Centro de Ciências Humanas e da Educação, Universidade do Estado de Santa Catarina.

Macedo, R. P.; Marques, C. S. da E.; Ulbricht, V. R. 2014. *Inovação e gestão de ambiente imersivo governamental e corporativo*. In: Proceedings of World Congress on Communication and Arts. [S.l.: s.n.]. v. 7, p. 55–59.

Masi, D. D.; Manzi, L.; Figueiredo, Y. 2003. *Criatividade e grupos criativos*. [S.l.]: Sextante Brazil.

Monteiro, Beany Guimarães. 2008. *Design e Inovação Social: práticas de atuação e uso do design em contextos locais*. VI Seminário de Metodologia de Projetos de Extensão. Universidade Federal de São Carlos, São Carlos.

Mozota, Brigitte Borja de. 2011. *Gestão do design: usando o design para construir valor de marca e inovação corporativa*. Porto Alegre: Bookman. 343p.

Pinheiro, I. R.; Merino, E. A. D. 2015. *Os 4 vetores da inovação: Um quadro de referência para a gestão estratégica do design*. Estudos em Design, v. 23, n. 2, p. 75–101.

Queiroz, Amanda, Souza, Richard Perassi Luiz de, Rech, Sandra Regina. 2012. *Por uma abordagem qualitativa dos dados: A pesquisa de tendências embasada na Grounded Theory*. E-periódico Moda Palavra, Ano 5, n.10, jul-dez, Florianópolis.

Robinson, Ken. 2012. *Libertando o poder criativo: a chave para o crescimento pessoal e das organizações*. Editora HSM, São Paulo. 780p.

Sanders, E. B.-N.; Stappers, P. J. 2008. *Co-creation and the newlandscapes of design*. Co-design, Taylor & Francis, v. 4, n. 1, p.5–18.

Sapper, S. L. et al. 2015. *Da ideia ao conceito de produto: o uso de técnicas criativas combinadas para auxiliar no processo de desenvolvimento de novos produtos de design*. Estudos em Design, v. 23, n. 1, p. 49–60.

Sebrae – *Serviço Brasileiro de Apoio às Micro e Pequenas Empresas. Design no Brasil*: relatório 2014 do setor de design. Brasília: Sebrae, 2014.

Silva, Edna Lúcia da; Mebezes, Estera Muszkat. 2005. *Metodologia da pesquisa e elaboração de dissertação*. 4. ed. rev. atual. Florianópolis: UFSC. 138p.

Souza, Richard Perassi Luiz de. 2005. *Roteiro da Arte na Produção do Conhecimento*. EDUFMS, Campo Grande. 23–37p.

Tarachucky, L.; Gomez, L. S. R.; Merino, E. A. D. 2014. *A utilização dos métodos cocriativos para a criação de marcas territoriais – o caso de alvito*. Universidade Federal de Santa Catarina.

Velada, A. R. R. et al. 2008. *Avaliação da eficácia da formação profissional: factores que afectam a transferência da formação para o local de trabalho*. 192p.

Textiles, Identity and Innovation: Design the Future – Montagna & Carvalho (Eds)
 ISBN 978-1-138-29611-4

Genderless clothing issues in fashion

B. Reis
FibEnTech, Faculty of Engineering, University of Beira Interior, Covilhã, Portugal

M. Pereira
FibEnTech and UNIDCOM, Faculty of Engineering, University of Beira Interior, Covilhã, Portugal

S. Azevedo
CEFAGE-UBI, Faculty of Social and Human Sciences, University of Beira Interior, Covilhã, Portugal

N. Jerónimo
LabCom.IFP-UBI, Faculty of Social and Human Sciences, University of Beira Interior, Covilhã, Portugal

R. Miguel
FibEnTech, Faculty of Engineering, University of Beira Interior, Covilhã, Portugal

ABSTRACT: This research aims to study and clarify ideas of genderless clothing, to contribute later in depth and contemporary knowledge about this subject. This study focuses on society's current issues and aspects, regarding fashionable gender issues, the most common types of human bodies, and addressing key consumer issues and facts. Sociologists draw a distinction between gender and sex. The purpose is to determine if genderless fashion will be a paradigm or a trend. The object of study in this research is to analyse genderless clothing in all its complexity, which encompasses an analysis of the phenomenon in many areas of study such as sociology, fashion design and consumer behaviour. Attributing consequently, an interdisciplinary character in this research. Considering historical-social, socio-economic and cultural factors and how genderless clothing fits into the contemporary market.

1 INTRODUCTION

There is already some research about genderless clothing, but these existing studies are superficial and underdeveloped. A brief research verified that this subject has always been present for the last decade, for at least one season per year (of the two that exist today).

Despite being a contemporary topic, there are no scientific and reflective works addressing genderless fashion focusing simultaneously on sociology, fashion design, and marketing, hence the relevance and contribution in producing knowledge about this phenomenon that is gaining an ever-increasing importance.

For Simmel (2008), the first issue is considering society as a field of tensions and interactions, that fashion itself, its expression, its variations, its rhythms, its mechanisms, its ambiguity, its meaning, and in its specific place within social reality is a privileged manifestation, because it is always present as a factor of both socialization and individualization.

The reality of genderless clothing is also a national reality, influenced by the international context of androgynous fashion, about which there is little research.

The implications of this research can be useful because it will help to better understand genderless clothing and what remains to be done in relation to this type of clothing. The problem is fitting the same garment in both men and women. Thus, we formulate the hypothesis of a garment that adapts to the body of men and women, regardless of gender.

Although there are already some brands that have made some capsule genderless collections, it is also necessary to analyse these collections and how the market research was undertaken.

In this sense, this research aims to fill that gap, not only in fashion design and patterns, but also on literature. The results can be useful for the market players to know that this target exists in the market.

2 INTERDISCIPLINARY RESEARCH IN GENDERLESS CLOTHING

2.1 *An interdisciplinary research*

For this context of genderless clothing, it is important to approach not only the fashion design research field, but also the fields of research like sociology and marketing (studies about the consumer), as well as several sub-themes inserted in these scientific fields.

According to Kawamura in 2011, several academic disciplines in the social sciences and humanities have contributed to our understanding and analysis of clothing and human behaviour. Consequently, one must regard fashion studies as an interdisciplinary area of knowledge that has emerged from theories and research findings that cross traditional disciplinary boundaries, thus becoming possible to integrate and incorporate multiple and different methodological strategies.

2.2 *The gender in sociology*

For Schouten (2011), studies on gender, in the sociology of gender, are not residual scientific categories, but rather a specific mode of observation and analysis of social reality. It attends to the ways of involving people in social relations and their ideas, depending on whether they are male or female. Gender must be studied because of the great importance of this principle in the organization of social life, but also because it is at the root of many situations of disparity and injustice.

According to Giddens (2010), at the sociological level, a distinction is made between sex and gender. Sex refers to biological differences between the male and female bodies, whereas gender concerns psychological, social and cultural differences between men and women.

The socialization of gender consists of learning the gender roles with the help of agents such as the family, the school, and the media. It is believed that gender socialization begins as soon as a child is born. Children internalise the norms and expectations that are supposed to correspond to their biological sex. In this sense, they adopt "sexual roles" and the masculine and feminine identities (masculinity and femininity) that accompany them. Usually, we do not even notice them – precisely because they are too diffuse. They have been introduced from birth.

Our conceptions of gender identity, as well as the sexual attitudes and inclinations associated with them, are formed so early in life that we consider them, most of the time, as certain in adulthood. But gender does not exist simply; All of us, as some sociologists claim, "doing gender" is in the social interactions that we establish daily with others humans (Giddens, 2010).

Anthony Giddens in 2010, says that before we review these contradictory approaches, it is necessary to make an important distinction between sex and gender. In general, sociologists use the term sex to refer to the anatomical and physiological differences that define the male body and the female body. By contrast, gender is understood as the psychological, social, and cultural differences between males and females.

Gender is associated with socially constructed notions of masculinity and femininity. It is not necessarily a direct product of the biological sex of an individual. The distinction between sex and gender is fundamental since many differences between men and women are not of biological origin. Sociological interpretations of gender differences and inequalities have assumed contrasting positions on this issue of sex and gender.

However, some authors argue that the aspects of human biology – from hormones to chromosomes, from brain size to genetics – are responsible for the congenital differences in behaviour between men and women. These differences, they say, are visible in all cultures, implying that natural factors are responsible for the gender inequalities that characterize most societies.

Dorfles (1984) says that the phenomenon of clothing and the fact that it lies on the very deep dimensions of our character, our humour, our way of being in the world, it always seemed of considerable importance, not only for oneself but also for others. In other words, the phenomenon of clothing, or indeed the fashion phenomenon, is in contact and is also a consequence of the experiences of our daily life (and all that includes). This is quite significant not only in individual terms but in general societal terms.

This same author (Dorfles, 1984) also mentions that fashion is not only an insignificant and superficial phenomenon, but the mirror of habits, the psychological behaviour of the individual, profession, political orientation, and taste. Thus, turning fashion fashionable, considering its social, aesthetic and cultural weight.

2.3 *Gender issues in fashion*

In 1905, Simmel (2008), said that fashion expresses, in the most visible and concrete way, the essentially dialectical and dynamic reality of society, made of inevitable conflicts between individuals and the multiple and different social formations, between individuals and groups or classes.

In a brief review of fashion history, there are notable male detailing in women's clothing in the nineteen-twenties and later in the nineteen-sixties.

According to Seeling (2000), everything began with Paul Poiret (1879–1944). Considered the first designer, he provoked a scandal with his skirts-trousers. In 1900, Poiret dressed Sarah Bernhardt (actress) in pants, for his first male role in L'Aiglon, thus being the first woman to wear pants in public. In 1913, Gabrielle Chanel presented her first sports models in jersey fabric. Of broader and relaxed cut, this material allowed freedom of movement.

The World War I freed most women from the choice of what to wear; there were not many options: work clothes, uniforms and mourning suits. In 1931, the mayor of Paris demanded that Marlene Dietrich left town for daring to take to the streets with a man's suit. In 1966, the women's tuxedo of Yves Saint Laurent, provoked a new scandal, nevertheless remaining forever connected to its name (Seeling, 2000).

Monneyron (2006) also addresses the question of whether, or not, a unisex fashion exists, also referring to the historical facts of Poiret, Chanel and Yves Saint

Laurent, mentioning later the dysfunction of sexual dimorphism between woman and man. With this brief approach, in the present research, one can address a contemporary representation of androgyny in clothing, its popularity and what impact it has on society today.

Considering the assessments of Wilson (1985), she mentions anti-cultural clothing and the way it expresses revolt, and she also refers scandalous clothing, which expresses only ambiguity. At first sight, the androgyny of rock stars, such as David Bowie, scandalised. Thus, there were boundaries of boldness when a man began to wear make-up or a woman to shave the head. Yet, the same author (Wilson, 1985), refers to fashion journalist Suzy Menkes (*The Times*, May 1, 1984) who wrote about the 1984 fashions of "androgynous undergarments" and the masculinity of women's clothing, suggesting that they were "the maximal affirmation of fashion about the sexual revolution".

In this way, the "*inverted clothing*", which is making way for the unisex departments "*breaking genders*", in the most sophisticated fashion stores, is just another madness of fashion, and the market that these styles appeal to is the market of affluent heterosexual couples, for whom androgynous clothing symbolizes not an attack on the gender, but only a reassertion of middle-class unity.

Part of the strangeness of clothing is that they link the biological body to the social being and the public to the private. This makes clothing difficult because it forces us to recognize that the human body is more than a biological entity. It is a cultural organism, even a cultural artefact, and its own borders are not well defined (Wilson, 1985).

2.4 *Pattern design in genderless clothing*

To understand the pattern of genderless clothes, firstly it is necessary to understand the pattern proceeded for the feminine and masculine gender, considering the tables of measures and sizes used and their respective types of bodies. Thus, it is necessary to take an ergonomic and anthropometric approach in this investigation.

Amstrong (2010) mentions which is the ideal standard figure, depending on whose measuring standards are based on and who is listening to whom. That is, the figure has evolved from consumer feedback to the buyer, from buyer to manufacturer, and from manufacturer to modeler.

Besides doing an analysis of the forms of the bodies, it also does the postures that they have. In this research, it is necessary to verify which anthropometric studies of the human body exist, as well as the following three, and how they will later influence the methods and research work in this ongoing investigation.

Sheldon (1940) apud Iida, (2005: 104) conducted a study of 4000 American students and concluded that there are three basic physical types: ectomorph, mesomorph, and endomorph, all with individual characteristics. Most people do not belong to a unique type, most show mixtures between them, as shown in figure 1.

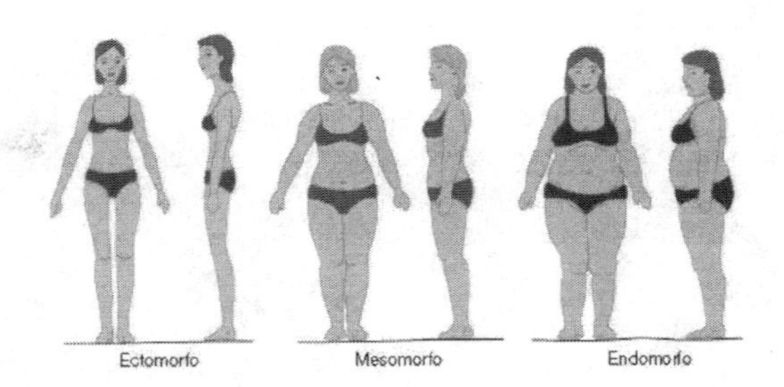

Figure 1. Types of physical bodies: ectomorph, mesomorph, and endomorph. Source: Sheldon (1940) *apud* Iida, (2005, p. 104).

However, in this work, it will be necessary to address other studies, regarding the diversity and changes in the types of human bodies, whether male or female.

3 FASHION CONSUMER

3.1 *Introduction*

Following a sociological and fashion design approach, it is essential to mention how genderless clothing will avenge the market. For this, it is necessary to see which brands and designers produce genderless clothing and what strategy each one uses.

Bohdanowicz and Clamp (1994) state that it is important to know the impact of the social and economic environment on the fashion marketing process, since few other industries are so powerfully and directly affected by social and economic trends as the fashion industry.

The concept of a '*satisfying exchange relationship*' is also of major importance. Marketing involves the exchange – usually of money, but also of time and effort, in return for goods. Sometimes people may be willing to invest more time or money in acquiring something that is highly desirable because they feel it has a greater value derived from its status as a fashion item.

A fashion item is worthless when it is no longer in fashion, as value is implicit in the item being fashionable. At the heart of the marketing strategy, concept is the appreciation of the need to satisfy customer's needs, wants and demands.

Besides, the factors that differentiate fashion marketing from the marketing of other types of goods or services can be broadly classified into three areas: strong influence of environmental pressures, time constraints, and the role of buyers.

3.2 *Genderless clothing consumption and the market*

To Pinto (2015), the development of genderless clothing is a fashion business attempt, a way to reach different audiences, responding to the needs of an increasingly fluid market niche.

According to Monneyron (2006), it raises an issue, which is also pertinent to answer in the present investigation, if and in what way the clothes or the dress determines our behaviour. Consumer behaviour will be something to consider in this investigation.

English (2007), in his conclusions of one of his chapters of "Cultural history fashion, 20th century", leaves several questions in which one adapts in this context: "As with the evolution of modern marketing strategies, in a product that has more commercial value than the product itself?" this is one of the questions that ends up fitting into the present investigation.

In fashion, to produce genderless clothing, it is necessary to verify the shapes of the men and women bodies and their measurements. For the modelling of clothing practiced for current stereotypes doesn't apply to many people and is only produced for a niche market.

Furthermore, it was possible to establish an association in physical differences between men and women. Knowing that there is a niche market for genderless clothing, thus enabling the creation of a business model for this target.

3.3 *Consumer identity*

The identity ambivalence referred by Davis (1992), where he discusses the relationship between fashion, culture, and identity, through the elements of clothes comes to share in the work of ambivalence management as much as does any other self-communicative device.

The most common identity ambivalences found their way into those representations of the self that address core sociological attributes of the person, the so-called master statuses, meaning the age, gender, physical beauty, class, and race. Ambivalence over how to enact the gender roles, in the social class identification, and the sexuality (Davis, 1992).

Therefore, it is important for this ongoing research and investigation to consider these mentioned attributes as raw material for consumer identity. However, it is significant to focus on which social identity the consumers are inserted in, not forgetting to mention how some consumers' exclusivity (considering that we live in fast-fashion and mass cultures society) achieve the sense of individuality, essential to the ideia of consumer identity.

3.4 *Consumer behaviour*

In Fashion-ology, according to Kawamura (2005), the consumption pattern changed with the advent of mass consumption, along with mass production. A clear division was established between the activities of production and those of consumption. With the industrial revolution came the consumer revolution which represented a change in tastes, preferences and buying habits.

The consumers have attributes and factors that are essential for this investigation. Some of these essential factors and attributes are indicated by Easey (2009). The role of the consumer's behaviour in fashion marketing provides a range of concepts to help fashion marketers think about their customers.

In the fashion consumer's decision process, that involves types of consumer decision, types of customer, a model of consumer behaviour; there is also a psychological process (consumer attitudes, consumer motivation, consumer personality) and the sociological aspects of consumer behaviour (social groups, opinion leadership, the family, social stratification, geodemographic, and the diffusion of innovation) (Easey, 2009).

Kawamura (2005) also says that the cultural meaning of consumer goods is shifting. Meaning that is constantly flowing to and from its several locations in the social world, aided by the collective and individual efforts of designers, producers, advertisers, and consumers. Contemporary culture has been associated with an increasingly materialistic or fetishist attitude, and the symbolic dimension of consumption is increasingly becoming important.

3.5 *Apparel attributes*

A study by Pereira et. al (2009), about the relationship between apparel attributes and advertising on consumer buying behaviour, refers that the consumer behaviour approaches and the tendencies in contemporary advertising display the reflux of its former pedagogical and constructive dimension.

Advertising's classical model – the famous copy strategy – insisted on a message that praised the functional or psychological benefits of one product. Also, the product's attributes play an important role in marketing communication for both the consumer and the marketer alike. The study concludes that advertising in fashion magazines influences the following attributes: used materials, style, durability, cut, treatment, quality, satisfaction, necessity, and function.

4 OTHER STUDIES

This ongoing research led a previous investigation, showing that there are other studies in this field we should consider.

4.1 *Unisex apparel as a response to gender stereotypes*

According to Pinto (2015), the development of genderless clothing is a fashion business attempt, a way to reach different audiences, responding to the needs of an increasingly fluid market niche. Still, in a study by the same author in 2015, the theme of unisex clothing as a response to male and female stereotypes is a clarification on the creation of unisex clothing in response to male and female stereotypes.

The author (Pinto, 2015) also mentioned that androgyny and the unisex are present in the main current trends of fashion, confusing and tearing the tenuous barrier between the masculine and the feminine. A historical analysis is made, a brief synthesis of the body related to Fashion and asks the question: "Unisex, a paradigm or trend?" (Pinto, 2015). However, he concludes his research without giving a concrete answer to the question. Given the complexity of the topic and the need for a greater depth of research carried out indicated in the suggestions for future research.

4.2 *The trans as a fashion phenomenon*

Previously, Filipe Godinho, in his 2012 research entitled "The trans as a fashion phenomenon", encompassed various scientific fields such as Fashion Design, Psychology, Philosophy, and Sociology. The author, with his study, intended to reflect on the motivations that make this subject a reference in the most varied kinds of manifestations. In his state-of-the-art review, he made a contextualisation of the trans phenomenon, then a brief historical-cultural report of the theme of androgyny since ancient times, based on the vestiges found. The author (Godinho, 2012), also makes a social reflection, taking as a base of analysis the perspective of philosophers, psychologists, and sociologists who have dedicated or still dedicate part of their career in the case studies related to the object of study.

4.3 *The gender issue*

In January, 2017, the printed American National Geographic addressed the more identitary side of existing types of genres, the special issue about gender revolution. The gender issue prepared a glossary from the book "The Teaching Transgender Toolkit", and it is possible to verify that there are not only the masculine and feminine gender, but it's clear that there are at least 13 types of gender classifications (officially), being: agender, androgynous, cisgender, gender binary, gender conforming, gender dysphoria, gender expression, genderfluid, gender identity, gender marker, gender non-conforming, genderqueer, and nonbinary. This National Geographic issue defines what types of gender are and what differentiates them. Beyond these 13 types of gender classification, we cannot forget to consider on this ongoing investigation the genderless and the unisex concepts.

4.4 *A case study of women wearing men's clothing*

According to Coutinho (2011), a case study belongs to a type of qualitative plan, which, in addition to understanding and describing in depth the phenomenon investigated, is an intense study of a single subject or group about an emerging problem. The expected results are to understand explanatory factors of behaviours; identify personalities/socialization traits; test and discover concepts; coding cases and discovering patterns.

In a case study conducted in 2017 by Reis (in press), it requested three women (not considering their gender) to wear men's clothing of male gender. Asking some questions about what they were wearing, regarding their appearance and the aesthetics of the piece they wore; how they felt when wearing that piece; whether they would be used daily or not; if they forgot the challenges they had dressing that piece of clothing they bought; and if they were confused when tightening the piece of clothing. In this study, we considered some characteristics such as age, height, stature (physical) and what sizes of clothes they currently wear.

In the final considerations of the study we verify that, with the given answers, it is possible to have a niche market. That is, although there are some anatomical differences between women and men, it is possible to have only one garment for both, thus not constituting an antagonistic subject. The study showed that there were pieces that fit perfectly.

Some negative aspects are the background of the people who wore the clothes in question not being considered (what they do, what they studied, and their level of education); only women were considered (referring to sex as a human being). The authors of the study did not consider the sizes of the pieces of clothing in relation to the people who wore the pieces of clothing.

After this case study, the author felt the need to put the term unisex in his research. The term unisex is defined in the dictionary as an adjective of designed or suitable for both sexes; not distinguishing between male and female; undifferentiated concerning biological sex.

5 FINAL CONSIDERATIONS

Recognising the importance of the theme, the present research is considered relevant, being a current and multidisciplinary topic addressing several fields of knowledge. Considering historical-social, socio-economic and cultural factors, it is possible to understand how genderless clothing fits into the contemporary market.

It is most likely that gender is a social construction that is acquired throughout our experiences in everyday life.

Although we live in a mostly binary system, there is a diversity of other genders besides feminine and masculine. This is due to the fact many people are looking for their own individuality and identity, even when it goes against the mass culture.

In a final consideration, with these latest research examples, according to Dorfles (1984), many political, economic and cultural events influence, directly or indirectly, fashion and customs and can themselves become "fashionable".

However, it is still not possible to determine if genderless fashion and clothing is just a trend or will be a paradigm, yet this issue will remain in this research in the future.

Considering the above-mentioned study, despite its underplayed aspects, it is necessary to consider them in future studies in this field, being a major contribution to the progress in this ongoing research. In the future of this research, we will include testing patterns in clothes, be it women's or men's garments.

REFERENCES

Amstrong, H.J. (2010). *PATTERNMAKING for fashion design*. 5th ed. PEARSON, New Jersey.

Bohdanowicz, J., Clamp, L. (1994). *FASHION MARKETING*. By Routledge, New York.

Coutinho, C.P. (2011). *Metodologia de Investigação em Ciências Sociais e Humanas: Teoria e Prática*. Edições Almedina, Coimbra.

Davis, F. (1992). *Fashion, Culture, and Identity*. University of Chicago Press.

Dorfles, G. (1984). *A MODA DA MODA*. Edições 70, Lisboa.

Easey, M. (2009). Fashion Marketing. Third Edition. WILEY-BLACKWELL, A John Wiley & Sons, Ltd., Publication.

English, B. (2007). *"A cultural history of fashion in the 20th century: from the catwalk to the sidewalk"*, Berg Publishers.

Geographic, N. (2017). *Gender Revolution: Special Issues the Shifting Landscape of Gender*. January 2017. Vol. 231. N° 1. Official Journal of the National Geographic Society.

Giddens, A. (2010). *SOCIOLOGIA*. 8ª edição, Fundação Calouste Gulbenkein.

Godinho, F. (2012). *Trans como fenómeno de moda*. Universidade Técnica de Lisboa.

Iida, I. (2005) *Ergonomia: projeto e produção*. São Paulo: E. Blucher.

Kawamura, Y. (2005). *Fashion-ology*. U.S.A., Berg.

Kawamura, Y. (2011). *DOING RESEARCH IN FASHION AND DRESS – AN INTRODUCTION QUALITATIVE METHODS*. U.K., Bloomsbury Academic.

Monneyron, F. (2006), *"50 respuestas sobre la moda"*, GG moda, Barcelona.

Pereira, M., Azevedo, S. G., Ferreira, J., Miguel, R. A. L., Pedroso, V. (2009). The Relationship between apparel attributes and advertising on consumer buying behaviour. *MPRA Paper No. 11908*, posted 3 December 2008 16:08 UTC.

Pinto, D. (2015). *Vestuário Unissexo como Resposta aos Estereótipos de Género*. University of Beira Interior. Final version in pdf available by the author.

Schouten, M. J. (2011). *UMA SOCIOLOGIA DO GÉNERO*. Edições Húmus, Lda.

Seeling, C. (2000). *MODA – O Século dos Estilistas*. Könemann.

Simmel, G. (2008), *"Filosofia da Moda e Outros Escritos"*, Edições Textos & Grafia, Lisboa.

Wilson, E. (1985). *ENFEITADA DE SONHOS. MODA E MODERNIDADE*. Edições 70, Lisboa.

Textiles, Identity and Innovation: Design the Future – Montagna & Carvalho (Eds)
© 2019 Taylor & Francis Group, London, ISBN 978-1-138-29611-4

U.MAKE.ID fashion sourcing platform project

M. Pereira
FibEnTech and UNIDCOM, Faculty of Engineering, University of Beira Interior, Covilhã, Portugal

B. Reis, L. Pina & R. Miguel
FibEnTech, Faculty of Engineering, University of Beira Interior, Covilhã, Portugal

P. Rafael
Consultant in the Project U.MAKE.ID, Covilhã, Portugal

ABSTRACT: Technology, profitability, reliability and globalization are words that are increasingly present and that are important in the fashion industry. In this work, the sourcing in fashion industry is important to understand, not only, how the digital economy works, but how the fashion industry sourcing platforms can support the interests of companies. This study, aims to show the capabilities of the U.MAKE.ID platform and how this project can be a real platform for the fashion industry players whether they are factories, brands or suppliers. The U.MAKE.ID platform is a recent project that has been developed to serve some of the players in fashion industry. Consideration has been given to the needs and difficulties that each one of the players probably have in the management of the production process.

1 INTRODUCTION

When fashion brand tries to find possible factories or suppliers to produce their new products, besides the price, there are some other critical and important aspects used to do the sourcing in fashion, that should be mentioned and questioned if they are the best solution or not for the brand business interests.

Since the 1990s, the fashion industry has become increasingly international due to important phenomena such as the removal of the multi-fibre agreement in 2005, the economic crisis of 2008, and the saturation of mature and traditional markets. These phenomena have disrupted the traditional models of the industry, leading to greater importance of the final customer due to lower demand (Abecassis-Moedas, 2006) and stronger competition at the international level against global competitors (MacCarthy & Jayarathne, 2009).

In the same line of thoughts, and on the other side of the supply chain, there are the factories and suppliers that aim to have a more agile communication process with their actual and possible new clients to create a good, fast and easy business relationship.

The objects of this theoretical study are the sourcing in fashion; the B2B commerce and the digital economy that are a relevant and current topic and they all are connected. It is identified by the acronym B2B but its meaning is business-to-business, that is the commerce established between companies.

Nowadays the communication and the technologies go around with hands together so that the brands, the manufactures and the suppliers will do their jobs with more flexibility.

To help to understand the complexity and the magnitude of digital economy and understand the real importance of it in fashion industry, it was made in the next chapters a presentation about some existed theories regarding the digital economy, based on sourcing platforms with a business-to-business model and directed to the fashion field.

To finalize this theoretical study, it will be made an exposure about U.MAKE.ID platform project and its potentialities for fashion brands, factories and suppliers, regarding how this project can help to improve the business relationships between the players and to minimize the challenges in current way of sourcing.

2 LITERATURE REVIEW

2.1 *Digital economy in fashion industry*

According to Lu, in Reed (2014), the globalization has changed the traditional definition of international trade, through both the Internet and various forms of data.

It has been said that change is just part of nature. Therefore, it is natural to witness our environment changing at an unprecedented pace, particularly in what we regard today as digital economy that brings a plethora of challenges due to the disruptive nature of digital technologies but also brings plenty of promising opportunities for those who are aware and prepare to embrace the digital evolution (Kowalkiewiez et al. 2017).

The internet has been used to deliver interactive commerce and pontificated experiences by the designers, brands and retailers in a way to satisfy a consumer desire for profounder brand engagement.

The fashion industry is delineated by different, complex supply networks, both in terms of fragmentation of production activities and geographical dispersion of the actors involved (Macchion et al. 2015).

Moreira (2011) says that normally to achieve a good and viable purchase, the first consideration made by the buyers is the purchase efficiency in terms of cost, followed by the distance to the market.

When a firm competes in any industry, it performs a several discrete but interconnected value-creating activities, such as operating a sales force, producing a component, or delivering products, and these activities have points of connection with the activities of suppliers, channels, and customers (Azevedo et al. 2007).

Globalization of markets has pointed out the opportunities either in term of finding new selling markets or in term of finding new markets for the supply and production (Hernandez & Alejandro 2016).

Also, according to Kowalkiewiez et al. (2017) the prevalence of digital technologies is forcing organizations to re-imagine the way their business models are configured. The way their business processes are designed, and the way they work in what is known today as a digital economy. Smart companies leapfrog industry borders and sell additional products and services into existing markets, leaving established market leaders perplexes and behind.

2.2 *B2B digital platform*

Black & Edwards (2015), confirms that the traditional fashion design cycle, at the same time as being contested by new digital approaches, platforms and services maintains a stronghold unlike other sectors within the creative industries.

To the fashion system, the dynamics of the sector and the potential for innovation in terms of new business models expressed by firms, outline the framework within which defining strategies and processes of internationalization (Runfola & Guercini 2009).

On an industry-wide scale, many B2B sites exist where buyers, sellers, distributors, and manufacturers can offer products, submit specifications, and transact business. This popular form of online B2B interaction is called supply chain management (SCM), or supplier relationship management (SRM) (Shelly & Rosenblatt, 2012).

According to Lewis (2014), when a fashion designer decides to start a creative business, one of the last things on your mind is logistics. But, as everyone knows deep down, the boring bits are what keeps a business thriving. And when it comes to clothes, accessories or anything related to fashion, nailing your production process might not be the most glamorous task, but it's certainly the most vital.

As Pina et al. (2017) refers, one of the main goals of any B2B e-commerce is to sell more, but every company needs to define and prioritize its objectives so that the B2B strategy works. A market that is ever more open, turbulent and with high levels of negotiation demands every day more the junction of competition and cooperation.

According to Lu, in Reed (2014), from the business perspective of the fashion industry, the consumer is the leader in the supply chain, followed by retailing, apparel manufacturing, and then fiber/textile production. In this process, each step happens on a global scale, not simply within one country.

Because it allows companies to reach the global marketplace, B2B is especially important to smaller suppliers and customers who need instant information about market prices and availability, because, on an industry-wide scale, many B2B sites exist where buyers, sellers, distributors, and manufacturers can offer products, submit specifications, and transact business (Shelly & Rosenblatt, 2012).

Regarding Strugatz (2012), that interview Macala Wright that is digital marketing consultant, founder and chief executive officer of *"Why This Way"* and editor in chief of *"FashionablyMarketing.me"*, cites that the economic climate is a contributor to the rise of trade based websites, regarding that many companies no longer have the boundless resources to travel to Europe, Asia and the U.S. to seek and source new products. Because of recent global trends, for example high street retailers are under increasing pressure to sustain the low-cost business model whilst at the same time fulfilling the needs of an increasingly demanding consumer (Ryding, Vignali et al. 2015).

2.3 *Sourcing in fashion*

Regarding fashion brands and their goals and needs, is always relevant to understand how, where and who make the sourcing of new manufacturers or suppliers and how they preserve a healthy and professional relation with the ones that they already have.

Is crucial to know their methods of sourcing and try to understand how they work and which faults they're try to manage and solve and how a B2B sourcing platform could be a plus for their business.

For Christopher et al. (2004), the fashion supply systems are characterised by three critical lead-times: time-to-market, time-to serve and time-to-react. All three of these factors stress the importance of agility in fashion supply networks. Agility does, however, necessitate radical changes in organisational structures and strategies and a move away from forecast-driven supply. Market sensitivity, virtual integration, networked logistical systems and process alignment all become fundamental prerequisites to achieving the ultimate agility, a quick response capability.

Also, according to Christopher et al. (2004), agile supply chains are more likely to be information-based.

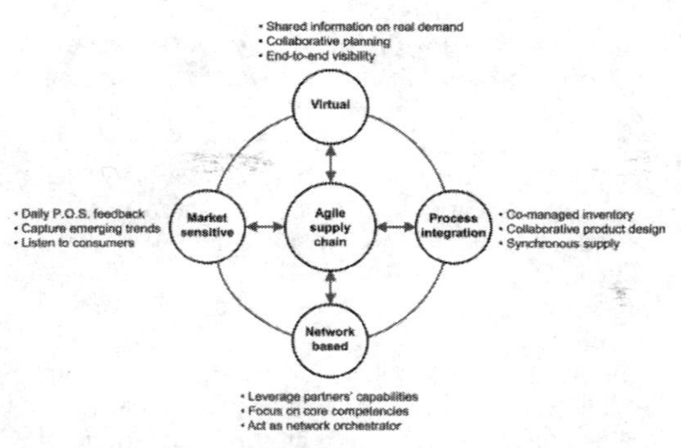

Source: Based on the model originally developed by Harrison, Christopher & van Hoek (1999)

Figure 1. Caption of foundations for agility in a fashion business.

By their very nature, fashion markets are instable and hard to forecast. Therefore, the need for agility. It has been suggested by Harrison et al. (1999) that an agile supply chain has several characteristics. Specifically, the agile supply chain is divided in: market sensitive (it is closely connected to end-user trends); virtual (it relies on shared information across all supply chain partners); network-based (it gains flexibility by using the strengths of specialist players) and process aligned (it has a high degree of process interconnectivity between the network members).

The figure 1 suggests that there are several practical ways in which these four key dimensions can be brought into play to create an agile supply chain for organisations competing in fashion industries. Considering each of these four dimensions in turn, many of observations can be made.

When observing fashion brands or fashion manufacturers, they are constantly looking for different methods of business approach because that can ease the sourcing process and consequently improving the function of the supply chain management to achieve the goal of increase their profits. That is, from a perspective of making the sourcing process more agile among the factories, the suppliers and brands in the fashion business.

3 U.MAKE.ID

Regarding Pina et al. (2017), the U.MAKE.ID project consists on a platform that works as an instrument for the improvement of processes which translates into economy due to the collaboration of users, presenting itself as a tool that can ease the productive process and streamline proven competences as well as a tool for internationalization of goods, services and knowledge.

3.1 *Project of a digital platform with a B2B economical model*

The project has innovated in the way that approaches the market, in the aggregation of several functionalities and in the improvement of processes having also three extra functionalities: the search engine of fashion brands as factories; a marketing and social and economic relations platform and the creation, development and production software.

Observing at the functionalities mentioned above, the pursued objective is the creation of a platform of integrated services that can be able to reduce the barriers to communicate with suppliers, building connections that can reduce the distance between conception and production of the different kinds of products, through the simplification of a group of processes from the sourcing to the production, simplifying the communication and organization between the different players.

The U.MAKE.ID platform will enable fashion brands/designers to understand in a simple and intuitive way all the steps of the productive process, since conception to the final product, allowing them to build their project in their own terms, indicating what the needs are so that from these they can find the relevant suppliers for their specific needs, turning the sourcing easier and also enabling a centralized and shared organization between the different fashion players inside the platform.

This project has a main goal to ease and streamline the access to industry for apparel, footwear, jewellery and to the necessary means to manufacture, with the objective of liberating the creative power of the ones that want to access these means. Consequently, reducing significantly, the costs of sourcing while organizing the resources already available in these industries in fashion. To better understand the way in that the platform works were defined six main goals that have been the basis for its development:

- Being a point of access to the fashion industry making the industrial sourcing and the productive process simple and accessible allowing brands to build a personalized chain of production having into account its needs;
- Contributing for the increase of entrepreneurship in the fashion cluster through the simplification and improvement of the access to the industry;
- Generating more efficiency and gains in the productivity throughout the whole production process because of the simplification on the access to the industry and because of the availability of affordable software for creation, development and production that can be shared between the different users connected to the productive process;
- Promoting the European fashion industry, its values and its "savoir faire" to all the society stakeholders;
- Go into the direction of the increasingly need of a sustainable fashion supply chain that allies innovation and presentation of new technological developments to functionalities that allows a better approach to products and more sustainable means of production;
- Being the reference and the most convenient way to find the "Best European Factory" (http://www.umakeid.com/).

3.2 *Advantages of a B2B sourcing platform*

The textiles and clothing sector, which is commonly defined as "supplier dominated" in the dynamics of technological innovation (which tend to come from the suppliers of machinery and industrial goods), in recent years, seems capable to offer a particularly fertile ground for innovation in terms of approaches to the market and production management models (Runfola & Guercini 2009).

The complexity of the markets and networks has been increasing as well as the products/services and even the partnerships tend to be more strategic, so, cooperation became fundamental (Carvalho & Encantado, 2006).

To remain competitive in the international arena, companies must incorporate local styles through coherent management of the new product development process, making this process particularly challenging (Caniato et al. 2014, Ganesan et al. 2009 & Sandberg, 2010).

Below are the main advantages in the usage of a B2B e-commerce platform such as U.MAKE.ID, captivating new clients (this is probably the main goal between the companies that decide to adopt a B2B strategy). Especially in times of crisis, captivating new clients means direct opportunities of new income.

Reactivate old clients, most established companies have a list of inactive clients for what is pertinent to try to work with them again. Contact old clients or build promotional actions focusing these clients can bring fast results. Augmenting the relationship with the clients, in other words, the bigger the level of contact with the clients and the bigger the overture to them the bigger are the chances of improving the commercial relationship. Companies that can establish this kind of communication sell the most.

The relationship requires the existence of a process of interaction in the sense first defined, and through which one can assume that the network of social relations impact the way in which the actors assume their behaviour and therefore how the economic institutions are developed (Blois, 2004).

Ease to buy and flexibility in the reposition of stocks, this is, as easier is to make business with a company more the clients will buy from it (if other factors like the core of the business and stock resemble themselves).

Increase the number of times that a sale is, as important as finding new costumers is also making them to buy more times in the same timeframe. This means a greater cash flow and an increase in profit since the cost of marketing is much less.

Sell to small costumers or fulfil small orders, in other words, usually companies focus on large clients and large orders leaving the small costumers and orders aside. But what if these small clients could make their orders through these platforms? The costs could be reduced and relationship between cost/benefit will be improved.

Also, the geographic covering, meaning, the clients geographically distant imply a higher cost due to situations related with time or transportation. A B2B platform will work in the same way for a client on the other side of the city or a 1000 km away. Without a geographic barrier, a company can now look for costumers that before weren't economically viable selling to regions that could never approach before.

Work efficiently all the products available, that means that many salesman focus their efforts only on the products that sell the most and that are easier to sell. Companies with a larger number of products will always have problems in getting their costumers to know everything that is available and so, sell every product. Building communication and marketing strategies to get the clients to know all the products will bring significant gains to the company that will in this way diversify the sales and consequently increase the average value of the products.

Increase the share of products with the greater margin of profit, that is usually, the more complex products are, more difficult is to be able to convince the clients of their advantages. However, many of these products are the ones that can bring the highest profit margin for the company. With a complete exposure of the product capabilities and characteristics in a platform, it will be easy for both parts to understand the product and its advantages.

In the end, but not least the efficient launching of new products, is that the new product launchings, require significant costs, such as printed materials, create events, customer visiting plans, etc. With a B2B e-commerce platform this task can have a great aid from the digital world. It is much faster and cheaper to create videos, virtual experiences and digital technical descriptions and spread this work through the internet than doing everything in person.

4 FINAL CONSIDERATIONS

In this initial approach the contributions of this paper are that U.MAKE.ID can create a new global dynamic in the fashion business and in technology, giving a positive impact on firm's business performance.

We can verify that, U.MAKE.ID can help brands, manufactures and suppliers in the sourcing process. Speeding up the search of what it is that adaptable (or not) to each company having into account the needs of each one in particularly.

According to Shah (2017), in his research about emerging paradigms of managing digital business, enterprises will not be satisfied with digital technology based assurances.

To make sure that the platform can be a positive and necessary tool, on a daily bases work to designers, technicians, buyers and other players. The launch of a Beta version is expected for the next months and after that a test is mean to be done.

The aim with the test is to verify the platform functions, with fashion designers/ brands and industry professionals, for which is intended the distribution of two different questionnaires. The first one will be distributed after using the platform and it aims to understand the way how sourcing is explored by companies nowadays and what are the advantages of using a platform such as U.MAKE.ID. The second questionnaire it will be distributed before the platform being used and the objective is to understand the impact in the users and their adaptation, connexion and interplay with the platform never discarding relevant comments and pertinent tips so that a so that a positive upgrade can be achieved having into account the user's expectations.

ACKNOWLEDGEMENTS

This investigation is supported by the project leaded by PICTÓNIO company and sponsored by Project 003385 "U.MAKE.ID" Financed by the Regional Operational Program Center, within the scope of Portugal 2020 – I&DT PROJECTS IN COPROMOTION ENTERPRISES and also co-financed for European Union.

REFERENCES

Azevedo, S. G., Ferreira, J. & Leitão, J. 2007. The role of logistics' information and communication technologies in promoting competitive advantages of the firm, *Munich Personal RePEc Archive.*

Abecassis-Moedas, C. 2006. Integrating design and retail in the clothing value chain: an empirical study of the organisation of design. International Journal of Operations & Production Management 26 (4): 412–428

Black, D. & Edwards, M. J. 2015. What's digital about fashion design? *Fashion, Technology and the Digital Economy. Project Report, AAM Associates and LCF, UAL.* London UK.

Blois, K. 2004. The Market Form concept in B2B Markets, in Håkansson, H, Harrison, D., Waluszewski, A., (ed), *Rethinking Marketing. Developing a New Understanding of Markets*, Wiley, Chichester

Caniato F., Caridi M., Moretto A., Sianesi A. & Gianluca Spina. 2014. Integrating international fashion retail into new product development. *International Journal of Production Economics*. 147 (Part B): 294–306

Carvalho J. C. De & Encantado L. 2006. *Logística e Negócio Electrónico. SPI-Sociedade Portuguesa de Inovação,* Porto.

Christopher M., Lowson R. & Peck H. 2004. Creating agile supply chains in the fashion industry. *International Journal of Retail & Distribution Management*, Vol. 32 Issue: 8, pp. 367–376

Ganesan, S., George, M., Jap, S., Palmatier, R.W. & Weitz, B. 2009. Supply chain management and retailer performance: emerging trends, issues, and implications for research and practice. *Journal of Retailing* 85 (1), 84–94.

Hernandez-Vargas José G. & Alejandro O. Z. 2016. Methodology and Strategies for Companies in the Process of Internationalization. *Journal of Global Economics* (4): 224. doi: 10.4172/2375–4389.1000224

Kowalkiewiez, M., Safrudin, N. & Schulze, B. 2017. The business consequences of a digitally transformed economy. *Shaping the Digital Enterprise*, Springer International Publishing Switzerland.

Lewis, P. 2014. Fashion entrepreneurs: How to find a factory to make your products. The Guardian – https://www.theguardian.com – Acceded at 24/03/2017.

Lu, S. 2014. The 21st Century Digital Economy and the Fashion Industry: A Macroview, University of Rhode Island, Summary by Kristen Reed.

MacCarthy, B.L. & Jayarathne, P.G.S.A. 2009. Fast fashion: achieving global quick response in the internationally dispersed clothing industry. NUBS Research Paper Serie.

Macchion L., Moretto A., Caniato F., Caridi M., Danese P., & Vinelli A. 2015. Production and supply network strategies within the fashion industry, *Int. J. Production Economics,* Vol. 163 (173–188)

Moreira, R. C. 2011. Estudo comparativo das cadeias de abastecimento na Indústria do vestuário – Sonae SR – Fashion Division, Dissertação de mestrado Faculdade de Engenharia da Universidade do Porto, Porto.

Pina, L., Reis, B. Rafael, P., Pereira, M. & Miguel R. 2017. U.MAKE.ID: U.MAKE.ID – A Digital Sourcing Platform For the Fashion Business: A Theoretical Study. *Annual Conference of the EuroMed Academy of Business*

Runfola A. & Guercini S. 2009. Business networks and retail internationalization: a case analysis in the fashion industry. Published at the 25th IMP – Marseille

Ryding, D., Vignali, G., Carey, R., & Wu, M. 2015. The relative significance of product quality attributes driving customer satisfaction within the fast fashion market: a UK perspective. *International Journal of Business Performance Management*, 16 (2–3): 280–303.

Sandberg, E. 2010. The retail industry in Western Europe—trends, facts and logistics challenges. Report of Department of Management and Engineering Logistics Management, 1–58.

Shah, V. S. 2017. Emerging Paradigms of Managing Digital Business: In Association with Factoring Incremental Risks. Conference Paper. DOI: 10.18178.

Shelly, G. B. & Rosenblatt, H. J. 2012. Systems Analysis and Design, 9th Ed., Course Technology, Cengage Learning.

Strugatz, R. 2012. B2B Fashion's Online Time. *WWD* – Acceded at 15/04/2017.

Textiles, Identity and Innovation: Design the Future – Montagna & Carvalho (Eds)
© 2019 Taylor & Francis Group, London, ISBN 978-1-138-29611-4

Making textiles talk: An experimental e-textile workshop

S. Uğur Yavuz & N. Cohen
Faculty of Design and Art, Free University of Bozen-Bolzano, Italy

ABSTRACT: Textile making is one of the oldest handcrafts around us, embracing a unique spectrum of fields and aspects within it – from plant and animal care to yarn processing and fabric making. It encompasses a truly unique history which could in many ways tell the story of mankind. However, today we see textiles as mere consumption goods, which are ready to wear and use, most often detached from culture, context or environment. This workshop reacts on this reality and tries to offer a different perspective. Making a reference to the traditional "sewing circles" which for centuries were bringing together women to sew, embroider, knit, weave together while sharing daily life experiences and stories. The aim is to reinterpret and give new life to this ancient ritual through embedding electronics into textiles within a collaborative working platform. The outcome is a collective work, in which textiles become interfaces bridging the digital world with traditional textile making skills. The workshop is an example of how e-textiles can work as an accelerator of interest in textile craft and a trigger for finding new tangible interaction modalities. Besides, this type of workshops can create new value and meaning for crafting textiles, empowering craft tradition and leading to an emergence of a new or revised definition of craft.

1 INTRODUCTION

Craft is defined as "a dialogue concerned with development of a body of work created over a significant period of time, with incremental shifts of knowledge, which build cumulatively" (Follet & Valentine, 2007). In our digital era, this cumulative knowledge generator has been evolving into new forms, by keeping its tangible, hand related aspects but gaining new dimensions through the use of digital tools. "New Craft" that is altered with digital technology shows us unique and interesting features which are the results of combination of diverse peculiarities of tangible and intangible practices (Nimkulrat et al, 2016). Internet has brought an opportunity for craft practitioners to collaborate with others, discover new methods, materials and create unique results that are openly shared with others in order to add on, modify and transform into new forms and applications.

As materials and production techniques evolve, textiles evolve, too. E-textiles that contain conductive yarns such as silver or stainless steel in order to create electrical circuits or sensors (Berzowska, 2005) are one of the results of this ever-evolving technological development happening in textile field. Although they are mostly associated with high-end user goods, like sportswear, healthcare garments, military uniforms etc, they have also a strong impact in DIY communities and contemporary craft practices. Textiles' expressive feature can be altered through embedding digital technologies, making them sensing and reacting, interacting with their user. This thrilling possibility of 'making textiles behave' encouraged many practitioners to merge traditional textile making techniques with e-textiles to create responsive surfaces that can be defined as "contemporary craft" (Satomi & Perner-Wilson, 2011). Today, there are many communities, makerspaces, online platforms dealing with e-textile craft through sharing and expanding this emerging tacit and explicit knowledge with others (Kobakant, 2017; High-Low Tech, 2014; eCrafts Collective, 2013; Waag Society, 2017).

In this paper, we aim at discussing the evolving role of digital technologies in crafting textiles and presenting a workshop conducted at UNIBZ, Faculty of Design and Art, as an exemplifier of a collaborative design activity that bridges between craft and digital technology. This workshop is replicable in other contexts and can give different results based on the participant's background, interest, skills, and knowledge, as well as the materials and techniques available. In conclusion, the paper reflects on how traditional collaborative textile making rituals can be applied in design workshops by integrating e-textiles and electronics in order to investigate the creation of new value and meaning for crafting textiles.

2 CRAFTING E-TEXTILES IN COLLABORATIVE SETTINGS

Textile making is one of the oldest handcrafts, which embraces a broad knowledge of techniques that have been invented, developed and cumulated through centuries. It was not only a craft done by special artisans but also a common practice that was known by most of the people in each family. Besides its personal applications, it was done as an activity that brought people together in a ritual setting, in which people could share their experiences, skills and ideas. For instance, sewing-circles were collaborative sewing sessions where women got together and shared different sewing/quilting techniques, stories and opinions about daily life. This community based making activities today can be seen again in makerspaces, Fablabs and other maker communities where digital fabrication is the medium to ease the making process. This type of "community of practices", like communities of artisans, makers, craftsmen, designers, can be defined "both as sources of boundary and contexts for creating connections" (Wenger, 1998, p 104) and "a mutual engagement in a shared practice" (Wenger, 1998, p 214). Textile making has always had this strong ritual of mutual engagement throughout history, and is recently reappearing by the agency of e-textiles and digital technology. Although textile making and digital world seem like distinct realities, they have interesting similarities, as also seen in Sennett's (2008) analogy between weaving and Linux software. While textile craft communities carry the cumulative tacit knowledge and skills from generations to generations, open source platforms and programs that are one of the important results of digital era can do the same in an intangible way by being open to public for sharing, improving and building new outcomes. This analogy affirms the e-textile & craft cohesion that we are witnessing today.

One of the earliest examples of e-textile craft practices was done by High-Low Tech (2014) research group at the MIT Media Lab. They conducted workshops and introduced tutorials for embedding electronics into DIY textile artifacts, by making the high technology easy to integrate with handmade textiles. Kobakant (2017) website was another initiation to share new knowledge about creating sensors, actuators and circuits with conductive yarns and e-textiles through open source tutorials. Besides the online platforms, eCrafts Collective (2013) organized series of workshops bringing people together to merge electronics with traditional embroidery, crochet, beading, felting, weaving techniques. E-Textile Swatch Exchange platform is also another community that introduces swatch samples of e-textiles that are realized during yearly organized workshops (Hertenberger et al., 2013). On the other hand, FabTextiles (2017) from Barcelona Fablab and The TextileLab (2017) Amsterdam are some other examples of physical spaces in which digital fabrication and e-textiles are explored in a form of community practice. As seen in these examples, today we witness a revival of crafting textiles with a new approach which incorporates digital technology into the making process.

3 MAKING TEXTILES TALK _ E-TEXTILE SEWING CIRCLE WORKSHOP

Within the TANA (Tangible Narrators) research project, conducted in UNIBZ, Faculty of Design and Art, a prior workshop was organized in Covilhã, in Portugal, at the UBI – Textile Department. The aim was generating new ideas for smart textiles as medium to tell stories and narrations about their makers, history, etc. (Ugur Yavuz et al, 2016). In this workshop, "World Café" (Brown & Isaacs, 2005) method was applied in order to facilitate open discussion and create scenarios in a collaborative way. While this workshop was mainly about idea generation, the subsequent workshop "Making Textiles Talk", mentioned in this paper, was organized at UNIBZ, in Bolzano, Italy, in order to create ideas through making and direct exploration of materials.

"To develop narratives for smart textiles we can learn from the traditional textile craft communities who, for generations, have successfully embedded their cultural narratives within their fabric, creating textiles that are functional as well as coded with meaning and purpose" (Tharakan, 2011, p. 197). Through embedding electronics, craft textiles' expressive feature can be altered into a new dimension, in which it is possible to make them 'talk' and tell their stories. Each textile work expresses its maker's story in an implicit or explicit way, "they carry a story and an identity" (Tharakan, 2011, p. 197). Starting from this phenomenon, "Making Textiles Talk" workshop aimed at creating textile samples that could express their maker's story or give the textiles a voice to tell something to others.

The workshop refers to the traditional "sewing circles" which bring people to sew, embroidery, knit, weave together while sharing their experiences and stories. The aim is to give life to this ancient ritual through embedding electronics into textiles in a collaborative working platform. The outcome is a collective work, in which textiles become interfaces bridging the digital world with traditional textile making skills.

3.1 *Workshop setting*

The workshop was done with 6 design students (product & communication design) and conducted by a product design researcher and a textile designer, who is an expert in embroidery. The participants were chosen among BA and Master students of Faculty of Design and Art, based on their interests in textile design and previously made textile related projects. The workshop took place at Faculty of Design and Art, at UNIBZ, for two days. The main tool, which was created ad-hoc for this workshop, was a table constructed as a canvas that allowed students to craft their work on a collaborative setting. A semi-transparent textile was fitted on a wooden frame that was carried by a metal table leg structure (Fig. 1).

The workshop started with a basic introduction on e-textiles and another presentation on traditional textile making techniques. Some samples of conductive yarns, and their applications as simple switches or sensors were shown to the participants in order to make them understand various possibilities that could emerge in the workshop. The participants were

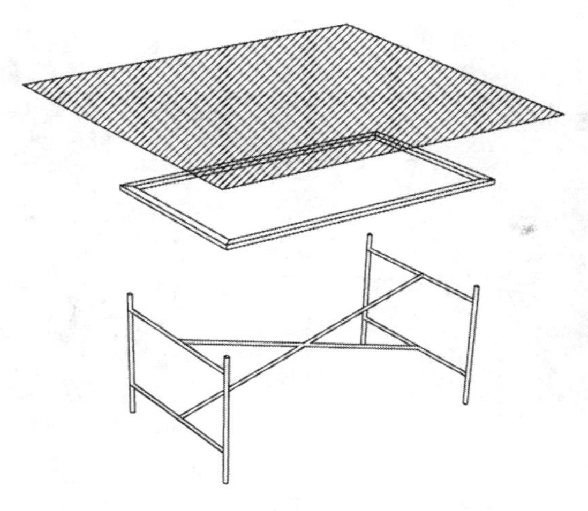

Figure 1. The wooden frame is covered with a semi-transparent textile that is fixed like a canvas.

Figure 2. Second day working on the main table.

Figure 3. Working on weaving loom with conductive and non-conductive yarns.

asked to create pieces of textiles that could work as switch/sensor to release a sound as output. To ease the prototyping process, we used Bare Conductive Touch Board (Bare Conductive, 2017) as a platform to turn conductive textiles into touch sensors to play different audio files. Besides conductive yarns and conductive ink, we provided them many other conductive elements, such as metal buttons, safety pins, metal foils, etc.

The participants first explored the materials that were given them in the workshop. Based on the materials, they started tinkering on how and what kind of interaction they could create and what kind of sound they would associate with this interaction. First day was dedicated to trials with materials and idea generation. The participants made small samples and created simple circuits of LEDs and batteries by creating textile switches. Through this activity, they gained knowledge about how a circuit can work and function on a soft surface. In the end of the day, participants shared their trials and ideas with others in order to decide the work plan for the next day of the workshop, where all the pieces done by participants would be applied to the main table cloth. The participants chose varied themes and techniques: sewing, weaving, felting, and embroidery.

On the second day, participants started to work on the main table cloth (Fig. 2) and created sound files that were put into the Bare Conductive Touch Board which was used as the main processor. On this last day, the participants mainly worked on constructing their pieces on the table, and connected them with the programmed processor. The workshop finished with a testing session, in which each piece was tested by the participants around the table.

3.2 *Results*

The participants chose different techniques to mix with embroidery. For instance, felting and weaving were explored as alternative techniques to create separated pieces that were later integrated onto the table cloth.

The participants preferred to try the techniques that they were more comfortable with. For instance, after making several trials with embroidery, one group decided to use a small weaving loom to create a woven piece consisted of different conductive parts that would later be attached to the main table cloth (Fig. 3). They used various yarns, (synthetic, cotton, and conductive) in order to create unique color mixtures. Each color patch was associated with a different melody, that was played by a MIDI interface in computer. By touching each patch a different sound was created.

As a second trial, the woven piece constructed with various patches was attached to Bare Conductive Touch Board and a speaker in order to play MP3 sound files that were uploaded into the board. Each patch was associated with a word that was a part of a sentence: "Textiles can talk to us" (Fig. 4).

A participant worked on playful aspect of sound and touch, in which she created a labyrinth game on the table cloth. The labyrinth's walls were embroidered with conductive yarn, and it was supposed to be played with both fingers aiming at arriving to the exit without touching the boarders. In case the finger touched the boarders, it was giving a warning sound.

Figure 4. Woven piece tested with Bare Conductive Touch Board and speaker.

Figure 5. Workshop table with all results attached on it.

As an alternative technique, felting was also used by mixing wool and conductive fibers. Felted conductive surfaces were sewn onto the table cloth with felted conductive balls. Different sound outputs were assigned to different parts of the felted surface by casually bouncing felted conductive balls onto it. This work was also exemplifying the playful aspect of touch and sound in a game setting.

One of the participants created a story for children by embroidering symbolic items onto the table cloth. By placing one item to another (working as simple switches), sound outputs were played in order to empower the story that was told through visual symbols embroidered with conductive and non-conductive yarns. The story had also an educative perspective in which children could learn simple principles through interacting with the textiles and sound outputs.

Unlike the other works, one participant worked on a glove which was integrated with touch sensors on different parts of fingers. This glove could translate various Italian gestures into short phrases played as a sound output. We integrated the glove on to the table becoming an extension of the final table cloth. The phrases were recorded with the participant's own voice and the sound file was integrated into the processor. The glove was tested by the participant. Although the workshop was inclined to create textile works on a surface base setting, some results like the glove went out of the limits of the table, extending it towards wearable applications.

Besides wearables, another participant worked on a three-dimensional textile switch which was constructed with metal wires covered with fluffy fibers. She made a spiral form with the wires, and by pressing the different size spirals, she could play various spring like sounds. These three-dimensional objects were sewn onto the main table cloth and connected with the processor to play the assigned sound files. This work was an example of how sound could be associated with hand gestures, as an embodied interaction and let the table grow in three-dimensional way, as the glove did.

Finally, all results were sewn on the main table cloth as a collective work (Fig. 5). While doing this work, participants helped each other to conclude the overall work, therefore being around a table increased the collaboration aspect of the workshop in which individual works became a part of a common platform. Each work was tested by participants as a conclusion session of the workshop.

3.3 *Reflections*

"The traditional textile craft processes are known for its slow and lengthy processes be it spinning, weaving, knitting or printing." (Tharakan, 2011) Merging textile making and electronics in a short workshop led to difficulties in finishing the work in time with a good quality. Participants spent more time on making, rather than the idea generation. Through their focus on making, they could learn different features of the conductive yarns and its possibilities by trials and errors.

Textile making embraces a tacit knowledge that is learnt through an "experiential and subjective" action (Staff, 2012). This workshop produced diverse results based on the participants' subjective contribution and exploration. The diversity between results show that while crafting e-textiles, participants could have chance to express their personal interests and stories through their works.

The workshop showed us that tactility and interaction with textile surfaces led to exploration of new interaction modalities and gesture-based interplays. Most of the participants created their ideas based on experimental interaction with the material, such as, bouncing the metal springs or throwing the balls on the surface. Besides, the glove was also an example of how gestures and sound could be linked within an idea of a wearable technology translating Italian gestures into phrases.

The second day of the workshop was dedicated to the collaborative work around the main table cloth. This activity worked as an engine to create spontaneous conversations, knowledge exchange and augmented the sense of collaboration. Although each participant worked on separate pieces, being on the same table led them to learn about other techniques and different ways of thinking with material.

4 CONCLUSION

Through this workshop presented in the paper, we aimed at opening a discourse on how a collaborative design activity could bridge between textile craft skills and digital technologies. The workshop was an experimental attempt which needs to be replicated and tested in different places and with participants who have different backgrounds. The results shown in this paper could change based on the participant's skills, knowledges, backgrounds, materials and techniques that were used. While this type of workshops could bring light to new ways of interaction modalities, it can also be used as a method to create new value and meaning for crafting textiles and to empower craft tradition, leading to an emergence of a new definition of craft. It is not anymore the same as before, but altered and evolved by digital skills and possibilities. Craft skills can survive or be revived through embracing new characteristics and behaviors, transforming into a new phenomenon by the integration of digital technology.

ACKNOWLEDGEMENT

The workshop was conducted together with textile designer and embroidery expert Dipl. Des. Kira Kessler and the participants of the workshop were students from Free University of Bozen-Bolzano, Faculty of Design and Art, Bachelor in Design and Master in Eco-Social Design: Chiara Zardi, Stefania Zanetti, Emanuela Frisari, Ylenia Steiner, Audrey Solomon, Lena Voegele.

REFERENCES

Bare Conductive. 2017. *Bare Conductive. Available at http://www.bareconductive.com accessed* 02.06.2017.

Berzowska, J. 2005. Electronic Textiles: Wearable Computers, Reactive Fashion, and Soft Computation, *Textile*, 3:1, pp. 2–19.

Brown, J. & Isaacs, D. *The World Café: Shaping Our Futures Through Conversations that Matter*, San Francisco: Berrett-Koehler Publishers.

E-crafts Collective. 2013. About. Available at https://ecrafts collective.wordpress.com/about-2/ accessed 02.04.2017.

Fabtextiles. 2017. Available at https://fabtextiles.org accessed 02.06.2017.

Follett, G. & Valentine, L. 2007. New Craft Future Voices, *The Design Journal*, 10(2), 1–3.

Gray, C., & Burnett, G. 2009. Making sense: An exploration of ways of knowing generated through practice and re ection in craft. In L. K. Kaukinen (Ed.), *Proceedings of the Crafticulation and Education Conference* (pp. 44–51). Helsinki, Finland: NordFo.

Hertenberger. A. et. al. 2014. 2013 E-textile swatchbook exchange: the importance of sharing physical work, ISWC '14 *Adjunct Proceedings of the 2014 ACM International Symposium on Wearable Computers: Adjunct Program.* – New York : ACM New York, 2014.

High-Low Tech. 2014. *High-Low Tech.* Available at http://highlowtech.org accessed 02.06.2017.

Kobakant. 2017. *About.* Available at http://www.kobakant.at/?page_id=475 accessed 02.06.2017.

Nimkulrat, N., Kane, F, & Walton K. 2016. *Crafting Textiles in the Digital Age.* Bloomsbury Academic.

Satomi, M. & Perner-Wilson, H. 2011. Future Master Craftsmanship: Where We Want Electronic Textile Crafts To Go. *Proceedings of the ISEA2011. The 17th International Symposium on Electronic Art.* 14–21 September, Istanbul.

Sennett, R. 2008. *The Craftsman*. New Haven & London: Yale University Press.

Staff, P. 2012. Drawing and Sewing as Research Tools. *The art of research II Process, Results and Contribution.* Keuruu.

Tharakan, M. J. 2011. Neocraft: exploring smart textiles in the light of traditional textile crafts. *Proceedings Ambience '11 Conference Borås, Sweden.*

Ugur Yavuz S, Cohen N, Salvado R, Araújo P (2016) Wandering With Textiles, *Proceedings of Designa 2016 International Conference On Design Research,* Covilha, Portugal.

Waag Society. 2017. *TextileLab.* Available at https://waag.org/en/project/textilelab-amsterdam accessed 02.06.2017.

Wenger, E. 1998. *Communities of Practice: Learning, Meaning, and Identity*. Cambridge: Cambridge University Press.

Textiles, Identity and Innovation: Design the Future – Montagna & Carvalho (Eds)
© 2019 Taylor & Francis Group, London, ISBN 978-1-138-29611-4

Footwear customization: A win-win shared experience

N. Oliveira & J. Cunha
Department of Textile Engineering, University of Minho, Guimarães, Portugal

ABSTRACT: The current paradigms of fashion design, production and consumption are in profound transformation. In the age where experience is more and more an element of differentiation and promoter of success, it is important to reflect on the potentialities of co-design and emotional design in the creation of interactive footwear customization experiences. Based on literature reviews, it has been determined that with the aid of new productive methods and support technologies it is possible that the creative footwear process itself becomes a shared experience with mutual advantages between industry and the consumer.

1 INTRODUCTION

1.1 *Theoretical framework*

The relationship of the consumer with the products is the subject of an endless number of studies in the most diverse fields of knowledge. Not being a particularly new subject, it apparently has not been exhausted in interest and content.

In fact, back in 1970, Toffler anticipated that "The turnover of things in our lives thus grows even more frenetic. We face a rising flood of throw-away items, impermanent architecture, mobile and modular products, rented goods and commodities designed for almost instant death." (Toffler 1990, p. 58). The author extrapolates this deprived relation to other dimensions of individuals' social life, suggesting that the relationships and social interactions themselves are moving to the disposable. This frantic transience relates to the craving for novelty, supported by a superindustrial society. Toffler argues that it is necessary to establish personal tactics towards social strategies, in order to dampen the changes in future society, introducing change into a chain of predictable, rather than random events (Toffler 1990).

The fashion industry is part of a market that tends to overproduce objects at a price that is often attractive to consumers, but with catastrophic impacts on the environment and societies (Gardetti & Torres 2013).

Mass production is seen as an outdated and too painful method, with other production processes more appropriate to the current reality from the point of view of sustainability (Anderson-Connell et al. 2002). On the other hand, personalization can prove to be a complex and expensive process for industry and consumers (Gilmore & Pine 1997; Boër & Dulio 2007).

Thus, when reflecting on these subjects, there are several issues to be discussed. Is it possible, in any way, to balance mass production and personalization? How can new technologies and consumption habits contribute to this eventual balance? Is it possible to promote win-win situations for all players in this value chain?

The answer to these questions may lie in rethinking the relationship between the industry with the consumer, the designer with the user and the consumer with the product. In this way, co-design, or collaborative design, emerges as an alternative design process to massification, which, although highly studied from various points of view related to management and design methodology, is relevant to study from the perspective of the mutual experience between industry and the consumer in a specific fashion sector, the footwear.

Brands such as *Reebok*, *Nike* and *Converse* are references to take into account in this subject and to be explored in this article.

1.2 *Objectives*

The general objective is to reflect on the transformations of the current creative scene in footwear design, specifically about customization processes, highlighting the contribution of co-design as a methodology of mutual benefits for the industry and for the consumer. To this end, efforts will be made to present a reflection on co-design, emotional design and customization, defining advantages and disadvantages, and establishing a direct relation with customization.

1.3 *Methodology*

In order to develop the present study and to reach the proposed objectives, it was required to establish the appropriate methodology for each of these objectives. Thus, firstly the problematic to be explored was determined to allow to define the subject of study, that is, the key idea to present. After this definition, a literature research in specialty themes was carried out, underlying the study of customization and co-design applied to footwear design, with a strong association to emotional design, based on books, scientific articles and proceedings. Through documentary analysis, it was attempted to: identify the main authors of each theme, who present current and legitimated work in the subject in study; find definitions related to the concepts in analysis; determine the advantages and disadvantages associated; and define areas of application, in particular in footwear sector. Then, it was aimed to argue the thesis presented based on studies and investigations of other authors.

2 THEORETICAL FOUNDATION

2.1 *Customization and co-design*

Given the increasing competition in the global market, brands have faced ever-greater challenges regarding the development of new products. Clients tend to be more informed and knowledgeable, raising the bar of demands on the requirements that products must accomplish to deal with the needs of their intended audience. In addition to this great challenge, brands have to control production prices to keep products affordable to their consumers (Zhang & Tseng 2007).

Concepts such as mass customization emerge as alternative processes to the massification of products, attending to the increased importance of the consumer's role in the creative process. According to Tseng & Jiao (2001, p. 285), mass customization is defined as the "production of goods and services to better meet individual needs of customers with the efficiency close to mass production".

The consumer eager for novelty, individuality and distinction compared to others, finds in customization the way to satisfy his needs (Ribeiro et al. 2009).

According to Gilmore & Pine (1997), four types of approach to customization must be considered: cooperative or collaborative – the company establishes contact with consumers to identify their needs and the best solutions to meet them; adaptive – the consumer customizes the product based on a varied range of possibilities offered by the brand; cosmetic – a standard product is presented to different consumers with minor changes; transparent – the company designs individualized products without the consumer necessarily having to intervene in the process.

The customization process comprises an active participation of the clients that collaborate directly with the designers in the development of products, in order to satisfy their needs. It is a process that includes three chronologically sequential phases: learning, correspondence and evaluation. In the learning phase, the company defines the clients and studies them to gather the necessary information to set goals to be achieved; in the correspondence phase, the company establishes direct collaborations with clients and uses their knowledge to develop proposals that fit their needs, these clients tend to be representative of the market segment that the brand intends to achieve; in the evaluation phase, the company evaluates the effectiveness of previous phases to verify the suitability of products to customers (Murthi & Sarkar 2003).

The first approaches to co-design appear in northern European countries, to democratize design, allowing consumers to engage in the process more actively by sharing their knowledge and experience. This approach gives consumers greater power by making them a decisive element in the creative process (Steen et al. 2007; Sanders & Stappers 2008; Fischer 2011; Vines et al. 2013). Several innovative approaches to design have been developed over the past two decades with the main purpose of including the user in the process. This type of collaborative design can be applied in the most diverse areas of design (Stewart & Hyysalo 2008; Markussen 2013). Within this current of thought are inserted several themes and methods of design, i.e., design thinking, sustainable design, ethical design, activist design, emotional design, slow design, among others (Verganti 2003; Sanders & Stappers 2008; Dubois et al. 2014; Kang 2017).

2.2 *Emotional design*

Emotional Design emerged in 1999 during the first conference on design and emotion, organized by the Department of Industrial Design at Delft University of Technology (Kurtgözü 2003). This emotional feature, allied to the development of products, can be justified by the new demands and needs resulting from cultural and social changes, from the relation between individual and object, boosted by the capitalist industrial production that considered this a competitive commercial advantage (Jordan 1998; Hekkert & Mcdonagh 2003; Tonetto & da Costa 2011).

According to Demir et al. (2009), emotional design consists in creating with intent to evoke or avoid an emotion.

One of the main representatives of this design current is Donald Norman who, in his book *Emotional Design: Why we love (or hate) things*, sought to explain that emotion is the result of three levels of human brain processing: visceral, behavioral and reflexive. The author argues that these three levels can serve as a guideline to processes for the development of design products from three different strategies: design for appearance (visceral design – emphasis on the automatic relations of worship, desire, attraction, normally triggered by sounds, colors, shapes, and other sensory and aesthetic features that may involve an object), design for ease of use (behavioral design – refers to the use and objective performance of a product's functions, how the user understands and operates the product and how it behaves), or design for meanings and reflection (reflective design – refers to use from the subjective point of view, intangible aspects, encompassing cultural and individual particularities, affective memory and meanings attributed to the product and its use) (Norman 2004). The concept of emotion does not gather a consensus between the scientific community and the different fields of knowledge, therefore there is no universal definition due to the difficulty of distinguishing it from other affective phenomena, such as: feelings, humor and attitudes (Scherer 2005).

Thus, for the present study it was understood to assume and agree with Norman (2004) vision, who understands that emotion belongs to the emotional system (evaluator, assigns positive and negative valences quickly and efficiently, allowing a decision to be made) and the cognitive system (interprets and understands situations, giving feelings to behavioral and sentimental actions). Still according to Norman (2004), most research on emotions focuses on the negatives (stress, fear, anxiety, and anger), but positive emotions are as important as negative ones, positive emotions are crucial to curiosity, learning, and to stimulate creative thinking.

2.3 *Applicability of co-design*

The products developed in co-design processes tend to follow a modular construction strategy. This modularity consists of decomposing products into structures that can be regrouped in different ways, resulting in several possible combinations, giving rise to products with significant differences. High degrees of modularity allow greater customer involvement, resulting in more customized products. However, the associated production costs and the flexibility requirements of the productive structure tend to increase (Jiao et al. 2007; Ribeiro et al. 2009).

Some authors question if all people have sufficient creative capacities to intervene in a co-design process, suggesting that only a few have this competence. According to them, it should not be assumed that all consumers are capable of conceiving and co-design could result in an increase of consumer power and generate entropies to the creation processes, shaking hierarchies (Von Hippel 2005).

The co-design process will depend on the level of knowledge, passion, interest and creativity of the intervenients, and can differ over the individuals' day. Therefore, it is necessary to establish effective communication and creation processes adapted to the kind of consumer and products (Sanders & Stappers 2008).

According to Gustafsson et al. (2012), the collaboration of consumers in the creative process has increased in popularity, triggering that more and more companies adopt co-design measures. The conclusions obtained by the authors suggest that there is a direct relation between co-creation and innovation and that these can be combined, but the relevance of each of the dimensions of the communication varies when dealing with radical or incremental product innovation. Thus, they suggest that radical innovations need to limit inputs that are too focused on the particular product or solution.

2.4 *Footwear and co-design*

The footwear sector is witnessing a transition from mass production to mass customization, giving rise to new business models that are under development and introduction on the market. In fact, collaborative design begins to find more and more followers, apologists for client integration in the product design process (Pandremenos et al. 2010).

The *EUROShoE* (Extended User Oriented Shoe Enterprise) research project considers that the emphasis of the mass customization lies in the design process and in the creation of a database containing the developed models. This provides all the information for the client to fill out a form with the specifics intended and which assists in the final decision of the model (Semenenko & Krikler 2004; Boër & Dulio 2007).

In the sports segment, textile materials are predominant and frequently combined with leather and other materials, namely polymeric, which enhances the development of footwear that combines aesthetic and functional dimensions (Shishoo & Buirski 2005). Within this universe, *Reebok* was one of the most important brands to join the online customization platforms (*Reebok Custom*), allowing the consumer to intervene in the creation process by defining the elements he wants to have in his model. *Nike* also focused on developing a process of innovation and marketing based on the best practices of value creation with its consumers. To this end, the brand developed the *Nike iD* website platform, which is a strong bet on footwear customization, allowing the consumer to create models that can only be purchased online. Similarly, the *Converse* brand launched in 2006 the *Converse One* platform that allowed consumers to intervene in the creation of their customized model. Other examples of footwear brands that have embraced this type of approach to co-design are *Adidas*, *Puma*, *Vans*, *Timberland*, *JG*, *O'Neill*, *Footjoy*, *Steve Madden*, among others (Berger & Piller 2003; Boër & Dulio 2007).

Within the scope of the European research project *DOROTHY*, a group of researchers developed a model to aid the footwear design process in which the client plays a collaborative role. This interactive model presupposes the development of a tool, an online software called *ShoeSM* (Shoe Design Support Module) that allows to establish a communication channel between designers, customers and the production department. In this way, when a designer creates a model he decides which components will be modular, that is, changeable or customizable. Then uploads the template, which can be 2D or 3D, and waits for approval from the production department. When the proposal is approved, the production department forecasts the costs and time required to manufacture the model. Then the proposal is sent to customers who provide feedback regarding the proposal in question and interact with the designer to improve the product, this in an initial phase of testing and analysis of acceptance of the product by a group of standard customers (Pandremenos et al. 2010).

3 STATEMENT

3.1 *Co-design as a break-even point*

Among the collaborative design processes, mass customization is distinguished from product personalization because it is not intended to develop a fully made-to-order product but rather to tailor a mass product to a group of consumers with common needs. Therefore, it is considered that there is no homogeneous market, but a heterogeneous market (Gilmore & Pine 1997).

According to Tseng et al. (2013), in the design field there is no established definition of personalization and customization, these terms being often used indiscriminately as synonyms. For the authors, personalization refers to increasing the personal relevance of the individual on the required product, with a more particular and proactive intervention, while customization shows a more passive and limited participation in which the client makes choices between a predefined set of options.

Mass customization appears as a balance between the demand for customized products and the cost-efficiency goals of companies. In a market that tends to be increasingly heterogeneous, mass customization seems to be a positive response to consumer needs and organizational goals (Wang et al. 2013). However, in spite of all the advantages and potential of co-design-based application, customization tends to make production and distribution processes more complex, which may at first appear to be a difficult adaptation for companies to produce custom models in particular for those who are specialized in producing few models in high quantities. However, it is possible if some simple adaptations are made, such as the creation of a parallel production line of a smaller and more sustainable size for the company. This trend towards customization is increasingly practiced and consumers who buy custom footwear tend to establish loyalty bonds with brands and buy more (World Footwear 2017).

3.2 *Contribution of technologies*

Some technologies have been developed to aid the customization process of mass products, namely, body scanning, 3D simulation software and digital printing (Anderson-Connell et al. 2002).

For footwear in particular, scanner technology has also been developed and applied which allows to get the specific features of an individual's feet to be adapted to suit the shoe lasts and comfort of footwear models. This information can be used in the production of new shoe lasts or in the search for the best correspondence of the existing ones. Examples of this type of technology are the *Corpus.e*, *Infoot*, *Vorum*, *FotoScan*, *UCS* and *Formalogix* scanners and the *FotoFit* and *Shoe Selector* software (Boër & Dulio 2007).

In creative processes that contemplate the direct intervention of the client, as co-creator, he has the possibility of interacting directly and actively in the creation of the product, indicating the desired requirements or concretely designing the product. Today, with the dissemination of information technology and the Internet, most of these processes are carried out through online digital interfaces, which, in addition to bringing the customer closer to creation, also enhance a more pleasant and seductive shopping experience. In this way, the customer becomes better acquainted with the product, which enhances brand confidence and loyalty. The most commonly used software models in customization processes are based on product configurators, through which different components and attributes of a product are available, which can be combined according to the customer's wishes, resulting in a more customized product. Thus, through the input provided by the client – options taken based on the needs, tastes and preferences – it is possible to materialize an output – variation and adaptation of a product – that corresponds more assertively to the one desired by the client (Wang & Tseng 2011; Tseng et al. 2013; Sandrin et al. 2017).

In today's era of easy and common cybernetic interaction, it becomes increasingly feasible for the customer to have control and mastery on the final creation of the product, becoming more and more an effective decision maker of creation. In turn, the designer is increasingly driven to the development of product platforms. This allows all products created to be customized at any time and according to the preferences and wills of the customer, even after the sale of the product (Tseng et al. 2013; Dubois et al. 2014). In this way, it is understood that co-design is increasingly dependent and related to technological evolution, leading to the development of software oriented to configurations based on the choices of the customers. This reality drives organizations to incorporate more elements of business intelligence into their management processes and to develop specific communication channels for this type of customer interaction (Freuder et al. 2003; Junker & Mailharro 2003; Tiihonen & Felfernig 2010).

The new technologies supported by the Internet are key elements for the development of co-design-based creative processes, and increasingly accepted by consumers, since the consumption of products and services through the Internet is a very present reality, more and more common and with significant growth forecasts. This change of behavior comes to redefine the consumption patterns of contemporary society (Ribeiro et al. 2009).

3.3 *Conscious consumption*

Currently, the footwear market, like the fashion market in general, is undergoing restructuring. Gone are the days when mass production was absorbed by the market. The consumer of the present tends to be more informed and knowledgeable, being, more and more, an active partner. In fact, consumers play an increasingly important role in the design process, interacting with creatives, in order to tailor the product to their real needs. This is due to the growing demand not for large quantities, but for a wide variety of products. To match this reality, brands are adapting, following market segmentation policies, focusing on very specific small target groups. This leads to the need for companies adjust and develop appropriate technologies and processes across the value chain, from creation, production, distribution, sales and after-sales service (Georgoulias et al. 2009; Pandremenos et al. 2010).

The demand for highly differentiated products, which meet the specific needs and expectations of each individual and consequently with high added value, has been a strong trend for the footwear sector throughout Europe since the beginning of this millennium (Brenton et al. 2000; Merle et al. 2010; Park & Curwen 2013; Curwen & Park 2014). The current paradigm of consumption is in a stage of change. Increasingly the consumer tends to make purchasing decisions that transcend the characteristics of the product itself. All the inherent involvement of the brand and the production process are directly influencing factors of consumer decision making, resulting in the growth of the ethical and sustainable market (Carrigan & de Pelsmacker 2009; Eifler & Diekamp 2013). According to Atkinson (2004, p. 2) "It is a truism to say that the mode of production and even the mode of retailing affect the relationship we have to objects". In this way, the secret of a brand's success can lie in the design of customizable products, capable of ensuring efficiency and effectiveness in the variety of possible combinations and communication with consumers (Gilmore & Pine 1997).

The traditional paradigms of the market dictate that the consumer is at the end of the value chain, and that he tends to be limited to choose and to acquire the products that are offered by the most diverse brands (Anderson-Connell et al. 2002). However, some alternative currents that reposition the consumer begin to proliferate, giving him a more active role in the act of creation, which results in an empowerment and increased involvement. This is due to the profile of the current consumer being increasingly educated, informed, demanding and globalized, as previously never verified (Ramaswamy 2008). However, this active participation of consumers results in an increase in expectations, that is, a consumer faced with a customized product has higher expectations than if he were purchasing a standard product, coming from the massification (Wind & Rangaswamy 2001). According to Franke et al. (2010), self-produced products give the customer a significantly greater predisposition to pay. This effect is mediated by feelings of achievement and moderated by the outcome of the process, as well as by the contribution perceived by the individual in the process of self-development. Thus, this process of creation has the ability to evoke emotions and states of mind in the consumer regarding their involvement with the product.

3.4 *Conscious production*

The needs of customers and consumers are so variable and imperative for the market that they guide it not for the development and production of large quantities of the same product but rather for a large variety of products in smaller quantities. This current reality could imply higher costs to the producer and customer, however, companies try to control these through the development of modular products (Zhang & Tseng 2007).

Co-design arises with special importance in the customization strategy, as it allows the consumer to find and develop products that more effectively meet their needs, resulting in increased satisfaction. This concept challenges traditional practices, giving new roles to consumers, who no longer have passive creative performances, based on an interactive cooperative relationship. This results in a new paradigm where products are no longer only developed by the internal critical mass of the company, but combining the direct collaboration of consumers (Prahalad & Ramaswamy 2004).

The landscape of design is changing, suggesting that more and more designers adopt approaches that are focused on individuals and not on products themselves. It is in this logic that co-design emerges as a collaborative process that takes advantage of the value of the creator and the consumer as a work team (Sanders & Stappers 2008; Brown & Katz 2011; Fischer 2012; Kang 2017). This growing trend of adopting co-creative approaches in design has been studied by several researchers who support a holistic view of design, which embraces and advocates collaborative design processes (Dubois et al. 2014; Kensing & Blomberg 1998; Sanders 2000; Verganti 2003). As determined by Berger & Piller (2003), the adoption of this type of methodology leads to the flexibility of productive capacity, the reduction of stocks, the increase of responsiveness, the advanced entry of the payment of the product, improvement of organizational resilience, the agility of information, the increase of value creation for the consumer, improving the capacity for innovation and development of new products.

3.5 *Future perspectives of co-design*

The forecasts of several researchers indicate that it may be possible to change the way in which products are distributed. The technological and social development lead to a new experience for the customers to obtain the products, moving from an exclusively physical distribution to including also online distribution, namely through 3D printing. This technology reveals an enormous potential for a greater satisfaction of the real needs of the consumer, being able to allow the client to take an even more active role and to admit a greater customization of the products. This may even allow to achieve truly personalized solutions, withdrawing the costs associated with the production of variations of the product family, as it may be the customer to assemble these elements. This possibility may correspond not to a threat to the creative's work but rather to an opportunity for companies and brands to attach greater importance to the ability of designers to develop modular products. That is, 3D printing technology can result in the increased relevance of the designer's role to the company, which will focus less on production and more on creation. This technology has the potential to revolutionize society and the way of doing design, giving greater relevance to customer participation, contributing to the growing of DIY (do it yourself) universe. This possibility opens the way to a new type of customization – individualized or open-ended customization, a new paradigm that motivates the customer to participate, create, learn, gain experiences and obtain products and choose production processes that meet individual needs. (Gandhi et al. 2013; Tseng et al. 2013).

4 CONCLUSION

4.1 *Conclusion*

The co-design process, understood as a client-based co-creation process, is seen as an extension of the mass customization process, based on a more open product architecture, on-demand production systems and digital systems that enhance participation in product design (Tseng et al. 2013). As a consequence of the collaboration, the role of the designer is changing in the current panorama, this includes the role of facilitator of the realization of the products idealized by the consumers (Sanders & Stappers 2008). Thus, the design ceases to be held only for the sake of the people, passing the designer to be encouraged to work with people.

The right definition of customer needs and requirements is the key to customization and creating pleasing and engaging experiences. The intervention of customers in the design process proves to be an effective methodology and an enhancer of the products success, as it allows to effectively include their real preferences.

This way of thinking design splits responsibilities between the designer and the client, ending with the control, which until then was almost exclusive to the designer, giving greater importance and power to the client. So, the client ceases to play an active role only at the end of the value chain, being included in the initial phase of product creation.

Being the main purpose of design meeting consumer's needs, co-design enhances the achievement of this goal. In this way, it becomes imperative for this new paradigm to know and understand the motivations, desires, expectations and needs of the consumer, as an individual and as a group. Thus, all opportunities for interaction between business and consumer should be seen as opportunities in the creation, sharing and obtaining of knowledge and information. Co-design makes it possible to satisfy customer needs more efficiently and it is a competitive advantage, but also reduces waste and production and delays. However, it must be taken into account that customization, although in mass, is a highly complex process for an organization.

The intrinsic variables of production and costs of the product can be difficult to determine and operationalize. In this way, the key to success in today's competitive marketplace can be the design of customizable products that can ensure efficiency and effectiveness in the variety of possible combinations, production and communication with consumers.

The fact that the customization tools available on the Internet are increasingly accessible to consumers, allows them to create products in an increasingly sophisticated way, being this the communication channel with the greatest potential for the practice of co-design.

All these dimensions contribute to making the creative process itself a shared experience between the industry and the consumer, focusing on closer relations in which the involvement, interactivity and communication become key drivers. There is thus a sharing of benefits, responsibilities and experiences, resulting in potentially beneficial situations for both parties involved.

4.2 *Limitations and final considerations*

The present article expresses a position that despite being sustained in scientific literature it requests a deeper study with more concrete data. In this way, it is assumed that research needs to be continued, particularly with regard to multidirectional communication channels and new processes, technologies and production techniques that contribute to the greater effectiveness and efficiency of the co-design models applicable to the customization of footwear.

ACKNOWLEDGMENTS

This work is supported by FEDER funding on the Programa Operacional Factores de Competitividade-COMPETE and by national funds through FCT – Foundation for Science and Technology within the scope of the project POCI-01-0145-FEDER-007136 and UID/CTM/00264.

REFERENCES

Anderson-Connell, L. J., Ulrich, P. V & Brannon, E. L. 2002. A consumer-driven model for mass customization in the apparel market', *Journal of Fashion Marketing and Management: An International Journal*, 6(3), pp. 240–258.

Atkinson, P. 2004. Post-industrial manufacturing systems: the impact of emerging technologies on design, craft and engineering processes, in Burnett, G. (ed.) *Challenging craft: International Conference 8–10 Sept. 2004*. Aberdeen, Scotland: Gray's School of Art, Robert Gordon University.

Berger, C. & Piller, F. 2003. Customers as co-designers, *Manufacturing Engineer*, pp. 42–45.

Boër, C. R. & Dulio, S. 2007. *Mass Customization and Footwear: Myth, Salvation or Reality?* Springer London.

Brenton, P., Pinna, A. M. & Vancauteren, M. 2000. Adjustment to Globalisation: A Study of the Footwear Industry in Europe. CEPS Working Document No. 151, October 2000.

Brown, T. & Katz, B. 2011. Change by design, *Journal of Product Innovation Management*, 28(3), pp. 381–383.

Carrigan, M. & de Pelsmacker, P. 2009. Will ethical consumers sustain their values in the global credit crunch?, *International Marketing Review*, 26(6), pp. 674–687.

Curwen, L. G. & Park, J. 2014. When the shoe doesn't fit: female consumers' negative emotions, *Journal of Fashion Marketing and Management: An International Journal*. Emerald Group Publishing Limited, 18(3), pp. 338–356.

Demir, E., Desmet, P. M. A. & Hekkert, P. 2009. Appraisal Patterns of Emotions in Human-Product Interaction, *International Journal of Design*.

Dubois, L.-E., Le Masson, P., Weil, B. & Cohendet, P. 2014. From organizing for innovation to innovating for organization: how co-design fosters change in organizations, *Aims 2014*, (June 2014).

Eifler, C. & Diekamp, K. 2013. Consumer Acceptance of Sustainable Fashion in Germany, *Research Journal of Textile and Apparel*, 17(1), pp. 70–77.

Fischer, G. 2011. Beyond Interaction: Meta-Design and Cultures of Participation, *23rd Australian Computer-Human Interaction Conference (OzCHI'11)*, pp. 112–121.

Fischer, G. 2012. Meta-design and Cultures of Participation: Transformative Frameworks for the Design of Communication, in *Proceedings of the 30th ACM International Conference on Design of Communication*. New York, NY, USA: ACM (SIGDOC'12), pp. 137–138.

Franke, N., Schreier, M. & Kaiser, U. 2010. The "I Designed It Myself" Effect in Mass Customization, *Manage. Sci.* Institute for Operations Research and the Management Sciences (INFORMS), Linthicum, Maryland, USA: INFORMS, 56(1), pp. 125–140.

Freuder, E. C., Likitvivatanavong, C., Moretti, M., Rossi, F. & Wallace, R. J. 2003. Computing Explanations and Implications in Preference-based Configurators, in *Proceedings of the 2002 Joint ERCIM/CologNet International Conference on Constraint Solving and Constraint Logic Programming*. Berlin, Heidelberg: Springer-Verlag (ERCIM'02/CologNet'02), pp. 76–92.

Gandhi, A., Magar, C. & Roberts, R. 2013. How technology can drive the next wave of mass customization'. McKinsey & Company.

Gardetti, M. A. & Torres, A. L. T. 2013. Sustainability in Fashion and Textiles – Values, Design, Production and Consumption, *Management of Environmental Quality: An International Journal*, 24(4).

Georgoulias, K., Papakostas, N. & Chryssolouris, G. 2009. Novel business models formanufacturing firms, *2009 IEEE International Technology Management Conference (ICE)*, pp. 1–7.

Gilmore, J. H. & Pine, B. J. 1997. The four faces of mass customization., *Harvard business review*, 75(1), pp. 91–101.

Gustafsson, A., Kristensson, P. & Witell, L. 2012. Customer co-creation in service innovation: a matter of communication?, *Journal of Service Management*, 23(3), pp. 311–327.

Hekkert, P. & Mcdonagh, D. 2003. Design and Emotion, *The Design Journal*, 6(2), pp. 1–3.

Von Hippel, E. 2005. Democratizing innovation', *Journal für Betriebswirtschaft*, p. 63–78.

Jiao, J., Simpson, T. W. & Siddique, Z. 2007. Product family design and platform-based product development: A state-of-the-art review, *Journal of Intelligent Manufacturing*, 18(1), pp. 5–29.

Jordan, P. W. 1998. Human factors for pleasure in product use, *Applied Ergonomics*, 29(1), pp. 25–33.

Junker, U. & Mailharro, D. 2003. Preference Programming: Advanced Problem Solving for Configuration, *Artif. Intell. Eng. Des. Anal. Manuf.* New York, NY, USA: Cambridge University Press, 17(1), pp. 13–29.

Kang, J.-Y. M. 2017. Customer interface design for customer co-creation in the social era, *Computers in Human Behavior*, 73, pp. 554–567.

Kensing, F. & Blomberg, J. 1998. Participatory Design: Issues and Concerns, *Computer Supported Cooperative Work (CSCW)*, 7(3), pp. 167–185.

Kurtgözü, A. 2003. From Function to Emotion: A Critical Essay on the History of Design Arguments, *The Design Journal*, 6(2), pp. 49–59.

Markussen, T. 2013. The Disruptive Aesthetics of Design Activism: Enacting Design Between Art and Politics', *Design Issues*. MIT Press, 29(1), pp. 38–50.

Merle, A., Chandon, J.-L., Roux, E. & Alizon, F. 2010. Perceived Value of the Mass-Customized Product and Mass Customization Experience for Individual Consumers, *Production and Operations Management*. Blackwell Publishing Inc, 19(5), pp. 503–514.

Murthi, B. P. S. & Sarkar, S. 2003. The Role of the Management Sciences in Research on Personalization, *Management Science*. INFORMS, 49(10), pp. 1344–1362.

Norman, D. A. 2004. *Emotional Design: why we love (or hate) everday things*. New York, NY, USA: Basic Books.

Pandremenos, J., Georgoulias, K., Chryssolouris, G., Jufer, N. & Bathelt, J. 2010. A shoe design support module towards mass customization, *2010 IEEE International Technology Management Conference (ICE)*, pp. 1–8.

Park, J. & Curwen, L. G. 2013. No pain, no gain?: Dissatisfied female consumers' anecdotes with footwear products, *International Journal of Fashion Design, Technology and Education*, 6(1), pp. 18–26.

Prahalad, C. K. & Ramaswamy, V. 2004. Co-creation experiences: The next practice in value creation, *Journal of Interactive Marketing*, pp. 5–14.

Ramaswamy, V. 2008. Co-creating value through customers' experiences: the Nike case, *Strategy & Leadership*, 36(5), pp. 9–14.

Ribeiro, L., Miguel, R. & Duarte, P. 2009. Aplicação experimental do co-design e da modularidade no design de moda, *Congresso Brasileiro de Pesquisa e Desenvolvimento em Design*, (November).

Sanders, E. B. N. 2000. Generative Tools for Co-Designing, in Scrivener, S. A. R., Ball, L., and Woodcock, A. (eds) *Collaborative Design*. Springer, pp. 3–12.

Sanders, E. B. N. & Stappers, P. J. 2008. Co-creation and the new landscapes of design, *CoDesign*. Taylor & Francis, 4(1), pp. 5–18.

Sandrin, E., Trentin, A., Grosso, C. & Forza, C. 2017. Enhancing the consumer-perceived benefits of a mass-customized product through its online sales configurator: An empirical examination, *Industrial Management & Data Systems*, 117(6).

Scherer, K. R. 2005. What are emotions? And how can they be measured?, *Social Science Information*. SAGE Publications, 44(4), pp. 695–729.

Semenenko, A. & Krikler, R. 2004. Knowledgw-based shoe design process, in *5th Workshop on Integrated Product Development*. Magdenburg, pp. 22–24.

Shishoo, R. & Buirski, D. 2005. *Textile use in sport shoes, Textiles in Sport*. Cambridge: Woodhead Publishing Limited.

Steen, M., Kuijt-Evers, L. & Klok, J. 2007. Early user involvement in research and design projects–A review of methods and practices, *Paper for the 23rd EGOS Colloquium*, pp. 1–21.

Stewart, J. & Hyysalo, S. 2008. Intermediaries, Users and Social Learning in Technological Innovation, *International Journal of Innovation Management*, 12(3), pp. 295–325.

Tiihonen, J. & Felfernig, A. 2010. Towards recommending configurable offerings, *International Journal of Mass Customisation*, 3(4), p. 389.

Toffler, A. 1990. *Future Shock*. 4ª ed. New York, NY, USA: Bantam Books (Bantam books).

Tonetto, L. & da Costa, F. 2011. Design Emocional: conceitos, abordagens e perspectivas de pesquisa, *Strategic Design Research Journal*.

Tseng, M. M., Hu, S. J. & Wang, Y. 2013. Mass Customization, *CIRP Encyclopedia of Production Engineering*.

Tseng, M. M. and Jiao, J. 2001. Handbook of industrial engineering, in Salvendy, G. (ed.) *Handbook of industrial engineering: Technology and Operations Management*. New York, NY, USA: John Wiley & Sons, Ltd, pp. 284–289.

Verganti, R. 2003. Design as brokering of languages: Innovation strategies in Italian firms, *Design Management Journal*, 14(3), pp. 34–42.

Vines, J., Clarke, R. & Wright, P. 2013. Configuring participation: on how we involve people in design, in *Proceedings of the IGCHI Conference on Human Factors in Computing Systems*, pp. 429–438.

Wang, Y., Kandampully, J. & Jia, H. 2013. "Tailoring" customization services: Effects of customization mode and consumer regulatory focus, *Journal of Service Management*, 24(1), pp. 82–104.

Wang, Y. & Tseng, M. M. (2011) Integrating comprehensive customer requirements into product design, *CIRP Annals – Manufacturing Technology*, 60(1), pp. 175–178.

Wind, J. & Rangaswamy, A. 2001. Customerization: The next revolution in mass customization', *Journal of Interactive Marketing*, 15(1), pp. 13–32.

World Footwear 2017. *How design is passing on to consumers*. Available at: /www.worldfootwear.com/news/how-design-is-passing-on-to-consumers/2494.html (Accessed: 1 June 2017).

Zhang, M. & Tseng, M. M. 2007. A Product and Process Modeling Based Approach to Study Cost Implications of Product Variety in Mass Customization', *IEEE Transactions on Engineering Management*, pp. 130–144.

Textiles, Identity and Innovation: Design the Future – Montagna & Carvalho (Eds)
© 2019 Taylor & Francis Group, London, ISBN 978-1-138-29611-4

Competitiveness in the textile industry due to electronic commerce

D.V. Leal
EACH, Escola de Artes Ciências e Humanidades, Universidade de São Paulo, Brazil

G.L. Toledo
FEA, Faculdade de Economia, Administração e Contabilidade, Universidade de São Paulo, Brazil

D.K. Junior
EACH, Escola de Artes Ciências e Humanidades, Universidade de São Paulo, Brazil

ABSTRACT: The purpose of this article is contextualize the electronic commerce, as a tool of the competitive position on the textile industry strategy. Comprehend the present worldwide textile production scenario is an essential part to point out the gains and the peculiarities of the impact caused by the online commerce establishment on this chain. In addition to that, have been raised details on the challenges of embrace this channel of negotiation, communication and marketing, in deals as B2B, more specifically on the textile industry.

Keywords: Competitiveness; Textile industry; Electronic commerce.

1 INTRODUCTION

The textile industry, in all its aspects, has a great meaning in the most varied countries where it is present. The events in recent years caused strong impacts and changes in the segment, contributing to the current textile scenario.

With the specific data survey done about the emergence of new players that has great weight on the international market (Asian countries, particularly China), this paper sought to investigate the textile industry competitive edge, in the face of the aggravation of the international competitiveness.

This way, the study seek to situate the textile worldwide field, pointing out distinct aspects related to the industry, and aims to assess new international trading forms, as a tool for the textile competitive position, thinking specifically in strategy and technology.

This paper also analyzes the contribution of the electronic commerce Business-to-Business (B2B) on the management, seeking to collect data that prove it's positive usability on the competitiveness, without forget to understand the barriers that the online business face, as risky deals, specific bureaucracies of the textile market, online and traditional channel conflicts, company and costumer data safety, and commercial interests.

2 METHODOLOGY

This article's investigation has qualitative focus. Has as main characteristic the use of the natural environment. In other words, the study is an applied work, since it is realized using practical expertise, directed to the solution of specific problems. (Richardson, 1989).

As for the goals, the research has exploratory and descriptive nature. Descriptive when it concerns on describe the features of a certain specific event. Exploratory when identifies through bibliographic research, factors that contribute to the occurrence of intended events (Gil, 1996).

The paper was organized in five steps described below: the first step concerned in describe the international panorama of the textile industry. The second one approached the main markets and the international textile commerce. The third step was characterized by the theoretical aspects survey, related to the textile industry competitiveness. The fourth step contextualized the electronic commerce appliance on the textile industry. Lastly, on the fifth step, were submitted the conclusions reached with basis on the preceding stages.

3 INTERNATIONAL PANORAMA OF THE TEXTILE INDUSTRY

The production increase is due to various changes suffered by the productive structure of the textile chain (spinning, weaving, knitting, finishing, beneficiation and confection). Parallel to this, the integration of world markets, with the progressive reduction of international market tariff barriers, resulted on the reduction of the textile products' prices. This enabled the

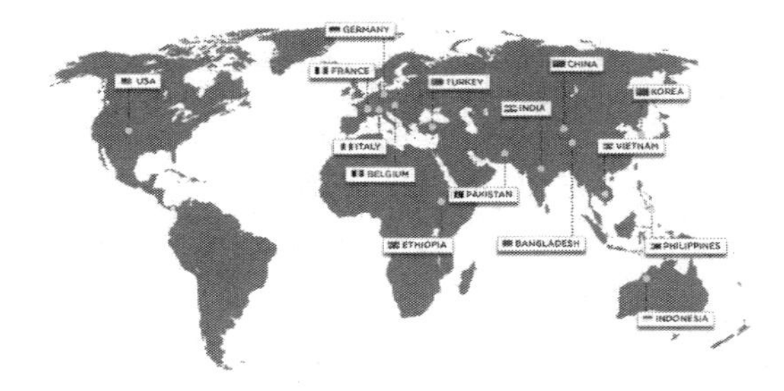

Figure 1. Country-map Textile Industry 2016.

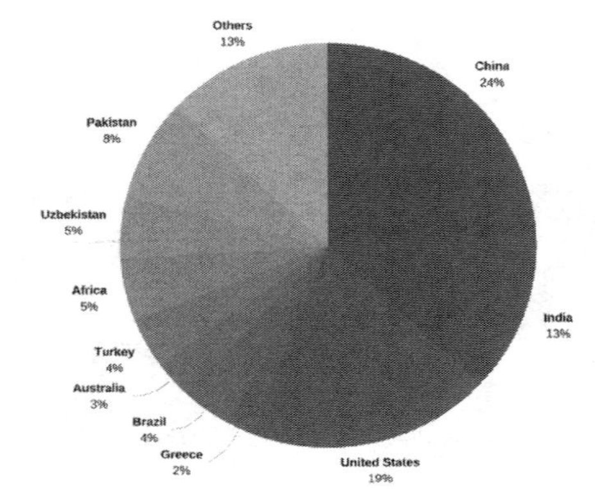

Figure 2. Distribution of world cotton fiber production 2015.

increasing importance of the international commerce, because of the competitive power of some peripheral countries, as India, Thailand, Pakistan, Indonesia, South Korea, Hong Kong and Taiwan (Costa & Rocha, 2009).

This changes have influenced the competition standards on the textile industry. The North American and European companies started thinking its production focusing on the steps with greater added value on the chain, as design, quality, flexibility, production organization and *marketing*. In the seeking for product differentiation and achievement of specify market niche – that are interested in branded products and specify meanings, and not only low prices – this world fashion guides aimed their market competitive strategy by the desires and financial limits of the final costumer (Jenkins & Barbosa, 2012). It is important to point out also the mixture of the reached technological advances on the textile industry with the use of cheap labor of some peripheral countries. The main boost to this displacement is the search for production cost reduction.

The textile industries on the developed countries started to invest on new technologies of conception, process, sales and product, changing their strategy of compete for items of lower added value, originating mainly for Asia, and proceeded to specialize in more profitable materials, possible by the new productive process and by the new developed chemical fibers (Kaplinsky & Messmer, 2008).

In the aim of increase its proximity with the bigger textile costumer markets, this companies developed techniques turned for the reduction on the time for design, production and trading of the textile items, in a way to allow that the production followed the *fast fashion*'s request pace, since it started to prevail on the fashion market (UNCTAD Handbook, 2013).

3.1 *Main international textile markets and commerce*

On the last 20 years it has completed changed the world textile production market. It was found the displacement tendency of a significant share of the production of articles and manufactured items of the developed countries (United States, European Union and Japan), to emerging countries at Asia, Eastern Europe, North Africa and Caribbean. With highlights for China, India, Pakistan, Indonesia, Taiwan, South Korea and Thailand, the main boost for this migration, in addition to the factors mentioned on item 3, is the search for production costs reduction, related mainly to labor (Costa & Rocha, 2009).

Responsible for a meaningful share of the current total amounts, the Asian continent concentrates some of the main world textile producers. Among the worldwide fifteen bigger textile producers, eight are in the Asian continent, and two of them are in the first positions: China and India (Kaplinsky & Messmer, 2008).

China holds a highlight position on the world textile and clothing production. The most populous country on the planet is responsible for almost 25% of the worldwide textile production (Jenkins & Barbosa, 2012).

The author Rangel (2008), defends that the high competitiveness on the Chinese textile chain can be justified by the over-abundant and low-cost labor. With respect to raw materials – cotton and polyester – it also has a privileged position. In addition to that, it produces internally last generation textile machinery.

It is also important in this connection to highlight two meaning changes on the textile market competitiveness standard. The first one involve the Agreement on Textiles and Clothing (ATC) signed in 1995, in which the nations compromised to eliminate the non-tariff barriers for the commerce on a deadline of 10 years. The second change was the removal of the shipping conference on the marine navigation. This way, the shipping prices got lower, raising the textiles globalization and affecting decisively the exports with low unit cost for amount of transported load (Lall & Weis, 2005).

At this point, the extremely aggressive politics adopted by Asian countries to the achievement of external markets was a meaning factor for China's expansion on the international textile commerce. The strategy is the price decrease due to the shipping of big amounts of standardized products, not necessarily of

10 Exporters and Importers of Textiles 2015 (Billion Dollars and Percentage)									
	VALUE	SHARE IN WORLD EXPORTS/IMPORTS				ANNUAL PERCENTAGE CHANGE			
	2015	1980	1990	2000	2015	2010-2015	2013	2014	2015
EXPORTERS									
China	109	4.6	6.9	10.4	37.4	47	12	5	-2
European Union (28)	64	-	-	36.7	22.1	3	4	4	-14
India	17	2.4	2.1	3.6	5.9	25	13	5	-6
United States	14	6.8	4.8	7.1	4.8	5	3	3	-3
Turkey	11	0.6	1.4	2.4	3.8	24	10	2	-13
Korea, Republic of	11	4	5.8	8.2	3.7	-3	1	-1	-11
Chinese Taipei	10	3.2	5.9	7.7	3.3	-4	-1	0	-6
Hong Kong, China	9	-	-	-	-	-7	2	-9	-7
Pakistan	8	1.6	2.6	2.9	2.9	13	7	-3	-9
Japan	6	9.3	5.6	4.5	2.1	-2	-12	-7	-3
IMPORTERS									
European Union (28)	68	-	-	35.2	22.1	3	5	6	-18
United States	30	4.5	6.2	9.8	9.6	13	4	5	5
China	19	1.9	4.9	7.8	6.1	8	9	-6	-6
Viet Nam	18	-	-	0.8	5.8	67	17	13	50
Bangladesh	10	0.2	0.4	0.8	3.2	49	9	14	48
Hong Kong, China	9	-	-	-	-	-9	0	-10	-9
Japan	8	3	3.8	3	2.6	11	-3	1	-8
Mexico	7	0.2	0.9	3.6	2.1	2	3	4	2
Turkey	6	0.1	0.5	1.3	2	24	5	5	-13
Indonesia	6	0.4	0.7	0.8	1.8	35	4	0	-2

Figure 3. 10 exporters and importers of textiles 2015 (Billion dollars and percentage).

low quality. This fast growth resulted on the maintenance of high tariffs for import, adoption of non-tariff barriers (occupational and environmental policies) and establishing anti-dumping actions by the countries that receive the imported products (Farooki & Kaplinsky, 2012).

3.2 *Structure and pattern of concurrence on the textile industry*

Been able to be tagged on markets of *commodities*, the textile market has as attributions segmented markets, patterned and differentiated products. The concurrence on the *commodities* market occur by the costs and prices reduction, with the result the patterned products offer. According to the high expansion of economies and investments, the Asian countries have dominated the world textile market, resulting also on the emergence of new players (China and Vietnam). It is relevant in this process to observe the market of high fashion niches, in which the developed countries' companies have aimed their strategies. They assist the most demanding portion of the population, with high purchasing power (Lall & Weis, 2005).

The resulting of it points out that the textile companies' strategy, should concern with key elements as marketing, the distribution and trading channels. It is important to say that in this market there are demands related to style, design, identity and quick response to fashion changes and on the costumers' profiles (Melo et al., 2007).

4 ASPECTS RELATED TO THE COMPETITIVENESS ON TEXTILE INDUSTRY

The major fashion brands are always seeking more responsible and efficient supply chains. With this, changing efforts in each one of the textile chain steps

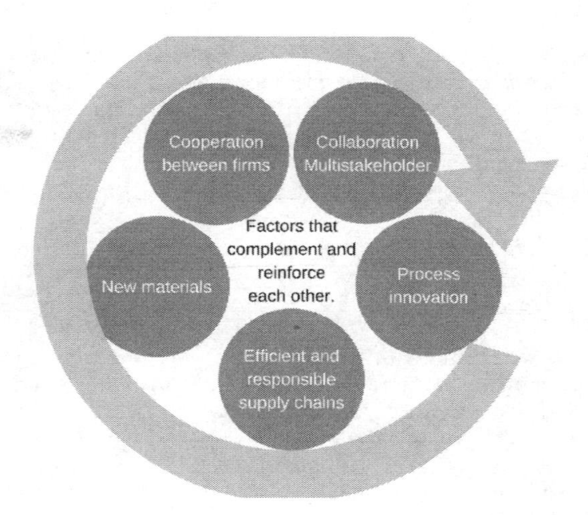

Figure 4. Competitiveness Aspects. ABIT 2008.

are conceived, in order to identify if it is the management model, the strategy or even the process, that need to suffer modifications, so they can assist more effectively, ethically and innovatively the market demands (Oliveira & Epaminondas, 2014). There are five major questions to be raised: new materials, new process and technologies, more efficient and responsible supply chains, cooperation between companies and multi stakeholders collaboration.

Five points has been emphasized for the competitiveness positioning of supplier companies on the supply chain of the textile and apparel sector: business competitiveness (management capacity on the business associated to the use of information technology and communication), products development capability (design innovation, full product development, closely to fast prototyping), product and process quality, delivery punctuality and socio-environmental responsibility (ABIT, 2015).

4.1 *Innovation as a source of competitiveness*

There is a need of attention of the managers to the importance of the innovations, considering them a competitive success critical component (Govindarajan & Trimble, 2010). Sparrow (2010), that defends a culture turned to innovation need, and Bowonder et al. (2010), associate the innovation based strategies to the achievement of competitive advantage.

Tidd et al., (2008, p. 85), understand that innovation is "more than simply conceive a new idea; it is the process of developing its practical use". The authors understand that the innovation can mean new ways of serving the already established and mature markets, affirming that tendencies change and the consumers themselves demand changes in diminishing periods of time. The point here is not just to create something new, but also to improve something that already exists, creating market competitiveness.

The increased profitability happens due the innovation, with the surging of new products and services, allowing the reach to other market slices. This way, new products and services are important mechanisms to keep the competitive company. Therefore, one of the manager major roles is to identify the innovative costumers (those one that are the first to acquire the product or service) because they are the ones that set the success or failure of the innovation (Venâncio et al., 2014).

4.2 *Innovative clusters on the textile industry*

The capability of creating a prosperous environment to the companies' innovation is extremely important for its survival (Weenen et al., 2013), it increases the productivity efficiency and effectiveness, enabling a fast answer to the costumers' needs, creating competitiveness and reaching new markets (Bae & Chang, 2012). The innovation can be seen in a technological perspective (Linton, 2009), once that is the technology what facilitates the products' acceptance or not, by the costumers' satisfaction and needs, and the technology competitiveness in relation to the previous (Dantas & Moreira, 2011).

The innovation generation depends increasingly of new technological expertise generated not only by internal activities. The process' complexity resulted in a huge growth on the use of external systems (Zeng et al., 2010). The companies can't innovate separately, in a way that they must complete their capacity of creating internal knowledge with external innovation sources, in order to reach market success, concluding that innovation is like a cycle that evolves interactions between silence and coded knowledge (Yam et al., 2011).

On the recent years, it's becoming apparent in the world textile industry a meaningful change on the strategy, changing from a guidance to the production to a guidance to the market, with offers of textile and apparel conceived and manufactured accordingly the final costumers' preferences and demands. Because of the large part of the textile producers create for the external market, the textile industry needs to focus on the globalization and on the internationalization, optimizing costs, adapting to the market changes, developing the capacity of innovate and fast responding the costumers, being always able to answer the now market demands.

5 ELECTRONIC COMMERCE APPLIANCE ON THE TEXTILE INDUSTRY AS AN INNOVATION

The increase of the concurrence orientated the restructuring of the textile companies' strategies. Instead of a vertical production model, we now have a fragmented model, with autonomous steps that adjust to the new conditions and use the advantages offered by the market. In this scenario, the electronic commerce comes up as an alternative lower investment to the textile negotiation, communication and trading, making it possible fast worldwide presence and democratizing the commercial acting. According to Porter (2003, p. 17), "the introduction of a new important technological innovation can allow that a company reduces the costs and simultaneously intensify the differentiation, and maybe reach both strategies". The introduction of a new technology, and we talked here about online platforms, is understood as steps towards the adoption and acceptance of a technology on the market.

Electronic commerce platforms with deals as B2B (Business-to-business) provide advantages to sellers and buyers that meet through this channel and deal in real time. According to Albertin (2004), one of the advantages to the buyers is to make the companies buying decisions faster, simpler and more efficient. With this kind of deal, the companies can use the system of research and negotiation with different sellers at once. As for the sellers, according to Pinna (2001), the major benefit is the easy access to the costumers all over the world, simplifying the prices, orders and deal approval process.

Due the benefits, with the costs reduction, more agility on the searching and contact procedures with suppliers and buyers, more control on the tendering process and, consequently, the reduction of mistakes on this process, makes the electronic commerce B2B almost a need on the competitive markets (Felipini, 2015).

Thinking on the textile industry, the great productivity brought by the electronic commerce is directly connected to the world representation that it provides. Expand the selling to beyond the national borders is related to two important competitive aspects: supply chains and new process and technologies. With this, the B2B commerce is now a natural tendency of the textile companies, when it involves meaningful changes on the commercialization, promotion and distribution, further allowing, with its evolution, be an important point in the textile production and export/import map.

5.1 *Electronic commerce B2B*

The electronic commerce Business-to-Business (B2B) relation is characterized by involving transitions between companies using the internet. This also includes the information flow inside the organization and also the communication between suppliers and buyers (Chaim, 2000). To follow up an electronic commerce strategy, the companies must understand that the internet is a tool and that before adopting a technological solution, it's essential to make a product, costumers, competitive situation, resources and company operations evaluation to better understand how all those elements fit in an electronic commerce strategy (Hutt & Speh, 2010).

In relation to the B2B market, the connections between the suppliers and buyers are constructed on the searching for beneficial partnerships for both sides. The platform presents itself as the image of the sales representative and takes a crucial role on the building and sustentation of the relationships and partnerships with the costumers, becoming a major value creator. This kind of relation becomes even more important in high value, complexity and risk deals (Brito & Vasconcelos, 2004). The companies pursue the capacity of take the lead, innovating and anticipating themselves on the search for solutions, on the value creation and on the general perception as a supplier that highlights among its competitors.

Using the online infrastructure, a company can provide higher operational speed and costs reduction (Mahadevan, 2003). For that matter, the e-business is more than simply trading products and services for money on an online network. It's a technology that allow the companies to raise up the transitions process' accuracy and efficiency and makes possible the exchange of information between clients and suppliers, benefiting all the chain members (Trepper, 2000).

5.2 *The challenges of applying the electronic commerce in the industry*

When the textile industry adopts as part of its strategy the investment in solutions for the electronic commerce, it end up facing some barriers and challenges. One of them is to open space for channel conflicts between physical sales and online, maybe presenting a scenario of disorder to the clients. An adopted solution for this problem would be organize the electronic commerce in a decentralized way, having multi online representations, capable of solving specific problems, customized service and other bureaucracies, and a centralized logistic base, platform performance, marketing, costumers' service and distribution (Zeng & Pathak, 2003).

Besides that, its relevant to think that virtual stores are susceptible to attacks (Andrade, 2011), what prejudices the realization of certain sales by the textile industry, by its high values and bureaucracies (Albertin, 2004). Monteagudo (2006) observes that the insecurity mood that lots of people still fell on internet makes that several companies don's begin their activities on the electronic commerce, what delays a greater growth of this kind of commerce.

Lastly, one of the biggest concerns of the textile companies towards the online commerce B2B is the vulnerability in relation to the intern data exposure on the web, because they can be given more easily to the concurrence (Albertin, 2004). For that, it's necessary to organize the platform in a way that, to have access to certain data (prices, technical information and specific data), and provision of information about the clients is mandatory.

Some of the key variables for the failure on the implementation of an information technology tool in a company are the technology, cultural, organizational and structural barriers, lack of strategy vision, lack of training and products and sales information complexity (Rigato, 2013). By the other side, if there is applied resources on this strategy, it's necessary attitude on breaking paradigms, searching the innovation, personal and organizational development (Zeng & Pathak, 2003).

5.3 *Proposal for the electronic commerce improvement on the textile industry*

Buying by the internet is riskier than the traditional, because of the absence of opportunity of inspect the product physically (Kim & Forsythe, 2009). When it becomes on textile products, this factor gets worse because of some attributes as quality, weight, texture and color, that are important for taking decisions and at the same time they are hard to be shown by the screen. The written descriptions are in general insufficient for the product analysis (Grewal et al., 2004). Fashion goods are usually classified as experience products, because the information acquisition about the product is hard during the shopping, and the quality analysis just is possible after its use (Ha & Stoel, 2012).

For textile goods, on which is very important to touch and feel, the attributes presentation in a website are critical incentives to promote the navigation (Park et al., 2012). Besides that, as the B2B shopping are in large amounts and high values, it becomes fundamental to think in a solution for the attribute "touch".

A proposed solution for this problem would be send to the more active costumers, periodically, catalogues with little textile samples of the next company's launches. Although it is delegate more duties to the creative and logistic sectors, susceptible to future complications, the catalogues can be positively on the selling realization. In addition, it will be a facilitator for listing the important clients, allowing long-standing customers loyalty and reaching new ones.

6 CONCLUSION

In a general way, on the textile market, the dislocation of a meaningful part of the production for multiple countries, linked to a lower cost production, resulted in a competitive positioning by part of the concurrence, changing strategies. The innovations on the technology area, quite often came not only to bring improvements to the processes but also to create a more competitiveness environment.

The bibliographic review emphasized the use of the online commerce on the propitious moment to the textile exportation and importation scenario, because of its capacity of worldwide and continued presence, operating around the clock and innovation on the communication between the deals' participants. With a good strategy, there's space for all at the textile commercial chain, when we think about virtual commerce. But for that, it's necessary to size the chain conflicts, balance the commercial interests and guarantee the data security, for the clients and the company. It's also necessary to think on solutions that facilitate risky deals, that involves high values and bureaucracies.

We can conclude that the technology advance that came with the internet and the electronic commerce can contribute for the textile market B2B development. The high potential of the information technology tools, in terms of costs, speed and research is a relevant channel for the segment competitiveness. The use of the electronic commerce as a capacity of revolution for the process of the organizations, providing productivity gains, procedural improvements, operational costs reduction and mainly on the innovation of the offered service.

REFERENCES

ABIT – Associação Brasileira da Indústria Têxtil. 2008. BNDES: A cadeia têxtil e de confecções – uma visão de futuro. *Rio de Janeiro: Apresentação realizada no BNDES.*

ABIT – Associação Brasileira da Indústria Têxtil. 2015. São Paulo.

Albertin, A. L. 2004. *Comércio eletrônico: modelo, aspectos e contribuições de sua aplicação.* São Paulo: Atlas.

Andrade, R. 2001. *Guia prático de E-Commerce.* São Paulo: Angra.

Bae, Y. & Chang, H. 2012. Efficiency and effectiveness between open and closed innovation: empirical evidence in South Korean manufacturers. *Technology Analysis & Strategic Management,* 24(10): 967–980.

Bowonder, B., Dambal, A., Kumar, S. & Shirodkar, A. 2010. Innovation strategies for creating competitive advantage. *Research Technology Management,* 3(53): 19–32.

Brito, L. A. L. & Vasconcelos, F. C. 2004. *A heterogeneidade do desempenho, suas causas e o conceito de vantagem competitiva: proposta de uma métrica.* 107–129.

Chaim, R. M. 2000. Comércio eletrônico ou canal de vendas eletrônico? *Perspect. Cienc. Inf., Belo Horizonte,* 1(5): 69.

Costa, A. C. R. & Rocha, E. R. P. 2009. *Panorama da cadeira têxtil e de confecções e a questão da inovação.* Rio de Janeiro.

Dantas, J. & Moreira, A. C. 2011. *O Processo de Inovação – Como potenciar a criatividade organizacional visando uma competitividade sustentável.* Porto: Lidel Editora.

Farooki, M. & Kaplinsky, R. 2012. The Impact of China on Global Commodity Prices. *New York, Routledge.*

Felipini, D. 2015. O comércio eletrônico B2B. *E-commerce.Org.* São Paulo.

Gil, A. C. 1996. *Como elaborar Projetos de Pesquisa.* São Paulo: Editora Atlas.

Govindarajan, V. & Trimble, C. 2010. *O outro lado da inovação: a execução como fator crítico de sucesso.* Rio de Janeiro: Editora Elsevier.

Grewal, D., Iyer, G. R.; & Levy, M. 2004. Internet retailing: enablers, limiters and market consequences. *Journal of Business Research,* 7(57): 703–713.

Ha, S., & Stoel, L. 2012. Online apparel retailing: roles of e-shopping quality and experiential shopping motives. *Journal of Service Management,* 2(23): 197–215.

Hutt, M.D. & Speh, T.W. 2010. *B2B: Gestão de Marketing em mercados industriais e organizacionais.* Porto Alegre: Bookman.

Jenkins, R. & Freitas, A. 2012. Fear for manufacturing? China and the Future of Industry in Brazil and Latin America. *The China Quarterly.*

Kaplinsky, R. & Messer, D. 2008. Introduction: The Impact of Asian Drivers on the Developing World. *World Development,* 2(36).

Kim, J. & Forsythe, S. 2009. Adoption of sensory enabling technology for online apparel shopping. *European Journal of Marketing,* 9/10(43): 1101–1120.

Lall, S. & Weiss, J. 2005. China's competitive threat to Latin America: an analysis for 1990-2002. *Oxford Development Studies,* 2(33).

Linton, J. D. 2009. *De-babelizing language innovation. Technovation,* 29(11): 729–737.

Lu, S. 2016. WTO Reports World Textile and Apparel Trade in 2015. *Department of Fashion & Apparel Studies, University of Delaware.*

Mahadevan, B. 2003. *Making Sense of Emerging Market Structures in Business-to-Business E-Commerce.* California: Management Review.

Melo, M. O. B. C., Cavalcanti, G. A., Gonçalves, H. S. & Duarte, S. T. V. G. 2007. Inovações Tecnológicas na Cadeia Produtiva Têxtil: análise e estudo de caso em indústria no nordeste do Brasil. *Revista Produção Online UFSC, Santa Catarina,* 2(7).

Monteagudo, R. 2006. *O E-Commerce e a I-Empresa.* São Paulo.

Oliveira, P. H. & Epaminondas, M. E. R. 2014. Conhecimento, Inovação e Estratégia competiti-va: um estudo no setor atacadista da moda. *Revista Eletrônica de Estratégia & Negócios,* 1(7): 82–104.

Park, E. J., Kim, E. Y., Funches, V. M. & Fox, W. 2012. Apparel product attributes, web browsing, and e-impulse buying on shopping websites. *Journal of Business Research,* (65): 1583–1589.

Pinna, R. 2001. Sopa de letrinhas. *Revista TI.* São Paulo.

Porter E. 2003. *Estratégias Competitivas Essenciais.* Rio de Janeiro: Campus.

Rangel, A. S. 2008. *Uma agenda de competitividade para a indústria paulista.* São Paulo: Instituto de Pesquisas Tecnológicas do Estado de São Paulo.

Richardson, R. J. 1989. Pesquisa Social: Métodos e Técnicas. *Colaboradores José Augusto de Souza Peres.* São Paulo: Editora Atlas.

Rigato, C. A. 2013. *Uso da internet na pós-venda de serviços business-to-business.* Tese de doutorado, Universidade de São Paulo, SP, Brasil.

Sparrow, P. 2010. Cultures of innovation. *Management Today.*

Tidd, J., Bessant, J. & Pavitt, K. 2008. *Gestão da Inovação.* Porto Alegre: Bookman.

Trepper, C. H. 2000. *Estratégias de E-commerce.* Rio de Janeiro: Campus.

Unctad. 2013. *Handbook of Statistics, United Nations.*

Venâncio, D. M., Andrade, D. & Fiates, G. G. S. 2014. Inovação em serviços: Um estudo bibliomé-trico da produção científica no Portal Capes até setembro de 2013. *Revista Eletrônica de Estratégia & Negócios, Florianópolis,* 1(7).

Weenen, T., Pronker, E., Cimmandeur, H. & Claassen, E. 2013. Barriers to innovation in medical nutrition industry: a quantitative analysis of key opinion leader. *Pharma Nutrition,* 1(3): 79–85.

Yam, R. C. M., Lo, W., Tang, E. P. Y., & Lau, A. K. W. 2011. Analysis of sources of innovation, technological innovation capabilities, and performance: An empirical study of Hong Kong manufacturing industries. *Research Policy,* 40(3): 391–402.

Zeng, A. & Pathak, B. 2003. Achieving information integration in supply chain management through B2B e-hubs: concepts and analyses. *Industrial Management & Data Systems,* 9(103): 657–665.

Zeng, S. X., Xie, X. M., & Tam, C. M. 2010. Relationship between cooperation networks and innovation performance of SMEs. *Technovation,* 30(3): 181–194.

Textiles, Identity and Innovation: Design the Future – Montagna & Carvalho (Eds)
© 2019 Taylor & Francis Group, London, ISBN 978-1-138-29611-4

Charles Frederick Worth and the birth of Haute Couture: Fashion design and textile renewal in the times of conspicuous consumption

Maria João Pereira Neto
CIAUD, Lisbon School of Architecture, University of Lisbon, Portugal

ABSTRACT: The paper focuses on Charles Worth, the pioneer of the Haute Couture industry as creator and entrepreneur, but also as someone who understood the role of progress in textiles, determinant for its global success. The intention is to demonstrate how Worth from 1860 reached an inauspicious development of the fashion industry and textile world which only had counterpoint in the golden years of Haute Couture of the 20th century. We emphasize the pioneering role of a man who takes a prominent place in the universe of nineteenth-century luxury, integrating the ability to assert the power of a name through personalized labels. As the first international fashion "dictator", Worth defined a style for a new social class and established the basis for fashion industry that endures. He proposed a new exquisite elegance: an age of elaborate ornamentation with social acceptability and identity even for those so called new world that with Worth may compete in elegance with western elites and royalty.

1 THEMATIC AND BIOGRAPHICAL FRAMEWORK

Being fashion a barometer of culture, concerns and interests, that shaped a time period or time we sought with this text to highlight an entrepreneur and a creator who had the strategic vision of creating a global business in a century where modernity and progress. The changes he introduced in the then disorganized confectionery business laid the foundations for the contemporary creative clothing industry and helped redefine the very modern concept of haute couture and the way we view textiles and fashion. Charles Frederick Worth, was born 1825 in Bourne, South Linconshire, England in October 1825, the son of a solicitor who left the family in a very fragile economic situation. At age 11, he was embarrassed to look for work: first in a printing office and shortly thereafter he went to London as an apprentice at the big retailers Swan & Edgar in Picadilly n° 10.

Accordingly to Worth biographers like Dianna de Marly (1990): London "was the centre of the fashionable world in Britain, particularly where male fashion was concerned (...) Very few ready made clothes were sold at his time, apart from loose garments like cloaks and mantles. A lady desirous of a new dress would go first to select the material at a haberdasher's or mercer's, and then take it to have up into a gown in the latest style. These milliner's were open for very long hours, particularly during the Season, and customers who wanted a ball gown in a hurry would expect an placed in a morning to be ready by that evening (...) Worth grew up in the establishment, maybe he lived on the premises, sleeping under the counter (..)". His main task was to unpack the news from Paris. Soon he understood that especially the emerging female audience was thirsty for news and also for expert advice. His presence in London was of utmost importance, not only by his first and decisive contact with luxury, but most of all by learning to know the tastes of elite who bought and ordered the most sophisticated products in the large and exquisite warehouses of Victorian London. Nor should you forget the proximity of the National Gallery where one can deepen an artistic passion by studying the styles and intricate details of the period costumes painted by the old masters.

The art of the past will became for Worth the inspiration for his future as costume designer. In 1846, at the age of 21 and with considerable experience in the world of textiles and his trade, Charles Frederick Worth went to Paris the center of luxury since the reign of Louis XIV, with a hand full of nothingness and a heart full of dreams, he didn't received any education or formal guidance in fine arts, but he had a major experience in selling textiles to be made in clothes. Soon Charles Worth will stands out as a fine salesman at *Gagelin – Opigez & Cie*, the prestigious novelty warehouse located at 83 rue Richelieu, celebrated for the excellence of silks used by the royal court tailors not only by intricate fabrics and embroidery, but also by the famous shawls of Cashmere, a fashion accessory very popular in the first half of the nineteenth century, as well as cloaks and covers. In these large warehouses he progresses to the position of head of sales in the section of shawls and covers.

It is here that he meets Marie Augustine Vernet with whom he came to marry in 1851.

As a salesman at Gagelin-Opigez & Cie, he promoted all the most innovative ways of capturing the attention of customers and potential buyers, using his wife as the living model for shawls and hats he sold and then went on to produce. Marie was Worth's inspirational muse and always instigated him to follow his own path, being also the first living mannequin.

According to De Marly (1990) "Marie was the catalyst which turned Worth from a textile salesman into a real dressmaker".

Breward (2003) and De Marly (1990) refer that Worth understood that he was capable to sell everything what she was wearing so, he became more and more concerned about Marie's overall appearance. During his London experience he must be aware of the quality of English tailoring techniques which gave both men's suits and women's riding habits an excellence of fit unequalled in the rest of the world. Worth began to tailor costumes to complement the shawls produced by *Gagelin Opigez & Cie*, initially of enormous simplicity, but which by the excellence of its cut, its sobriety and elegance attracted an ever-increasing customer to the shop.

The first dresses he prepared for Marie were made of white muslin and utterly plain, but the fit was so good that customers become addicted to his clothes. It was following this immense success that Gagelin allowed Worth to open a section for women's dresses in his *magasin*, which according to the consulted sources (De Marly, La Haye, Coleman) assumes his official entry into the world of clothing. Clients who sought the creations of Charles Worth, the first man in the world mainly dominated by women, expected him to define the costume design that was always made to order and by measure what constitutes an immense innovation. However, this partnership with Gagelin was not always peaceful, as young Worth's ideas clashed with a more traditional view of his employers: "they were the most distinguished silk mercers in the whole Paris and not common dressmakers. But it was at the service of his employers that Worth's creations were the creations of Worth were displayed at the 1851 Universal Exhibition in London, which earned him not only a huge reputation but also, later, the possibility of opening his own business in one of the most prominent arteries of Paris in 1858. At the London Universal Exposition, the Gagelin House was awarded a gold medal "for excellent embroidered silk, wrought up in dresses of elegant style" (Marly, 2009). According to some of the authors consulted, and especially one of Worth's earliest biographers his son Jean Philippe (1856–1926) quoted by Diana de Marly, this medal was only made possible by the impact of his father's extraordinary creations – with the most intricate borders in magnificent silks. Later with the Exposition Universelle of 1855 in Paris, Worth's employers entered his dresses again in a public demonstration of pride for his works.

2 THE HOUSE OF WORTH

In 1858 Charles Frederick Worth had already acquired the knowledge and the creative talent to start its own business, so he joined in a partnership with Otto Bobergh (1821–1881) a fellow colleague from another çompany, who provided the capital required. They arranged the first floor at Rue de la Paix, in central Paris, near the Opera House, with an initial staff of 20 workers. The political situation was favourable to all luxury and extravagance proposed by Worth.

It was the time of the outcome of a new social order. This street was soon to become the most fashionable address in the world. Worth and Bobergh furnished all the premises along the lines of a private residence to ensure the maximum comfort to their clients, and by doing so, these elite fashion firms became known as Houses (Maisons).

It was Worth's idea that as clients were out in the country, or in vacations during summer, it could be interesting to present a new Spring/ Summer collection, earlier, at the beginning of the year. It was a major innovation.

Since the beginning of his career, Worth tried to be associated to an artist and not only as a business man. And doing so, he tried to look like a master painter like Rembrandt, with a velvet beret, fur trimmed coat and a floppy neck tie.

But we need to go back to 1848 when the Prince louis Napoleon (1808–1873), the first elected president of the French Republic, after the fall of King Louis Philippe, is proclaimed emperor. He and his wife since 1853, Eugenia de Montijo (1826–1920) a spanish aristocrat (To be knowned as Empress Eugénie), are conspicous consumers and assumed all the excesses of luxury, which they trnsport to the court whose opulence exudes to the streets of Paris, a city effervescent of change and modernity, celebrated by painters and poets.

In 1859, one year after the inauguration of Worth's and Bobergh's Couture House, the princess Paulina of Metternich (1836–1921) arrives in Paris as the wife of the Austrian ambasador under the élan of distinction, elegance and good taste that will become legenadries.

It's supposed to be her who presented Worth to the Empress through a wonderful floral tulle ball dress used for a ball at the Tuilleries, under the own request of Marie and Charles Worth. Soon the Empress becomes the most fervent fan and costumer of Worth, failling to be faithful to the royal dressmakers who had been responsible for the wedding dress and trousseau. The Empress Eugénie thus becomes Worth's main ambassador and will be the most sought and most chosen seamstress for all the major events of Parisian society.

Thus, from 1864, Worth becomes the official couturier of the Empress. He has created all her wardrobe, from court to gala dresses. For The inauguration of the Suez Canal in 1869 Eugénie felt the need to order more than two hundred and fifty suits. The Worth phenomenon was lauched. With so much demand, Worth and Bobergh had more than 1200 seamstresses, assuming its founder not only as an artist and creator, but also as someone who could make small adjusments through the smallest and exquisited details to the end of the work, as a real work of art. He was a remakable entrepreneur, and also a remarkable man ahead of his time.

In the 1860s Worth envisioned the possibility of creating a new silhouette that would change the course of fashion in the following decades until the end of the nieteenth century. Worth has reinforced the circular crinoline, so popular since the 1830s, by introducing a narrower, rather heavily armed skirt behind it: rustling, creating a new S shaped silhouette, that will change women body.

Preferred and exclusive couturier of Eugenia Empress and her court, and after the success and revolution proposed by the wedding dress of Princess of Wales, Charles Worth has become a symbol of luxury and privilege of a conspicuous consumer class that was not visible since the 18th century.

The silhouette proposed by Worth extended to the other real houses like those of Austria being the Austrian empress Elizabeth, one of its most faithful clients. Worth also knew how to take advantage of the potential of the international market, such as the US, whose newly wealthy families came to Paris in search of new fashions and new forms of consumption, and mainly for respectability and credibility, as we can read in 19th century authors like Edith Warthon.

Great bankers like the Rostchilds were also his customers, as well as American and Australia's *noveau riches*. Worth had the strategic vision of opening what are now called corners in some of the world's largest retail but body-adjusted warehouses: mass-produced garments but personalized. His season collections were prepared in Paris, and soon, in a few weeks, were been sold in the other part of the world: the originals, or copies, under his control. His signature became a mark with real and symbolic value. With the end of the Second Empire and the Prussian Franco War, Worth found himself forced to close his business, and the building was used for a hospital. The times were a little difficult with the war and the fall of the monarchy, and he had to adjust to new clienteles and new products with lines of mourning and maternity and even sport. As Veblen and Bell said it was the heyday of the new paradigms of consumption of an emerging new class and a new social order that integrates the conspicuous consumption of goods as a way of asserting his power.

Attentive to market developments and to the emergence of new classes, Charles Worth will seek and secure new customers outside France at the same time encouraging the French textile, silk industry to the new laws of the market.

By the end of the 1880s the structure of a modern couture house with two seasonal collections was consolidated, as was the extension of the brand to the franchising of patterns and fashion designs. The consequences of the War and the fall of the monarchy in 1871 led to the end of the partnership between Worth and Bobergh, however Worth's popularity did not falter.

Despite the end of the business partnership with its Swedish partner, finally came the House of Worth in all its brilliance continuing to impose itself on the world of fashion even after the death of its creator in 1895, continuing to reinvent itself in the twentieth century, first with one of his sons and then with Poiret having closed its doors in the 1950s.

3 THE HOUSE OF WORTH AND TEXTILE WORLD

We highlight Worth's role as a great fashion and entrepreneur. He was the first to know how to withdraw from the client the power of choice over creation, assuming himself the role of trendsetter and the ostentatious styles of the elites. He was the main responsible for the revolution in the process of creating Fashion. As Lipovetsky (2010) said under its aegis fashion reaches the modern age, it has become a creative and also an advertising enterprise.

He was responsible for a whole series of small/large changes in the fashion industry and many of them remain current, more than a hundred and fifty years years later.

The technical innovations he used in terms of detail, fabric and colour denote extraordinary knowledge and skill: at the level of the complexities of engineering and geometry applied to the world of textiles.

Worth had forged good working relationships with textiles manufacturers whilst at Maison Gagelin-Opigez et Cie and invited them to make special production runs of unique fabrics for his collections. He worked especially closely with the silk weavers in Lyon and collaborated with them on the designs, including reviving historical patterns. Worth was also responsible for ordering the production of fine, lustrous silk satin which has, ever since, became a mainstay fabric for luxurious evening gowns (De La Haye, et al. 2015).

According to authors such as Watt, Veblen and Bell, the innovations of the so called Industrial revolution created the means to produce more objects and goods than those who usually were required by consumers. "Therefore new markets were eagerly sought and the new consumers seduced into believing that they needed mass produced decorative items" (Watt 2004), London's Great Exhibition of 1851 is referred to be the first in a row of major international venues to display the wonders of technical as artistic achievements to the public.

Accordingly to Watt (2004) textile designers took the full advantages of these major opportunities to display not only new designs but also the new techniques, assuming a continuous of technical improvements which combined with the popularity of historical revival styles in fashion (mainly dresses) but also in home furnishings, produced some spectacular textiles for both body and furnishings.

This applied to all types of textiles from the most inexpensive printed cottons to the most exquisite woven silks. There was also a dramatically change in the textile dying technology with some discoveries like aniline dyes – the first completely synthetic dyes, by W. H. Perkin in 1856.

Prior to his discovery, "all textile dyes were derived from natural sources – plants, insects and minerals. The first aniline dye was a manmade re- creation of the colouring agent in the madder rot, which produced numerous shades of red. Different aniline dyed coloured fabrics were shown at the London International Exhibition of 1862 were the textiles were displayed next to the sticky black coal tar waste from which the dyes were derived(...). Soon there was a major competition between French and British textiles manufactures and designers, fuelled by the impact of the major international exhibitions" (Watt 2004).

However France remained the leader country in costume and interior fashions, while England "sought to capture a larger share of luxury goods market. French textile designers were better trained and earned more money than their counterparts elsewhere in Europe (...) Printing was the main technique used by late 19th century textile artists to the ease of translating a concept into a finished product. As mechanical printing techniques continued to improve, the range of fabrics that could be successfully printed increased and texture fabrics such as cotton velveteen, became very popular as furnishing fabric in the later decades of the century" (Met Museum 2017).

3.1 *Worth and the French textiles*

Perhap's one of the most important contribution to the world of fashion and textile industries of the House of Worth, was the type of fabrics that was chosen. Following the 1870 and the fall of the Second Empire,"Charles Worth became an even more important client for the textile and trim producers of Lyon and its environs. Worth used pre-existing yard goods and worked with manufacturers to come up with lavishing patterns for new materials. Charles Worth had begun his designing career by following the expansion of women's skirts in the 1850s, when they were supported by layer upon layer of petticoats. In the late 1850s Worth draped yards of fabric over the skirts' increasing width, as the newly devised crinoline cage, permitted expansion without increased bulk. Many Worth's dresses from this period, sadly, were frothy, cloudlike confections in silk tulle that have melted into oblivion." (Coleman 2014). However it's major impact, can be seen in works by artists such as Franz Xaver Winterhalter like the famous portrait of Empress Elizabeth of Austria of 1865.

As Coleman (1989) said Worth introduced hooped dresses with flatter fronts in the early 1860s however it has evident that he was careful when he choose not to diminish the amount of material needed; he only pushed the fabric to the back of the dress. "From decade to decade Worth clearly championed a progressive attitude to fashionable changes, experimenting with the flat fronted crinoline and gored skirts(...), the mermaid – like: 'princess – line 'devoid of waist-seam in the 1870s, and the revival of gigot sleeves in the 1890s" (Breward 2003). These apparent less expansive styles posed a curious economic challenge. Having been trained in great and specialized warehouses, Worth recognized the danger of weakening trades that may contributed to the success of his own business. Therefore he had to either incorporate large quantities of material into his garments or support the production of costlier luxury goods. In order to maintain a high level of consumption, the House moved material throughout much of the 1870s and 1880s from draped overskirts to trains, bustled backs, and a variety of combinations of these styles. So," just as the Empress Eugénie's patronage of the French textile industries had been crucial before 1870, so also was Charles Worth's business vital for the looms of Lyon and Paris that created spectacular luxury materials afterward" (Coleman 2014).

Many of Worth's early garments were made of unpatterned silks-tulles, taffetas, reps, and satins-or nominally patterned fabrics featuring stripes and small floral sprays-in. "Since the 1870s, perhaps as a move to fill the immense void left by the departed French Court, the House increasingly employed more expensive textiles usually associated with household furnishing in its garments." (Coleman 1989, De Marly 1990)

As we know, Worth, in an innovative and perhaps provocative way, used grand-scale floral motifs generally designed for wall coverings in garments whose skirts were often not long enough to include a full repeat of the pattern.

Such luxury and exquisite fabrics, exhibiting an immense richness of material and the highest level of technical skill, were a major feature of the House of Worth's models until the first years of the twentieth century. "With the exception of machine-made laces, Worth's trims and embroideries matched the ground fabric to which they were applied" (Coleman 2014). According to the same sources, the consensus among Worth's clients was that these costly outfits were worth the price, and a real and priceless status symbol.

Charles Worth and his house did not merely purchase materials; they are also known to have worked closely with textile manufacturers. From such concerns as *A. Gourd et Cie*, *J. Bachelard et Cie*, and *Tassinari et Chatel*, the Worths either commissioned specific designs or ordered pre-existing patterns which were often been displayed in important exhibitions as for instance the 1884 London's International Health Exhibition, where the textile world and fashion were displayed.

Many of the fabrics that were found in late nineteenth, or early twentieth-century Worth's garments, feature subjects that were especially popular within the House such as feathers, stalks of grain, stars, butterflies, carnations, iris, tulips, chestnut and oak leaves, scallops and scales, and bowers of roses. Every Worth's gown was a real masterpiece by its own right, overflowing with rich silk materials, trimmings of intricate embroidery, ruffles, frills and floating fine silk tulle. White, cream and gold were some of his favourite colours (De La Haye 2015). However, lavish textiles were not only used for evening wear in Worth's designs as his wonderful day dresses of cut and uncut voided velvet attests. Some ensembles are nowadays shown in great Museums like the MET and the V&A which provides extraordinary examples of Worth's art and practice of incorporating elements of historic dresses in his designs.

As we can see thorough V&A website, or even in the archive organized by De la Haye (2015), we still have wonderful and well preserved examples of the large scale of the pomegranate and floral motifs that follows the Louis XIV's style in textile patterns, and also other kind of fashions, like the 18th century. The bodices and overskirts, made in one piece and worn over separate skirts, are known as polonaises, and they are modified versions of an eighteenth – century style popularized by the French Queen Marie Antoinette. These kind of day dresses would have been appropriate for walking or making social calls and worn by women of the upper classes. The back of these polonaises were expertly draped to accommodate the bustles worn by fashionable women in the 1880s all over the world.

4 LEGACY: WORTH AND COUTURE

"(...) The men believe in the Bourse, and the women in Worth" said the 19th century journalist Felix Withehurst (Marly 1990).

Charles Worth was not only a designer, a creator but also an arbiter of taste. He dictated the styles that his clients wore and this was unprecedented. It was also remarkable how he organized his confections as a true production line in which each employee had a specific function as well as production goals, labour standards and distribution, space organization: a taylorist organization long before the term has been created. Worth's elaborate costumes were created in standardized and interchangeable parts allowing not only faster results but also enormous savings in fabric and great quality in the details and finishes, comparable, or surpassing the old Parisian master tailors and the great court dressmakers.

Still in the 1860s he adopted the paper molds and trained the modellers who worked with him, customizing services, while producing in series. Worth used the sewing machine for all tasks with the exception of the smaller and more intricate and delicate details, being receptive and promoting any innovations that would allow the adornments in a fast, efficient and refined way. He encouraged the French factories to produce textiles and silks of the highest quality, collaborating with the industrialists in the design and production of innovative products and patterns anticipating contemporaneity.

He was, however, a kind of demiurge in encouraging the strained use of corset so extremely tight that it sacrificed women's bodies, deforming them, molding them and imprisoning them to enhance the beauty of his own creations. Crinolines and the new S-shaped silhouette shaped and deformed the woman's body for decades, until the liberation proposed by names like Poiret. Charles Worth was the first couturier to identify his clothing through an internal custom labelling.

This was a symbol of status, a certificate of power, of extreme quality and, as Bourdieu (1975) said quite as a magic power, whose signature, or miraculous *griffe* is a manifestation of power, a recognized abuse of power. Worth recognized this magical value. The power of clothing and garments that goes beyond its functionality and / or beauty through its high costs. Inflated by the simple fact that they are inaccessible to most people, only accessible to a few chosen. Worth innovated the process of producing luxury clothing, creating the Haute Couture system.

By adding symbolic value, cultural capital and prestige to clothing, Charles Worth defined the fashion system for the coming ages. According to some major historians of the modern fashion and textile world (Breward 1995, 2003, Tarrant 1994) in contrast to what happened in nineteenth-century England, where industrialists were not great innovators, but if they were able to adjust to the new needs and above all the inventions of others, making it feasible, accessible and reproducible in France, Charles Worth was able to work with what already existed. Before Worth, dresses were specifically created for a single woman, for each woman. They were produced by a professional or amateur, almost always woman, seamstress or by the woman herself.

Although some coarse fabrics could be produced at home, most of them were bought or marketed through commercial establishments in the interior of the country, usually from where the fabrics came, and / or in warehouses selling a little of everything from food to tools and animal food, up to lace, or in the context of nineteenth-century modernity, in large, more or less sophisticated urban warehouses; *les grands magasins.*

The possibility has also arisen that, in some places, especially those specialized in fabrics and accessories such as the popular shawls, can be made some adjustments to the bodies, in which may have been the first forms of ready-to-wear, especially for women: Worth did it first at Gagelin. One could also buy the fabric, associate it with a style and then order the fabrication of the piece to someone outside the system.

Worth promoted a new sequential system of Design, Production, Exhibition and then the Order which was a huge innovation. Before him there was no such idea of associating a costume with his creator. As we have seen, it was also a pioneer in presuming the alteration of the imprisoned female silhouette, to be a beautiful and happy consumer. He was also the first couturier to sign his works as an artist: with a personalized label. He developed the notion of haute couture house, as a public place, open to all with *francs and faith*, but also a private place and surrounded by luxury and discretion: it should be remembered that Worth once was at the same time the official couturier of the Empress Eugénie and of the lovers of the Emperor dressing them, with refinement, for the same social events. New customers from the new worlds could also order by mail and receive their precious orders of chests full of personalized clothing in the space of a week or two.

He understood the concept of the fashion press and its immense power, knowing better than anyone until then to use it in his behalf. In a time when travel is becoming fashionable and made easy, linking continents and countries through steamers and trains, Worth has become a designer and brand on a global scale, wearing not only the aristocrats but also the new masters of the world, a new and powerful clientele (the dollar princesses) that is expanding with industrialization. Customers of modern life could also have confidence in the past through a strong craft and artistic component that made them unique.

Accordingly to Breward (2003) "the sketchbooks which recorded his design ideas" across plus than thirty years of activity and career show "in abbreviated form Worth's preference for fabrics that emphasized sculptural effects and surfaces over peripheral decoration or deeper textures; tulles, satins, silks and brocades rather than moirés and velvets". This magnificent sketchbooks draw attention to "the refining of the body's proportions and a process of skilful pattern-cutting, repositioning dress design as a form of engineering rather than a mere synthesis of designing of existing elements".

However, what distinguishes Worth from other congeners, making him unique, was not only a radical change in fashion design, but the fact that it introduced the idea of seasonal collection, presenting twice a year new and complete sets of clothing, definitively imposing a calendar on creativity, artistic innovation and consumption: with the introduction of each new collection it transforms into old, outdated and out of fashion the previous collection and, thus giving a very limited useful life to the clothes and their accessories, imposing new rules new ways of living / taking over consumption. His success was great and lasted beyond his life, and above all, he encouraged others to follow in his footsteps and to promote similar businesses.

In 1868 he founded the *Chambre Syndicale de la Haute Couture* with his sons to organize and regulate the growing number of creative designers and haute couture houses while at the same time attempting to prevent abusive copies of his creations.

Haute Couture assumes the role of social status and good taste, or rather the dominant taste, novelty and luxury from a city, Paris, which is increasingly becoming the epicentre of fashion and modernity, in an almost identity logic.

All costumes were tailor-made, custom-made and entirely original. A personalized and absolutely exclusive service that made each piece extraordinarily expensive, and together with the label and added the value of Worth's name, affirmed the elitist status of the pieces and their owners, which in an obsessively classist and conspicuously consumerist society, was understood as necessary and, above all, symbol of social status and identity.

As Wilson (2003) quotes in this "rapacious world beauty becomes the passport to social nobility (...) Appearance replaced reality. Whoever wished to crash high society could, provided they looked the part". Worth's over ornamented crinolines of the fashionable women were a very important contribute to this display, these swaying, trembling bells (...) created the illusion, not the reality of modesty (...). They were decorated with a stylish rifling from previous periods a promiscuity that reflected the promiscuity of a society in which bourgeois morality *clothed* the rapacity and animal energy of young capitalism."

Worth's costumes were a kind of artificial works of art, like some paintings of that time, tried paradoxically to re – invent nature. One of his clients, Madame Octave Feuillet, a wife of a fashionable novelist wrote:

"He had decided upon a dress of lilac silk covered with clouds of tulle in the same shade in which clusters of lilies of the valley were to be drowned. A veil of white tulle was to be thrown like a mist over the mauve clouds and the flowers, and finally, a sash with flowing ends should suggest the reins on Venus's chariot" (Wilson 2003).

By doing this, he became immortal.

ACKNOWLEDGMENTS

This research was only possible with the support of CIAUD– Research Centre in Architecture, Urbanism and Design, Lisbon School of Architecture of the University of Lisbon, and FCT – Foundation for Science and Technology, Portugal.

REFERENCES

Bell, Q. 1976. *On Human Finery.* London: The Hogarth Press.

Bordieu, P. & Desault Y. 1975. "Le couturier et sa griffe: contribuition à une theorie de la magie". In *Actes de la recherché en sciences Sociales: 7–36* (1)1

Breward, C. 1995. *The culture of fashion.* Manchester: Manchester University Press.

Breward, C. 2003. *Fashion*, Oxford/New York: Oxford University Press.

Coleman, A. 1989. *The Opulent Era: Fashion of Worth, Doucet and Pingat.* London: Thames & Hudson.

Coleman, E. (2014) http://fashion-history.lovetoknow.com/fashion-clothing-industry/fashion-designers/charles-frederick-worth [31 of July of 2017]

De La Haye, A. & Mendes, V. 2015. *The House of Worth portrait of an archive*, London: V&A Publishing.

De Marly, D. 1990. *Father of Haute Couture – Worth*, New York/London: Holmes & Meir.

International Exhibition 1862: Official Catalogue of the Fine Art Department (1862) https://books.google.pt/books?id=HzBbAAAAQAAJ&pg=RA1-PA5&hl=pt-PT&source=gbs_toc_r&cad=3#v=onepage&q&f=false [31 of July of 2017]

International Health Exhibition (1884) https://archive.org/details/gri_33125008618163 [31 of July of 2017]

Lipovestky, G. 2010. *O Império do Efémero – a moda e o seu destino nas sociedades modernas.* Lisboa: Dom Quixote.

Lysack, L. 2008. *Come buy, come buy shopping and the culture of consumption in Victorian women's writing*. Athens: Ohio University Press.

Met Museum (2017) Nineteenth-Century European Textile Production. http://www.metmuseum.org/toah/hd/txtn/hd_txtn.htm [31 of July of 2017]

Rzepa, H. (2017) http://www.ch.ic.ac.uk/motm/perkin.html [31 of July of 2017]

Tarrant, N. 1994. *The development of costume*. London: Routledge.

Veblen, T. 1965. *A teoria da classe ociosa*

Watt, M. 2000. "Nineteenth- century European Textile production". In *Heilbrunn Timeline of Art History*. New York: The Metropolitan Museum of Art

Wilson, E. 2003. *Adorned in dreams, fashion and modernity.* London/New York: I.B. Taurus

Worthparis (2017) Worthparis – Paris Haute Couture. http://www.worthparis.com/paris-haute-couture [31 of July of 2017]

3. Innovation

3.1. Technical and smart textiles

Textiles, Identity and Innovation: Design the Future – Montagna & Carvalho (Eds)
© 2019 Taylor & Francis Group, London, ISBN 978-1-138-29611-4

Shape memory alloy knitted fabrics: Functional fabrics that work

Kevin Eschen & Julianna Abel
Department of Mechanical Engineering, University of Minnesota, Minneapolis, MN, USA

ABSTRACT: The recent rise of soft robotics, soft computing, and portable devises has highlighted the potential of functional fabrics. Functional fabrics are a novel approach to create pliable, lightweight structures with intrinsic active functionalities, enhancing diverse engineering applications by providing actuation, sensing, energy harvesting, and communication. The added functionality of such fabrics is accomplished by integrating active fibers into designed textile geometries. Active fibers can be created from various material systems including shape memory alloys (SMA), shape memory polymers (SMP), electro-active polymers (EAP), and carbon-nanotubes (CNT). Shape memory alloy knitted actuators, one type of functional fabric actuator, use shape memory alloy wire as an active fiber within a knitted textile. Utilizing the hierarchical nature of knitted architectures for tailored designs through changes in the knit grid, knit pattern, and the knitted loop itself, a variety of actuation deformations can be achieved. SMA knitted actuators can provide extension, contraction, scrolling, corrugation, and twisting upon thermal actuation. SMA knitted functional fabrics excel at traditional actuation performance metrics such as actuation displacements, actuation forces, and performed mechanical work, while achieving complex, customizable three-dimensional deformations. These multifunctional characteristics within a fabric architecture renders the potential for applications in many fields, including medical devices, rehabilitation, aerospace, and defense.

1 INTRODUCTION

Passive fabrics have protected from harsh weather conditions, granted comfort, and served as a platform to express individualism through color and patterns throughout modern humankind (Mccabe, 2010). Passive fabrics are created with a one-dimensional fiber that is arranged into a two- or three-dimensional stable fabric architecture using traditional textile manufacturing techniques such as knitting, weaving, and braiding. Textile research has been committed to understanding the underlying mechanics of the fabrics that we wear every day and to improving the manufacturing process. Extensive research on knitted fabric geometric properties (Munden, 1959), analytical and numerical modeling, and experimental characterization (Smirfitt, 1965) of passive fabrics has led to a transformation of the design and manufacturability of fabrics. However, the passive nature of fabrics has remained constant until recently.

Functional fabrics are a novel approach to utilize the unique geometric properties of textile architectures in combination with the enhanced functionality of active material fibers. Due to the direct integration of active fibers, functional fabrics have intrinsic active capabilities, compared to traditional fabrics with attached conventional electronic or electro-mechanical devices. Extensive active materials research has propelled the realization of various functional fabrics. For example, research on carbon-nanotubes (Foroughi, 2017), piezoelectric fibers (Lemaire, 2015), electroactive polymers (Maziz, 2017), photosensitive fibers (Abouraddy, 2005), and shape memory alloys (SMA) (Abel, 2013) have demonstrated promise for functional fabrics. Through the integration of multifunctional fibers within textile architectures, the characteristic properties of both the structure and the active fiber are leveraged to generate actuation, sensing, energy harvesting, and communication capabilities. The integration of these capabilities within pliable, lightweight fabrics has generated interest among various industries including medical devices, rehabilitation, aerospace, and defense.

The change of energy domains is the defining feature of active materials and enables their multifunctional use. Electroactive polymers and piezoelectric fibers transform mechanical energy into electrical energy and vice versa. While these materials perform this transformation efficiently, have high strain resolutions, and can operate at high frequencies, the supported stresses and strains are relatively small. These characteristics make electroactive polymers and piezoelectric fibers ideal for sensing and energy harvesting applications. The low cost and easy accessibility of these materials (Huber, 1997) have bolstered research efforts in the fields of sensing and energy harvesting. Reseach on the integration of piezoelectric and electroactive polymers has resulted in unique functional fabric applications (Scilingo, 2003), (Hasegawa, 2008), (Ahn, 2015). A theoretical foundation for sensing functional fabrics has been established through experimental characterization and conceptual research (Xie, 2016), (Zhang, 2005), (Wang, 2014), (Bashir, 2014). Additionally, mathematical model models to calculate sensing parameters, such as resistive models for fabrics, have been established (Xie, 2014), (Li, 2010).

While the early research on functional fabrics was primarily occupied with creating fabric sensors and energy harvesters, an increased interest in functional fabric actuators, especially those leveraging the knitted architecture, can be reported across material systems. Functional knitted fabric actuators have been implemented by incorporating active fibrous materials including SMA (Abel, 2013), electroactive polymers (Maziz, 2017), and carbon nanotubes (Foroughi, 2016). The variation of the knitted architecture geometry and the choice of fiber accommodates specific application cases. For example, electroactive polymer knitted actuators have been investigated for medical applications because they provide forces that are safe for tissue and can achieve actuation through novel chemical mechanisms such as ionic liquids (Maziz, 2017). Additionally, SMA springs have been attached to traditional fabric and were investigated for their potential to provide mechanical counter-pressure to the body (Holschuh, 2012) and act as dynamic structural systems (Yuen, 2014). While these implementations of functional fabrics address challenges in their specific fields, SMA knitted actuators have demonstrated their potential to excel at traditional mechanical performance metrics such as the maximum mechanical work upon thermal actuation (Eschen, 2017). This chapter discusses the enhanced capabilities of functional SMA knitted fabrics and provides an overview of promising SMA knitted fabric applications.

2 MATERIALS AND FABRIC ARCHITECTURE

Shape memory alloy knitted actuators provided distributed three-dimensional actuation motions as a result of the material properties of the shape memory alloy actuator wire and the mechanical deformations induced during the knitting manufacturing process.

2.1 *Shape memory alloy material properties*

The most prominent member of the shape memory alloy material family was discovered when the Naval Ordnance Laboratory (NOL) developed a nearly equatomic alloy composed of nickel (Ni) and titanium (Ti). The combination of the element symbols of the alloying elements and the acronym of the laboratory resulted in the name Nitinol, by which these shape memory alloys are commonly known. The intrinsic shape memory effect of Nitinol and other shape memory alloys are attractive features for a variety of industrial and research applications. Through thermal actuation, the shape memory effect of Nitinol is initiated, resulting in temperature-dependent material properties and the ability to recover material strains of up to 8% (Shaw, 1995). These macroscopic effects occur because of different configurations in the crystalline lattice structure of shape memory alloys. A martensitic solid-to-solid phase transformation is initiated by heating the shape memory alloy above its specific austenite start temperature, and comes to completion when the material reaches its austenite finish temperature. The phase transformation from martensite to austenite results in temperature-dependent material properties, such as the material stiffness, which is significantly higher in the austenitic phase. The transformation temperature of the material can be tailored through heat treatments and changes in the alloy composition, e.g., through adding ternary elements to the alloy (Eckelmeyer, 1976). Upon cooling, the reverse phase transformation back to the martensitic lattice structure occurs, and the material changes back to its martensitic material properties.

The shape memory effect leverages the temperature-dependence of the material properties to provide actuation capabilities. While shape memory alloy materials are available in many different actuator forms (e.g., bars, springs, ribbons), the most common form is a wire. A typical shape memory application begins with an undeformed SMA wire at a temperature for which the material is in the martensitic phase. At the same temperature, the wire is deformed through the application of mechanical loads. After unloading, the deformed wire is thermally actuated, and the phase transformation to austenite occurs. The material properties change, and induced strains of up to 8% are recovered. Subsequent cooling of the wire to the martensite temperature leads to a full restoration of the martensitic wire material properties so that the wire is in the exact same state as it was in the beginning of the process (Sastri, 1968). Shape memory alloy wires are a unique actuator form that can be integrated directly into fabrics.

2.2 *Hierarchical active knitted architecture*

Functional SMA knitted fabrics accomplish various customizable actuation deformations dependent on the geometric and dimensional properties of the architecture. To fully describe the kinematic actuation motions and the performance of SMA knitted actuators, a multi-scale, hierarchical description of the architecture is required. Textile architectures are common described at three different levels of detail and accuracy, namely the microscale, the mesoscale, and the macroscale (Figure 1). Microscale descriptions of knitted fabrics are concerned with fiber interactions or, for monofilament fibers such as SMA, the physics at grain resolution. The mesoscale is an intermediary scale between the micro- and macroscale, which examines the fabric at a unit cell level. The unit cell of a textile is commonly defined as its smallest repetitive geometric element. Lastly, a macroscopic investigation of textiles is involved with the performance and

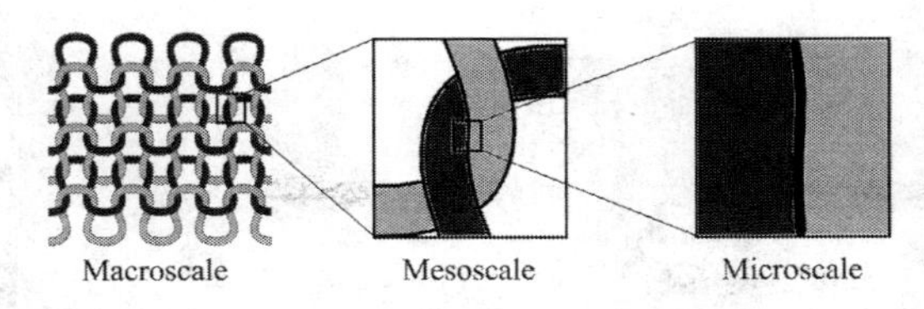

Figure 1. Multiscale Fabrics: Fabrics are commonly described and investigated at the macro-, meso-, and microscale.

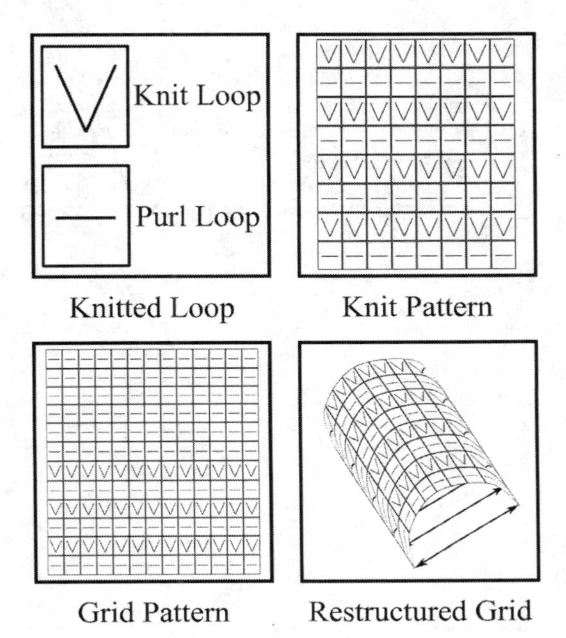

Figure 2. Hierarchical Functional SMA Knitted Fabric Architecture: Functional SMA knitted fabrics can be described on four hierarchical levels, the knitted loop, knit pattern, grid pattern, and restructured grid.

behavior of the complete textile. The division of fabrics into multiple scales allows an investigation of various levels of granularity. While this multiscale approach fits textiles in general, the hierarchical nature of functional SMA knitted textiles is not fully captured through this description.

Additional layers of hierarchy are added on the macroscale to characterize the range of kinematic actuation motions and establish a formalized functional SMA knitted fabric classification (Abel, 2013). This hierarchy includes four levels of detail, which are from the lowest to the highest level: the knitted loop, the knit pattern, the grid pattern, and the restructured grid.

The first level of the active knitted architectural hierarchy, the knitted loop is concerned with the two fundamental knitted loop configurations, the knit loop and the purl loop (Figure 2, top left). The knit and purl loops are equal in the geometry but interlace with the previous row of knitted loops from opposite directions. The strains induced in the knitted loop during manufacturing are partially recovered during thermal actuation. The recovery of manufacturing strains and sliding of interlacing adjacent loops provide the foundational mechanism governing the large deformation of active knitted actuators.

The second level, the knit pattern, describes the homogenous assembly of knitted loops within the textile (Figure 2, top right). Various combinations of knit and purl loops in both, the horizontal and vertical directions result in knit patterns that achieve different homogeneous actuation deformations. Common knit patterns, such as stockinette, garter, or rib, that provide homogeneous actuation motions can be described with this level of architectural hierarchy.

The combination of various knit patterns in a single fabric is described on the third level of hierarchy, the grid pattern (Figure 2, bottom left). While knit patterns are homogenous in their assembly of knitted loops, grid patterns form a heterogenous structure with variable mechanical properties and actuation deformations distributed throughout the functional fabric. Grid pattern architectures expand the range of kinematic motions afforded by knitted SMA actuators and enable controlled assembly of complex shapes.

The highest level of the hierarchy, the restructured grid, describes the post-processing of planar knit patterns and grid patterns into three-dimensional, spatial architectures (Figure 2, bottom right). This level can be achieved through a post-knitting connection of knitted loops, course-wise restructuring, grid cell merging, or re-ordered grids (Abel, 2013). Restructured grids further expand the range of attainable actuation motions.

3 SMA KNITTED FUNCTIONAL FABRIC MECHANICAL CAPABILITIES

The ability to modify the system parameters of functional SMA knitted fabrics enables a wide range of enhanced capabilities, such as complex two- and three-dimensional deformations (Abel, 2013), localized and distributed actuation (Anderson, 2017), and variable stiffness and adaptive actuation forces (Eschen, 2017). This section discusses the enhanced functional SMA knitted fabric capabilities and introduces applications that can benefit from them.

3.1 *Complex three-dimensional deformations*

The kinematic actuation motions of SMA knitted actuators are governed by the hierarchical composition of the knit and purl loops within the fabric. The basic deformation modes are a function of the selected knit pattern, which represents the second level in the active knit hierarchy.

The most basic knit pattern, the stockinette pattern, is an assembly consisting of uniquely one type of knitted loop. As knit and purl loops are equal in geometry, but opposite in direction, stockinette patterns can be created from both types of knitted loops, with the only difference of flipped front- and backsides of the fabric.

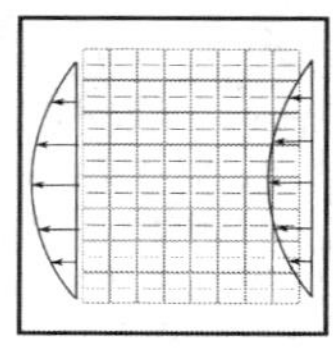
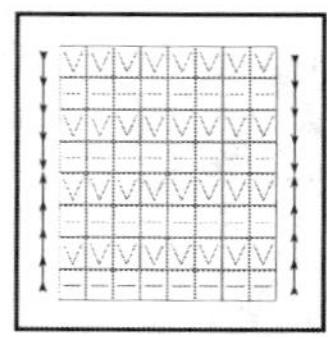
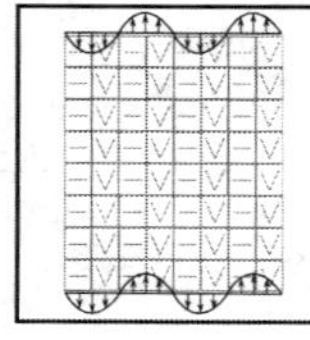

Stockinette Pattern Garter Pattern Rib Pattern

Figure 3. Knit Pattern Schematic: Planar knit patterns produce homogenous actuation deformations. The stockinette knit pattern bends out of plane, the garter knit pattern produces uniaxial contractions, and the rib pattern forms a corrugation deformation upon thermal actuation.

Stockinette knit patterns produce asymmetric actuation deformations resulting in arching or scrolling of the fabric into a cylindrical shape (Figure 3, left). Such actuation deformations render potential for deployable structure applications, e.g., in aerospace and extra-terrestrial structures that require tight packaging for travel and expand into their intended shape upon actuation when place in their desired environment.

Knitted fabrics with interlacing rows of knit and purl loops are called garter knits (Figure 3, middle). These knitted architectures produce macroscopically planar actuation contraction of up to 50%, which can be primarily attributed to recoverable bending deformations and relative sliding of the interlacing loops (Eschen, 2016). The large actuation displacements and forces of these lightweight, pliable structure shows potential for applications in aerospace, where the currently used hydraulic and pneumatic actuators contribute significantly to the weight of the airplane.

The rib pattern uses a similar architectural approach as the garter pattern. However, the interlacing knit and purl loops are not along the rows of the fabric, but along the columns of the fabric (Figure 3, right). The columns of purl and knit loops bend into opposite directions upon actuation, causing an effect that can be geometrically described as accordioning or corrugation. These patterns produce significant actuation forces along the corrugation direction and are promising for variable surface topology applications, e.g., in fluid flow control and optimization (Bil, 2013) (Abel, 2010).

As discussed in Section 1.3, more complex functional SMA knitted fabrics can be produced by utilizing the higher hierarchy grid patterns and restructured grids. Functional fabrics that utilize multiple knit patterns within a grid pattern provide the previously discussed actuation deformations spatially distributed throughout the fabric structure. Restructured grids, the highest level of the hierarchy, accomplish additional enhanced actuator capabilities, which are impossible to realize on the knit pattern level. For example, the planar, contractile garter pattern can be turned into a powerful compression sleeve by connecting the ends of the architecture. Such functional SMA knitted compression fabrics are promising for rehabilitative medical devices, e.g., through the application of circumferential pressure to certain parts of the body (Macintyre, 2006) and (Granberry, 2017), and within counter-pressure space-suits (Holschuh, 2012).

Even though the focus of this section has been laid on planar knitted fabrics that actuate out-of-plane, it is noteworthy that the hierarchical architecture of functional SMA knitted fabrics can also be applied to circular knits. The circumferential pressure generation from restructured garter pattern functional fabrics can similarly be achieved with a circular knitted rib pattern, which accomplishes the pressure generation through circumferential corrugation and decreases the effective diameter of the knitted sleeve.

Utilizing the available hierarchical levels for the creation of unique functional SMA knitted fabrics enables the realization of fabric actuators capable of nearly any conceivable deformation upon a thermal impulse. SMA knitted actuators are capable of producing complex motions required to realize a variety of engineering applications.

3.2 *Distributed and localized actuation*

Shape memory alloy wire is a multi-functional material that transforms thermal energy into mechanical energy using the shape memory effect. While the shape memory effect is an excellent alternative actuation mechanism, only a few engineering applications can be controlled by a change in environmental temperatures. One way to avoid reliance on environmental temperature changes is to employ resistive (Joule) heating. Resistive heating is a simple method to accurately control the temperature in SMA wires based on electrical power input and resistivity measurements (Ma, 2004). For straight SMA wire, resistance feedback power control has been used in experimental characterization procedures, with thermal accuracy far beyond the requirements of general actuator design applications (Furst, 2012). While the resistance feedback power control is an established actuation mechanism for plain SMA wire, new challenges arise when applying the same methodology to functional SMA knitted fabrics. Specifically, the interlocking adjacent loops within the knitted structure enable electrical shorting across the complex knitted resistive network.

An analytical model representing the functional SMA knitted fabric as an interconnected network of resistive elements has been developed and experimentally verified (Anderson, 2017). This model predicts the diffusion of the current away from the primary electrical path and determines the number of courses around the primary path that experience temperatures above the austenite start temperature due to resistive heating (Figure 4). Predictive capabilities of the primary path enable distributed and localized actuation of functional SMA knitted fabrics by resistive hearing. Through distributed and localized actuation, complex motion control of functional SMA knitted fabrics is feasible. An example of the impact of this development can be derived from investigating the simple contractile SMA knitted actuator architecture. Without localized actuation and under constant applied mechanical

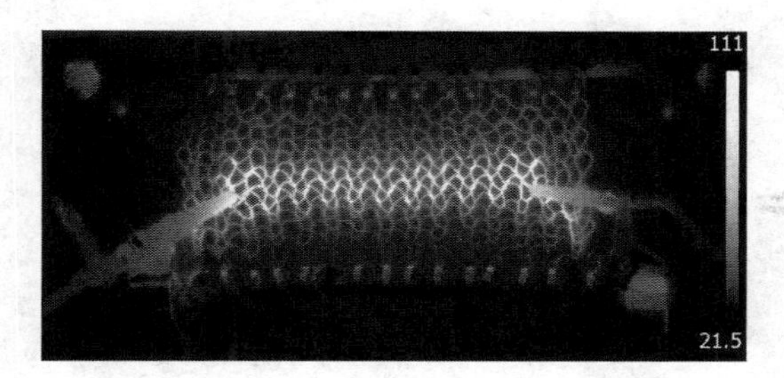

Figure 4. Locally Actuated Functional SMA Knitted Garter Fabric: Resistance feedback power control enables local actuation of functional SMA knitted fabrics, e.g. for continualization of actuation displacements in garter patterns. Thermal imaging shows the actuated rows and the wire temperatures in °C.

a)

b)

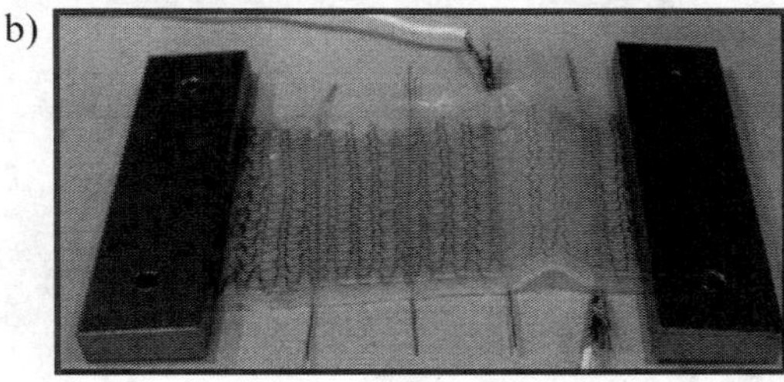

Figure 5. Localized Actuation: The actuation of localized areas in functional SMA knitted fabrics can be used to "switch" areas on and off, as shown in this rib knit pattern fabric. In a), the most centered rib is actuated, whereas a border rib is actuated in b).

loads, contractile SMA knitted actuators provide a single designed actuation displacement as the actuator is limited to either no or complete actuation contraction. Using advanced localized actuation, nearly continuous actuation contractions can be realized providing a real alternative to traditional pneumatic and hydraulic actuator systems.

While this approach utilizes localized actuation to allow for the continualization of a functional fabric property, it can also be used for the actuation of discrete spatially restricted sections of the functional SMA knitted fabric. For example, when using the rib pattern in functional SMA knitted fabrics, single ribs can be resistively heated and actuated (Figure 5). Such behavior may be used in unmanned aerial vehicle (UAV) fluid flow control, as the presence of surface dimples (Lo, 2016) and full-span corrugations (Milholen, 2005) can lead to superior flight properties in different flight situations, e.g., climb or travel.

3.3 *Mechanical actuation performance*

The variety of obtainable actuation deformations increases the difficulty of defining universal actuation performance metrics for functional SMA knitted fabrics. While the garter knit pattern provides uniaxial actuation contractions that can be characterized in simple uniaxial tensile testing apparatus, complex restructured grids demand similarly complex performance measurement systems. Because the performance of higher hierarchical structures, such a grid patterns and restructured grids can be seen as the combined performance of knit patterns, the focus of the mechanical actuation performance description is set to functional SMA knit patterns.

Planar functional SMA knit patterns are usually described with classic actuation performance metrics such as actuation displacements, actuation forces, and mechanical work. However, the direction of useful actuation is different amongst knit patterns. For example, the uniaxial contractions of functional garter knitted fabrics are entirely planar, while functional rib knitted fabrics deform out-of-plane and perform significant work in that direction.

The performance of functional garter knitted fabrics is dependent on the geometric loop parameters and the number of rows and columns within the knitted architecture. Actuation contractions, defined as the scaled difference between the martensite and austenite knit lengths, up to 50% can be achieved (Eschen, 2017). For relatively small functional garter knitted fabrics with dimensions of approximately. 10 cm × 10 cm, geometric loop parameter dependent actuation forces on the scale of 1 N–10 N have been experimentally verified. The attainable actuation displacements of any functional garter knitted fabric can be scaled with the number of rows, while the forces scale with the number of columns in the fabric (Evans, 2006).

While the functional garter knitted fabric architecture provides large actuation displacements with useful, but not extensive actuation forces, the functional rib knitted fabric architecture serves opposite actuation requirements (Figure 6). The corrugation deformation produces relatively small linear actuation displacements on the scale of millimeters to a few centimeters. The actuation forces, however, can reach magnitudes of 20 N–200 N for single functional rib knitted fabrics (Abel, 2010). Besides the customizability of forces through changes in geometric loop parameters on the lowest hierarchical level, it is possible to significantly increase the obtainable actuation forces through nesting of functional rib knitted fabrics into a restructured grid. Nesting of functional rib knitted fabrics alters the actuation performance similar to parallel spring systems, with the actuation displacements remaining those of a single actuator, but obtainable forces equal to the sum of the obtainable forces of the single actuators (Abel, 2010).

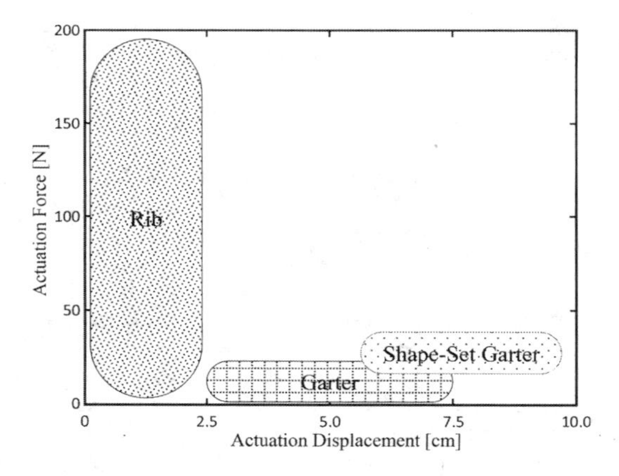

Figure 6. Design Space for SMA Knitted Actuators. Rib knit pattern SMA actuators provide moderate displacements under substantial applied loads, whereas garter knit pattern SMA actuators produce large displacements under moderate loads.

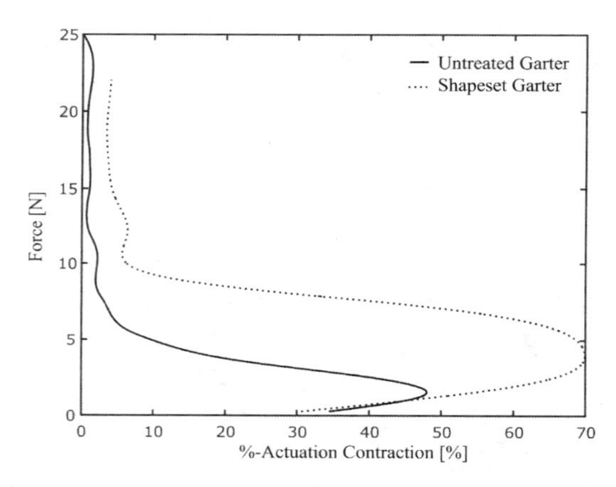

Figure 7. Effect of Shape-Setting: Performing a global shape-setting procedure in a pre-strained configuration can significantly increase the achievable %-actuation contraction and actuation forces of garter knit pattern SMA actuators.

The thus far describe functional SMA knitted fabrics are composed of SMA wire with an entirely linear memorized shape. Through thermal treatment, heating of the wire above its recrystallization temperature, a new memorized shape can be imprinted to the SMA material. This process is called shape setting and has demonstrated the potential for increasing the functional SMA knitted fabric mechanical actuation performance (Figure 7) (Yi, 2017). Shape setting can be performed locally on distinct segments of a structure and globally on the entire fabric. Global shape setting in a strained configuration along the wale direction can be used to shift the point of maximum %-actuation contractions to significantly higher forces. Additionally, this shape setting procedure results in a distinct cut-off force, at which the achievable actuation displacement drops from a value close to the maximum to a negligibly low number. This behavior could be exploited as a safety mechanism for applications that require sudden release at a specific critical load.

4 CONCLUSIONS

This paper provides a review of the current state of the art for functional SMA knitted fabrics. The intrinsic shape memory effect of SMA wire leveraged as a fiber within the hierarchical knitted architecture enabled enhanced functionalities including complex three-dimensional actuation deformations and localized and distributed actuation. Across various knit patterns, grid patterns, and restructured grids, intriguing mechanical actuation performance can be reported, with pattern-specific, unique strengths such as large actuation displacements and forces, as well as complex deformations and motions. Different methodologies to further improve the functional SMA knitted fabric capabilities are discussed. Resistive heating to realize localized actuation and shape set to increase and tailor the achievable actuation forces will amongst other methodologies help to realize relevant applications with functional SMA knitted fabrics that work.

REFERENCES

Abel, J., Luntz, J., and Brei, D., "Hierarchical architecture of active knits," *Smart Materials and Structures*, vol. 22, no. 12, p. 125001, 2013.

Abel, J., Mane, P., Pascoe, B., Luntz, J., and Brei, D., "Experimental investigation of active rib stitch knitted architecture for flow control applications," *Proc. SPIE 7643, Active and Passive Smart Structures and Integrated Systems 2010*, 76430H (April 09, 2010), vol. 7643. SPIE, 2010.

Abouraddy, A. F., Shapira, O., Bayindir, M. Arnold, J., Joannopoulos, J. D., and Fink, Y., "Fabrics that See," *Optics and Photonics News*, vol. 17, no. 12, pp. 21–21, Dec. 2006

Ahn, Y., Song, S., and Yun, K.-S., "Woven flexible textile structure for wearable power generating tactile sensor array," *Smart Materials and Structures*, vol. 24, no. 7, p. 075002, 2015.

Anderson, T., and Abel, J., "Spatially Distributed Actuation of Shape Memory Alloy Knitted Composites," *Proc. of the International Conference on Shape Memory and Superelastic Technologies*, May 2017.

Bashir, T., Ali, M., Persson, N.-K., Ramamoorthy, S. K., and Skrifvars, M., "Stretch sensing properties of conductive knitted structures of PEDOT-coated viscose and polyester yarns," *Textile Research Journal*, vol. 84, no. 3, pp. 323–334, Feb. 2014.

Bil, C., Massey, K., and Abdullah, E. J., "Wing morphing control with shape memory alloy actuators," *Journal of Intelligent Material Systems and Structures*, vol. 24, no. 7, pp. 879–898, May 2013.

Eckelmeyer, K. H., "The effect of alloying on the shape memory phenomenon in nitinol," *Scripta Metallurgica*, vol. 10, pp. 667–672, Aug. 1976.

Eschen, K and Abel, J., "Effect of Geometric Design Parameters on Contractile SMA Knitted Actuator Performance," *Proc. of ASME 2017 Conference on Smart Materials, Adaptive Structures, and Intelligent Systems*, 2017.

Eschen, K and Abel, J., "Architectural Settling Mechanisms Contractile SMA Knitted Actuators," Technical Presentation. *ASME 2016 Conference on Smart Materials, Adaptive Structures, and Intelligent Systems*, 2016.

Evans, J., Brei, D., Luntz, J., "Preliminary Experimental Study of SMA Knitted Actuation Architectures," *Proc. ASME International Mechanical Engineering Congress and Exposition*, Nov. 2006.

Foroughi, J., Spinks, G. M., Aziz, S., Mirabedini, A., Jeiranikhameneh, A., Wallace, G. G., Kozlov, M. E., and Baughman, R. H., "Knitted Carbon-Nanotube-Sheath / Spandex-Core Elastomeric Yarns for Artificial Muscles and Strain Sensing," *ASC Nano*, 2016.

Furst, S. J., and Seelecke, S., "Modeling and experimental characterization of the stress, strain, and resistance of shape memory alloy actuator wires with controlled power input," *Journal of Intelligent Material Systems and Structures*, vol. 23, no. 11, pp. 1233–1247, Jul. 2012.

Granberry, R., Abel, J., and Holschuh, B., "Active Knit Compression Stockings for the Treatment of Orthostatic Hypotension," SIGCHI Symposium on Wearable Computers, September, 2017.

Hasegawa, Y., Shikida, M., Ogura, D., Suzuki, Y., and Sato, K., "Fabrication of a wearable fabric tactile sensor produced by artificial hollow fiber," *Journal of Micromechanics and Microengineering*, vol. 18, no. 8, p. 085014, 2008.

Holschuh, B., Obropta, E., Buechley, L., and Newman, D. "Materials and Textile Architecture Analyses for Mechanical Counter-Pressure Space Suits Using Active Materials," *technical manuscript presented at AIAA Space 2012*, AIAA 2012–5206, 2012.

Huber, J. E., Fleck, N. A., and Ashby, M. F., "The Selection of Mechanical Actuators Based on Performance Indices," *Proceedings: Mathematical, Physical and Engineering Sciences*, vol. 453, no. 1965, pp. 2185–2205, 1997.

Li Li, Wai Man Au, Kam Man Wan, Sai Ho Wan, Wai Yee Chung, and Kwok Shing Wong, "A Resistive Network Model for Conductive Knitting Stitches," *Textile Research Journal*, vol. 80, no. 10, pp. 935–947, Jun. 2010.

Lo, K. H., Zare-Behtash, H., and Kontis, K., "Control of flow separation on a contour bump by jets in a Mach 1.9 freestream: An experimental study," *Acta Astronautica*, vol. 123, pp. 229–242, 2016.

Ma, N., Song, G., and Lee, H.-J., "Position control of shape memory alloy actuators with internal electrical resistance feedback using neural networks," *Smart Mater. Struct.*, vol. 13, no. 4, p. 777, 2004.

Macintyre, L., "Designing pressure garments capable of exerting specific pressures on limbs," *Burns*, vol. 33, no. 5, pp. 579–586, Aug. 2007.

Maziz, A., Concas, A., Khaldi, A., Stlhand, J., Persson, N.-K., and Jager, E., "Knitting and weaving artificial muscles", *Science Advance*, Vol. 3, No. 1, 2017.

Mccabe, M., Potter, E., Simon, C., Potter, E., and Baggerman, C., "Smart Fabrics Technology Development," 2010.

Milholen, W. E., and Owens, L. R., "On the Application of Contour Bumps for Transonic Drag Reduction" in *AIAA Paper, 43rd Aerospace Sciences meeting and Exhibit*, 2005.

Munden, D. L., "The Geometry and Dimensional Properties of Plain-Knit Fabrics". *Journal of the Textile Institute Transactions*, 50(7), pp. T448–T471, 1959.

Sastri, A. S., Marcinkowski, M. J., and Koskimaki, D "Nature of the NiTi martensite transformation", *physica status solidi* (b), 25(2), pp. K67–K69, 1968.

Scilingo, E. P., Lorussi, F., Mazzoldi, A., and Rossi, D. D., "Strain-sensing fabrics for wearable kinaesthetic-like systems," *IEEE Sensors Journal*, vol. 3, no. 4, pp. 460–467, Aug. 2003.

Shaw, J. A., and Kyriakides, S. "Thermomechanical aspects of NiTi". *Journal of the Mechanics and Physics of Solids*, 43(8), Aug., pp. 1243–1281, 1995.

Smirfitt, J. A., "Worsted 1 x 1 Rib Fabrics Part I. Dimensional Properties". Journal of the Textile Institute Transactions, 56(5), pp. T248–T259, 1965.

Wang, J., Long, H., Soltanian, S., Servati, P., and Ko, F., "Electro-mechanical properties of knitted wearable sensors: Part 2 Parametric study and experimental verification," *Textile Research Journal*, vol. 84, no. 2, pp. 200–213, Jan. 2014.

Xie, J., and Long, H., "Equivalent resistance calculation of knitting sensor under strip biaxial elongation," *Sensors and Actuators A: Physical*, vol. 220, Dec. 2014.

Xie, J., Long, H., and Miao, M., "High sensitivity knitted fabric strain sensors," *Smart Materials and Structures*, vol. 25, no. 10, p. 105008, 2016.

Yuen, M., Cherian, A, Case, J. C., Seipel, J., and Kramer, R. K., "Conformable actuation and sensing with robotic fabric," *2014 IEEE/RSJ International Conference on Intelligent Robots and Systems*, pp. 580–586, 2014.

Zhang, H., Tao, X., Wang, S., and Yu, T., "Electro-Mechanical Properties of Knitted Fabric Made from Conductive Multi-Filament Yarn Under Unidirectional Extension," *Textile Research Journal*, vol. 75, no. 8, pp. 598–606, Aug. 2005.

Textiles, Identity and Innovation: Design the Future – Montagna & Carvalho (Eds)
 ISBN 978-1-138-29611-4

*Sun*Plum*Lime*Berry, a collaborative textile design and manufacturing illuminated wool carpet project

S. Heffernan
College of Creative Arts, Massey University, Wellington, New Zealand

ABSTRACT: This paper presents an illuminated wool carpet project collaboration with a bespoke carpet tufting company, CRONZ (Carpets and Rugs of New Zealand) and a textile design academic. It aims to establish new conduits between design research, technology and New Zealand (NZ) wool through the development of speculative designs. The background introduces the reasons for both parties to invest in the research. From an academic perspective, the design research and findings are discussed. CRONZ based in Christchurch, NZ, is interested in new design and technology led functional textile innovation. The role designers have in bringing traditional textile industries, high tech developments and user aspects is illustrated. The multi-disciplinary project research started with discussions, yarn sourcing, and painting. It then involved iterative sampling, outsourcing of wool dyeing by Woolyarns Ltd, lighting configuration trials and carpet tufting production. The reinvestment of traditional material processes is reinterpreted in light of manufacturing imperatives.

1 INTRODUCTION

What if a carpet could feature not just pattern and texture, but also light? Curiousity about surplus waste stream material and technology coming together to create innovative design to bring new add on values for carpets led this work. How can novel material products improve the interior environment? Is there a significant benefit for the addition of speculative design to the carpet production cycle? What would performance tests reveal? What are the key performance attributes and how can the aesthetic be improved through the application of non-heavy metal free dyes? What are the insights of architects and business entrepreneurs? What are the health and well-being benefits? Together ways were devised to give new life to surplus industry materials.

The outcome of the multi-disciplinary project in the intersection of design, science, engineering, manufacturing and is a light transmitting shaped wool carpet *SunPlumLimeBerry.* The project built on an existing relationship which began during a 2012 postgraduate collaboration between Rebekah Harman and CRONZ, that was funded by government and supervised by the author and Jessica Payne. A mutual relationship of trust, knowledge and expertise sharing developed between CRONZ personnel and the academic team.

The carpet was manufactured using eco rated coloration and rug tufting techniques combined with LED technology and tufted by both an electronic and hand held tufting gun (Fig. 1). The integration of LEDS enabling a sensory linear light effect to be emitted on the surface of a rug was a first for CRONZ, indeed any NZ carpet tufters. The LED system enabled pressure to be exerted from walking without incurring damage to the lights or lighting system.

Figure 1. *SunPlumLimeBerry* eco dyed wool carpet displaying daylight illumination effect. 2 × 4 m × 1.2 m dimensions. Photograph by Belle Gwilliam.

Before discussing the development of the illuminated carpets, the viewpoints of the company, wool

Figure 2. Production in progress at CRONZ, near Christchurch, showing the mechatronic tufting system used to tuft the llop pile carpet. Photograph by author.

industry background and product performance is outlined.

1.1 *CRONZ – carpets and rugs of New Zealand*

Niche style carpet production in New Zealand includes the use of the latest technology to produce highly engineered materials. CRONZ is a small family owned and operated business led by the husband and wife team of John and Helen Wyma. They produce boutique carpets and rugs for high-end domestic and commercial clients. Typically, their products are one-off works commissioned by architects and designers. Customers can specify every parameter in a rug: size, shape, structure, color, pattern and texture. CRONZ have developed software specific to their needs and have the facility to combine hand tufting with a unique mechatronic control system to ensure quality bespoke structured woollen surfaces are manufactured (Fig. 2).

2 BACKGROUND

2.1 *Wool*

Over the last few decades the use of carpets in our homes has significantly decreased. The decline in demand for coarse carpet wool is a major dilemma for the NZ wool industry and is the focus of various reports and government funding initiatives. The contribution of NZ wool exports has declined to 2% down from 30% in the 1950s. In 2016, wool fibre exports for the June 2016 year were $760M (Burtt, 2016). NZ produced 10% of the world wool production, making it the world's fourth largest producer. Wool carpet and rug exports added $115.5M (Burtt, 2017). This research considered how we could alleviate the impact of the decline in coarse wool demand by introducing novel ideas.

Steven Parsons, Market Development and Innovation Manager, Wools of NZ, Ilkley, UK, remains optimistic and highlights the advantages of wool carpets emphasising the use of rapidly renewable resources, environmental and the human habitat advantages. He confirms the positive consumer experience woollen carpets offer (Parsons, 2017).

2.2 *Competition from synthetic carpet*

Parsons considers there is a pressing need for higher value wool products as the competition from nylon carpets increases. Parsons belives this is possible through good design drawing on the inherent characteristics of wool while demonstrating environmental integrity and sustainability. Nylon is added to wool carpet to meet a price point. The addition of cheap oil-based fibre allows yarn strength to be maintained with lower quality, cheaper fibre. The nylon adds no benefit to the product, it only increases the visual appearance of wear as nylon is shiny and wool is dull. What ergonomic health benefits do wool carpets offer?

2.3 *Ergonomic considerations*

Researchers identified carpets provide cushioning and support for the feet, which can have benefits in both dynamic walking and static (standing) conditions (Cantoral-Ceballos & Nurgiyatna & Wright & Brown-Wilson & Scully, 2014) These researchers found a higher pile height and compressible fibre significantly reduced leg and lower back discomfort when standing at workstations. This information led decisions to produce a carpet strong enough to be walked on while still performing in a waterproof manner to protect against spillages causing overheating.

2.4 *Fresh perspectives from Philips and DESSO*

In a different approach, industrial designers at Eindhoven University of Technology and DESSO, an international flooring company developed an intelligent and interactive floor rug. They explored how people interact with their environment and intelligent floor rug systems. The project proved the need for a multi-disciplinary approach and required making and doing throughout the research and benefitted from a multi-disiplinary approach within academia and with personnel at the carpet manufacturing company (Deckers & v.d Stouw & Peutz, 2012). Two years later, an engineering news bulletin announced a partnership between Philips and DESSO testing carpets with incorporated illuminated signs (Neroth, 2014). It hailed the innovation as adding an extra dimension to interior design and space planning to transform the way people react with information and environment. A further light transmitting carpet development is articulated in a commercial partnership between Philips and Tandus Certiva, a global broadloom carpet leader. The marketing brochure asserts the carpets exhibit the capacity to transform the way people interact with a space (Tandus Certiva, 2017). Thus, a carpet may become a dynamic canvas, appealing to our senses and our natural attraction to light. Such high quality carpets have the potential to complement the aesthetics of a building or the creation of memorable experiences in hotels. Furthermore, advantageous practical considerations include the enhancement of safety for children and elderly at night, to nightclub security to fire exit lighting.

3 DESIGN AND MAKING PROCESS

3.1 *Research questions*

Some strategic practical research questions in this project included the following: how much yarn is required to produce the carpets? Do the planned colors correlate with CRONZ color palette? What is the range of the CRONZ color palette? How can novel materials improve the interior environment? Is there a significant benefit for the addition of speculative design to the carpet production cycle? What would performance tests reveal? What are the key performance attributes and how can the aesthetic be improved through the application of non-heavy metal free dyes? What were the insights of architects and business entrepreneurs? Keywords included aesthetic, ambiguity, dialogue, documenting, empathy, intuition, questioning, reflecting and tactility.

3.2 *Wool yarn*

A surplus quantity of forty kilograms of strong export wool yarns was sourced. The quality spun yarns were rejected from export due to the presence of black dust flecks or tussock seed vegetable matter, a problem widely known as *vm* in the wool industry. However, in this project the *vm* did not prove to be a problem and is not recognizable in the final works. This then raises questions about standards and the sustainability of an industry.

3.3 *A parallel fashion focused project*

Parallel to this carpet project my fashion colleague Sue Prescott developed an illuminated, interactive fashion collection, Global Nomad. Using a back and forth discourse we developed key aspects of the shared project boundaries and continued to meet and discuss our individual projects during the design development, reflection, refinement stages and shared the final photography session.

3.4 *Design and color*

The design process exploring new carpet functions began with a series of paintings abstracting arid river landscape scenes viewed from above. Dye, yarn and pile selection was made implementing principles of sustainability in local manufacturing enterprises (Fig. 3). Colours were chosen to ensure the material looks good when the lights are off and for a symbolic relationship to the concept: the impact of the sun, rope as an arid landscape colour, lime as a contrast for flourishing growth, and plum, wine, and berry were skillfully blended in the tufting process forming the straight edge to represent the impact of high temperatures.

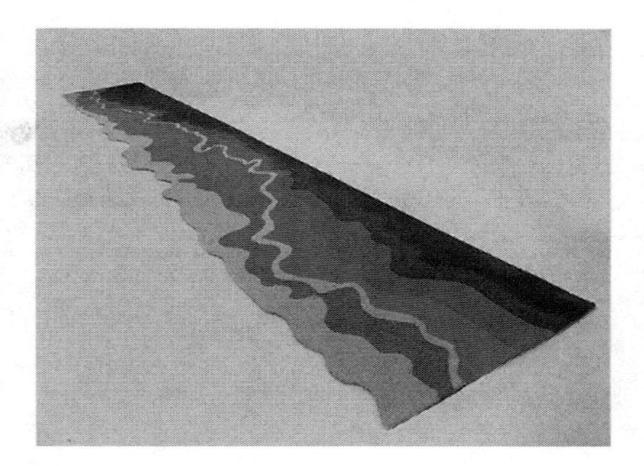

Figure 3. *Sun*Plum*Lime*Berry full view profiling the undulating edge. Photograph by Belle Gwilliam.

The consideration of a range of factors to determine lively, rich color choices was discussed. I developed palettes for review by Prescott to ensure our projects were connected not only by the use of surplus wool yarn and material, interactive illumination, but also by color. The paintings were converted to Illustrator files for transfer to CRONZ's software.

To interpret the designs a range of coloration approaches were considered rather than using pre-dyed yarn, but discounted as being high risk and perhaps detracting from the planned sensory illumination effects.

This led to consultation with CRONZ to discuss their color palette. The importance of the determination of the equivalent color is because dye recipes of the CRONZ colors are held by the dyers, Woolyarns, Wingate, Wellington, who use eco dyes in their industrial process. The sun, wine, lime, berry, plum and rope colors were perfect matches. Lime is not within their typical commercial color spectrum, so a Pantone color chip was selected for the lime dye development by Woolyarns. Fast approaching deadlines meant time was not available to sight the dye color samples and one colour, rope was stronger in green grey hues than anticipated. The variation perhaps due to compositional change of spring water due to weather fluctuations.

3.5 *Design process*

A practice-based reflective process incorporated thinking-through-making to create knowledge and insights. The relationship between the manufacturing and thinking created opportunities to express knowledge through the making of the works. Experimentation and reflective thought examined the decisions made and re-visited the pathways taken. The process is characterized by acquiring knowledge through design, aesthetic and technical experiment sampling, refining and evaluating. In the context of the carpet it coupled design with industrial sampling and manufacture broadening the experience. Iterative tufting and yarn sampling using mechatronic and hand held tufting guns was undertaken in conjunction with lighting configuration trials. In this way, design was coupled with industrial sampling and manufacture to broaden the user experience. Collaboration ensured decisions were made based on a range of relevant practical, functional and aesthetic perspectives.

3.6 *Carpet structure*

A range of dyed and undyed cut pile and loop pile samples were developed using four or six strands per tuft to allow for an expression of a sense of direct image transformation from the paintings. Color blending samples using plum, wine and berry colored yarn were successfully developed. CRONZ completed a range of loop and cut structure profile, texture and color blending pile samples. The curves within the design were identified as complex for normal mechatronic tufting capacity. Hand tufting some design components was identified as essential to maintain the design integrity. One undulating, uneven long edge on the carpet was considered to be possible and within the normal mechatronic tufting and finishing process.

3.7 *Illumination*

Together we trialed lighting options and the innovative incorporation of motion sensitive LEDs in tufted carpet evolved through a collaborative process of technical trials and reflection and evaluation. CRONZ undertook several iterations in a second range of trial samples to test the strength of light transference. An art silk/wool blend was introduced to ensure stronger light transmission effects. The apertures between the pile loops increase the intensity of emitted light. Initially I envisaged the LEDS could be placed along the wiggly white line of the design or, alternatively, multiple rows along the line. It was necessary to consider if the wiggly line needed to be further stylized to facilitate tufting practicalities. Additionally, allover light effects were considered and sampled.

The first technical light consideration was the use of three small LEDS on a strip, then, other ideas were sampled, for example, we identified the need for the LEDs to withstand pressure from walking while still protecting against spillages without overheating. Uli Thie, Massey University, provided electro-technical support preparing the lights and sensory devices for insertion.

At the CRONZ factory I designed a system to implement lights to integrate the LEDs, by embedding the lights in the tufted pile connected to sensor devices. Installed underneath the tufted wool, the enclosed unit LEDS are built strong enough to be walked upon and are protected within the laminated structure of the carpet. John Wyma planned both the carpet tufting and backing meticulously. He ensured the carpet illumination has water proof performance characteristics. The final backing material is tough and applied to ensure high endurance and longevity.

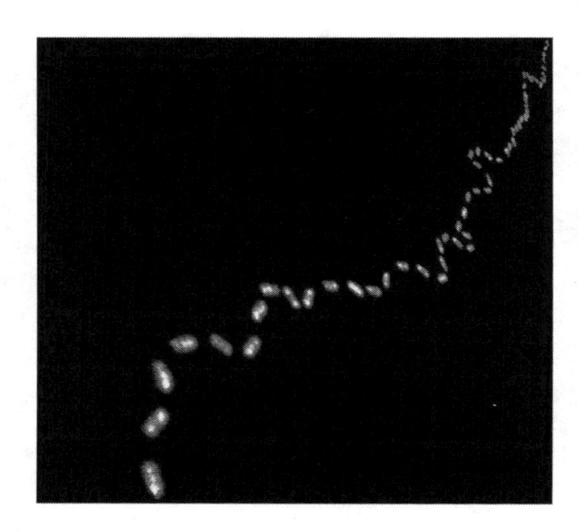

Figure 4. *Sun*Plum*Lime*Berry sensory illumination effects revealed in conditions of darkness. Photograph by Belle Gwilliam.

4 THE PARALLEL PROJECT

Parallel to this project my colleague Sue Prescott, coordinator of the fashion programme at Massey University, used surplus wool yarn and fabric from industry sources to create a fashion/interiors collection. Prescott's *Flight and the Global Nomad Tent,* is an outfit that transforms into a portable sanctuary/shelter using furniture and mobile forms. The outfit separates to form five parts: a dwelling, a lamp, a sleeping mat, pillow and a sleeping bag. A sensorial, rich environment encloses and protects or expands into multiple shelter versions. To consolidate one's space whilst travelling *Flight* allows either a private or community space. In 2017 Prescott received a Red Dot award for Global Nomad.

5 DISCUSSION

The collaboration made innovation possible, drawing on a range of expertise, allowing a strong relationship to develop between a textile design researcher and industry in this project.

The team considered the total experience of the carpets for the consumer and combining aspects of form, texture, color, wool and interaction with light transmitting materials. Another positive performance characteristic is the wool carpets absorb noise unlike hard floors.

This research identified a range of benefits from health and well being way finding for children and elderly, to ambiance enhancement in hotels, corporate signage, to night club ambiance and security to fire exit signage. The novel interior carpet appears both luxurious and resilient with the potential to breathe new life into the surplus wool waste stream and carpet industry. Such carpets could produce a persuasive unique signature for corporate clients or a comfortable, positive ergonomic experience at a workstation.

The carpet provides cushioning and support for the feet, which can have benefits in both dynamic walking and static (standing) conditions. This higher pile height of compressible wool fibre will significantly reduce leg and lower back discomfort when standing at workstations. The integrated LEDs enable pressure to be exerted from walking without incurring damage to the lights.

The total experience of the carpets for the consumer and combined aspects of form, texture, color, wool and interaction with light transmitting materials is considered. Based on other carpet testing using similar wool structures, positive performance factors were assumed for structural performance, durability, warmth, odour, resilience and sound. No nasty chemical encapsulated products were used in the wool processing and dyeing. The vm present at the outset is no longer visible, raising questions about potential uses of waste wool.

6 CONCLUSIONS

Today, wool innovation often occurs in the intersection of design, science, engineering, manufacturing and technology. Technology coupled with traditional methods of making and interpretation can provide a real way to enable the powerful subtlety of these approaches to embrace, navigate and form with complexity of process.

The novel interior carpets appear both luxurious and resilient with the potential to breathe new life into the surplus wool waste stream and carpet industry, especially for corporate clients. With the increasing use of electronics in everyday products, it seems this approach could improve the perception and potential of NZ wool. This collaboration with industry communicates new market opportunities for wool and shows the feasibility and value of design research outside the academic environment.

In 2016, the carpet *Sun*Plum*Lime*Berry was selected by a jury of architects for exhibition at the BORDERS Festival *Future Landscapes* Exhibition at Palazzo Ca'Zanardi in Cannareggio, Venice for seven weeks from October until November and were very well received (Fig. 5). This exhibition ran parallel to the Architecture Biennale in Venice. In-kind and funding support was received from NZIA, a professional institute of architects and a wool brokerage company revealing a level of recognition and endorsement for the carpet innovation.

Figure 5. Detail of *Sun*Plum*Lime*Berry in Future Landscapes Exhibition Palazzo Ca'Zanardi, Venice 2016. Photograph by author.

REFERENCES

Burtt, A. 2016. Compendium of New Zealand Farm Facts 2016 Version. Available online at: http://www.beeflambnz.com/Documents/Information/nz-farm-facts-compendium-2016%20Web.pdf.

Burtt, A. 2017. New Zealand Wool Export Statistics. Available online at: http://www.beeflambnz.com/information/export-statistics/wool-export-statistics/.

Cantoral-Ceballos, J. & Nurgiyatna, N. & Wright, P. Brown-Wilson, C. & Scully, P. 2014. Intelligent Carpet System Based on Photonic Guided-Path Tomography for Gait and Balance Monitoring in Home Environments. *Research Journal of Textile and Apparel* 15(1): 279–289. Publisher.

Deckers, E. v.d, Stouw, B. Peutz, J. 2012. An Intelligent and Interactive Carpet Role of Design in a textile Innovation Project. *Research Journal of Textile and Apparel* 16(4): 14–22.

Neroth, P. 2014. Light emitting carpets mark the way forward. *Engineering and Technology* 8(12): 16.

Parsons, S. Snippets on innovation and wool blog. Available online at: https://woolblog.com/author/premierwool/2017.

Tandus Certiva, 2017. Luminous Carpets marketing brochure. Available online at: https://www.tandus-centiva.com/product-solutions/luminous-carpets.

Textiles, Identity and Innovation: Design the Future – Montagna & Carvalho (Eds)
© 2019 Taylor & Francis Group, London, ISBN 978-1-138-29611-4

Smart textiles in the performing arts

Aline Martinez, Michaela Honauer, Hauke Sandhaus & Eva Hornecker
HCI Group, Faculty of Media, Bauhaus-Universität Weimar, Germany

ABSTRACT: We present the Sonification Costume, an interactive costume created for modern dance performances, that makes movements perceivable through sound. We created it in the context of studying textile-based sensors, and here describe the exploration of different production techniques, which led to a knitted whole-body suit with seamlessly integrated stretch sensors. Self-reflection and a user study reveal that the costume concept is interesting to dancers and how we can further improve textile-integrated sensors.

1 INTRODUCTION

The power of smart textiles is a growing topic in social and cultural contexts. In general, smart textiles are technology-enhanced fabrics or garments. Experts distinguish between embedding technology features directly in the fabric (e-textiles), or attaching computational devices onto clothes and bodies (wearables). Ryan (2014) shows that the first visions of smart garments are over 100 years old (p.27), and gives an overview of existing projects throughout her book. In the last two decades, dozens, maybe even hundreds of wearables and e-textiles projects have emerged. Many of these are design or artistic applications, with some meant to be worn by a performer who is trained to wear this type of clothing. In the following, we present the *Sonification Costume* – a knitted whole-body performance suit created for making a dancer's movements perceivable through sound (fig. 1). We categorize it as e-textile since technological components are seamlessly integrated and almost all are built from textile materials. Further, we call it an *interactive costume* because this dance costume is reactive to its wearer's movements. We here explain the idea behind using a whole-body suit for transforming movements into sound, and describe how the costume was crafted.

We share the material explorations made for figuring out the best do-it-yourself (DIY) technique of integrating conductive thread that enables sensing of positions or motions, and finally, we discuss the results of our user studies and feedback we received from two dancers who wore the costume.

This costume was created in a student project that had the goal of exploring self-made textile-based sensors and is part of a larger research project on interactive costumes for the performing arts. Our approach feeds from several disciplines ranging from fashion design over media arts and computer sciences to the social sciences. We strive to answer the following research questions: How can e-textiles support the performing arts and enhance the expressiveness of performers? How do we best produce textile-integrated sensors? How do wearers adapt to interactive dance costumes? This paper contributes a report on experiences in using interactive costumes for dance performance, knowledge on crafting stretch sensors, and insights in how dancers engage in wearing technology-enhanced costumes.

Figure 1. The sonification Costume. Photo by Tim Vischer.

2 BACKGROUND AND RELATED WORK

2.1 *Mass-market products and DIY community*

In the 1990s, Yamaha and Mattel released the first interactive wearable. Besides the lack of precision of

Mattel's Power Glove (en.wikipedia.org/wiki/Power_Glove), Yamaha's Miburi (en.wikipedia.org/wiki/Miburi) was used by artists in music and art performance. Only two decades later, commercial companies like Ralph Lauren (press.ralphlauren.com/polotech) and Google in cooperation with Levi's (Jacquard Project, Poupyrev et al. 2016) invest in smart textiles and wearables. Still, most products on the market tend to be consumer-focused applications (e.g. integrated buttons to control a smartphone) rather than tools directed at artists. These do however exist, and have been researched and developed over the last two decades by makers and DIY artists alike, who share their findings and processes online, embracing open-source principles. For instance, the works of Mika Satomi, Hannah Perner-Wilson (kobakant.at/DIY) and Afroditi Psarra (afroditipsarra.com) were fundamental to developing the project presented here. Schools and universities have also started giving more importance to the subject, and now offer courses related to e-textiles, bridging the gap between different disciplines (fashion and interaction design, engineering).

2.2 *Knitting and textile stretch sensors*

Knitting is the method of creating a fabric from multiple loops of yarn. Knitted fabric can be stretched, allowing the production of tight, but flexible, textiles. This freedom of movement is important for realizing a dance costume (Bicât 2012). Because of this ability to stretch, initially designers of circus costumes, and later of dance costumes, explored knitted textiles (vam.ac.uk/content/articles/d/dance-costume-design). Handcrafted knitted sensors integrating conductive yarn are being developed by the maker and DIY community, often from a combination of different materials. Some projects constitute interactive installations (Persson 2013) or discuss benefits of interactive knitted applications (Bredies & Gowrishankar 2014). Afroditi Psarra's Soft Articulations costume measures a dancer's motions, generating sound output (afroditipsarra.com/index.php?/on-going/softarticulations). It integrates handmade bend sensors crafted from resistive fabric and velostat.

2.3 *Interactive dance and sonification*

Many examples of interactive dance performances can be found in the literature. These range from improvisation (Huang et al. 2015, Moura et al. 2007) and modern dance (Johnston 2015, Latulipe et al. 2011), over oriental dance (Hsu & Kemper 2015) and tango (Brown & Paine 2015), to breakdance (Hosomi et al. 2007). Technologies are diverse. Some use on-body sensors to track motion and others use camera-based tracking, some provide visual feedback via projections and others with LEDs attached to the garment. Sensor technologies can be categorized according to Mulder's (1994) classification of three systems for tracking human movements: inside-in, inside-out and outside-in. With inside-in, sensor and source are on the body. Inside-out, the sensor is on the body and captures external sources from the environment (e.g. an accelerometer moving in the earth's gravitational field). Finally, outside-in, external sensors track artificial or natural sources (e.g. markers).

Previous work with sound output involved non-real-time sonification (Naveda & Leman 2008), use of external sensors such as video streams (Alaoui et al. 2015), external markers (James et al. 2006, Winkler 2002), external sensors (Siegel & Jacobsen 1998) or combinations (Camurri et al. 2000). Aiming to create music from a dancer's motion, Siegel & Jacobsen (1998) present a case study of interactive dance and list ten functional requirements for interactive dance interfaces. They conclude that as dance is visually aesthetic, and due the restriction of movements, it is impossible to create music by dancing. Thus, sensor devices are not music instruments.

In general, sonification can be understood as 'visualizing with sound'. Houri et al. (2011) use the term 'audiolizing' to describe body movements that are translated into sound. Sonification is frequently used to represent sensor data (for example geiger counter clicks) with non-speech audio. Hermann et al. (2011) categorize three sonification functions as alarms, alerts, and warnings; status, process and monitoring messages; and data exploration. They add a fourth function: art and entertainment. Various costumes and sensors exist that can create music, or sound, from body movements. However, these devices are aimed at musicians, and not dancers. In collaboration with computer engineers and other artists, the British musician Imogen Heap created gloves that can make music (imogenheap. co.uk/thegloves). With devices such as this, a performer's movements are made with respect to the musical notes, harmonics and beats the musician want to achieve. In contrast, the sound output of a costume for dancers should depend on the dance poses and body movements of the performers.

3 THE SONIFICATION COSTUME

3.1 *The costume idea*

Interactive costumes for stage performances tend to utilize video images and light. In contrast, our project aims for a sonification of the dancer's motion. The costume is a whole-body suit knitted to incorporate the circuit within the fabric structure. Thus, the knitted sensors are connected with threads to the microcontroller placed in the back, enabling the dancer to move freely without damaging the connections. The costume has six strategically placed sensors (hands, elbows, knees) that are wirelessly connected to a computer that creates the sound output. We intend to make performer's movements perceivable through sound – in other words, we aim to *sonify* dance poses and body movements via the costume. Each movement stretches the fabric so that it generates a variation in the electrical current going through the garment, which we measure as change in resistance. The final output is a raw translation of the changing resistance current into sound.

Thus, sound output depends on the dancer's movements (i.e. interaction with the garment). We assume that during dance improvisation, both will influence each other, creating an interactive dialogue. Even in a rehearsed dance performance we don't expect that the sonified movements compose a music.

With this work we explore dance, bodily motion, and e-textiles as an interface for interacting with digital systems. Choreographers and dancers such as Pina Bausch, Martha Grahams, and Mimi Jeong were our starting point to understand the dancer-space relationship. Their choreographic works inspired our concept design. While in the Tanztheater Wuppertal, Pina Bausch's dancers fuse live sounds with the background music (theguardian.com/stage/2014/aug/24/pina-bausch-music-sweet-mambo-tanztheater-wuppertal), in Martha Grahams' Lamentation piece, the tubular costume is reshaped by her motion (en.wikipedia.org/wiki/Lamentation_(ballet)). In Mimi Jeong's Double U performance, the alliance between two dancers inspires us to think about the exchange between the dancer and the sonification of his/her motions (jeongmimi.net/portfolio/double-u).

3.2 *Material exploration*

During our material exploration, different textile techniques were tested. Weaving, knitting, crochet and stitching were done with normal, resistive and conductive thread. Eventually, we found that each technique was suitable for different styles of sensors according to the methods' qualities. The first step in finding the ideal sensor was to produce different sensors from textile materials (fig. 2). We crafted bending (f,i), stretch (a,b,c,e,g,h,k,l,m) and pressure sensors (d, j). All fabric sensors – stretch, bend, or push sensors – work by changing resistance when force is applied. This can be measured with an ohmmeter. After exploring these types of sensors, we found that knitted and crocheted sensors seamlessly integrate the sensor in the fabric structure, and thus are less obtrusive for the wearer's body.

As in the Jaquard project (Poupyrev et al. 2016), weaving has been shown to produce good touch sensors. Because the technique uses two different sets of interlaced threads, it is possible to determine where exactly someone touches the fabric. The loom holds the wrap set of threads and the weft is wound between it, producing the patterns. Different threads either from the wrap or the weft can be wound together. Touch identification is feasible due to the fact that conductive threads are placed at specific places from both sets, building a matrix system. When the fabric is touched, the computer receives the data of column and row touched.

Felting, on the other hand, results in good pressure sensors. In Yihyun Lim's work (archive.monograph.io/ylim/felt-me), the condensed mix of normal and conductive fibers produces a fluffy structure that changes in conductivity according to the pressure force applied to it. This changes because, from the pressure, the conductive fibers touch each other more or less, improving the currency's flow through this structure.

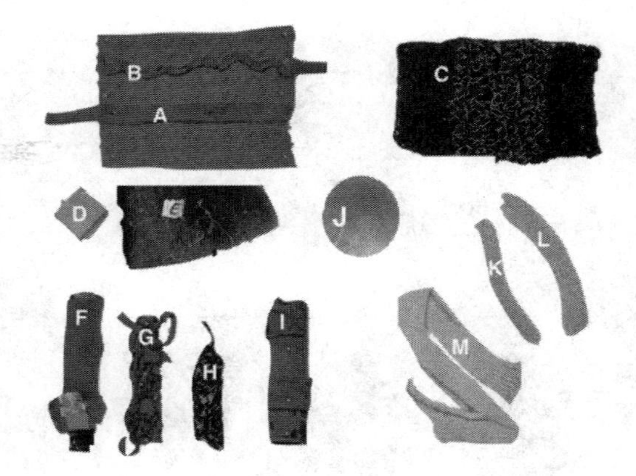

Figure 2. Textile sensors different in type and materiality.

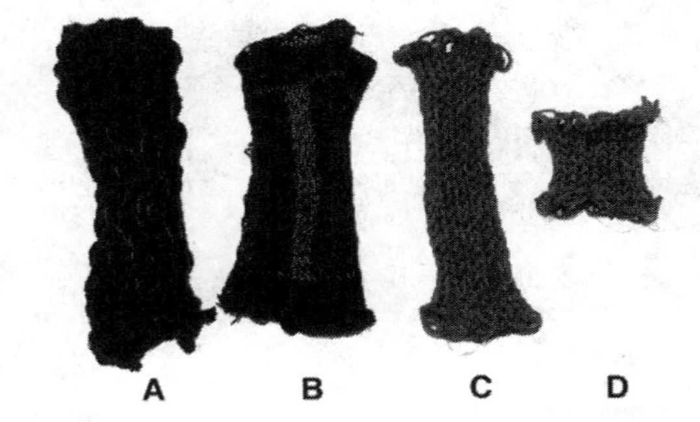

Figure 3. Knitted stretch sensors.

Knitting and crocheting result in a quality we were looking for: stretchability. While both are made by interlocking loops of yarns, their production method differs: in crocheting, each stitch is completed before proceeding to the next, while knitting keeps a large number of stitches open at a time. We used both techniques to produce stretch sensors. However, crocheted fabric does not quickly return to its former shape after being stretched and released. But knitted structures return easily to their former shape. Due to their stretchability, the motion of bending and opening joints can be sensed. Thus, when the knitted surface is stretched, the conductive thread is tensioned, changing its data flow.

After deciding to go for knitted stretch sensors, we crafted (fig. 3) and evaluated four knitted sensors (table 1). The sensors differ in thickness of wool, amount of resistive thread used, and size. Measurements were made with an Arduino microcontroller. The results in table 1 show the maximum and minimum value whenever the arms are bent or opened. Sensor A is made from thick wool, and differences were small (68), most likely because it consists of fewer resistive threads than the others, and also because the structure has less loops, and therefore less stretchability. Sensor B is a 2cm resistive line with more threads than sensor A. It produced the greatest

Table 1. Comparison of knitted stretch sensors.

	Grammage of wool	Characteristic of sensor – height × width (in cm)	Min. Value (in ohms)	Max. Value (in ohms)
A	50 g/45 m	circular 15 × 8	72	140
B	50 g/125 m	straight line 15 × 2	150	665
C	50 g/125 m	circular 23 × 3,5	190	420
D	50 g/125 m	circular 8 × 8	79	110

Figure 4. Textile-integration of a wrist stretch sensor (resistive thread in light gray) and circuit paths on the arms and legs (conductive thread in golden).

range of values, from 150 to 665. Sensor C, however, was the best. It accurately reveals the motion/position of the arm due to its circular shape. Sensor D is shorter but wider than the other sensors. This shows that height is more important than width for this type of sensor. Furthermore, during data collection, the circular sensor was worn in two ways: as sleeve and as stripe. We found that an attached tube sensor is more reliable. However, when the elbow is inserted in this tube, the data changes drastically, reducing reliability. Thus, another structure from normal wool should be knitted around the arm, placing the circular sensor as a stripe on the elbow.

3.3 *Realization*

3.3.1 *Textile interface*

From concept to implementation, the idea was to explore the possibilities for an interactive costume made from textile materials: thread and yarn. After testing materials and techniques for a feasible and aesthetic solution, the knitting technique was chosen to integrate the sensors. We used resistive thread in the knitting mesh for the dancer's joints, acting as stretch sensor (fig. 4). These sensors are connected with conductive thread manually stitched alongside the arms and legs to a pocket in the back (fig. 5); this houses the non-textile components of the system (a microcontroller, batteries and wireless communication module). Our intention was to make the functional textile parts of the dance suit obvious. Hence, the suit basically has been created out of black yarn, and the conductive/resistive threads are brighter, in silver/gray/golden color tones, what we found aesthetic because it visually puts highlights on the parts of the body our costume is sensitive to.

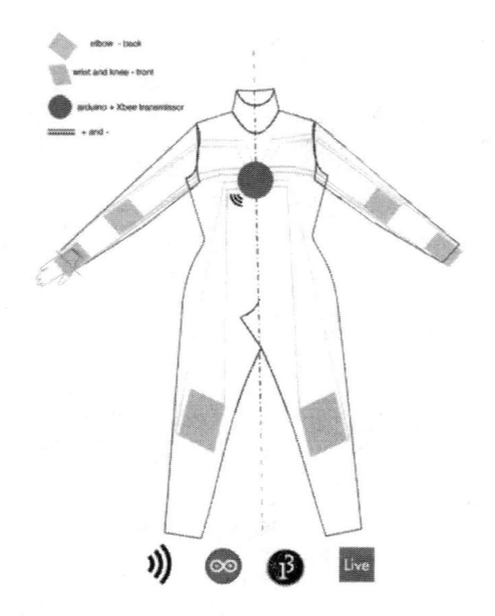

Figure 5. Schematic sketch of the costumes showing the position of all six stretch sensors and the microcontroller.

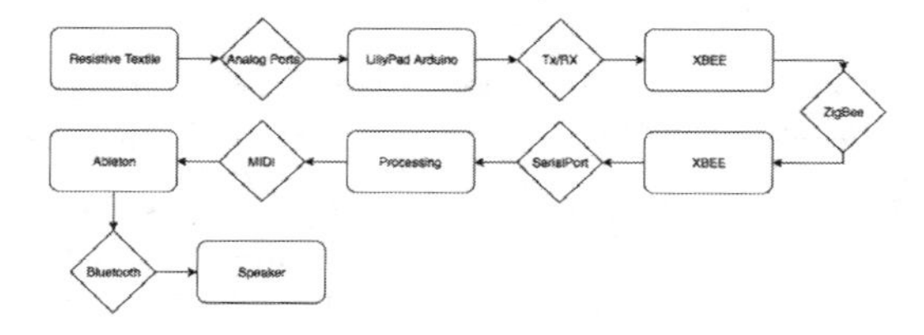

Figure 6. Schematic showing how the signals are processed.

3.3.2 *Technical implementation*

The softwares used are: Arduino for sensing motion signals and sending wifi data, Processing for receiving it, and Ableton for the sound output (fig. 6).

The microcontroller (Arduino Lilypad), placed at the back of the costume, reads six analog inputs (wrists, elbows, and knees). Different from digital buttons which only have an on/off state (bended or not), with analog signals, we can detect the degree of bending a joint. This information is sent wirelessly through an XBee module, also at the dancer's back, to another XBee module connected to a computer. We then encode this information with Processing (Java) to MIDI signals and forward the results to a commercial music sequencer (Ableton) to output the sound. The code enables an off-stage operator to calibrate the costume, change the sounds, or turn the costume on and off, all during performance.

3.3.3 *Sound output*

The final costume has two different sound sets and two different control modes. One sound set provides three different sounds with deviations for left and right limbs. The other sound set uses only one instrument for all limbs and combines harmonically in a minor gamut.

Figure 7. First user test.

Figure 8. Second dancer making sounds with specific postures.

The first control mode allows only constant movements to trigger sounds. The second control mode saves a null position of the performer and turns up the volume for the sensors individually if the dancer deviates from the original position.

Both modes sonify different aspects of dance and are equally valuable depending on the use case. Choreographies that put high emphasis on final positioning, e.g. in ballet, should not fade out the music if these positions are reached. Other dance styles that favour quick movements, e.g. improvisational or modern dance, could make more use of the constant movement mode.

4 EVALUATION

4.1 *Method*

To evaluate the costume, we conducted qualitative semi-structured interviews with dancers, and reflected on the crafting (self-observation) and staging process (indirect observation). The costume was used twice. First, by a dancer from children and youth ballet TPT Altenburg/Gera (fig. 7) that performed three dances with it (solo performance where she heard extra music via headphones and not the sonified sounds; in a group performance with sonified sounds to additional background music; and solo with sonified sounds and without additional music). At that time, the sensor signals were still noisy and unstable regarding their range. Before the second user test, and in hope for improved and clear signals, we added ready-made flex sensors from stiffer and plastic material and further stitched another layer of resistive yarn and sewed elastic thread into the fabric to keep the sensors in place. In a second test, a local dancer in Weimar performed two free improvisations with this improved costume, one for each of the two sound sets, without any background music (fig. 8). After each test, the dancers were interviewed with a set of questions, concerning wearability/usability, functionality, effect/perception, and concept/choreography integration of the costume.

4.2 *Results*

The dancers talked mainly about the two distinct costume characteristics, being knitted and making sound. As the two dancers performed very differently, interview responses differ. This was helpful to get user feedback and further improve the costume.

Wearability: Both dancers noted that the costume was comfortable and cozy. The technology housed in the back of the costume and the conductive/resistive threads did not irritate them nor restrict them in their movements, even the ready-made sensors were ok for the second test dancer. The knitted sensors did not feel different to the rest of the knitted costume. However, the dancers complained that the woolen fabric accumulates heat, which is inconvenient for long performances or during rehearsals. The second dancer was taller than the first and the costume did not fit well for her body. This was a bit uncomfortable, although she said it was no problem. In addition, observation revealed that the second dancer could not fully raise her arms up, as the costume restricted movement. During the crafting process, we had not thought about making the costume adjustable to different body sizes, as conductor lines and sensors are integrated into the costume's structure and the sensors need to be in the right place to deliver reliable signals. In addition, both dancers needed help to dress because the costume is tight; this enabled us to ensure correct sensor positioning.

Functionality: In general, both dancers experienced the costume as functional and reactive. They reported that they tried to exaggerate and adjust their movements to produce more sonified output. The first dancer noted a delay and could not hear how the costume reacted to her movements during the first two dances because she concentrated on the additional background music for her choreography. But its proper functioning was observed by us, even though the textile-integrated sensors, at that time, generated noise and consequently only had a small range. In contrast, the improved costume was experienced as fast and accurate by the second dancer. The integrated ready-made sensors yielded more precise bend or release information. Later, while still improving the technique for knit sensors, we found a way to create knitted sensors with more precise results (tube-like shape instead of flat design, fig. 3). During the entire crafting and staging process, we observed that the textile-based sensors sometimes did not function accurately. This was because we did not insulate them from skin moisture or protected them from light, which causes corrosion and decreases sensor's and conductor lines' quality after a while. We tried to alleviate this issue by integrating multiple layers and lines of conductive/resistive thread. Further, the taller dancer could not fully raise her arms as a functional boundary of the costume but, as we observed, she played with that restriction.

Effect: The costume potentially adds something new to the performance, and dancers thought it was new and futuristic. They found the concept and textile-integration of sensors very interesting. The first performer stated that the sound output does not disturb dancing because she knows the choreography "by heart". On the other hand, the second performer experienced this differently while improvising. She first tried out the sounds of each joint isolated and then attempted to combine them. This at first created undesired effects. Later, she became more experienced and was able to combine the sound "aesthetically" in a controlled manner. We observed this process as true interactivity as she was in a dialogue with the costume over time, and the system reacted properly to her inputs. Further, the second dancer was more interested in the first sound set and movement mapping because the abrupt movements and unnatural sounds allowed her more experimental freedom. From our observer perspective, the second sound set uses sounds that are combined more harmonically, and hence, encouraged more fluent and less abrupt movements.

Choreography: The first dancer, who only dealt with the first sound set and danced a given choreography, noted that the sounds generated by the costume did not fit the original meaning of the choreography. It suggested another meaning, altering the interpretation. The dance was "simple, happy", whereas the sound generated felt like "something bigger, maybe even mean"; she said this can fit to dance but would need to be staged consciously. She prefers happier harmonic sounds, but also noted that normally choreography and production are not made by the dancers, rather "they have to follow the rules of the choreographer and be professional". In comparison, the second dancer who improvised her own choreography was satisfied with the given scenario. Already during ideation, we assumed that our costume is more appropriate for performances that are not choreographed to a specific piece of music or have a higher degree of improvisation. The feedback from this dancer and our observations confirm this assumption. We experienced her performance as successful and further observed that the different sound modes and sound sets influenced how she danced; in particular, how output was mapped to input dramatically influenced her movements.

5 DISCUSSION

We here discuss the results of our evaluation on the background of our research questions. We show how functional requirements for interactive dance systems (Siegel & Jacobsen 1998) and costume design requirements (Honauer 2017) are successfully fulfilled, and point out what future research on interactive dance costumes can do.

5.1 *Main findings and lessons learned*

User tests and reflection on the creation process revealed insights in how wearable and easy to use the costume is, how stable and reliable it works, what dancers as wearers think of it, and how the costume concept could be integrated into choreography.

Materiality and Shape: Wool is an uncomfortable material for dance – although it adds interesting aesthetics, our tests dancers got into sweat quickly. Nevertheless, knitting appears an ideal technique to integrate textile-based stretch sensors into whole-body dance costumes – our test dancers did not feel them, but rather perceived ordinary fabric on their skin. This means the interface is unobtrusive (Siegel & Jacobsen 1998). Another argument for knitting is high stretchability which supports the sensors used. Despite the stretchy characteristics, one dancer was too tall and the costume restricted her movements because it was non-adjustable. According to Siegel & Jacobsen 1998, a costume must never restrict a dancer's movements. One could argue that stage costumes are always made for one person but we tested with two different dancers. Or, on the background of Oskar Schlemmer who experimented with costumes that restrict and direct performers movements (Scheper 1988), the costume could be staged for similar explorations and purposes. Further, the dancers needed some help for dressing, but since all technical components are integrated into the textile, the interface can easily be mounted onto the dancer (Siegel & Jacobsen 1998).

Functional Requirements: According to the functional requirements for interactive dance systems listed by Siegel & Jacobsen (1998), our costume was only in parts stable enough and ready for stage performances. Especially correct measurements, fast and accurate system responses, and robustness were major issues we struggled with. For instance, we experienced that long traces of conductive yarn can be problematic as they lose resistance, and uninsulated conductive/resistive threads suffered from corrosion of the material because of exposure to light and humidity. Thus, insulation and short paths of conductor lines are essential when creating smart dance costumes. Only when functionality and robustness are ensured, costumes become durable and reliable for stage performances (Honauer 2017). For the second user test, we attached ready-made bend sensors because the sensor measurements were not very clear during the first dance test. From the software side, we later were able to modify the textile sensor signals to balance noisy measurements, and from the hardware side we later found a more appropriate textile-based solution (circular instead of flat shape). Further, we measure angles of the body's limb joints in pose mode, we managed to transmit data wirelessly, our software/hardware setup is adaptable to other costume concepts or multiple costumes, and is not too expensive (Siegel & Jacobsen 1998).

Acceptance by Dancers: Our test dancers liked the costume concept and its aesthetic appearance. Like us, they believe in the potential of this costume concept for the performing arts. But this clearly depends on a dancer's personality and taste. The effect of the costume's sounds differed – the first dancer seemed to be irritated because the sonification did not fit the music normally used for the choreography. In contrast, the second dancer who danced without additional background music, enjoyed this more aggressive sound set (over the harmonious tones), as it was giving here more experimental freedom. We believe that dancers will like the Sonification costume more if they have improvisational freedom and a less predefined choreography.

Integration of Choreography: In general, an interactive costume can be applied for dance performances if it meets all necessary design requirements (Honauer 2017, Bicât 2012). But our case study showed that dances without a given choreography are higher rated by performers, and integrating the Sonification costume into existing choreographies might be hard. Thus the interaction concept of a costume needs to be developed hand in hand with the choreography, or at least, in our case the choreography is to be developed for the Sonification concept. Likewise, our second test dancer needed to make herself familiar with the costume's operating mode – this argues for early rehearsals with a costume prototype in order to best adapt choreographic work to the costume's interactive features (Honauer 2017).

5.2 *Future work*

Our project exemplarily shows a collaborative work of design and technology tested with potential users. Further research is to be conducted on several levels – the perspectives of different stakeholders are to be considered to satisfy the interdisciplinary character of smart performance costumes (Honauer 2017).

Dance and Choreography: Only a choreographer willing to explore the possibilities of interactive dance has the practical knowledge to stage a performance with the Sonification Costumes. Likewise, it could be interesting to stage multiple Sonification costumes or to enable blind or visually impaired people to experience dance. Further, dancers need to learn early-on how to control the costume's features (Honauer 2017), and future studies can determine how much time is needed in which conditions.

Textile, Fashion and Costume Design: Textile integrability is the characteristic we were mostly looking for. Fashion and costume designers might be able to explore other techniques (e.g. crocheting or weaving) suitable for integrating textile-based sensors. But this demands knowledge in electronics and hardware design, which currently is not part of a fashion or costume designer's workflow. We could overcome this by education, or by creating a closer collaboration with professions that have these competencies. Also, washing and repairing interactive costumes is not trivial (Honauer 2017), and needs to be explored further with experts of garment or textile design.

Interaction Design: Our work showed value of experimenting with more materials and techniques for future scenarios. Interaction Designers might find solutions not only through conceptual work, and might come up with good designs by actively exploring fabrics and combining them with conductive fabrics and yarns. Future research should ask if interaction designers need basic knowledge in dance costume design because this seems to be crucial to successfully stage dance performances (Bicât 2012). This is similar with a designer's experience with dance – Latulipe et al. (2011) give insights in how to design interactive dance performances, but future research could focus more on the processes of designing interactive dance costumes.

Engineering and Computer Sciences: Developing textile based circuits or even microcontrollers is still a challenge but further investigations might be able to discover new approaches (Buechley & Eisenberg 2009). Hence technologists need to be involved in the creation process. Furthermore, modular approaches, as already common in software and hardware development, appear promising for interactive costumes. Modularity, developed in collaboration with designers, could make interactive costumes more robust, functional, and easier to maintain.

Audience: The literature provides a few insights in the spectator's perception of interactive stage effects in general (Reeves et al. 2005), or explains how projects involved spectators (Friederichs-Büttner et al. 2012). But investigation in the perception of interactive dance costumes is still missing. On the artistic and human perception fields, our work leads to the question of how the Sonification concept opens new ways to the perception of dance. When we are able to stage robust and reliable interactive costumes, audience evaluations can provide more insights in the effects this type of costume has.

6 CONCLUSION

We presented the design of an interactive dance costume. A small user study and reflection of the design process reveal insights in challenges that emerge when creating and staging garments with textile-integrated stretch sensors for the performing arts. Our prototype showed that it is possible to seamlessly integrate hardware, but textile sensors must be stable and deliver reliable signals. The implementation process required a lot of experimentation and interdisciplinary teamwork. Designers and engineers needed a mutual understanding to develop design and technology together. At the end, working with the potential end-users – dancers in our case – was important to figure out how the textile product feels like, if it is appropriate for the intended purpose, and how intuitive its use is.

In light of all this, we believe that textile-integrated sensors are a great chance for dance applications, but future research should consider choreographic work as well as the audience's perception to reach an encompassing understanding of how to design interactive dance costumes. The results we presented here may also be helpful for designing interactive costumes with other types of textile-based sensors and actuators, or for the design of interactive clothing in general because our research collects basic insights in how to craft e-textile products.

ACKNOWLEDGEMENTS

We thank the children and youth ballet Altenburg/Gera, in particular Charlotte Streu who performed in our first prototype, and the teacher Claudia Kupsch. We thank Claire Dorweiler for performing in our final costume, Tim Vischer for his help and taking photographs, and Kevin Jahnel for producing the second sound set.

REFERENCES

Alaoui, S.F., Bevilacqua, F. & Jacquemin, C. 2015. Interactive Visuals as Metaphors for Dance Movement Qualities. *ACM Trans. Interact. Intell. Syst.* 5, 3, Article 13 (September 2015), 24 pages.

Bicât, T. 2012. *Costume and Design for Devised and Physical Theatre*. The Crowood Press, Wiltshire, UK.

Bredies, K. & Gowrishankar, R. 2014. Interactive Knitting Patterns as Boundary Objects in e-Textile Design. In *Proc. of Ambience '14&10i3m*. Tampere, Finland. 1–6.

Brown, C. & Paine, G. 2015. Interactive Tango Milonga: designing internal experience. In Proc. *of the 2nd International Workshop on Movement and Computing* (MOCO '15). New York, USA: ACM. 17–20.

Buechley, L. & Eisenberg, M. 2009. Fabric PCBs, electronic sequins, and socket buttons: techniques for e-textile craft. In *Personal Ubiquitous Comput.* 13, 2. London, UK: Springer Verlag. 133–150.

Camurri, A., Hashimoto, S., Ricchetti, M., Ricci, A., Suzuki, K., Trocca, R. & Volpe, G. 2000. EyesWeb: Toward Gesture and Affect Recognition in Interactive Dance and Music Systems. In *Computer Music Journal* 24, 1 (March 2000). Cambridge, USA: The MIT Press. 57–69.

Friederichs-Büttner, G., Walther-Franks, B. & Malaka, R. 2012. An unfinished drama: designing participation for the theatrical dance performance Parcival XX-XI. In *Proc. of the Designing Interactive Systems Conference* (DIS '12). New York, USA: ACM. 770–778.

Hermann, T., Hunt, A. & Neuhoff, J. 2011. *The Sonification Handbook*. Berlin, Germany: Logos Publishing House.

Honauer, M. 2017. Designing (Inter)Active Costumes for Professional Stages. In Schneegass, Stefan & Amft, Oliver (Eds.), *Smart Textiles – Fundamentals, Design, and Interaction*. Cham, Switzerland: Springer International Publishing. Chapter 13, 279–302.

Hosomi, S., Tsukamoto, M. & Nishio, S. 2007. A system for controlling LED blink in wearable fashion. In *Proc. of the 2007 intern. Conf. on Wireless communications and mobile computing* (IWCMC '07). New York, USA: ACM. 665–670.

Houri, N., Arita, H. & Sakaguchi, Y. 2011. Audiolizing body movement: its concept and application to motor skill learning. In *Proc. of the 2nd Augmented Human International Conference* (AH '11). New York, USA: ACM. Art. 13, 4 pp.

Huang, H., Huang, H.-C., Liao, C.-F., Li, Y.-C., Tsai, T.-C., Teng, L.-J. & Wang, S.W. 2015. Future circus: a performer-guided mixed-reality performance art. In *Adjunct Proc. of the 2015 ACM Intern. Joint Conference on Pervasive and Ubiquitous Computing and Proc. of the 2015 ACM Intern. Symposium on Wearable Computers* (UbiComp/ISWC'15 Adjunct). New York, USA: ACM. 551–556.

Hsu, A. & Kemper, S. 2015. Kinesonic approaches to mapping movement and music with the remote electroacoustic kinesthetic sensing (RAKS) system. In *Proc. of the 2nd International Workshop on Movement and Computing* (MOCO '15). New York, USA: ACM. 45–47.

James, J., Ingalls, T.; Qian, G., Olsen, L., Whiteley, D., Wong, S. & Rikakis, Th. 2006. Movement-based interactive dance performance. In *Proc. of the 14th ACM intern. Conf. on Multimedia* (MM '06). New York, USA: ACM. 470–480.

Johnston, A. 2015. Conceptualising interaction in live performance: reflections on 'encoded'. In *Proc. of the 2nd International Workshop on Movement and Computing* (MOCO '15). New York, USA: ACM. 60–67.

Latulipe, C., Wilson, D., Huskey, S., Gonzalez, B. & Word, M. 2011. Temporal integration of interactive technology in dance: creative process impacts. In *Proc. of the 8th ACM conference on Creativity and cognition* (C&C '11). New York, USA: ACM. 107–116.

Moura, J.M., Barros, N., Branco, P. & Marcos, A.F. 2010. NUVE: in between the analog and virtual body. In *Proc. of the fifth intern. Conf. on Tangible, embedded, and embodied interaction* (TEI '11). New York, USA: ACM. 409–410.

Mulder, A. 1994. *Human movement tracking technology*. Retrieved March 18, 2017 from http://www.xspasm.com/x/sfu/vmi/HMTT.pub.html

Naveda, L. & Leman, M. 2008. Sonification of Samba dance using periodic pattern analysis. In *Proc. of 4th Intern. Conf. on Digital Arts* (ARTECH 2008). Porto, Portugal: Portuguese Católica University. 16–26.

Perner-Wilson, H., Buechley, L. & Satomi, M. 2010. Handcrafting textile interfaces from a kit-of-no-parts. In *Proc. of the fifth intern. Conf. on Tangible, embedded, and embodied interaction* (TEI '11). New York, USA: ACM. 61–68.

Persson, A. 2013. Stretching Loops – Exploring interactive textiles expressing Imitation for recording and replaying interactions. In *Crafting the Future, 10th European Academy of Design Conference*. Göteborg, Sweden.

Poupyrev, I., Gong, N.-W., Fukuhara, S., Karagozler, M.E. & Schwesig, C. & Robinson, K.E. 2016. Project Jacquard: Interactive Digital Textiles at Scale. In *Proc. of the 2016 CHI Conference on Human Factors in Computing Systems* (CHI '16). New York, USA: ACM. 4216–4227.

Reeves, S., Benford, S., O'Malley, C. & Fraser, M. 2005. Designing the Spectator Experience. Designing the Spectator Experience. In *Proc. of the SIGCHI Conference on Human Factors in Computing Systems* (CHI '05). New York, USA: ACM. 741–750.

Ryan, S.E. 2014. *Garments of Paradise – Wearable Discourse in the Digital Age*. Cambridge, USA: The MIT Press.

Scheper, D. 1988. *Oskar Schlemmer – das triadische Ballett und die Bauhausbühne*. Berlin, Germany: Akad. d. Künste.

Siegel, W. & Jacobsen, J. 1998. The Challenge of Interactive Dance: An Overview and Case Study. In *Computer Music Journal* 22, 4 (Winter 1998). Cambridge, USA: The MIT Press. 29–43.

Winkler, T. 2002. Fusing movement, sound, and video in Falling Up, an interactive dance/theatre production. In *Proc. of the 2002 conference on New interfaces for musical expression* (NIME-02). Singapore, Singapore: National University of Singapore. 1–2.

Textiles, Identity and Innovation: Design the Future – Montagna & Carvalho (Eds)
© 2019 Taylor & Francis Group, London, ISBN 978-1-138-29611-4

Functionalization of cotton fabric with chitosan microspheres containing triclosan

L.G. Magalhães, C.S.A. de Lima & S.M. da Costa
EACH – USP, School of Arts, Sciences and Humanities, University of São Paulo, São Paulo – SP, Brazil

A.C.S. Santos
Department of Microbiology and Immunology – UNESP, State University of São Paulo, Botucatu – SP, Brazil

S.A. da Costa
EACH – USP, School of Arts, Sciences and Humanities, University of São Paulo, São Paulo – SP, Brazil

ABSTRACT: Medical textiles are structures designed for use within the medical field and can be classified into surgical textiles, extra body systems and hygiene and health products. Chitosan is a natural polymer with interesting properties such as bactericidal activity, biocompatibility and non-toxicity. In this work, an antibacterial finish for hospital fabrics was developed from the synthesis of chitosan microspheres with the incorporation of the triclosan drug. The finish was impregnated in 100% cotton fabric with twill structure. The fabric was characterized by mass per unit area, yarns linear density and yarns density determination tests. Optical microscopy analyzes were performed and showed that the microspheres were impregnated between the fibers of the fabric. The material, therefore, presented potential for application as a finish for fabrics used in the medical area.

1 INTRODUCTION

The constant search for improvement in the quality of life and the aging of the population has stimulated research on textile raw materials – whether in the form of fibers, filaments, fabrics or nonwoven – for medical applications. Due to the innumerable physical characteristics of fibers and fabrics, such as three-dimensional structure and variety in length and fineness, textiles have been gaining more and more space in medical applications (Ferreira et al. 2014, Quin, 2017).

Medical textiles represent a segment of technical textiles and are structures developed for use within the medical field. They play an important role in this field, and can be found in many applications, from surgical masks to artificial organs. They are divided into three categories (Araújo et al. 2001):

- Surgical textiles (includes all products and materials – implantable, non-implantable and protective – used in surgeries);
- Textiles for extra body systems (artificial organs);
- Hygiene and health products (surgical, protective clothing, bedding, etc.).

The use of natural polymers in the production of these materials in the hospital, hygiene and health area is very promising due to the properties that these materials present in relation to the synthetic ones. Chitosan is a biocompatible, biodegradable and non-toxic polymer (Heinemann et al. 2008). It consists mainly of β-(1$\rightarrow$4)-2-amino-2-deoxy-D-glucopyranose units (Fig. 1). Chitosan is a natural cationic polysaccharide derived from the removal of the acetyl groups (C_2H_3O) from chitin, which is a polysaccharide found in the exoskeleton of crustaceans and in the cell wall of fungi. This deacetylation reaction is carried out in an alkaline medium (Jayakumar et al. 2010, Elgadir et al. 2015, Hosseinnejad & Jafari, 2016).

Deacetylation of chitin to produce chitosan is considered one of the fundamental parameters that influence in its properties, mainly in its bactericidal action (against gram-positive and gram-negative bacteria) (Eichhorn et al. 2009).

Chitosan has molecular formula $(C_6H_{11}O_4N)_n$, is classified as a weak base, and has three reactive functional groups, which are: one amino and two hydroxyls (primary and secondary). The amino grouping is predominant and has great influence on the properties of this copolymer.

The protonation of this group, which occurs in acidic medium, contributes to the appearance of positive charges in the chitosan and its bactericidal activity, whereas in the alkaline pH the deprotonation of the NH_3 grouping occurs, causing the polymer, consequently, to lose its positive charges, its bactericidal

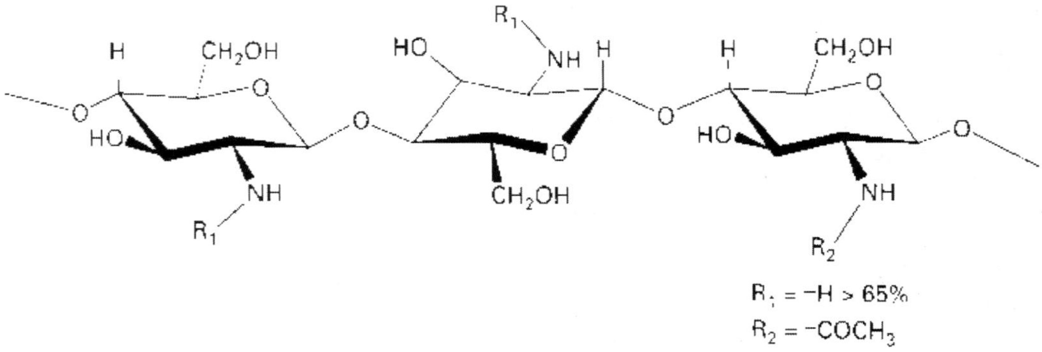

Figure 1. Chitosan Chemical Structure (Eichhorn et al. 2009).

capacity and its solubility (Joshi et al. 2009, Elgadir et al. 2015, Hosseinnejad & Jafari, 2016).

In addition to its antibacterial capacity, other properties, such as viscosity, solubility, metal ion chelation, reactivity, elasticity, elongation, tensile strength and moisture absorption, make this polymer progressively employed in medical, food and agricultural fields, and others (Eichhorn et al. 2009, Elgadir et al. 2015).

Chitosan has the advantage of variability of the ways in which it can be prepared, like gels, fibers, nanofibers, membranes, flakes, films, scaffolds, microcapsules, microspheres, etc. (Matté & Rosa, 2013, Jayakumar et al. 2011).

Microspheres are irregular spherical shaped microparticles containing active substance in a dispersed polymer base and have biodegradable and biocompatible polymers as the main raw material. The use of microspheres in the industry had its development starting in 1960 and with the advance of researches it began to be applied for several purposes, for example, protection of active agent against external agents, control release of active agent, treatment of industrial effluents and finishes Textiles. (Boury et al. 2009, Matté & Rosa, 2013).

In this research, chitosan was used as triclosan drug envelope (TCS, 2,4,4-trichloro-2'-hydroxyphenyl ether). This drug is widely used as an antimicrobial agent in personal hygiene products (Dayan, 2007).

The objective of this work was functionalization of cotton fabric with chitosan microspheres containing triclosan for applications in medical hygiene and health textiles.

2 MATERIALS AND METHODS

2.1 *Materials*

Chitosan used in this study is a commercial product (Sigma Aldrich, St. Louis, USA) obtained from the purification of chitin extracted from shrimp shells with at least 75% deacetylation; triclosan was of pharmaceutical grade. All solvents used in this study were of analytical grade. The cotton fabric used was with 100% cotton composition and twill structure.

2.2 *Methods*

2.2.1 *Synthesis of Chitosan Microspheres*

The method used for the production of the microspheres was as described by Hui et al. (2013).

To prepare the chitosan and triclosan gel, 0.05% (m/m) of triclosan was diluted in 5 mL of ethanol and 1 g of chitosan was added to a Becker containing 20 mL of 2% acetic acid solution (v/v). Both solutions were mixed and the resulting mixture was kept under magnetic stirring overnight until complete homogenization.

After complete dissolution of the polymer, the chitosan and triclosan solution was added in small quantities to a solution of 150 mL of liquid paraffin containing 2% (v/v) Span-80 surfactant, at a constant temperature of 55°C under mechanical stirring for 20 minutes. The mixture was then cooled in an ice bath to −10°C and the pH adjusted to 9–10 with 10% (m/v) sodium hydroxide solution. 10 mL of glutaraldehyde (25% aqueous solution) was then sprayed, and the mixture was further stirred for another 40 minutes to stabilize the microspheres. Then, to separate the microspheres, the solution was transferred to falcon tubes, which were placed in the centrifuge for 40 minutes at 4000 rpm and 5°C. At the end of this process, the oily part of the solution was discarded, the microspheres were separated, washed and filtered with petroleum ether and water, and left dispersed in distilled water overnight and then filtered again with water and absolute ethanol After being cleaned, the microspheres were deposited in Petri dishes and dried under vacuum in a desiccator containing silica.

2.2.2 *Characterization of the fabrics used for impregnation of the finish*

Fabrics without and with antibacterial finish were evaluated by yarn density tests, based on ISO 7211-2, determination of mass per unit area of fabric (ISO 3801), and linear density of yarns (ISO 7211-5).

Figure 2. Moist chitosan microspheres.

Figure 3. Chitosan microspheres after vacuum drying.

2.2.3 *Impregnation of microspheres on textile substrates*

Impregnation process of the chitosan microspheres onto 100% cotton twill fabric consisted on the preparation of 250 mL of a bath containing 20 g/L Phobol resin (Huntsman, USA).

In a 50 mL becker it was weighed 1 g of microspheres and added 10 mL of distilled water. The dispersion was kept under magnetic stirring for 10 minutes.

Thereafter, the microsphere dispersion was added to the resin bath and the mixture remained under stirring for another 10 minutes.

The impregnation was carried out in a pad, adjusted with pressure of 1 Bar and speed of 0.031 m/min in order to obtain a pick-up of approximately 80%.

Once the microspheres were impregnated, substrate was dried in a steamer for 5 min at 110°C and cured for 1 min at 130°C.

2.2.4 *Optical microscopy of fabrics with antibacterial finish*

Antibacterial finished fabrics were analyzed by light microscopy on a Leica DME microscope (Wetzlar, Germany) at magnification 10x.

3 RESULTS AND DISCUSSION

3.1 *Synthesized microspheres*

The material obtained by the synthesis reaction had a yellowish coloration while still moist and it was possible to observe the microspheres were formed (Fig. 2).

After vacuum drying (Fig. 3), the sample reduced in volume, got brownish color (due to oxidation, probably) and appearance of powder.

3.2 *Characterization of the fabrics used for finish impregnation*

Determination of mass per unit area test was performed with cotton substrate samples before and after the impregnation of the chitosan microsphere with Phobol resin. Results are shown in Tables 1 and 2.

Table 1. Results of mass per unit area of 100% cotton twill.

Tecido sarja - algodão 100%			
Corpo de prova	Massa (g)	Média (g)	Gramatura (g/m²)
1	2,05		
2	2,07		
3	2,03	2,05	204,74
4	2,04		
5	2,05		

Table 2. Results of mass per unit area of 100% cotton twill after application of Phobol resin.

Tecido sarja - algodão 100% - PHOBOL			
Corpo de prova	Massa (g)	Média	Gramatura (g/m²)
1	2,10		
2	2,08		
3	2,11	2,08	207,78
4	2,06		
5	2,04		

After impregnation of the microspheres a small increase in fabric weight was noted, about 1.5%, probably due to the addition of both the resin and the microspheres to the substrate, in order to raise the amount of grams per square meter.

Linear density of the yarns, both weft and warp obtained before and after impregnation of the finished fabric are given in Tables 3 and 4 (warp) and Tables 5 and 6 (weft).

According to the data obtained, it was observed that the warp yarns shown a small decrease of linear density after the impregnation of the resin containing the microspheres. Since this difference was minimal, it was probably related to the irregularity of the specimens used and not to the impregnation of the finish. In the case of weft yarns, impregnation of the finish

Table 3. Warp yarn linear density.

Tecido sarja - Título – Urdume			
Amostras	Comprimento	Massa	Título
	(mm)	(mg)	(Tex)
1	105	3,6	34,28
2	90	3,7	41,11
3	102	3,6	35,29
4	107	3,8	35,51
5	95	3,6	37,89
		Média	36,82

Table 4. Warp yarn linear density after microspheres impregnation.

Tecido sarja - Título - Urdume			
Amostras	Comprimento	Massa	Título
	(mm)	(mg)	(Tex)
1	100	4,1	41
2	105	3,7	35,24
3	91	3,3	36,26
4	94	3,2	34,04
5	103	3,4	33,01
		Média	35,91

Table 5. Weft yarn linear density.

Tecido sarja - Título – Trama			
Amostras	Comprimento	Massa	Título
	(mm)	(mg)	(Tex)
1	110	3,4	30,91
2	110	4	36,36
3	110	3,8	34,54
4	111	3,7	33,33
5	108	3,4	31,48
		Média	33,33

Table 6. Weft yarn linear density after microspheres impregnation.

Tecido sarja - Título - Trama			
Amostras	Comprimento	Massa	Título
	(mm)	(mg)	(Tex)
1	105	3,7	35,24
2	80	2,7	33,75
3	109	4	36,70
4	103	3,9	37,86
5	94	3,9	41,49
		Média	37,01

implied in the increase of 10% in the average linear density, going from 33.3 Tex to 37.0 Tex.

Finally, the density test of the 100% cotton twill fabric (Table 7) was performed only with the control (unfinished) fabric, since the amount of warp yarn

Table 7. 100% Cotton Twill fabric yarn density.

Tecido sarja - algodão 100%		
Corpo de prova	Fios de urdume/cm	Fios de trama/cm
1	32	16
2	32	16
3	32	16
4	33	16
5	33	16
Média	32,4	16

and weft per centimeter present in the substrate does not change by application of the resin with the microspheres. The analyzes showed that the fabric had 32.4 warp yarns/cm and 16 weft yarns/cm.

3.3 *Impregnation of particles on textile substrates*

Phobol resin applied to the cotton samples for impregnation of chitosan microspheres made the fabric less softer, causing it to partially lose its malleability.

The resin used was hydrophobic, which, according to the application, can confer an interesting waterproofing property for medical textiles and be another form of prevention against hospital contamination (Mitchwell et al. 2015).

Visually the substrate showed no variation in its coloration, but the applied microspheres, found on the surface due to its brownish hue, could be observed.

Microspheres impregnation in the fabric showed acceptable fixation. It was noted that smaller microspheres were housed between the fibers of the tissue and therefore well secured. However, as larger microspheres were concentrated on the surface of the yarns and more susceptible to detachment from the fabric.

3.4 *Optical microscopy of fabrics with antibacterial finish*

The images obtained by optical microscopy made it possible to observe the fixation of the particles in the innermost areas of the tissue, in order to show why some of the impregnated microspheres were more fixed than the others. It was also possible to analyze the morphology and size of the microspheres.

It has been observed in Figure 4 that some areas of the fabric have concentrated a greater amount of microspheres, and their fixation between the fibers of the tissue is remarkable. The microspheres accumulated in this area of the substrate had on average 100 μm of diameter, presented well defined spherical shape and regular surface.

Figure 5 shown a larger microsphere, with a diameter of 194.11 μm and isolated from the others. Also with spherical aspect and smooth surface.

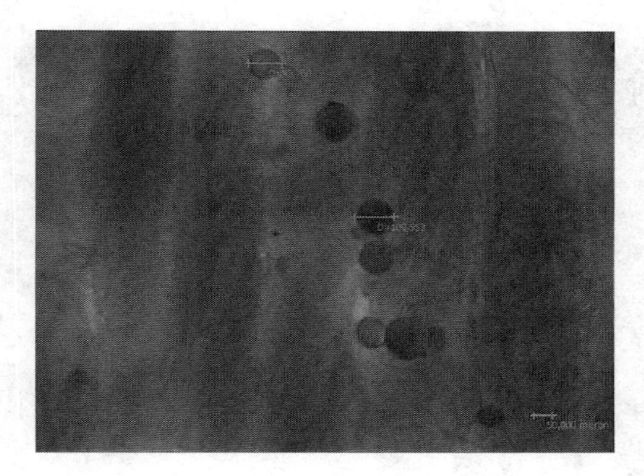

Figure 4. Microspheres images obtained by optical microscopy (10x magnification).

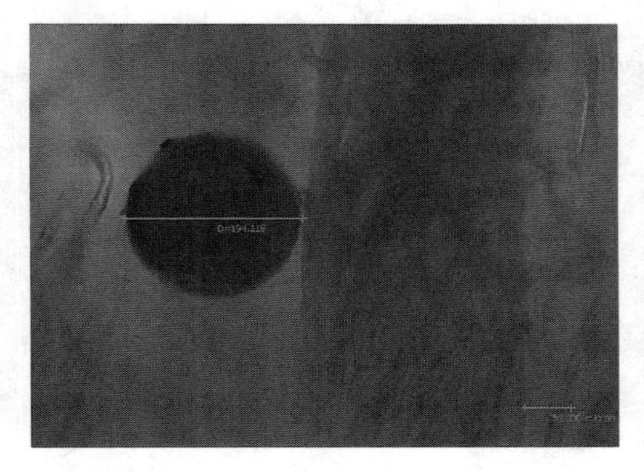

Figure 5. Isolated microsphere image obtained from optical microscopy (10x magnification).

4 CONCLUSION

Chitosan microspheres with triclosan were synthesized and applied onto 100% cotton fabric with twill structure. Application of the finish did not cause significant changes in the physical characteristics of the fabric. By means of optical microscopy, it was possible to verify a good distribution and fixation of the microspheres between the fibers of the substrate. Therefore, it can be concluded that the material has potential for application as a textile finish and for the development of new studies. Further tests are still required.

ACKNOWLEDGMENT

Authors would like to thank FAPESP and CAPES for the financial support for the research development.

REFERENCES

Araújo, M.; Fangueiro, R.; Hong, H. *Têxteis Técnicos: Materiais do Novo Milénio*, Ed. Williams, Ltda, Ministério da Economia, v. 2, p. 167, 2001.

Boury, F. et al. Method for preparing particles from na emulsion in supercritical or liquid CO2. US 2009/0087491 A1. 22 dez. 2005, 2 abr. 2009.

Dayan, A.D. Risk assessment of Triclosan (Irgasan) in human breast milk. In: Food Chemical Toxicology, v. 45, p.125–129; 2007.

Eichhorn, S.J. et al. Handbook of textile fibre structure. 1.ed. Boca Raton: CCR Press, 2009. 2v.

Elgadir, M.A. et al. Impact of chitosan composites and chitosan composites on various drug delivery systems: a review. Journal of Food and Drug analysis. n. 23, p. 619–629, 2015.

Ferreira, I.L.S. et al. Aplicação de Materiais têxteis na área da saúde. In: Congresso Científico Têxtil e de Moda, 2., 2014, São Paulo.

Heinemann, C., Huinemann, S., Bernhaedt, A., Worch, H. & Hank, T. Novel Textile Chitosan Scaffolds Promote Spreading, Proliferation, and Differentiation of Osteoblasts. Biomacromolecules v.9, 2913–2920, 2008.

Hosseinnejad, M. & Jafari, S.M. Evaluation of diferente factors affecting antimicrobial properties of chitosan. International Journal of Biological Macromolecules, Gorgan, p. 467–475, 2016.

Hui, P.C.L. et al. Microencapsulation of Traditional Chinese Herbs – PentaHerbs extracts and potential application in healthcare textiles. In: Colloids and Surfaces B: Biointerfaces, v. 111, p. 156–161, 2013.

International Organization for Standardization. ISO 3801: Determination of mass per unit length and mass per unit area. Switzerland, 1977. 4p.

International Organization for Standardization. ISO 7211-2: Determination of number of threads per unit length. Switzerland, 1984. 6p.

International Organization for Standardization. ISSO 7211-5: Determination of linear density of yarn removed from fabric. Switzerland, 1984. 4p.

Jayakumar, R. et al. Biomedical aplications of chitin and chitosan based nanomaterials – A short review. Carbohydrate Polymers. v. 82, p. 227–232, 2010.

Joshi M., Ali, S.W. & Purwar R. Ecofriendly antimicrobial finishing of textiles using bioactive agentes based on natural products. Indian Journal of Fibre & Textile Research. v. 34, p. 295–304, 2009.

Matté, G.M. & Rosa, S. A tecnologia da microencapsulação através das microesferas de quitosana. Revista Iberoamericana de Polímeros. v. 14, n. 5, p. 206–218, 2013.

Mitchell, A., Spencer, M. & Edmiston JR, C. Role of healthcare apparel and other healthcare textiles in the transmission of pathogens: a review of the literature. Journal of Hospital Infection, v. 90, p.285–292, 2015.

Quin, Y. *Medical Textile Materials*. Cambridge (GB): Woodhead Publishing, 2017. 264p.

Textiles, Identity and Innovation: Design the Future – Montagna & Carvalho (Eds)
© 2019 Taylor & Francis Group, London, ISBN 978-1-138-29611-4

Smart textiles in architecture

A.C. d'Oliveira, G. Montagna & J. Nicolau
CIAUD, Lisbon School of Architecture, University of Lisbon, Lisbon, Portugal

ABSTRACT: The overall objective to introduce the Smart Textiles into architecture goes directly to a new paradigm shift into green buildings or sustainable buildings. Inhabitants, want to be part of the changes, their participation makes the projects successful. The smart textiles can be a new integrant part into office spaces, into commercial buildings or even at private homes. Fast introductions and less expensive actions can support new conductors of electricity or even into new supports of lighting fixtures under architecture interiors with the smart wall textiles. The use of these new type of walls, for example, music systems can be introduced into spaces, without cables or construction support, which take time and money. Either, with the new smart textiles, will be possible to support the heating, with simple curtains or fabric wall paper, adapting the temperature with the interior needs, without a new air-conditioning system. This research will introduce new possible actions into architecture with sustainable measures and with a possible strong support from the users.

1 INTRODUCTION

1.1 *Smart textiles & architecture*

The combination of these two areas, textiles and Architecture can a be a possible investment using the nanotechnology concepts into a new paradigm.

With the growing interests of the building sector, and the intense levels of differentiation that the market arrives, a new introduction of intelligent technologies into the textiles offering to the end user's smart actions into spaces. The wall covering components could work with an aesthetic function or could include a new type of comfort, creating a new possible option or improving the traditional installations exposing a possible more efficient system (Aliprandi, 2015).

The fabrication processes for smart textiles with conductive fibers with an integration of soft sensors as pressure, temperatures or others (Capineri, 2014).

1.2 *New paradigm & green buildings*

With the fast change that the world is passing, a new paradigm to buildings is part of global action days. Green buildings are consuming adaptions into materials and trends, as a market as the market demand.

For example, in Germany, a new sports center implemented a sophisticated membrane under the construction where they applied a layer that enables to achieve large column-free spans where the building shapes and allows for high indoor daylight quality. They used three different membrane materials as the fluoropolymer-coated PTFE (Polytetrafluoroethylene) fabrics, the PTFE- coated glass-fiber fabrics and the PVC-coated (Polyvinylchloride) polyester (PES) fabrics.

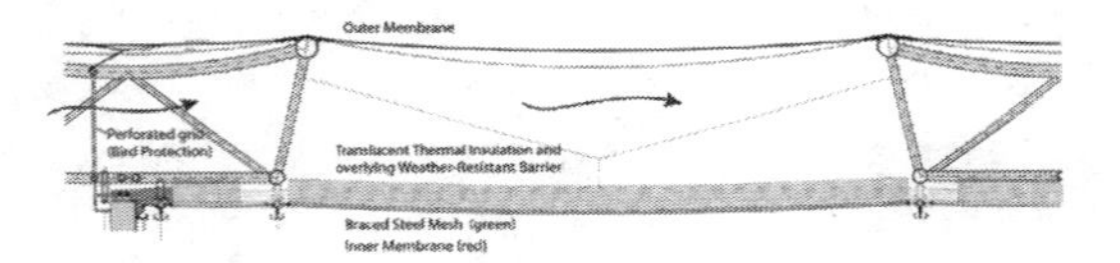

Figure 1. Cross section roof construction (fab_architects) (source: Capineri, L, Elsevier)

These materials created benefits to reduce the energy demand for heating and cooling. The building created benefits with a translucent Thermal Insulation, and the Fiber spun fabric made of glass or polyester and synthetic resins or aerogel insulation material, which has the best insulating properties and light transmission (Cremers, 2016) currently.

Studies like this in Germany are positive to other possible cases. Choosing the right materials and the right combinations can create benefits to the reduction of energy demand and more resources will be saved in the future. More experiences like these are needed to avoid more mistakes under construction (Cremers, 2016).

1.3 *Scientific research on structures*

A new tendency can occur under the building's rehabilitation with the support of new advanced textile materials for the construction of tensile structures.

The own structures can be introduced a high performance when they combine innovative sensors to constant monitoring the structures measuring, deformations, operating temperatures or other parameters. An example on this is MULTITEXCO a European Research Project funded by FP7- SME where was

Figure 2. Halar® High Clarity printing (Courtesy of P.A.T.I. S.p.A.) (source: Toniolo, P, Elsevier).

developed a scientific and technologic research based on the construction sector and the textiles. Problems like unexpected stresses in the fabric, extreme wind or snow, friction between support and material or elevated temperatures called misuse have a detail under the investigation. The high cost of these type of analysis makes the market slower but they although used sensors as state of the art and with general low cost compared to other techniques such as optical fibers or videogrammetric monitoring (Heyse, 2016).

1.4 *Offices, Commercial spaces & private houses*

The interventions under offices spaces, commercial areas or even at private houses can have a change with the new technology that will create new habits.

With the development of a new fluorinated polymer to meet the need of an excellent transparency at the high thickness, a thermoplastic resin that at the end, can in the future substitute glass in Architectural applications.

With a product that has a high transparency and that has an excellent outdoor stability, hydrolysis, and UV aging, with a fire resistance approval with NFPA 701, UNI EN13501 certified the benefits are intense for construction. This new material has a very high vapor barrier properties, an advantage in an excellent printability, a good mechanical property, and an enormous thermal properties level can transform architecture, and as the design imagined today.

This possible substitute for glass, with a high range of applications under facades, domes, umbrellas for residential and public facilities such as stadiums, airports, trains stations, commercial centers or others, can be the begging of a new market. The architecture will need to prepare a new possible knowledge to implement under the objects (Toniolo, 2016).

1.5 *Contemporary art*

The new creative cities, combined with smart cities are part of our lives nowadays. The American artist, Dave Cole, with his teddy bear in 2004, made by rolls of fiberglass, created an impact on the future of textiles imposing and attracting people into contemporary art using an idyllic American childhood

Figure 3. David Cole fiberglass Teddy Bear (source: www.brown.edu).

Figure 4. Toshiko Mori house, external stairs (source: Florida house design Toshiko Mori).

ideal, at first glance. Fiberglass Teddy Bear, an enormous stuffed pink bear visible through the windows of List Art Building, appears like a charmingly silly commemoration of youthful innocence (Quinn, 2010).

The cover for Architecture using textiles, Toshiko Mori's is an example of the use of fibers as a hallmark of work. The introduction of industrial strength filaments has featured in Mori's commercial and residential projects, where he introduced calling systems stretch systematically through space to suspend and support architectural details and whole structure.

Mori has built some buildings in the Gulf of Mexico, where seasonal hurricanes challenged her to design buildings robust enough to withstands extreme weather conditions and to sleek enough to embody the lightness and transparency that her work has.

One of the most famous projects from Mori was the external staircase with a single length of a fiberglass textile. Was created "landings at the top and bottom of the staircase by smoothing each end of the textile into a flat surface and folding the tips of the fabric upward to form a banister for each level. The middle length was pleated like a fashion garment as Mori wrapped it thirty times to create the staircase's steps and treads." And Mori was focused on this idea 'The staircase is structurally sound yet weighs less than 300 lbs... which is extremely light for a prefabricated staircase. The fabric performs flawlessly as a staircase, and it can be maintained as easily a boat deck' (Quinn, 2010).

An intense transformation can be part of this paradigm shift under interventions connected with office spaces, where the textiles will probably increase the options but decreasing in some traditional and old application systems.

Under the commercial area, and the knowledge improvement about smart components applied on materials to re-draw on spaces and brands that a new generation of architects and engineers will need to be a focus on unique systems that our generation was a focus on before.

Smart homes, smart spaces or the home automation are increasing solution to citizens. A house can be brilliant, but there are different needs from even nowadays, a prolonged internet use can have an association with prolonged sedentary behavior and all of the associated health consequences.

The internet was designed to connect us with others. It has certainly done this, but it's also true that the web can be a cause of disconnection in our lives. The idea of adapting the tendencies to the smart spaces need to go directly to a health system that creates benefits to citizens and not the opposite.

The smart textiles combine with nature will be an intelligent necessity and a quick action into the territory.

Figure 5. World premiere in Munich: EVALON® Solar cSi – the world's first solar roofing membrane with semi-flexible PV modules consisting of crystalline silicon solar (source: alwitra.de).

Figure 6. Adaptive Systems by Maria Mingallon (2010) (source: http://www.mariamingallon.com/fcas.html).

2 APPLICATIONS

2.1 *Powered surfaces*

The materials that are under study can have a characterization by different levels and characteristics.

When we talk about Energetic systems, we connect directly with LED efficiency systems. They can be integrated into buildings and reduce exponentially the used energy, years ago, under on simple building. They transformed the electrical system during the last decade, and they had grown with quality and they are now an integrant part of an intervention with quality.

When we talk about facades, we could connect with different possible benefits as the photovoltaic modules with crystalline cell structure, with extracted cells from the raw material silicon that was recognized by Alexandre-Edmond Becquerel in 1839. During the time, the new manufacturing technologies made possible to produce modules which could be integrated into design concepts or even into sophisticated integrated systems in architecture.

Some other possible types of powered surfaces like the Piezoelectric, the Illuminated solutions, the Luminescent that can be Electroluminescent or Photoluminescent or even light transmitting are an integrant part of our future introductions into architecture and future intense researchers.

The called Responsive system works with a network of microchips that work with information-transmitting action will change building.

Color-changing will be actin into the visual design surfaces where materials change their appearance, and where the people can be in front of the decisions, can be an integrant part of the action directly connected with personal desires.

The Shape changing, goes directly to some specific external stimuli. Piezoelectric materials can produce a voltage that represents transformation or deformation of processes. The reverse can have happened to change the original shape. And this is a new possibility for buildings and a new creative action into Neighborhoods.

2.2 *Faster & Less expensive & NEW*

The new possible applications under smart walls can result in the future in a use of a new method that made the inhabitants more participative under decisions and opportunities to choose better and in most intelligent and productive type of action.

The information transmitted into a textile, removing the need for an infrastructure as a TV, or an IPAD or a LAPTOP, or even a PHONE, and transferred for clothes, or a wall paper or a chair can be a new point of view under the electronic textiles called – e-textiles.

2.3 *Opportunities*

By definition textiles that can interact with their environment. They can sense and react to environmental conditions and external stimuli from mechanical, thermal, chemical or other sources. Such textiles are multifunctional or even "intelligent" which is fulfill by some sensors incorporated in the textiles. The embedded sensors are sensitive to various parameters such as temperature, strain, chemical, biological and other substances.

Smart Biomedical garments and clothing can act as "second skin" and detect, in seconds, vital signals of the wearer's body or changes in the wearer's environment.

The new integration of polymer optical fibers (POF), with their outstanding material properties, into technical textiles has not seriously been considered, until now under Architecture and this could be another possibility.

A health monitoring and the incorporation of optical fibers in geotextiles leads to additional functionalities of the textiles with the monitorization of mechanical deformation. The strain, temperature, humidity, pore pressure, detection of chemicals, measurement of the structural integrity and the health of the geotechnical structure (structural health monitoring) can be inserted into a new integrant part a rehabilitation strategy to cities that want to grow or transform into revitalized places and creative spaces.

Another example is the Saxon Textile Research Institute (STFI) in Germany has developed a technology that integrates optical fibers into geotextiles so that the sensing fiber can adapt onto the textile and the integration procedure does not affect the optical and sensing properties of the fibers. The use of special coating and cable materials are of crucial importance to protect the fragile single mode silica fibers against fiber breakage during the integration into the textiles and the installation under construction sites.

3 NEW OBJECTS

At the moment, there are a strong group of textiles and polymers used in building and construction. Technical characteristics and requirements of textiles used for building and construction. The fiber reinforced polymer composite materials for building and construction are Developing and testing textiles and coatings for tensioned membrane structures.

Figure 7. Detection of cracks in a masonry wall by a textile-embedded distributed POF OTDR sensor after several seismic shocks applied to the building (source: Krebber, K, http://dx.doi.org/10.5772/54244).

The textiles applications and polymers in construction can adopt an update in tensile structures for architecture and design.

The possible role, the properties and applications of textile materials in sustainable buildings; Learning from nature: Lightweight constructions using the 'technical plant stem.'

The role of textiles in providing biomimetic solutions for construction; Textiles for insulation systems, control of solar gains and thermal losses and solar systems; Sustainable buildings: Biomimicry and textile applications; Challenges in using textile materials in architecture: The case of Australia; Innovative composite/fiber components in architecture.

4 CONCLUSIONS

Textiles, polymers and composites are increasingly being utilized within the building industry. This pioneering text provides a concise and representative overview of the opportunities available for textile, polymer and composite fibres to be used in construction and architecture.

Textiles, polymers, and composites are increasingly intensifying the utilization within the building industry. This pioneering text provides a concise and representative overview of the opportunities available for textile, polymer and composite fibers to be used in construction and architecture.

The main types and properties of textiles, polymers, and composites used in buildings. Key topics include the types and production of textiles, the use of polymer foils and fiber reinforced polymer composites as well as textiles and coatings for tensioned membrane structures. A selection of applications within the building industry it is under an intense research to improve the market at the macro scale.

Some teams and international contributors are working under textiles properties, polymers and composites for buildings. Some of this research started to be important references for architects, fabric manufacturers, fibre-composite experts, civil engineers, building designers, academics, and students.

A concise and representative overview of the opportunities available for textile, polymer and composite fibers to be used in construction has to be a new knowledge to a future expert to grow into the market and with the market.

The insight into how high-tech textiles already influence our daily lives as well as potential applications in modern buildings will be continuous investigations on our next five years with the introduction of new technologies and features a thorough discussion of technical characteristics and requirements of textiles used for buildings and construction (Pohl, 2010).

Smart textiles are materials and structures that sense and react to environmental conditions or stimuli, and their integration into protective clothing has led to the development of products with greatly enhanced protective capabilities in hazardous situations. Smart textiles for protection provides a comprehensive analysis of smart materials used in producing protective textiles, and explores a wide range of end-use protective applications.

The considered smart materials and technologies. Are focus on smart textiles for protection, this section goes on to discuss types of materials, surface treatments and the use of nanofibres and smart barrier membranes. The application of sensors, actuators and computer systems in smart protective textiles is explored, followed by a review of biomimetic approaches to design.

There is ongoing new researches in some specific applications of smart textiles for protection, in smart technology for personal protective equipment and clothing, in smart protective textiles for older people and smart high-performance textiles for protection in construction and geotechnical applications are all discussed in depth, as is the use of smart textiles in the protection of armored vehicles and in protective clothing for fire fighters and first responders (Chapman, 2012).

New spaces, and new cities, maybe the new Climate Smart Cities could be better welcome with solutions like create benefits for a citizens' participation under this new type of introductions.

ACKNOWLEDGMENTS

This research was only possible with the support of CIAUD – Research Centre in Architecture, Urbanism and Design, Lisbon School of Architecture of the University of Lisbon, and FCT – Foundation for Science and Technology, Portugal.

REFERENCES

Aliprandi, Stefano; Monticelli, Carol; Zanelli, Alessandra – Technical Textiles and Thin Insulation Materials. New Scenarios for the Energetic Retrofitting. ISSN 18766102. Vol. 78 (2015), p.501.

Capineri, L – Resistive Sensors with Smart Textiles for Wearable Technology: From Fabrication Processes to Integration with Electronics. ISSN 18777058. Vol. 87 OP – In EUROSENSORS 2014, the 28th European Conference on Solid-State Transducers, Procedia Engineering 2014 87:724–727 (2014), p.724.

Chapman, R. Smart textiles for protection, Elsevier Science, 2012.

Cremers, Jan, [et al.] – Analysis of a Translucent Insulated Triple-layer Membrane Roof for a Sport Centre in Germany. ISSN 18777058. Vol. 155 (2016), p.38.

Heyse, P.; Buyle, G.; Beccarelli, P. – MULTITEXCO – High Performance Smart Multifunctional Technical Textiles for Tensile Structures. ISSN 18777058. Vol. 155 (2016), p.8.

Krebber, K. et al., Smart technical textiles with integrated pof sensors. Proc. of SPIE, 2008, 69330 (69330V) 1–15.

Pohl, G. Textiles, polymers and composites for buildings, Elsevier Science, 2010.

Quinn, Bradley – Textile Futures, Fashion, Design and Technology. Oxford, New York: BERG, Oxford, New York, 2010. ISBN 9781845208073.

Sauer, Christiane – Made of...New Materials Sourcebook for Architecture and Design. Gestalten, Berlin, 2010. ISBN 978 3 89955 289 8.

Toniolo, Paolo; Carella, Serena – Halar® High Clarity ECTFE Film – An Highly Transparent Film for New Buildings Structures. ISSN 18777058. Vol. 155 (2016), p.28.

Textiles, Identity and Innovation: Design the Future – Montagna & Carvalho (Eds)
 ISBN 978-1-138-29611-4

Technological clothing as an element of innovation: Survey of young Portuguese users

C. Carvalho & G. Montagna
CIAUD, Lisbon School of Architecture, Universidade de Lisboa, Portugal

H. Carvalho & A. Catarino
Textile Engineering Department, University of Minho, Guimarães, Portugal

ABSTRACT: Technological garments in general can be considered as a strategic niche market for the Portuguese and international textile and clothing industry, and its definition in conceptual and operational terms needs to be defined with some strategic urgency. The recognition of the general needs of its potential users is a primary necessity for its correct development and for the adaptation to the national and international market. This article aims to analyze the relationship between research, innovation and industry in a European country such as Portugal, considering the business transformation undertaken in recent years, by collecting consumer data and consumer desires for this type of product. This study, unique in its kind, can be considered representative of the national universe and expresses the vision of the Portuguese university population, in everything comparable to its international counterparts. To this end, about 600 questionnaires, filled by students from the north, central and southern Portugal, were collected and validated, which may allow for an own and updated reading of the needs and desires of the users closest to this type of product, analyzing the factors that bridge the industry and the market. The need for increased innovation is a reality of the Portuguese textile industry that needs the integration and monitoring of Portuguese university education institutions, which promote research in this type of clothing, and the textile industries.

1 INTRODUCTION

Technological clothing increasingly marks the daily life of society, covering different areas and uses. Term used as a generic definition for technical clothing, high performance clothing and/or smart clothing (wearable or smart garment), this definition refers to a type of clothing that, due to its characteristics of incorporating technologies, operates in areas as different as health, high competition sport, leisure, protection, workwear, among others (Figure 1).

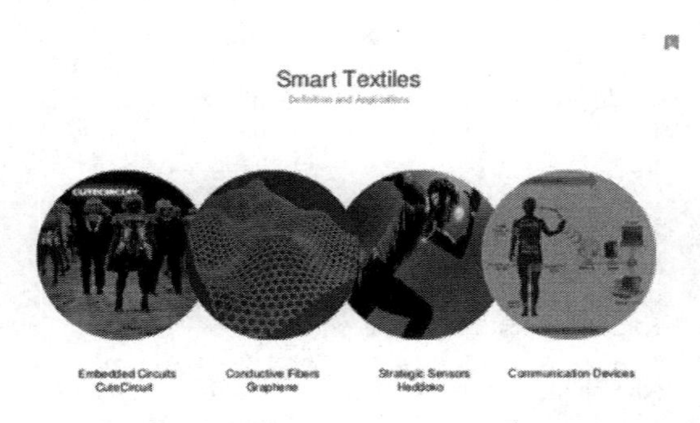

Figure 1. Slide Share by Barbara Trippeer (2016).

The industrial production of such clothing in such a way as to enable it to be marketed on a sustainable scale is an innovation exercise to be carried out by the industries in collaboration with the research centers to potential of this type of products. Although much discussed and investigated, this application still has little commercial presence. The lack of use by the industry of the effort already invested by the research centers in general for the development of these new products is notorious. This fact, on the one hand, highlights the lack of business opportunity, which could cover many areas of activity and generate synergies and profits that could be reinvested in the same research and improvement of innovation products. On the other hand, the failure to fill this market range with products that in many cases have already been identified as having great potential (and whose market could be considered as consistent) weakens not only the Portuguese business position in this type of product, but also the developed in the national territory, delaying and weakening the possibility of market penetration.

To be able to describe the market in terms of the needs associated with this type of clothing, enabling the development of innovation in the Portuguese textiles and clothing industries, it is fundamental to know the universe of possible users for the design of this type of pieces, as well as its domestic and international production and marketing. Riikka Matala (2005) states that, "by asking how technology [including clothing] can influence people's lives, women refer to technology as a medium and men refer to it as being a product."

Thus, a specific questionnaire was carried out to collect and organize specific information about the needs and use of this type of clothing, and was applied to about 600 students of higher education, which allowed for the outlines of trends and trends for the near future. The study aims to achieve, as proposed by McCann (2006), that aesthetic concerns and cultural needs can be balanced with the physical needs and specific characteristics of the user and the environment in which they operate.

2 STATE OF ART

According to Baurley (2004), "black and silver boxes that now contain computers and electronic objects will disappear, as technologies will be incorporated into new environments, such as interiors, buildings, furniture and clothing".

A first generation of smart clothing was created in the 1990s using conventional electronic materials and components and attempting to adapt the various elements by design so that the electronic object could be coupled to the garment in question. In the late 1990s, a collaboration between Philips® and Levi's® originated the ICD + Line (Marzano et al., 2000), which can be considered as the first successful step in the development of an adaptable piece that had ability to integrate any pre-existing device such as a mobile phone, a microphone or a remote control, for example. In the late 1990s, the constant and rapid growth introduced by the applications of multimedia technologies, and a greater awareness of the capabilities and reach of what a wearable could mean, made the scientific community realize the potential of this technology applied to different areas. Some of the first to realize the great potential of smart garment were the comfort/leisure areas and the health area.

There are already products developed for the health area, namely sweaters that monitor vital signs and that have the capacity to alert the medical team in case of danger of the patient. Examples are Health Vest®, Numetrex®, Sensatex's Smartshirt®, Vivometrics Lifeshirt®, Verhaert Mamagoose® (O'Mahony and Braddock, 2002), or the Portuguese Vital Jacket® product. This type of clothing is commercialized in a technological solution in which the processing system is associated with the textile product as if it were an accessory and not fully integrated into the textile structure.

Still with a great impact on health, but beyond this, the area of sport is particularly interesting for the development of projects of this nature, and Lihui Wang (2002) recalls the conceptualization of design is increasingly a central factor in the encounter with

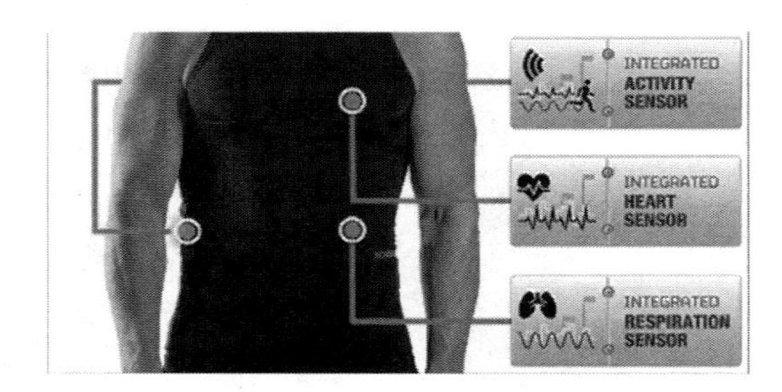

Figure 2. http://blog.softograph.com/wearables/ (2017).

the specific needs of users. The use of conventional or textile sensors, electrical conductive textiles, the evolution of integration techniques and the miniaturization of electronics, has revolutionized the design and production capacity of clothing for specific uses, increasing the user's own performance capacity, as well as their safety levels in performing specific tasks as shown on figure 2.

3 METHODOLOGY

For the development of this research, and to collect the data that allow to draw a panorama of what matters around smart and technological clothing in general, so that they can be taken over by industries that generate innovation products as creators of marketable products capable of responding to users' needs, four-part questionnaires were drawn up to allow the characterization of respondents and their preferences in terms of this type of clothing. The questionnaire included non-mutually exclusive multiple-choice questions. The obtained data were evaluated and later crossed between them. Thus, the results presented will consider the multiple responses of respondents, but also the crossing of the different choices. In this way it was intended to create some perception about the values attributed by the respondents to a set of situations and needs, and to reveal a network of meanings in many cases despised but allowing transversal readings.

The questionnaire had a first pre-test application and, after its formal review, was successively applied on a large scale. About 600 students from different universities and institutes of the national territory were interviewed, all of whom were students of different academic years and varied areas of learning.

Participants in this questionnaire were students from the north of Portugal through the School of Engineering of the University of Minho and the Polytechnic Institute of Cávado and Ave; In the center of Portugal, students from the Faculty of Architecture of the Technical University of Lisbon, students of the Higher Education Sciences Institute of Lisbon (ISCE) and students from the Lusíada University of Lisbon participated. Representing the south of Portugal, students from the University of Algarve participated in the surveys.

The data obtained were statistically treated and analyzed with SPSS software V.14.

4 RESULTS

Research in the field of technological clothing and applications for high-end clothing, such as smart, high-performance clothing and technical clothing, is now a reality at the international level and an important area of study for some universities and Portuguese companies. The research produced and its application to the development of new products and new areas of innovation of the national industry is of extreme importance at a time when the textile industry may have the capacity to embrace market niches that until recently were exclusive of other areas, such as health, well-being, biomedical sciences, rehabilitation or leisure.

In this sense, the questionnaire carried out for this purpose was developed in the university environment that today, thanks to the new forms of communication offered by the Internet and to the broader capacity of displacement of young people in general, but also thanks to the programs of exchanges of students inserted in higher education, is internationally comparable and could allow a parallelism of students' desires at the international level. Thus, it can be considered that this hearing can be considered representative of the use of this type of clothing by European university students, although regional differences can not be completely excluded.

The results presented here can be considered as developers the main lines and tendencies expected by young university students, who at the same time could represent the most informed and close users of the market in question. The study could thus allow a greater assertion in the development of innovation products by Portuguese textile and clothing industries with added technological value. Considering the potential of this type of market, and the need to apply the research produced in the academic world this study is of high strategic interest for the sector.

The following are some data considered important for the framing of the respondents and for their identification.

In the observation of the data provided by the respondents, about 85% are female and 15% are male, and about 85% are university students from mainland Portugal, ranging in age from 19 to 25 years.

Of the respondents, 45% say they have 2 to 4 hours a day to devote to their personal activities.

The following questions were then put to the respondents' consideration, the respondents assigning each question an order of importance ranging from a minimum of 1 (not important) to a maximum of 7 (very important):

Respondents were asked to respond as they preferred to occupy their free time and the answers below indicate the preferences in this sense of the respondents who responded having between 2 and 4 hours daily for their personal activities.

The indicators of activities most frequently selected by the respondents as first preference were crossed with the most selected indicators in the remaining questions, by the same users, to correlate the users' choices among the various issues. This process allowed us to better define the chain of values assigned to each choice by the respondents.

1. Preferred activities for leisure time:

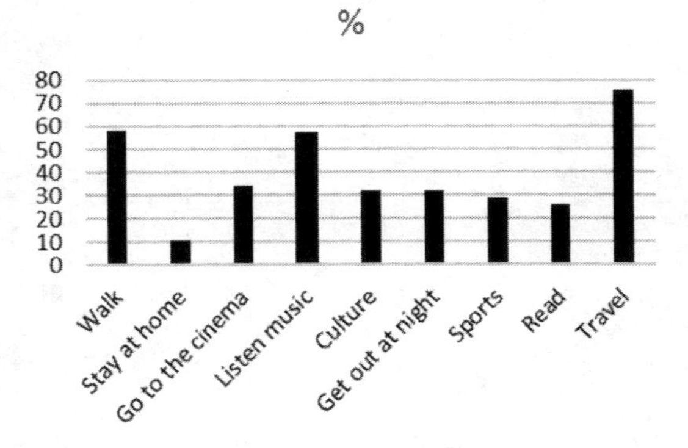

Respondents identified their habits as consumers, declaring that they are very fond of shopping (44%) and being essentially free in the choices they make and acquiring only what they really like (61%). This new generation of consumers affirms that they are satisfied with the products purchased (35%) and declares that they are indifferent when asked about the fact that they are loyal consumers of a brand (31%), even claiming that they do not necessarily buy at each station and that it disagrees completely with the purchase of brand-name products (43%).

The choices indicated by the respondents in terms of consumption habits were crossed with other indicators so that it was possible to indicate which other questions were selected by the subjects who answered the questionnaire and could be considered important in association or singularly. Below are the three major choices associated with the most chosen indicators.

2. Consumer habits of the respondents:

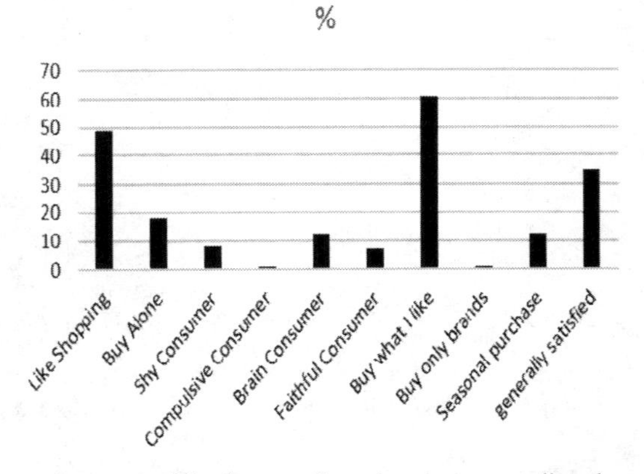

In terms of the degree of requirements regarding the products it acquires, general comfort was indicated as a priority by 43% of the respondents. The results are shown on next figure.

As previously described, the indicators selected here were correlated with the indicators selected by the same users in the other questions. The result is as follows:

3. Degree of consumer demand in the acquisition of general clothing:

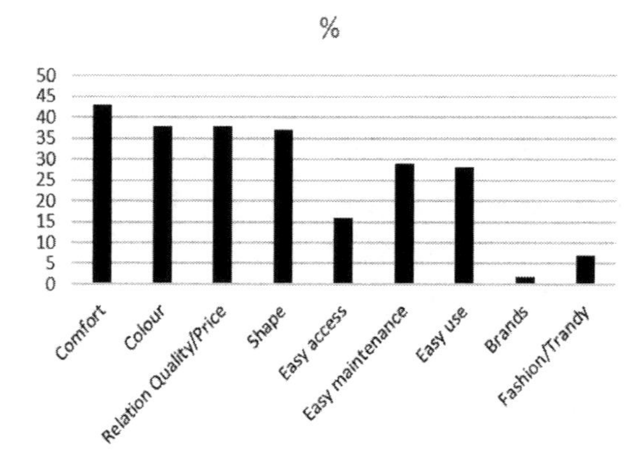

Respondents identified their preferences in terms of technological clothing and, in general terms, the areas of extreme sports (36%), rehabilitation (33%), health (32%) and personal protection (30%) appear as the areas of greater interest on the part of the interviewees, for the application of capital gains from this type of clothing.

4. Consumer interests of respondents in terms of technological clothing:

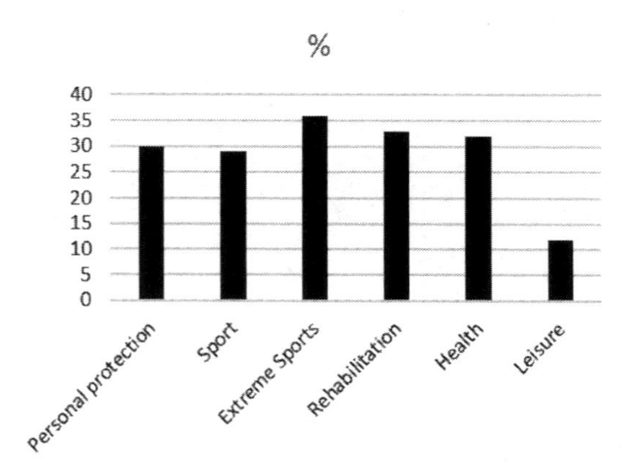

The application of the distinctive elements that could allow the creation of clothing considered as "technological" is privileged by respondents in garments such as trousers (36%), underwear (34%), t-shirts (33%) and sports jacket (20%). The same respondents declared themselves indifferent to the introduction of this type of technology in garments such as suits (6%), shirts (17%) or classic blazers (8%).

5. Garments with technological elements that users find interesting:

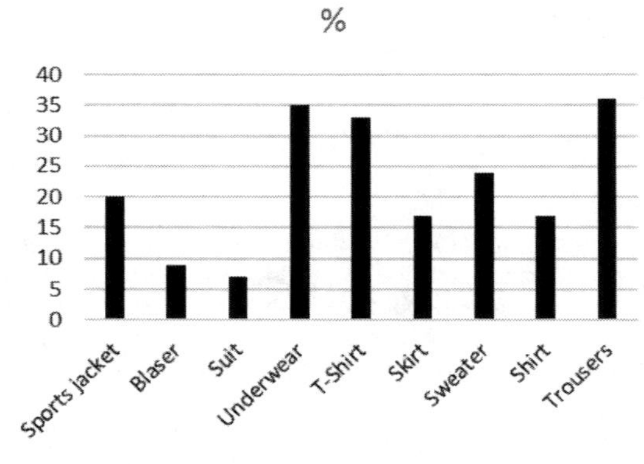

The students covered by the survey were still asked about the characteristics that possible garments of this type should incorporate. The comfort of the movements continues to be one of the most sought-after characteristics in the garment (66%), followed by the thermal comfort and the adaptability of the piece to the use by 48% of the respondents, and the impermeability of the piece with 38% of the preferences. With less importance but still with some expressiveness, information technologies (17%) and color adaptability (24%) were indicated as characteristics to be introduced in clothing.

6. Respondents were asked about the technological features they would like to see included in the garments they find most interesting:

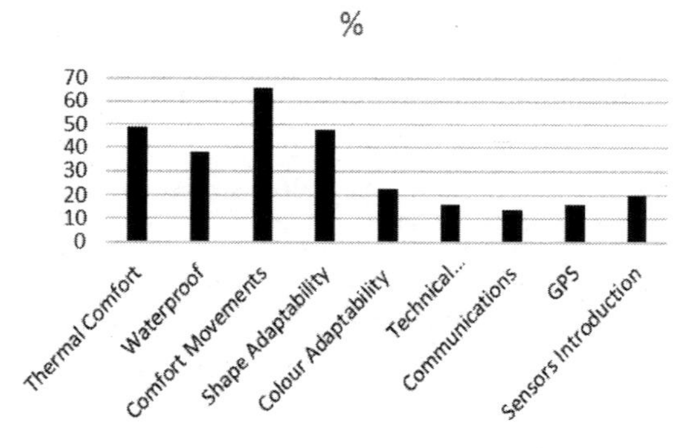

5 CONCLUSIONS

This study was carried out to legitimize the needs of the possible users of this type of clothing. The lack of existence of this type of study carried out in the national territory made it possible to create hypotheses, without a solid basis, on what the needs and desires of users of this type of clothing could be.

The importance of this type of product is nowadays indisputable. While on the one hand Portugal is undergoing a major restructuring of its industrial network to adapt to the ongoing change in international terms, on the other hand, companies and educational institutions already design research projects in this direction, producing high technological garments visibility and with industrial and commercial potential of international aptness.

University students can now be considered quite universal and international in terms of knowledge and manner of being, and for this reason they were the ideal users of this type of product and for this age group. The rapid dissemination of knowledge, information networks and student exchanges through community programs are just some of the examples of the elements that allow a greater availability of knowledge, which means that these students, besides being the nearest future users of this type of clothing, may also be the most informed and most objective in their evaluation and development.

It should be emphasized that this study refers to university students and that the answers should be understood as specific to this group of users, with all the inherent answers to them. The choice of certain options to the detriment of others may have to do with the intensive or less intensive use of certain articles or objects by the respondents, which does not mean that options completely rejected in this study cannot be considered important when applying the survey to other groups social policies. This study reveals some novelty in the choice of the respondents of the application of this type of technologies to the underwear. The choice of this option may have been made by the fact that young people view technology as useful but at the same time something personal and intimate without the need for outside or public demonstrations. This fact can also be corroborated by the choice that the respondents made about the use of this type of leisure technology, where leisure activities occupy much of their free time, but the application of these supports for free time was only appreciated by 7% of respondents.

It also seems interesting to underline what seems to be a tendency of the young consumers questioned in assuming as little or nothing related to the fashion brands, buying only the products they like independently of the brand and categorically refusing to buy articles just because they are branded.

Radical sports, rehabilitation and health have been identified as the most immediate applications of smart clothing, with the most preferred being the inclusion of technologies such as pants, underwear and T-Shirts. The preferred characteristics of the respondents, associated with this type of clothing, are the thermal comfort, the comfort of movements and the adaptability of form. Surprisingly, technologies such as sensors, GPS and communication were the characteristics indicated as the least desirable. This results in the fact that users prefer the application of new technologies and technologies called "smart" to enhance and improve the traditional functions of clothing, to the detriment of the inclusion of new functionalities that are usually found in other media.

ACKNOWLEDGMENTS

This research was only possible with the support of CIAUD – Research Centre in Architecture, Urbanism and Design, Lisbon School of Architecture of the University of Lisbon, and FCT – Foundation for Science and Technology, Portugal.

The authors are also grateful for the support and availability of the Universities and Institutes involved in the distribution of questionnaires and in the collection of information.

REFERENCES

Baurley, S. 2004. Interactive and experiential design in smart textile products and applications. *Personal and Obiquit Computing*, 274–281.

Marzano, S., Green, J., Heerden, C. V., Mama, J. & Eves, D. 2000. *New Nomads*, Rotterdam, 010 Publisher.

Matala, R. 2005. Technophobia in the Intelligent Clothing Design Process. In: REAL, E.C., ed. Pride & Pre-Design – The Cultural Heritage and the Science of Design CUMULUS LISBON 2005, 2005 Lisbon. IADE, 462.

Mccann, J. 2006. A Design Process for the Research and Development of Smart Clothes with Embedded Technologies with Potential to Enhance Quality of Life for Older People. Smart Homes and Beyond.

O'Mahony, M. & Braddock, S.E. 2002. Sports Tech, London, Thames & Hudson.

Wang, L., Shen, W., Xie, H., Neelamkavil, J. & Pardasani, A. 2002. Collaborative conceptual design – state of the art and future trends. *Computer-Aided Design*, 34, 981–996.

3.2. Sustainable textiles

Textiles, Identity and Innovation: Design the Future – Montagna & Carvalho (Eds)
© 2019 Taylor & Francis Group, London, ISBN 978-1-138-29611-4

Sustainability of textile and fashion supply chain: Case study approach

S. Azevedo, R. Miguel & P. Mesquita
University of Beira Interior, Covilhã, Portugal

ABSTRACT: This study aims to identify sustainability practices in the textile and fashion supply chain (SC) according to the Triple Bottom Line (TBL). To achieve this objective, a case study of a fashion retailer brand is performed. This study shows that the adoption of sustainability practices provides lower production costs and better management of resources. There is also a concern on the minimization of emissions and environmental impacts across the supply chain using natural, recycled or biodegradable fabrics, garments reform or recycle fabrics for the manufacture of new parts. Regarding social sustainability, the case study is interested in providing good working conditions and support projects that benefit the communities where the company operates. This study contributes for raising the attention of academics and professionals for the sustainability in the textile and fashion industry for the kind of practices that could be implemented to become textile and fashion' retailers companies more sustainable.

1 INTRODUCTION

Advances in industrialization and sophistication of industrial technology have contributed for improving the quality of life since the beginning of the industrial revolution in the eighteenth century. However, these advances exacerbated resource depletion, global warming and environmental destruction (Na & Na, 2013). The growing concerns on environmental issues has transformed the environmental protection in a matter of global interest (Nagurney & Yu, 2012; Thanikaivelan et al., 2005).

Sustainability has become a common goal and critical for organizations and is currently one of the fundamental issues of society.

In this context, fashion and textile manufacturers felt compelled to have a commitment to these issues, as the textile manufacturing processes produce large quantities of waste with a negative impact on environment (Lo et al., 2012).

The aim of this study is to identify sustainability practices of fashion SC based on its three dimensions (environmental, economic and social). To achieve this, the case study of H&M is performed.

2 LITERATURE REVIEW

2.1 *Sustainability in textile and fashion industry*

The Textile and Fashion industry provides numerous benefits to our everyday life and around the world but has also a negative side associated to the exploitation of workers, disposable fashion production, waste of resources and no encouragement for sustainable consumption (Bennie et al., 2010).

In Textile and Fashion industry the organizational changes began about 30 years ago, when the stable structure of long-standing traditional luxury fashion industry faced several challenged related to globalization, changes in its customer base and the entry of competitors in previously protected markets (Djelic & Ainamo, 1999).

Currently, this industry has an enormous production capacity at surprisingly low costs which make it an important contributor for the economies (Allwood et al., 2006; DEFRA, 2011). Moreover, according to the FashionUnited (2017) the global apparel market is valued at 3 trillion dollars, 3,000 billion, and accounts for 2 percent of the world's Gross Domestic Product (GDP) and employed 57.8 million of people in 2014.

Although various problems arise associated to Textile and Fashion industry, such as: abuses of working conditions, human rights are not respected, workers are hired at low wages for long hours; the creation of these jobs represent important benefits especially for women in the poorest countries, which have opportunities to obtain some income (Gardetti & Torres, 2013).

The Textile and Fashion industry being a sector with fractional and heterogeneous characteristics, it represenst one of the longest and most complicated industrial chains in manufacturing industry (European Commission, 2011). Regarding natural resources (eg. water), the Textile and Fashion industry is compelled by other industrial sectors in Europe to be more sustainable (Vajnhandl & Valh, 2014).

Based on the three pillars of sustainability, to be sustainable this industry must incorporate a set of concerns such as: the negative impacts on the ecosystem, use of raw materials from animals, human rights, labour standards, the welfare of workers, the use of non-toxic materials, cleaner production processes, how to recycle creatively fashion items, discarded materials and reuse of waste created (Eder-Hansen, et al., 2012). Sustainability is an important issue for the Textile and Fashion supply chain because of its characteristics, such as: i) the high use of resources; ii) transferring production to low-cost supply countries with weak environmental concern; ii) the fierce industry competition; and iii) the miserable working conditions in some regions. These characteristics make the sustainability issue obviously indispensable for this industry (de Brito et al., 2008; Faisal, 2010). Moreover, some studies (Arbogast et al., 2012; Battaglia et al., 2014) highlight sustainability initiatives as the most effective and direct strategy for this industry to become more competitive and expand their markets.

Kozlowski et al. (2012) claim that only now the Textile and Fashion industry begins to consider sustainable fashion production beyond a passing trend, the proof is the growing number of implementations and development of corporate social responsibility policies, stakeholder engagement, increased reports communicating the sustainable actions developed by the company, life cycle analysis of specific apparel products and greater transparency in the supply chain management.

To realize a sustainable production of Textile and Fashion industry some obstacles should be overcame such as: higher costs of environmentally friendly' materials; good working conditions in supply chains, design and supply activities based on fast fashion concept.

For Kozlowski et al., (2015) the main challenge for sustainability in this industry is its function as a conductor of a consumer culture. Other studies (Eder-Hansen et al, 2012;. Fletcher, 2010) support this idea arguing that controlling the rapid flow of demand for fashion goods is the main challenge for achieving sustainable fashion consumption.

Bhardwaj & Fairhurst (2010) believe that fast fashion is a concept that affects and will continue to affect the Textile and Fashion industry in the next decade. According to Niinimäki (2010) it is necessary to involve retailers since they have the power to inform consumers about their initiatives and influence their habits and purchasing decisions.

In addition to the rapid flow of demand for fashion goods and poor working conditions the great quantity of pollutants created by this industry is an important concern (Fulton & Lee, 2013). Many of the production phases along the supply chain of Textile and Fashion industry generate waste and contain solids, oils and toxic heavy metals such as copper and chromium, such as: fibre production, fibre processing and spinning, wire preparation, production of fabrics, bleaching, dyeing, printing, finishing and wastewater (World Bank Group, 1998). These substances can cause neurological, carcinogenic, allergies and affect fertility of workers who are constantly in contact with them (Gardetti & Torres, 2013). Because of these various chemicals and toxic substances and the large volume of water used in the different stages of production processes, the Textile and Fashion industry is considered the main industrial production wastewater (Hasanbeigi & Price, 2015).

According to Vajnhandl & Valh (2014), as most textile companies are SMEs, wastewater treatment and textile recycling becomes a big challenge to them because of their limited economic resources and less sophisticated technologies. Regarding the distribution and retail, environmental impact occurs mainly due to transport and materials and energy used in the packaging. Already in retail stores, the environmental impacts also include the emission of greenhouse gases, fossil fuel use, waste generation and the maintenance of buildings such as: electricity, heating, lighting and use of small volumes of water (Schönberger et al., 2013).

Due to health concerns, government restrictions on chemicals used in clothing has contributed to the implementation of sustainable practices in supply chains of many countries in Europe, translated into Ecological labels (eg. Global Reporting Initiative – GRI) or the compliance certification from the International Organization for Standardization (ISO 14001 and ISO 26000). These strategies have stimulated the supply and demand to reduce environmental impacts across the product life cycle and usually involves establishing limits use of substances which are harmful to the environment during production, use and handling of waste (Jørgensen & Jensen, 2012).

3 METHODOLOGY

In this study a qualitative methodology is used based on a case study. According to Rowley (2002), a case study approach is adequate when the boundaries of a phenomenon are not clear and there is also no control over behavioural events. Moreover, this methodology is considered adequate when the examination of the data is most often conducted within the context of its use (Yin, 1984), as it is the case.

Yin (2003) considers that case studies can be exploratory, descriptive or explanatory and single or multiple. In this study a descriptive case study is present.

As in all other type of researches, in the case study attention must be given to construct validity, internal validity, external validity, and reliability (Yin, 2003). Levy (1988) established construct validity using the single-case exploratory design, and internal validity using the single-case explanatory design. Yin (2003) suggested using multiple sources of evidence as the way to ensure construct validity. The current study uses multiple sources of evidence (newspaper news, company's website and sustainability reports), so, construct and internal validity is reached in this study. Yin (2003) states that external validity could be achieved from theoretical relationships and their generalizations. The objective of using the research case study is to identify the main sustainability practices and

Table 1. Objectives and research propositions

Objective	Propositions
To identify sustainability practices (environmental, economic and social) used by fashion SC.	P_1. Companies belonging to Fashion SC deploy sustainable practices. P_2. There is a concern in minimizing the negative impacts of fashion SCs on environment. P_3. The adoption of sustainability practices by companies belonging to fashion SC contributes to reduce production costs. P_4. The adoption of economic practices by companies belonging to fashion SC contributes to a better resources management. P5. The adoption of social practices by companies belonging to fashion SC contributes to sustainability.

corresponding relationships. Yin (2003) also considers that the development of a formal case study protocol provides the required reliability.

Regarding the data collection methods, Forza (2002) argues that data can be collected in a variety of ways, in different settings and from different sources. In case study research, Yin (2003) suggests that evidences may come from six main sources: documents, archival records, interviews (semi-structured, structured or unstructured), direct observation, participant-observation, and physical artefacts. Other sources of data can include informal conversations, attendance at meetings and events, surveys administered within the organisation films, photographs, and videotapes (Yin 2003).

3.1 *Objectives and research propositions*

This study aims to identify sustainability practices (environmental, economic and social practices) used by fashion supply chain (SC). Being so, the main research question is the following: "Which sustainable practices are deployed by the fashion supply chain?". Based on literature review some goals and propositions are suggested (Table 1).

4 CASE STUDY

In order to identify the main social, economic and environmental practices used by companies belonging to Textile and Fashion SC a case study of a fashion brand (H&M) is performed.

4.1 *Case study characterization*

H&M was founded by Erling Persson in 1947 at Vasteras in Sweden. Initially the brand only sold clothes for women and was named "Hennes" which means "them" in Swedish. It had as main objective to sell quality fashion at low prices (H&M, 2014).

In 1968 the founder Erling Persson bought the store named Mauritz Widforss that was a clothing and equipment store for hunting and fishing. After that, the store was renamed to Hennes & Mauritz, adopting later the initials H&M.

The goal of H&M reflects sustainability concerns since the company try to respect the three dimensions of sustainability (H&M, 2016).

The fashion industry is extremely dependent on natural resources, so the H&M seeks to change the way of doing fashion. That seems to be a major challenge but also a great opportunity to transform the linear model of production into a circular pattern and carry it on a large scale.

4.2 *Sustainability practices used*

The case study company (H&M) purchases products from independent suppliers who are long-term partners. From long time ago the company has been working on improving employee conditions, minimizing the negative impacts of its activities on environment, working closer with their supply chain members and also helping the communities where the company is located. So, the sustainability has been a concern of the H&M (H&M, 2016).

To facilitate the perception of sustainable practices across the supply chain, the challenges faced by each supply chain member and the respective sustainable practice adopted are described in Table 2.

Attending to the information in Table 1, it can be seen that H&M uses sustainability practices across its supply chain, so the proposition 1 (*P1. Companies belonging to Fashion SC deploy sustainable practices*) is verified.

4.3 *Environmental impacts of textile and fashion SC practices*

H&M has a commitment to the reuse, minimization and recycling having as future goal the elimination of any kind of waste. Once this target is reached, suppliers, employees and all connected to the business are encouraged to change their behaviours (H&M, 2016).

In 2015, were collected over 12,341 tons of clothes for discharge, representing more than 60% of the amount collected in 2014. In the same year, 1.3 million pieces were made with closed loop material. That represents over 300% more compared with 2014 (H&M, 2015).

According to Ceschin & Vezzoli (2010) the recycling of garments as raw-material in the manufacture of other products should be considered a priority. Several authors (Arbogast & Thornton, 2012; Nidumolu et al., 2009) argue that an organization is only considered sustainable if there is a commitment with the environment and a recycling program.

The waste reduction and minimization of impacts and emissions to the environment is an important goal for H&M. The company is aware of the negative

Table 2. Challenges and practices adopted by H&M SC.

Stages of the supply chain	Challenges	What has being done
Design	Create fashion without compromising the design, quality, price or sustainability.	Use of recycled and more sustainable materials. Use of organic materials as an alternative to traditional fabrics in the manufacture of conscious collections.
Raw-material	The intensive use of water and chemicals are concerns associated with the processing of raw materials.	Use of organic cotton, recycled materials and other more sustainable raw-materials as Tencel and Lyocell fibres.
Fabrics Transformation	The fabrics processing utilizes an intensive amount of water, energy and chemicals.	Promoting responsible use of water from how the cotton crops are irrigated to the way customers do the washing of clothes. Prohibiting the use of chemicals considered dangerous.
Clothing production	Has not their own production and sometimes it occurs in developing countries where law enforcement authorities are weak.	Vendors sign a commitment to comply with all the environmental and social requirements of the company's code of conduct Encouraging the long-term performance and more profitable contracts.
Transport	Transport accounts for 6% of greenhouse gas emissions in the life cycle of a garment.	H&M uses more environmentally friendly transportations such as ships and trains to transport about 90% of all its products.
Point of sale	Entry to new markets and sustainable expansion.	Use of renewable energy in stores, offices, warehouses and other places.
Customers	36% of the environmental impact is generated in the consumer use phase.	Educate the consumer to make washing clothes at 30° degrees instead of 60° degrees, and alert them for the clothes drying process in order to economize money and energy.
Recycling	Failure of the created recycling of waste.	System clothing collection that offers the consumer an easy solution so that the parts are not discarded in the trash.

Source: Own elaboration based on data from H&M Sustainability Report, 2016 (H&M, 2016a).

impacts of their production activities for the environment and across its supply chain. The negative impacts across H&M' SC are distributed in the following way: 12% of the negative impacts are generated during the production of raw materials, 36% in the production of fabrics and fibres, 6% during garment production, 5% is associated to packaging, 6% to transport and 26% is a result of the activities of consumers to wash and care for their garments (H&M, 2015).

Most of the waste produced is treated in the warehouses of the companies belonging to its supply chain. H&M collected more than 15,888 tonnes of garments (29% increase from 2015) and between 2013 and November 2016, almost 39,000 tonnes of garments were collected (H&M, 2016a).

Another feature of the fashion industry is that it uses very significant volumes of water in their manufacturing processes, which is a scarce natural resource. To improve the way the H&M uses water, a partnership with the World Wide Fund for Nature (WWF) was developed (H&M, 2015) and nowadays most of the stores already have an efficient system of water use.

Additional concern of H&M is related to energy consumption. The goal is that renewable energy sources are used whenever feasible and what is actually verified is that more than 96% of the energy used by the company is self-produced (H&M, 2016b). Also, the electricity consumption in H&M stores has been decreasing significantly since 2010 with a target of 20% reduction in 2020 (H&M, 2015).

One of the reasons that explains this positive results is the addition of LED lighting screens and video in stores, which aims to provide not only a more inspiring environment to consumer purchases but also contribute for decreasing the energy expenses. Moreover, company continuing to work with store rebuilds, and investing in new technology for lighting, heating, ventilation and air-conditioning (HVAC) systems. By 2030, every store will use 40% less energy per square meter than the stores constructed today. Also, 27% of all energy used in 2014 was sustainable, in addition, many stores are being equipped with energy metering systems to control energy consumption and allow its use more consciously (H&M, 2016).

Through supplier energy efficiency programs in Bangladesh, China, India and Turkey H&M has reduced greenhouse gases emissions. During 2016, suppliers achieved savings of nearly 30 million kWh as a result of H&M' improvement programs. These programs have also supported savings of nearly 8 million tonnes of natural gas and 1.5 kilo tonnes of coal. Combined, this adds up to over 200 kilo tonnes fewer greenhouse gases emitted. The company is also working to reduce the energy used in logistics, transport and warehouses. For example, H&M has a goal to ensure that 100% of their transport service providers (TSPs) are controlled by environmental programs (H&M, 2016).

Another measure that has being taken is to reduce emissions of greenhouse gases during the life cycle of clothing. H&M has been struggling to reducing emissions. In 2016 the aim of reducing emissions of greenhouse gases was achieved in absolute numbers, which reached 70,165 tons compared to 142,445 tons in 2015 (H&M, 2016).

According to UNEP (2011) so that fashion retailers become more sustainable, they have to manage activities associated to environmental impact, water conservation, reducing energy consumption and emissions and recycling plans.

The information published in "The H&M Group Sustainability Reports 2016" indicates that H&M is concerned with the minimization of the impacts and emissions to the environment. Being so, the Proposition 2 (*P2. There is a concern in minimizing the negative impacts of fashion SCs on environment*) is verified.

4.4 *Contribution of sustainability practices for production costs reduction*

A variety of recycled materials such as cotton, polyester and wool are used in all collections of (H&M, 2016).

Recycled material is made from a waste product, such as old textiles, production waste or used PET bottles. By using recycled material, the amount of waste that go to landfill, the use of virgin raw materials and the chemicals, energy and water used in raw material production can be reduced. H&M SC uses a wide range of recycled materials including recycled cotton, polyester, wool, cashmere and plastic. In the long term, the company wants to turn all products that can no longer be used into new materials and products (H&M, 2016).

Environmental initiatives focused on products and processes contribute for a reduction or mitigation of environmental impacts in manufacturing processes, thereby reducing costs (Sem et al., 2013). Thus, the proposition 3 (*P3. The Adoption of sustainability practices by companies belonging to fashion SC contributes to reduce production costs*) is verified, since the use of recycled materials contributes to the reduction of production costs.

4.5 *Impact of economic practices on resources management*

The H&M' concern on the emissions and waste derived from its activities, has contributed for the development of a comprehensive strategy to manage their resources in a more efficient way.

Chemicals are very useful contributing to a higher productivity, softening of fibres or cleaning clothes using less water. Most chemicals when used and managed properly, do not present any risks. Nevertheless, some chemicals are harmful to people and the environment when handled incorrectly. Being so, H&M is investing in alternatives solutions to minimize its use, replacing these chemicals by better alternatives and managing the resources and wastes in a more environmental way (H&M, 2015).

As regards wastewater, modernization of treatment technology will make more efficient the management of water so that the environmental impact is reduced on the ecosystems of the industry catchments (H&M, 2016).

Vajnhandl & Valh (2014) argue that it is crucial that the resources used by the fashion industry, such as water, be managed in a sustainable manner. H&M achieved a reduction of 56% water and 58% of energy in the production of his first collection of ecologic jeans, this represents more than half of the value used by the denim production "normal". This was only possible through innovative tools and sustainability experts that after studying the situation have suggested a better way to manage production resources (H&M, 2016). According to the H&M Group Sustainability Reports 2016, 82% of business partner factories are in full compliance with wastewater quality requirements. Moreover, some achievements in terms of environment impacts reached by H&M' SC are: use of 2.3 million m3 less water due to water saving programs that were implemented; 8,234 workers were trained in four regions (Bangladesh, China, India and Turkey) on water, energy and resource efficiency; 75 factories participated in cleaner production program in four regions; Half of H&M' denim products were ranked at highest level (green) using Jeanologia's tool; 51% of H&M group stores, offices and warehouses have water-efficient equipment.

Murray (2016) argue that only through the control and management of natural resources can the sustainability be achieved, especially in the economic dimension.

Despite reports available do not provide much information about economic sustainable practices, the above information is sufficient to demonstrate that a better management of resources contributes for implementing sustainable practices. By this way the proposition 4 (P_4. The adoption of economic practices by companies belonging to fashion SC contributes to a better resources management).

4.6 *Impact of social practices on sustainability of textile and fashion SC*

Social practices contribute to greater sustainability in the fashion SC.

H&M group and SC' partners have implemented from long time ago several practices to improve the employee welfare.

Some of these practices are the following: i) development of a Global Labour Relations training program for the H&M group; ii) 290 factories received workplace dialogue training; iii) 140 factories use Fair Wage method, which represents 29% of the total product volume; iv) gender split in Board of Directors is 50:50; v) a total of 3,411 workers received certification in 2015 and 2016 within the Enterprise Based Training (EBT); vi) all of commercial business partners must have a policy for recruitment that guards against anti-discrimination, harassment and abuse in the workplace. (H&M, 2016).

In a partnership with UNICEF, H&M Conscious Foundation develops a program that reaches the most vulnerable children and seeks to mitigate the inequalities. Today, it is rare to find any workers below the statutory minimum age in H&M' suppliers' factories. The H&M group has taken a clear stance against all use of child labour for many years. It is a minimum requirement for all factories producing for the H&M group, and represents their continuously monitor compliance (H&M, 2016).

Over 50% of people living in developing countries do not have access to toilets and clean water in schools and workplaces causing serious illness and negative impacts. In partnership with WaterAid H&M invests in clean water and sanitation in schools. The program aims to deliver water, sanitation and hygiene education in schools to improve health care, education and transform the future of children (H&M, 2015). The program also aims to prevent dropout of girls, which is a major cause of impact on the economy and prevents the advancement of women in society.

In partnership with CARE, H&M invests in the training and development of women and is dedicated to the promotion of actions to strengthen women role in developing countries to transform their future (H&M, 2015).

The main desire of the company is to demystify ideas and beliefs about what women can and cannot do and modify the systems that prevent women and girls to reach their potential and realize their dreams. Basic skills training in the areas of self-esteem, calculation of cost price, marketing, trading and sales are offered, as well as an incentive for setting up small businesses or improvement of existing companies (H&M, 2015).

The goal for 2017 is that 100 thousand women from poor communities have access to knowledge and skills or capital to allow enhance or create a company (H&M, 2016).

Based on the principle of equal rights, human dignity and solidarity between society, social sustainability seeks to match the standards of living so that all people can enjoy basic and necessary resources to live with dignity (Majid et al., 2016). Missimer et al. (2017) argue that benefiting society and helping social development is also a form of sustainability.

After the report of three examples of practices taken by H&M Conscious Foundation, it can be stated that the proposition 5 (P5. The adoption of social practices by companies belonging to fashion SC contributes to greater sustainability) is verified.

5 CONCLUSIONS

The concept of sustainability has changed along the time. Actually, is considered as part of day-to-day life of people and organizations. Due to environmental issues, society organizations are pressing companies for adopting a more environmentally friendly behaviour. Moreover, increased consumption, supply, ease of disposal of fashion and textile goods, short life cycle of products are challenges that the fashion industry are facing and becoming the competitiveness of the fashion companies more difficult to reach.

In this context, the fashion industry, have sustainability to create value for stakeholders, improve its image, to rêmain competitive and offer a product that meets consumer needs and preserves the environment.

There are many practices that can be applied by fashion supply chain with sustainability concerns and some of them are already being adopted such as the product and transport packaging containers (Delai & Takahashi, 2013).

To analyse whether the fashion industry adopts sustainable principles in the three dimensions of sustainability, a case study of H&M SC was performed in this study.

The main results showed that the adoption of sustainability practices provides numerous benefits for the fashion industry. The H&M Group, which was the fashion retailer brand studied in this research, has reached important benefits from adopting sustainability practices, mainly the reduction of production costs and better management of resources.

It was also noted that there is a concern to minimize emissions and environmental impacts throughout the supply chain, which features the environmental dimension. Using natural, recycled or biodegradable fabrics, garments reform or recycle fabrics for the manufacture of new parts are examples of efforts to achieve environmental sustainability (Na & Na, 2013).

Regarding the social dimension, H&M Group is interested in offering good working conditions. It supports some projects (already mentioned above) that benefit the communities where it operates. Therefore, based on the case study of H&M, it can be said that the fashion industry adopts sustainable behaviour based on the three dimensions of sustainability.

Delai & Takahashi (2013) say that sustainable development can only be achieved through the production that minimizes the use of the conscious consumer consumption and natural resources. For the authors, the fashion retailers have an important role in changing the consumer behaviour, because of its position between supply and demand.

Policies adopted in water policy, climate, innovative changes, and design will be the key to the fashion industry in the future (Bennie et al., 2010).

In terms of limitations, due to the difficulty in collecting data from primary source, the data collection was carried out through secondary sources, namely, sustainability reporting and the company's website, which could bias the information.

For future lines of research, it would be interesting to study more cases of companies operating in the fashion industry and check on their sustainable behaviour, as well as a comparative analysis between the sustainable behaviour of the fashion industry companies located in different continents.

ACKNOWLEDGEMENTS

The authors would like to thanks to the CEFAGE-UBI which has financial support from FCT, Portugal, and FEDER/COMPETE 2020, through grant UID/ECO/04007/2013 (POCI-01-0145-FEDER-007659) and to the FibEnTech – Fibre Materials and Environmental Technologies.

REFERENCES

Allwood, J., Laursen, S., & Malvido de Rodriguez N., C. B. 2006. Well Dressed? University of cambridge.

Arbogast, G. & Thornton, B. 2012. A global corporate sustainability model. Journal of Sustainability and Green Business, 1: 1–9.

Battaglia, M., Testa, F., Bianchi, L., Iraldo, F., & Frey, M. (2014). Corporate social responsibility and competitiveness within SMEs of the fashion industry: Evidence from Italy and France. Sustainability 6(2): 872–893.

Bennie, F., Gazibara, I., & Murray, V. 2010. Fashion Futures 2025 Global Scenarios for a sustainable fashion industry. Forum for the Future, 61. Retrieved from https://www.forumforthefuture.org/sites/default/files/project/downloads/fashionfutures2025finalsml.pdf

Bhardwaj, V., & Fairhurst, A. 2010. Fast fashion: response to changes in the fashion industry. The International Review of Retail, Distribution and Consumer Research. 20(1): 165–173.

Brockett, A.M., & Rezaee, Z. 2013. Corporate Sustainability. In S. O. Idowu & R. Schmidpeter (Eds.), Corporate Sustainability: 1–27. New York: Springer Heidelberg.

Ceschin, F., & Vezzoli, C. 2010. The role of public policy in stimulating radical environmental impact reduction in the automotive sector: the need to focus on product-service system innovation. International Journal of Automotive Technology and Management. 10: 321–341.

De Brito, M.P., Carbone, V., & Blanquart, C. M. 2008. Towards a sustainable fashion retail supply chain in Europe: Organisation and performance. International Journal of Production Economics, 114(2): 534–553.

DEFRA – Department for Environment Food and Rural Affairs DEFRA. 2011. Sustainable Clothing Roadmap Progress Report 2011. Retrieved from https://www.gov.uk/government/uploads/system/uploads/attachment_data/file/69299/pb13461-clothing-actionplan-110518.pdf.) Acessed in March, 15, 2017)

Djelic, M.-L., & Ainamo, A. 1999. The Coevolution of New Organizational Forms in the Fashion Industry: A Historical and Comparative Study of France, Italy, and the United States. Organization Science, 10(5): 622–637.

Delai, I., & Takahashi, S. (2013). Corporate sustainability in emerging markets: Insights from the practices reported by the Brazilian retailers. Journal of Cleaner Production, 47: 211–221.

Eder-hansen, J., Kryger, J., Morris, J., & Sisco, C. 2012. The NICE consumer: research summary and discussion paper. Danish Fashion Institute.

European Commission. 2011. Communication from the Commission to the European Parliament, the Council, the European Economic and Social Committee and the Committee of the Regions: A Renewed EU Strategy 2011-14 for Corporate Social Responsibility. COM(2011) 681 Final, 1–15.

Faisal, M.N. 2010. Sustainable supply chains: a study of interaction among the enablers. Business Process Management Journal, 16(3), 508–529.

Fashionunited (2017) global fashion industry statistics – international apparel, https://fashionunited.com/global-fashion-industry-statistics (acessed 23 april, 2017)

Fletcher, K. 2010. Slow Fashion: An Invitation for Systems Change. Fashion Practice. 2(2): 259–266.

Florea, L., Cheung, Y.H., & Herndon, N.C. 2013. For All Good Reasons: Role of Values in Organizational Sustainability. Journal of Business Ethics, 114(3): 393–408.

Forza, C. 2002. Survey research in operations management: a process-based perspective. International Journal of Operations & Production Management, 22 (2): 152–194.

Fulton, K., & Lee, S.-E. 2013. Assessing sustainable initiatives of apparel retailers on the internet. Journal of Fashion Marketing and Management. 17(3): 353–366.

Gardetti, M.A., & Torres, A.L. 2013. Sustainability in Fashion and Textiles. Greenleaf Publishing Limited, 44(0), 0–20. Retrieved from http://www.greenleaf-publishing.com/productdetail.kmod?productid=4010

H&M. 2014. Annual Report 2014. Retrieved from http://about.hm.com/en/About/Investor-Relations/Financial-Reports/Annual-Reports.html (Accessed: April, 2, 2017)

H&M. 2015. Conscious Actions Sustainability Report 2015. Retrieved from http://sustainability.hm.com/en/sustainability/downloads-resources/reports/sustainability-reports.html. (Accessed April 18, 2017).

H&M. 2016a. H&M Annual Report 2016, https://about.hm.com/content/dam/hmgroup/groupsite/documents/master language/Annual%20Report/Annual%20Report%202016.pdf (Acessed March, 20, 2017).

H&M. 2016b. The H&M Group Sustainability Highlights 2016, https://about.hm.com/en/media/news/general-2017/hm-sustainability-report-2016.html. (Acessed 28, march, 2017)

Hart, S.L., Milstein, M.B., & Ruckelshaus, W. 2003. Creating sustainable value. Academy of Management Executive, 17(2): 56–67.

Hasanbeigi, A., & Price, L. 2015. A Technical Review of Emerging Technologies for Energy and Water Efficiency and Pollution Reduction in the Textile Industry. Journal of Cleaner Production, 95: 30–44.

Jørgensen, M.S., & Jensen, C.L. 2012. The shaping of environmental impacts from Danish production and consumption of clothing. Ecological Economics, 83: 164–173.

Kozlowski, A., Bardecki, M., & Searcy, C. 2012. Environmental Impacts in the Fashion Industry. Journal of Corporate Citizenship, 45: 17–36.

Kozlowski, A., Searcy, C., & Bardecki, M. 2015. Corporate sustainability reporting in the apparel industry. International Journal of Productivity and Performance Management, 64(3): 377–397.

Levy, S. 1988. Information technologies in universities: An institutional case study. Unpublished doctoral dissertation, Northern Arizona University, Flagstaff.

Lo, C.K.Y., Yeung, A.C.L., & Cheng, T.C.E. 2012. The impact of environmental management systems on financial performance in fashion and textiles industries. International Journal of Production Economics, 135(2): 561–567.

Majid, I.A., & Koe, W.-L. 2012. Sustainable Entrepreneurship (SE): A Revised Model Based on Triple Bottom Line (TBL). International Journal of Academic Research in Business and Social Sciences, 2(6): 293–310.

Missimer, M., Robèrt, K-H, Broman, G. 2017. A strategic approach to social sustainability – Part 1: exploring the social system, Journal of Cleaner Production, 140(1): 32–41.

Murray, M. 2016. Partitioning ecosystems for sustainability, Ecological Applications; 26(2): 624–636.

Na, Y., & Na, D.K. 2013. Investigating the sustainability of the Korean textile and fashion industry. International Journal of Clothing Science and Technology, 27(1): 23–33.

Nagurney, A., & Yu, M. 2012. Sustainable fashion supply chain management under oligopolistic competition and brand differentiation. International. Journal of Production Economics, 135(2): 532–540.

Nidumolu, R., Prahalad, C.K., & Rangaswami, M.R. 2009. Why sustainability is now the kep driver of innovation. Harvard Business Review, 15: 57–64.

Niinimäki, K. 2010. Eco-Clothing, consumer identity and ideology. Sustainable Development, 18(3): 150–162.

Rowley, J. 2002. Using case studies in research. Management Research News, 25(1), 16–27.

Schönberger, H., Martos, JLG, and Styles, D. 2013. Best Environmental Management Practice in the Retail Trade Sector, European Commission Joint Research Centre Institute for Prospective Technological Studies, ISBN 978-92-79-30495-8.

Thanikaivelan, P., Rao, J.R., Nair, B.U., & Ramasami, T. 2005. Recent Trends in Leather Making: Processes, Problems, and Pathways. Critical Reviews in Environmental Science and Technology, 35(1): 37–79.

UNEP, United Nations Environment Program. (2011). The Chemicals in Products Project: Case Study of the Textiles Sector.

Vajnhandl, S., & Valh, J.V. 2014. The status of water reuse in European textile sector. Journal of Environmental Management, 141: 29–35.

World Bank Group. 1998. Textiles. Pollution Prevention and Abatement Handbook. Washington, D.C. 20433, U.S.A. ISBN 0-8213-3638-X.

Yin, R.K. 2003. Applications of case study research, 2nd ed., vol. 34, Thousand Oaks: Sage.

Textiles, Identity and Innovation: Design the Future – Montagna & Carvalho (Eds)
© 2019 Taylor & Francis Group, London, ISBN 978-1-138-29611-4

The possible use of the fish leather in a lingerie confection: Approach to the Amazon

T. Figueredo
BELEM, Center of Natural Sciences and Technology, University of Para, Brazil

N. Silva
MANAUS, Faculty of Technology, Federal University of Amazonas, Amazonas

ABSTRACT: The use of alternative materials in the fashion industry has been tabling productive potential as this is an unsustainable Market of the social and environmental point of view, and from this perspective, this market presents potential for change and economic development, if it is directed to a conscious production, as people are becoming more informed about the environment and what comes to harm him. Therefore, this article aims to present the lingerie design, thinking of fish leather as a base raw material in the manufacture of parts, where the same ones aim to meet concepts of sustainability and design, in a way that the Project function and aesthetic satisfy the public to whom they are directed. The presented research was based on bibliographic studies (websites, articles, blogs) and survey, and had as outcome to show the importance of idealization and creative thinking to arrive at concise results of the production and environmental perspective.

1 INTRODUCTION

The fashion in Brazil and in the world, is an area that presents challenges to society, since this is one of the industrial sectors that most pollutes the environment, and at the same time, it is one of the industries that most turn a profit to those who invest in it, and to the actors of society that make the economy of the sector move. This is due to a strong culture of consumption, which reinforces the idea of the old and the new, the beautiful and the ugly, the replacement, concepts constantly employed by professionals of the area.

It is this economy that has been questioned, as it passes over people and nations to explore and continue forming this over - consumption, that is damaging the environment in which it dwells. However, it is known that it is possible to achieve a balance between need, production, profit, and consumption, and thus, one can modify a system that does not have ecological support to have continuity.

One option found for this problem is the use of less harmful materials to the environment, which can be used for the production of a new product or modifying an existing one. In this case, we chose to use the leather fish of the Amazon region in the production of underwear and designed the product to the local public and reusing the input of fishing industries that do not work this by-product.

1.1 *Materials and methods*

This article aims to present possibilities of using fish leather in the manufacture of underwear, aiming to reuse an input that has been little explored in the Amazon region, and so that the underwear market is benefited in the concentration area study, so that aggregates are tangible and intangible concepts to the material presented.

The presented methodology is experimental, based on literature (articles, websites, blogs) and survey.

2 DEVELOPMENT

2.1 *The fishing and tanning industry in Brazil*

The practice of tanning happens in Brazil for many years, since indigenous tribe inhabit the territory. The use of fish meat for food, bones and scales for crafts and skin to produce vestment have always been constant practices in these societies. Today, the tannery is directly related to the fishing industry, which went through necessary changes in recent years so that now satisfies the consumer of fish, the fisherman and the environment.

The fishing activity is a constant reality of residents of the coastal areas of the country, who survive of this exercise, but even in a period of development of important technologies, some small and medium enterprises and artisans do not work the by-products of fish, which causes unwanted environment Impacts. One of these by-products is the skin of the fish, which when transformed into leather, can generate income for communities that survive of these industries (Tilapia leather process 2010).

In Rio Grande do Sul, for example, the practice of tanning has been happening because "the Group of Management Studies in Aquaculture (GEMAq) developed the ecological tanning design of animal skins to add value through the craft making "(Tilapia leather process 2010, 0:45 min) having the tilapia leather *(Pseudocrenilabrinae)* as base raw material in the manufacture of parts produced by the local community, which is a common fish in the southern region of Brazil.

In the state of Pará, the Almira Alice Fonseca Martins researcher observed how the pescada amarela *(Cynoscion acoupa)* fur were wasted in the city of Curuçá, by the local exporter of seafood industry, and saw their potential for creation and production of fashion accessories. Since then, Almira met with the local community to know the possibility of developing your project, seeking to reuse the skin, adding concepts of sustainability, and generate income to the community, aiming to expand the local production chain. However, the researcher did not continue the project, because the lack of government incentives and the lack of community interest were deterrent factors (Pinto & Moraes 2014).

In Manaus, fish farming is a strong export industry of fish a pirarucu *(Arapaimagigas),* aruanã *(Osteoglissidae)* and dourada *(Carassius aurata),* however, the leather tannery of these species does not happen, getting to return to the rivers more than 10 Tons per month of this input. According to Melo (2016), the researcher José Jorge da Silva Rebello (90), together with the National Institute of Research in the Amazon (Inpa), has been conducting more than 20 years of studies on tannery practice in the region, transforming the fish skins in a Good quality leather. José says that it is a dream to see the technology that he developed being applied in an industrial production, because he believes in the economic potential of the project (information obtained in an interview with Tanner and researcher José Jorge da Silva Rebello (90), by journalist Kelly Melo to the online page of the local newspaper 'The Critic').

And, in Rio de Janeiro, the director of the Institute-E (a non-profit institution that aims to promote sustainable fashion in Brazil), Nina Almeida Braga, in partnership with the international brand, Osklen, developed a project on the use of leather of pirarucu *(Arapaima gigas)* to produce clothes and fashion accessories. The initiative started from the idea of ??producing a conscious fashion in the environmental aspects and possible in the social aspects, so that the community involved in the manufacturing process is directly benefited by the work. According to Simone Cipriani, founder of *Ethinical Fashion Initiative* (institution that works the ethical aspects of the fashion), working with sustainable fashion is a process in development of the current textile industry, and should be careful while in this new perspective, the cultural appropriation of these communities is a dangerous issue and should be treated with caution (Cunha, 2015).

According The Support Service to Micro and Small Enterprises (Sebrae [2010–2017a]), the tannery sector moves in Brazil, around US $ 21 billion per year, and sustains a market of 450 tanneries, most of which are small companies employing between 4 and 99 employees.

2.2 *Underwear market in Amazon region*

According to research conducted by the Institute of studies and industrial marketing (2015), the Brazilian clothing market showed low production in 2014, however, the sportswear market, *jeanswear,* underwear and sleep and swimwear, showed significant growth in annual production in the country (Rodrigues 2015). Also according to the Institute of Studies and Industrial Marketing (2015), it was estimated that in 2015 there was a 1.5% increase in the volume of private parts of the country, including panties, bras, sweaters, pajamas and styling; most notably The bras, "which are increasingly diversified and attractive" (interview with Marcelo V. Prado, director of the Institute of Studies and Industrial Marketing) recording average expanding above the remaining underwear (Cajano 2015).

According to the study, in the north of the country, focus only 1% of companies in the underwear segment, getting the Southeast with 53.2% of national production; Northeastern region with 19.9%; South region with 17.9% and Central West with 8.3% of the total. The study also points out that, for female consumers of this product, the comfort regard is above the price regard and the brands that are getting emphasis in this scenario shown that characteristic. (Cajano 2015).

2.3 *The project*

The Support Service to Micro and Small Enterprises (Sebrae [2010–2015]), released another survey, also conducted by the Institute of Studies and Digital Marketing (2015), which identifies the characteristics of Brazilian underwear consumers. Most of them are in the age group of 25 and 34 years; Their monthly income is between R$ 2,041.00 and R$ 7,650.00 and they are concentrated in classes B2 and C; the cash payment was made in 66% of purchases and the average spend on lingerie, for these consumers was R$ 93.20 (The Support Service to Micro and Small Enterprises, [2010–2015]).

The underwear developed for this study were not produced in practice; however, they were designed for a small-scale production, where the use of fish leather is the great advantage of the product; and the use of alternative fabrics such as organic cotton, which also serves a less aggressive feature to the environment.

The cultivation and production of organic cotton uses less chemicals and chemical inputs, reducing damage to the soil, with an ordinary cotton production accounts for 25% of worldwide use of pesticides (Santos 2016).

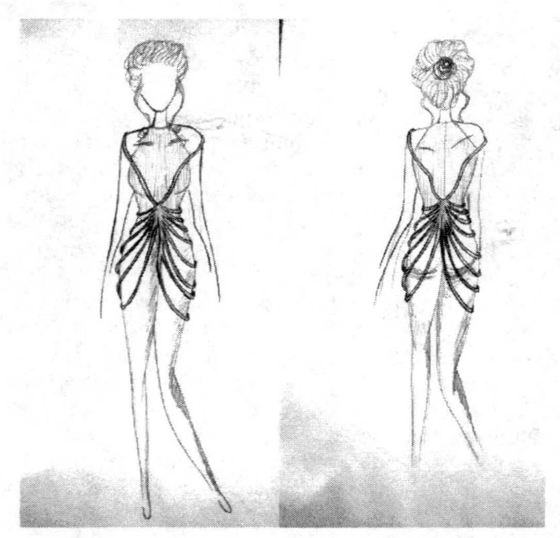

Figure 1. (Authors 2017). Lingerie 1: The Golden Embrace.

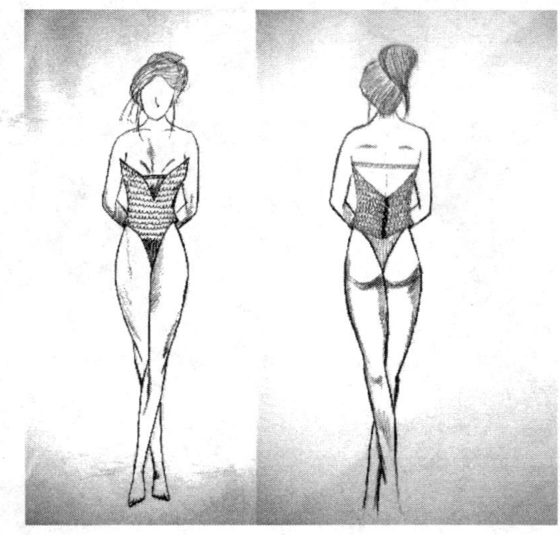

Figure 2. (Authors 2017). Lingerie 2: The Touch of Shining Scale.

In this project, we sought add to the idealized parts concepts such as mysticism, since the raw material base of lingerie is fish leather, and this icon of Brazilian fauna consists legendary stories of the Brazilian riparian peoples; sustainability, in the environmental aspects of production, since the parts would be produced with recycled inputs; simplicity in application of few materials to manufacture; innovation, in the case of underwear, which must meet the youth/adult audience, which buys products that call them attention by aesthetics, and comfort. And lastly but not leastly, sensuality, necessary and natural concept when it comes to lingerie.

In the Figure 1 (Authors 2017), the lingerie was designed with the *body* model, where it was thouto use organic cotton in nude colored, hid with fiber of fish leather, in this specific case, with the use of leather of dourada *(Carassius aurata)*. The piece is a whole where the filament is tailored to the *body* by the center and it is dressed, with dispensing (invisible zipper) on the back, in the vertical direction.

In the developed outline in Figure 2 (Authors 2017), the corset would be produced with the leather of pescada amarela *(Cynoscion acoupa)* Finished made with fish scales. In this model, the filament of the leather makes a visual connection between the front and the back side of the accessory, and the fabric that covers the female intimate part was thought with transparent synthetic material (pure black tulle), sewn inside the corset, Making the piece a unique one.

In the third model (Fig. 3) is formed by the lingerie bra and panties, made with leather aruanã *(Osteo glissidae)* and filament with the use ofleather of pescada amarela *(Cynoscion acoupa)*.

The bra has lock on the front through a node, the underwear is worn and must have elasticity; pantyhose (Lycra in nude color) is secured in the panties by the filament produced along the lines of pescada amarela

Figure 3. (Authors 2017). Lingerie 3: The Relationship of Skins.

leather, which makes the model make a visual contrast through the skins of different species, but that are connected visually.

According to the Amazon Portal site ([2010–2017]), the pirarucu *(Arapaima gigas)* "(. . .) is not only the largest scale fish in the world, but also a symbol of Amazonian waters. The pirarucu is a fish of great personality." And it is precisely this personality that the lingerie is looking to transpire in your design, through a pirarucu leather *body (Arapaima gigas)* and the finishing done with his long scales. The transparent skirt would be produced with pure tulle, in the nude color, fastened with buttons by the waist. The mirror mentioned in the name of the piece, refers to the cutting of the product, which is the same on the front and on the back.

Figure 4. (Authors 2017). Lingerie 4: The Mirror of the Most Beautiful.

Figure 5. (authors 2017). Lingerie 5: The Arowana tail.

And here, again, the lingerie of figure 5 (Authors 2017) was designed with the use of leather of aruanã *(Osteoglissidae)* as a raw material base, because he has the most darkened skin and its leather has the same shade.

In this proposal, the leather would be used only in pantyhose, which has a design that remind scales, in order to refer to the proposed of use fish leather; and also, the panty and bra would be produced with black Lycra. The idea is that the leather filament make connection between the pantyhose and the rest of lingerie, as we see in the drawing.

The idea of using fish leather also assume that this material, compared to the bovine leather is more durable and has great aesthetic potential. According Nobrega (2015), the fish leather is three times more resistant than cowhide, has high durability and resistance to textile processing and dyeing.

According to Pratini (2000), the generation of sketches as a design step becomes essential, and from then on, an innovative product in the functional and/or aesthetic aspects can be born, to satisfy the needs and desires of the consumers for whom the same will be destined.

But not only as a need for product and project, the ideas presented here have as a primary driver a sustainable product proposal, since the idea is that the input used in the production is obtained from existing industries that do not adequately destine fish skins slaughtered for consumption, and also the social aspects of production, all the participants involved must be properly fitted in the process, so that there is not excess of work or production, and that the income generated from the project will benefit the state where It will be produced through the generation of employment.

3 CONCLUSION

In analyzing this study, we can see that the idealization of the pieces started from the problem presented, where fish skin has been devalued in the Amazon Region of Brazil by industries that do not work the by-product of fish. Noting that the Amazon is still an area that has a vast scenario in differentiated raw material to produce some products, in this case, of lingerie.

Unsurprisingly the use of fish leather has proved to be a different material in textile production and in the manufacture of fashion in Brazil. With a growing search for the use of alternative materials in this industry, which presents a high degree of disposal and pollution, the leather is still a resilient and aesthetically pleasing fabric, and therefore, its application continues to be valued.

And yet, the choice of using fish leather in a confection of lingerie came from the idea that in the north of Brazil this is a relatively unexplored market, and therefore promising as the research presented by the Institute of Studies and Industrial Marketing (2015) they say that there was an increase in production and consumption of underwear in Brazil, still in 2015.

Therefore, adding the differential to use the leather of native fish of the study area ??concentration, and other materials to compose the pieces, proved to be an interesting alternative from an environmental point of view and design, so that the concepts applied to lingerie were thought to create a relationship between the material, the product, and consumers of the Amazon region in Brazil.

At this point, the idea come from the principle that the parts are produced using recycled raw materials, rather than virgin raw material (fish for use the leather), and the actors involved in the process - the angler and his family, fishing industries and the refrigerating company, the worker and the machines involved in the production are suitably employed in the process, so that it generates employment and income with a small local production, seeking to target any and all waste generated.

The five pieces presented here have not been produced in practice, and therefore the experimental research proves issues of comfort, aesthetics and functionality addressed throughout the project. But the idea that you can reuse natural inputs in a fashion production, while the production of lingerie in the Amazon region is still unexplored and presents investment potential, makes it encourage such production.

And, it is a matter addressed in this project, in a practical way, the importance of idealization phase in the process, since the five pieces were not produced in fact, but to get to the differential in the design of lingerie, this is Makes a fundamental step, flowing creative thinking.

REFERENCES

Brito, G.A. 2013. Environmental impacts generated by tanning. 9th Fashion Colloquium, Fortaleza. Available: http://www.coloquiomoda.com.br/anais/anais/9-Coloquio-de-Moda_2013/ARTIGOS-DE-GT/Artigo-GT-Moda-e-Sustentabilidade/Impactos-ambientais-gerados-pelos-curtumes.pdf.

Cajano, P. 2015. Intimate fashion sector foresees growth of 1.5% in 2015. Available: http://www.investimentosenoticias.com.br/noticias/negocios/setor-de-moda-intima-preve-expansao-1-5-in-2015.

Cunha, R. 2015. The arapaima leather and the challenge of mixing tradition and innovation in the fashion industry. Available: http://www.stylourbano.com.br/o-couro-do-pirarucu-eo-desafio-de-mesclar-tradicao-e-inovacao-na-industria-da-moda/. Epochtimes.

Leite, RP 2012. Fish leather as a sustainable option for fashion. Textile and Clothing Industry – Textile Industry – Year IX. Available in: http://textileindustry.ning.com/forum/topics/couro-de-peixe-como-op-o-sustent-vel-para-a-moda.

Melo, K. 2016. Technology of harnessing fish leather in the Amazon is wasted. Manaus. Available: http://www.acritica.com/channels/governo/news/tecnologia-de-aproveitamento-do-couro-de-peixes-da-amazonia-e-desperdicada. The critic.

Nóbrega, L. 2015. Fish leather and its benefits in the textile and clothing industry. São Paulo: University of São Paulo, USP. Available: http://www.teses.usp.br/teses/disponiveis/100/100133/tde-08092015-131941/en.php.

Pinto, W. & Moraes 2014 yellow Hake is fashionable. Beira do Rio. Journal of the Federal University of Pará. Year XXX N°130. Available: http://www.jornalbeiradorio.ufpa.br/novo/index.php/2014/152-2014-08-01-17-25-17/1623-2014-08-04-16-43-57.

Portal Amazônia. Amazon from A to Z. [2010–2017]. Fish of the Amazon. Available: http://portalamazonia.com.br/secao/amazoniadeaz/interna.php?id=957.

Pratini E. 2000. Of the sketches on paper to sketch in space: the design of a gestural system for 3D modeling. Rio de Janeiro: 4th Sig Radi. Available: Http://papers.cumincad.org/data/works/att/13d8.content.pdf.

Rodrigues, K. 2015. IEMI: growth potential in national retail production. ADS Corporate Communication. IEMI Market Intelligence. Available: Http://www.textilia.net/materias/ler/moda/moda-destination-market/iemi_potential_of_growth_in_national_production_of_tourism.

Santos, E. 2016. Brazilian experience with organic cotton will be disseminated in Mercosur. Embrapa. Available: http://www.brasil.gov.br/economia-e-emprego/2016/03/experiencia-brasileira-com-algodao-organico-sera-difundida-no-mercosul.

Support services for micro and small businesses, Sebrae [2010–2015]. Intimate Fashion: Profile of Consumption and Behavior of Purchase. Available: http://sebraemercados.com.br/moda-intima-perfil-de-conjuice-and-behavior-of-purchase.

Support services for micro and small businesses, Sebrae [2010–2017a]. How to Set Up a Fish Leather Tanning Company. Available: https://translate.google.com/?hl=pt#pt/en/Como%20montar%20uma%20empresa%20de%20curtume%20de%20couro%20de%20peixe

Support services for micro and small businesses, Sebrae [2010–2017b]. The leather tannery market. Available: http://www.sebraemercados.com.br/o-mercado-de-curtume-do-couro-de-peixe/.

Tilapia leather Process 2010. Managements Study Group on Aquaculture (GEMAq), State University of Western Paraná, Toledo, Brazil. 2010. 7:46 min. Available: https://www.youtube.com/watch?v=sCxaXluKai.

Textiles, Identity and Innovation: Design the Future – Montagna & Carvalho (Eds)
© 2019 Taylor & Francis Group, London, ISBN 978-1-138-29611-4

Traditional techniques of Portuguese tapestry as the basis of a sustainable construction for fashion

C. Morais, G. Montagna & M. Veloso
CIAUD, Lisbon School of Architecture, University of Lisbon, Portugal

ABSTRACT: Tapestry is an ambiguous procedural form, able to become a shelter item, a religious good or even a type of connotative mural. The significance of carpet and tapestry depends on how the structure is constructed although the functionality of both can be reversible, since there are tapestries to cover the floor and carpets to decorate walls. It is not usual to talk about their techniques and their long-lasting way of yarns interweaving to re-produce all kinds of products. In Portugal there are some regional qualities of significant cultural value, different from their true origin (almost always oriental) by the insertion of other types of materials or by the yarns density variation. This handcraft weaving, made in horizontal or vertical loom, when associated with creativity and cultural training of the producer, can provide not only linear and two-dimensional objects (carpets or tapestries) but also wearable objects of three dimensions field. In fact, it is in the pattern making matter and in the zero waste planning that this paper purposes to demonstrate a versatility of producing garments on looms. In addition to control waste and add durability through the quality of materials, it's a manual labor which respects the slow production system (Slow Fashion). Therefore, it let the consumer to have customized drawings and textures, being a focus of differentiation and innovation in fashion system.

1 TAPESTRY

1.1 *Its evolution in the Iberian Peninsula*

The etymology of the tapestry remains uncertainly and shows a duality between function and exhibition, horizontality and verticality, with different meanings and cultural contexts (Shanahan, 2014). Its existence is related to historical, social and geographical factors, which wide use range from the Islamic prayer rug to the European decorative mural.

The two main methods of tapestry production tend to be made around high-warp loom (or vertical loom) and the low-warp loom (or horizontal loom), whose directions are associated to the position of the loom worker.

In ancient times, the weaving, whether made in low or high-warp was part of the typical artisanal system, often limited to the local raw materials that came from agriculture and the livestock of their producers.

Although it was developed for necessity, by people who lived in Mesopotamia, Egypt, Greece, Rome, Persia, India, and China, it was a labour also inseparable from art, still playing at the present time an important role in the manufacture of carpets, generally with geometric designs.

On the other hand, Europe become a great centre of realistic designs. During the middle Ages, the panels' production took on a great importance in the interior decoration of castles and medieval churches, improving as well the thermal comfort.

The Bayeux tapestry from 11th century, which represents the conquest of England by the Normans, is an example of the earliest tapestries still in use today. They all started to be embroidered on canvas, but the evolution of this work let woven techniques grew quickly. Flanders tapestry was spread by Flemish cartoonists, while French was influenced by Turkish people that worked in the weaving centres of Aubusson and Beauvais (Morais, 2005).

In Portugal, this activity presented two distinct realities: a larger growth in population centres and a smaller and limited development in country places. Here it was a subsidiary activity of the agriculture and established at self-consumption. Although the Portuguese factory was presented in Antwerp, in 1555, with the famous *Pastrana* tapestries, the kingdom of Portugal continued to be rural, with unprofitable productive processes (Medeiros & Lopes, 2000), comparing to neighbouring countries.

The weaving had only begun to flourish thanks to the role that *Marques de Pombal* played in the establishment of small factories. Belonging to the crown these manufactures had fiscal benefits and financial subsidies, as happened with the Royal Factory (1772) in Portalegre for weaving wool and cotton, or the Royal Factory (1764) built in Covilhã to support local manufactures in dyeing and fabrics finishing procedures.

There are few writings to prove the national tapestries production. However, the inventory made by the researcher Maria José Mendonça, in 1983, for the

Portuguese Institute of Heritage, most references are dated from the 16th and 17th centuries, from Flanders and Brussels (Huylebrouck, 1986). Even the mythical tapestries of Pastrana were executed in Tournai, France, in the 15th century, whose copies for the Portuguese State were later made in 1957 at the Real Fábrica de Tapices de Madrid. On the other hand, according to Fernando Baptista de Oliveira (1995), the first Arraiolos embroidered carpet comes from the 17th century (by the hands of the Arabs who had fled to the south of Portugal). So it is probably that embroidered tapestry had appeared in our country later than the woven tapestries.

The Industrial Revolution brought down this type of goods to a smaller amount, valuing those of fast consumer, where the *Technique* prevails over the *Art* (Mumford, 1952). In developed countries manufacturing processes went alongside mechanical efficiency, overcoming the artisan work, let him reduce his personal contribution and lower the quality standards. The use of machine and the progressive development of textile industry created two different paths for tapestry: manual or semi-industrial tapestry and the industrial utilitarian tapestry (Morais, 2005). In England, the success of textile and carpeting industry had boosted the production and exports of this type of product. Since mechanization began to be part of everyday life the artist had to handle both manual and industrial processes in order to choose the best method to adopt for the product he wanted to produce (Morais, 2005). During this period, some artists tried other systems, such as un-tufting guns, while they were working in mechanical weaving companies. The teaching of arts also changed and the new concept that valued simultaneously the useful and the beautiful (Design) was developed and applied on fabrics technology and upholstery classes at the Bauhaus school (1919–1925, Weimar and 1925–1932, Dessau).

The International Exhibition of Modern Decorative and Industrial Arts in Paris (in 1925) had already positioned carpets and tapestry in a total decoration plan, based on an eclectic style. However it was after 2nd World War that signed works reinforced the role of modern tapestry as a figurative mural. Its verticality started to be a mean of communication and a form to differentiate anonymous artisan works, valuing the author's individuality in the market (Shanahan, 2014). In the meantime, the foreign market opening created a wide cooperation between Eastern and Western countries. The Eastern countries such as China, Turkey and India became major centres of manual production while European countries leader in the mechanical carpets production, apart from those weaving shops that has continued to produce small series for elite markets. Till the half of the 20th century Portugal followed this reality, having as its great impetus the French artist Jean Luçart.

1.2 *Portuguese tapestry techniques*

According to the master research of the author (Morais, 2005), most of the Portuguese manufacturing techniques comes from the wisdom of oriental people that were established in the Iberian Peninsula during the Reconquest and, over the time, they got aesthetic and material changes. Among them we have:

Table 1. Different types of stitches and methods of Portuguese weaving and tapestry.

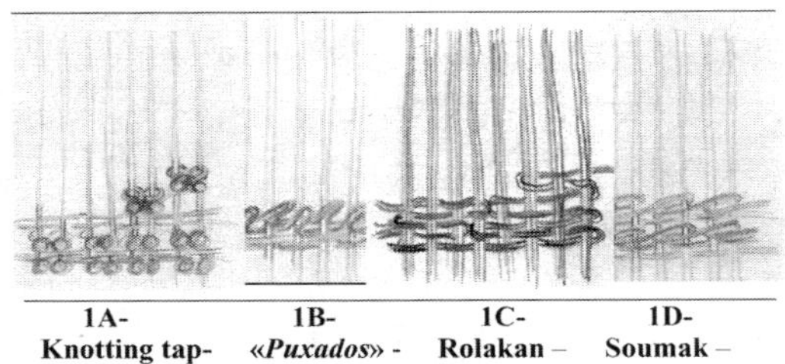

1A- **Knotting tapestry** – A shag structure, made with Turkish knots, (symmetrical knots) between two weft weaving threads. Beriz basic weaving.	1B- ***«Puxados»*** - *Looped piled tapestry*, with Tibetan knots. Similar method of Castelo Branco	1C- **Rolakan** – *Slit-tapestry* from *Gobelins*, but made on two warps threads simultaneously.	1D- **Soumak** – *tapestry* from Central Ásia,very identical from Portalegre tapestry.

- The weaving of *knotting tapestry*, as it is produced in Beiriz (Table 1, 1A);
- *Rollakan tapestry* from the north of Portugal, similar to Slit-tapestry or Eastern Kilim, the base tapestry structure from Beauvais, Gobelin and Aubusson (Table 1, 1B);
- The *looped pile tapestry*, made in 3 or 4 pedal looms, such as the one from Castelo Branco (Table 1, 1C);
- The *Soumak weaving* (Slit-tapestry variant) with wrapping stitches, such as the *Portalegre tapestry* (Table 1, 1D);
- The weaving of plain tapestries, such as *rag carpets*;
- The embroidered tapestry, made on canvas with needle, visible in the *Arraiolos tapestry* (Slavic cross stitch) and the *Madeirense tapestry* (open and full stitches);
- Tapestry and *tufted carpets*, using the Tufting gun technique;
- Art tapestry made of fabrics;
- Contemporary tapestry, made of mixed techniques.

In the present work the structures made in loom are more relevant for the proposal work, namely the *knotting tapestry* (beriz), *looped pile weaving*, *rollakan* (slit-tapestry) and *soumak*.

All of them allow numerous variations of tightening and can provide different textures (plain or piled).

2 FROM TAPESTRY TO THE CLOTHING

2.1 *Introduction*

Clothing has a functional dimension (Miller, 2007, p. 27) which, when combined with the contemporaneous phenomenon of Fashion, becomes an "aesthetic medium for the expression of ideas, desires and beliefs that circulate in the Society" (Wilson, 1985). For some researchers, Fashion does not equate to other arts, such as Architecture, for the reason of frivolity which is implicit in it (Hollander, 1993) perhaps stemming from Kant's theory that characterizes Fashion as a vanity title for those who follow it as a behaviour without inner value (Kant, 1963). However, other theories value the *Couture Art* because it depends on the synergy between client and creator (Martin & Koda, 1995) a method where ideas flow with clear advantage for artistic creation (Marietta, 2013).

Among Fashion trends, High Fashion has been the example of an esthetic experience of possession and status, with unique and exclusive characteristics, made of materials and fabrics of great quality. Techniques of production and manual application such as embroidery and lace, often performed by artists and craftsmen, are Design elements present in all parades.

Heavy structures, such as tapestry, are not often seen in the garment construction, but from time to time there are slight reminiscences. *Take me Home* collection (Autumn-Winter 2011/2012) by Maria Gambina was an example where we could show *Arraiolos* embroidery technique in fashion clothes.

The perception of these textiles structures can express a communicative ability, since, according to Andrew (2008) the tapestries, besides having colors, textures and patterns, also evoke feelings and humor for who observe them.

Producing garments directly in looms is a constructive possibility but not has been explored, perhaps due the difficulty of its process or due the time it would take. Therefore, the proposal of the present work is not well-suited to the fast fashion system. The methodology respects a slow Fashion process (Slow Design) that emphasizes quality and differentiation.

2.2 *The market*

In spite of the fact that is possible, both in High Fashion and Ready-to-wear, making complex silhouettes and designs with high technological textile materials, not always the customer satisfaction is matched.

A desire for novelty is a usual feature of Fashion market but nowadays it is also necessary to link that with ethical laws to make the production chain clearer and improve fairness at work. These issues are essential requirements for future generation, as well those present, named Lohas and Millennials generation (Cortese, 2003; Everage, 2002).

2.3 *Sustainable methodologies*

The Fashion has grown up as an object of expression from western culture. The environmental concerns and the necessity of reducing the consumption, by a minority, opened up opportunities never seen in Fashion, such as business around *vintage clothing* or even around the design of unique pieces. Nowadays Fashion has been a goal of other manifestations, integrating sustainability concepts, social and cultural changes which originate abilities to slow the system. A reuse of materials and their transformation into products of high added value, has gradually developed an ecological awareness from who produces and for who consumes the products (Schulte & Lopez, 2007). According to Slow Food movement leader Carlo Petrini, "It is useless to force the rhythms of life". This concept, created and worked on an ethic of life that seeks to balance the human being with the local, social and environmental mean, proposes the production of quality goods that will last longer and leads the consumer to wake up to a less materialistic attitude (Strauss & Fuad-Luke, 2008).

In clothing field, it is possible to slow down the garment system, analysing all stages of each product life cycle; In the stage of pre-production through the maximization of materials and selection those that can be reused or recycled; at the production stage implementing design measures that provide durable and repairable parts; at the distribution and retail stage through local transport policies and implementing practices of clothing rental; in the utilization stage with measures of personalization in order to prolong the affective feeling of the user; in the end-of-life stage, through measures of reusing and materials valorization, etc.

Zero-Waste is a methodology that focuses on reducing waste from cutting and sewing garments (Gwilt & Rissanen, 2011), even though other examples may highlight this concern along the product chain. In fact, one of the first examples was presented publicly emphasised the reduction of time production, such was visible in APOC design from Issey Miyake and Dai Fugiwara, in 1998, whose process were part of the design itself. But, apart from reducing the pace garment supply chain, measures of maximizing resources or increasing the products functionality also become constant and attractive at their projects, promoting the delay of garments into the landfill (Gwilt, 2014). The important thing is that total project should be nonlinear process and designer can rethink the silhouette and the pattern making or *drapping* their pieces before drawing them, in contrast to the conventional fashion design (Morais, 2013).

The zero waste pattern making approach can be worked in 2 or 3 dimensions but the result is well differentiated. The first method has ancient references, visible in the Japanese kimonos or in the Greek chiton, where the simple and straight lines of the garment have become objects of study for a more complex planning. Here several components can be integrated into an only piece or rather integrating several pieces of garment into a single cutting plan. In this type of work, called "Jigsaw puzzle" and explored by designers Holly McQuillan, Mark Liu and Timo Rissanen (Rissanen, 2013) the negative spaces of the block drafts are part of the final construction.

In addition to two-dimensional planning, there is a three-dimensional planning called 'Subtraction Cutting', created and developed by Julian Roberts (2013)

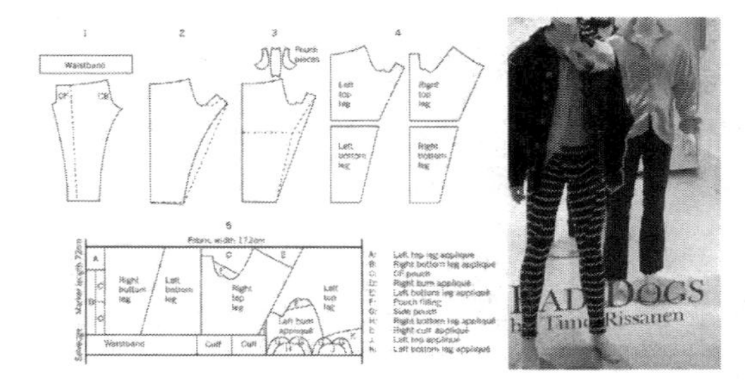

Figure 1. Design of leggings demonstrating function of shape library building (Rissanen, 2013).

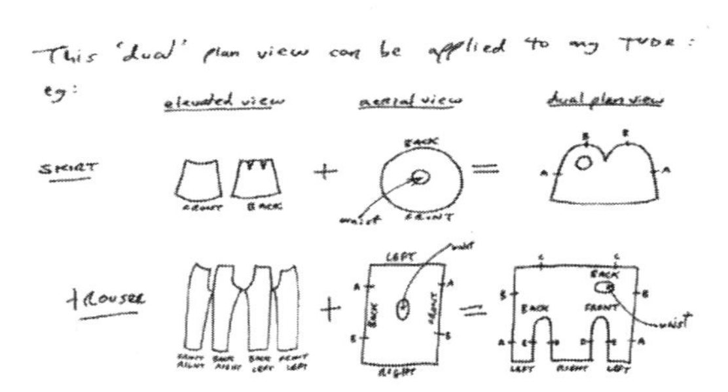

Figure 2. Examples of skirt and trousers Patternmaking planning. (Roberts, 2013, p. 30).

which integrates Design and production just in time. The creator idealizes an outward vision of a silhouette and conceives it by improvisation, supervising its development and fabric falling (Morais, 2013). This method does not let reduce waste as also limit the designer activity by the fabric size he works with, often forcing him to change the initial idealized forms (Lindqvist, 2013).

The customization of clothing and the production of individualized Couture are activities that integrate the humanist vision of design. Although the first judgment has an industrial economic approach (for Ready-to-wear) and the second has a more elitist economic approach (for High-Fashion), both respond to the purpose of better satisfying the client.

3 PROPOSAL

3.1 *Methodology*

This proposal demands an alignment between art and technology, integrating concepts of sustainability, for articles of differentiated value.

In order to do this aim, we intend to expose the fusion of tapestry techniques in clothing production, assuming a methodology that reduces waste and allows a customizable fit. Associating the dress modeling with its reproduction in a loom for fabrics also requires a break of conventional construction rules. The proposal methodology can be experienced in two ways.

Figure 3. *Purple zero-waste dress* by Line Sander Johansen, 2008 (Gwilt, 2014, p. 83).

Figure 4. Prototype of the human figure shape to weave garments (master work of Mariana Veloso).

The first way can be made through a previous planning of two dimensions garment modeling which later will be transferred to the loom (see Fig. 3, Line Sander Johansen work); The second way can be made through a three-dimensional planning, using a loom with a bust shape where clothes can be woven on it with tapestry techniques (see Fig. 4, work by Mariana Veloso).

4 CONCLUSIONS

A link between blocks pattern making and three-dimensional modeling (*drapping*) in clothing construction has been explored in order to minimize errors and better conceive a desired silhouette. Cutting fabric on dummy fashion and taking adjusts in sketches is a current method for ready-to-wear garment or independent productions. However, turning that into a loom production process is an innovative slower system that revitalizes the *Art* and the *Technique*.

This type of tapestry work, connected with the fact that it can provide customizable clothing while expressing cultural diversity of the country, is such conscious as an artistic solution. In addition to provide durability on clothing, it can reduce waste (during its manufacture) and enables differentiated designs and textures, making it also an excellent means of communication.

ACKNOWLEDGMENTS

This research was only possible with the support of CIAUD – Research Centre in Architecture, Urbanism and Design, Lisbon School of Architecture of the University of Lisbon, and FCT – Foundation for Science and Technology, Portugal.

REFERENCES

Andrew, S. (2008). Textile Semantics: Considering a Communication-based Reading of Textiles. In *Textile* (Vol. 6, pp. 32–36). United Kingdom: Berg.

Cortese, A. (2003, July 20). They Care About the World (and They Shop, Too). *The New York Times*.

Everage, L. (2002, October 1). Understanding the LOHAS Lifestyle. *The Gourmet Retailer Magazine*.

Gwilt, A. (2014). *A Pratical Guide to Sustainable Fashion*. London, Uk: Bloomsbury Publishing.

Gwilt, A., & Rissanen, T. (2011). Shapping Sustainable Fashion-Changing the Way we Make and Use Clothes. London, Uk: Earthscan.

Huylebrouck, R. (1986). Portugal e as Tapeçarias Flamencas. *Revista Da Faculdade de Letras Do Porto*, 3, 165–198.

Kant, I. (1963). *Analytic of the beautiful*. Indianapolis .: The Bobbs-Merrill Company.

Marietta, M. (2013). Action not Words. In *Power of Making: the importance of being skilled* (Daniel Charny). London: V&A Publishing and the Crafts Council.

Martin, R., & Koda, H. (1995). *Haute Couture* (MetPublications). New York.

Medeiros, C. L., & Lopes, F. (2000). *A Tecelagem Tradicional-Motivos e Padrões*. Lisboa: Livros e Leituras.

Miller, S. (2007, March). Fashion as Art; is Fashion Art? *Fashion Theory. The Journal of Dress, Body & Culture*, 11(1), 25–40.

Morais, C. (2005). *A Tapeçaria e Suas Técnicas de Produção*. Escola de Engenharia, Universidade do Minho, Guimarães.

Morais, C. (2013). *A Sustentabilidade no Design de Vestuário*. Universidade de Lisboa, Lisboa.

Mumford, L. (1952). *Art and Techniques*. New York: Columbia University Press.

Rissanen, T. (2013). *Zero-waste Fashion design:- A study at the intersection of Cloth, fashion design and pattern cutting*. University of Technology, Sidney.

Roberts, J. (2013). Subtraction Cutting by Julian Roberts. Retrieved from http://www.julianand.com/

Schulte, N., & Lopez, L. (2007). Sustentabilidade Ambiental no Produto de Moda (pp. 1–7). Presented at the I Encontro de Sustentabilidade em Projecto do Vale do Itajaí, UFSC, Vale do Itajaí, Brasil.

Shanahan, M. (2014, October 3). Tapis/Tapisserie: Marie Cuttoli, Fernand Léger and the Muralnomad. *Konsthistorisk Tidskrift/Journal of Art History*, pp. 228–243.

Strauss, C., & Fuad-Luke, A. (2008). The Slow Design Principles –A New interrogative and reflexive tool for design research and practise. In *Changing the Change Design, Visions, Proposals and Tools Proceedings*. Torino: Allemandi Conference Press. Retrieved from http://www.allemandi.com/university/ctc.pdf.

Wilson, E. (1985). *Adorned in Dreams: Fashion and Modernity*. London: Virago Press.

Textiles, Identity and Innovation: Design the Future – Montagna & Carvalho (Eds)
© 2019 Taylor & Francis Group, London, ISBN 978-1-138-29611-4

Sustainable textiles – A review on strategies for product lifetime extension

V. Januário & R. Salvado
Faculty of Engineering, University of Beira Interior, Covilhã, Portugal

S. Uğur Yavuz
UNIBZ – Free University of Bozen, Bozen, Italy

ABSTRACT: Concerning the fashion system, the beginning of the 21st century is marked by the proliferation of Fast Fashion and over-consumption. The way people consume today is pushing at the boundaries of the planet and those who are living in it. Therefore, disruptive and radical solutions are urgently needed towards a more sustainable world. This paper presents a literature review on strategies to extend fashion textile products lifespan, exploring some examples of businesses that are already implementing it, focusing on advantages and disadvantages both to costumers and companies. This strategies for themselves won't lead to radical change in the current system, however combining them in a totally new fashion system will be key to a promising sustainable model.

1 INTRODUCTION

Although sustainability is on the agenda of the major fashion brands, its main focus has been on the material and production level (Vuletich 2012), which may not be a sufficient answer to the complex environmental and social problems the planet is facing.

The same brands that are getting attention for being "green", are the ones responsible for the current consumption patterns, specially in what concerns to the exponential growth of waste, an answer to the planned obsolescence of products that are always chasing passing trends. According to the "Time out for Fashion" research, published in November 2016 by Greenpeace, clothing production doubled between 2000 and 2014 and the average person buys 60% more items of clothing every year and keeps them for about half time compared to 15 years ago. Moreover, 95% of the clothes that are thrown away could be re-worn, reused, repaired, or recycled but are part of the millions of tones of textiles that end up in landfills every year around the world (Greenpeace 2016).

The challenge for fashion brands and designers in order to achieve a more sustainable practice is to focus in all the phases of product lifetime instead of the current focus in the production and sales one (Aus 2011). The aim is to develop products with a longer and meaningful lifespan, an approach that will lead to innovation in the production, use and discarding phase. As Armstrong et al. (2015) pointed, the most promising design strategy is the one that merges products with service elements, Product-Service System (PSS). This strategy is not a new idea, but the potential of services to support long-term engagement with fashion products to enhance sustainability is very promising (Armstrong et al. 2016).

To make a new sustainable model in the fashion field the quality of products is determinant, only a shift to a more highly skilled, craft-based production will permit the extension of clothing lifetime through repair and reuse (Cooper 2005). Moreover, in a model claiming for a reduced demand of products it would provide new employment opportunities (Cooper 2005) and therefore brings credibility to the business model. Solutions such as co-creation, customization and halfway products, for the production phase, services such as repairing, renting and upgrading, for the use phase, and upcycling or recycling for the disposal phase are being suggested as key to the new model (Niinimäki & Hassi 2011).

At the same time, deeper relationships between products and consumers are needed to achieve deeper attachment and care for products. As Niinimäki (2011a) clarifies "if the designer can connect the design outcome deeply with a consumer's emotions, identity construction, aesthetic needs and personal memories, that is, values and lifestyle, the design process can achieve a deep product satisfaction and product attachment."

It's impossible to ignore the contrast between the long-term perspective from sustainability and the change promoted by fashion for purely semiotic or symbolic reasons (Koefoed and Skov undated) yet according to Walker (2006), that doesn't mean that these two will necessarily come into conflict, in the opposite, fashion may even be a key element towards a more sustainable way of living. Even if it seams an antithesis to all concepts we have about fashion, the reality is that fashion is one of the best tools to shape lifestyles and cultural values (Fletcher 2008), hence the best tool to change current behaviors to more sustainable ones.

2 RESEARCH QUESTION

Fashion is expressed and worn by people as a material product, but fashion is much more about feelings than about the product itself. Fashion has the power to create a sense of individuality and belonging at the same time, as two different narratives in the same character. So, the main question of this work is:

How should a new sustainable fashion system be designed in order to engage the consumer with long-term products without loosing the sense of fashion individuality and belonging?

Fashion is the biggest cause of discharge of textile material, so the main objective of the work is to understand how we can adapt fashion products to the changing needs of people, and how we can address sustainability in the fashion industry once is proven by several studies (Niinimäki, 2011a and Joy et al. 2012) that even people with a stronger commitment with sustainability and sustainable habits don't feel the same when it concerns to fashion.

To answer this question a bibliographic review in sustainable fashion and textiles will be presented, focusing especially in the contribution of participatory design during all phases of product lifespan, to promote engagement, value and meaning.

3 DESIGN STRATEGIES TO REDUCE TEXTILE WASTE

Kate Fletcher (2008) argues that the present model of fashion consumption is too passive, the result of two simple actions: to buy and to have. On the contrary, to foster lasting relationships, it's necessary for the actors to remain active, so, the fashion consumption model should forecast a way to engage the consumer with the garment, what suggests a change in the way fashion is consumed, not only produced. Fletcher (2008) suggested that open design (participatory design, co-design, etc) should be key. Putting the consumer in the middle of the design process, products will be suited to individual needs and therefore will probably fulfill consumer's needs longer.

Other authors like Van Nes (2003), Mugge, Schifferstein & Schoormans (2004) also see in some strategies of product design such as "Do-It-Yourself", "Do It With Others", or Product Personalization, for instance, a way to improve CPA – Consumer-Product attachment, once all the efforts stimulated by the product bring significant synergies for the people involved. The product motivates the "live creature" to create its own experience, and as long as the experience is considered valuable, the product is considered valuable itself and become memorable overtime. Memories and history can only be part of a product through its use, something that a new replacement product is absent, so when creating a product is important to enable participation and experimentation in order to facilitate memories and therefore to cool down the replacement (Gulden 2013).

In this section will be presented design strategies to extend product lifespan during production, use and discard phase.

3.1 *Textile design and production*

The current fashion business model is driven for a faster ever-changing production phase (Alastair and Hirscher 2013) in order to keep up with new trends and consumers desires.

As Niinimaki (2013) pointed, "most clothes purchases are driven by a need to fulfill emotional desires or vanity needs rather than a real need for new clothing", this type of purchase rarely contributes to a truly relevant experience capable of creating attachment between the product and the consumer. In the opposite experiences that involve investment of something from the consumer, establishes a bond that last much longer (Cramer 1999).

In this sub chapter, there will be presented some design strategies that could be applied in the production phase to create scenarios that facilitate this connection and therefore postpone textile discarding.

3.1.1 *Co-creation*

To create emotionally durable design, a deep understanding of consumer needs and values is crucial, once the main objective is to design a product that will be meaningful for a long period of time and difficult to detach (Niinimäki & Hassi 2011). To better answer to this costumer's needs and values, they should participate in an open-ended design process and in an open-source design system, to feel engaged within a collaborative action. Consumers should become co-creators, changing the current relationship between consumer – designer – fashion product (Kozlowski 2012). With an active role, they will more easily develop a sense of attachment with the product, emotional fulfillment and satisfaction.

There are already some innovative fashion businesses using co-creation as strategy. "Be the Buyer" is an example of it. The company Modcloth gives their customers the chance to vote for their favorite models to be produced and even share some tips with designers. Such kind of program allows customers to be more involved, feel heard and valued. At the same time the company offers their clients exactly what they wish (consumer value creation 2015).

But this strategy is not only applied by new brands. For instance, Burberry, an iconic luxury brand, also saw in co-creation a way to be closer and connected to their costumers. Together with salesforce.com they re-invented the interactions between store's sales, services, people and customers as a mini-community. The Burberry platform unleashes mutual emotions and shares important data with both the brand and the customer. Once again, the customer feels engaged with the brand, he may be asked to give suggestions on a new product or participate remotely on a fashion show. The brand will win by getting the feedback the costumer shares directly with it or in social media with others (Gouillart 2012).

3.1.2 *Customization*

The customization strategy allows models that combine products and services to be capable of fulfill consumer's, manufacturers' and other stakeholders' expectations and satisfaction. There are interesting solutions, specially linked with on-line purchases of customization. This kind of solutions, so far seem focused to sell more products, and not to really offer a role in decision-making to the costumer (Niinimäki 2011b) once they are deeply related with small changes (e.g. colors, style) but, its potential towards CPA (consumer-product attachment) and long-term textile products it's undeniable.

Personalized or customized products favor the creation of a deeper emotional connection with the consumer. Allowing the consumer to add personal meaning to the product facilitates the feeling of attachment (Chapman 2009).

There are several mass customization options in the web. These options use fast, flexible and digital manufacturing technologies associated with computer-aided design (Niinimäki & Hassi 2011).

Appalatch (https://appalatch.com/) and Unmade (https://www.unmade.com/) are two brands dedicated to 3D knit sweaters made on demand. The costumer is able to design his own sweater defining its style, pattern and color. He will feel unique with a product that it's all he desired; the experience of being part of the process has great potential to guarantee a long-term relationship. On the other hand, the brand avoids stocks while is always aware of consumers' preferences.

3.1.3 *Open Source*

Open source fashion is another way of co-creating and the result of the new digital era. In this kind of strategy, the designer sell his skills trough patterns and instructions, but it's the user that does the construction itself. The consumer leaves his passive role to become an active maker, with a leading role in decision-making. For this reason, this kind of design strategy in not led by large companies or retailers (Niinimaki and Hassi 2011).

A great example to this model is the project "The living wardrobe" from Jo Cramer, where the wearer is encouraged to interact with the garment over time to repair, modify and update it. The user will not only get a product that fulfill his needs, but will also be empowered with new skills and knowledge (Cramer 2014).

3.1.4 *Halfway products*

Another strategy, that in what concerns to users' involvement is very like the last one, is the concept of halfway products. In this one, products are provided in kit-bases, enabling the users to be the builders and develop inner knowledge about the product, which may be a key factor in order to be able to repair it in the future, if needed. With this method, the user adds his preferences, creativity and even memories to the product, and experiences a deeper connection with the product as result of his effort (Niinimäki 2011b). On the other hand, a kit-based product is designed for disassembly and reassembly of modules, which offers great possibilities for evolutions in time: upgrading, modifications and repair, as well as less laundering (Fletcher 2008).

Make(able) is an example of an open collaborative project that encourages everyone to become active makers of their own wardrobe. They perform half-way clothing workshops that offer customers an easy start with sewing, while creating real unique design pieces for themselves. The customer/maker will be able to customize and finalize unfinished designer products they have, becoming an active participant of the design process.

3.2 *Textile use*

It is estimated that the use phase of a product is responsible for 75–95% of its total environmental impact (Sherburne 2009). Moreover, this is the phase of product lifetime we have less information about both use, care and consumer-product relationship, although this is a critical phase to understand product replacement (Mugge et al. 2010).

For the use phase, the current work will focus, especially, in services strategies that added to products may improve it and therefore extend its lifespan.

Services in the design fashion field aren't really a new thing; in the past made-to-measure was a regular option. With this service, the costumer used to be an active part of his garment since its design, the fact that he may have waited quite some time for it and pay a higher price creates naturally feel of connection. For the same reasons these products were well taken cared and repaired when needed (Cramer 2014).

Nowadays, there is a several number of consumers returning to the engaging process of that business model due to the possibility of having a different product, customized and made especially for him. With internet and digital technology, this kind of business has evolved and spread globally (Cramer 1999), which is a win for both costumers and brands. To the firsts ones is offered a perfect fit, both physically and emotionally, which predicts a longer product life (Niinimäki & Hassi 2011); to the brand these techniques enable savings in materials and stocks, once production depends on existing orders (Niinimäki 2009).

Even if services added to the product at the design phase are promising for its life extension, only a shift to a business model that combine services with products will allow a deeper knowledge of the consumer and monitoring their changing needs. Changing product value to use value will create new opportunities laid in product quality and durability (Mont 2002), with consumer satisfaction at the center of focus. We are talking about adding services that will promote product lifespan trough upgrading, updating or repairing. Most of all, these services postpone the psychological obsolescence of the product (Niinimäki 2011a). On the other hand, also services as renting or leasing could offer innovative relationships consumer-products-brands.

3.2.1 *Upgrading*

Services that offer the possibility of upgrading a product have great impact in extending product lifetimes through satisfaction. As said before the main cause for fashion discard is related to changing needs and aesthetics preferences, so a service that offers the possibility to redesign or to modify the product allow to reverse this premature obsolescence. Furthermore, these services offer new business opportunities on a local and global level (Niinimäki 2011b). This kind of services can work together with strategies like halfway products and open source to get more value.

A very interesting project that mixes the open-source strategy and upgrading is Recyclopedia, from Otto van Busch, where step-by-step tutorials on how to update the garments dying in the wardrobe are suggested. A library of documented methods of updating clothes could be downloaded and printed from the platform, by anyone and developed as a DIY project (Busch 2008).

3.2.2 *Repair*

Repair services were very common in the past, but once fast-fashion spread all over, poor quality of materials and cheapness of products have made it no longer worth to mend. Moreover, in the new generations there is a lack of skills to carry minor amendments, so that is an increasingly less frequent act, and most often limited to small tasks like sewing a button (Holroyd 2013). Even with a pay service for repair people feel reluctant with this solution due to a social stigma (McLaren & McLauchlan 2015), feeling fear of being seen as poor.

In order to go against all these stereotypes some projects emerged. The Reknit Revolution (https://reknitrevolution.org/) run by Amy Twigger is one of them. In this project, the mending stays visible by choice in any kind of knit pieces. The mending is converted in an updating or upcycling, and once the consumers sees it like a new piece, the repair process is worth it and less pejorative.

Other activities like Love Your Clothes (LYC), a public event that happened in Scotland in 2014 run by McLaren & McLauchlan, tried to educate the consumers to engage with older products trough repair and how this strategy can empower not only the garment but the owner itself. This project has become a platform (http://loveyourclothes.org.uk) to help raise awareness to do better clothes choices – buy, care and repair, refashion and upcycling, and discarding.

3.2.3 *Renting and leasing*

Renting and leasing are used-oriented services characterized by a lack of personal ownership – companies retain product ownership and consequently its responsibility, (Style 2015) while costumers only pay for the use of the product. This kind of strategy "allows products to be used more intensively, which reduces the number in circulation and the use of the old, inefficient models, and removes a supplier's incentive to curtail life spans" (Cooper 2005), at the same time suppliers must design better products, take care of them longer and find creative solutions for product's end of life.

Product renting involves sequential use by different consumers which may contribute to a less careful behavior and therefore lead to higher impacts, while leasing involves a single consumer that is only paying for the use of the product (Tukker 2015). These strategies are well disseminated in products like cars or videos, but in fashion and textiles are only now having a substantial growth, once fashion products are intimate, and issues like hygiene are raised.

A great example on fashion leasing is Mud Jeans (www.mudjeans.eu), a dutch company that offers a leasing contract for jeans made of organic and recycled cotton. On the first purchase the customer pay a €20-member fee, then he can choose his jeans for a €7,50 monthly fee with one year duration. In the end of this year he has three options: keep the jeans, switch for a different model (keeping paying the €7,50 fee), or send it back to the company and receive a €10 voucher to use on a next purchase. The company is responsible for the recycling of jeans, so this strategy allows them to stay the owners of the raw material, getting them after use. The user is paying for the use and is able to change the product according to his changing needs.

A totally different approach, that doesn't have sustainability in its value proposition but instead is offering accessible high fashion, is Chic by choice (www.chic-by-choice.com). This Portuguese company rents luxury dresses from well-known brands and designers to special occasions, such as wedding, cocktails, proms, etc. The costumer only pays for the use of the dress, the company is responsible for its repairs and laundry.

3.3 *Improvement of old textile*

One of the main concerns with the current fashion system is the excess of textile waste created daily all over the world. It's known that if clothes were worn for another nine months than what they are, making a total lifetime of about three years, a 20-30% reduction on carbon, water and waste footprint would be reachable (McLaren & McLauchlan 2015). So, end-of- life strategies are as much important as design and use ones.

Reuse, recycle and upcycle are being integrated in new and old business models, and are key towards a sustainable fashion system.

3.3.1 *Re-use*

Re-use is, above all, the redirection of discarded products, that are still in great shape, to new owners, without needing any repair intervention (Goldsworthy 2012) – second-hand products.

Even if this concept of second-hand isn't new in retail, there are some social stigma related to it. So, changes in consumer behavior are still needed to support new business models.

Brands are focusing in the re-use concept to strengthen the relationship they have with existing customers and, also, to reach new markets. Take-back schemes have been implemented in-stores, supported by resell/reuse platforms. This kind of approach is mainly chosen by premium and high fashion brands with higher quality products.

Reuse should be a key theme in brand strategy towards sustainability, that is the first step towards brand's responsibility for the end-of-life of their products (Hvass 2014; 2015).

One of the first retail brands opening its own second-hand store was Filippa K, a Swedish company with a philosophy built around core values like style, simplicity and quality. They have been focusing along the years to produce trends free, quality clothes, which brings great opportunity for long-term clothes as well as for reuse (Piegsa 2013).

Back in 2008 they opened an exclusively Filippa K. second-hand retail store in collaboration with Judit's Second Hand, a successful consignment second-hand store in Stocklom. They are currently developing a system that allow all Filippa K. stores to accept take-back products and working to spread the concept to other second-hand stores (Hvass 2015).

3.3.2 *Upcycling*

The main goal of upcycling is to provide a solution to waste, "by optimizing the lifetimes of discarded products from an inefficient system" (Han, Tyler, and Apeagyei 2015), in other words, to give value to waste (Richardson 2011).

In what concerns to fashion design, upcycling means creating products with greater value from textile waste, a concept that gives both designers and consumers the opportunity to increase satisfaction (Han, Tyler, and Apeagyei 2015) (Sung, Cooper, and Kettley 2016), while being environmental friendly.

There is a different approach to upcycling if done from consumers' or brands' point of view. For this work it is important to distinguish them, since they have different results regarding attachment.

For either players design is the agent of transformation (Goldsworthy 2012), however individual upcycling provides for more active consumer involvement, whether in design alone or also, and especially, in its physical transformation. What, as in previously presented strategies, favors the attachment consumer-product due to the experience coming from the work itself as well as the resulting learning. From the environment point of view is also the best approach, once avoids transportation and could save more embodied energy (Sung 2015).

The main advantage that companies and brands have, to upcycle, face to consumers is the ease of access to other textile waste in addition to the used products, i.e. production waste, returned defected products or even leftovers from stores (Aus 2011).

Upcycling, as brand initiative, is usually related to reinvigorate the old into something new. With a great design component, this method offers the costumer a uniqueness feeling, supported by one-of-a-kind product, very like to a customized one (Piegsa 2013) which presents a huge impact in CPA.

It should be noted that the uniqueness offered by upcycling, and so valued for attachment, has an inconvenience, the fact that it does not offer choice in terms of size, and hence cannot respond all consumers (Aus 2011).

Upcycling is already used as a post-production process in some market niches, a good example of it is Alabama Chanin. A company with a simple philosophy: "use local materials and local skills to create garments that can meet the ambitious biography of 'Hand-sewn in America'"(Hemmings 2010). With a production that does not aim mimic the mechanical processes, the garments are hand sewn by local women, inspired by the local quilting tradition, with use fabrics from local thrift stores and more recently from the Salvation Army (Goldsworthy 2012).

3.3.3 *Recycling*

Recycling is the last strategy presented in this work, because it should also be the last strategy for textiles. Even if in the first recycling it is possible to obtain a quality product, the truth is that after repeating the process for sometimes the quality is lost to the point of being impossible to do it again (Goldsworthy 2012). On the other hand, recycling also implies high environmental impacts resulting from re-manufacturing processes and transport. This doesn't mean that recycling isn't a strategy to consider reducing and give new value to textile waste, on the contrary.

4 CONCLUSION

This paper has presented a set of strategies that could be introduced in all products stages (design, production, use and discard) in order to extend its lifespan and prevent textile waste. These strategies are one of the best contributions to a more sustainable fashion and textile system since they not only consider the reduction of environmental impact, but because, above all, they represent an opportunity for behavior change.

Looking at the current fashion system, the main problem to combat towards sustainability is the reduction of waste, and recycling in not a sufficient answer to the scale of the problem. A holistic and systemic approach that places the consumer at the center of the solution seems to be the right way to go. All stakeholders in the fashion system must recognize that consuming is part of live in the developed world, and that just asking to stop consumption is utopian, so, even if the presented strategies may not lead to more sustainable practices alone, they have a great potential to strengthen relations between consumers and their products, and therefore avoid premature obsolescence.

For brands and designers, it is imperative to think of products that allow the creation of emotional bonds, the best way to get it is by combining this products with services. Services that require more effort and commitment from the consumer and that gave them a co-creation role have more value. Moreover, these services also represent changes and innovation in business.

To conclude, it may seem difficult to implement these strategies in a system that is not yet ready for such profound changes, however there is already a niche market working for also a niche target that uses them as value prepositions, some examples of them were presented in this work. This proves that consumers' behavior can be re-educated, and that innovative and value-differentiated proposals can be an attractive to change.

REFERENCES

Alastair, Fuad-Luke, and Anja-Lisa Hirscher. 2013. "Open Participatory Designing for an Alternative Fashion Economy." *Sustainable Fashion: New Approaches* 459 (June): 174–82. doi:10.1038/459915a.

Armstrong, Cosette M., Kirsi Niinimäki, Chunmin Lang, and Sari Kujala. 2016. "A Use-Oriented Clothing Economy? Preliminary Affirmation for Sustainable Clothing Consumption Alternatives." *Sustainable Development* 24 (1): 18–31. doi:10.1002/sd.1602.

Aus, Reet. 2011. *Using Upcycling in Fashion Design. Design.*

Busch, Otto von. 2008. *FASHION-Able: Hacktivism and Engaged Fashion Design*. Edited by Art Monitor. Gothenburg: Johan Öberg.

Chapman, J. 2009. *Emotionally Durable Design*. Objects, Experiences & Empathy. London: Earthscan.

Cooper, Tim. 2005. "Slower Consumption – Reflections on Product Life Spans and the 'Throwaway Society.'" *Journal of Industrial Ecology* 9 (1–2): 51–67.

Cramer, Jo. 1999. "Made to Keep: Product Longevity Through Participatory Design in Fashion." *Design Principles and Practices: An International Journal* Volume 5 (Issue 5): 437–46.

Cramer, Jo. 2014. "Wear, Repair and Remake: The Evolution of Fashion Practice by Design." *Conference, Shapeshifting*, no. April: 1–18.

Fletcher, K. 2008. *Sustainable Fashion & Textiles*. London, UK: Earthscan.

Fuad-Luke, A. (2009) *Design Activism: beautiful strangeness for a sustainable world.* London: Earthscan.

Goldsworthy, Kate. 2012. "Design for Cyclability: Pro-Active Approaches for Maximising Material Recovery." *Making Futures Journal* 3.

Gouillart, Francis. 2012. "Co-Creation: The Real Social-Media Revolution." *Harvard Business Review*, 2–5.

Greenpeace. 2016. "Timeout for Fast Fashion."

Gulden, Tore. 2013. "Modelling of Memories through Design." In . doi:10.13140/2.1.4880.0647.

Han, S, D Tyler, and P Apeagyei. 2015. "Upcycling as a Design Strategy for Product Lifetime Optimisation and Societal Change." *Product Lifetimes And The Environment*, no. June: 130–37.

Hemmings, J. (2010) Alabama Chanin: Hand sewn in America. *Embroidery* 61 (7), 16–21.

Holroyd, Amy Twigger. 2013. "Not the End of the Journey: Exploring Re-Knitting as A 'craft of Use.'" *Making Futures Journal* 3.

Hvass, K.K. 2015. "Business Model Innovation through Second Hand Retailing, a Fashion Industry Case." *Journal of Corporate Citizenship* 2015 (57): 11–32.

Kant Hvass, K. (2014) Post-retail Responsibility of Garments – A fashion industry's perspective. Journal of Fashion Marketing and Management. 18 (4), 413–430.

Koefoed, O., Skov, L. (undated). Sustainability in Fashion. *Openwear: Sustainability, Openness and P"P Production in the worl of fashion.*

Kozlowski, Anika. 2012. "Sustainability Driven Innovation and Fashion Design."

McLaren, A., McLauchlan, S. (2015) Crafting sustainable repairs: practice-based approaches to extending the life of clothes. *Proceedings of the PLATE Conference*. 17–19 June, Nottingham, UK.

Mont, O. 2002. Clarifying the concept of product-service system. *Journal of Cleaner Production*. 10 (3), 237–245.

Mugge, R., Schifferstein, H., & Schoorman, J. (2010) Product attachment and satisfaction: Understanding consumers' post-purchase behavior. *Journal of Consumer Marketing*. 27 (3), 271–282.

Niinimäki, K., 2009. Developing sustainable products by deepening consumers' product attachment through customizing. *Proceedings of the World Conference on Mass Customization & Personalization*. 5–6 October, Helsinki, Finland.

Niinimäki, K. 2011a. *From Disposable to Sustainable: The Complex Interplay between Design and Consumption of Textiles and Clothing*. Edited by Aalto University. Helsinki. https://aaltodoc.aalto.fi/handle/123456789/13770.

Niinimäki, K., 2011b. Sustainable Consumer Satisfaction, in: Vezzoli, C., Kohtala, C., Srinivasan, A., (Eds.), *Product-Service System Design for Sustainability*, LeNS publication. Greenleaf, Sheffiel.

Niinimäki, K., Hassi L. 2011. Emerging Design Strategies in Sustainable Production and Cosnumption of Textiles and Clothing. Journal of Cleaner Production. 19 (16), 1876–1883. doi: 10.1016/j.jclepro.2011.04.020.

Piegsa, Edith. 2013. *Green Fashion*. doi:10.1007/978-981-10-0111-6.

Richardson, Mark. 2011. "Design for Reuse: Integrating Upcycling into Industrial Design Practice." *International Conference on Remanufacturing*, no. Ipcc 2006: 1–13.

Sherburne, A. (2009) Achieving Sustainable Textiles: A Designer's Perspective. *Sustainable Textiles: Life Cycle and Environmental Impact*. Cambridge, UK: Woodhead Publishing Limited/The Textile Institute.

Style, Just. 2015. *Sustainable Fashion. The Guardian*. doi:10.1038/459915a.

Sung, Kyungeun. 2015. "A Review on Upcycling: Current Body of Literature, Knowledge Gaps and a Way Forward." In *International Conference on Environmental, Cultural, Economic and Social Sustainability*, 17:28–40.

Sung, Kyungeun, Tim Cooper, and Sarah Kettley. 2016. "An Alternative Approach to Influencing Behaviour: Adapting Darnton ' S Nine Principles Framework for Scaling up Individual Upcycling." *2016 Design Research Society 50th Anniversary Conference*, no. 2007: 1277–90. doi:10.21606/drs.2016.391.

Tukker, Arnold. 2015. "Product Services for a Resource-Efficient and Circular Economy – A Review." *Journal of Cleaner Production* 97. Elsevier Ltd: 76–91. doi:10.1016/j.jclepro.2013.11.049.

Van Nes, N. 2003. *Replacement of durables: Influencing product lifetime through product design*. Rotterdam: Erasmus University.

Vuletich, Clara. 2012. "'We Are Disruptive': New Practices for Textile/Fashion Designers in the Supply Chain." *10th European Academy of Design Conference – Crafting the Future*, 1–14.

Walker, S. 2006. Sustainable by Design: Explorations in Theory and Practice. London: Earthscan. https://consumervaluecreation.com/2015/04/05/co-creation-in-the-online-fashion-world-modcloth/acessed 04/05/2017.

Textiles, Identity and Innovation: Design the Future – Montagna & Carvalho (Eds)
© 2019 Taylor & Francis Group, London, ISBN 978-1-138-29611-4

Sustainability through traditional processes: Strategy of "Salva a Lã Portuguesa" for revival of natural wool

M. Pacheco & E. Bazaraite
CERIS, Instituto Superior Técnico, University of Lisbon, Lisbon, Portugal

ABSTRACT: Technological development of this century must step forward by looking back: traditional processes end up offering a higher level of sustainability and market homogeneity. A case of wool is an example of malicious cycle of neglecting the sheep breeders of small scale, high percentage of synthetic fibres in clothing industry that implement complicated industrial processes for deriving a product. What is more, traditional processes, accessible and easy to do, are being forgotten as the industrial products take place in all the daily aspects. By disappearance of wool industry in Portugal, sheep breeders were subjected to selling the wool to foreign industries; however, the breeders of small scale, not able to produce sufficient quantities for reasonable sales, would end up destroying the wool. Portuguese wool is sold abroad for later receiving wool products back to be sold in Portugal. On the other hand, a necessity to produce with our own hands, serving both as therapy and a way to fight against consumerism, has become more and more popular during the last decades. People look for handmade activities yearning for capacities to both plant their food and make their clothing. The paper portrays the current context of sheep breeding and wool market in Portugal, and describes the approach of "Salva a Lã Portuguesa" project to the importance of local sheep breeds, the traditional wool treatment processes and social aspects of such activities. The project "Salva a Lã Portuguesa" intends to offer natural wool supply of local Portuguese sheep breeds, giving access to the knitters, spinners and felt makers to the high quality raw material that justifies the manual process of creation. Working with small quantities of wool would enable a production of yarn types, differentiated by sheep breeds and variation of spinning processes.

1 INTRODUCTION

An extremely quick technological development after the industrial revolution must be seen with its pros and cons. The comfort of rapid production, access to all kinds of foods, textiles, and building materials *inter alia*, lead to globalization and a lack of variety throughout the global market.

The whole problem of wool embraces several changes of the 20th century society: stratified economy of the first and the third world, technological development and invention of new cheaper textiles, a change of work routine, leading more people to pursue a career out of home and leaving home crafts behind – this way small scale sheep breeders had no use for their wool considered as "bad wool" for the industry. Sheep shearing became a mere necessity to keep sheep healthy, and wool was thrown away, burnt or sold for an absurdly low price.

After industrial revolution wool production was confronted by two heaps of scale: first it passed from manual to local industry and second from local industry to global industry, leaving small scale local producers with no empowerment to sell their raw material. Every change had its criteria of wool quality, differentiating good and bad wool. "Bad wool" meant unacceptable raw material for industrial process, however it continued to be suitable for artisanal purposes. During the first half of the 20th century these artisanal crafts continued alive, but the great industrial heap, introducing cheap and accessible textiles and clothing of low quality, led to abandoning natural wool because it was expensive or it required long hours of work at home.

Why we find important to "Save the Wool"? First, wool is already an existent resource as the sheep are still grown for meat and milk and wool comes "in a package". Throwing wool away means abandoning raw material that has various uses: textile industry and civil engineering. And if already existent wool is not used, other materials must be produced to satisfy these needs – it's an unsustainable cycle of production.

Second, different sheep breeds have different qualities of texture, colour and comfort. This variety must be preserved as part of the natural world valuing all the species. This variety can be appreciated not just by seeing, but also by using wool textiles in everyday life (D.G.A.V., 2013: 125). Variety of wool qualities is a result of centuries of genetic development. Today there are more than 15 native sheep breeds in Portugal, and each of them tells a piece of genetic history.

Third, wool is an amazing material for its qualities, mainly its thermal and moisture properties (Laing, 2016: 26). It has been lately announced as the top

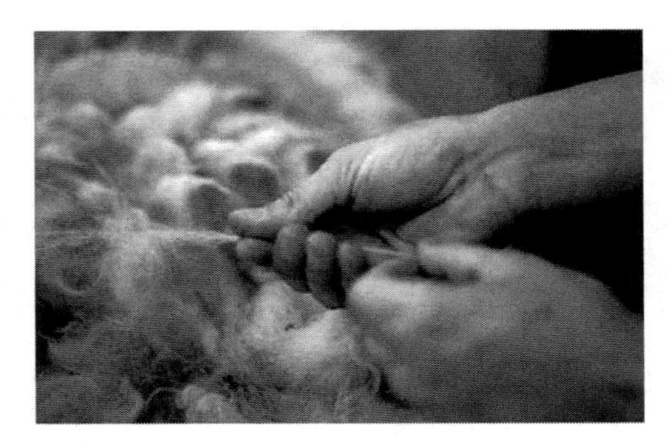

Figure 1. After shearing of Churra Galega Mirandesa, checking the length of the fibers.

choice for uniforms of firemen and soldiers. The reason is a great heat resistance: "Subject wool to a flame and it will hardly burn, only char" (Grove, 2013: 71). What is more, "wool manages to moisture, it's open and porous and it doesn't stick to the skin", this means that wool properties make it "the coolest fibre to wear even on the hottest day" (Grove, 2013: 73).

Fourth, wool and textiles are part of local identity, and promoting local crafts can bring an empowerment to rural communities. As ruins of castle or Palaeolithic drawing on stone is conserved for future generations, a fragile rural environment requires care as it is about to disappear, taking away a millenary knowledge of survival on this planet. This care must be sustainable – the crafts cannot be turned into a mere tourist attraction, but they must be useful for producing durable items of high quality, able to be used throughout the future generations.

A project "Salva a Lã Portuguesa" ("Save the Portuguese Wool") aims to use locally grown wool giving an opportunity to small-scale sheep breeders to sell their production. Our aim is to offer the knitters and designers yarn differentiated by the sheep breeds, enriching the market with more options, and avoiding wool waste. A promotion of manual wool spinning intends to empower people to produce their own yarn and maintain the craft alive. All in all, we aim for intelligent use of high quality resources, sustainable production – natural, regenerative and local, and revival of wool textile crafts as part of local identity.

2 A NECESSITY TO PRESERVE NATIVE BREEDS

By the end of the 20th century many wool factories in Portugal were closed down, leaving the breeders with their wool unused. Portuguese wool production started to be exported to Spain or even further. The local interest in wool products lowered as well, leaving the sheep breeders with wool necessary to be sheared every year without receiving or receiving very little for the wool itself.

Considering *wool problem*, that embraces various issues of cultural identity and industrial effectiveness, "European Group of Wool" was founded in the late 1990's with two main goals: 1) to preserve the identity of traditional landscape and local culture; 2) to strengthen production systems that respect the environment and improve the wool industry. At the same time, "Journeys of Merino sheep breed" were organized in Portugal, discussing issues related to sheep farming in Portugal, native breeds, the characteristics of herds and the use of wool in relation to its variety, quality and quantity (Várzea Rodrigues, 1999).

Fifteen years later a problem of lack of interest in wool and its commercialization led to a worrying situation: there are only few experts in wool classification and typification and technicians able to prevent changes to the genetic basis of sheep, and this situation endangers the quality of wool that is being produced now.

Due to various ethnic groups of sheep races, the quality of wool produced in Portugal is heterogeneous. Commonly, Merino breed is considered to have the highest quality in relation to other wool produced in Portugal. This kind of wool, characterized by wavy fibres and soft touch, is produced in the centre and in the South of the country, and is chosen for industrial processes and clothing production. The wool that is considered lower quality in the perspective of industry, because of its straight fibre and rough touch (e.g. Churro breed) is left for the local and regional multi purposes like artisanal production, crafts, decoration, from the rustic style to contemporary design (Várzea Rodrigues, 1999: 78).

The different characteristics of wool fibres and the industrialization of the process of treating wool and spinning into yarn, originated the classification of "good wool" (Merino ethnic races), which are suitable for industrial process, and "lower quality wool" (due to its fibre characteristics) that are difficult to adapt to the industrial criteria. The latter was left outside of wool market, leading the breeders to getting rid of it by throwing away or simply burning.

Portugal has been using a system of wool gathering, classification and storage since 1940. This system, only applied for the Merino breeds, is under control of sheep producers' associations, divided by the geographic spread of breeds.

A problem of the decreasing value of wool is obvious in the numbers published in the D.G.A.V. reports. The percentage of gathered wool of all produced wool varies notably, being just 9% (when statistics is around 15/20%). The last decades revealed even higher descent of the gathered wool (up to 50%) in relation to wool production (Várzea Rodrigues, 1999: 79). It means that the majority of producers do not deliver wool in the associations and do not participate in wool auctions that happen few months after sheep shearing, opting for direct sell to an intermediate person, even receiving a lower price for kg without evaluating wool quality. The price difference in wool auctions

Figure 2. A fleece when clipped off a sheep and laid out, resembles the shape of the animal. It is then rolled up for transportation, and is unrolled before sorting the wool.

Figure 3. Older generations are still able to process wool at all its stages. It is important to learn with them before it is too late.

and direct sell can be up to 50% (Várzea Rodrigues, 1999: 82).

Still, a necessity to increase milk production of the herds led some producers to crossbreed Merino female sheep with males of exotic breeds (such as Frísia, Manchego, Awassi, Assaf and Lacaune). This had negative effects on wool quality.

In Portugal a size of flock of sheep is directly related to geographical environment. Big flocks are grown in extensive grasslands common in Alentejo region, and small flocks are grown in the mountainous regions like Trás-os-Montes or Serra da Estrela (Ribeiro, 2001: 122–127).

Portugal has a high number of native breeds, representing centuries of genetic evolution of adapting to the environment. According to the most recent inventory, there are 15 native Portuguese breeds (D.G.A.V., 2013) divided into 3 main ethnic breed groups: Merino, Bordaleiro and Churro (Várzea Rodrigues, 1999: 79).

The Merino ethnic group is divided into white Merino and black Merino of Alentejo, most common in the South of river Tagus' region; and Merino of Beira Baixa, spread in the North – interior region of the river Tagus.The Bordaleiro ethnic group is geographically distributed on the Northern side of river Tagus and is divided into Bordaleira of Entre Douro and Minho, Mondegueira, Saloia, Serra da Estrela and Campaniça (although some authors consider Campaniça as a different ethnic group as the results of genetic studies vary).

Churro ethnic group is divided into Churra Badana, Churra da Terra Quente, Churra do Campo, Churra do Minho, Churra Galega Mirandesa and Churra Galega Bragançana, most common in the Northern part of Portugal, and at the South – Churra Algarvia, genetically different due to its geographic separation.

3 BACK TO THE WOOL

During the last decade a global renaissance of interest in knitting and spinning inspired several start-ups. Besides various public meetings of knitting and spinning enthusiasts, small-scale wool industry was restarted. A small-scale productions seek to give use for wool of various sheep breeds, offering new quality of products for knitters and spinners, while bigger scale factories started producing pieces of clothing and home ware embracing the local identity and employing factory workers that left the closing factories in the late 90s and only got an opportunity to apply their knowledge again. This is a case of "Burel factory" and "Ecolã" factories in Manteigas, Portugal.

In United States the renaissance of wool crafts, embracing all the phases of wool treatment, took place already in 70–80s, leading to edition of high numbers of handbooks about the home scale wool works (Baines, 1977; Chadwick, 1980; Dixon, 1979; Kroll, 1981 inter alia). Until nowadays, there are still no extensive handbook in Portuguese regarding wool works, and interested enthusiasts search for information through the direct communication with the rural. Fortunately, there are still people who know how to treat wool from their own experience derived from the necessity of survival. Generational gap between the ones who know and the ones who want to learn exists. However, luckily, the knowledge is now acquired and has more chances to be passed to the future generations.

Small scale initiatives like "Saber Fazer", led by Alice Bernardo, "Qlã" by Fátima Gavinho Rodrigues, "Retrosaria" by Rosa Pomar, "Lhana" by Isabel Sá, are seeking the solutions that combine industrial, semi-industrial and manual processes of yarn production, bringing to market the products with local identity. Besides that, tools for carding, spinning and knitting are being reproduced following the ancestral techniques, which are still functional in the age of electrical tools. Several associations, like "Associação ALDEIA", based in the rural regions and promoting local traditional knowledge, started to give short courses of introduction to wool cycle, including sheep shearing, spinning, dyeing, knitting and weaving. These initiatives bring double benefits – knowledge

Figure 4. A good shearing is crucial to obtain the best of wool. Experienced sheep shearers are able to keep a fleece intact enabling better sorting.

Figure 5. Sheep breeders of very small flocks of sheep take more care of each animal. This leads to better quality of meat, milk and wool.

sharing, building bridges to connect generations and appreciation of rural society that has been fading to oblivion during last decades. In the century of high-speed the importance of slow-living concept is growing.

"Obrador Xisqueta" got even further. Besides the wool works, yarn and knitted clothing production, a shepherding school has been founded, bringing more attention to this ancestral profession necessary to keep sheep breed variety.

4 "SALVA A LÃ PORTUGUESA": CHALLENGES

A project "Salva a Lã Portuguesa" (Portuguese of "Save the Portuguese Wool") seeks for the solutions to produce yarn differentiated by breed. As different breeds have different wool textures, the wool can be used for producing a great range of clothing and home ware products.

"Salva a Lã Portuguesa" project aims to participate in all the phases of wool cycle. establishing a network that dialogues with sheep breeders, technicians, shearers and wool collectors, delivering wool to the delivery centres, as well as National Associations (ANCORME, ACOS, ACRO, ACOME *inter alia*) promoting a quality of breed.

"Salva a Lã Portuguesa" seeks for a direct connection with shepherds that have small flocks of sheep of pure breed. Small quantities of sheep in a flock enable a better care of each animal and attention to the wool quality. However, the field work done by "Salva a Lã Portuguesa" shows that this is not always a case: as the wool has been a forgotten raw material during 2 decades, the shepherds have a scarce knowledge of how to take care of wool while it's still on sheep, what is a correct shearing (because during this period shearing was a mere necessity to maintain sheep's health) and how to store wool correctly (fleece rolling).

Until the 80's, Portugal had a well-developed industry of wool processing with an experience of more than one century, concentrated in Castelo Branco and Covilhã regions. During the 90's most of them closed, the machinery was sold out to dismantle or simply abandoned.

As mentioned before, big part of Portuguese wool is currently exported and the small wool quantities that stay in the country is gathered in a few wool-processing factories that treats quantities of wool bigger than 1 ton, meaning that wool from different breeds and regions are joined and mixed to achieve the necessary quantities for industrial treatment. All in all, "generic" Portuguese yarn is produced, depreciating sheep races, local specificities and traditions and consequently shepherd's' pride on an exemplary flock is lost.

"Salva a Lã Portuguesa" is now working to find a solution that can work between big scale industry and small scale artisans groups, bringing to life old machinery that are still functional and able to solve the tasks of the scale of "Salva a Lã Portuguesa" wool cycle.

"Salva a Lã Portuguesa" starts its wool cycle in the sheep farm, following shepherds and their small breeds during the year, until the day of shearing. The ability and skills of the shearer are crucial to provide a complete fleece that facilitates the further scouring of wool parts, allowing choosing best quality fibres.

This wool is then taken in Covilhã to be washed, dyed, carded and spun in small factories, creating yarn from wool that was facing obliteration during the last two decades.

Simultaneously, in the "Salva a Lã Portuguesa" atelier in Lisbon, part of wool is treated in traditional manual way, using wooden tools such as carders and spinning wheels. The aim is simple: to teach the artisans and retain traditional ways of wool craft, as well as to produce special yarn 100% traditional and handmade.

Though a project could consider working only with the old people still able to card, spin and twist wool in perfection, it seems to be a momentary solution. Apparently these people are more useful teaching, while younger can work for production, gaining experience throughout the years.

Figure 6. Artisanal yarn produced by "Salva a Lã Portuguesa", spun by spinning wheel, white and black Merino from Alentejo.

5 CONCLUSIONS

To solve "wool problem" in Portugal and a lack of appreciation of it, require looking back to the traditional wool cycle joining it with contemporary industrial processes, from production and processing to trade. The final product, both artisanal and industrial, should be adapted to each market, to its quality requirements and processing conditions in a sustainable production practice.

One of the reasons why industrial process mixes all the different kinds of wool is because only big quantities of wool make the industrial process economically beneficial. As there are smaller numbers of sheep of certain breeds, they are either rejected, or they are simply mixed with other, producing a generic type of yarn simply designated as "wool". This fact led the project "Salva a Lã Portuguesa" look for small-scale options of semi-industrial wool process that would enable keep the wool of different breeds separated.

At this moment the solution of "Salva a Lã Portuguesa" leans onto the use of the surviving machinery of the old wool industries. As an alternative to the artisanal wool process and to the mass industry, it is necessary to establish an "in between" process, turning back to a national and local-scale industry that has almost disappeared in recent decades.

An answer to wool problem should point out to the issue of scale that has no aim to grow global. It means that a controlled and secure contraction of the global industrial process should be considered, respecting ecological limits of the planet and ensuring that the needs of current generations are met without compromising future needs.

ACKNOWLEDGMENTS

The authors of this paper wish to thanks the Salva a Lã Portuguesa Association for founding this project.

REFERENCES

Baines, P. 1977. *Spinning wheels: spinners & spinning*. London: B T Batsford Limited.

Chadwick, E. 1980. *The Craft of Hand Spinning*. London: A Batsford Craft Paperback.

D.G.A.V.(Direcção Geral de Alimentação e Veterinária). 2013. *Raças Autóctones Portuguesas*. Lisboa: D.G.A.V.

D.G.V. (Direcção Geral de Veterinária) 1997. Primeiras Jornadas das ovelhas Merina da Beira Baixa e Churra do Campo e da Cabra Charnequeira. In *Colectânea Sociedade Portuguesa de Ovinotecnia e Caprinotecnia 1997–1998*, vol. 8 nº1. Lisboa: S.P.O.C./D.G.V., 5–97.

D.G.V. 1998. Jornadas das ovelhas Serra da Estrela e Churra Mondegueira e da Cabra Serrana. In *Colectânea Sociedade Portuguesa de Ovinotecnia e Caprinotecnia 1997–1998*, vol.8 n°1. Lisboa: S.P.O.C./D.G.V., 99–276.

D.G.V. 1999. Jornadas das ovelhas de Raça Merina. In *Colectânea Sociedade Portuguesa de Ovinotecnia e Caprinotecnia 1999–2000*, vol.9 n°1. Lisboa: S.P.O.C./D.G.V., 5–101.

Dixon, M. 1979. *The Wool Book. A guide to spinning, dyeing and knitting.* Middlesex: The Hamlyn Publishing Group Limited.

Grove, S. 2013. Wooly wonkas. In *Monocle*, issue 63, vol. 07, London, 71–76.

Kroll, C. 1981. *The whole craft of spinning: from the raw material to the finished yarn.* New York: Dover publications.

Laing, R. & Swan, P. 2016. Wool in Human Health and Well-Being. In Fangueiro, R.; Rana, S. (Eds.), *Natural Fibres: Advances in Science and Technology Towards Industrial Applications From Science to Market*, vol. 12. Dordrecht: Springer, 19–34. [Accessed in https://link.springer.com/chapter/10.1007/978-94-017-7515-1_2, in 1st June 2017].

Matos, C. et al. 2010. *Caracterização das pastagens do Vale do Guadiana: contributos para a definição de boas práticas agrosilvopastoris*. Mértola: Associação de Defesa do Património de Mértola.

Ribeiro, O. 2011. *Portugal, o Mediterrâneo e o Atlântico*. Lisboa: Tema livre.

Várzea Rodrigues, J. et al. 1999. Produção de lã e de lã fina em Portugal. In *Colectânea Sociedade Portuguesa de Ovinotecnia e Caprinotecnia 1999-2000*, vol.9 n°1. Lisboa: S.P.O.C./D.G.V., 75–88.

Textiles, Identity and Innovation: Design the Future – Montagna & Carvalho (Eds)
© 2019 Taylor & Francis Group, London, ISBN 978-1-138-29611-4

Design of sustainable textiles through biological systems and materials – innovative narratives within the circular economy

G.S. Forman & C. Carvalho
CIAUD, Lisbon School of Architecture, University of Lisbon, Lisbon, Portugal

ABSTRACT: One of the greatest challenges of this century is the endeavour to guarantee commercial profit without placing excessive demand on natural resources and without disrupting Earth and all ecosystems. Climate change, resources scarcity, sustainable energy, waste and the design of resilient sustainable goods are critical issues. The textile industry is one of the most pollutant sectors in the world; fashion fast paced systems plus the superfluity of mass production and high volume of products at low prices has great significances both at the social and environmental levels. The issues within the industry are usually related to the little effective planning and designing of products and the general crisis in disposing textile waste. From a Circular Economy perspective, the understanding of material sources and flows is paramount and long-term view is key whilst analysing products and their attached values. Many varied studies concluded the application of scientific and technical advances in life science to develop commercial products is route to sustainable design. The intensification of biotechnologies as the foundation of design projects expresses approaches often involving living organisms at all scales, used to build, influence and improve the objects around us. Exploring biology as a vehicle to design not only offers a new vision for the development of products but also feeds the interest in learning the peculiar aesthetic values of biological resources.

1 INTRODUCTION

1.1 *Textile challenges*

Textile industry is a demanding sector, continuously challenged by a complex society. Challenges are vast, ranging from transformation paradigms to the great expectations of brands by consumers (Bendell & Kleanthous, 2006). General issues are connected to an economic crisis threatening commercial success; limited budgets, tight deadlines, policies, legislation limitations, obsolesce, etc. are amongst the main concerns (Fletcher, 2008). On the other hand, the sector's negative impacts on the ecossystem are varied; non-renewable energy, contaminations and waste are but a few of the sustainability problematic aspects (Bendell & Kleanthous, 2006); Fletcher, 2008; Klein, 2000, El-Hagar, 2010; Malik, Ghromann, & Akhtar, 2014). Due to the natural resources exploitation and constrains as well as related physical and political access, environmental issues have direct repercussions in the price volatility.

Furthermore, the fast business model effects and the relocation of manufacturing to the East in search for low-cost labour facilitates mass production and low priced products high volume (often of low quality), which has implications at both environmental and social levels (Koring & Arold, 2007).

Modern society is characterized by the overconsumption of goods. To operate in, and around, this fast culture, one of evolution, adaptation and change, it is crucial to understand in depth the real kinds of value attached to products and to grasp their lifecycle and different speeds of usage (Fletcher, 2008; Ellen MacArthur Foundation, 2012; Earley & Goldsworthy, 2015). Effective strategies to design resilient products and deal with the textile waste, which is suffering a crisis in Europe (European Union, 2010; WRAP, 2012), are paramount (Niinimaki & Hassi, 2011). Several challenges are forcing many changes, and one must "develop novel guidelines, methods and procedures for design and innovation towards sustainability" (Ellen MacArthur Foundation, 2012; Garcia-Serna, Perez-Barrigón, & Cocero, 2007).

Textile sectors not only have to deal with the reduction of ecological burden (with the depletion of resources having direct repercussions at the natural and economical capital levels) but also they must, ultimately, provide for wellbeing, equality and social improvement through their products.

1.2 *Overconsumption and textile waste*

Textile industries related waste is of varied origins (Fletcher, 2008; WRAP, 2012). The first bulk of waste occurs during the production phase (use of non-renewable energy as well as vast amounts of water and its contamination, by the use of toxic substances). The second bulk occurs in a pre-consumerism stage, not only with the fabric remains from pattern cutting but also with textiles distribution and packaging. Finally, the third bulk of waste happens in the end of life of the product, when the consumer no longer needs it and throws it to the bin to be incinerated. The non-biodegradability of certain materials plus the adding of toxic substances (as dyes or finishing products) are added concerns (Fletcher, 2008; Greenpeace, 2011, 2012, 2014).

The amount of waste generated is deeply linked to apparel low prices and super fluidity of mass production with the consequent overconsumption of textiles. This has repercussions in the clothing-recycling field, which has seen a decrease in the past years (European Union, 2010) as cheaper apparel lacks in quality and deteriorates quicker (Koring & Arold, 2007). Rapid consumption patterns are losing a lot of value to landfills and the search for alternatives is imperative; Several reports urge for the need of fresh policies that facilitates waste management and the importance of looking at waste no longer as a burden but rather as a valued resource (WRAP, 2012; Teijin, 2016; FOE Europe, 2013).

According to the European Union (2010), there are still varied strategies to explore regarding waste management and the effective planning of ecofriendly products as well as to encourage consumers' responsible behavior (European Union, 2010). Reduce, reuse, repair and recycling activities must be prioritised so the demand and extraction of raw material resources can be mitigated (European Union, 2010) FOE Europe, 2013; Niinimaki & Hassi, 2011). However, albeit reducing the environmental burden of textile industries and creating jobs, there is evidence of a decline in the collection of garments, as apparel has seen a decrease in material quality over the years (Koring & Arold, 2007). Indeed, numerous studies do indicate that there is a general crisis in disposing waste (European Union, 2010; Greenpeace, 2011, 2012, 2014; H&M, 2016) – no reuse/recycle targets, a lack of effective services in managing the different textile wastes, general misconceptions (about innovative strategies or technology) (Tomás, 2016) and policy or legislation limitations (H&M, 2016).

Additional issues in recycling textiles are largely determined by poor design – nature of the garments, the employment of unsafe substances (Greenpeace, 2011, 2012, 2014), low quality materials (Koring & Arold, 2007) or clothing often manufactured with blended fibres not allowing for their re-yarn or reprocessing into new suitable material, that could be use for the production of new apparel (up-cycling) (Fletcher, 2008; Teijin, 2016). Textile products that are designed without their end of life in mind are usually down-cycled (frequently shredded into products of lower value or when the original yarn is transformed into a lower value textile (Beasley & Georgeson, 2014).

These are all aspects that contribute to the low levels of reuse/recycle in Europe and the loss of valuable materials that could otherwise return to the industry (European Union, 2010; H&M, 2016). According to the Waste and Resource Action Program (WRAP) (WRAP, 2012) reports from the year 2012, around 350.000 tonnes of used clothing ends up in the landfill per year, in the United Kingdom only; this corresponds to an estimated £140 million worth in used garments (WRAP, 2012). A study from the entity The Friends of the Earth (FOE Europe, 2013) gives account of 5.8 million tonnes of textiles discarded every year, across the EU, according to the European Commission; only 25% of these post-consumer textiles is recycled by charities and industrial enterprises whilst the remaining 4.3 million tonnes goes to landfill or energy from waste (FOE Europe, 2013).

Low prices and reckless unlimited consumption has repercussions regarding the amounts of waste that is generated. Textile industry and related Design departments are still not fully prepared to create responsible, intelligently planned goods integrating culture, communities, environment and economy as well as the broad and long-term perspective on materials, innovative technologies and what each step of products lifecycle really entails (Fletcher, 2008; Bendell & Kleanthous, 2006; Greenpeace, 2011, 2012, 2014).

1.3 *Material sources and flows*

As noted by the Ellen MacArthur Foundation (2013), the drivers for change into a new economic model are largely determined by price and supply risks, natural systems degradation, economical losses and structural waste, material and technological innovation, amongst many others. A shift from the conventional linear economy to an alternative economy starts in accepting the limitations of the former and the opportunities of the latter.

The Circular Economy (CE) is restorative and regenerative by design and it is defined by two sets of principles – technical and biological materials. It is a framework where safe materials are disassembled and a) recycled as technical nutrients or b) composted as biological nutrients in two distinct closed-loop systems. In other words, technical materials are designed to circulate with minimal loss in quality (recycled in a cascading system) whilst biological materials are designed to re-enter the biosphere safely (Ellen MacArthur Foundation, 2012).

Regarding the sustainable imperative within the textile industry, studies reveal that textiles are one of the sectors that can greatly benefit from the cascading systems that define the CE (WRAP, 2012; Ellen MacArthur Foundation, 2012; H&M, 2016), creating value (from waste) and reducing the environmental footprint. From a CE perspective, the understanding of materials flows is paramount and the long-term view is key whilst analysing products/materials and their attached values (values of today and values of the future).

Deeper knowledge on both technical and biological materials offer great opportunities for intelligent product designing, particularly in the current environment of several technological advances in many

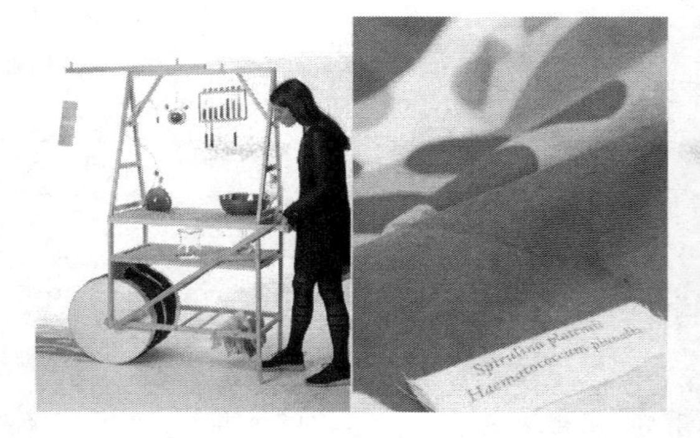

Figure 1. *Algaemy*, from *Blond&Bieber*. Printing textile withalgae-produced dyes.

distinct fields of research (3D printing, biotechnologies, etc.). The possibilities of innovative materials and technologies are endless (MoMA, 2012; Myers, 2012; Ginsberg, Calvert, Schyfter, Elfick, & Endy, 2014; van Wijk & van Wijk, 2014).

Figure 2. Varied leather-like material samples and textures; *Mycoworks* present an alternative textile made entirely by mushrooms.

2 MERGING SCIENCE AND DESIGN

2.1 *The biological approach – designing with nature*

Many varied studies, including the reports from the Organization of Economic Cooperation and Development (OECD), concluded that the application of scientific and technical advances in life science (biological technologies) to develop commercial products is route to sustainable design (OECD, 2013; Science Gallery, 2013). Regarding biological materials, the scope of science currently allows for completely new radical ways of producing materials (MoMA, 2012); innovative methods of manufacturing through natural resources such as, for example, bacterium. So, how can textiles enable for a more sustainable future?

With increasingly more technological innovation and scientific breakthroughs as well as collaborations between scientists and designers one has noticed in the past years the intensification of biotechnologies as the foundation of several design projects (MoMA, 2012; Myers, 2012; Ginsberg, 2009).

The symbiosis science/design plays a vital role as it strengthens the results for the quest of effective sustainable design of materials and products. Projects and their approaches often involve living organisms at all scales – animals, plants, microorganisms (bacteria, fungi, algae), cells, etc. that can be used to build, influence and improve the objects around us (e.g. Exhibitions "En-vie" or "Grow your own") (Baumel, 2009; Myers, 2012; Alive/En-Vie, 2013; Science Gallery, 2013; Lee, 2009; Blond & Bieber, 2014, Mycoworks, 2016).

The application of classic or modern biotechnology to the design process goes "beyond mimicry to integration" as mentioned by Myers (2012). Biological systems not only serve design by consulting organisms and ecosystems to apply their underlying design principles to its innovations, as in *Biomimicry* (Benyus, 2002), but also are essential components of the projects, boosting their properties, both at the function and aesthetic levels. Furthermore, biology based technologies also refer to the replacement of industrial and mechanical systems with biological processes (Ginsberg et al., 2014; Mycoworks, 2016).

On the other hand, nutrients from biological sources have the capacity to re-enter the water or soil without depositing hazardous synthetic substances or materials and toxins.

2.2 *Microbial activity as part of the creative process in sustainable design*

There are currently several textile designers exploring different microorganisms species to understand the possibilities of creating textiles that are innovative and sustainable (Lee, 2010; MoMA, 2012; Ginsberg et al., 2014; Mycoworks, 2016). Bacteria, fungi or algae are living organisms capable of functioning as bio-factories, manufacturing materials such as cellulosic fibers (Lee, 2010; Mycoworks, 2016) and textile dyes and ink printing material (figure 1) (Santos, 2010; Blond & Bieber, 2014; Forman, 2016).

A significant feature of microbial activity is the characteristic rapid development, adding the economical and quick-to-produce factor. Additionally, one can consider the physical aspects and practicability of designing with micro living organisms; they can operate within a new view of scale and resource – "designing local" and "designing light" (Fletcher, 2008) – and function or be manipulated within (or according to) natural atmospheric conditions (e.g. *New Balance Bio-Logical* breathing textiles technology, using bacterial cells that expand and contract relative to moisture in the air) (Lacey, 2016).

Furthermore, microorganisms such as fungi may even be of help diverting waste as they feed on discarded matter (such as wood dust, corn husks and pistachio shells, in a carbon-free process) to produce by-products such as mushroom leather, for example, as seen in figure 2 (Mycoworks, 2016).

As the understanding of microorganisms grows deeper, their environmental aspects can be manipulated in order to explore or enhance their possibilities and functionalities.

Bacteria producing cellulosic fibers from a Kombucha tea batch (Lee, 2009) or the production of several dyes from the bacteria found on soil around certain plants (rizhosphere) (Santos, 2010) provide for an innovative narrative for textiles design field in regards to sustainability.

This is achieved by confronting scientific data and methodology with the opportunity of finding and highlighting product design sustainability issues and the need for the effective change (Forman, 2016).

3 CONCLUSION

Our current economy is based in a linear model of "take, make, dispose", a model that does not ensure consumers or businesses on the reduction of both nature depletion and ecological footprint; it is limited and dependent, it relies on non-renewable energy and resources. However, this century is characterised by price volatility due to resource scarcity as well as physical and/or political access. The unpredictable atmosphere, plus the competitive environment that define industries, is slowly pushing businesses to operate – shifting to a more circular model of manufacturing, given commercial profit is not compromised. A diversity of initiatives and approaches to sustainability are not only needed but also possible.

CE principles, as guidelines for the design of textile products, provide alternative narratives towards sustainability in the textile industry, in general (WRAP, 2012), also emphasizing the importance of the nature of both technical and biological materials regarding waste.

The intensification of biotechnologies as the foundation of design projects expresses bio-approaches involving living organisms at all scales in use to build, influence and improve objects.

Exploring biology as a vehicle to design not only offers a new vision for the development of sustainable products but also feeds the interest in learning the peculiar aesthetic values of biological resources. On the other hand, it instigates the deeper consideration on the importance of conserving the biodiversity on Earth, as well as its immense potential in the design field.

Besides the provocateur element inviting debate, some of the exhibited projects or on-going assignments where biological resources were incorporated in the creative process of design, attempt to use local environmental conditions to inform the performance and aesthetics of fashion and apparel. Not only the creation of a product that is biodegradable is achieved but also, in most cases, natural biological systems to produce goods are seen as inspirational and as effective bio-machines.

Furthermore, the radical innovative nature of some of the projects that assimilate biological systems or materials creates a rupture on the meaning of traditional interactions of body/clothing. This will additionally help to question and debate (Ginsberg, 2014) on emergent topics (some in need of urgent resolving) whilst disseminating the sustainable imperative.

ACKNOWLEDGMENTS

This research was only possible with the support of CIAUD – Research Centre in Architecture, Urbanism and Design, Lisbon School of Architecture of the University of Lisbon, and FCT – Foundation for Science and Technology, Portugal.

REFERENCES

Alive/En Vie. (2013). *gallery*. Retrieved May 8, 2013 from This is alive: http://thisisalive.com/gallery/attachment/000626/

Baumel, S. (2009). *Sonja Baumel*. Retrieved August 5, 2010 from Sonja Baumel: http://www.sonjabaeumel.at/work/bacteria.

Beasley, J., & Georgeson, R. (2014). *Advancing resource efficiency in Europe Indicators and waste policy scenarios to deliver a resource efficient and sustainable Europe*. European Environmental Bureau. Brussels: European Environmental Bureau.

Bendell, J., & Kleanthous, A. (2006, August 31). *Deeper Luxury – quality and style when the world matters*. Retrieved October 3, 2009 from www.org.uk: http://www.wwf.org.uk/deeperluxury/_downloads/DeeperluxuryReport.pdf

Benyus, J. M. (2002). *Biomimicry: Innovation inspired by nature*. William Morrow.

Biomimicry Institute. (2010). *Biomimicry Institute – Ask Nature*. Retrieved January 23, 2010 from Biomimicry Institute – Ask Nature: http://www.asknature.org/article/view/why_asknature.

Blond & Bieber. (2014). *Algaemy*. Retrieved September 3, 2014 from Algaemy: http://www.blondandbieber.com/algaemy.

Braungart, M., & McDonough, W. (2002). *Cradle to Cradle – remaking the way we make things*. New York: North Point Press.

Braungart, M., McDonough, W., & Bollinger, A. (2007). Cradle-to-Cradle design: creating healthy emissions – a strategy for eco-effective product and system design. *Journal of Cleaner Production*, 15, 1337–1348.

Earley, R., & Goldsworthy, K. (2015, June). *Designing for Fast and Slow Circular Fashion Systems: Exploring Strategies for Multiple and Extended Product Cycles.*

Retrieved July 26, 2015 from Academia: https://www.academia.edu/13232768/Designing_for_Fast_and_Slow_Circular_Fashion_Systems_Exploring_Strategies_for_Multiple_and_Extended_Product_Cycles.

El-Hagar, S. (2010). *Sustainable Industrial Design and Waste Management: Cradle-to-Cradle for Sustainable* Development. Academic Press.

Ellen MacArthur Foundation. (2012, January 25). *Towards the circular economy vol.1: an economic and business rationale for an accelarated transition.* Retrieved March 15, 2015 from Ellen MacArthur Foundation: http://www.ellenmacarthurfoundation.org/publications/towards-the-circular-economy-vol-1-an-economic-and-business-rationale-for-an-accelerated-transition.

European Union. (2010). *being wise with waste: the EU's approah to waste management.* European Union. Luxembourg: Publications Office of the European Union.

Fletcher, K. (2008). *Sustainable fashion & textiles – design journeys.* (1st Edition ed.). London, United Kingdom: Earthscan Ltd.

FOE Europe. (2013). *Less is More – Resource efficiency through waste collection, recycling and reuse of aluminium, cotton and lithium in Europe.* FOE EUROPE. Vienna: Global 2000.

Forman, G. S. (2016). *The importance of biotechnology application in the textile industry – natural textile dyes.* Lisboa: Dissertação de Doutoramento, Universidade de Lisboa – Faculdade de Arquitectura.

Foxley, G. (2011, December 25). *The Culture Shift of 2011, and What It Means for Business.* Retrieved March 30, 2012 from Triple Pundit – People, Planet, Profit: http://www.triplepundit.com/2011/12/culture-shift-2011-business/

Garcia-Serna, J., Perez-Barrigón, L., & Cocero, M. J. (2007). The new for Design towards sustainability in chemical engineering. *Chemical Engineering Journal* (133), 7–30.

Ginsberg, A. D. (2010, July 10). *Redesigning the Tree of Life.* Retrieved March 31, 2012 from The Synthetic Kingdom: Designing Evolution Research into synthetic biology & design futures: https://synthetickingdom.wordpress.com/2010/07/10/redesigning-the-tree-of-life/

Ginsberg, A. D., Calvert, J., Schyfter, P., Elfick, A., & Endy, D. (2014). *Synthetic Aesthetics – Investigating Synthetic Biology's Designs on Nature.* Massachussets: MIT Press.

Greenpeace. (2011, July 13). *Dirty Laundry.* Retrieved September 3, 2011 from Greenpeace: http://www.greenpeace.org/international/en/publications/reports/Dirty-Laundry.

Greenpeace. (2011, August 23). *Dirty Laundry 2: Hung Out to Dry Unravelling the toxic trail from pipes to products.* Retrieved September 3, 2011 from Greenpeace: http://www.greenpeace.org/international/en/publications/reports/Dirty-Laundry-2.

Greenpeace. (2012, March 20). *Dirty Laundry: Reloaded How big brands are making consumers unwitting accomplices in the toxic water cycle.* Retrieved May 29, 2012 from Greenpeace: http://www.greenpeace.org/international/en/publications/Campaign-reports/Toxics-reports/Dirty-Laundry-Reloaded/

Greenpeace. (2012, December 4). *Toxic Threads: Putting Pollution on Parade.* Retrieved February 5, 2013 from Greenpeace: http://www.greenpeace.org/international/en/publications/Campaign-reports/Toxics-reports/Putting-Pollution-on-Parade/

Greenpeace. (2012, November 20). *Toxic Threads: The Big Fashion Stitch-Up.* Retrieved February 5, 2013 from greenpeace: http://www.greenpeace.org/international/en/publications/Campaign-reports/Toxics-reports/Big-Fashion-Stitch-Up/

Greenpeace. (2014, January 14). *A Little Story About the Monsters In Your Closet.* Retrieved January 21, 2014 from Greenpeace: http://www.greenpeace.org/international/en/publications/Campaign-reports/Toxics-reports/A-Little-Story-About-the-Monsters-In-Your-Closet/

Greenpeace International. (2012). *Dirty Little Secret.* Retrieved April 3, 2012 from Greenpeace International: http://www.greenpeace.org/international/en/multimedia/videos/Dirty-little-secret-/

H&M. (2016). *Comments on An EU action plan for the Circular Economy* Dec. 2, 2015. Hennes & Mauritz. Hennes & Mauritz.

Klein, N. (2000). *No Logo: Taking Aim at the Brand Bullies.* Knopf Canada.

Koring, C., & Arold, H. (2007). *Rreuse.* Retrieved May 3, 2016 from www.rreuse.org: http://www.rreuse.org/t3/fileadmin/editor-mount/documents/Leonardo-SH-Sector/LSH004-European-Sector-Analysis-Report-english.pdf.

Kruschwitz, N. (2012, June 21). *Leverage Points for Creating a Sustainable World.* Retrieved November 4, 2012 from MIT Sloan – Management Review: http://sloanreview.mit.edu/article/leverage-points-for-creating-a-sustainable-world/.

Lee, S. (2009). *Biocouture.* Retrieved August 3, 2010 from Biocouture: http://www.biocouture.co.uk/newimages/Biocouture_press.pdf.

Malik, A., Ghromann, E., & Akhtar, R. (2014). *Environmental deterioration and human health: Natural and anthropogenic determinants.* London: Springer.

MoMA. (2012). *BioDesign Preview.* Retrieved January 3, 2013 from MoMA – Museum of Modern Art: https://www.moma.org/momaorg/shared/pdfs/docs/publication_pdf/3167/BioDesign_PREVIEW.pdf?1349967238.

Mulvaney, D. (2011). Green Politics: An A-to-Z Guide. California: SAGE Publishers.

Myers, W. (2012). *Bio Design: nature, science, creativity.* London: Thames & Hudson.

Niinimaki, K., & Hassi, L. (2011). Emerging design strategies in sustainable production and consumption of textiles and clothing. *Journal of Cleaner Production* , 19, 1876–1883.

OECD. (2013, July). *Biotechnology and its role in sustainable design.* Retrieved November 2, 2014 from CRODA: http://webcache.googleusercontent.com/search?q=cache:d-nZiw4QEo4J:www.croda.com/download.aspx%3Fs%3D1%26m%3Ddoc%26id%3DD4481F0E-29C9-485B-ABDD-17F48E5EB14E+&cd=2&hl=en&ct=clnk&gl=pt.

Santos, G. C. (2010). *A Biotecnologia desde a Antiguidade até ao século XXI: Corantes texteis naturais.* Lisboa: Dissertação de Mestrado, Universidade Técnica de Lisboa – Faculdade de Arquitectura.

Science Gallery. (2013). *Grow your own – life after nature.* Retrieved 2013 from Science Gallery: https://dublin.sciencegallery.com/growyourown.

Steffen, A. (2006, August 31). *The Future Laboratory, Conscience Consumers and the New Austerity.* Retrieved October 3, 2012 from World Changing – Change Your Thinking: http://www.worldchanging.com/archives/004876.html.

Stoddar, R. (2014, June 3). *Who are these purpose-driven consumers?* Retrieved October 7, 2014 from oliverrussel: http://www.oliverrussell.com/purpose-driven-brands-consumers-social-impact-2.

Teijin. (2016). *Evolutionary Step in Recycling.* Retrieved May 19, 2016 from www.teijin.com: http://www.teijin.com/solutions/ecocircle/.

Textiles Environment Design Research. (2014). *Polyester recycling.* Retrieved July 1, 2014 from TED Research:

http://www.tedresearch.net/media/files/Polyester_Recycling.pdf.
Tomás, C. (2016, January 10). 230 toneladas de roupa vão para o lixo. *Jornal Expresso*.
van Wijk, A., & van Wijk, I. (2014). *3D Printing with Biomaterials – towards a sustainable and circular economy*. Amsterdam: IOS Press/Delft University Press.
World Economic Forum, E. M. (2016). *The New Plastics Economy – Rethinking the future of plastics*. Retrieved May 5, 2016 from Ellen MacArthur Foundation: http://www.ellenmacarthurfoundation.org/publications.
WRAP. (2012). WRAP, *Valuing our Clothes: the true cost of how we design, use and dispose of clothing in the UK*. Retrieved May 1, 2016 from WRAP: http://www.wrap.org.uk/sites/files/wrap/VoC%20FInAL%20online%202012%2007%2011.pdf.

Textiles, Identity and Innovation: Design the Future – Montagna & Carvalho (Eds)
© 2019 Taylor & Francis Group, London, ISBN 978-1-138-29611-4

Textile sustainability: Reuse of wastes from the textile and clothing industry in Brazil

C. Jordão & A.C. Broega
University of Minho, Guimarães, Portugal

S. Martins
Londrina State University, Londrina, Brazil

ABSTRACT: The negative impact of the textile industry in the in the world is not new. The textile waste management is a today's problem but will be even bigger for the future generations if we do nothing. But, though very insipid, there are already new approaches to solve the problem, and two examples come from Brasil. In this work we are carrying out as a qualitative research (case study's) about tow textile waste management organizations: the "Clothing Bank of CS-RGS" and the "Reusable Fabric Bank of SP". Both unities offer a textile waste management systems with a sustainable redirection of textile waste or material that would no longer be used, allowing the extension of the Life Cycles and add value to it. The result shows that these systems do not prioritize in a balanced way, the 3 dimensions of sustainability (environment, economic and social-cultural).

1 INTRODUCTION

In front of the scenario we are experiencing in this contemporary society, an increasing concern worldwide emerges in relation to the negative impacts caused by the waste that is daily discarded, product of exacerbated consumption or as a result of high industrial discards, damaging, directly or indirectly, local and global systems in the social, economic and environmental spheres.

Thus, the concept of sustainability emerges as an attempt to minimize these ecological and social crises and as a condition for the maintenance and perpetuation of the resources for future generations. However, it is necessary to have a broad awareness of the societies for this issue and radical changes in the way of production, consumption and lifestyles (Vezzoli, 2010).

In the face of the negative impact of the textile and clothing industry in the environmental and social spheres, innovative initiatives are emerging that seek solutions to these problems through the development of systems that are built under the pillars of sustainability.

Through this panorama, the present work develops an exploratory study on projects of textile sustainability in development in Brazil, with focus on management and reuse of clean waste for further development of new products or to ensure the correct destination of waste that cannot be reused. The main objective of the study is to analyze which sustainability dimensions are prioritized by these systems, aiming at a better understanding of the systemic structure that is being outlined in the eco-efficient projects of the Brazilian textile scenario.

Based on the data analyzed, it is expected that the work may have relevance as a source of information and research for future implementation of innovative sustainable projects that aim to reuse clean waste from the textile industry. Also, it will contribute to reduce environmental impacts, add more value to the products, promote social equity and awaken society to a truly sustainable consumption.

2 STATE OF ART

2.1 *The pillars of sustainability*

The concept of sustainability has expanded in the course of its evolution. Initially addressing issues restricted to environmental impacts, the concept runs through its concern for a debate of social problems and the economic and cultural environment.

In this sense, Elkington (2012) introduced the theory of the three pillars of sustainability in 1997 in the first publication of his classic work "Cannibals with Forks: The Triple Bottom Line of 21st Century Business", although since 1995 he had already created the terms "People, Planet & Profits" to popularize the idea of multidimensional value creation.

The concept of Triple Bottom Line- Profit-Planet-People will be used in this article to analyze sustainability under different approaches. In addition, the cultural pillar will be used, even though it was not contemplated by Elkington, but it is fundamental when thinking about sustainability in contemporary society.

2.1.1 *Economic pillar (Profit)*

For Elkington (2012, p.112), the economic pillar is the simplest to be understood by the companies, as it really refers to the main support base of organizations. Thus, one must calculate how to maintain economically sustainable operations through economic capital, understood by the total value of its assets subtracted from its obligations, also evaluating the capital understanding that encompasses physical capital (technology and factory), human capital and capital intellectual.

In this view, Boff (2015) criticizes the standard model of the three pillars, in terms of economic development. For the philosopher, development and sustainability obey different logics and that are self-standing: "... one privileges the individual, the other the collective; one emphasizes competition, the other co-operation; one the evolution of the fittest, the other the coevolution of all together and interrelated" (Boff, 2015, p.45).

2.1.2 *Environmental pillar (Planet)*

Elkington (2012, p.116) defines eco-efficiency as the provision of goods and services at competitive prices, capable of meeting human needs, offering quality of life while having the capacity to gradually reduce the ecological impact and intensity of use of the product life cycle.

The environmental dimension for Vezzoli (2010) can be perceived as part of sustainable development concerned with not overcoming the resilience of the biosphere and geosphere, ie the planet's ability to absorb human actions without the irreversible consequences of environmental degradation.

When it comes to Braungart and McDonough (2013), they reflect on the need to imitate natural systems in order to create a more inspiring relationship through strong partnership with nature. Designing systemic structures, whose products, by-products or services can feed ecosystems and allow re-circulation of technical materials, instead of dumping or discarding them erroneously, is now a key issue for the survival of future generations.

2.1.3 *Social pillar (People)*

Social capital needs to be seen by organizations that claim to be sustainable beyond human capital, embracing reflections that address more than the society's health of and the potential for wealth creation (Elkington, 2012, p. 123).

Kablin (2010) affirms that the concept of sustainability is widening in the sense that it aims to use as substratum the adoption of ethical stances, since the sustainability of the model itself can be equated in the responsibilities of the entrepreneurs, the citizen and the State.

2.1.4 *Cultural pillar (Culture)*

The cultural pillar is not seen in the *Triple Bottom Line* by Elkignton (2012), but initially presented by Hawkes (2001). It was added in this analysis, since many recent studies point to the need to see sustainability also in the eyes of the cultural question, which is the fundamental basis of the values of each society, building the different cultural identities.

Hawkes (2001, p.25) highlighted the cultural pillar in his book *The Fourth Pillar of Sustainability: Culture's essential role in public planning,* in which he affirms that there is a need to insert the cultural pillar vitality in order to bring societies welfare, creativity, diversity and innovation, emphasizing that culture must be treated as a basic requirement for the health of societies.

2.2 *A Socio-environmental impacts of the textile industry in Brazil*

In Brazil, the textile and confectionery sector has a history of almost 200 years, which allowed positioning the country in the world reference frames as the fourth largest confectionery production park in the world and the fifth largest textile producer. (ABIT-Brazilian Textile Industries Association, 2015).

However, despite its prominent position, according to the Brazilian Textile Industry – Brazil Textile Sector Report 2015, prepared by ISIM (Institute of Studies and Industrial Marketing) with the institutional support of ABIT (Brazilian Textile Industries Association), the textile and confectionery sector in Brazil faces a scenario of uncertainties due to the current macroeconomic instability, but despite this fact still represents with a strong impact on the Brazilian economy.

Analyzing the publication "The Power of Fashion" of ABIT (2015), which presents the scenarios and creates an agenda of competitiveness for the sector from 2015 to 2018, it is observed a decrease in the production of the industry for the fourth consecutive year. In 2015, the industry earned US$ 36.2 billion, a significant drop compared to the 2014 billing of US$ 53.6 billion (ABIT, 2016).

In Brazil, it is estimated that 175,000 tons of solid waste from the confectionery industry will be produced each year, with trousers, socks and shirts being used among others that could be used by other industries. However, it is found that 90% of the total of these wastes are discarded incorrectly (*Sinditêxtil*, 2012).

The date presented by ABIT, show numbers that reflect a scenario of macroeconomic threats and a future with many uncertainties; however, the textile and confectionery sector still represents an effective impact on the Brazilian economy and this leads the industry to seek Marketing's adaptations and changes.

3 RESEARCH METHOD

The research method used was the case study in order to investigate in greater depth a contemporary phenomenon, in this case, the phenomenon of sustainability in the textile sector, within its real life context. Since it is a technically unique situation, two units of analysis were mapped, defined by the similarity criterion of their systemic structures, since both, although with different business models, per-form the collection, separation and commercialization/allocation service of leftover fabrics from the textile and garment industry, which would end up with landfills or other destinations that could have a negative impact on the environment and society.

The units that served as the object of analysis of this study were the Bank of Clothing, located in the city of Caxias do Sul in the state of Rio Grande do Sul, and the Reuse Fabric Bank of São Paulo capital, which will then be explained.

In the second phase of the development of the empirical work, after categorizing the data, the information was submitted to a tool called Sustainability Design-Orienting Toolkit (SDO), developed by Carlo Vezzoli and Ursula Tischner, whose objective is to guide the design process for the solution of Sustainable systems.

It is an open-source software with a copyleft license, which has the option of being used online (www.sdolens.polimi.it) or downloaded and installed on a network of Local Internet (LAN) (VEZZOLI, 249, 2010). In the case of this work, the online version was used.

The tool is originally structured based on three dimensions of sustainability: (1) the environmental; (2) the socio-ethical and (3) the economic and political dimensions.

For each of the three dimensions of sustainability proposed by Vezzoli (2010, p.238, 239), six requirements are established, both for evaluating a given system and for guiding the design process.

The requirements of the environmental dimension are:

- Optimization of system life;
- Reduction of distribution and transportation consumption;
- Reduced resource use;
- Minimization/appreciation of waste;
- Conservation/biocompatibility and;
- Toxicity reduction.
- In the socio-ethical dimension, these are established requirements:
- Improvement of employment and work condi tions;
- Increased equity and fairness in relation to the authors of the system;
- Capacity Building/Promotion of Responsible /Sustainable Consumption;
- Integration of people with special and marginal ized needs;
- Favoring social cohesion and;
- Prioritization of local resources.
- Regarding the economic dimension, the requirements evaluated are:
- Market position and competitiveness;
- Profitability/value added to businesses;
- Added value for customers;
- Long-term business development;
- Partnerships and cooperation;
- Macroeconomic effects.

The SDO Strategic Analysis phase was used to identify the system's priorities in relation to the environmental, socio-ethical and economic dimensions, based on pre-existing requirements analysis.

As the SDO program allows analyzing up to two case studies, the Clothing Bank was established as the existing system, because it has a more complex structure and the field research has been performed in greater depth in this unit of analysis. The Reuse Fabric Bank was used as case study 1, so that comparisons could be established between the two systems in order to highlight the qualities and opportunities for improvements in the existing system (BVCS).

For each criterion, the SDO system sets a project priority level to A (High), M (Medium), B (Low), or N (No). When the existing system (Clothing Bank) is compared to case study 1 (Reuse Fabric Bank), improvements classified by SDO are evaluated on a scale of levels, ranging from a radical improvement (++), incremental improvement (+), without significant changes (=) or depreciation (−). The definition of both the priority level and the evaluation of the improvements were carried out by the research investigators.

4 RESULTS ANALYSES AND DISCUSSION

4.1 *Cases study*

4.1.1 *Analysis of the clothing bank of Caxias do Sul (BVCS)*

The Clothing Bank of Caxias do Sul (BVCS) in Rio Grande do Sul was founded in October 2009 as a centralizing body for textile waste from the industries of the state of Rio Grande do Sul. The BVCS receives the waste from the industries, sorts through types of materials and organizes leftover fabrics for later disposal to registered communities that will reuse these materials. What cannot be reused by the communities goes to other industries that make reuse for upholstery or is destined for co-processing, closing the product cycle. According to Juarêz Paim, BVCS's manager, the project came up initially to solve a the textile industrial pole problem of Rio Grande do Sul, which had no place to correctly dispose of its waste. However, the need to contribute to social inclusion through the generation of work and income through training courses was soon perceived.

4.1.2 *Analysis of reuse fabric bank of São Paul (BTR, from Portuguese, Banco de Tecido de Reuso)*

The structure of the São Paulo Reuser Fabric Bank (BTR) was created in January 2015 by Lu Bueno, with the objective of prolonging the life of stopped, stocked fabrics, leftovers of cuts of clothes or of rolls and flaps.

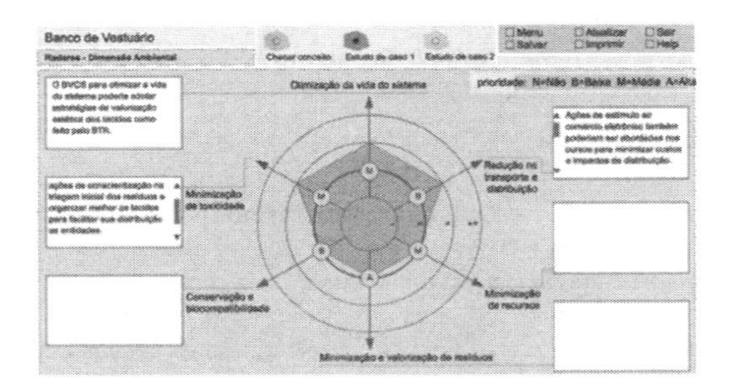

Figure 1. Improvement radar of the existing BVCS system compared with the BTR-environmental dimension.

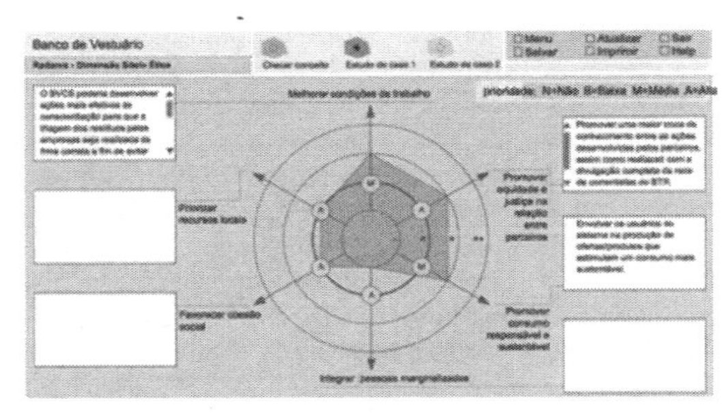

Figure 2. Improvement radar of the existing BVCS system compared with the BTR-socio-ethical dimension.

The purpose was giving these products new use, preventing it from ending up in landfills. Additionally, fabrics, that are no longer in use, are brought to the units and are organized and sanitized. After this process, they are put up for sale, and marketed per kilo to the current value of R$ 45. In case a person deposits fabric in the Bank, she receives credits for each kilo deposited and later can get new ones with these credits.

4.2 *SDO- sustainability design orienting toolkit application*

4.2.1 *Environmental sustainability radar and improvements*

The diagram of figure 1 shows the existing system, in this case the Clothing Bank of Caxias do Sul, the result of the analysis process of the system's environmental priority for each criterion established by the SDO. The visualization shows that the BVCS system has a low environmental priority regarding the issues related to the transportation and distribution of waste, since they do not have the responsibility of collecting the materials from the industries (are the industries that deposit their waste there). Another point of low priority is related to the consumption of non-renewable energy, which does not apply to the case of the existing system.

In addition, there is a high priority of the system regard to both minimizing and valuing waste. This awareness is conveyed to all partners and to all training courses and workshops provided by the BVCS.

When the radar is superimposed on the BTR case study analysis, improvements can be seen in three aspects: (1) optimization of system life, (2) reduction of transport and distribution, and (3) toxicity minimization, as can be seen in the shaded area of the figure.

4.2.2 *Socio-ethical sustainability radar and improvements*

The diagram of figure 2 is the analysis result of the socio-ethical dimension, in which four items with high priority can be observed in the BVCS system. These are highlighted: (1) promote equity and justice, (2) integrate marginalized people, (3) pro-mote social cohesion, and (4) prioritize local re-sources.

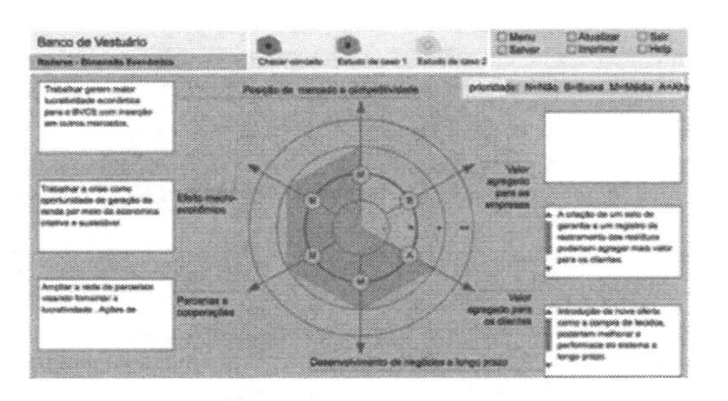

Figure 3. Improvements radar of the existing system BVCS system compared with the BTR-economic dimension.

When compared the BTR radar with the diagram's cross-referenced, can be seen that the radar points to some improvements in the system, in terms of: improving working conditions, promoting equity and fairness among partners, and promoting responsible and sustainable consumption; mainly, due to this system, to guarantee high visibility to the actions practiced by the different actors in its structure.

4.2.3 *Economic sustainability radar and improvements*

The diagram shown in figure 3 demonstrates that, among the requirements analyzed, only one was evaluated as being of high priority (added value to the client). The other four were categorized as medium and one as low priority, which is the low added value to the company that still cannot obtain its financial independence, needing the public financial support (information from Caxias do Sul's City Hall). Overlapping the radars, when comparing the BVCS with the BTR, there is an improvement of the system, regarding to: market position and competitiveness, macroeconomic effects, partnerships and cooperation, long-term business development and added value for customers. This fact can be justified by the organizational structure of BTR, which promotes actions aimed at guaranteeing greater financial autonomy of the system.

5 FINAL CONSIDERATIONS

The application of the tool SDO – Sustainability Design Orienting Toolkit, aimed to support the planning and management of eco-efficient systems, allowing helping a more specific understanding, so that the priorities of the sustainable dimensions could be mapped. In the BVCS case, there was a strong inclination towards the environmental and social dimensions appreciation.

Although the project was initially conceived, there is no point in offering solutions to legal problems in the environmental sphere. It ended up gaining more social contours due to the "training courses", which are offered free of charge by the BVCS and by the work of inclusion and social cohesion. The strong partnership network, established by the BVCS, reinforces the idea of cohesion necessity between different partners, who, despite having different specific objectives, are united in a cooperative work. Such thinking reinforces Vezzoli's (2010) by pointing out the connection between socioeconomic actors as one of the primordial elements in the development of sustainable systems.

However, there was a fragility regarding the economic dimension, since the BVCS, although in evolution, still depends on financial resources of Caxias de Sul's City Hall to continue with its work, not presenting solutions to take advantage of the system's profitable potential. Thus, it is not fully sustainable in terms of economic maintenance and system autonomy.

Because it is a private initiative structure and focused on the profitability of the system, the São Paulo Reuse Fabric Bank applies more efficient techniques aimed to guaranteeing the maintenance of its structure and, despite being a recent business, already presents evidence of its Eco-efficiency.

In the BTR structure, it was possible to observe a greater balance between the three dimensions of sustainability analyzed. In addition to environmental and socio-ethical benefits, it can also be economically sustainable by adopting fair trade practices.

The BTR system presents perceptible improvements when compared to the BVCS, since it can show the value of the products coming from the reuse fabrics throughout its chain. It also reveals the work of its network of partners, highlighting the work of designers, always aiming to value initiatives of fabrics' reuse, showing that sustainable fashion and the role of designers contribute to the improvement of eco-efficient systems.

It is believed that the present work has succeeded in achieving its objectives in order to present to us clearly the sustainable dimensions that are being prioritized.

The research thus confirms the initial assumptions that the systems, which manage the textile and clothing industry clean waste in Brazil, do not work in a balanced way all the pillars of sustainability (environmental, socio-ethical and economic pillar). In addition, despite a greater mobilization, they are still incipient when it comes to the resolution of the complex problems related to the impacts of this sector in Brazil.

ACKNOWLEDGEMENTS

"This work is supported by FEDER funds through the Competitivity Factors Operational Programme – COMPETE and by national funds through FCT – Foundation for Science and Technology within the scope of the project POCI-01-0145-FEDER-007136"

REFERENCES

ABIT. Cadeia Textil e de Confeção: Visão de Futuro 2030. São Paulo: CNI/SENAI/ABIT, 2015.

ABIT. O Poder da Moda – Cenários- Desafios e Perspectivas. Agenda de Competitividade da Indústria Têxtil e de Confecção Brasileira 2015-2018. São Paulo: ABIT, 2015.

Boff, L. Sustentabilidade: o que é: o que não é. São Paulo, Ed. Vozes, 2015.

Braungart, M., & Mcdonugh, W. Cradle to cradle: criar e reciclar ilimitadamente. São Paulo: Editora Gustavo Gili, 2014.

Elkington, John. Canibais com Garfo e Faca. (Edição Histórica de 12 anos).São Paulo, M.Books, 2012.

Hawkes, J. The Fourth Pillar of Sustainability: Culture's Essential Role in Public Planning; Commom Ground Publishing Pty Ltd in association with Cultural Development Network (Vic), Victoria, Australia, 2001.

Kablin, I. Desenvolvimento Sustentável : um conceito vital e contraditório. In : Zylbersztajn D. E. Lins C. Sustentabilidade e Geração de Valor- A transição para o século XXI. Rio de Janeiro, Elsevier Editora, 2010.

Sinditêxtil-SP – Sindicato das Indústrias de Fiação e Tecelagem do Estado de São Paulo. Projeto com Responsabilidade Social. Disponível em junho de 2012 <http://www.sinditextilsp.org.br/index.php/materias/item/840-sinditêxtil-sp-lança-projeto-com-responsabilidade-socioambiental> Acessed in: 4th of June 2016.

Vezzoli, C. Design de sistemas para a sustentabilidade: teoria, métodos e ferramentas para o design sustentável. Tradução Mauro Rego. Salvador, Edufba, 2010.

Textiles, Identity and Innovation: Design the Future – Montagna & Carvalho (Eds)
 ISBN 978-1-138-29611-4

In direction to sustainable textile roofs in architecture and design

C. Alho
CIAUD, Lisbon School of Architecture, Universidade de Lisboa, Portugal

ABSTRACT: This research paper intends to show the evolution of sustainable textile roofs concepts and the emerging direction to Sustainable Architecture and Design. The purpose of the study takes into consideration study cases based on literature review about important architects and theorists all over the World in order to select the basic principles of sustainable urban development defined in the Agenda XXI. The research methodology is based on relevant research and literature. The results show that sustainability is one possible direction defined for architecture as a response to solve the needs for future generations. The conclusions evidence how Sustainable Architecture and Design should stand for a basic integrative attitude to introduce in all levels of the Architectural Process in order to achieve sustainable development.

1 SUSTAINABILITY

Sustainability has become a widely applied Concept – so much, that the meaning lost precision and definition; today, it probably acts more like a symbol of a necessary Civilizational change, i.e. a different perception of Human activities and Values, in relation with an environment conscious attitude and accounting.

2 THE CONCEPTS AND ORIGINS

The Concepts and its Origins, may have resulted from the confrontation of two trends of thought regarding mankind's existence. One which claimed the priority of human development at any cost, even with the depletion of the natural environmental resources. The other that claimed safe guarding of the environment and resources.

Since both problems are global ones, as they go beyond the borders of the state and request solutions that could hardly be offered separately by any of them, a debate became possible round the table of the United Nations. Such debate has been tackling in turn a number of the global issues, for several decades now. A compromise around the concept of sustainability has emerged, thus, development must be sustainable in order to be acceptable, while concern for the environment must not remove the central need for development. The philosophy of sustainability is simple: no resource should be consumed at a pace higher than its speed of regeneration.

In the present context, sustainable development implies a central focus on people therefore it must be a dynamic system like human society. Thus sustainability is fundamentally a question of balance maintained over time.

Figure 1. Textile cover of ephemera structure.

Sustainable development is not a new idea, the protection of farmland forests, fisheries, cultural landscapes, historical areas and its aesthetics, etc., dates back at least to the eighteenth century.

> *"Then I say earth belongs to each generation... generation during its course, fully and in its own right, no generation can contract debts greater than may paid during the course of its own existence" Thomas Jefferson September 6, 1789* (ORTEE, 1992)

More recently in the (1960s, 70s), at the beginning of the environmental revolution, damage to the natural system became a main concern. However terms such as sustainability or sustainable development became more often mentioned during the last decade, due to scientific evidence that excessive exploitation of natural resources in parallel to the ever-growing polluting agents would lead to irreversible environmental destruction.

Figure 2. 'Tubaloon' textile sculpture by Snohetta for the Kongsberg Jazz Festival; Photo © Snohetta.

Figure 3. 'Tubaloon' textile sculpture by Snohetta for the Kongsberg Jazz Festival; Photo © Snohetta.

The Paris biosphere conference, held in 1968, sounded the alarm of environmental degradation and the need for action.

> "*The conference discussed science and management issues posed by the pervasive environmental and resource problems that were becoming increasingly evident in the world at that time. It concluded that countries must develop greater capabilities for undertaking cross-disciplinary research linked to policy and management issues for environmental conservation, and to what now would be called sustainable resource use*". (CBRA, 2002)

In 1972, a report published by the so-called Club of Rome, which shocked the world, predicting risks of the forthcoming collapse of life on earth. The 'club' consist of scientists, economists, businessmen, international high civil servants, Heads of State, and former Heads of State who pooled their different experiences from a wide range of backgrounds to come to a deeper understanding of the world problematic, working with that new device of modernity, the computer.

In 1972, a report published by the so-called Club of Rome, known as "Limits to Growth," brought the message that the world was heading for disaster because of population growth and industrial expansion, exhaustion of stocks of natural resources, environmental destruction, and food shortages.

The "Limits", report was published in June 1972 just prior to opening of the United Nations Stockholm Conference on the Environment "Only One Earth".

In the (1980s and 90s), a greater importance was also given to these problems, broadening the scope of scientific actions.

In April 1987 a report known as "Our Common Future" or Brundtland Report, was published by (WCED/UN) where Development and Environment *state of the art* relations were balanced and evaluated. The Report basically suggested a different type of economic growth, where the carrying capacity of natural resources had to be respected (along with an ethical imperative of leaving future generations the possibility of using those resources) in order to obtain and improve "Quality of Life"; furthermore, it claimed that the maintenance of a conflictual *status* between Development and Environmental policies was a dead-end street – rather they should be seen as two interlocked faces of the same coin.

However the most important summit was held at Rio de Janeiro, Known as the Earth Summit, this event was the largest and most celebrated international conference ever organized by the UN, with an unprecedented attendance up to that time: 178 countries were present, 120 heads of state, 800 journalists and more than 30,000 people.

2.1 *The concept of environment*

The concept of Environment was also evolving, at the same time – from an almost identity with Nature and the physical quality of its components affecting Mankind, to the perception and evaluation of the surrounding Universe, through social, economical, philosophical and cultural *criteria*, focused on the more subjective goals of "Quality of Life" and "Sustainable Development"; this broadening of scope conflicted with the ruling logic of sectorializing Human Knowledge and *Praxis* – but, on the other hand, provided a powerful instrument for an integrated approach to understand Reality, improving the quality of assessment and decision- making.

3 SUSTAINABILITY IN ARCHITECTURE

In the field of Architecture, sustainability is now also becoming *mainstream*; but the seeds were already there for the last decades- mainly after the oil crisis of the 70s:- Passive Solar, Bioclimatic, Green and Eco-Architecture had often claimed for the need of a better relation with Site, physical Environment, Resources, Human scale and cultural diversity, though their somewhat marginal approach didn't secure a significative number of followers; also, without major impact, trends like Critical Regionalism and a few of the Post-Modern movements and architects reacted against the cultural "Global Village" normalization approach, that ruled under the Modern Movement, pointing out the importance of local input and scale, towards a more Humane Architecture.

Governments, specially of the industrialized northern countries, have supported the climate conscious approach that some of these trends proclaimed, on a saving energy policy basis; but up till now, failed to influence significatively architects and public opinion – besides the first buildings formal inconsequence and certain lack of quality, the Consumerist Way of Life that the industrialized World also sustained and publicized, and the civilizational blind faith on Techno-scientific solutions to dominate Nature and mechannicaly solve problems, prevented a wide acceptance of an environmental attitude in the Architectural process.

The individual building's approach also revealed unsatisfactory; some of the fundamental problems needed to be tackled with and solved at an Urban scale (sometimes even regional and global), the importance and weight of the Urban Areas to the Environmental Agenda only this last decade to be fully recognized: – major events like ECO92 in Rio, or HABITAT2, in Istambul, gave top priority to the "ecological footprint" of urban settlements, and a multitude of international and national organizations among cities were launched, in search of the lost quality of urban environment combined with new opportunities for Development.

Under the influence of the 3Rs environmental policy (Reuse, Recycle, Rehabilitate), the Life-cycle analysis of building materials and construction systems made also its entrance in the Architecture *lexicon*, becoming one of the trademarks of the sustainable building process.

But the major break-through, that opened the eyes of professional and public opinion towards a sustainable Architecture, came with the (semi-)conversion of some of the best known High-Tech Architects (such as Renzo Piano, Sir Norman Foster, Richard Rogers, to name a few) and other influential architects like Tadao Ando or Thomas Herzog, stating explicitly in their most publicized Works the embracing of this *Cause*.

The co-existence of high-rise urban buildings with some almost new-Vernacular small buildings blending with Nature, both claiming to be Sustainable Architecture, brought a bit of confusion and controversy to the field:- amidst this wide variety of philosophical and formal approaches, what was in fact a Sustainable Architecture?

Clearly, all those architects became aware of the potential impact of Architecture in the physical environment, both the negative and positive contributions towards a more sustainable local and global World. A better resource management, namely the energy, water and air cycles that went in and out the buildings, was an obvious path; but the unavoidable problem related with this quest was the concept of "Quality of Life" that remain an underlined objective- which couldn't escape easily the current consumerist Way of Life (values and references).

Though positive, most of those sustainable attitudes were partial ones, failing multi-scale coherent, integrated approach, where local specifications and global references blend in an adaptable and diverse Architecture, responding at different (but complementary) scales to the philosophical, territorial, spatial and constructive levels, and to the ecological imperatives of the before/after periods on the Building's life.

A very significative number of architects and theoricists choose the Ecological Principles as the reference to follow, in order to achieve the desired sustainability in Architecture- even here with a wide range of attitudes: -the Ecosystems used as a methaphor or analogy to understand and manage the functioning of a Building/urban structure; the reproduction of Natural systems scientific organization for formal vocabulary invention or *data* assessment; the bioregional approach in order to understand and incorporate the local and contextual ecological structure in Site analysis and bioclimatic behaviour; the global understanding of environment/mankind relationship, to better integrate natural and cultural heritage... Sinergetic, Homeostatic or Simbiotic behaviours, even new concepts like Negentrophy (*McHarg, Ian*), observed in some organisms/ecossiystems, translated to the architectural methodology, revealing the high potential of this field research to make a more holistic, interactive, responsable Architecture – and probably more alive. *("Life is right,...")*.

If one follows the original concept applied in the Bruntdland Report, and besides the optimal resolution of the binomial relation between Resources Management and Quality of Life, Sustainability requires also other fundamental aspects:- Continuity, which translates better (rather than durability) in the dynamic adaptation of a building (or urban fabric) to the continuous changing ways of Life (and spatial requirements/references), and specially, Ethical responsibility towards next (and also past) generations, to incorporate local and civilizational information, seen as the essencial resource to understand the Past and to provide alternative paths to build the Future.

Therefore to continue its development towards sustainability, return to the basic principles stated more than twenty years ago, at "Earth Summit – Rio 92" and the most import document out of that conference " Agenda XXI" with its twenty seven principles, ten of which (listed below), are most relevant for the practice an any intervention in the built human environment: (BEQUEST 2001).

Principle 1: Human beings are the centre of concerns for sustainable development. They are entitled to a healthy and productive life in harmony with nature.

Principle 2: The right to development must be fulfilled so as to equitably meet developmental and environmental needs of present and future generations.

Principle 3: In order to achieve sustainable development, environmental protection shall constitute an integral part of the development process and cannot be considered in isolation from it.

Principle 4: All States and all people shall cooperate in the essential task of eradicating poverty as an indispensable requirement for sustainable development, in order to decrease the disparities in standards of living and better meet the needs of the majority of the people of the world.

Principle 5: States shall cooperate in a spirit of global partnership to conserve, protect and restore the health and integrity of the Earth's ecosystem. The developed countries acknowledge the responsibilities that they bear in the international pursuit of sustainable development in view of the pressures their societies place on the global environment and the technologies and financial resources they command.

Principle 6: To achieve sustainable development and a higher quality of life for all people, States should reduce and eliminate unsustainable patterns of production and consumption and promote appropriate demographic policies.

Principle 7: Environmental issues are best handled with the participation of all concerned citizens, at the relevant level. At the national level, each individual shall have appropriate access to information concerning the environment that is held by public authorities, including information on hazardous materials and activities in their communities, and the opportunity to participate in decision-making processes. States shall facilitate and encourage public awareness and participation by making information widely available. Effective access to judicial and administrative proceedings, including redress and remedy shall be provided.

Principle 8: In order to protect the environment, the precautionary approach shall be widely applied by States according to their capabilities. Where there are threats of serious irreversible damage, lack of full scientific certainty shall not be used as a reason for postponing cost-effective measures to prevent environmental degradation.

Principle 9: National authorities should endeavour to promote the internalisation of environmental costs and the use of economic instruments, taking into account the approach that the polluter should, in principle, bear the cost of pollution, with due regard to the public interest and without distorting international trade and investment.

Principle 10: Environmental impact assessment, as a national instrument, shall be undertaken for proposed activities that are likely to have significant adverse impact on the environment and are subject to a decision of a competent national authority.

4 CONCLUSIONS

Some authors consider Sustainable Architecture an impossibility, an *oxymoron* (contradiction in terms) if a strict meaning is applied to the concept; in the context the definition was presented above, I would rather consider it a redundancy, because a responsible Architecture should always incorporate those fundamental aspects referred, regardless of programatical, economical, formal or other conditioning aspects in the process of Architecture design and implementation. However that is still not yet the case for the majority of the architectural approaches all over the World, and so , rather than another trend or formal style, Sustainable Architecture should stand for a basic integrative attitude to introduce in all levels of the Architectural Process.

Textile roofs gains a new importance in front of the new paradigm and it seems to appear in the camping leisure, two different and opposite forms in evolution.

In an upward direction, where this proposed project is integrated, expresses sophistication and the constant institutionalization, earning new contours pointing until the new classifications of "Luxury". In descending direction, the provisory lodging becomes in a form of permanent residence for a considerable part of population with modest incomes.

Thus, made these considerations of this case study, it seems us pertinent to take off the following conclusions:

- Tourism is to move, answering the new eco-logical and environmental concerns and also to a dramatic problem created by the development of masses tourism, that aggravate from Second World War until present, leaving entire cities structures that are not used during half of the year, what configures a clear problem of sustainability.
- That, will make the industry of tourism, on the XXI century, go thought substantially structuralized changes, and surely, make it more responsible, environmental and socially speaking.
- The proposal of echo-camping-resort that configured our case study is surely a new reply to a new program of tourism, looking for creative standards of quality and to propitiate a bigger contact of the customers with nature, as well as minimizing the negative impact of tourism.
- Thus, it makes sensible to conclude that the lodging in tourism of nature, in the case of eco-camping, is probably the evolution on the direction of sophistication and comfort, in the physical and architectural type of lodging, as well as in the number and quality of the leisure equipment.
- The notion of "luxury" is changing, and the close link to nature, related to unpolluted areas and harmonious natural environment is one of the great luxuries today.
- Solar and aeolian alternative energies had come to be and are in a primitive period of training development and will go to prosper and reach efficiency standards that at this moment still are considered utopian.
- As well, architecture and urbanism will go to follow development standards supported on "low" and "high" technology concepts. In the first case, simple and economic traditional constructive processes

will be retaken and improved. In the second case, new, lighter and sophisticated materials will be developed, as well as, new equipments in order to achieve and improve technology.

- The 'economic' and 'efficacy' notions will walk along with energy efficiency, comfort and human well-being and in harmony with nature preservation.
- Eco-camping-resort appears as a valid option to the construction of a new conventional tourist enterprise, and search to create permanent employment.
- In conclusion, changes in society creates new human and physical conditions on the built environment witch defines emerging concepts for eco sustainable resorts in rural areas of Europe.

ACKNOWLEDGMENTS

This research was only possible with the support of CIAUD – Research Centre in Architecture, Urbanism and Design, Lisbon School of Architecture of the University of Lisbon, and FCT – Foundation for Science and Technology, Portugal.

REFERENCES

Alho, C. 2000. *Authenticity Criteria for the Conservation of Historic Places*. Ph.D. Thesis. Salford, U.K.: University of Salford.

Denscombe, M. 1998. *The Good Research Guide for small-scale social research projects*. Buckingham, Philadelphia: Open University Press.

Emmanuel, R. 2005. *An Urban Approach to Climate Sensitive Design: Strategies for the Tropics*, Spon Press – Taylor & Francis Group.

Geyer-Allely E. 2002. *Sustainable consumption: an insurmountable challenge*. UNEP, Industry and Environment Review, n°. January–March

Hinks, J. 1996. *Research Methods. M.Sc. In Information Technology In Property and Construction*. U.K.:Unpublished Collection, University of Salford.

Stern, N. H. 2007. *"The Economics of Climate Change: The Stern Review"*, Great Britain Treasury Edition: illustrated, reprint, Cambridge University Press.

Yin, Robert K. 1994. *Case Study Research: Design and Methods* (2nd ed.), Sage Pub., California.

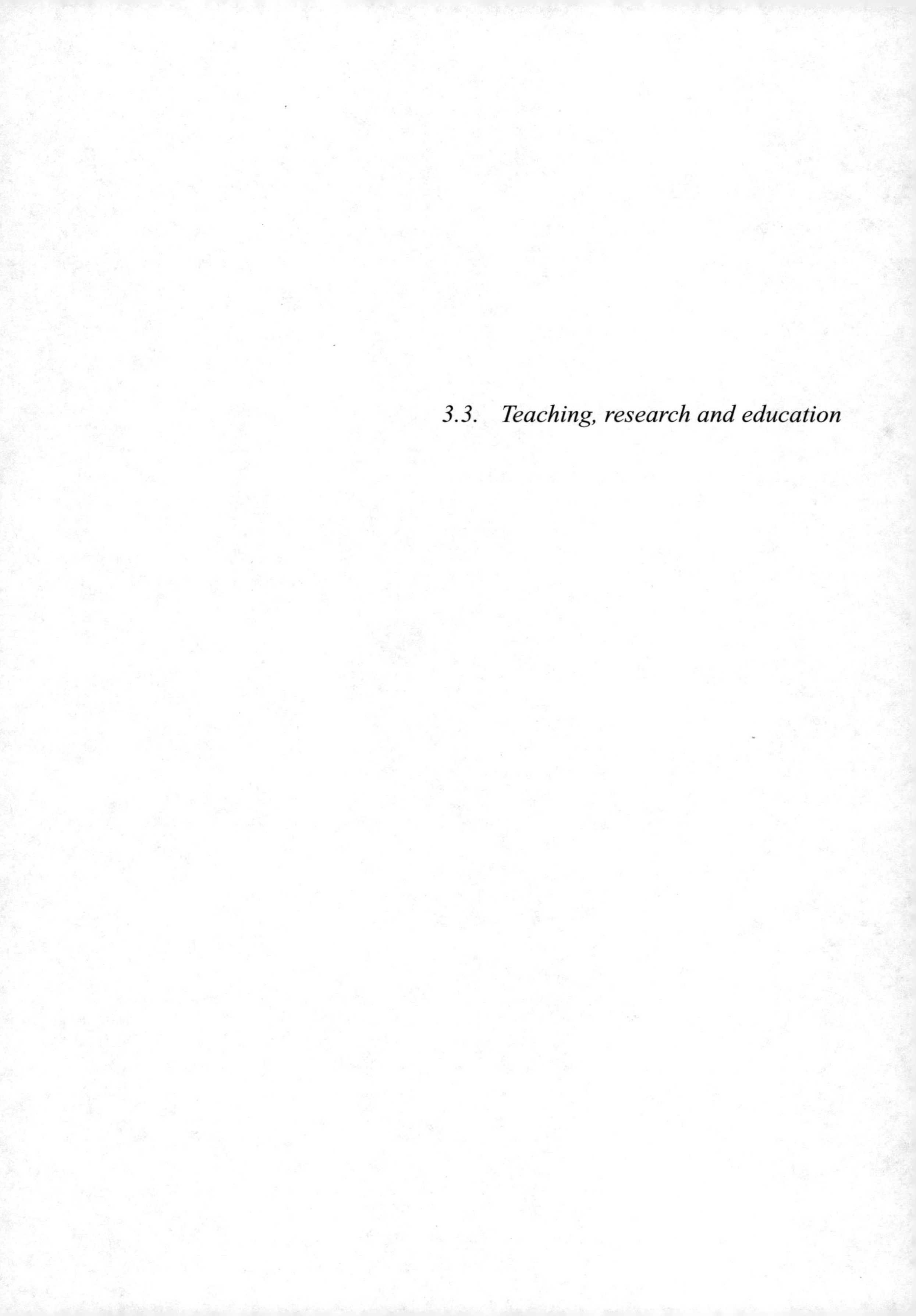

3.3. Teaching, research and education

Textiles, Identity and Innovation: Design the Future – Montagna & Carvalho (Eds)
© 2019 Taylor & Francis Group, London, ISBN 978-1-138-29611-4

The tactual experience of textiles in fashion design education

I. Simoes, F. Silva & N. Nogueira
CIAUD, Lisbon School of Architecture, University of Lisbon, Portugal

ABSTRACT: Because the selection of the right fabrics is paramount for clothing to materialize as imagined, this paper raises some questions regarding teaching textiles to Generation Z fashion design students. If topics presented through formal lectures and demonstrations were replaced by hands-on activities making use of tacit knowledge, the correlation between fabric properties and behaviour would become embedded more easily. The internalized information acquired largely through tactual experience would certainly enhance the design process (from conceptualization to prototyping) and ultimately the experience of wearing.

1 INTRODUCTION

Over the past decade, educators acknowledged the need to rethink the approach to teaching design. For some this requirement became obvious because of the important part design plays in "the transition to a sustainable society" (Irwin 2015, p. 91); for others because of "the ever-shifting student population" (Faerm, 2013), particularly those born from the early 1980s to the mid-1990s (known as Generation Y) and those born from the mid-1990s to early 2000s (known as Generation Z), as their focus and expectations are significantly different from their predecessors' (Ashdown, 2013; Simoes & Silva, 2016).

Not aiming to devalue the relevance that social and environmental issues have, in themselves, and for preparing future designers, this paper focuses on the challenges of teaching textiles to Generation Z fashion design students: what makes this generation unique, what are the best ways to motivate them to know about textiles, why are they seemingly unable to relate their own experiences as dressed bodies with the process of designing clothing?

The questions pursued by us derive from our experience as fashion design educators. The importance that the selection of fabrics has for clothing to materialize as imagined, in addition to the implications that fabrics have for the whole design process, makes it crucial to come up with alternative approaches, grounded on the understanding that "we can know more than we can tell," as Polanyi puts it ([1966] 2009, p. 4). Even if our field is neither textile science nor textile design, the huge part played by fabrics from conceptualization to prototyping, encouraged us to ask how Generation Z students could develop a cross-disciplinary thinking largely founded on the tactual experience of textiles, with the aim of enhancing the design process and ultimately the experience of wearing.

2 THE ONE-DIMENSIONAL, FLATTENED OUT WORLD

With the new millennium, many students entering college choose Fashion Design above all other study options, because, unlike their 1990s peers, "they find the fashion world better than any other job" (Hameed, 2017). This shift in career preference should mean that teaching Fashion Design is easier, as students would be more willingly involved in learning the implicated skills. However, we have been observing that generation Z students – born from the mid-1990s to early 2000s –, today's undergraduates, seem to be unable to relate their own experiences as dressed bodies with the process of designing fashion products.

The reasons for this apparent lack of tacit knowledge – i.e., what we know how to do unthinkingly, e.g., how to get dressed and undressed, how to dress for cold or warm weather – may lay in the fact that fast fashion, a phenomenon that came to the fore in the late 1990s, is the current consumption practice. For one thing, the quick manufacturing of clothes based on current, high-cost luxury fashion trends, sold at an affordable price encourages a 'throwaway' attitude among users (Joy et al., 2012). The constant need for stimulation explains, to some extent, why both fast fashion and users embrace obsolescence. Youngsters, in particular and unsurprisingly, are the perfect 'victims,' as they are "continually evolving temporary identities" (Joy et al., 2012, p. 276). New clothes are available every few weeks and correspondingly discarded, a pace that results in long-run indifference (as though old clothes cannot be affectionately regarded) and excludes from consciousness much of the details (as though clothes are merely perceived as inkblots).

For another thing, the huge number of fast fashion stores that opened in Portugal since the mid-1990s forced the majority of retailers and wholesale suppliers specializing in imported, specialty fabrics to close their doors, as the demand for everyday custom-made clothing decreased substantially. To claim that there is a negative correlation between the two occurrences is not easy to substantiate, if not impossible: according to *Visão Sete*, the disappearance of long-established shops from the main cities is attributable to high rent increase and real estate development (visao.sapo.pt). But the truth, whatever it might be, signals not only the inherently mutable morphology of commercial activity but also the impact of globalization.

Also not easy to substantiate, if not impossible, is the negative correlation between the end of the retailers and wholesale businesses referred to before and the appearance of various chain stores all over Portugal since the early 2000s, selling low quality, low-cost fabrics – in Feira dos Tecidos, for example, fabrics are mainly made of polyester or other synthetic fibres and their prices range from 2.95€/meter to 14.95€/meter –, or alternatively, the parallel between these fabric chain stores and the phenomenon of fast fashion. In truth, fast fashion and consequently the low quality, low-cost fabrics sold at chain stores are grounded on visual stimulus rather than tactile stimuli, causing the sensory and sensual qualities to disappear from artefacts (Pallasmaa, 2005), namely fabrics and clothing.

To top it all off, as true digital natives, Generation Z spends approximately nine hours a day on entertainment media, excluding screen time spent at school or for homework (Trifecta Research, 2015). The problem is that computer imaging tends to flatten our magnificent, multi-sensory, simultaneous and synchronic capacities of imagination by turning the design process into a passive visual manipulation, a retinal journey (Pallasmaa, 2005).

Thus the one-dimensional, flattened out environment that Generation Z students are subject to unable them to identify intuitively – i.e., from experience – what makes or doesn't make fabrics suitable for a particular style, much less to put into exact words what they unconsciously know about fabrics.

On the other hand, the almost exclusive use of computers and other electronic devices "creates a distance between the maker and the object, whereas [...] model-making puts the designer into a haptic contact with the object or space" (Pallasmaa 2005, p. 12).

3 THE SENSE OF TOUCH

Western culture has always favoured vision over other sensory modalities. Classical antiquity tradition, for instance, correlated vision with fire and light, and considered touch as the most earthly of the senses and therefore the most unlikely sense to inspire the intellect. In fact, the history of touch is somewhat detached from the history of skin and until the 16th century the skin was regarded as a by-product of the hardened flesh, when in contact with air, and therefore not having sensory properties.

However, touch is truly a complex sensory modality. Unlike the other sensory modalities, it is not located at one specific point in the body and it involves a myriad of informational channels that directly support both notions of bodily awareness and exploratory action. The skin, now commonly established as the organ for our tactual experiences, is fully innervated by different nerve endings – each of them specialized in the discrimination of a particular set of stimuli (Linden, 2015). But the excitation of those nerve endings does not suffice for touch to happen: other sensory perceptual apparatus need to come into play, namely proprioception, the awareness of our bodily position, and kinesthesis, the awareness of our bodily movements. The information captured by all of these different perceptual mechanisms brings about the sensory experiences that we generally denominate as touch.

Due to the intricacy of the perceptual mechanisms that constitute touch and the stimuli they perceive, most authors dealing with this subject tend to define different dimensions manifested in the sense of touch. One of the most frequent distinctions is the passive touch/active touch, even though the multitude of sensory transducers engaged in each of these categories makes it is difficult for the scientific community to reach a consensual definition. For example, Gibson (1962) defines passive touch as merely the stimulation of the skin's receptors located in the surface of the skin and in tissues immediately underneath, while active touch – because exploratory in nature and therefore self-generated – as the intertwining of proprioception, kinesthesis, and cutaneous stimuli, along with a sense of purposiveness. Fulkerson (2004) describes passive touch more thoroughly, naming it also tactile or cutaneous sensation, characterizing it as being body-directed, involving no control, and tending to occur in parts of the body we have little direct exploratory capacity; conversely, active touch, also called haptic touch, is a combination of cutaneous, kinesthetic and other motor feedback stimulation, body controlled, object directed and exploratory in character, mainly present in the hands, which are specifically designed for engagement and exploration.

Regardless of the definitions that better describe the two dimensions of touch, the truth of the matter is that touch is a special sensory modality, as it requires a unique connection with the surrounding world. To know the world through touch is to reach out and actively engage with it (Fulkerson, 2004), and therefore touch involves a commitment that no other sense modality does. In addition, touch gives every terrestrial animal a "constant background of stimulation" (Gibson 1962, p. 480), functioning as the operative interface that integrates every sense of experience (vision included) with the sense of self – "the very locus of reference, memory, imagination and integration" (Pallasma 2005, p. 11). Touch is also of paramount significance to the physiological development of human beings. Linden (2015) describes studies of touch-deprived infants in understaffed orphanages or in isolated incubators where the deficit of nourishing touch, during the early stages, invariably resulted in compromised cognitive, motor, and immune systems.

In brief, tactual experience – the term Fulkerson (2004) uses to refer to the various aspects present in the sense of touch – is a combination of different dimensions whose complex intertwining makes possible the engagement with/exploration of the environment, the notions of embodiment and self and the proper physiological development.

4 THE TACTUAL EXPERIENCE OF TEXTILES

But in what respect is touch essential for fashion design? Why do tactual experiences need to be instigated amongst Generation Z fashion design students as a means for obtaining knowledge about textiles?

Fashion, and for that matter clothing in general, has been frequently described by 20th century theorists as a cultural phenomenon, where clothes act as signs of social, taste and gender distinctions. This particular line of thought – one that fails to reflect on the corporeal sensations of the wearer – tends to perpetuate the preference for vision over touch. But as Fulkerson (2004) highlights, there is a wide variety of features obtainable through touch, namely thermal, weight, density, and texture qualities, just to name a few; Li & Wong (2006) add that touch can also provide a good basis for the estimation of fabric mechanical properties, such as bending, shearing and tensile. All these aspects are crucial for the physical experience of the wearer, and consequently for a fashion designer to consider.

Moreover, to overlook the fact that dressing is inevitably feeling the fabrics in direct contact with the skin, is to discard the profoundly emotional experience that wearing clothes can entail. Interestingly, a team of neuroscientists used fabrics to assess the role of the skin's C-tactile fibres in the affective components of touch (Essick et al., 1999); for that matter different fabrics (velvet, cotton and plastic mesh) were stroked across the subjects' hairy skin, concluding that the hedonic qualities of touch can be psychophysically evaluated and that valid and reliable estimates are obtained. Thus, it is not hard to understand Candy's (2005) description of her intense childhood memories of the feelings elicited by her school uniform, donned after summer vacations, where qualities like weight, texture and shape of the garments brought out an affective anticipation of wintry emotions. By experiencing different fabrics through touch, taking into account their subjective/emotional aspects, like pleasantness/unpleasantness, can bring forth "new ways to understand and to design for human experience" (Candy, 2005).

This is not to say that aesthetic appeal of fabrics in fashion design should be rejected. On the contrary, experiencing different fabrics by means of pressing and rubbing it through the fingers and thumb – a feeling commonly referred to as 'fabric hand' – allows to predict the suitability of the fabrics for the designed clothing, preventing discrepancies between conceptualization and prototyping.

The physical and visual properties of fabrics can enable innovation in the context of fashion design – both in couture and ready-to-wear –, as the ground breaking creations by designers from different generations of the twentieth and twenty-first centuries can corroborate: the work by Mariano Fortuny (1900s), Madeleine Vionnet (from 1912 to 1939), Paco Rabanne and Pierre Cardin (1960s), or more recently Issey Miyake, Rei Kawakubo and Junya Watanable (all since the 1980s), Anrealage (since 2003) and Iris Van Herpen (since 2007) pivoted around the aesthetic and mechanical properties of fabrics and fabric manipulations.

What needs to be done in regard to Generation Z fashion design students – mesmerized by all the visual allure of luxury fashion delivered through fast fashion –, is to remove fashion from the one-dimensional realm of electronic devices and bring its tangible, multidimensional character to the fore. What better way to start than to experience the structural, emotional and sensory aspects of different textiles through touch?

5 TEACHING TEXTILES TO GENERATION Z

In order to rethink the way fashion design is taught, educators must, before anything else, remember that Generation Z students, today's undergraduates, are the first to be truly digital natives. And so they have never known life without Internet, cell phones and other electronic devices; even though they were born a few years before the boom of social media they don't remember "a world in which letters were printed and sent, much less hand written" (Palfrey & Gasser 2016, p. 4). They have lived their entire lives with instant access to a colossal amount of data about any topic that crosses their minds – without questioning its accuracy, losing interest quickly (Williams, 2015) – and have never known the challenge of researching using analogical tools – such as encyclopediae, archives and specialized literature. They avoid libraries because reading takes too long and it implies concentration – an impossible deed since their attention span is believed to have shrunk to eight seconds (Finch, 2015). They are technologically savvy and prefer spending their time playing videogames while communicating with their peers via social media; they prefer texting (using acronyms, emoticons and emojis) over talking; they typically communicate in spurts of shorter, but more frequent, bursts of information (Lenhart, 2010).

If we, educators, are not prepared to understand and face the characteristics of Generation Z, we risk increasing the gap between the way we teach and the way this generation learns. Obviously to avoid this barrier requires thinking of and embracing alternative learning approaches. Actually researches on teaching Generation Z students highlight that we should not use "the same draconian methods employed from our former college professors" (Faerm, 2013). To oppose this "on the stage" form of teaching (Ford, 2015), we should remember that, in fashion design education, many courses deal with fabrics, even if they focus on different aspects, such as Materials, Drawing, Design Studio, Draping and Sewing. Could it be that if we were more willing to deal with the 'big picture' and less with micro details, students would be able to make connections and interrelations concerning different aspects of textiles more easily?

We should also remember that students have a long experience as dressed bodies, thus they "can know more than [they] can tell" (Polanyi [1966] 2009, p. 4). An easy starting point could be to encourage discussions among them, using their clothes as examples of the visual, tactual, structural and emotional aspects of fabrics. By providing the adequate means for students to express their tacit knowledge on the subject – i.e., showing them how to combine their present tactual sensations with the imprints of their life experience (Linden, 2016) –, they would be able to relate more easily fabric appearances with corresponding commercial names, as well as fabric hand with fabric properties and qualities.

Hands-on activities are undoubtedly more efficient than listening to formal lectures and watching demonstrations, as students would engage actively with their bodies rather than depending on their passive capacities to retain 'abstract' information. Implementing exercises involving solely the students' tactual experiences, thus demanding their "disengagement from a broader field of attraction [namely, visual] for the sake of isolating or focusing on a reduced number of stimuli" (Crary 2000, p. 1) – normally excluded from consciousness –, could demonstrate their attention-worthiness. By nurturing an active learning environment the design process and ultimately the experience of wearing would become enhanced.

6 CONCLUSIONS

The questions pursued by us derive from our experience as fashion design educators. By drawing parallels between fast fashion and the disappearance of specialty fabric stores, between fast fashion and the emergence of chain stores selling low-cost, low-quality fabrics, we question if the current consumption reality impedes Generation Z students to relate their own experiences as dressed bodies with the process of designing fashion products.

Thus, to teach Generation Z students demands that we understand and accept their characteristics and, consequently, rethink the way fashion design is taught.

Because fabrics are transversal to many courses, alternative approaches must be implemented, particularly those based on hands-on activities.

We believe that drawing on the students' past and present tactual experiences of textiles would remove the one-dimensional realm of fashion, glimpsed through the screens of their electronic devices, and bring its tangible, multidimensional character to the fore.

ACKNOWLEDGMENTS

This research was only possible with the support of CIAUD – Research Centre in Architecture, Urbanism and Design, Lisbon School of Architecture of the University of Lisbon, and FCT – Foundation for Science and Technology, Portugal.

REFERENCES

Aldrich, W. 1996. Fabric, form and flat pattern cutting. Oxford: Blackwell Publishing.

Ashdown, S. 2013. "Not craft, not couture, not 'home sewing': teaching creative patternmaking to the iPod generation." *International Journal of Fashion Design, Technology and Education*, 6 (2): 112–120.

Candy, F. 2005. "The fabric of society: an investigation of the emotional and sensory experience of wearing denim clothing." *Social Research Online*, 10 (1). http:// socresonline.org.uk/10/1/candy.html [Accessed 20/August/2014].

Crary, J. 2000. Suspensions of Perception: Attention, Spectacle and Modern Culture. London: MIT Press.

DeKay, M. 1996. "Systems thinking as the basis for an ecological design education." *Proceedings of the 21st National Passive Solar Conference*, April 1996. https://works.bepress.com/mark_dekay/11/ [Accessed 6/March/2016].

Essick, G., James, A. & McGlone, F. 1999. "Psychophysical assessment of the affective components of non-painful touch." *Neuroreport*, 13 (10): 2083–2087. https://www.ncbi.nlm.nih.gov/pubmed/10424679 [Accessed 24/April/2017].

Faerm, S. 2013. "Why art and design higher education needs advanced pedagogy." *Miscmagazime*, November 2013. https://miscmagazine.com/why-art-and-design-higher-education-needs-advanced-pedagogy/ [Accessed 17/July/2016].

Finch, J. 2015. "What is Generation Z, and what does it want?" *FastCompany*, April 5, 2015. https://www.fastcompany.com/3045317/what-is-generation-z-and-what-does-it-want [Accessed 24/April/2017].

Ford, T. 2015. "5 Tips for Teaching Generation Z in College." https://blog.tophat.com/generation-z/ [Accessed 24/April/2017].

Fulkerson, M. 2004. The First Sense: A Philosophical Study of Human Touch. Massachusetts, England: The MIT Press.

Gibson, J. 1962. "Observations on active touch." *Psychological Review*, 69 (6): 477–491.

Hameed, R. 2017. "The effects of fashion on teenagers." *Voice of journalists*. http://www.voj.news/the-effects-of-fashion-on-teenagers/ [Accessed 24/April/2017].

Irwin, T. 2015. "Redesigning a design program: how Carnegie Mellon University is developing a design curricula for the 21st century." *The Solutions Journal*, January-February 2015: 91–100.

Jenkins, R. 2015. "15 aspects that highlight how Generation Z is different from Millennials." http://www.business2community.com/social-data/15-aspects-that-highlight-how-generation-z-is-different-from-millennials-01244940#6ABkhm20PH6lzA02.97 [Accessed 24/April/2017].

Joy, A., Sherry, J., Venkatesh, A., Wang, J. & Chan, R. 2012. "Fast fashion, sustainability, and the ethical appeal of luxury brands." *Fashion Theory*, 16 (13): 273–296.

Lenhart, A., Ling, R., Campbell, S. & Purcell, K. 2010. "Teens and Mobile Phones." *Pew Research Center*. http://www.pewinternet.org/2010/04/20/teens-and-mobile-phones/ [Accessed 24/April/2017].

Li, Y. & Wong, A., eds. 2006. Clothing Biosensory Engineering. Cambridge: Woodhead Publishing in association with The Textile Institute.

Linden, D. 2015. Touch: The science of the sense that makes us human. UK: Penguin Random House.

Palfrey, G. & Gasser, U. 2016. Born digital: How Children Grow Up in a Digital Age. Philadelphia: Basic Books.

Pallasmaa, J. 2005. The Eyes of the Skin: Architecture and the senses. Chichester: Wiley-Academy.

Polanyi, M. (1966) 2009. The Tacit Dimension. Chicago and London: The University of Chicago Press.

Simoes, I. & Silva, F. 2016. "Collaborative learning in fashion education." *Fashion: Exploring Critical Issues 8th Global Meeting*, September 2016.

Trifecta Research 2015. "Generation Z media consumption habits." http://trifectaresearch.com/wp-content/uploads/2015/09/Generation-Z-Sample-Trifecta-Research-Deliverable.pdf [Accessed 24/April/2017].

Visão Sete 2012. "O fim de uma era." February 9, 2012. http://visao.sapo.pt [Accessed 24/April/2017].

Williams, A. 2015. "Move Over, Millennials, Here Comes Generation Z." *The New York Times*, September 18, 2015. https://www.nytimes.com/2015/09/20/fashion/move-over-millennials-here-comes-generation-z.html [Accessed 24/April/2017].

Textiles, Identity and Innovation: Design the Future – Montagna & Carvalho (Eds)
© 2019 Taylor & Francis Group, London, ISBN 978-1-138-29611-4

Teaching sustainable fashion and textiles

Theresa Beco Lobo
Unidcom-Iade-Universidade Europeia, Lisbon, Portugal

ABSTRACT: As educators teaching sustainable fashion and textiles, we must examine the apparel and textile industry's major impact on the environment. The paper will explore new ideas from proponents of the slow fashion movement and advocates of sustainable practices. It will examine different perspectives by professors of the design, manufacturing and non-profit community. This paper relates the results of their teaching and research in this subject. In conclusion, as educators and as consumers of fashion, we need to educate and reach out to the consumer and convince them to demand quality. The professors should encourage fashion students to research sustainable materials and fabrics, and to examine the social and environmental impact of the fashion industry. While mastering sustainable fashion issues is not easy, our experience has shown that students can accomplish well thought out projects that reflect a strong eco-consciousness and sense of social responsibility.

1 INTRODUCTION

The paper relates the study of sustainability in fashion and explores experiences in teaching eco fashion. The paper asserts that both design and merchandising students in a fashion program need to be educated about the need to foster sustainability, while being encouraged to find ways to make fashion more ecologically friendly. Two professors from IADE in Lisbon, Portugal, are part of a program in the Department of Fashion to better understand and foster efforts to make fashion more sustainable. This paper relates the results of their teaching and research in this subject.

2 TEACHING EXPERIENCES

First was considered the idea of teaching an eco-fashion course in 2008. It was identified and pursued the leading innovators of sustainable textiles and fashion in Portugal. The students were informed about the environmental impact of their future creative decisions in the fashion industry and they should think about sustainable practices and principles. The master program class was focused on creating green apparel that has a reduced impact on the environment but is still appealing to the consumer. The class learned that fashion and sustainability appear to be contradictory concepts. Therefore, it was critical to teach students to think about the life cycle of their products from production to disposal. These students were eager to learn that both short term and long-term views on sustainability issues are necessary to grow the future of environmentally conscious textiles and fashion. In today's volatile times, it is inevitable that a "power shift" will take place where consumers will demand responsible behavior from brands.

Sourcing organic materials and gaining the support of eco-designers for the course was a remarkable experience for the students. The designer invited Isilda Pelicano owner of JANS CONCEPT label, to act as design critic. Isilda was a former partner in the acclaimed Portuguese-based eco fashion label. She had created her signature material from recycled bicycle inner tube tires. This material was often mistaken for leather. She was well known for her one-of-a-kind couture designs. Isilda selected student designs that created new textiles or experimented with natural dyes that used less toxic processes.

The student designs were edgy and hip, unlike the drab, hippy garb from the 60s.

The professor's innovative ideas were rewarded when they receive a generous donation of organic cotton, and cotton bamboo blend fabrics from Montebelo Textile, a Portuguese manufacturer of circular knit, warp knit and woven textiles. In 2007, the Textile Outlook International Report deemed Montebelo Textile as a pioneer in the organic cotton movement. (Textiles Intelligence, 2007).

The students also utilized hemp and hemp silk blend of Fibnatex of Portugal, peace silk, and organic cotton denim fabrications from vendors. Some used perch fish skin leather as trim detail in their designs.

Other students recycled second hand clothing into new garments or experimented with natural dyes for new colorations.

Between classes, the professor mentored a student grant research project called "Deconstruction-Reconstruction." The project was based on the principle of 'zero-waste' in the design studio use of muslin. The students researched recycling fibers in response to the large amount of scraps wasted each day in the fashion design department.

Throughout the project's duration, the students collected muslin scraps from design studios and tracked muslin waste, with the goal of creating a fabric-recycling program. Students from various areas of study, including fashion, fibers and metals, collaborated to create one-of-a-kind garments and accessories that showcased their creative abilities while minimizing waste. The students concluded their collaboration with an exhibition made possible by the undergraduate research grant.

Isaac Mizrahi, internationally acclaimed designer, as a leader in the design business for more than twenty years, Isaac has been awarded four CFDA awards, including a special award in 1996 for the groundbreaking documentary Unzipped. In 2008, he became creative director of Liz Claiborne, Inc., where he has the opportunity to wield enormous influence on popular taste as well as the means to explore sustainable practices in mass production. Isilda visited IADE's fashion department as a lecturer and class critic. Isaac's appearance inspired the students to employ his hand-made techniques in their projects.

In comparison to Carlos, Theresa came late to the teaching of sustainability. As a merchandising professor and a veteran of over thirty years in the Portuguese fashion industry, she had a strong interest in this area of study, but was unsure how to bring the subject of eco-friendly fashion into the classroom on any but an informational basis.

From a mostly business (merchandising) point of view, Theresa had concerns that the fashion companies (as well as those in other industries) striving to become eco-friendly had less than altruistic motives. While a proponent of keeping the fashion industry profitable and thus viable, Theresa sought examples of companies that were able to do both: reduce their carbon footprints and still make enough money to sustain and grow their businesses.

From a historical perspective, it was disturbing to witness the destructive path pursued by the fashion business during the past several decades: less creativity, more emphasis on price while often sacrificing quality and durability, and the "rush" by so many retailers and wholesalers to embrace the new god called "fast (and disposable) fashion".

By this point, several other professors in the department had become interested in the subject of sustainable fashion and were looking for ways to incorporate the subject into their classes. A generous grant from Cotton, Inc. enabled the entire department to devote two design classes and one merchandising class (Product Development) to the study of sustainable and recycled clothing. The grant enabled the department to take ten students and three professors to MOMAD-Salón Internacional de Moda in Madrid, Spain in September 2014. The students attended seminars on eco-friendly and sustainable fashion. They also met vendors who were engaged in the green manufacturing and fair trade of fabrics, components, and finished garments. To conference participants, this exposure was extremely valuable in increasing awareness of the progress and challenges associated with reducing the carbon footprint of the fashion business.

Back at the university, a new semester started with the research IDEA-LAB, about 10 design graduation students and 5 marketing graduation students. Information from the MOMAD trip was dispensed to the other students in this team. Design and merchandising students (approximately 15) in the two classes they made a visit to the Agratil – Industria Textil, Lda. This cotton company in Maia, North of Portugal. In the highly intense one-day visit, they were exposed to Agratil's efforts to develop more sustainable fabrics and additional ways to recycle cotton fabrics and fibers. Students learned about the steps needed to grow and process certified organic cotton, and other research underway to reduce the amount of pesticides and harmful chemicals used in traditional processes of growing and processing standard cotton.

The factory Diniz & Cruz had 3 designers' students and 2 marketing students in an internship, who worked in this company during a semester. It simulated building a line of denim jeans for a store in Lisbon.

While students did not make actual garments, they did merchandise a line and show the product on design boards. As well, they were required to create inspiration boards for their products, complete a specification package with all details and measurements and cost sheets for the products. Finally, they designed a marketing piece to inspire and educate store associates on how to sell and market this new private label line of jeans (Elzakker, B. van, 1999). The class challenges included a requirement that fabrics be either sustainable (could be partly made of organic cotton or other materials), contain at least 75% cotton, and/or consist of at least partially recycled materials (fabrics, findings, trim, or packaging).

After outlining the basic requirements, the professor allowed the researchers teams flexibility and room for creativity to visualize their sustainable line of jeans. An early assignment was to prepare a detailed Market and Trend Report that communicated the results of their research into the subject of fashion and sustainability. It was crucial that they do sufficient research to separate fact from fiction on the subject and to have a feel for what is possible and new in developing eco-friendly fashion. The results were very exciting and rewarding.

Some students put emphasis on the fabrics themselves, selecting blends that used a combination of organically grown cotton with regular cotton and other fibers. Other students use partially recycled fabrics, trim, or findings. Still others put their emphasis on recycled packaging, hangtags, and hangers. Lastly, some researchers put their emphasis on in-store events, a good example being marketing efforts that included an offer for customers to bring in old jeans for recycling in exchange for a discount on a new pair of the store's private label jeans.

Further emphasis was placed on plans for sourcing components that could be found closer to the sewing factory. This issue spoke to the environmental costs of transporting goods from factories to stores. The assignment had to take into account an important element not readily apparent in a design-only class, i.e., all products had to be able to be sourced and manufactured at price points that would be viable in a private label program for a mid-price department store chain in medium sized towns around Lisbon, Portugal.

As a result of this program, students devised very creative approaches to completing their assignment, learned a more realistic view of the eco-friendly fashion movement, and realized the importance of a consumer message that communicates the sustainability theme. A semester-end presentation of the projects by a Lisbon fashion industry critic and a representative from Citeve – Centro Tecnológico das Indústrias Têxtil e do Vestuário de Portugal yielded high compliments on the work created and the message presented.

As part of the cotton project semester program, the professors also introduced the students to major environmental organizations concerned about eco fashion. They invited the executive director of ANJE – Associação Nacional de Jovens Empresários and the director ATP - Associação Têxtil e Vestuário de Portugal is to Support and Promote Sustainable practices, facilitate collaboration, raise Awareness and provide the Tools and Resources needed to reduce poverty, reduce environmental damage and raise standards in the fashion industry. In 2004, on a grey January day in London, a group of determined fashion designers and business people sat around a table to discuss the challenges they faced and what they could do about it. They saw that some of the challenges they faced could be resolved by a unified approach, better communication and dissemination of information, pooled resources and shared practices. An ethical fashion network would open markets for fair trade, organic and values led producers, and make it easier for fashion businesses to source and produce in a sustainable way.

Over the course of the next year networking events and round table meetings took place around Portugal – and in 2014 the Ethical Fashion Forum was formally founded as a unified, not for profit organisation with 20 founder member businesses.

By 2014, GREENFEST EVENT, Estoril.," an annual, large-scale sustainability summit and a call to action. The GREENFEST organized a sustainable fashion show during Lisbon's Fashion Week. The show enlisted 32 top designers to create one-of-a-kind pieces made from sustainable materials.

The show demonstrated the use of renewable, reusable, non-polluting fabrics such as organic cotton and wool, bamboo, corn-based fibers and biopolymers, which were designed into everything from couture to street wear (Hoffman, L., 2003).

Students learned how the organization's initiative on Future Fashion spurred the fashion industry to transition to more sustainable practices. The GREENFEST published the Future Fashion White Papers to promote innovative thinking within the industry. Furthermore, Hoffman assembled a sustainable textile library, providing resources for designers and professionals.

Hoffman hopes that newly minted environmentalists will become informed consumers able to guide the market toward better green products (Hoffman, L., 2003).

For the professors second year research on sustainable fashion, they team-taught a studio called Idea.Lab.. Students did independent research and contacted designers who addressed design and sustainability issues in their collections. They also interviewed emerging designers with exhibits at the Modtissimo; a trade shows for the boutique market and better retailers. The premise of the investigation was for students to explore the growing street wear scene and to utilize sustainable, recyclable or organic cotton fabrics (Slater, K., 2003). The approaches ran the gamut from recycling cotton garments to sourcing sustainable fabrics to creating collections made with new cotton fabrics. The semester resulted in deep learning for the students, as evidenced by how their design outcomes reflected their understanding of this ethical challenge.

The ten projects, juried by both an industry critic and a Citeve representative, yielded a publication featuring their sustainability collections. A selection of the projects was exhibited on campus and at the department's annual fashion show.

In planning for the semesters following the Cotton, Inc. grant class, the professor sought ways to keep the subject of sustainability in his Product Development class and to expand the subject into other classes. He was gratified to find that students easily identified with and showed great enthusiasm for this area of fashion. For Product Development, the classes continued to include a focus on sustainable fashion. It required students to include research on eco-friendly fashion in their Market and Trend Report, and allowed them the flexibility to include elements of sustainability into their private label jeans line.

Another team of Fashion Forecasting, proved to also be a natural fit to introduce discussions and studies on the effect of the green movement on current and future fashion.

At the end of the semester, after a period of research focus on future fashion (10 years from now and beyond), students were required to write a forecast on macro issues likely to impact the fashion industry in ten years. The environmental message is a natural subject of interest for this report, and many of the students chose this driver as the major focus of their prediction paper. The student predictions were creative, insightful, and perhaps gave a glimpse of our world to come.

3 ECO-FASHION, THE PAST AND THE PRESENT

In the past few years, sustainable clothing and eco fashion came into mainstream consciousness with much media attention and hype. Major campaigns were being waged for leadership in the organic food and clothing markets. Megastores like H&M in Lisbon, and El Corte Ingles in Porto, took notice and introduced organic food and clothing. While organic clothing evolved out of the organic agricultural movement, sustainable clothing is a product of the environmental movement. In other words, organic clothing has its roots on the farm and sustainable clothing, in the lab. The last approach places emphasis on reuse and recycling of manufactured products.

One market leader in environmentally conscious practices is Patagonia, Inc., who was the first outdoor apparel retailer to use post-consumer recycled fleece (PCR). By using PCR fleece, Patagonia has prevented more than 86 million plastic bottles from being landfilled.

In the early 90s, the company commissioned a study of the environmental impact of its raw materials and found that oil-based synthetics, such as polyester and nylon, caused less harm to people and the environment than standard cotton. Patagonia was one of the first companies to introduce organic cotton and recycled polyester. In 1994, Patagonia converted its entire cotton line to organic cotton. Dedicated to using processes and materials that have as little negative impact on the planet as possible, Patagonia partnered with Wellman Inc., the developer of the PET recycled bottle polyester fibers known as EcoSpun (Sustainable Technology Education Project, 2008).

4 FAST FASHION

Fast fashion became an industry catchphrase with the growth of Zara, owned by Inditex of Spain. It is currently the world's largest clothing retailer, operating 3900 stores in 70 countries around the world. Although the first Zara store opened in 1975, the real buzz started with the first Oporto, Portugal in 1989 (Keeley, G. & Clark, A. 2008).

Many consumers, especially teens, find themselves shopping in retail chains known for their trendy "fast fashion," such as H&M, Target, and Zara. The prevailing concept is that low-cost garments can be readily discarded. Environmentally, such consumerism is disastrous.

The focus on speed, large volumes and greed contributes to global warming. In the report, "Well Dressed," Cambridge University researchers interested in changing consumer behavior, characterized the term, sustainable clothes, as an oxymoron (Rosenthal, E., 2008).

The Zara model is incompatible with sustainable fashion because organic and sustainable fashions typically take more, not less, time to produce. Raw materials such as organic cotton need extra growth and processing time and components such as recycled materials are not abundantly available. Eco-friendly dyeing methods are more time-consuming than commercially available quick dyeing methods (Capell, K., 2008).

Fast fashion and sustainability reflect inherently conflicting priorities. If getting fashion to the store faster and cheaper is the focus, sustainability suffers. In addition, fast and cheap fashion is usually regarded by the consumer as disposable, and runs counter to the concept of clothing that lasts. Fast fashion is often the first to end up in landfills.

5 SLOW FASHION

The concept of slow fashion is, of course, not new. Prior to mass production, all fashion was slow. Garments were produced one at a time with limited fabrics and hand-made. The life spans of these garments were much longer – usually until they wore out. Restyling or passing garments along to another family member was quite common. Overtaxed landfills were not a concern. Fashion has been both slow and restricted for the elite. Since the late 1800s, the wealthy desired to dress in the style of the European couture. However, they had to wait for these hand made garments to be available.

As with so much of what happens in the world of fashion, a backlash is happening. Today, fast fashion is being challenged by a slow down: less quantity, more quality. The over consumption of cheap, low quality products is looking dated. Kate Fletcher, an eco fashion pioneer, says the solution is Slow Fashion (Fletcher, K., 2008).

"Slow fashion is not time based, but is quality based. It's a shift away from an agenda that's about the quantity of stuff." She also states that while Slow Fashion will never go mainstream, it can still have the power to change behavior (Fletcher, K., 2008).

Many stores are promoting recycled clothes and several, like Urban Outfitters, offer reconstructed clothes alongside new ones. Some high-end fashions are very expensive garments.

For example, the customized garments of Isaac Mizrahi are not likely to end up in a landfill. These exquisite pieces have what the fashion industry refers to as "emotional durability," a personal connection that makes it likely they will not only remain in one's wardrobe, but get passed on to the next generation (Vitale, Ami, 1999).

For many in the industry, the very nature of the fashion business makes the promotion of buying less and wearing it longer an alarming thought. There exists powerful incentive to encourage consumers to buy more and more.

In this way, their own jobs are assured. Also, because much of what goes into the manufacturing of clothing on a mass scale is, by its very nature, the antithesis of ecological friendliness, fashion industry insiders are often overwhelmed with the idea of building a sustainable business model. As many in the eco and sustainable movement are quick to say, it's unrealistic to think that we can ever make fashion completely sustainable. According to Janet Hethorn and Connie

Ulasewicz in their book, Sustainable Fashion, Why Now, the definition of sustainability is when "you take out of a system the same amount of energy as you put in, with no pollution or waste." (Hethorn, J. & Ulasewicz, C., 2008).

Nonetheless, fashion industry insiders can take actions to promote the shift to slow fashion.

- The time seems to be right for the growth of small, independent designers, manufacturers, and even retailers (including internet retailers) to emerge and succeed. The need for high quality, well-designed, unique items that justify "investment purchasing" would point to the need for hands-on original design, component parts (including fabrics) that are not available in large quantities, and individualized specialized manufacturing. Considering that many consumers today (both men and women) are more confident about their fashion choices and like to establish an individual "look," this growth of smaller fashion businesses would seem to be very appealing to these customers.
- Along the same lines, a renewed interest in vintage fashion opens up additional opportunities for retailers to resell vintage items, for designers to secure and re-style vintage items, and for growth of a market in component parts to be recycled into new garments.
- We need a re-birth of couture-like designers to create original, quality styles for a larger market. However, their fabric and component choices need to be more sustainable and able to be worn for many seasons and years.
- All retailers and manufacturers must renew their efforts to improve the sustainability of their businesses through conservation of energy, wise use of natural resources, reduced waste, and sourcing closer to manufacturing and end-destination site. If everyone makes progress in this area, the industry can make exceptional progress toward a more ecologically friendly future.
- Not to be ignored in this conversation is the need for a more rapid move toward fair trade practices in manufacturing and the importance of ethics in the subject of fashion production.

6 CASE STUDIES: (1) LOCAL PRODUCTION MODEL, AND (2) ANSWERING THE NEED FOR SUSTAINABLE FASHION ON A LARGER SCALE

Today, the global market in textile production causes many products to be transported several times over long distances between processors before they reach a user.

Transporting a product uses resources and creates pollution, both seen as externalized costs. Some argue for the creation of a model of local production, and global branding.

Isaac Mizrahi is an outstanding example of local production. Isaac introduced a series of U.S.-based community revitalization projects that utilized local craft traditions to produce hand-made couture garments. He enlisted local women who were former factory workers, retired teachers, widows, housewives and secretaries to sew one-of-a-kind, handmade garments for her fashion line.

Isaac employs a labor intensive, time-honored stitching tradition. He approach 'Slow Design' embraces the long-term view over a short-term gain by using age-old techniques to create products that celebrate strong design principles. He meets the challenge by using local ingredients harvested and put together in a socially and environmentally responsible way. In light of today's economic and environmental challenges, it may be wise to develop strong regional manufacturing centers closer to home that would help reduce the negative impacts of global production.

The Nature Conservancy commissioned Mizrahi to work with salmon skin. Mizrahi's design features leather paillettes—small disks akin to sequins that are sewn into a supporting layer of fabric to create a rich, fluid surface. Previous Mizrahi projects include cutting paillettes from aluminum soda cans, a cheeky experiment in using trash to create luxury.

By contrast, Martha Stewart Company (a public company) is a large-scale producer of women's sleepwear and robes and they have recently introduced an eco-friendly sleepwear line made from certified organic cotton and other eco-friendly components such as recycled hangtags and buttons. For a large corporate retailer or manufacturer to enter the eco-friendly field, the challenges are daunting. While small runs and handmade products meeting organic standards can be managed, production in large quantities for multiple outlets can be overwhelming.

The requirements for oversight of each step of the process, the need for speedy deliveries, and the larger quantities of fabrics and other component parts can be difficult to accomplish.

A desire to enter this field combined with the needs of a large customer led Jaclyn Industries' sleepwear division to develop a collection of organic cotton sleepwear.

According to Elaine Crossley, Martha Stewart Company's, there were many challenges. They had to find a source for the cotton and it was only logical to seek that source in China since their manufacturing facilities were located there. Knowing that it takes three years to convert farmland that has been used to grow conventional crops to organic farmland, they had to seek a grower who was already operating an organic farm. Agents in their Shanghai office, who then had to locate a quality inspection service certified in the organic field to conduct the required number of farm and factory inspections to insure compliance, accomplished this.

They found this service in the Netherlands, an extra expense considering the distance from source to farms and factories involved with the program.

7 CONCLUSION

The interest among consumers today for an "organic" and "environmentally friendly" lifestyle is indeed growing. Consumer interest and focus is certainly stronger when it comes to foods and cleaning products; an online 2008 survey of 2000 adults found that between 40% to 56% of males and females between the ages of 18 and 64 make a distinction (when purchasing) between natural and organic foods and beverages. In spite of the lack of government oversight of product claims and establishment of a clear definition of natural, these consumers understood the difference. Similarly, a 2007 McKinsey survey of 7751 people in eight countries found that 87% of consumers worry about the environmental and social (fair trade) impact of the products that they purchase (Mintel International Group, Ltd., 2008).

According to Ethical Fashion Forum EFF 2008 survey found that consumers also considered the environment when deciding upon their clothing purchases.

When asked to rate the motivations for their purchases, eco-friendliness was given by 50% of respondents, an increase of 5% over the responses in a 2006 survey. The results also showed that eco-friendliness was more important to these consumers than brand name. Although a 2007 survey by Chain Store Age found that 25% of those surveyed had not purchased any green products except food and household cleaners, the McKinsey survey mentioned above found that 33% of their respondents had either already purchased green products or were ready to do so (Murphy, S., 2009).

Many large and small fashion companies are moving to reduce their carbon footprints in every aspect of their businesses. It is widely known that H&M's green program was initiated by the CEO Lee Scott and then filtered down in many different areas of the corporation: Changes to incandescent lights, newer more efficient delivery truck engines, education of all employees in creating a more sustainable workplace, and insistence that all suppliers in countries like China comply with strict environmental laws.

There are, however, hurdles that all fashion companies must face in their attempts to reduce the carbon footprint:

- Longer lead times, as much as an additional two months from conception to store delivery
- The challenge of color consistency when vegetable and fruit dyes replace more traditional ones
- Not all natural fibers make fabrics that are commercially viable, and must be evaluated for problems such as pilling.
- Costs for producing organic fashion can add 20% to 30% to total costs.

Progress towards overcoming these challenges will surely come, albeit in unexpected ways.

According to Mark Penn and E. Kinney Zalesne, authors of the book Microtrends: The Small Forces Behind Tomorrow's Big Changes, "The biggest trends today are micro: small, under the radar patterns of behavior, which take on real power when propelled by modern communications and an increasingly independent minded population. In the U.S. one percent of the nation or three million people, can create new markets for a business, spark a social movement, or produce political change." (Penn, M. & Zalesne, E.K., 2009).

In conclusion, as educators and as consumers of fashion, we need to educate and reach out to the consumer and convince them to demand quality.

We need to promote a standardized rating system for textiles and clothing that shapes the next generation of farmers, producers and consumers.

Many organizations and manufacturers are working towards establishing an evaluation system for textiles. The Leadership in Energy and Environmental Design (LEED), developed by the U.S. Green Building Council, can serve as a model. We encourage fashion students to research sustainable materials and fabrics, and to examine the social and environmental impact of the fashion industry. (Haight, S., 2008) Additionally, it is not enough to design ecofriendly, because the garment must appeal to the consumer, by fitting well and being stylish.While mastering sustainable fashion issues is not easy, our experience has shown that students can accomplish well thought out projects that reflect a strong eco-consciousness and sense of social responsibility. We must do more to nurture future green designers and product developers, and make a difference to save the planet.

REFERENCES

Anderson, K. Sustainable Fashion. [TC]2 Bi-Weekly Technology Communicator, 2008, November 26.

Capell, K (2008, October 29). Zara thrives by breaking all the rules. Business Week, Retrieved October 13, 2008 from http://www.businessweek.com/magazine/content/08.

Elzakker, B. van, Organic cotton production, in D. Myers and S. Stolton (eds), Organic Cotton: From Field to Final Product, London: Intermediate Technology Publications, p22, 1999.

Fletcher, K. Sustainable Fashion and Textiles : Design Journeys.
London, England and Sterling,VA: Earthscan, 2008.

Haight, S. High minded; a handful of designers make use of eco-friendly elements in their latest collections. Women's Wear Daily, p. 12-13, 10/28/2008.

Hethorn, J. & Ulasewicz, C. (2008). Sustainable fashion, why now? New York, NY: Fairchild. p. 255, 2008.

Hoffman, L. (Ed.). Future Fashion White Papers. Lincoln Park, NJ: Greg Barber Co., 2007.

Houck, A. (2007, September/October). Intelligent design: from classroom to catwalk, Sustainable design has arrived. Veg News Magazine, p. 40–45.

Keeley, G. & Clark, A. (2008, August 12). Retail: Zara bridges Gap to become world's biggest Fashion retailer. Guardian.co.uk, Retrieved October 15, 2008 from http://www.guardian.co.uk/business/2008/aug/12/retail.spain.

Mintel International Group, Ltd., (2008, December). The evolving natural lifestyle.

Murphy, S. A look ahead: the supply chain in 2009. Supply Chain Management Review, Retrieved January 20, 2009.

Natural

Penn, M. & Zalesne, E.K. (2009, March 13). Value is the new green. Wall Street Journal, Retrieved March 18, 2009 from http://online.wsj.com/article/SB123689912898512981.html.

Rosenthal, E. 'Fast clothes' versus 'green clothes.' New York Times, 2007, January 24 Retrieved August 3, 2008 from http://www.nytimes.com/2007/01/24/business/worldbusiness/24ihtclothing.4332207.html.

Slater, K., Environmental Impact of Textiles: Production, Processes and Protection, Cambridge: Woodhead Publishing, p. 27, 2003.

Sustainable Technology Education Project (2008). See http://www.stepin.org/, Retrieved August 12, 2008.

Textiles Intelligence (2007, July-August). Organic cotton small but growing.

Textiles Outlook International, Issue 130. 2007, July–August Retrieved July 21, 2008 from http://www.textilesintelligence.com/tistoi/index.

Vitale, Ami, Design for a Living World, Isaac Mizrahi, New York: Assouline, pp. 16–20, 2009.

Textiles, Identity and Innovation: Design the Future – Montagna & Carvalho (Eds)
 ISBN 978-1-138-29611-4

Methodology to study the creativity of Portuguese fashion designers

J. Barata, R. Miguel & S. Azevedo
University of Beira Interior, Covilhã, Portugal

ABSTRACT: The instable and unpredictable economies have challenged all types of companies to be concerned with creativity and innovation matters. Nowadays innovation is known as the result of a system, involving the creation (creativity), the management of process, the distribution of products and the final use. In this work the 'design' is perceived as an important contribution for differentiation as it is associated with creativity and then innovation. The Portuguese Textile and Apparel Industry presents models and trends so that until 2020 these transforming entities are prepared to collaborate in a competitive way by offering products and services with added value. This is the best way to ensure the survival and the opportunity to target external markets. These indications are mostly directed to the employers. This study argues that workers, especially fashion designers, are a great driving force of the companies' results, therefore its competitiveness. And so it is appropriate to study the creativity of fashion designers working in Portuguese Industries. The strategy to be adopted is interdisciplinary and the creativity is considered as being directly related with the surrounding contexts/variables.

1 INTRODUCTION

The following methodology was developed to study and understand how both physical, labour environment, sociocultural and geopolitical contexts affect the creative behaviour of the fashion designers that work in the Portuguese Textile and Apparel Industry.

By understanding these connections, we are aiming to develop scientific contents to be used by the industries, notifying for a behave that enables the creativity of its workers; fashion designers in particular.

The objects of this study are creativity, innovation and Portuguese Fashion Industry; here we mind the professional activity of the fashion designers and their creative predisposition. The creative outputs may lead to innovation and thus to a product or service with increased value. These facts contribute to the company's competiveness in the market.

This document presents the theoretical and empirical methodology, a literature review on creativity, innovation and the industry. In the end it is presented a framework about relevance and originality of this study.

2 THEORETICAL AND EMPIRICAL METHODOLOGY

The strategy adopted when designing this methodology was to have a multidisciplinary concept about creativity and its study. The methodologies to be applied are mixed (qualitative and quantitative) in order to achieve the goals of this study.

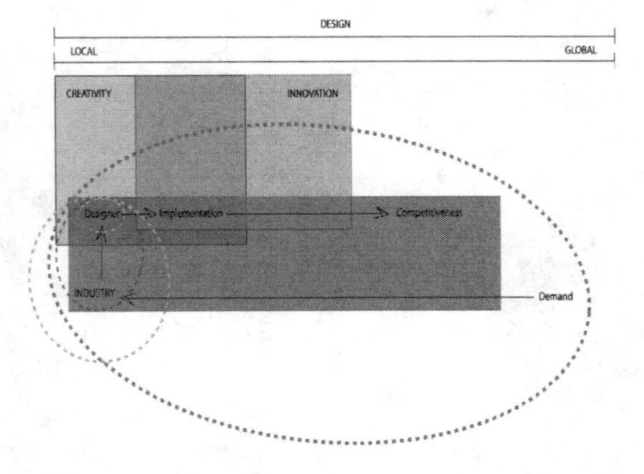

Figure 1. Work frame and onion style method. Font: authors.

The first step is a deep literature review about creativity, innovation and the textile and apparel industry, this review aim, among the deep knowledge in these areas of study, to detach the precise contexts to be study.

Data collection will approach an onion style method. The figure 1 shows the areas of study in rectangles (creativity, innovation and industry) and the dashed lines represent the onion method mention above. From small to larger circles, from the workplace and company's structure and environment to local and global geopolitical and sociocultural conditions, every relevant context is to be study.

The next stage of this analysis is to collect data among the sample companies in order do deduce its results.

3 LITERATURE REVIEW

3.1 *Creativity*

Etymologically, creativity is related to the term 'create' and comes from the Latin word 'creare' and from the Greek 'krainen' (Bahia, 2008), meaning "to exist, to come out of nowhere, to establish relationships that were not established by the universe of the individual" (Pereira, Mussi, & Knabben, 1999, p. 4).

The creative output can be reviewed in an anthropological approach. It is associated to a personal contribution to a common future through the development of the creative potential, attempting to adapt.

The creation seems to be genetically associated with the survival skill, and so, it can be analysed in the most diverse areas of human innervation. Adapting and problem solving process are intrinsically related with the species survival (Bahia, 2008; Gurgel, 2006; Pereira et al., 1999).

In a world of constant technical, social, political and economic transformations, the adaption skills seem more important than ever before.

These subjects can be referred to both individual and group/organizational matters. In an industrial context, this subject is highly pertinent (Amabile, 1983, 1996; Gurgel, 2006; Hennessey & Amabile, 2010; Martins & Terblanche, 2003) once "Post-industrial organisations today are knowledge-based organisations and their success and survival depends on creativity, innovation, discovery and inventiveness." (Martins & Terblanche, 2003, p. 64).

3.1.1 *Approaches on creativity*

3.1.1.1 Divine Approach

Ancient times saw creativity as attributed to inhuman elements to the gods or demons who were responsible for the creative act (Gurgel, 2006; Oliveira, 2012; Valente, 2015; Wechsler, 1998), "something magic which is out of our control" (Von Stamm, 2008, p. 10). And this notion lasted for centuries, the existence of a 'genius' that channelled clairvoyance to some men with privileged access.

3.1.1.2 Psychodominance Approach

By the 16th century, creativity began to be regarded as a form of madness, "This [Freud] saw in the unconscious the origins of both neurosis and creativity, having conducted case studies of eminent people, such as Leonardo da Vinci, considering his creative work as a sublimation of repressed complexes" (Alencar, 2003, p. 1)

Gestalt psychologists have indicated that the occurrence of creativity would be associated with a insight moment, when the creative person suddenly discovers an outcome to the problem (Lima, 2006). "Another theory that contributed to the study of creativity was Gestalt theory. It brought up the research on the insight moment, that is, on the moment when suddenly a solution to a problem appears" (Alencar, 2003, p.1).

3.1.1.3 Cognitive approach

During the Renaissance the human being was the focus of attention and this concept was renewed. Here the creative ability was linked with the human condition, as a result of a cognitive process (Oliveira, 2012). The 'Attitudes Research Project' conducted by Guilford (1949 to 1969) focused the cognitive abilities as study-base criteria to understand the creative process and the creative being (Oliveira, 2012; Wechsler, 1998).

3.1.1.4 Psychometric approach

The psychometric approach represents an evaluation, measurement and quantification of cognitive processes. From a set of variables, these analyses operate thought biographical and personality inventories, predisposition for creativity and behavioural tests (Amabile, 1996; Lima, 2006).

The 'Torrance Tests of Creative Thinking' (TTCT®) are a set of instruments used to understand the creative skills. These tests use verbal and figurative batteries (Wechsler, 1998).

3.1.1.5 Socio-interactionistic approach

Since the 70's until the present time, creativity is observed as a result of a very complex interaction between the creative being and the environment. This is a multifaceted approach, wider than the mere concern on studying the individual features. "The underlying theme of this review is the need for a systems view of creativity. We believe that more progress will be made when more researchers recognize that creativity arises through a system of interrelated forces (. . .)" (Hennessey & Amabile, 2010, p. 571). Here we understand the requirement to add social and organizational psychology approaches to this study.

Considering creativity as a result of multidimensional interactions, we must have in concern four factors, (1) the individual, (2) the process, (3) the environment and (4) the product (Alencar, 2003; Gurgel, 2006; Oliveira, 2012; Parolin, 2003; Valente, 2015).

For the present study we took in consideration some models from the Socio-Interactionistic approach. Amabile's Componential Model (1983,1996) presents a theoretical componential model that includes some concepts as the influence of task motivation, the relevant techniques in the area and also refers to the processes of creativity. The model presents the phases for the creative outcome: (1) identification of the problem or task, (2) preparation, (3) generation of outcomes, (4) validation of these outcomes and communication, and (5) final result (Alencar, 2003; Amabile, 1983, 1996; Jiménez, 2016; Lima, 2006; Oliveira, 2012).

Csikszentmihalyi's domain-field-person model (1988, 1999, 2014) has also been used to understand creativity as resulting of a system. The domain is responsible for understanding of creativity once it may define what is novelty compared to what is old. "Domain is the area of creativity with which the individual's thinking, product and action are confronted (. . .)" (Lima, 2006, p. 17). The field represents the social context that is responsible to legitimate the new outputs, "which evaluates innovations and decide which are valid and should be retained" (Jiménez, 2016, p. 43). Regarding the person/subject, the model invokes the personal background, considering the connections the creator establishes in terms of emotions and motivation.

3.1.2 *Creativity and industry*

"In the 21st century, given the growing supply of products on the market, it's noticeable the urgency of companies to differentiate themselves by aiming to the extension of their competitive capacities" (Tironi, 2014, p. 3).

On the belief that thought creative and innovative outcomes companies are better prepared to face unpredictable economies, one can see the increasing focus on the designers' creative processes. This concernment can be read in many academic publications, and for years many of these did not concern all the contexts surrounding the creative being (Amabile, 1983, 1996; Hennessey & Amabile, 2010; Lima, 2006; Swann & Birke, 2005).

Amabile (1983, 1996) is a pioneer on recognizing not only the workplace environments but also the interpersonal dynamics alongside with the influence of task motivation when studying creativity.

3.2 *Innovation*

Etymologically, 'innovation' derives from the Latin word 'innocatione', meaning renewal. Nowadays innovation is known as the result of a system, involving the creation (creativity), the management of process, the distribution of products and the final use.

Commonly the terms 'creativity' and 'innovation' are mentioned together, "This is reflected in the now widely accepted definition of innovation equalling creativity plus (successful) implementation." (Von Stamm, 2008, p. 1).

In this framework, the adaptation of the human species initially mentioned regarding the survival instinct appears again, concerning the business context; innovation is a form of survival that "has become the rule and not the exception: there are new demands, new customers, new products, new processes, new management techniques, new markets, etc." (Gurgel, 2006, p. 68).

When it comes to generating profit, creativity is not enough, and so there is the need to monetize these new and creative ideas into an economic feedback.

3.2.1 *The models of innovation*

Understanding innovation in its dynamic quality follows the models and processes of innovation: (1) Linear Model, (2) Interactive Model and (3) Systemic Model.

The (1) Linear Model is related to the classic and neoclassic growing and development theories. The model is a combination of two approaches: science push and demand pull (Gurgel, 2006; Lacono, Almeida, & Nagano, 2011; Lobosco, Barbosa, Moraes, & Antonio, 2011).

Science push indicates socio-economic growth results from the technologic innovation and the academic research. Demand pull "(. . .) considers that innovation is stimulated by the market's needs (. . .)" (Lobosco et al., 2011, p. 409).

The (2) Interactive Model is related to an evolutionist current and it proposes interaction rather than a linear process. This model is built in looping feedbacks between its elements (academic research, markets, procedures, products, distribution channels and users); these feedbacks have to do with the individual contributions and its needs (Gurgel, 2006; Lobosco et al., 2011).

In the (3) Systemic Model is understood that companies can not innovate outside the business community and without direct and indirect relations with the partners. These partners are other companies, academic and research entities or the global economic framework (Lobosco et al., 2011).

In these terms one can say that if new and creative ideas are related to the survival of the species, innovation has to do with the implementation of these outputs. Implementation activities are scientific, technologic, organizational, financial and commercial steps towards innovation. According to Oslo Manual (OECD, 1997, 2005), companies may innovate in four terms, product, process, organizational and marketing. This implementation process has to be regarded since the ideas' sellection, trough product's development and its destribution (Europeia, 1995; OECD, 1997, 2005; Von Stamm, 2008).

3.3 *Design as innovation*

Design is a difficult term to be defined with precision, even the dictionaries differ according to contexts, "in addition to a wide range of opinions of what 'design' refers to, there is also potential for further confusion due to national differences. (. . .) design is a word used in many countries, its meaning varies." (Von Stamm, 2008, p. 17).

Here 'design' is used as a process, as a conscience decision-making activity aiming the final result (product or intangible service). "(. . .) design is defined as the shaping (or transformation) of ideas into new products and processes; and innovation is defined as the exploitation of ideas (. . .)" (Hollanders & Cruysen, 2009, p. 5).

Extensively, the interest in design in the last decades has to do with innovation goals and thus with adding value to the outcomes. These considerations are linked to the companies' and brand's competitively (Ferreira, 2012; Hollanders & Cruysen, 2009; Von Stamm, 2008).

In the economic context, Ferreira (2012) refers to vertical and horizontal strategies. The vertical approach is related to quantifiable factors and the product's or services' improvement wile the horizontal one refers to atributes and qualities percived by the consumer. In this frame, the intangible attributes have to do with style and reputation.

In the current study 'design' is perceived as an important contribution for differentiation as it is associated with creativity and then innovation.

Design "(...) contributes to the expansion of available ideas (...)" (Hollanders & Cruysen, 2009, p. 6) and innovation to a "(...) chance of successfully commercialising these ideas." (Hollanders & Cruysen, 2009, p. 6). Thus, design "allows us to better understand the structure and evolution of a national reality (...) in an knowledge and innovation based economic era" (Ferreira, 2012, p. 2).

3.4 *Textile and apparel industry*

The term in title represents a complex and dynamic context of industries that aim for the design, production and distribution of apparel products. These industries respect several cycles and trends but they have in common the following steps: identification of the trends, textile production, production of artefacts, distribution and selling (Jones, 2005; Lima, 2006; Neves & Branco, 2000).

The Portuguese Textile and Apparel Industry (ATP in Portuguese) is an association with more than fifty years old and with the mission of "Representation, intervention and adequate defence of the interests of the textile and apparel industry" (ATP – Associação Têxtil e Vestuário de Portugal, 2015, p. 7). "Regarding the apparel industry, Portugal is the 5th largest producer, with 4% of apparel production and 4% turnover" (ATP – Associação Têxtil e Vestuário de Portugal, 2014, p. 35).

In 2016, according to the online newspapers 'Journal of Business' (Jornal de Negócios in Portuguese) and 'Observer' (Observador in Portuguese), the exports from the Portuguese textile and apparel industry have overcome the 5.000 million's milestone (Agência Lusa, 2017; Larguesa, 2017).

The strategic plan "Textile 2020: Planning the Development of the Textile and Apparel Industry until 2020" ATP (2014) set this goal in 2014.

4 THE STUDY OF CREATIVITY AND THE ESTRATEGIC PLAN UNTIL 2020

The plan mentioned above presents a set of guidelines, trends and strategies for companies to prepare themselves to contribute to the commonly desired competitiveness. The changings within Portuguese apparel and textile industry point to adapt for the instable macroeconomic scenarios.

One of the structuring axes for this goal is trough creativity and innovation, which are mostly directed to the employers. This study argues that workers, especially fashion designers, are a great driving force of the companies' results, therefore its competitiveness. And so it is appropriate to study the creativity of fashion designers working in Portuguese industries (Agis, Bessa, Gouveia, & Vaz, 2010; ATP – Associação Têxtil e Vestuaário de Portugal, 2014, 2015).

Moreira da Silva (2010)	**(ATP – Associação Têxtil e Vestuário de Portugal, 2014)** **(Agis et al., 2010)**	
(1) the researching problem must specifically belong to the subject area of analyses	"Representation, intervention and adequate defence of the interests of the textile and apparel industry" (ATP – Associação Têxtil e Vestuário de Portugal, 2015, p. 7)	
(2) the methods must become a model that can be applied in future investigations in the area	"To ensure, by the end of the current decade, a prosperous economic activity, based in an industry of excellence, assisted by innovation, creativity and service, elements of differentiation capable of offering added value, making it competitive to strengthen its exporting virtue in an increasingly competitive and selective context." (ATP Associação Têxtil e Vestuário de Portugal, 2014, p. 47).	✓ Pertinence of the study & ✓ Innovation and contribution to academic and industrial knowledge
(3) the topic should be relevant to society		
4) the design professionals should be part of the research	"Specialization, segmentation, creativity, service, technological innovation in products and processes, and commercial strength are the key to the future and the grand trend for the Textile and Apparel Industry..." (Agis, Bessa, Gouveia, & Vaz, 2010, p. 94).	

Figure 2. Relevance and originality of the study. Font: authors.

According to Moreira da Silva (2010), there are four essential conditions for academic studies in design, (1) the researching problem must specifically belong to the subject area of analyses, (2) the methods must become a model that can be applied in future investigations in the area (3) the topic should be relevant to society and (4) the design professionals should be part of the research. See figure 2 for a better understanding of how these references come together to justify the relevance and originality of the study.

This study was not found in the primary review or for this framework within Portuguese textile and apparel industries. The strategy to be adopted is interdisciplinary and creativity is considered as being directly related with the surrounding contexts/variables.

5 CONCLUSION

In the instable economies' markets, innovation is the best way to ensure one's competitiveness and survival. Innovation is the successful implementation of a creative output. Hence, creativity is an important process to be study in a wider perspective, as it is a reflex of many contexts.

The methodology was designed to deeply study and understand how these contexts influence the designer's predisposition to be creative.

Horizontal economic strategies aim for the symbolic principles and the products or services with added value in the consumer's perspective.

Design is observed with high relevance in this document, as it has to do both with the initial ideas (creativity) and the global conception of the projects (innovation).

The textile plan until 2020 set an economic goal regarding exports and in 2017 the industry has overlapped these values, representing prosperous indicators. This study aims to contribute for this economic growth and for a sustainable creative production.

REFERENCES

Agência Lusa. (2017). Exportações de têxteis e vestuário crescem 5% em 2016. Retrieved May 19, 2017, from http://observador.pt/2017/02/09/exportacoes-de-texteis-e-vestuario-crescem-5-em-2016/

Agis, D., Bessa, D., Gouveia, J., & Vaz, P. (2010). *Vestindo o Futuro Microtendências para as indústrias têxtil, vestuário e moda até 2020.*

Alencar, E. M. L. S. (2003). Contribuições Teóricas Recentes ao Estudo da Criatividade. *Psicologia: Teoria E Pesquisa*, 19(1), 1–8.

Amabile, T. (1983). *The Social Psychology of Creativity*. (R. F. Kidd, Ed.), *Statewide Agricultural Land Use Baseline 2015* (1st ed., Vol. 1). Springer Series in Social Psychology. http://doi.org/10.1017/CBO9781107415324.004

Amabile, T. (1996). *Creativity in Context*. Oxford: Westview Press.

ATP – Associação Têxtil e Vestuaário de Portugal. (2014). *Têxtil 2020: Projetar o Desenvolvimento da Fileira Têxtil e Vestuário até 2020*. Vila Nova de Famalicão.

ATP – Associação Têxtil e Vestuaário de Portugal. (2015). *ATP 50 Anos*. Vila Nova de Famalicão.

Bahia, S. (2008). Promoção de ethos Criativos. In M. de F. Morais & S. Bahia (Eds.), *Criatividade e educação: conceitos, necessidades e intervenção* (pp. 229–250). Braga: Equilibrios.

Europeia, C. (1995). *Livro Verde sobre a Inovação*.

Ferreira, M. C. (2012). *Design como indicador de inovação: estudo sobre as atividades de design na economia portuguesa*. ISCTE-Business School/Instituto Universitário de Lisboa.

Gurgel, M. F. (2006). *CRIATIVIDADE & INOVAÇÃO: Uma Proposta de Gestão da Criatividade para o Desenvolvimento da Inovação. Dissertação – Universidade Federal do Rio de Janeiro, COPPE.*

Hennessey, B. A., & Amabile, T. M. (2010). Creativity. *Annual Review of Psychology*, 61(1), 569–598. http://doi.org/10.1146/annurev.psych.093008.100416

Hollanders, H., & Cruysen, A. Van. (2009). *Design , Creativity and Innovation: A Scoreboard Approach. Innovation.*

Jiménez, M. G. (2016). *Employee Creativity and Culture: Evidence from an examination of culture's influence on perceived employees' creativity in Spanish organizations.* Universitat Ramon Llull.

Jones, S. J. (2005). *Fashion design: O manual do estilista*. Barcelona: Gustavo Gili.

Lacono, A., Almeida, C. A. S. De, & Nagano, M. S. (2011). Interação e cooperação de empresas incubadas de base tecnológica: uma análise diante do novo paradigma de inovação. *Revista de Administração Pública*, 45(5), 1485–1516. http://doi.org/10.1590/S0034-76122011000500011

Larguesa, A. (2017). INE confirma exportações têxteis acima dos 5.000 milhões em 2016. Retrieved May 19, 2017, from http://www.jornaldenegocios.pt/empresas/industria/detalhe/ine-confirma-exportacoes-texteis-acima-dos-5000-milhoes-em-2016.

Lima, E. (2006). *Um Contributo para Potenciar a Criatividade no Design de Vestuário*. Universidade Técnica de Lisboa.

Lobosco, A., Barbosa, M., Moraes, D., & Antonio, E. (2011). Inovação: uma análise do papel da agência USP de inovação na geração de propriedade intelectual e nos depósitos de patentes da Universidade de São Paulo. *Rev. Adm. UFSM*, 4(3), 406–424. Retrieved from http://cascavel.ufsm.br/revistas/ojs-2.2.2/index.php/reaufsm/article/view/3164

Martins, E. C., & Terblanche, F. (2003). Building organisational culture that stimulates creativity and innovation. *European Journal of Innovation Management*, 6(1), 64–74. http://doi.org/10.1108/14601060310456337

Neves, M., & Branco, J. (2000). *A Previsão de Tebdências para a Indústria Têxtil e do Vestuário.* TecMinho.

OECD. (1997). *Manual de Oslo: Diretrizes para a Coleta e Interpretação de dados sobre Inovação Tecnológica. OCDE, Eurostat e Financiadora de Estudos e Projetos.* http://doi.org/10.1787/9789264065659-es

OECD. (2005). *Oslo Manual – GUIDELINES FOR COLLECTING AND INTERPRETING INNOVATION DATA. Communities* (Vol. Third edit). http://doi.org/10.1787/9789264013100-en

Oliveira, R. S. R. de. (2012). *Um Programa de Treino da Criatividade Estudo exploratório com alunos do 1° Ciclo*. Universidade da Madeira.

Parolin, S. (2003). A criatividade nas organizações: um estudo comparativo das abordagens sociointeracionistas de apoio à gestão empresarial. *Caderno de Pesquisa Em Administração*. Retrieved from http://www.ead.fea.usp.br/cad-pesq/arquivos/v10n1art2.pdf

Pereira, B., Mussi, C., & Knabben, A. (1999). Se sua empresa tiver um diferencial competitivo, então comece a recriá-lo: a influência da criatrividade paea o sucesso estratégico organizacional. *23° Encontro Anual de Associação Nacional Dos Programas de Pós-Graduação Em Administração*, 1, 1–10.

Swann, P., & Birke, D. (2005). *How Do Creativity and Design Enhance Business Performance?: A Framework for Interpreting the Evidence. Think Piece.*

Tironi, M. R. (2014). "Conexões" Design Estratégico E Economia Criativa: Inovação Além Do Design De Moda. *10° Colóquio de Moda*, (7ª edição), 1–10.

Valente, L. F. (2015). *O livro do artista; um caminho para a criatividade*. Universidade de Lisboa. http://doi.org/06-09-2014

Von Stamm, B. (2008). *Managing innovation, design and creativity* (2°). Chichester: John Wiley & Sons Ltd.

Wechsler, S. M. (1998). Avaliação multidimensional da criatividade: uma realidade necessária. *Psicologia Escolar E Educacional (Impresso)*, 2(2), 89–99. http://doi.org/10.1590/S1413-85571998000200003

Textiles, Identity and Innovation: Design the Future – Montagna & Carvalho (Eds)
© 2019 Taylor & Francis Group, London, ISBN 978-1-138-29611-4

The teaching of materials in fashion design: The importance of holistic platforms

A. Miranda Luís & L. Santos
CIAUD, Lisbon School of Architecture, University of Lisbon, Portugal

ABSTRACT: The teaching of materials is a particularly important topic in design faculties, especially after the implementation of the Bologna Process. There has been a lack of attention regarding the teaching of materials in fashion design, an essential subject on a good design practice. There is not a notorious change in the teaching methodologies applied to this subject, essential for the instruction of future fashion designers. This paper presents a first analysis in a study of interconnected platforms on material communication (both physical and digital), since it represents an important interconnection between industry and academia, that has the ability to promote technological innovation in fashion design. The aim of this work is to generate an improvement in the area of material knowledge and communication, promoting teaching methodologies that enhance creativity and performance of fashion design students, during their academic and professional life.

1 INTRODUCTION

> "Design is the process by which abstract ideas assume concrete form and thus become active agents in human affairs. One of the critical parameters in any discussion of designed artifacts is material: what something is made of and how the material employed affects the form, function and perception of the final design. In a broad sense, the story of materials, their discovery and subsequent manipulation constitutes a significant thread in the history of civilizations and often provides a common point of reference for cultural discourse in general." (Doorman, 2003).

The evolution in technological production, in technological education and in the specific area of materials for design promoted the study and evaluation of the programmatic contents of design graduations. These concerns resulted in a greater responsibility by university professors that now are asked to have a profound knowledge on materials and production techniques (Luís, 2017).

In 1999, a new law on education – The Bologna Process – introduced a new paradigm in teaching, where a new educational strategy based on specific learning goals, sought for a deeper skill acquisition by design students. This new way of gaining knowledge was focused on the enhancement of design practice during student training, enabling a more self-guided learning process (Luís, 2017).

According to data released by the European Union, higher education should be restructured in order to boost innovation. This growth in innovation would promote economic competitiveness not only for each individual country, but also for the whole European Union (Dias C. et al. apud Murphy, 2013).

Cristina Dias suggests that education and training in design should predict and be aware of market needs, having to adequately anticipate the development of professional skills that each designer should acquire in order to promote innovation (Dias C. et al. 2014).

This focus on innovation is increasingly present in the fashion practice, which seeks to respond to the needs and expectations of an informed and demanding final consumer. Textile plays a nuclear role in consumer adoption of a fashion product, because "garment attractiveness can be determined by both its surface quality and shape in relation to the textile(s) utilized", says Akiwowo K. (2016).

According to Sinclair (2015), in a near future, daily life will embrace clothes that use new varieties of micro fibers and nanotechnologies, able to generate ultra-lightweight, breathable, quick drying and supportive garments that are able to offer UV protection and excellent fit to body contours [fig. 1]. These are only some examples of the central importance innovation will have in the field of fashion. That innovation can only be effective if means of communication are developed to maintain a continual flow between industry and academia, allowing new experimentations and responses to the actual needs of consumers.

According to Edma de Paula, "The design education programs around the world have some differences that may vary from program to program, from country to country, according to different factors, but all of these programs share a strong commitment to create a Depth knowledge of design and an enthusiastic partnership

Figure 1. The *Oricalco* shape-memory shirt uses a Nitinol (Nikel and Titanium) memory metal that can be manually distorted, then regain its shape when heated. http://advancedtextilessource.com/2016/01/22/fashion-grounded-space-inspired/ [21 April 2017].

between the educational program, students and local society" (Paula E. et al. 2007).

Consumers are tired of the standardization imposed by the fast fashion phenomenon and will be looking for customized products, rooted in local alternatives. We will see the re-rise of local, *in-loco* production, and a re-affirmation of new regional cultural languages, expressed through fashion and textiles. According to Sinclair (2015), individually designed garments that are tailored to specific consumer needs will be at the forefront of fashion in a near future. For Aman Advani, a cofounder of the brand 'Ministry of Supply', that has permanently installed a 3D-knitting machine at its flagship store in Boston [fig. 2]:

> "This isn't a niche product, (. . .) this is step one of a longer route to a sustainable and strong production method that's here to stay. At this point we are pretty convinced that we can get one-quarter, maybe even one-third, of our production done in-store, or even to serve the online market within the next couple years." (Bain M. 2017).

That is why, a material and technological communication is essential for the fashion product and industry to evolve. A new space of communication, the *cyberspace*, which emerges from the global interconnection of computers, users and information, allows this information to naturally flow in the post-modern era (Lévy P. 1997), and should be empowered in the field of the fashion project, both educational, academic and commercial.

Therefore, in the area of materials, and more specifically in its teaching, technical innovation will be of great importance. It will allow the development of

Figure 2. Ministry of Supply's 3D knitting machine in store. https://qz.com/949026/brands-including-adidas-uniqlo-and-ministry-of-supply-see-the-future-of-fashion-in-on-demand-3d-knitting/ [21 April 2017].

design solutions more adapted to the needs of consumers and will trigger a new form of teaching, where web platforms will act as a bridge between information and physical materials, functioning as true textbooks. This technological evolution and holistic thinking, not only related to the new materials, but also to the information and dissemination of the same, will increasingly contaminate education and as such, the methodology applied by professors in the teaching of textile materials, in fashion design degrees.

According to this new paradigm, new learning experiences must be a participatory and collaborative process. Daniela Barros and the research team of which she is member are concerned with the cognitive and communicational dynamics of the generations that grew with computers, mobile phones and the internet's connectivity, multi-directionality, sharing, collaboration and interactivity: "For new times, new media and their approaches" (Barros D. et al. 2011).

Therefore, the digital emerges as the new natural form of communication, and materiality is going to be increasingly transmitted recurring to computers. This generates some sensorial and emotional questions because the textile surface is more than a functional gesture, is an emotional experience, triggered not only by sight but also by touch. According to Quinn (2010):

> "a textile's surface is more than just a façade — it's a curious layer of aesthetics and identities, and a contentious site of exploration and resistance. (. . .) Because surfaces are interpreted by sight and touch, physical contact with the surface initiates a complex multisensory, emotional and cognitive experience that provides a uniquely individual interpretation of the world."

Here resides the importance of defining a strategy of implementation for this new teaching methodology, which aims to contribute to a richer skill acquisition in the field of materials. This approach will enhance the education of fashion designers, promoting creativity and innovation.

The interaction between different kinds of platforms (both physical and digital), on industry, academia and education levels can help to ensure that "new technologies are applied in a correct way" (Papanek V. 1995) and, therefore, build effective forms of communication and learning in the field of materials.

2 MATERIALITY IN DESIGN TEACHING AND PRACTICE

In Portugal, higher education institutions have undergone a process of change and evolution at various levels. According to Dora Agapito and the research team behind the publication "Design In Portugal", on April 25th, 1974 there were about four Universities in Portugal, that "went to a network of university and polytechnic institutions distributed by various locations around the country". Currently we have in public education, about 15 universities, 5 university schools, 15 polytechnic institutes and 9 polytechnic schools. The private university network covers around 14 universities, 33 university schools, 2 polytechnic institutes and 56 polytechnic schools (General Directorate of Higher Education, 2015) (Agapito, 2015), which proves that teaching design in Portugal has evolved in all areas of knowledge and consequently in the area of fashion design. In that specific area, there are 5 public universities dedicated to fashion design education, and 2 public universities with a special focus on the textile area. According to Agapito, graduates in fashion design have been growing since 2002: about 30 in 2003/2004 and 127 in 2012/2013 (Agapito, 2015).

The raw material and the knowledge about it are fundamental to the entire process of developing a design project. It is then, of an extreme importance for the future designer, to create a strong connection between practice and theory (Louis, 2017).

The fashion design project is no exception, and the deep connection between the garment and its materiality, strictly related to the technological advancements of a certain age, is essential (Udale J. 2008). Like Breward (2003) states, "without the basic coordination of needle and thread there would be no raw material for the formation of such processes".

In fashion design, the creation of products is deeply connected to experimentation, and so, with the characteristics of a certain material. For example, complex forms, connected to a material's behavior are explored recurring to the *moulage* technique, that allows the creation of prototypes, directly in a bust, with the anatomical measurements of a human being [fig. 3] (Lima J. et al. 2016).

A new generation of consumers show necessities that go beyond the context of quality in product performance and use, seeking objects that bring them not only functional benefits, but also emotional ones (Jordan P. 2000). To interact with a textile surface is an extremely emotional experience because it connects with us in a sensorial way. Textiles are a nostalgic, recurring piece in our life, that touch and involve us from birth to death (Auch M. 2016). Akiwowo K. (2016) corroborates this thought by stating that sight and touch are the most important senses when a consumer is confronted with a garment, in store.

Figure 3. The moulage technique allows the creation of prototypes, directly in a bust, with the anatomical measurements of a human being. https://www.instagram.com/p/2x-ADgE3hu/ [21 June 2017].

Because of this complexity, the research in the field of materials – especially of textile materials – is of extreme importance, existing a gap in these studies, namely in the field of understanding the tactile properties of clothing and its material production (Breward, 2003).

Thus, we believe that at a time where we are experiencing changes at the most various levels (economic, social...), in which sustainability and technology are subjects concerning all scientific areas, the importance of studies on materials becomes notorious. These studies can empower a rigorous and conscientious improvement on design innovation, not only in fashion design teaching, but also in the future of fashion design practice.

Education has undergone significant changes, however, its programs, how they are taught, especially in the areas of technologies and materials, may not be keeping up with the overall sociocultural changes.

According to Nigel Cross, researcher and professor in design, the exploration process in this field should be centered on the articulation of three factors: People, Processes and Products. Therefore, these three factors should also be the focus in design teaching.

People, Processes and Products are interdependent and interconnected by culture. Materiality appears in the middle of these interconnections, deeply rooted to the human condition.

About this fact Lévy P. (1997) says that "It is impossible to separate the human from its material environment, or the signals and images through which he gives meaning to his life and world" (author's translation), meaning that studies and communication regarding materials are vital to any activity that deals with the human being, fashion design being one of those activities.

Sudjic D. (2009) states that "Design is the language that a society uses to create objects that reflect its purposes and its values." Meaning that culture acts as a natural filter that influences what products are made and what materials are used. Therefore, an efficient communication is necessary to produce relevant design responses to contemporary society.

For Cooperstein S. et al. (2004), it is important to create a new way of teaching and learning where professors are instructors, with a global knowledge of teaching and work tools, showing students how they can investigate and produce their projects. This teaching philosophy is aligned with the main goals of the Bologna Process, a new educational paradigm for the new millennium. Teaching shifts to an exploratory and self-taught approach by the student, who must receive from the instructor-teacher the necessary tools for the correct development of the process. In this sense, there is an intrinsic need for platforms (physical, digital and/or mixed) that disseminate access to materials, essential for the development of the fashion project.

According to Almeida (2009), the way design is being taught in Portugal followed the principles that should now have been implemented in all universities, according to the Bologna Process. "Artistic" Schools were the first to systematize the requirements for the process implementation. According to the author, design education in Portugal followed the system of reproducibility based on 'who knows how to do, knows how to teach'. This strategy shows the extreme importance that material knowledge, held by the educators, has in fashion design teaching. On the other hand, teaching on materials needs to be supported by physical samples that can efficiently trigger the learning process.

An efficient bridge between the communication and accessibility of materials is essential to a design education that finds in materiality the essence of its products. The relationship between materiality and final product must be carefully understood because it can result in wrong transcriptions between product idealization and production. In fashion design, this technical maturity is essential because the use of bidimensional representations like fashion *croquis* tend to capture inefficiently the complexity and materiality of a garment [fig. 4] (Souza P. 2008). The designer must have in-depth knowledge of techniques and materials so the intended result is achieved.

A digital tool has the potential to be powered by the *cyberspace* community, which ranges from professionals of a specific area (educators, investigators and industrials) to students that intend to be part of it, promoting an open and constant dialogue that fuels the growth of information on that same tool. Only through an efficient communication can a platform like this present itself as an efficient instrument in a society where the information is reinvented on a daily basis (Lévy P. 1997).

Figure 4. The communication gap between fashion croquis and the real product. https://www.bellanaija.com/2013/01/president-barack-obama-officially-sworn-in-as-the-44th-president-of-the-united-states-of-america-bn-scoop-photos-from-the-swearing-in-ceremony-inaugural-ball/ [21 June 2017].

Digital material platforms like 'material connexion', 'virtual material', 'archiproduts' and 'materianl', can be an easy way to access and research materials, allowing students to contact with new materials and check their applicability. But most of these virtual platforms are not specific to the textile/fashion industry, lacking material solutions on those fields, especially on textile research and innovation. However, an effort has been made in the development of structures that can help to preserve and divulge the textile industry and its historical textile solutions. It is in this context that fashion and textile museums, like the Textile Industry Museum, are important tools that are expected to support present and future fashion designers (Museu Têxtil, 2017).

Another issue with these virtual platforms is the difficulty to communicate the accurate sensorial information on a certain material. Thus, physical material samples are needed to complement the digital information. These physical samples allow each student and designer to verify the characteristics of a certain material, and to understand its applicability to the project in hand.

The existent digital material platforms allow an easy material research but lack of a cross dialogue with each other, which would enable the creation of a transversal cataloging system, standardizing the shared information about a certain material. Only through an uniformed cataloging can the materials be efficiently communicated and linked through platforms, both physical and digital.

One example of a cross platform strategy is the PANTONE system that was created to standardize the world communication of color. Each color is identified through a code, used on both digital and printed representations of that color [fig. 5].

Many other attempts on material communication have been made. Daniel Charny, professor and investigator at the Kingston University (United Kingdom),

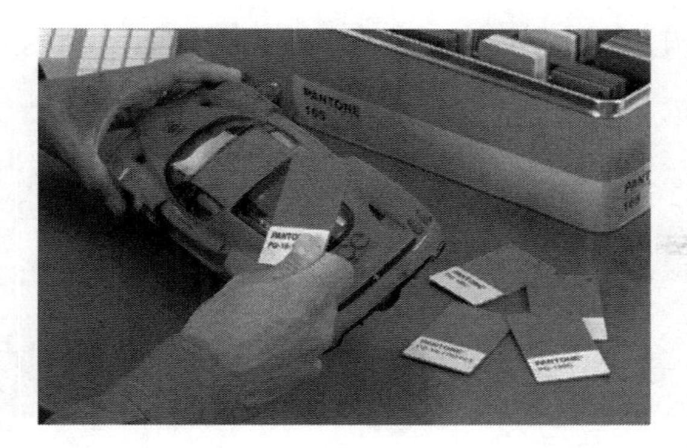

Figure 5. The PANTONE system refers a color in both physical and digital ways. https://www.pantone.com/pantone-numbering-explained [21 June 2017].

advocates in his investigation the development of innovation in design education through the creation of mobile libraries and exhibition modules: "the maker library network" (Charny D. 2015). To Charny, innovation can only happen in the physical presence of the innovative agent, and only by creating efficient connections between industry, academia and education, can a real creative flow emerge.

The "Institute of Making – Materials Library", (University College London), also in the United Kingdom, developed another example of material communication: a diverse material library/database. The institute, which believes that innovation resides in the knowledge and manipulation of materials, frequently organizes seminars where the theme of materials is debated. They allege that only through the access to a physical sample can a designer understand the real potential of a certain material (Institute of Making – Materials Library, 2017).

3 MATERIAL PLATFORMS

> "Men do not understand one another by relying on signs for things, nor by causing one another to produce exactly the same concepts, but by touching the same link in each other's spiritual instrument." (Goodman apud Cassirer, 2003).

According to Ramalhete P. (2012), some authors such as Dobrzanski and Ashby and Johnson, debated on the fact that designers are the professionals with the best capacity to leverage the evolution of technical and product aesthetics, by channeling the development and application of new materials. They also consider that, the development of new materials and their use by designers, can only be achieved through a rigorous holistic knowledge of materials and their handling.

According to the Portuguese Society of Materials (Materials, 2015) since 1983, conferences have been held at LNEC (National Laboratory for Civil Engineering) to discuss the use and application of materials. The congresses are held every two years and already have the partnership of several educational institutions such as Instituto Superior Técnico, University of Aveiro and University of Minho. The materials' conferences are organized by SPM – Portuguese Society of Materials, which hosted on April 2017, the conference entitled "Materials 2017" (SPM, 2017).

None of the events on the theme of materials, that take place in Portugal, is focused on materials for design application. The potential that could come from a partnership between industry and designers is forgotten by those entities.

In the case of fashion and particularly of textiles, there is a stigma regarding the passage of knowledge between industry and academia. Textile fairs are held, such as Modtissimo in Oporto, however the opening of companies to collaborate with education and academia is always limited by merely commercial interests.

But not all the examples are knowledge-hermetic. In Vila Nova de Famalicão, the Center for Nanotechnology and Smart Materials – CENTI, is the first material center that, through its strategic vision, seeks to establish a link between various industries and research centers. This center develops new technologies applicable to textiles, but due to commercial secrecy needs, information is only transmitted after the due tests and registrations.

Outside Portugal, in fashion schools such as the Faculty of Education in Passo Fundo, Brazil, the physicality of materials is understood as a nuclear need to the creative process. The access to physical materials provides "support to the practical activities of the Fashion Design course" that requires "materials for research such as fabrics, accessories, equipment, documents and catalogs of trends and information on the area of textile confectionery, fashion, clothing and related areas". The school also provides a material library, busts for *moulage* and clothing fitting, fashion and textile trend bureaus, and diverse technical instruments for fibers and fabrics studying (About Materials Laboratory – Materioteca of the Faculty of Education of Passo Fundo, 2017).

In São Paulo, Brazil, educational institutions have been promoting workshops under the theme of 'Design and Materials', with a special attention on Fashion Design. In 2016 they held for the first time, in parallel to the workshop, the first conference on materials and design, at the University of Anhembi Morumbi – São Paulo, which proved to be extremely important for the debate on textile materials in the fashion field.

4 FINAL CONSIDERATIONS

An efficient communication on materials is becoming a general concern on both industry, academia and education. Multiple strategies are being tested on different levels, seeking to implement material libraries (both digital and physical), exhibitions, conferences, workshops and other forms of dialogue.

Work on this field has already begun and it is now essential to create a form of unified discourse that can strengthen the way materials are known, accessed and transmitted.

In the specific case of Fashion Design education, the access to materials is essential to the student's development, in an area where the product lives intrinsically connected to the material from which it is constructed. That is why, a holistic discourse about the textile material is extremely important, demanding a platform that can cover all the product's characteristics: functional, utilitarian and emotional (Jordan P. 2000). Only through this overall vision, the future designer can be aligned with the fashion market, growingly demanding and informed. As said by Akiwowo K. (2016):

> "(...) the 'wearability' of a garment may not only be defined by its use, appearance, and comfort characteristics. Instead, a more holistic understanding can be perceived due to an increasing general awareness of materials, properties, and life cycle considerations, such as fibers and their acquisition, manufacturing methods, distribution information, product use/functionality, disposal, recyclability, and other environmental aspects."

This new platform must be developed recurring to a diversity of supports (physical and digital), and be the result of academic-industrial relationships that can uphold a constant dialogue on the fields of innovation and consumption.

ACKNOWLEDGMENTS

This work was carried out with the support from the Research Center of Architecture, Urbanism and Design from the Faculty of Architecture of the University of Lisbon (CIAUD), and the FCT, Foundation of Science and Technology.

REFERENCES

Agapito, D. E., 2015. Design em Portugal. 1° edição ed. Loulé, Portugal: sílabas& desafios.
Akiwowo, K. 2016. "Garment ID: Textile Patterning Techniques for Hybrid Functional Clothing". In Kane, F., Nimkulrat, N., Walton, K. (eds.) *Crafting Textiles in the Digital Age*. London/New York: Bloomsbury Academic, pp. 171188.
Almeida, 2009. *O design em Portugal um tempo e um modo*. Lisboa: Faculdade de Belas Artes.
Auch, M. 2016. "The Intelligence of the Hand". In Kane, F., Nimkulrat, N., Walton, K. (eds.) *Crafting Textiles in the Digital Age*. London/New York: Bloomsbury Academic, pp. 171–188.
Bain, M. 2017. "Brands see the future of fashion in customized 3D-knitted garments produced while you wait" https://qz.com/949026/brands-including-adidas-uniqlo-and-ministry-of-supply-see-the-future-of-fashion-in-on-demand-3d-knitting/ [21 of April of 2017].
Barros, D. et al. 2011. *Educação e tecnologias: reflexão, inovação e práticas. Universidade Aberta*. Lisboa. E-book. http://www.intaead.com.br/ebooks1/livros/pedagogia/18.Educa%E7%E3o%20e%20Tecnologias.pdf [2 of July of 2017].
Breward, C. 2003. *Fashion*. Oxford: Oxford University Press.
Charny, D. 2015. The Institute of Making. http://www.instituteofmaking.org.uk/. [18 of May of 2017].
Cooperstein, S. et al. 2004. *Beyond active learning: a constructivist approach to learning*. USA: Emerald Group Publishing Limited.
Dias, C. et al. 2014. "Teaching digital technologies in industrial/product design courses in Portugal". In: *International Conference On Engineering And Product Design Education*.
Dias, C. et al 2013. "Ensino superior do design e indústria em Portugal: estudo do ensino e práticas do design industrial". In: *Segundo Encontro Nacional De Doutoramentos Em Design, Porto*. http://ciaud.fa.utl.pt/index.php/pt/sobre-2/2013-06-17-13-39-61 [5 of January of 2017].
Jordan, P. 2000. *Designing Pleasurable Products*. Boca Raton: Taylor & Francis Group.
Lévy, P. 1997. *Cibercultura*. Lisboa: Instituto Piaget.
Lima, J. et al. 2016. "O uso da moulage como ferramenta pedagógica no ensino do design de vestuário" in Italiano, Isabel et al. (eds.) *Pesquisas em Design, Gestão e Tecnologia de Têxtil e Moda: 2° semestre de 2014*. São Paulo: EACH/USP, 55–61.
Papanek, V. 1995. *Arquitetura e design*. Lisboa: Edições 70.
Paula, E. et al. 2007. "Revista convergência". http://convergencias.esart.ipcb.pt/artigo.php?id=78 [26 of January of 2017].
Ramalhete, P. 2012. "Metodologia de seleção de materiais em design: base de dados nacional". Phd Thesis. University of Aveiro.
Souza, P. 2008. "A moulage, a inovação formal e a nova arquitetura do corpo" in Pires, Dorotéia (ed.) *Design de moda: olhares diversos*. Barueri, SP: Estação das Letras e Cores Editora, 337–345.
Sudjic, D. 2009. *The Language of Things*. London: Penguin Books.
Udale, J. 2008. *Textiles and Fashion*. London: AVA Publishing.

WEB REFERENCES

Congresso da Spm [online].
http://www.spmateriais.pt/portal/index.php/about/atividades/congresso [Consulted on: 26-01-2017].
Institute of Making – Materials Library [online].
http://www.instituteofmaking.org.uk/materials-library [24 of June of 2017].
Museu da Indústria Têxtil. [online].
http://www.museudaindustriatextil.org/txx.php?tp=3&co=50&cr=3&LG=0&SID=&mop=52 [24 of June of 2017].
Sobre Laboratório de Materiais – Materioteca da Faculdade de Educação de Passo Fundo [online].
http://www.upf.br/Faed/unidade/laboratorios/materioteca [26 of January of 2017].
SPM – Sociedade Portuguesa de Materiais. [online].
http://www.spmateriais.pt/portal/index.php/about/atividades/congresso [26 of January of 2017].

Textiles, Identity and Innovation: Design the Future – Montagna & Carvalho (Eds)
© 2019 Taylor & Francis Group, London, ISBN 978-1-138-29611-4

Design – Non-fabrics – A practical case for the Castelo Branco Embroidery

A.M.P. Fernandes & C. Carvalho
CIAUD, Lisbon School of Architecture, University of Lisbon, Portugal

ABSTRACT: The present article is part of the development of an investigation in the Textile area having as main objective to validate new textile bases for the Castelo Branco Embroidery. For this purpose, 11 textile bases were developed with raw silk cocoons and the construction of bases based on nonwoven technology as the raw material. In order to validate these bases, a questionnaire survey was carried out on a pre-defined group of embroiderers with experience in the area, formulating the following hypothesis; Is it possible through the recycling of the silk cocoon to develop new textile bases to apply in the embroidery of Castelo Branco while complying with the technical conditions of the respective embroidery? In order to test the hypothesis, a methodology of mixed non-interventionist (action-research) research to attest the results was used.

1 INTRODUCTION

Castelo Branco Embroidery is one of the cultural icons of Portugal. Considered as one of the heritage elements of the county of Castelo Branco it is mostly recognized as an artisan art that prides the region and many more who work in it. (ADRACES, IMC\MFTPJ, CMCB, IPCB, 2007).

Recognition is given to its fascinating features which capture the eye of those who appreciate it: the enormous variety of stiches and techniques, the composition of the drawings, the associated symbolism, the rich color palette, as well as the valuable raw materials it employs: the linen and the silk. (ADRACES, IMC\MFTPJ, CMCB, IPCB, 2007).

In the light of the numerous and diversified references that exist on the Embroidery of Castelo Branco the authors Viana (1942), Moura (1949, 1996), Pinto (1992, 1993), Magalhães (1995) and Rosa (2007) refer to a whole historical and geographical context, its identity, as well as to present, its technical and aesthetic characteristics.

The technical execution of the Castelo Branco Embroidery does not only depend on the quantity of motifs, compositions or colors that it employs but also on the various types of fills used in the drawings.

According to Clementina Carneiro de Moura (Moura, s.d.), the current terms will be used, which are those that tradition and habit have adopted, more or less, in all regions of the country.

According to the works consulted, Figure 1, the Castelo Branco Embroidery is characterized by the preferential and predominant use of the Castelo Branco stitch. The main feature of this stitch is to be able to fill large surfaces saving raw material, since the raw material expenditure from the inside out is minimal and the quality of the embroidery is maintained by the pressure made in the frame of the fabric in question.

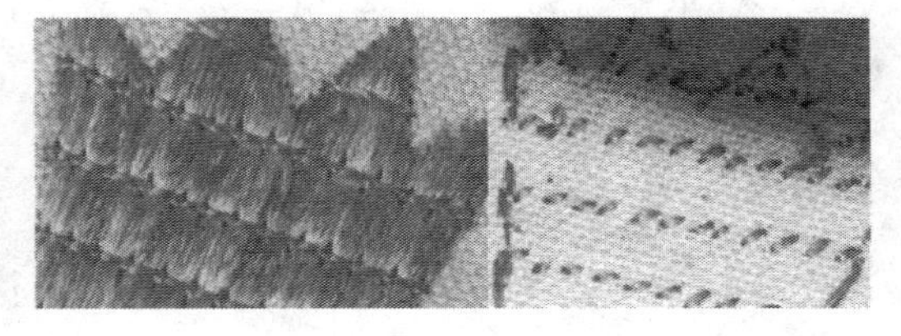

Figure 1. Castelo Branco stitch, with stripes – back (ADRACES; IMC/MFTPJ; CMCB; IPCB, 2007, p.31).

As a rule, the transverse stitches that lock the first ones are perpendicular and straight, but, exceptionally, they form nozzles or even intersect, giving rise to a grid. Also, the small points, called "prisons" because they hold these transverse, have been made in various ways. They can be spaced and alternating those of a row with those of the next row, or so little and together, that offer the look of a top stitch.

In the publication of Clementina Carneiro de Moura (Moura, sd, page 242) we can read: "There are certain oriental embroideries in which the stitches float so loose on the fabric that it is necessary to fix them with other stitches crossing, but in such a way as to avoid harming the glossy aspect of the job. It is not a new stitch, but the adaptation of the satin stitch to wider surfaces" (Needlework, 1900).

In addition to the technical execution of the stitches applied to the Castelo Branco Embroidery as a way to fill and bypass the respective drawings, there are more factors that contribute to the technical and aesthetic quality of the piece.

Figure 2. Tavares Proença Bedspread (Dim.: 2,50 × 2,00 m) Base in tow linen embroidered with linen thread (Pinto, Embroidery of Castelo Branco – Catalog of drawings, Bedspreads – I, 1992.

The linen and silk fabrics used in the embroidery have increasingly come to include a mechanical process.

Various shapes may be used to obtain fabrics but in the case of fabrics for the Castelo Branco Embroidery it is obtained from the interlacing of two sets of yarn known as warp and weft. The warp being the lengthwise direction of the fabric and the weft the direction of the width of the fabric; The resulting fabric is known as flat fabric.

The fibers used in the execution of the Castelo Branco Embroidery are, linen, silk and cotton for the base (filament only used in the fabric web) and for the embroidering, silk and linen, both with connections to the region of Beira where the city of Castelo Branco is located.

In historical embroidery, Figure 2. there are some examples of silk quilts embroidered in silk, and linen embroidered with linen, but the most common will be the use of linen for the base and the silk thread for embroidery.

After reflecting on the history of the motifs and the traditional materials used, the suggestion is to develop and analyze new fabrics.

The analysis of any textile material depends on the micro and macro structures where information can be obtained for analysis of the relationship between the structures of the material applied and its properties.

As such, starting from different types of studies, different types of structural models can be defined which will allow for an analysis of the performance and the behavior of the textile fibers in the application of the new textile bases based on technology of non-fabrics.

Figure 3. Autor.

In this research the non-fabrics are based on the needle-punching through the mechanical interlacing of the fibers originating from a carding from a process of dry threading.

The behavior of the structures being developed and analyzed do not only depend on the on the properties but rather of a set of parameters which may influence the structure of a needle-punched non-textile to apply to a typical Castelo Branco Embroidery

After analyzing the existing textile bases in the current embroidery, the proposal is the development of new textile bases Figure 3, based on non-fabrics to present to the Castelo Branco Embroidery market. The suggestion then is a new textile base taking the cycle of production of the silk to exhaustion where the raw material of the silk cocoon is used to the maximum.

The production technique of the new base rests on the process of needlework for nonwoven production according to the author Winterburn (1972) in which it says that the needlework fabrics "are produced for very fragmented market segments and some of them of great sophistication". Therefore the conditions of this statement are met since the Castelo Branco Embroidery has a very segmented public.

2 METHODOLOGY

In order to reach the central objective in the development of new textile bases based on wastes of silk

Table 1. Synthesis for analysis – elaboration of questionnaire-action.

Concept	Dimension	Components	Indicators
Embroiders with experience in creating embroidery on a medium that is not traditional.	Production of Castelo Branco Embroidery on new bases textile bases	Difficulties in production. Technical components	Stabilize the fabric. Difficulties in embroidery. Behaviour of embroidery on the non-fabric

cocoons applying the technology of the nonwoven by the process of needlepoint, resort to a mixed methodology. In order to attest to the objective of the present study the following hypothesis was formulated:

Is it possible, through the recycling of the silk cocoon, to develop new textile bases to apply to the Embroidering of Castelo Branco obeying the technical conditions of the respective embroidery?

Starting from the initial question, a questionnaire survey was formulated that reflects what is intended to be investigated. The question is a proposition on the starting point. Whether or not it will be accepted will depend on the results of its analysis and testing, through data collection and analysis techniques, which will lead to its validation through a focus group of seven embroiderers. This allows for the direct observation of technical components on the behavior of the textile surface presented in non-woven fabrics.

3 EXPERIMENTAL PROCEDURES

In order to analyze the embroidery motifs on non-woven textile surfaces, the intention was to analyze and test the possibility of producing the Castelo Branco Embroidery on new textile bases, the composition of which is silk. To synthesize and analyze the bases developed in needle punched fabrics by the embroiderers. The concept was identified to the dimension of its components as well as the indicators that would give the starting point for the development of this investigation -action.

In order to prove the validity of the samples, suggestions and other details that may be important for the reproduction of the nonwovens were collected, this having been done after initially having presented the objectives of the session and explaining the necessity as well as the importance of carrying out these experiments. Each of the embroiderers was given a technical sheet so that each one would be guided to produce a typical element of the Castelo Branco Embroidery and as such obtain answers for each presented nonwoven sample. For this purpose, the survey model was presented, where 11 textile surfaces developed in the laboratory for the execution of the embroidery are indicated, describing the stitch by stitch analysis of each base.

4 PRODUCTION DIFFICULTIES

4.1 *Stabilizing the fabric on the frame. Figure 4-A*

When the embroiderers are asked about the difficulties (or not) in stabilizing the fabric in the frame, the unanimous opinion is that the T1 sample does not have this problem. The same is true for T5, T6 and T10, although in these cases difficulties are also noted in the rolling of the fabric onto the beam and difficulties in attaching the fabric to the side (in the case of T5). One of the 7 respondents pointed out a difficulty in T6, when attaching the fabric to the side.

Difficulties in stretching the fabric was noted in T2, T3, T4, T9, T10 and T11 samples. However, samples T4 and T9 are the ones that are referenced (6 in 7 embroiderers) as those that show the greatest difficulties in stretching the fabric onto the frame. One of the 7 embroiderers had no difficulty in stabilizing the T9 on the frame.

The difficulty of attaching the fabric to the side is not often given as a result, which may lead one to suspect that the ends of the samples, with the initial changes introduced (reinforcement of the ends), fit the process.

4.2 *Difficulties embroidering. Figure 4-B*

As for the difficulty in embroidering on this fabric, it is suggested that the most difficult fabrics are samples T1, T2, T4 and T5. In the first case, considered by 3 of the 7 embroiderers surveyed, in the case of T2 by 6 of the 7 and in samples T4 and T5 for 5 of the 7.

Samples T3, T6, T9, T10 and T11 are indicated as being those that do not present difficulties in embroidering. 6 of the 7 respondents referred to T11 as the least difficult fabric.

4.3 *The Greatest difficulties found in embroidering. Figure 4-C*

The greatest difficulties encountered by the embroiderers when embroidering the fabrics. It is immediately visible that all the answers indicate the option, "the silk thread is loose, slack".

We can therefore highlight the fabrics that, by agreement, were the most difficult to embroider.

In the T2 and T5 sample, 5 of the 7 respondents report that loose and slack silk were the major difficulties they encountered when embroidering these fabrics.

In T1, T2, T4 and T10 fabrics, there is still reference to the option that the silk thread does not flow, it often snags although the number of embroiderers who pointed it out is insignificant.

4.4 *The Behavior of the embroidery on the fabric. Figure 4-D*

Regarding the collection of data from the appearance of the embroidery behavior compared to the fabric

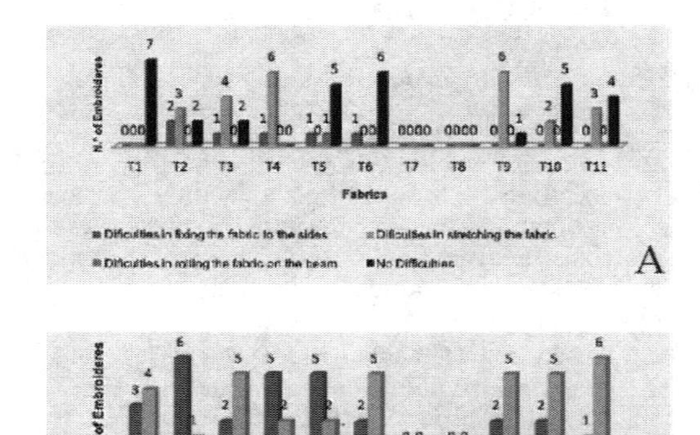

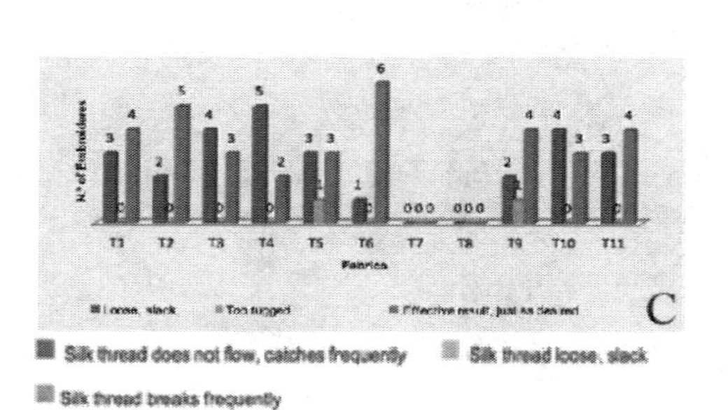

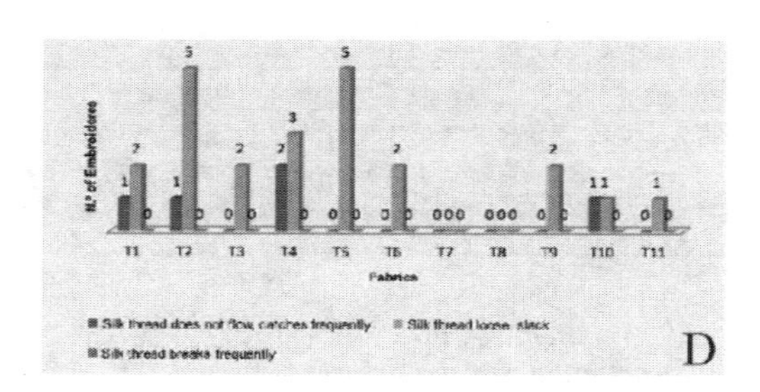

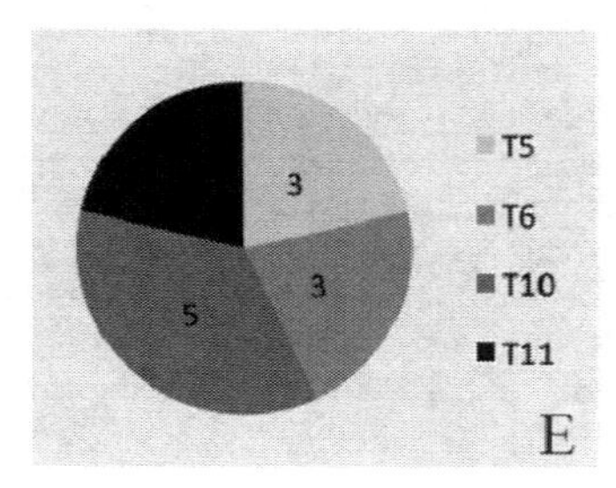

Figure 4. Graphical analysis.

used (when removed from the frame) the responses indicate 6 of the 7 respondents considered that the result is effective, as desired, with respect to T6. The same happens with the T2, where 5 of the 7 embroiderers share the same opinion. The remaining 2 responded that the embroidery (compared to the fabric used) becomes loose, slack, when removed from the frame.

Not all samples have an effective end result. Thus, T3, T4 and T10 show that most the respondents opted for the loose, slack option, as the behavior of the embroidery against the fabric.

In these non-woven fabrics, care and finesse must be taken. Figure 4.

5 FABRICS THAT ARE TECHNICALLY MORE VIABLE

In Graphic E, Figure 4, only the samples indicated by the respondents are identified as being preferred, the embroiderers understood that the finished samples to be an esthetic and technically perfect (or almost perfect) final work, were those samples that most resemble the traditional Embroidery of Castelo Branco or else the ones that best adapt to the embroidery production.

Among the 9 bases tested, each embroiderer was responsible for choosing two of the preferred samples (the sum of the answers is double the respondents).

The choice of T10 fabric as the preferred sample is visible. Of the 14 responses, 5 indicated T10 as the preferred sample. The remaining references were made to T5, T6 and T11.

5.1 *Testing*

Tests were carried out in the laboratory of the needle-punched non-fabric to determine the tension and extension strength (NP EN ISO 9073-3 of 1989).

In order to carry out the determination of the tension strength and the extent of rupture in the developed and presented non-fabric (needle-punched) to the embroiderers the resistance after the intervention was tested in the laboratory where the standard NP EN ISO 9073-3 of 1989 was used to test the resistance of the bases to laboratory scale.

Three longitudinal and cross-sectional samples we cut for testing. The principle of the longitudinal and transverse direction is to apply forces to the samples with a certain length and width to determine the con stant ratio of the elongation. The samples were cut with an approximate width of 50 mm and 0.5 mm and a length that ensured a spacing of the 200 mm.

After the test of the determination of the tension and extension of rupture, new samples were given to the embroiderers for direct verification on the touch and the possibility of embroidering with technical rigor.

5.2 *Analysis of results through direct observation*

In order to answer the starting question about the possibility of developing new textile bases based on nonwovens for the embroidery of Castelo Branco, it was again applied the qualitative interventionist methodology as an active research at this point of the work. An interview with direct observation in the analysis of the development of new embroidery was carried out where some issues that crossed with the previous

Table 2. Summary of the analysis of values in the laboratory by strip test method.

Samples| Fabric	Newton	%LO	Joule
S1	324.9	15.84	1.974
F1	325	5.795	1.312
S2	76.9	126.7	6.379
F2	125.9	79.99	9.136
S3	40	106.6	2.852
F3	52.05	113.3	3.701
S4	281.8	9.14	1.907
F4	111.6	14.98	1.154
S5	84	23.32	2.173
F5	95.8	30.8	3.107
S6	198.6	19.97	2.702
F6	834.7	12.46	13.78
S7	1.36	43.31	0.04484
F7	5.76	24.15	0.2053
S8	9.4	68.3	0.8628
F8	41.1	14.15	0.650
S9	195.2	14.13	3.356
F9	168.6	14.13	2.837
S10	153.6	9.995	1.781
F10	93.9	7.48	0.8762
S11	266.5	74.99	15.34
F11	494.1	93.33	28.28

results were developed and that allowed us to reach conclusions.

Through the direct observation of the embroiderers, the possibility of embroidering in any of the selected samples has been verified since some restrictions such as washings, finishes, the purpose of the pieces, etc. are taken into account. It was once again verified that the non-fabrics offer resistance to the creation and the embroidery, not losing the stability of the fabric nor affecting the technical rigor of its embroidery. It was verified that aesthetically, irregular non-woven fabrics, without large structures, with more texture, can be perfectly coordinated transparent bases. From the various analyses, samples T5, T6, T10 and T11 were selected as the preferred ones to apply and to develop new projects related to the emblematic Embroidery of Castelo Branco.

It has been found that the malleable characteristics of this type of non-woven fabrics allow the production of parts with great ease. It was also highlighted the ease in which the embroiderers identified through the touch the use of raw material of silk being the excellence of Embroidery of Castelo Branco.

It was also found that in some fabrics they lost lightness when applied to embroidery, mainly produced with the application of closed, very filled stitches. Try to use open points, less filled and contours, without distorting the reading, symbolism.

Embroidery:

It can be applied on the back of nonwovens, considering that most of this differentiation cannot be determined.

An attempt was made to introduce linen fibre to embroider on these bases where it was immediately verified through the analysis and direct observation tests on these new bases that the use of this fibre damaged the presented base and was not a viable solution because its characteristic was different from that of silk.

6 CONCLUSION

Based on the study proposal presented, it was verified through questionnaire analysis and direct observation the validity of these bases created as new textile supports for the application of Castelo Branco embroidery. The result presented here was surprisingly positive since these embroiderers had never worked on this type of support and when they tested it they were obtaining surprising answers. Analyzing the results obtained by the presented graphs, it was known the possibility of embroidering with efficiency and rigor the embroidery in these presented bases. On the bases analyzed as shown in figure 4 the embroiderers refer to the fabric 10 as being the sample chosen as technically perfect for embroidering. The remaining positive references are left to the bases Fabric4, Fabric6 and Fabric11.

It was concluded that in this type of bases it is possible for embroidery production not neglecting the technique that characterizes it so much.

For future production of new tissues based on the needling technique some technical adjustments should be introduced for future production in order to streamline and improve the final product intended.

From the analyzes carried out by the embroiderers on the use of new materials, there was also a strong and consistent opinion about the development of this type of bases to carry out new work in the area of Embroidery of Castelo Branco. In summary, the use of innovative solutions for the revitalization of Castelo Branco Embroidery applying raw materials and technologies other than the traditional ones can confer added value products enhancing the niche of the Embroidery Market of Castelo Branco which allows the region of Castelo Branco an enhancement of identity of the Castelo Branco Embroidery.

ACKNOWLEDGMENTS

The authors of this paper wish to thanks the Centre for Research in Architecture, Urbanism and Design (CIAUD) of the Lisbon School of Architecture of the University of Lisbon and FCT for founding this project.

REFERENCES

ADRACES; IMC/MFTPJ; CMCB; IPCB. (2008). *Colchas de Castelo Branco – Percursos por Terra e Mar*. Castelo Branco: ADRACES, IMC/MFTPJ, Câmara Municipal de Castelo Branco, IPCB.

ADRACES; IMC/MFTPJ; CMCB;IPCB. (2007). *Motivos – Descrição dos principais motivos do bordado.* Em ADRACES, IMC/MFTPJ, CMCB, & IPCB, Caderno de Especificações Técnicas do Bordado de Castelo Branco (p. 12). Castelo Branco: Parceria "Ex-Libris".

ADRACES; IPM/MFTPJ; CMCB;IPCB. (2005). *A situação actual do Bordado de Castelo Branco – Diagnóstico situacional.* Vila Velha de Ródão: ADRACES, MFTPJ, IPCB, Cârama Municipal de Castelo Branco.

Associação de Ensino e Pesquisa de Nível Superior de Design do Brasil. (2008). *Anais do 8º Congresso Brasileiro de Pesquisa e Desenvolvimento em Design. Design de Produto com Valor de Moda* (p. 380). São Paulo: Associação de Ensino e Pesquisa de Nível Superior de Design do Brasil.

Belino, N.J. 2006. *Optimização de um sistema automático de controlo da isotropia de mantos pré-agulhados por análise de texturas e recuperação de imagem baseada em conteúdo. Universidade da Beira Interior*, Departamento de Ciências e Tecnologias Têxteis. Covilhã: Universidade da Beira Interior.

Cardoso, J.R. 1944. *O problema da Sericultura Nacional e a Exposição de Colchas de Castelo Branco in Subsídios para a História Regional da Beira Beixa* – Vol. I. Castelo Branco: s.n.

Cardoso, M.H. 2011. *Novas bases têxteis para aplicação do Bordado de Castelo Branco em vestuário contemporâneo.* (A. M. Fernandes, Entrevistador).

Catálogo da Exposição de Bordado de Castelo Branco. (s.d.). (prefaciado por António Realinho) . Câmara Municipal da Vila de Luxemburgo: ADRACES.

Espaço e Desenvolvimento. 2007. *Estratégias para a Valorização e Promoção do Bordado de Castelo Branco.* Lisboa.

INDA – Association of the Nonwoven Fabrics Industry. (s.d.). Obtido em 20 de Julho de 2011, de www.inda.org

The Textile Institute. (s.d.).14 de Julho de 2011, de www.textileinstitute.org.

Vaz, P. 2004. *Tradição e modernidade. Quem tem passado tem obrigatoriamente futuro?* Em P. Vaz, O tradição tem futuro? O sector têxtil e do vestuário português na mudança de paradigma de desenvolvimento (p. 109).

Viana, E. 1942. Catálogo de Exposição – *Colchas de Noivado. Bordados de Castelo Branco.* Lisboa : Estúdio do SPN.

Winterburn, S. 1972 .*Needled-felted Fabrics.* Manchester.

Textiles, Identity and Innovation: Design the Future – Montagna & Carvalho (Eds)
© 2019 Taylor & Francis Group, London, ISBN 978-1-138-29611-4

The relevance of different players on the design project: Garment, identity, motivation and social innovation

J. Oenning
Department of Textile Engineering, University of Minho, Portugal
Department of Fashion Design, Federal University of Technology, Brazil

J. Cunha
Department of Textile Engineering, University of Minho, Portugal

J.B. Garcia Jr
Vale da Seda Institute, Brazil

ABSTRACT: This article is the result of one of the research steps of a Textile Engineering PhD investigation project, that starts with the premise that design for social innovation represents a challenging discontinuity of the traditional project methods, because it articulates an ability to enounce collective and individual interests. The motivation matrix, a new relationship compilation, is used with the aim to generate ways of social innovation. It is a co-design and visualization important tool, in which each player defines expectations on its system involvement, allowing the comprehension of the connections between the different system players and what makes them linked to each other.

1 DESIGN AND SOCIAL INNOVATION

The design project, active in different areas such as product, graphic or with interest in social relationships, has been shown specially on the last 10 years a powerful innovation driver (Hillgren et al. 2011). The current situation is shaped in a context of reflection, in which designers are invited to rethink their roles in projects and in what perspectives they wish to develop their ideas, offering new solutions to social problems and collaborating for a more assertive vision of sustainable futures. To achieve such achievements, it is necessary to understand which primary concepts portray Social Innovation in its generic field of prospecting. It is not a term with a closed definition, but the authors converge towards the same meaning.

Social Innovation is an innovative solution to a social problem, more effective, efficient, sustainable or fair than existing solutions, creating value for society as a whole and not just for private individuals. (Hillgren et al. 2011). Geoff Mulgan (2006), refers to social innovation as innovative service activities that aim to meet social needs and that stand out in the field of organizations, whose purposes are responses to social needs.

In addition to these explanations, social innovation emerges in a process of change and recombination between some agents such as: technology, traditional crafts, historical heritage, public policies and people, whether they are researchers, individuals or communities that act to resolve their problems or create new opportunities. The different players participate in the scope of the project at different times, each with its own goal, aiming at the common good (Manzini, 2013).

The applied Social Innovation course indicates significant changes in the way the society faces emerging social problems and the concern with the collective, where players demonstrate clearly the interdisciplinary character with methodological approach in continuous development (Moulaert et al. 2013).

For Young Foundation (2010, pg 16) Social Innovation can be a product, a production process, or a technology, such as a principle, an idea, a social movement or a combination between them, and it searches for answers to the social problems: "to identify and deliver new services that enhance individuals or community life quality; to identify and implement new integration processes on the workforce, new competencies, new jobs and new forms of participation as diverse elements, that contribute to improving the position of individuals in the workforce."

Innumerous are the examples of well-succeeded social innovation projects: microcredit startups, fair trade efforts; community tribunals, online and self-help health groups (Mulgan 2006).

Naturally, there were historical precedents in the social transformation and collective action. The syndicates, the human rights organizations, the socio-political movements towards independence: a wide variety of organizations in different contexts, and different sociopolitical ages approached problems and social wishes and fought for a better world (Mulgan, 2006).

However, over time, it is clearly understood that the research cannot be sole responsibility of the social sciences, nor can it be entrusted only to theoretical and empirical analysis, but rather has a strong orientation towards collective and systemic action (Manzini, 2013).

This systemic action of innovation by players networks who participate directly or indirectly on the solution development, starts in an identified social problem, with the collect of needed tools to find a suitable solution, develops a mapping to indicate which players are able to interact with the system in the sense of exploring abilities and knowledge (tacit and explicit) useful for social construction (Morelli 2007).

The understanding of these concepts is necessary to obtain a solid base on how to work on players competencies and abilities, in order to create a suitable prognostic to the social problem, respecting the singular and collective motivations of each one.

In the design for social innovation field, the work in nets shows much more effective and involves individuals, startups, non-profit organizations, local and global institutions that put their knowledge in a collective way for a variety of social and individual problems (Manzini, 2013).

Therefore, the present research project started by surveying some possibilities of social innovation that, starting with the Vale da Seda Institute in interaction with other players that could generate promising scenarios with results in collective and individual terms. Through the interacting the design with handcrafted units of manufacture, based on a system of co-creation and collaboration on a local scale, a method of networking was established with the objective of generating significant outputs for all those involved.

2 PLAYERS NETWORK

The local innovation systems are built by structuring a network of players who participate directly or indirectly in the development of solutions. The definition of these players should be made by a critical way, and explore the systems of interests and knowledge useful for future results. The investigation project works with the acting of three players: Instituto Vale da Seda (Silk Valley Institute), an association of Christian women named "In Casa de Talentos" and volunteer designers.

2.1 *Silk valley*

Silk is considered an ecological fiber due to the production processes and technologies that are not very aggressive to the environment with absence of the use of insecticides and involving small amounts of fertilizers for the cultivation of mulberry trees, whose leafs serve as food for silkworm.

In the state of Paraná, sericulture began in the late 1950s, with a boom in productivity in the 1970s and 1980s. About two thousand farmers work in this activity in the State, with mulberry cultivated area of 4000 ha. Silkworm cocoon production is characterized by activities that requires discipline from farmers and usually in this region, learning is passed on over generations of the same family (Cirio 2014).

It is an economical viable and low-risk activity, because there is a guarantee of sale, since it is maintained by the system of integration between farmers and the silk filature that provides silkworms and mulberry trees seedlings and latterly buys the cocoons. BRATAC, the only silk reeling company operating in Brazil, is located near the Silk Valley region and absorbs all silk cocoon produced. It produces high quality raw silk to wait on the most demanding markets worldwide (Cirio, 2014).

Brazil is an important producer of high quality raw silk, currently is the second largest exporter of raw silk in the world. The state of Paraná answers for 91.95% of this export and 88% of the Brazilian production of silk cocoon. From 2008, with the global economic crisis, these numbers have declined, with direct reduction on number of silkworm producers. These circumstances motivated the development of projects that aims to improve the silk production chain as a whole, emphasizing sustainable regional development, such as the Vale da Seda Project developed by Vale da Seda Institute (Berdu Jr & Pereira, 2013).

The Silk Valley, located in the northwestern region of Paraná, is the region that accounts for the biggest production of silkworm cocoon in the western part of the world. It is limited by the watershed of the Pirapó River, and the average area planted with mulberry trees in the municipalities that constitute this region is the double of other regions in Paraná.

2.1.1 *Vale da Seda Institute*

The Vale da Seda Institute is a project developed and installed in the Technological Incubator of Maringá. Created in 2009 by the entrepreneur João Berdu Garcia Junior, it is: "a private initiative of cultural nature, non-profit, with patrimonial, financial and administrative autonomy, with the purpose of promoting the regional identity of the Silk Valley, as a way to enhance silk production chain in the State of Paraná, contributing to sustainable regional development" (Vale da Seda 2010).

All these attributes came from a concern with the increase in the spectrum of silk chain performance in the municipalities of this region, since today only cocoon production itself generates revenue for those involved. Taking advantage of the high potential of value added that silk offers, made with great competence by French and Italian companies, can benefit small companies and individual entrepreneurs who have an interest in this activity (Vale da Seda, 2010).

The Vale da Seda Institute works on several fronts, and one of them, which converges with the interests of Social Innovation, is networking. Always focusing on projects that contribute to the strengthening of the regional identity, the increase of silk productive chain and added value in the State of Paraná, the Institute aims to put in contact small companies of different levels of performance in the market, which in cooperation would create new products, in a sustainable and socially responsible way. Fair trade is one of the guidelines of the current projects, proposing a profit increase among players involved, guaranteeing fair remuneration for producers and artisans, with strong economic benefits and social impacts (Pereira et al. 2009).

One of the mechanisms that Vale da Seda Institute uses to validate all the concepts of social innovation involved is the application of a holographic label to indicate authenticity and origin of products. This label has QR code technology that leads to on line certification of authenticity, origin, composition and commitment to sustainable regional development. Recently, information about the carbon footprint that each garment will generate throughout its useful life and as well, the carbon sequestration made by the mulberry trees necessary for the production of each silk piece is also available via QR code (Vale da Seda 2010). It is necessary to grow mulberry trees without pesticides to rear silkworms and produce the silk cocoons, because the mulberry leaves are the only food for the silkworm. Recent studies show that there is a positive correlation in the use of silk and the mitigation of carbon footprint (Giacomin et al., 2017).

In addition to the holographic label, another important initiative to get link between players, that are interested to connect with silk, proposed by Instituto Vale da Seda, is a trademark registered by the National Institute of Industrial Property (INPI): Renda Paraná. It aims to identify articles made with 100% silk yarn (made with by-products of silk reeling in Brazil), developed in knitting or crochet, with industrial or manual techniques of production and produced by individual entrepreneurs or artisans.

2.2 *NGO – Association of Christian women: In Casa de Talentos*

The Association of Christian women named "In Casa de Talentos", is a non-governmental organization, founded in 2004 in the neighborhood of *Cinco Conjuntos* in the city of Londrina, in the state of Paraná, south of Brazil, by a woman from the community. The objective is to provide opportunities for women who are not in a position to enter the formal labor market, for different possible reasons, or who are in situations of social vulnerability, that they have access to training for work and income generation in the areas of clothes making and handicraft development. In addition, it aims to offer an environment of human development and social interaction, where the users can experience the exchange of personal and professional experiences.

The NGO is located in a Londrina's community cultural center and is linked to the County Program of Incentive to Culture, which is supported by the pillars: symbolic expression; right of citizenship and potential for economic development.

The NGO's work structure is currently based on three distinct groups: a) Women working in painting, kitchen towels production, knitting, crochet and macramé; b) cutting and sewing operations and c) beauty salons. All groups are supported by NGO women collaboration group, who teach the students on classes and guides the production of articles for sale.

In the case of cutting and sewing, knitting and crochet, the NGO currently has five collaborators. The president and founder of the NGO is the manager and she also teaches in the cutting and sewing classes for the community. The income of the collaborators comes from sale of ready-made products (cushions, rugs, towels), repairs and adjustments of pieces that people from the community brings to adjust, sales of products developed (kitchen towels finished with handicraft techniques) and also part of the value of tuition fees from offered courses.

The NGO has an agreement with Londrina City Hall, that offers free vacancies on the courses for associated women in the Program of the Municipal Department of Social Assistance of the City of Londrina called *CRAS – Centro de Referência de Assistência Social*. This program "includes a set of actions developed with vulnerable families aiming at the development of their potentialities, as well as meeting the basic needs and inclusion in other public policies" (Prefeitura de Londrina 2017). One of these collaborating female student came from this CRAS program.

In addition to these groups, the NGO has an annex, the second-hand clothing bazaar, which has a collection of clothing donated mainly by service entities. The collaborators and NGO students customize some pieces of this collection, before trading. All the collaborators live in the *Cinco Conjuntos* neighborhood, where the NGO is settled, are married and have incomplete or complete high school degree of education.

2.3 *Volunteer designers*

The designer, in a social innovative project, has the possibility to act in many fronts, for being quite a dynamic process and often unpredictable, when it relates to a group of players working for a common denominator. Its role is to mediate the group of players involved in the offered network with mutual and individual goals, and to ease the rise of ideas and all participants' initiatives. The responsibilities in terms of creativity and design knowledge (to conceive and perform projects) become possible to the conception and implementation of new solutions (Manzini, 2013).

The context of the Social Innovation design project requires a large positive shift of thinking to the local context in which it is embedded, and focuses on local solutions centered on human being. This

Criterion	Justification
designers with a college degree	Competencies and abilities acquired through professional an quality training
designers working in the formal labor market	experiences in design projects applied in the labor market that could be passed to other groups
voluntary collaboration	paid work becomes impracticable, since there is no funding for the research project.

Board 1. Criteria and justifications for choosing the volunteer designers.

requires exploring their intuition skills, recognizing patterns, building ideas together, in an environment of co-creation and collaboration. (Brown & Wyat,t, 2010).

In sum, the designs foster a discontinuation on the way of performing tasks and proposes new sustainable methods to work around solutions (Manzini, 2008).

The group of designers who compounded the network's project in question, aims to offer training for the people involved in the process of social innovation in a voluntary way. The knowledge approaches are contextualized on the scope of work in the industry, either being molding, sewing, design and product marketing. To select the professionals, some criteria was established and reasonably justified, represented on board 1.

3 BUILDING THE NETWORK – MOTIVATION MATRIX

The concept of creative community, named by Manzini (2008), lays down the elements already existed supplied by individuals, institutions or available infrastructure, can recompile and improve the solutions to new ways of life.

Whenever a group gathers to discuss problems and solutions in social innovation, they take their own vision within, as well as their expectations. Before debating possible solutions for given questions, it is necessary to build a new collective vision, taking in consideration all possible relations amongst players and the best way of exploring every single one of them.

One of the ways to work this new compilation of relations, to ensure development in social innovation, is to make use of the motivation matrix. It's defined for co-design and visualization tools. In it, each player will define their engagement expectations in the system. The motivation matrix's purpose is to comprehend the connections between the system players and what makes them bundled to each other (Morelli, 2007).

This is possible because of data collecting techniques along with users carrying information, to improve the construction of a system, product or working process. In other words, to define what

gives to...	Vale da Seda Institute	NGO	Volunteer Designers	Investigation Project
Vale da Seda Institute	- to find and advance to new perspectives of development of activities and services with the silk of Paraná - to open and finalize the research in new areas - to deepen the involvement with Social Innovation	-qualification and experience in manufacturing of clothes never experienced - knowledge of the silk market - qualification in design and sewing - outreach of work in the fashion market - opportunity for co-creation and collaborative work	- knowledge of the silk market - contact with new raw materials - new competence for fashion market	- field of action for the development of social innovation - concepts and principles to be explored - support for research and experimental design - supply of raw material
NGO	- space for the development of Renda Paraná - to fortify values through co-creation - reach of Social Innovation in an experimental way	- to get knowledge - to expand the service portfolio - to find new solutions for generating income - to improve the management model	- to show new ways to design fashion products - opportunity for professional and human development - new abilities in the fashion design process	- social problem to be worked - environment of collaboration and co-creation - competence and abilities to be trained - support for the development of the investigation project - challenges for product development and management
Volunteer Designers	- exchange of experiences - qualification in manufacturing processes for product development	- qualification in needy areas - co-creation and collaboration - strengthening of project culture in design	- to acquire experiences in areas never experienced - to test capabilities - to collaborate with disadvantaged people	- support to achieve the proposed objectives - motivation for players involved - strengthening of social innovation concepts for the project
Investigation Project	- expansion of performance in social innovation - development of the trademark "Renda Paraná" - to gain visibility in other cultures - valorize craftsmanship	- support for the development of a new business model - to create new capacities in design and manufacture of fashion products - help establish new possibilities for generating income	- to acquire new experiences in the areas of fashion design and social innovation - opportunity to think over running business activity - exchange of experiences	- to generate meaningful outputs for all players - to achieve the proposed objectives - to generate future prospects

Board 2. Motivational Matrix built from Morelli's model (2007).

motivation each person has while being part of the system, expressing what the needs are or what to expect from it.

This matrix is a way of investigating the solution since it takes the point of view of each stakeholder along with its own interests (Morelli, 2007).

As such, the model created on board 2 related all players who were involved in the System and it displays what each individual expects to receive from self and others involved.

3.1 *Motivation matrix discussion*

The information processed through the Motivation Matrix suggests a challenge analysis process. In the context of collaboration, it is necessary to build a vision of the future, mapping and projecting strategies to reach these visions, and to respond in a proper manner to situations of exclusions, situations of necessity and to situations of desire for betterment human condition (Moulaert et. al 2013). For larger objectivity, the first action was to measure which one of them is emphasized amongst players, to elect the order of priorities and to define a plan of execution to approach them.

The first task, defined as key factor to develop solutions for social innovation in the system's context mentioned, is the need for qualification of the NGO collaborators. This aspect is highlighted in the research step, since every time a player (Vale da Seda Institute, Volunteer Designers and Investigation Project) links to the NGO, the term "qualification" comes out.

To establish a working plan to supply given expectations, some methodologies were put in place, with the intent to meet the desires from the NGO, as well as Vale da Seda Institute and Volunteer Designers. The central focus on the methodology is based on precepts of social innovation, which defines that the work must be developed in a collaborative way among all players, highlighting the relations between researchers-players and the social world in which they are going to perform. The decision making steps are communal and the work development itself is inclusive and cooperative. This analysis can take advantage from a holistic methodology in which allows theoretical perspectives to interact with empirical observations from an ethical positioning.

Some qualitative technical researches were laid to allow the outline relevant information into this phase of the project. Initially two were approached: the Observation and Semi-Structured Interview, hereinafter

3.2 *Data collection through observation and semi-structured interview*

The method of observation is a research technique that aims to understand the people and activities who perform within their cultural contexts of action, where the researcher is the own research instrument. Some criteria need to be defined as initial objectives that will guide observation, to understand the people and the field of observation, and establish strong social interactions between the researcher and the subjects (McKechnie 2008).

In addition to this, the other qualitative research technique used was the semi-structured interview. This type of interview consists of organizing the questions by topic guides, and the interviewer introduces the subject of the research, and allows the person interviewed to be free to talk about that, having the opportunity to talk about their experiences in richest possible way. The interviewer can do small occasional interferences for the deepening of aspects that may be relevant to the object studied, while maintaining an openness in the communication process (Bogdan & Biklen, 1994).

Dresses and Skirts	**Pants, Shorts and Jeans**	**Shirts and Blouses**	**Knitting, crochet and Macramé**
Adjustments of circumferences, strap, hem	Waistband, hem, zipper, side seams	Buttons, cuffs, sleeves, collar	Kitchen towel hem developed with handicraft techniques

Board 3. List of activities developed in the NGO in the areas of clothing and handicraft.

With a data collection tool based on the observation method (a check list to define the work structure installed in the NGO's manufacturing area) it was collected some data that could be used to build a strategic qualification plan. As a result, it was understood that the work structure, in spite of simple, meets the many requirements of a small manufacturing work unit, since it has industrial machines, overlocks, tables, chairs and supplies for the development of activities in the area of sewing and handicraft. In addition of this observation, guidelines were drawn that formed the basis of the semi-structured interview, and these questions were directed to the NGO president. Regarding human structure, it was understood that the NGO is currently formatted in two distinct groups: the collaborators working in the crafts making paintings, kitchen towels, knitting, crochet and macramé and the collaborators working in the cutting and sewing. These activities are sources of income generation, both in the sale of the products produced by them and in the courses offered to the community. Another important moment of the semi-structured interview was to understand about competences and abilities that employees already have. Therefore, all activities developed by them are listed in board 3.

It was realized that, almost in their totality, the existing sewing work modalities, are summarized in operations for small adjustments of pieces brought by the community. It is concluded that the routine of work is to operate the sewing machine in the simplest possible way, using the primary techniques to solve the problems. There is a concern for a good finish, but they don't know techniques that can help them achieve this goal.

In addition, another qualitative research technique used was the semi-structured interview. This type of interview consists of organizing the questions by topic guides, in which the interviewer introduces the subject of the research and leaves the interviewee free to discuss it, having the opportunity to talk about their experiences in a free and richer way in the information. The interviewer can make small occasional interferences for the deepening of aspects that may be relevant to the study object while maintaining an openness in the communication process" (Silva 2014 apud Bignetti, 2011, 4).

The perceptions, thoughts, cultures and knowledge of the community are potentialities seen as opportunities for action in the design process and all these elements, if well worked, culminate in an effective result. This context is in two complementary interests that the collaborators had the opportunity to signal:

1. development and trading of fashion products through a collective brand (creation, modeling, confection and sale) to a specific public with great demand, identified by them, and that meet the aesthetic, physical and psychological needs. The raw material would be the fabrics coming from second-hand clothes that are received monthly on donations, in good conditions, processed on other products;
2. to attend new brands of designers that are entering the market and wish to develop their collections on small scales, and can't achieve the minimum production imposed by traditional industry.

This observed scenario allowed to conclude that it is necessary, firstly, to list tools that allow the design as a social practice can be applied. After this, and based on these chosen tools, it is possible to elaborate and propose a strategic plan of training for these women to work in network, building a process of local development, including new skills and meanings for the generation of labor and income.

These tools are focused to answer the possible needs that all the players will have throughout the project, for the making of fashionable products that will be developed in the context of the established network. This list indicates in which areas the new competence and abilities will develop:

- Textile Structures: twill; satin; taffeta
- Draping: construction of skirt, blouse and dress bases; construction elements (pleats, ruffles, full circle); interpretations of molds
- Union seams: french seam; english seam; hand sewing techniques
- Construction of hem: scarf hem; Italian hem
- Finishes: invisible zipper application; application of bias; sewing machine needle clamp for finishes
- Craft techniques (improvement): knitting; crochet

At a later stage, two other tools will be approached for qualification: Project in Design and Marketing for Final Consumer.

4 DISCUSSION AND CONCLUSION

The preliminary study of the motivations that involve the net participants, through field research methods that allow investigate and collect suitable data, contributes to the design to establish strategies oriented to social innovation allowing local development.

The observation and the semi-structured interview, as research tools, showed up as excellent alternative to collect data in a social innovation project, because it puts the researcher as a journey facilitator. This positioning allows an open dialog, barrier free, and it promotes the access to the people and to the information intended to be collected. If, in the past, the social players were merely information passive inputs, now, there is an open space for collaboration, making co-producers and co-designers in social innovation projects.

On the collaborative projects, there is a paradigm that the collective interest is always ahead, guiding the main actions, in the pursuit of the common goods. However, the social innovation predicts a look towards the individual interests as well, because they give meaning and generate perseverance of each player into the project, allowing the common good possible.

The lack of this initial look towards each player's motivation can generate a negative impact to the goal development. An example is an important social innovation project, launched in 2010 and led by the Vale da Seda Institute, one of the net players, mentioned in this article, which is closing its activities in the current year (2017), just because the lack of motivation among women in the cooperative. This project, 'Artisans Brasil' gave birth to Cooperativa dos Produtores de Artesanato de Seda, a cooperative society formed by 46 women living in the rural area of Nova Esperança, one of the most relevant regions on silkworm production in the state of Parana. The project guided and trained women living in the countryside, on knitting, weaving and craft dyeing, to produce scarfs made of silk, obtained from the silkworm cocoons produced in Nova Esperança. These products were exported to the Artisans Du Monde, a French chain of fair trade retail stores and also on handcraft stores in the 12 cities that hosted Soccer World Cup in 2014, in Brazil. Engaged on fair trade system, one of the main goals was to create jobs and fair income to the woman from Nova Esperança, in a viable and sustainable way, that would also contribute to the region development (Berdu Jr 2011).

The cooperative wasn't able to keep the artisan's motivation after the failure of a hopeful activity, and that frustrated all the initial expectations.

The application of the Motivational Matrix, intends to avoid these consequences, since it refers to each one of the actors in order to understand and reveal what are the expectations and desires that drive them in the development of the project's tasks. This process mapped several interest actions, and the most featured in the beginning, was the training process that encloses and integrates all the actors. This was done by observing how many times the word "training" comes up over the matrix, and for most of them, it is linked to the non-profit organization player. The understanding of this action generated a look at the field of action in apparel and handicrafts existing in the NGO. Based on these data, new approaches in terms of skills and abilities were outlined so that a strategic plan for training could be built later in the next stage of the ongoing research project.

ACKNOWLEDGMENTS

This work is supported by FEDER funding on the Programa Operacional Factores de Competitividade-COMPETE and by national funds through FCT – Foundation for Science and Technology within the scope of the project POCI-01-0145-FEDER-007136 and UID/CTM/00264. The first author would like to thank for the support of the Federal University of Technology – Paraná, especially the Fashion Design Department and the Postgraduate Board.

REFERENCES

Berdu Jr., J. G. 2011. *Artisans Brasil – Artesanato de Seda no Comércio Justo e Solidário.* Available at http://tecnologiasocial.fbb.org.br/tecnologiasocial/banco-de-tecnologias-sociais/pesquisar-tecnologias/artisans-brasil-artesanato-de-seda-no-comercio-justo-e-solidario.htm (Accessed on 19 july 2017).

Berdu Jr., J. G. & Pereira, M. F. 2013. Silk Production in Latin American Regions – Constraints & the Way Forward. In: *Sericologia 53(1):* 15–23.

Bogdan, R. C. & Biklen, S. K. 1994. *Investigação Qualitativa em Educação: Uma Introdução à Teoria e aos Métodos.* Porto, Editora Porto.

Brown, T. & Wyatt, J. 2010. Design thinking for social innovations. *In: Stanford Social Innovation Review.* Leland Stanford Jr. University, 2010.

Cirio, G. M. 2014. Sericicultura. Dezembro 2014. SEAB – Secretaria de Estado da Agricultura e do Abastecimento. DERAL- Departamento de Economia Rural. Available at http://www.agricultura.pr.gov.br/arquivos/File/deral/Prognosticos/sericicultura_2015.pdf (Accessed on 29 july 2016).

Giacomin, A. M. & Berdu Jr, J. G. & Zonatti, W. F. & Silva-Santos, M. C. & Laktim, M. C. & Baruque-Ramos, J. 2017. Brazilian Silk Production: Economic and Sustainability Aspects, *In: 3rd International Conference on natural fibers, Braga 2017.*

Hillgren, P. & Seravalli, A. & Emilson, A. 2011. Prototyping and Infrastructuring in design for social innovation. *Co-Design Vol. 7, Nos. 3–4*, September–December 2011, 169–183.

Manzini, E. 2008. *Design para a inovação social e sustentabilidade: Comunidades criativas, organizações colaborativas e novas redes projetuais.* Rio de Janeiro: Editora E-papers.

Manzini, E. 2013. Making Things Happen: Social Innovation and Design. *Design Issue 30*: 57–66.

McKechnie, L. E. F. 2008. *The SAGE Encyclopedia of Qualitative Research Methods Participant Observation.* Thousand Oaks: Lisa M. Given.

Morelli, N. 2007. Social Innovation and New Industrial Contexts: Can Designers "Industrialize" Socially Responsible Solutions?. *Design Issue 23*: 3–21.

Moulaert, F. & MacCallum, D. & Hillier, J. 2013. Social innovation: intuition, precept, concept, theory and practice. In: Moulaert, F. & MacCallum, D. & Mehmood, A. & Hamdouch, A (eds), *The International handbook on Social Innovation collective action, Social learning and Transdisciplinary research.* 13–24. USA: Massachusetts.

Mulgan, G. 2006. The Process of Social. *Innovation. Innovations: Technology, Governance, Globalization*, volume 1, issue 2: 145–162.

Pereira, M. F. & Galeti, N. A. & Uchida, K. K. & Berdu Jr, J. G. 2009. Criação de sustentabilidade via princípios de comércio justo: o caso da Artisans Brasil. *In: A Economia em Revista Volume 17 Número 2.* Dezembro de 2009: 57–65.

Silva, J. S. G. 2014. Estratégias em design orientadas para a inovação social com enfoque no desenvolvimento local. *Tese de doutorado Programa de Pós-Graduação em Design do Departamento de Artes & Design da PUC-Rio.* Rio de Janeiro, Abril de 2014.

Vale da Seda, 2010. *Instituto Vale da Seda.* Available at http://www.valedaseda.org.br/?page=quem (Accessed on 9 march 2016).

Young Foundation (2010). *Study on Social Innovation, Study for BEPA in cooperation with Social Innovation Exchange (SIX),* Brussels: EC.

Textiles, Identity and Innovation: Design the Future – Montagna & Carvalho (Eds)
© 2019 Taylor & Francis Group, London, ISBN 978-1-138-29611-4

Graphic design and textiles: Visual appropriations in Sebastião Rodrigues' work

E. Rolo
CIAUD, Lisbon School of Architecture, University of Lisbon, Lisbon, Portugal

ABSTRACT: Talking about textiles implies, most of the times, also talking about decorative motifs over these textiles' surface and, in some cases, these visual motifs are created by a graphic designer. However, due to the visual richness of textiles, its relationship to graphic design can also be the reverse: textiles can influence design. We focus our study in this inverse relation, especially in Sebastião Rodrigues's work, identifying cases in which the designer used textiles to find visual themes for graphic projects. These appropriations concentrate mainly on two aspects: one, ethnographic, referring to popular embroidery motifs, and another, more erudite, in which the designer refers to the tapestry complex patterns. Simultaneously to this, we also approach Sebastião Rodrigues' proximity (and possible influence) to Marimekko textiles, a reference in modern Finnish and world design.

1 GRAPHIC DESIGN AND TEXTILES

The relationship between graphic design and textiles can be considered from several perspectives. The first one is based on typography and on the name of one of the main graphic design elements: the text – which etymological origin comes from *texere* (to construct, to weave) and *textus* (weaving, woven fabric, structure) (Dicionário Etimológico 2008). This origin is justified, according to Bringhurst (2004), with the homogeneous and flexible character that the written page presented and which continued with the typographic composition.

"The typesetting device, whether it happens to be a computer or a composing stick, functions like a loom. And the typographer, like the scribe, normally aims to weave the text as evenly as possible. Good letterforms are designed to give a lively, even texture, but careless spacing of letters, lines and words can tear this fabric apart." (Bringhurst 2004, p. 25).

Besides this approach, talking about textiles usually implies talking about the presence of decorative elements in these textiles. Concerning the decoration of textile surfaces, according to McDonald, F. (2011), essentially, five different technologies are considered: dyeing, "batik", printing (woodcut, copper engraving, stencil, screen printing, digital printing) painting and embroidery. All these techniques are more or less ancient and have undergone major changes with the mechanization brought by the Industrial Revolution. All of these intend to give visual richness to the textiles and thus contribute to the immense ethnographic heritage that this area encloses.

Regarding the morphology of the textile visual motifs, we can distinguish, according to Meller & Elfers (1991), five distinct groups: floral, geometric, conversational, ethnic and "artistic movements and period styles". It should be mentioned that some of the motifs represented in fabrics have meanings and connotations, which go beyond mere surface decoration. Examples of this are most of African textiles, which, this way, strengthen even more their ethnographic and ethnological value.

Graphic design is a discipline with some affinities with the textile area, because it can be the graphic designer to conceive the patterns to be reproduced on the fabrics' surfaces. Before the discipline existed with the present concept, artists or 'decorators' could perform the same task, as we can observe in the Arts and Crafts, the Bauhaus and Modernism movements, for example. However, the artistic manifestations in textile decoration, in particular in the tapestries and embroidery, are not rare in previous moments.

On the other hand, the relationship between the graphic designer and textiles can exist in the opposite direction, meaning that, instead of being the designer to design and deposit his creation in the textile support, the textile can work as an inspiration for the designer, who appropriates its visual elements to create in other design areas.

In the Portuguese graphic design history, there is the notorious case of a designer who collected and used visual motifs with ethnographic origin, and often from the textile area. This designer was Sebastião Rodrigues, who regularly conceived works based on popular visual themes present in *crochet*, in the typical embroideries from Castelo Branco or in the "lenços dos namorados" typical from Minho. However, the designer was also inspired by the Calouste Gulbenkian Museum tapestries, using this collection essentially in projects for the Calouste Gulbenkian Foundation.

2 SEBASTIÃO RODRIGUES AND GRAPHICAL APPROPRIATIONS FROM TEXTILES

Sebastião Rodrigues (1929–1997) was distinguished as a reference graphic designer in the panorama of the discipline history in Portugal. One of his graphic production main characteristics is the specificity of its visual resources. The designer frequently collected visual elements from ethnography, folk art and even archaeology, with which he kept building a kind of "library" of iconographic symbols. He made recurrent use of these graphic elements, inserting them into various projects in several moments of his career.

And I keep these forms. Loose sheets, notebooks... and later on, when I need, I review them and it often serves my job." (Um dia na vida de Sebastião Rodrigues 1969).

The term "recurrences" was introduced by José Brandão in the exhibition "Sebastião Rodrigues Designer", in 1995. However, then, this study wasn't deepened. In the thesis "Olhar | jogo | espírito de serviço: Sebastião Rodrigues and graphic design in Portugal" (Rolo 2015) it was possible to identify the various groups of these recurring elements: eyes, birds, hearts, rosettes, flowers, the sun, stars, faces or geometric patterns.

Some of these recurring motifs come from textiles, such as figures of birds, hearts and flowers. To better understand this appropriation, let us analyse some works that represent this dimension of the designer work.

In the book cover "Poesias Completas" by António Gedeão (Sá da Costa publisher, 1982) (Figure 1), we can notice the influence of ethnography on the represented graphical elements. On the cover background, we can see the representation of crochet instructions. As the main element, we can see a stylized heart, which refers to the popular textiles of Minho (especially "lenços dos namorados") (Figure 2), but it can also refer to some popular paintings and wood engravings.

In the poster "Portugal, País de Turismo" (1971) (figure 3), we can observe a traditional motif of Castelo Branco's embroidered bedspreads (Figure 4): a bird that the designer redesigned and simplified in order to transform it on the poster central element. Textures are also visible in the background, which, by its formal characteristics, we guess also comes from textiles.

This use of resources derived from popular art, by Sebastião Rodrigues, approached the "Estado Novo" ("New State" – the Portuguese dictatorship) propaganda philosophy, which was focused on folk art and folklore, and had in its basis a "folkloristic politics". This policy's purpose was to legitimize the dictatorial regime, to bring it closer to the people and, at the same time, to develop the cult of traditions and stimulate regionalisms. All thi s had the greater purpose of "fight against (...) the penetration in our country of

Figure 1. *Poesias Completas* cover. António Gedeão, Sá da Costa Publisher, 1982. [14 × 21 cm]. Author. "(...) For many years I have been using in my work shapes from popular origin that I collect in village museums... and at fairs throughout the country.

Figure 2. "Lenços dos Namorados", traditional Portuguese embroidery. *MatrizNet*. [online]. 2011. Consult. em 15/06/2013. Disponível em www: <http://www.matriznet.dgpc.pt/Matriz Net/Home.aspx>.

Figure 3. Portugal País de Turismo poster. 1971, Secretaria de Estado de Informação e Turismo. Calouste Gulbenkian Foundation archive.

any disturbing and dissolving ideas of national unity and interest" (Alves 2007, p. 21).

Another use of textiles in Sebastião Rodrigues' work is related to the appropriation of the tapestry graphic elements, namely from the Calouste Gulbenkian Foundation collection.

In the "Arte da Pérsia Islâmica" catalogue (Calouste Gulbenkian Foundation, 1985) (Figure 5), textile motifs are used on the dust cover and endpapers. The justification of this use is easily understood if we analyse the book content, in which the Persian tapestries occupy a prominent place.

Similarly, in the work "Calouste Gulbenkian, Coleccionador" (Calouste Gulbenkian Foundation, 1969) (Figure 6), the textile motifs are used as

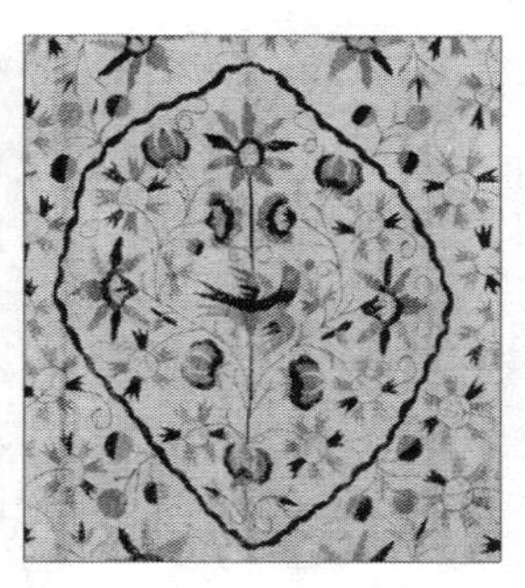

Figure 4. Castelo-Branco embroidery. MatrizNet.

ornamentation on the cover and on the dust cover. The pieces that constitute the basis of these graphics are part of the Calouste Gulbenkian Foundation collection, which justifies the use.

This type of graphics is used in these works in a directly way, without secondary significances. They only intend to enhance the editorial object and, at the same time, highlight the artistic piece represented.

We see in these cases a different use than the one that occurs with the appropriation of popular art. Here, the use of the graphic motif is directly related to the work content, in the case of popular motifs, the use may even be considered arbitrary, with no relation (or at least not direct) with the content.

3 GRAPHICAL APPROPRIATIONS AND TRANSFORMATION INTO GRAPHIC MATERIAL

In the context of this study, it is important to mention that Sebastião Rodrigues did not merely reproduce the elements he collected in diverse sources. On one hand, there was room for drawing and refinement of shapes: starting from more or less rough fast drawings, without much concerning with precision, the designer transformed them insistently, employing geometric drawing methods (Figure 7). His preoccupation with perfection and detail became an obsession, and because of that, his work process was very time-consuming.

On the other hand, the designer also used photography. In these cases, the transformation was essentially done by changing colour shades, transforming coloured patterns into monochromatic, more appropriate to the type of surface where they were used – generally, on more sober works, for institutions such as museums or the Calouste Gulbenkian Foundation.

In the case of the "Portugal País de Turismo" (Portugal, Country of Tourism) poster we can observe the transformation of an embroidery in a Sebastião Rodrigues' graphic motif, through the drawing simplification. The main graphic motif was isolated and geometrized, in order to form the thin figure, with very rational and perfect forms that we can observe in figure 8.

Figure 5. A arte da Pérsia Islâmica. 1985, Calouste Gulbenkian Foundation. [265 × 318 mm|64 pp]. Author's photography.

Figure 6. Calouste Gulbenkian Coleccionador. 1969, Calouste Gulbenkian Foundation. [21 × 28 cm|237 pp]. Source: Author's photography.

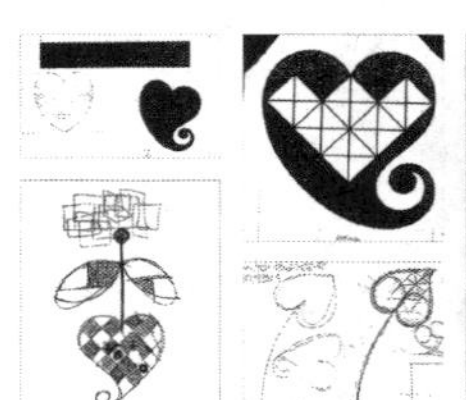

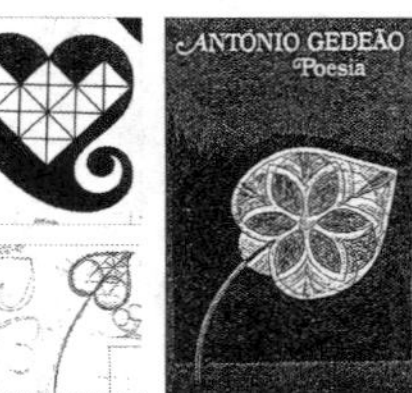

Figure 7. Sketches for Poesias Completas cover. 1982. Calouste Gulbenkian Foundation archive.

Figure 8. Motif of Castelo Branco typical embroidered quilt and its transformation into a graphic design element. Author/Calouste Gulbenkian Foundation archive.

Summarizing, we can say that Sebastião Rodrigues simplified, debugged, geometrized and perfected forms, in order to make them stable, graceful and with extreme perfection and cleanliness. However, even with the rationality degree he instilled, the designer had the gift of never making them "boring" or uninteresting. On the contrary, their graphics always evidenced liveliness and "joy", a luminous harmony, a modern class and a light stability. Therefore, his works acquired a serious appearance but without being heavy or dated.

Figure 9. Marimekko patterns. Designer Maija Isola and Vuokko Nurmesniemi. Watson-Smyth 2011.

Figure 10. *Almanaque* nr. 2 cover. November 1959. Source: Author's photography.

Figure 11. Covers of some numbers of the Almanaque magazine. Source: Author's photography.

4 SEBASTIÃO RODRIGUES AND THE MARIMEKKO TEXTILES

Another curious relation of Sebastião Rodrigues' work with the textiles refers to the Marimekko – a reference brand in the textile history. Created in 1951 by Viljo and Armi Ratia, it would profoundly mark the history of Finnish design and the whole course of modernist design.

Marimekko textiles are characterized by vibrant colours and simple and bold patterns, namely by the designer Vuokko Nurmesniemi's large stripes and the designer Maija Isola's big floral patterns (Figure 9). According to Meller and Elfers (1991), this type of pattern was called "Supergraphics", due to its affirmative and contrasting character.

These textiles arouse in line with the spirit of their time, in which graphic design was also governed by the same visual characteristics we can observe in the works of Paul Rand, Saul Bass (USA), Tom Eckersley (United Kingdom), Franco Grignanni, George Giusti (Italy) or Olle Eksell (Sweden), just to name a few examples.

A curious aspect of Sebastião Rodrigues' biography is the fact that in December 1958 he married Eila Rautakorpi, a Finnish woman. This fact led him to go on a train trip to Finland in 1959. According to Robin Fior (Fior, 2005, p. 207), the number 2 of Almanaque magazine cover was inspired by this trip, by this country and, in particular, by Marimekko textiles. We shall not forget though that these monthly covers had also an association with the respective Zodiac Sign.

In this cover (November 1959) (Figure 10) we can see the symbol of the Scorpio (representing the letters "M" and "U") in which a pattern of rectangles of coloured and geometric shapes is applied. According to Robin Fior, these shapes intend to represent the designer's travel, evidencing hidden connotations in each represented train compartment. These connotations culminate in the triangular shape, which leads us to the idea of home. Nevertheless, it is in the use of colour and in the type of contrasts, that Fior establishes a parallelism with Marimekko patterns. This cover evidences an explosion of colour, which we can also observe in some other works of the same period, for example, in the brochures for SNI (from 1959 to 1965) (Figure 11) and also in some posterior Almanaque covers (Figure 12). All this drawing simplification also invokes the collages, both by the character of cut outs and by the live and well-matched colours.

5 FINAL REMARKS

Textiles and their visual component, especially in figurative representations, are an important and interesting

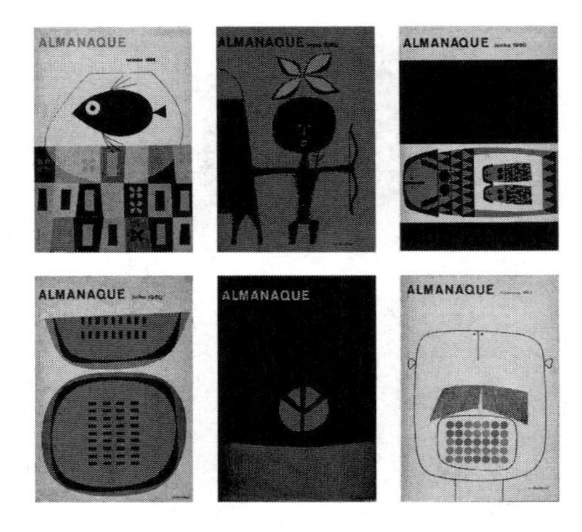

Figure 12. Salão dos Novíssimos Brochures. 1959–1965. SNI. [12,5 × 21 cm]. Source: Author's photography.

artistic heritage in the mankind history, both on a popular and ethnographic level and on a more erudite level.

In Sebastião Rodrigues's work we can identify two types of approach to textiles: one through the use and simplification of popular art and ethnographic motifs (present in traditional embroidery), and another through the use of textile patterns by their photographic manipulation.

Among the large amount of textiles, tapestries have occupied a prominent place over time, and examples of this are the rich Oriental tapestries, namely the Persians. In Portugal, we must highlight a different kind of tapestries – related to the Estado Novo architecture – which were integrated into architecture with a decorative purpose and assumed a prominent place in the most magnificent buildings. In this kind of tapestry, the reproduced graphic motifs are directly connected with the same period painting, but, as for their reproduction in a loom, the pictorial motifs have to be simplified and interpreted, we can declare that these are also graphic design objects.

We think it's interesting to focus on this character of drawing simplification and despoiling, which was common among the 50's and 60's designers and not only in textiles. Sebastião Rodrigues integrated that generation and notably simplified his graphics too.

In the case of more complex and traditional patterns, the shapes are not modified, but the colours are usually altered, overlapping tones, in order to soften the graphics and/or to approach them to the editorial object nature.

Another important aspect of this short study is connected to the Marimekko textiles and the assumption that they have been vehicle of transmission from Finnish to Portuguese design, through Sebastião Rodrigues. Gifted with a timeless and strong language, the Marimekko textiles are a reference in the world design panorama. Sebastião Rodrigues's travel to Finland, in 1959, allowed his contact with the visual richness of these objects, and consequently, some his works show characteristics that reveal this influence, especially in terms of colour and contrasts.

With this brief reflection we can realize that the affinity between graphic design and textile is a long-standing, bilateral and profitable reality. In the designer Sebastião Rodrigues's case, we consider of particular interest the connection to the popular textiles. We consider this visual heritage extremely rich and we believe that a compilation made at the present time would be very pertinent and serve as a basis for contemporary interesting graphic appropriations.

ACKNOWLEDGMENTS

This research was only possible with the support of CIAUD – Research Centre in Architecture, Urbanism and Design, Lisbon School of Architecture of the University of Lisbon, and FCT – Foundation for Science and Technology, Portugal.

REFERENCES

2008. *Texto* [Online]. Dicionário Etimológico. Available: https:// www.dicionarioetimologico.com.br/texto/ [Accessed 10/07/2017.

2013. *History of Marimekko* [Online]. Available: http://finnish design.com/history-of-marimekko/ [Accessed 8/07/2017.

2017. *Marimekko* [Online]. Available: https://www.marimekko.com [Accessed 10/07/2017.

AAV, M. 2003. *Marimekko: Fabrics, fashion, architecture,* New Haven and London, Yale University Press

AAVV 1959. *Almanaque.* Lisboa: Grupo de Publicações Periódicas.

AAVV 1995. *Sebastião Rodrigues, Designer,* Lisboa, Fundação Calouste Gulbenkian.

Alves, V. M. 2007. *"Camponeses Estetas" no Estado Novo: Arte Popular e Nação na Política Folclorista do Secretariado da Propaganda Nacional.* [Tese de Doutoramento], ISCTE.

Bringhurst, R. 2004. *The Elements of Typographic Style,* Point Roberts, Hartley & Marks Publishers.

Fior, R. 2005. *Sebastião Rodrigues and the Development of Modern Graphic Design in Portugal.* Tese de Doutoramento, University of Reading.

MatrizNet. [onine]. 2011. Consult. em 15/06/2013. Disponível em www: <http://www.matriznet.dgpc.pt/MatrizNet/Home.aspx>.

McDonald, F. 2011. *Textiles: A History,* Barnsley, Remeber When.

Meggs, P. B. & Purvis, A. W. 2006. *Meggs' History of Graphic Design,* Hobooken, New Jersey, J. Wiley & Sons.

Meller, S. & Elfers, J. 1991. *Textile designs: 200 Years of Patterns for Printed Fabrics arranged by Motif, Colour, Period and Design,* London, Thames and Hudson.

Rolo, E. 2015. *Olhar | jogo | espírito de serviço: Sebastião Rodrigues e o Design Gráfico em Portugal.* [Tese de Doutoramento em design], Faculdade de Arquitectura da Universidade de Lisboa.

Um dia na vida de Sebastião Rodrigues, 1969. RTP.

Watson-Smyth, B. K. 2011. The Secret History Of: Marimekko Unikko fabric. *Independent* [Online]. Available: http://www.independent.co.uk/property/interiors/the-secret-history-of-marimekko-unikko-fabric-2184168.html [Accessed 10/07/2017].

Textiles, Identity and Innovation: Design the Future – Montagna & Carvalho (Eds)
© 2019 Taylor & Francis Group, London, ISBN 978-1-138-29611-4

Sustainable fashion design: Social responsibility and cross-pollination!

F. Moreira da Silva
CIAUD, Lisbon School of Architecture, University of Lisbon, Portugal

ABSTRACT: The concept and practice of sustainable design is the response to the increasing of environmental and social pressures. Eco-fashion is a possible approach. From an environmental point of view, the clothes we wear and the textiles they are made from can cause a great deal of damage. Eco-fashion clothes are made using organic raw materials, such as cotton grown without pesticides and silk made by worms fed on organic trees don't involve the use of harmful chemicals and bleaches to colour fabrics are often made from recycled and reused textiles. Slow Fashion is a movement that is steadily gaining momentum and is here to stay. Slow Fashion represents all things "eco", "ethical" and "green" in one unified movement. Zero-waste design is a new way of thinking, a philosophy that forces us to challenge existing techniques and become a smarter designer. The present paper disseminates the theoretical contextualization of a research project aiming to demonstrate that to achieve a Sustainable Fashion Design, we have to work the three concepts at the same time, in a cross-pollination attitude, incorporating culture, identity, social responsibility and ethics, with a special focus on the Human Being.

1 INTRODUCTION

Recent concerns about the impact of global warming upon the world's climates and increased pollution of natural habitats have introduced the concept of environmental and ecological sustainability as a criterion against which to judge all human activity.

The concept and practice of sustainable design is the response to the increasing of environmental and social pressures. Nowhere are these issues more critical than in the creation of the world's built environments.

The built environment consumes energy and resources in its making and also in its day to day operation. The complex pattern of human habitats so formed are intimately connected to both natural and man-made energy and resource flow networks. The making of these habitats, therefore, requires knowledge of how the actions of individual designers, planners, builders and users of these habitats interacts with these energy and resource flows.

Thus, one of the greatest challenges facing all involved in this process is the creation of future products which not only continue to fulfill all of the traditional social, aesthetic and functional requirements of human settlements but also minimises its impact upon the natural environment by limiting what is taken from it and given back to it.

2 SUSTAINABLE DESIGN

Until very recently, the usual response to environmental problems was the reduction of pollution and post production waste.

Attention then turned away from such approaches, trying to increase awareness at the outset, allowing "cleaner" production, which, consequently, results in less waste and pollution in relation to the product that is being created

Then there was the perception of the major environmental impacts of the choices of materials and their use, and products disposal. The most advanced companies are beginning to see sustainable development as an opportunity rather than a threat: they recognize that prevention is better than cure

The concept and practice of sustainable design is the response of the design community to the increasing of environmental and social pressures.

During the past two decades we have witnessed a proliferation of terminology related to the incorporation of environmental considerations in the design.

According to Han Brezet and Carolien van Hemel (1997), experts that have been researching deeply on sustainable design during the last decade, eco-efficency improvement grows in time in four sequential levels:

1. product improvement;
2. redesign product;
3. innovation in function;
4. system innovation.

Product improvement: improvement of existing products as regards pollution prevention and environmental care; the products are designed and made accordingly.

Redesign of the product: the product concept remains the same, but the parts of the product are developed or replaced by others; has as common goals increasing the reuse of parts and raw materials, or minimize energy use at various stages of the product life cycle.

Innovation in function: it involves changing the way the function is fulfilled.

System Innovation: new products and services that require changes in infrastructure and related organizations.

To pass from level 1 to level 4, increasing amounts of time and complexity, which leads to higher levels of enhancements, are necessary for the sustainability. This model suggests that these complex innovations sustainable design can only be achieved over a significant period of time, say 10-20 years.

Until the beginning of this century most important aspect of production was cost efficiency. At the moment product's impact on the environment and more efficient use of resources becomes more important in entrepreneurship (Koklacova1 & Atstaja, 2011).

In every stage of production process design takes an important role, which implies not only functional and aesthetic, but also intangible value. At increasing demand for sustainable products, the manufacturers have to pay more attention to design and product life cycle which allows for improvement in the production and innovations in order to create more environmentally friendly items and for the sake of resource economy. It is possible if buyers, manufacturers and designers cooperate and create mutual understanding of business, manufacturing and design processes.

3 ECODESIGN

Ecodesign is product and service development in order to minimize the impact on the environment in the whole product life cycle at the same time ensuring functionality, quality, costs and aesthetic design.

Ecodesign is a strategic design management process that is concerned with minimizing the impact of the life cycle of products and services (e.g. energy, materials, distribution, packaging and end-of-life treatment). This involves assessing, prioritizing and then designing out problems, or designing new solutions. These solutions can range from specifying renewable materials, reducing the energy during usage to innovating the business model.

The aim of the ecodesign is to reduce consumption of the resources, to use eco-friendly materials, to optimize product manufacturing, distribution and usage as well as ensure the management at the end of the life cycle – recovery, recycling or disposal. (Trott, 2008).

4 SUSTAINABLE FASHION DESIGN

To develop a responsible awareness through the exploration of some emerging concepts in fashion, such as eco-fashion, slow fashion and zero-waste fashion, we have to focus on alternative methods of design and production of clothing giving special emphasis to the design of molds.

The present research has been comparing the different approaches to fashion design from a sustainable point of view.

The trans and interdisciplinarity of the different disciplines and areas which interact in fashion design industry lead to a different way of thinking and acting, using the cross-pollination of the new fashion perspectives.

4.1 *Eco-fashion*

Eco-fashion develops fashion products that take into account the environment, the health of consumers and the working conditions of people in the fashion industry.

Such is the case of the nonprofit *Sustainable Technology Education Project* (STEP), which aims to increase people's awareness of sustainable technology, enabling them to recognise the economic, environmental and social impacts of their own technology choices.

From an environmental point of view, the clothes we wear and the textiles they are made from can cause a great deal of damage. The pesticides that farmers use to protect textiles as they grow can harm wildlife, contaminate other products and get into the food we eat. The chemicals that are used to bleach and colour textiles can damage the environment and people's health.

Traditional fashion industry isn't very sustainable. People who make fashion products often have to work in terrible conditions.

Eco-fashion garments are made using organic raw materials, such as cotton grown without pesticides and silk made by worms fed on organic trees, not involving the use of harmful chemicals and bleaches to colour fabrics; they are often made from recycled and reused textiles.

Eco-fashion garments are high-quality garments that can be made from second-hand clothes and even recycled plastic bottles; they are made to last.

The people who make them are paid a fair price and have decent working conditions.

Eco-Fashion movement started at New York City's famed Fashion Week in February 2005 when the nonprofit *EarthPledge* teamed up with upscale clothing retailer Barneys to sponsor a special runway event called *FutureFashion*.

At the event, famous and up-and-coming designers showcased outfits made from eco-friendly fabrics and materials including hemp, recycled poly and bamboo. Barneys was so enthused that it featured some of the environmentally sensitive designs in its window displays for several weeks following the event, imparting a unique mystique to this emerging green subset of the fashion world.

Top Designers embraced Eco-Friendly Fabrics. One of the highlights of *FutureFashion* was a stunning pink-and-yellow skirt made from corn fiber by uber-cool Heatherette designer Richie Rich.

Eco-fashion Movement moved to the West Coast in June 2005 when San Francisco culminated its World Environment Day celebration with "Catwalk on the Wild Side," an eco-chic fashion show sponsored by the nonprofit *Wildlife Works* featuring top models and designs from the likes of EcoGanik, Loomstate, Fabuloid and others.

(http: //questpointsolarsolutions.com/?page_id=16877, accessed in June, 16, 2017)

One of the pioneers of the emerging eco-fashion movement is the designer Linda Loudermilk. Her "luxury eco" line of clothing and accessories uses sustainably produced materials made from exotic plants including bamboo, sea cell, soya and sasawashi. The latter is a linen-like fabric made from a Japanese leaf that contains anti-allergen and anti-bacterial properties.

Loudermilk also incorporates natural themes in each season's line—her most recent one being an oceanic motif. For designer Linda Loudermilk, fashion and ecology are inexorably linked. What Linda sees in nature, she applies to her designs.

As one of the first designers in the world to successfully take the idea of environmental consciousness into the elite realm of couture fashion, Linda Loudermilk has pioneered what she has trademarked as a Luxury Eco movement.

(https://www.yogitimes.com/article/linda-loudermilk-eco-luxury-high-end-fashion, accessed in 16 June, 2017)

4.2 *Slow fashion*

Slow fashion is a term coined around the year 2004 in London, some say by Angela Murrills, a fashion writer for the online news magazine Georgia Straight. The term became known since it is widely used in fashion blogs and on the internet. Inspired by the concept of "slow food", which originated in Italy in the 90s, the slow fashion adapted a few points to the scope of the model.

Others say it was first coined by Kate Fletcher, from the Center for Sustainable Fashion, when it was compared to the Slow Food experience.

In contrast to fast fashion – a fashion production system that prioritizes mass production, globalization, visual appeal, new appeal, dependence, concealment of environmental impacts on production, labor-based cost and cheap materials without taking into account the social aspects of production -, slow fashion emerged as a more sustainable socio-environmental alternative in the fashion world.

(https://www.ecycle.com.br/component/content/article/73-vestuario/5950-slow-fashion.html, accessed in June 17, 2017)

Slow Fashion is a movement that is steadily gaining momentum and is here to stay. Today's mainstream fashion industry relies on globalised, mass production where garments are transformed from the design stage to the retail floor in only a few weeks.

With retailers selling the latest fashion trends at very low prices, consumers are easily swayed to purchase more than they need. But this overconsumption comes with a hidden price tag, and it is the environment and workers in the supply chain that pay.

The fashion industry uses a constant flow of natural resources to produce 'Fast Fashion' garments. This industry is constantly contributing to the depletion of fossil fuels, used, for example, in textile & garment production and transportation. Fresh water reservoirs are also being increasingly diminished for cotton crop irrigation.

The fashion industry is also introducing in a systematic way, and in ever-greater amounts, manmade compounds such as pesticides and synthetic fibres, which increases their persistent presence in nature. As a result, some natural resources are in jeopardy and forests and ecosystems are being damaged or destroyed for such things as fibre production, leading to issues such as droughts, desertification and not least, climate change, that are affecting society at large.

Slow Fashion represents all things "eco", "ethical" and "green" in one unified movement. "Slow Fashion Values" are used to guide the entire supply chain.

So we can point out some core aspects which underline the importance of Slow Fashion, defining ten main principles:

1. Seeing the big picture – Slow Fashion encourages a systems thinking approach because it recognises that the impacts of our collective choices can affect the environment and people;
2. Slowing down consumption – reducing raw materials by decreasing fashion production can allow the earth's regenerative capabilities to take place. This will alleviate pressure on natural cycles so fashion production can be in a healthy rhythm with what the earth can provide;
3. Diversity – biodiversity is important because it offers solutions to climate change and environmental degradation. Diverse and innovative business models are encouraged;
4. Respecting People – participating in campaigns and codes of conduct can help to secure the fair treatment of workers;
5. Acknowledging human needs – designers can meet human needs by co-creating garments and offering fashion with emotional significance: the needs of creativity, identity and participation can be satisfied;
6. Building relationships – collaboration and co-creation ensure trusting and lasting relationships that will create a stronger movement;
7. Resourcefulness – Slow Fashion brands focus on using local materials and resources when possible and try to support the development of local businesses and skills;
8. Maintaining quality and beauty – encouraging classic design over passing trends will contribute to the longevity of garments: sourcing high quality

fabrics, offering traditional cuts and creating beautiful, timeless pieces;

9. Profitability – Slow Fashion producers need to sustain profits, and increase their visibility in the market to be competitive;
10. Practicing Consciousness – making decisions based on personal passions, an awareness of the connection to others and the environment, and the willingness to act responsibly.

(https://www.notjustalabel.com/editorial/slow-fashion-movement, accessed in June 17, 2017)

4.3 *Zero-waste fashion*

Zero-waste design is a new way of thinking, a philosophy that forces us to challenge existing techniques and become a smarter designer. Considering that roughly 15 percent of the fabric is discarded when a typical garment is made, the cumulative effect of leaving behind no waste has far-reaching environmental consequences. Zero-waste is about working within those constraints to invent beautiful new forms of fashion. It challenges the fundamentals of making clothing.

Typically, zero waste is associated with innovative grocery markets like this one or as a personal quest to reduce our trash footprint to zero. But the definition is expanding as more clothing manufacturers join the call to rein-in their wasteful practices.

Considering fast fashion is the second dirtiest industry in the world, next to big oil, reforming this toxic business is one of the best things we can do for the health of the planet — and our own.

Design is a growing body of research that draws on different branches or science and mathematics. It requires pattern-making know-how, a working understanding of sustainability principles. Anyone can drape a rectangle of fabric. But to make a zero-waste tailored ensemble for high fashion requires an entirely different level of skill. So, zero-waste requires smarter, more fearless designers who can see beyond drape and cut.

Featuring the work of international fashion designers, including Zandra Rhodes, Holly McQuillan, Timo Rissanen and Yeohlee Teng, with a common thread of how the fabric is cut and how much of it is wasted through the design process. They share a passion for reducing waste, without compromising style for sustainability. All their garments are created with little or zero fabric waste.

Zero-waste fashion research becomes an incubator for the fashion techniques of the future. They started out as a "cute" idea, but designs have become increasingly more complex and sophisticated with each passing season.

Zero-waste design isn't a new technology or material. It's mainly a new way of thinking—a philosophy that forces us to challenge existing techniques and become smarter designers. Technique-wise, it involves fitting all the flat pieces of clothing pattern like a jigsaw puzzle so no fabric is wasted.

Considering that roughly 15 percent of the fabric is discarded when a typical garment is made, the cumulative effect of leaving behind no waste has far-reaching environmental consequences. More than that, however, zero waste about working within those constraints to invent beautiful new forms of fashion.

Zero-waste design is probably not easy, but it's one of the more creative tools the fashion industry has to build a brighter future: fearless Designers adopt the Smart Zero Waste Fashion!

4.4 *Cross-pollination*

The intersection of different fields produces a fertile ground for new ideas, and the recombination of concepts previously separated at the core of creativity and novelty. Cross-pollination is a new way of generating innovative ideas.

The advances that emerge from a multidisciplinary work, grouping different approaches to the problem, different perspectives, are often of high value, superior to the best innovations obtained by partial approaches.

The cross-pollination of different areas stimulates the emergence of new fields, new points of view.

Hargadon and Sutton (1997) have demonstrated the condition under which teams are more likely to generate cross-pollination results. Fleming (2004) has shown that interdisciplinary teams produce more radical innovations than teams that act only within their disciplinary areas.

Cross-pollination is defined as the recombination of previously separated concepts. For example, biotechnology arose from the intersection of biology and organic chemistry. The collaboration of people in multiple disciplines allowed the discovery of DNA as a double helix.

The further development of molecular biology and its commercialization in the form of biotechnology has generated greater collaboration in the areas of chemistry, biology and physics.

Digital sound, which emerged at the intersection between computer science and music, is another example of a new interdisciplinary field of obvious cross-pollination. It was only through close collaboration between people with connections both within computer science and the music field that the first digital synthesizer was developed.

Nowadays we are aware that linear growth models for technology are useless, because they only describe one-dimensional progress, and not the recombination of ideas through mutual or reverse knowledge flows.

Last century we saw the emergence of thousands of new technologies, especially in science-based industries such as aerospace, plastics, personal computers, communication networks and optics. Any investigation of the history of these industries shows they were all based on a relatively small number of early technological breakthroughs; a clear example of exponential growth. (Gülzow, 2015)

More ideas mean better ideas, because more ideas are more likely to generate a really great idea, especially when it comes from working together for a more diverse group of creative. The key is to go beyond our

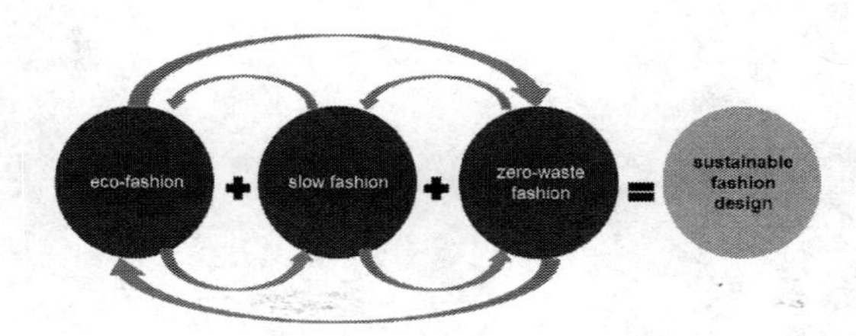

Figure 1. Sustainable fashion design through cross-pollination of three main concepts.

comfort zone and out of our normal circle of ideas generators.

In the present case, in addition to a cultural change, it will only be possible to achieve sustainable fashion design with a cross-pollination of the different types of approach and involving people from different areas.

Interdisciplinarity and cross-pollination allow us to obtain more holistic results, as a result of new strategies, integrating skills, leading to a Change of Mindset and a Change of Attitude.

5 CONCLUSIONS

The philosopher Karl Popper (1963) stated that we are not students of some matter, but students of problems, and that problems can go far beyond the limits of any subject or discipline. This statement has become increasingly relevant.

Today, many of the phenomena and problems that we try to understand and solve really go well beyond the boundaries of traditional academic disciplines. Modern technological developments and globalization increase the complexity of problems, and in response, we are increasingly aware that an integrated and holistic approach is needed.

If we don't want to 'hit the narrowing walls of the funnel', we must re-design the current unsustainable practices in society, including the fashion industry.

We can only achieve a sustainable fashion design policy if we bring together the different stakeholders involved in the process and if we make a cross-pollination of the three main concepts: eco-fashion, slow fashion and zero-waste fashion.

If this change is achieved, it will result in a gradual return to equilibrium, where societal behaviour is not in conflict with natural resources, and the fashion industry can carry on without compromising the health of the people and our planet, contributing to social-economic sustainability!

ACKNOWLEDGMENTS

This research was only possible with the support of CIAUD – Research Centre in Architecture, Urbanism and Design, Lisbon School of Architecture of the University of Lisbon, and FCT – Foundation for Science and Technology, Portugal.

REFERENCES

Brezet, H., van Hemel, C. 1997. *Ecodesign: A Promising Approach to Sustainable Production and Consumption*, United Nations Environment Programme, Industry and Environment, Cleaner Production.

Fleming, L. 2004, Perfecting Cross-pollination, Harvard Business Review, 82:22–24, September Issue, https://hbr.org/2004/09/perfecting-cross-pollination, accessed in January 12, 2017.

Gülzow, N., 2015. *Why degrees of cross-pollination determine the speed of innovation*, Journal Future Crunch.

Hargadon, A. & Sutton, R. 1997. Technology Brokering and Innovation in a Product Development Firm, in *Administrative Science Quarterly*, 42:716–749.

Koklacova1 S., Atstaja, D., 2011. Paper and Cardboard Packaging Ecodesing and Innovative Life Cycle Solutions, *Safety of Technogenic Environment*, Scientific Journal of Riga Technical University, 1, pp. 40–45.

Popper, K. 1963. *Conjectures and Refutations: The Growth of Scientific Knowledge*, Routledge

STEP [online] Available: http://stepin.org/ accessed in August, 03, 2017).

Trott, P., 2008. Innovation Management and New Product Development, 4th Edition, Edinburgh: Pearson Education Limited, pp. 426–432.

What is Ecodesign, 2011. [online] Available: http://www.ecodesigncentrewales.org/about-us/what-is-ecodesign/ accessed: June 25, 2017.

http://questpointsolarsolutions.com/?page_id=16877, accessed in June 16, 2017.

https://www.yogitimes.com/article/linda-loudermilk-eco-luxury-high-end-fashion, accessed in 16 June, 2017.

https://www.ecycle.com.br/component/content/article/73-vestuario/5950-slow-fashion.html, accessed in June 17, 2017.

Textiles, Identity and Innovation: Design the Future – Montagna & Carvalho (Eds)
© 2019 Taylor & Francis Group, London, ISBN 978-1-138-29611-4

Towards artistic children's education with textiles: A K10 challenge

M.J. Delgado & I.D. Almeida
CIAUD, Lisbon school of Architecture, University of Lisbon, Lisbon, Portugal

ABSTRACT: Making art with textiles has become part of the development process of creative potential that has led to the acquisition of different types of skills strongly correlated to the success of the textile and clothing industry. There are objectives defined in the curricula of basic education and pre-school education (K10 children) that tackle the artisanal process of transforming raw textile materials: these include acquiring knowledge of textile materials, highlighting the importance of handicrafts in culture and heritage, textile arts and the development of fine motor skills. The objective of this study is to learn about the practices of K10 children using textiles in schools, and to understand how these practices contribute to the development of their creative potential with regard to the area of textiles and clothing. Integrating the activities with textiles makes sensitization to the materials and techniques of the textile arts possible, and paves the way for greater sustainability of this sector in Portuguese society.

1 INTRODUCTION

Textile arts in Portugal date back to ancient times but it was in the 16th century that the field of handicrafts and textile fabrics made from wool, cotton, linen and silk grew considerably. The first domestic and family enterprises for wool wooling, fabric finishing and weaving sprang up in the weavers own homes (Sendim, 2005).

The beginning of textile industrialization triggered the transition from handicraft to mass production. In Portugal, as in other parts of Europe, this also occurred in the late eighteenth century and the textile industry grew significantly during the Industrial Revolution. The demand for cloth grew, so brokers had to compete with others for the supplies to make it. This created a problem for the consumer because the manufactured products cost more. The solution was to massify production with the type of machinery that connected multiple spinning so that up to eight threads could be processed at once, and led to the use of water-powered cotton mills and mechanical looms (Pereira, 2017).

However, even though mechanization played an undeniable role in increasing production and profits in the textile sector, resistance to the technological revolution – the default position of many Portuguese companies - was a major constraint to production and innovation. Changes were postponed until the mid-19th century, and only then was there a significant increase in the production and development of the textile, cotton and wool industry. The use of electrical power facilitated the establishment of spinning and weaving mills in towns and cities in the Minho region in northern Portugal. This occurred mainly in the Rio Ave valley, where there had been an ever-growing number of women workers from before the mid-nineteenth century onwards. As a result, the region became the country's textile hub. Between 1881 and 1917, the Portuguese textile sector concentrated the largest number of unskilled workers in companies engaged in the wool, cotton, jute, linen and stamping industries.

While the factories brought benefits in the way of automation and mass production, they also brought the social ills and hardships associated with child labour and poor workplace safety, which led to labour protests. By the beginning of the 20th century, the industry was still in growth mode, establishing its centre of operations in the centre and northeast of Portugal.

Although some periods of major competitive shocks have been overcome in textile and clothing companies, the current socio-economic conditions continue to pose great challenges for them. In pursuit of dynamic competitive advantages, as promotors of sustained growth, companies are now developing new activities and new ways of doing things. Technical expertise, creativity, technological innovation, new fashion business models and the internalization of companies play a fundamental role in their future projections. Nowadays, the textile sector is a hub of constant discovery through smart solutions. From Citeve (1990) to Cilan (1992) to Cenit (2009), several research and development centres are working towards bringing together smart technologies and clothing (ATP, 2016).

The textiles and clothing sector should not only be brought up to date as far as technology, techniques and design are concerned, but it must also invest in training and management. As globalisation occurred more quickly in clothing than in textiles, it is up to the latter to respond as soon as possible.

In this innovation paradigm, the textile sector is a critical area, one that is playing an ever-greater role in Portuguese business and the country's economy. It represents 10% of all Portuguese exports and 20% of Industry employment, making it dynamic and competitive in European and world markets (ATP, 2016).

Increased knowledge of skills in human resources, alongside a sustained dynamic textile production, highlight the importance of specialized training in the textile and clothing sector, not only in the fields of technology, innovation, processes and textile materials, but also in the areas of management and production planning. Hereafter, the design and creation of new products with high value, in addition to innovation and creativity, will depend on a renewed workforce whose professional training and qualification should have strong inputs of an artistic, scientific and technological nature, as justified by the purposes and contents of the curricula of higher education in this area.

Art has never been so much part of the textile world as it is today, and the reasons for this are founded in a constant exchange between schools, exhibitions, studios and shops.

In response to such demands, art education assumes an important position. The arts become part of the development process of creative potential, leading to the acquisition of different types of skills essential to the success of the textile and clothing industry.

The history of childhood education in Portugal reveals that a set of aesthetic experiences were included in the aesthetic education of children, and that they were based on guidelines drawn up by the Congress of Drawing of 1900 (Oliveira, 1996). They were implemented particularly through manual techniques using different materials, among which were textiles.

At present, there are objectives defined in the curricula of basic education and pre-school education that tackle the artisanal process of transforming textile raw materials. These include acquiring knowledge of textile materials, understanding the processes and technologies, awareness of biodiversity as well as environmental awareness, the importance of handicrafts in culture and heritage, textile arts and the development of mobility.

The integration of activities making use of textiles makes sensitization to the materials and techniques of the textile arts possible, and paves the way for innovative and qualified companies to maintain the sustainability of this sector in Portuguese society (ATP, 2016).

The objective of this study is to learn how textiles are being used in children's art education, and to understand how these practices contribute to the development of creative potential in these children with regard to the area of textiles and clothing.

2 OUTCOMES OF THE LITERATURE REVIEW

2.1 *The use of textiles in children's artistic education in Portugal: an historical overview*

The projects and reforms that characterized the various periods of children's education in Portugal were always associated with social values, political and religious ideologies, and the technological advances that have marked society. At the same time, mediation between family and school has always maintained a close relationship, combining affective values with the technical and scientific values of education (Magalhães, 1997).

i) Early modern period – ca. 1500 to mid-18th

Going back to the 16th and 18th centuries, sovereignty of the Companhia de Jesus, whose objectives were to develop the intellectual, physical and moral capacities of children and young people greatly influenced educational action in Portugal and Europe. This was a schooling plan based on the robust progression of development and knowledge (Gomes, 1995).

ii) Late modern period – after mid-18th to WW2

In the second half of the eighteenth century, with the extinction of the Companhia de Jesus in 1759 and inspired by Illuminist thought, the Marquês de Pombal, First Minister of King D. José I, undertook two major reforms – one in 1772 and the other in 1795 – that marked the domain of government action in education. These reforms influenced both the programs and the enlargement of the primary and secondary school systems, with "Minor Schools" opening throughout the country (Gomes, 1995). As far as higher education is concerned, a project was initiated with the general aim of drawing up a proposal to change the programmatic and methodological contents of the University of Coimbra, and to create the Faculties of Medicine and Mathematics. It was a modernization of education in Portugal that placed it alongside Europe's education systems of the time. However, any content related to the teaching of the arts was still non-existent.

Later, in the reign of D. Maria I (from 1777 to 1815), other changes were made in educational policy, which was marked by the return of religion to public education, along with approval for private schools to be opened. In 1815, and in the context of the social problems arising from the earthquake of 1755, the institution Real Casa Pia de Lisboa was founded. This institution promoted teaching females for the first time, with the opening of "girls' schools" (Oliveira, 1996).

Also in the reign of D. Maria I, measures were adopted that contributed to the progress of the sciences and the teaching of the arts in Portugal. The most representative in Lisbon include the constitution of the Royal Academy of Sciences of Lisbon (1779) and the Aula (1781), as well as the opening of the Drawing Class of Casa Pia and the Academy of Nudes, both in 1780.

In Oporto, the Public Classroom of Debux and Drawing was inaugurated in 1779 and it is considered the pioneering national institution in organized and systematic artistic teaching. Moreover, in 1860, high schools officially implemented the teaching of drawing, and only a few decades later, art education was included in trade curricula (Nóvoa, 1994; Lisboa, 2007).

The Liberal Revolution in 1820 was a starting point for major reforms in education. Those in power, recognizing how far behind public education was regarding the challenges of scientific progress, indicated the need for major educational changes; pointing out that these changes should be suitable to prepare students, at scientific and practical levels, for careers in agriculture, industry and commerce (Pulido Valente, cit. by Ó, 2009).

The General Regulation of Primary Education was an outcome of the first major reform of the constitutional regime. This occurred in 1836, and secured provision of free public primary education for all. A few years later, this reform also covered secondary and university education. It was "the first official document that systematized secondary education, integrating curricular, pedagogical and administrative aspects" (Ó, 2009, p.17). Also noteworthy is the work of Almeida Garrett who, defending the role of the arts in education, inaugurated the National Conservatory in Lisbon in 1836, to promote the teaching of the arts.

The year 1888 marks the creation of the women's lyceums. With the creation of these secondary schools, a curriculum for the training of young people was defined and implemented. The main goals were to develop specific competences that could lead either to the pursuance of university studies or to entering the world of the work. This was the end of "the false distinction between education and schooling, between theoretical and utilitarian teaching" (Coelho, cit. by O, 2009, p.16).

Between 1862 and 1886, schools for training teachers of primary education appeared for the first time. In Lisbon and Oporto, Commercial and Industrial Schools, and Industrial Design Schools were created, marking the beginning of a special technical, theoretical and practical education to prepare citizens to respond to the industrial growth needs of the country.

With regard to pre-school education, Law N° 1106 was passed in 1880 to encourage the creation of educational institutions that would support primary schools by taking in children from three to six years old and provide them with structured and consequential educational plans (Magalhães, 1997).

In addition to that, in 1882, the first Froebel Kindergarten, which followed the pedagogical currents of Europe was set up by public initiative. In the same year, the Association of Mobile Schools whose aim was to teach children literacy was created, with the wider objective being to combat illiteracy among the Portuguese population. (Cardona, 1997; Gomes, 1977, Magalhães, 1997). To achieve this goal, the João de Deus Method, one of the most effective types of material and texts ever designed to teach literacy and develop reading fluency, was created (Magalhães, 1997; Rodrigues, 2014). This association proposed "to establish nursery schools or school gardens for children from 3 to 7 years of age, where the spirit and doctrine of the educational work of João de Deus would be applied in its entirety, thus modifying the Portuguese type of nursery schools" (Gomes, 1977, p.51). The evolution in pre-school education is highlighted in the 1896 Children's Schools Regulation which, strongly influenced by the pedagogical conceptions of Froebel, Montessori, Décroly, among others, presented a program based much more on the integral development of children than on "instruction".

In the early years of the Republic (1911), and with the participation of pedagogues João de Barros and João de Deus, a comprehensive reform of the basic education system was designed, which included education for children between 4 and 7 years of age. Hence, the first kindergartens were set up. These were private schools for children's education, with innovative pedagogical methods to promote sensory education, the acquisition of hygiene habits, working methods and learning (Cardona 1997).

The multiplicity of reforms initiated during liberalism, and continued in the Republic, definitively marked the educational panorama in Portugal, with the highlight being the creation and autonomy of the University of Lisbon and Oporto (OEI – Ministério da Educação de Portugal, 2003).

The Estado Novo, which prevailed in Portugal from 1926 to 1974, made great changes in the educational system and was sustained by a new political and moral ideology. The official closure of public pre-school education in 1937 resulted in a sharp decline in early childhood education, (Cardona, 1997).

However, the history of children's education was always tied to family history and the history of women, and early childhood education was once again deemed the responsibility of the family and of an institution designated Work of Mothers for National Education. This institution, created in 1936, took on official responsibility for pre-school education and its role was "to stimulate the educational activities of the Family", "to ensure cooperation between the Family and the School" and "to better prepare the female generations for their future maternal, domestic and social duties" (Pimentel, 2011, p. 211).

In 1939, one single book for Elementary School learning was approved. The drawing activities at this level of education were not considered an autonomous discipline that had objectives, competencies and defined content (Oliveira, 1996). Rather, it was considered an activity that supported themes developed in other areas of knowledge through graphic records. In the context of aesthetic education for children and young people, the academic programs up to the 40s reflect the guidelines drawn up at the Congress of Drawing in 1900. These promoted developing a capacity for observation and reproduction through freehand drawing, as well as the practice of invention and geometric drawing, which were both associated with decoration.

By 1942, regulation of Schools of the Primary Magisterium had been adopted, and teachers were taught to train students to memorize the representation of the objects they drew from sight. The incorporation of this exercise in the student assessment at the end of the first grade studies was a pre-requisite for access to high school. Simultaneously, the classes of feminine courses promoted an aesthetic education, with the sole purpose of preparing those students for their future working life (Oliveira, 1996).

iii) Contemporary history – from 1945 to the present

In the field of artistic education, the influence of Herbert Read's theory is renowned. In his work Education Through Art, he takes up the foundations of Plato's doctrine defending that art should be the basis of education and that "every person is a proper type of artist" (Oliveira, 1996, s/p). The first guidelines for the introduction of art into the educational process, based on the merging of all forms of artistic expression, lead to an education of the senses, that is, an aesthetic education, which should form the basis of a broad conception of mans' formation.

This influence is reflected in the creation of several centres for children's art and children's art exhibitions, and through the work developed by the Calouste Gulbenkian Foundation (CGF). The CGF also designed and support (motivationally and financially) a number of pilot projects at pre-primary level such as the training and professional qualification of young people in art abroad. Similarly, the Children's Art Centers were implemented in partnership with the educational services of several museums. At this time, children's artistic education is marked by the action of Cecília Menano and João Couto, who founded the School of Art in 1949 to promote free expression in Cultural Education of children.

In the first cycle of the Technical School, the drawing program (1947) included the development of the following skills: observation, reproduction, interpretation, imagination and decoration, promoting manual work, free drawing, and geometric and decorative design. In 1948, Free Drawing became part of the official program of the lyceums, supported by the "Compendium of drawing for the first cycle of Lyceums" (Oliveira, 1996).

In 1959, the United Nations General Assembly adopted the Declaration of the Rights of the Child, and this provided the momentum for a turning point in the artistic education of children and young people in Portugal. However, up until the 1970s, artistic output in public schools was no more than choral singing, drawing, and handicrafts.

In the following years, some structural phenomena appeared that marked education and Art education in Portugal. Worth mentioning here is the relevance of the role played by Minister Veiga Simão's School System Reform Project (1971). This proceeded to definitively reintegrate public pre-school education into the educational system as a preparatory phase for school education (Gomes, 1977), and also led to the approval of 8 years of free basic school education. In tandem with this reform, which simultaneously tackled the artistic education of children, in 1971–72 the Training Course for Teachers of Art Education at the National Conservatory contributed to the aesthetic and artistic education of children (Oliveira, 1996).

In the period after April 25 1974, educational policies were redefined in accordance with the sociocultural orientations of the Revolution that were based on the principle of promoting equal opportunities. The expansion of the public network of kindergartens was to ensure the education of children from the age of three (Vasconcelos, 2015). In the field of art education, the prevalence of posters and murals inspired by the Revolution led to the promotion of graphic communication in schools, as well as to the introduction of cartoon work.

In the first Cycle of Basic Education, new curricula that integrate the artistic areas were drawn up; namely, Plastic Expression, Movement, Music, and Drama. While this emerged from the principles of integrative aesthetics, Visual education in the second and third Cycles of Basic Education, which extends to the 9th year of schooling, was oriented towards aesthetics and communication, with a strong focus on cultural values (Sousa, 2003).

Within this field of changes in educational policies, the Framework Law on Pre-School Education (OCEPS), published in 1977, reinforced the importance of this level of education and its integration into the basic education system.

In that period, artistic education at various levels of the education system was encouraged through the implementation of various projects and actions, in particular, the CAI and ACARTE (Service of Animation, Artistic Creation, and Education through Art), and with the development of artistic activities and projects for children, educators, and the community.

In 1986, with the publication of the Basic Law of the Educational System in Portugal (LBSE), and with the changes introduced successively in 1997, 2005 and 2009, a review of the structure of the educational system was undertaken to identify improvement opportunities and to ensure a higher quality education that would integrate aesthetic values into the curriculum (Sousa, 2003).

In the context of this evolution, the relationship between knowledge and expertise; between theory and practice; and between manual activities and artistic education, are some of the dimensions that are valued in this curriculum and are considered essential both for social integration and for the educational success of students. Art became part of the curricula at the Preschool and Basic Education levels, and is definitely accepted as a disciplinary area capable of promoting the development of the capacity for expression, the creative imagination and various forms of aesthetic expression, among others.

In the report of the study "Basic Knowledge of All Citizens in the 21st Century", the National Council for Education (2004) highlights the four pillars of education for the 21st century namely: (i) learning to know, i.e., acquiring the instruments of understanding; ii) learning to do, in order to act on the environment; (iii) learning to live together, in order to participate and cooperate with others in all human activities; and iv) learning to be, an essential means to integrate the previous three (UNESCO, 1996). In this report, the emphasis is on Basic Education, which is the level of education that, aimed at all citizens, is directed towards a curricular approach focused on the competences that lead to an active citizenship (Cachapuz, 2004).

2.2 *Artistic languages presently in the overall curriculum guidelines and in the program of the 1st cycle of basic education*

The Pre-primary Education Curriculum Guidelines in 2016 consider pre-school education as fundamental to the progress of each child's skills and learning in the transition to the first cycle, as well as "the first stage of basic education in life" (Silva, Marques, Mata & Rosa, 2016, p. 5).

In this sense, educational intentionality is characterized by the construction and management of a curriculum that includes strategies adapted to a social context involving different actors (children, educators, parents, family, professionals), and that facilitates communication and articulation with the community.

The relationships and interactions established between the different players in the educational process, from a systemic and ecological perspective, are essential for the development of this process. It is in this interaction with the various contexts that the child, establishing a continuity between play and learning, gets to know new forms and artistic languages that help them understand and represent the physical, social and technological world around them (Silva et al, 2016).

By playing and communicating with others, the child is not only appropriating new concepts through exploration and experimentation of the surrounding environment, but is also developing learning in the different content areas in an integrated and global way. The appropriation of multiple artistic languages allows the child to interact with others, express thoughts and emotions in a proper and creative way (Silva et al, 2016).

This approach to different artistic and cultural manifestations leads to the development of creativity, symbolic representation, and aesthetic sense. Therefore, children are encouraged to experiment, to perform and to create, to develop their capacity for expression, observation, appreciation and the transmission of ideas.

Artistic learning and making in the children's educational curriculum can be enhanced by the exploration and use of different materials and techniques. Activities carried out with various fabrics, papers, textures, paints, pencils, moldable materials, natural objects and recyclable materials, among others, lead to the development of creativity and meaningful learning.

Also, the multiplicity and diversity of materials that can be used, like waste materials for example, means they can be integrated into different activities and acquire new functionalities and meanings through their construction and deconstruction, and thus further help develop divergent and creative thinking.

Nevertheless, it is possible to even further enhance all this learning through understanding in the cultural and artistic context, with the implication of processes of observation, analysis and critical judgment. The approach to artisanship is a very important cultural aspect of children's education, not only for producing art but also in the acquisition of traditional knowledge concerning its use, the ethical dimensions, the family stories and traditions; all of which can be exploited by making the interconnection between different aspects of knowledge.

As Sousa affirms, "all experiences that take place in the field of arts education will be a source of knowledge, but also a source of connection and enrichment among peoples and cultures around the world" (2008, p.41).

3 RESEARCH METHODOLOGY

3.1 *The drawing up and application of the research instrument*

From the reading of both the first cycle teaching programs and the OCEPS, we came up with a questionnaire survey. The main goal was to seek information about teachers' perceptions of the adequacy of textile art in the education of children. The questionnaire was limited to 5 groups of differentiated but interconnected questions, as follows: i) sociodemographic characterization of the sample; ii) identification of techniques and materials related to the textile arts that are used in an educational context; iii) perception of the relevance of the use of these techniques and materials in the acquisition of skills and abilities essential for the development and training of children; iv) demonstration of the interdisciplinary dimension of textiles; and v) recognition of the importance of including activities with textiles in pedagogical practice.

This questionnaire was structured using closed questions, which included questions adapted from the Likert scale with four levels of response, and dichotomous questions. In this way, respondents would comment on the proposals presented, and classify them accordingly, with (1) being not at all relevant and (4) being very relevant. The option of I don't know/I can't answer (n/s; n/r) was also included.

3.2 *Data collection*

A total of 48 public and private schools from the Primary Education Cycle in the Lisbon area comprised

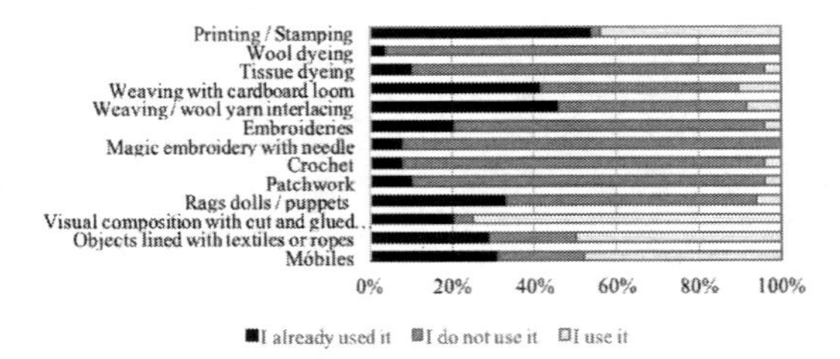

Figure 1. Techniques and textile materials used in an educational context.

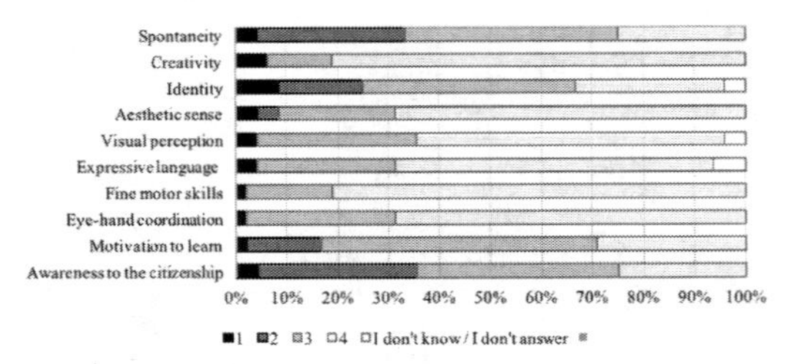

Figure 2. Teachers' perception of children's acquisition of skills and abilities through textile utilisation

the deliberately non-probabilistic sample. The teachers responded in all of them. Data were collected from several responses after 48 online survey submissions.

The results coming from descriptive statistical analysis were sufficient to define a very first exploratory definition of textile utilization in the artistic education with textiles.

4 RESULTS ANALYSIS

i) Profile of respondents

The sample consisted mostly of women (99%), and all the answers were considered valid. Regarding the age group, 50% of respondents were between the ages of 21 and 40, 10% were between 41 and 50 years and 40% were over 50 years.

With regard to academic training, 54% of the sample had a master's degree, 29% had a degree, 9% had a bachelor's degree, and 8% had a PhD. The doctoral degree belonged to teachers aged over 31 years. None of the teachers revealed any specialization or training in the field of fine arts.

ii) Techniques and textile materials used in an educational context

Although in this exploratory survey, the data were quantitative, the analysis was qualitative because statistical significance was not addressed. Therefore, the average values registered in the graphics may only be interpreted in a qualitative way.

Understanding the teachers' perceptions of experience using techniques and materials related to the textile arts in an educational context, as shown in figure 1, was an important driver for future improvement of textile utilization in an educational context, as the survey applied to the teachers' sample shows.

After analyzing these results (Figure 1), it is evident that the teachers, in general, do not use textile manufacturing and transformation techniques, such as weaving, embroidery, and dyeing of textile materials during their pedagogical practice. Only printing, and stamping on fabrics with rubber stamps are activities that approximately half of these teachers integrate into their classes.

The findings highlight that textile activities gain expression only when they complement or support other techniques, such as the creation of compositions, mobiles, and stamps, among others.

iii) Acquisition of skills and abilities through textiles

We also studied the relevance of using these materials and techniques in an educational context. Thus, based on the objectives defined in the programs and curricular guidelines, the respondents were presented with a set of recognized skills and abilities that can be developed through accomplishing activities based on textiles.

From our reading of the results presented in Figure 2, teachers' perception about the contribution of the materials and textile techniques associated with both sensitization and artistic education was deemed, by the majority of respondents, to be relevant or very relevant.

Materials, plasticity of materials, textures, and patterns, as well as techniques and technologies of textiles were also valued by the great majority of respondents as vehicles for promoting creativity, aesthetic sense, and visual perception. Also valued were the multiplicity of expressions revealed in embroidery, tapestries, crochet, and knitting, among others.

Just slightly more than 50% of respondents considered that the traditional matrix of knowledge and knowledge associated with textiles, materials and ancestral techniques lent added value to an identity and awareness of citizenship, as well as to the theme of encouraging learning.

Activities related to the field of textile materials were highlighted by more than 90% of the respondents as a privileged opportunity to promote the acquisition of skills overall, and those associated with motor skills, namely manual eye coordination and fine motor coordination.

iv) The interdisciplinary dimension of textiles

From the perspective of integration of the different areas of expression and communication, it has been interesting to learn how the activities with textiles in classes participate and contribute to the creation of interdisciplinary work or activities (Figure 3).

As shown in Figure 3, the art of textiles it is an activity that it is not promoted by these teachers. However, the specific materials and techniques of this art

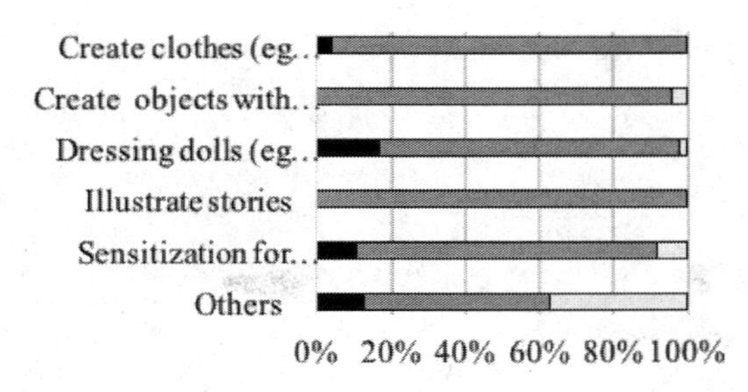

Figure 3. The interdisciplinary dimension of textiles in classes.

are used and reinvented to adapt to interdisciplinary contexts, such as creating garments for carnival or to illustrate stories, among others.

v) Inclusion of textiles in pedagogical practice

Finally, it should be pointed out that almost all the respondents (99%) were peremptory in their affirmative answer regarding the importance of including activities with textiles in their pedagogical practice.

5 DISCUSSION AND FINAL REMARKS

The literature review carried out in this study on children's artistic education evidenced that for several decades in Portugal, this dimension was not part of the curricular design of the educational system. Only in 1986, was art officially accepted as an important factor for the integral formation of the individual, with its inclusion in the educational system, namely in pre-school education and basic education (Sousa, 2003).

In the analysis of current school programs, the ability to respond to certain tasks, involving cognitive, social and emotional abilities, shows there is a need for guidance to support training in the development of skills.

Acquisition of these skills, which is not exhausted in each cycle of study, but progresses throughout the education and training of each individual, begins early in childhood and progresses throughout life, in a constant endeavour to expand knowledge and skills.

Competencies such as these are considered fundamental for the profitable promotion of a society oriented towards innovation and quality, and one in which knowledge and expertise are associated with knowing how to be and knowing how to do.

It is within this set of competencies that the textile arts fit with regard to contributing towards developing different forms of expression and communication, as well as motor skills; thus giving meaning to the educational effort assumed by the teachers interviewed. Simultaneously, the textile arts can enhance the development of aesthetic sensibility, abstract thinking, creativity, and self-esteem; all of which contribute to the integral formation and construction of citizenship.

The empirical study showed that when associated with other activities, the textile arts are one of the main pedagogical strategies of these respondents, by allowing the child to create, communicate and express themselves in different ways and in various contexts. The creation of carnival props and clothing, and other interdisciplinary activities highlights the close relationship of these materials and techniques with the emotional, sentimental and cognitive development of the child (Sousa, 2003).

Nevertheless, while textiles were understood as being a vehicle for children's artistic learning and sensibility through raising awareness of and bringing them closer to different artistic expressions, materials, techniques, and technologies, we realized that the activities associated with textiles, are currently very little implemented. It can even be observed that the specific activities of weaving and dyeing materials have fallen into disuse, despite the appreciation they received from almost all the respondents regarding their inclusion in pedagogical practices.

This discussion highlights the confrontation between the pedagogical intentionality of activities with textiles and practices that have been devalued. It is the devaluation of these practices that has closed the field of discovery for textiles as an art form.

In short, this study helps us to understand the discourses produced regarding children's education. In addition, it allows critical reflection on the nature of knowledge considered essential in the process of developing citizens in a society oriented towards the development and creation of innovative and qualified companies in the area of textiles.

Currently, with the strong resurgence of the textile and clothing sector in Portugal, the critical reflection that this study brings may contribute to the debate on the methodological orientations in K10 teaching.

ACKNOWLEDGMENTS

This research was only possible with the support of CIAUD – Research Centre in Architecture, Urbanism and Design, Lisbon School of Architecture of the University of Lisbon, and FCT – Foundation for Science and Technology, Portugal.

REFERENCES

ATP (Eds) (2016). *Fashion from Portugal. Diretório 2016.* Vila Nova de Famalicão: Multitema.

Cachapuz (2004). *Saberes Básicos de todos os cidadãos do Séc. XXI.* Lisboa: CNE. http://www.cnedu.pt/pt/publicacoes/estudos-e-relatorios/outros/793-saberes-basicos-de-todos-os-cidadaos-no-sec-xxi.

Cardona, M.J. (1977). *Para a história da educação em Portugal: O discurso oficial (1834–1990).* Porto: Porto Editora.

Gomes, J. (1977). *A educação infantil em Portugal.* Coimbra: Almedina.

Gomes, J. (1995). Luís António Verney e as Reformas Pombalinas do Ensino. In J. Gomes (ed.), *Para a História da Educação em Portugal-Seis estudos*. Porto: Porto Editora.
Lisboa, M. H. (2007). *As Academias e Escolas de Belas Artes e o Ensino Artístico (1836–1910)*. Lisboa: Edições Colibri.
Magalhães, J. (1997). Para uma história da educação de infância em Portugal. *Saber (e) Educar*, (2): 21–26.
Nóvoa, A. (1994). *História da educação*. Lisboa: Faculdade de Psicologia e de Ciências da Educação.
Ó, J.R. (2009). *Ensino Liceal (1863–1975)*. Lisboa: Faculdade de Psicologia e de Ciências da Educação.
OEI: Ministério da Educação de Portugal (2003). *Breve História do Sistema Educativo.* http://www.oei.es/historico/quipu/portugal/
Oliveira, E. (1996). In Pais, N. & Correia, L. (Coord). *Educação arte e Cultura*. Lisboa: FCG.
Pereira, A.S. (2017). *A indústria têxtil portuguesa*. Lisboa: Clube do Colecionador – CTT.
Pimentel, I. (2011). *A cada um o seu lugar, a política feminina do Estado Novo*. Lisboa: Editoras Temas e Debates.
Rodrigues, M.L. (Org) (2014). *40 anos de politicas de educação em Portugal*. Coimbra: Almedina.
Sendim, B. (2015). *Breve história da indústria têxtil. Da idade média à industrialização.* Famalicão: ATP-Associação Têxtil e Vestuário de Portugal. http://www.atp.pt/fotos/editor2/ATP_Brochura_Comemorativa_50_Anos.pdf.
Silva, I. Marques, L., Mata,L.,& Rosa, M. (2016). *Orientações Curriculares para a Educação Pré-escolar*. Lisboa: DGE.
Sousa, A. (2003). *A Educação pela arte e artes na Educação: Música e artes plásticas*. Lisboa: Instituto Piaget.
Sousa, M. R. (2008). *Música, educação artística interculturalidade a alma da arte na descoberta do outro*. Tese de Doutoramento em Ciências da Educação. Lisboa: Universidade Aberta.
Vasconcelos, T. (2015). *Percursos da Educação Pré-Escolar em Portugal: recuperar a memória. Encontro Educação Pré-Escolar.* www.fenprof.pt/Download/FENPROF/SM_Doc/Mid.../Fenprof_*jan2015.pptx*.

Textiles, Identity and Innovation: Design the Future – Montagna & Carvalho (Eds)
© 2019 Taylor & Francis Group, London, ISBN 978-1-138-29611-4

Author index